Lecture Notes in Artificial Intelligence   12886

Subseries of Lecture Notes in Computer Science

More information about this subseries at http://www.springer.com/series/1244

Hugo Sanjurjo González ·
Iker Pastor López · Pablo García Bringas ·
Héctor Quintián · Emilio Corchado (Eds.)

Hybrid Artificial Intelligent Systems

16th International Conference, HAIS 2021
Bilbao, Spain, September 22–24, 2021
Proceedings

 Springer

Editors
Hugo Sanjurjo González ⓘ
University of Deusto
Bilbao, Spain

Iker Pastor López ⓘ
University of Deusto
Bilbao, Spain

Pablo García Bringas ⓘ
University of Deusto
Bilbao, Spain

Héctor Quintián ⓘ
University of A Coruña
A Coruña, Spain

Emilio Corchado ⓘ
University of Salamanca
Salamanca, Spain

ISSN 0302-9743   ISSN 1611-3349   (electronic)
Lecture Notes in Artificial Intelligence
ISBN 978-3-030-86270-1   ISBN 978-3-030-86271-8   (eBook)
https://doi.org/10.1007/978-3-030-86271-8

LNCS Sublibrary: SL7 – Artificial Intelligence

This Springer imprint is published by the registered company Springer Nature Switzerland AG
The registered company address is: Gewerbestrasse 11, 6330 Cham, Switzerland

Preface

This volume of Lecture Notes on Artificial Intelligence (LNAI) includes accepted papers presented at the 16th International Conference on Hybrid Artificial Intelligence Systems (HAIS 2021), held in the beautiful seaside city of Bilbao, Spain, in September 2021.

HAIS has become an unique, established, and broad interdisciplinary forum for researchers and practitioners who are involved in developing and applying symbolic and sub-symbolic techniques aimed at the construction of highly robust and reliable problem-solving techniques, and bringing the most relevant achievements in this field.

The hybridization of intelligent techniques, coming from different computational intelligence areas, has become popular because of the growing awareness that such combinations frequently perform better than the individual techniques such as neuro-computing, fuzzy systems, rough sets, evolutionary algorithms, agents and multiagent systems, and so on.

Practical experience has indicated that hybrid intelligence techniques might be helpful to solve some of the challenging real-world problems. In a hybrid intelligence system, a synergistic combination of multiple techniques is used to build an efficient solution to deal with a particular problem. This is, thus, the setting of the HAIS conference series, and its increasing success is the proof of the vitality of this exciting field.

The HAIS 2021 International Program Committee selected 55 papers, which are published in these conference proceedings, yielding an acceptance ratio of about 68%.

The selection of papers was extremely rigorous in order to maintain the high quality of the conference and we would like to thank the Program Committee for their hard work in the reviewing process. This process is very important in creating a conference of high standard and the HAIS conference would not exist without their help.

The large number of submissions is certainly not only a testimony to the vitality and attractiveness of the field but an indicator of the interest in the HAIS conferences themselves.

HAIS 2021 enjoyed outstanding keynote speeches by distinguished guest speakers: Javier del Ser, who is a principal researcher in data analytics and optimization at Tecnalia, Spain, and is a part-time lecturer at the University of the Basque Country, Spain; Concha Bielza, who is a Full Professor in the Department of Artificial Intelligence at the Polytechnic University of Madrid, Spain; and Enrique Zuazua, who holds a Chair in Applied Analysis at Friedrich Alexander University, Germany, and a Chair of Computational Mathematics at the University of Deusto, Spain.

HAIS 2021 has teamed up with two journals, *Neurocomputing* (Elsevier) and the *Logic Journal of the IGPL* (Oxford Journals), for a suite of special issues including selected papers from HAIS 2021.

Particular thanks go the conference's main sponsors, Startup OLE, the Department of Education and Universities of the Basque Government, the Logistar Project of

DeustoTech, and the University of Deusto, who jointly contributed in an active and constructive manner to the success of this initiative.

We would also like to thank Alfred Hoffman and Anna Kramer from Springer for their help and collaboration during this demanding publication project.

September 2021   Hugo Sanjurjo González
                                                              Iker Pastor López
                                                         Pablo García Bringas
                                                              Héctor Quintián
                                                              Emilio Corchado

Organization

General Chair

Emilio Corchado   University of Salamanca, Spain

Local Chair

Pablo García Bringas   University of Deusto, Spain

Local Co-chairs

Iker Pastor López   University of Deusto, Spain
Hugo Sanjurjo González   University of Deusto, Spain

International Advisory Committee

Ajith Abraham	Machine Intelligence Research Labs, USA
Antonio Bahamonde	University of Oviedo, Spain
Andre de Carvalho	University of São Paulo, Brazil
Sung-Bae Cho	Yonsei University, South Korea
Juan M. Corchado	University of Salamanca, Spain
José R. Dorronsoro	Autonomous University of Madrid, Spain
Michael Gabbay	Kings College London, UK
Ali A. Ghorbani	UNB, Canada
Mark A. Girolami	University of Glasgow, UK
Manuel Graña	University of País Vasco, Spain
Petro Gopych	Universal Power Systems USA-Ukraine LLC, Ukraine
Jon G. Hall	The Open University, UK
Francisco Herrera	University of Granada, Spain
César Hervás-Martínez	University of Córdoba, Spain
Tom Heskes	Radboud University Nijmegen, The Netherlands
Dusan Husek	Academy of Sciences of the Czech Republic, Czech Republic
Lakhmi Jain	University of South Australia, Australia
Samuel Kaski	Helsinki University of Technology, Finland
Daniel A. Keim	University Konstanz, Germany
Marios Polycarpou	University of Cyprus, Cyprus
Witold Pedrycz	University of Alberta, Canada
Xin Yao	University of Birmingham, UK
Hujun Yin	University of Manchester, UK
Michał Woźniak	Wroclaw University of Technology, Poland
Aditya Ghose	University of Wollongong, Australia

Ashraf Saad	Armstrong Atlantic State University, USA
Fanny Klett	German Workforce Advanced Distributed, Learning Partnership Laboratory, Germany
Paulo Novais	Universidade do Minho, Portugal
Rajkumar Roy	The EPSRC Centre for Innovative Manufacturing in Through-life Engineering Services, UK
Amy Neustein	Linguistic Technology Systems, USA
Jaydip Sen	Tata Consultancy Services Ltd., India

Program Committee Chairs

Emilio Corchado	University of Salamanca, Spain
Pablo García Bringas	University of Deusto, Spain
Héctor Quintián	University of A Coruña, Spain

Program Committee

Iker Pastor López (Co-chair)	University of Deusto, Spain
Hugo Sanjurjo González (Co-chair)	University of Deusto, Spain
Aitor Martínez Seras	University of Deusto, Spain
Alfredo Cuzzocrea	University of Calabria, Italy
Amelia Zafra Gómez	University of Córdoba, Spain
Anca Andreica	Babes-Bolyai University, Romania
Andreea Vescan	Babes-Bolyai University, Romania
Andrés Blázquez	University of Salamanca, Spain
Angel Arroyo	University of Burgos, Spain
Antonio Dourado	University of Coimbra, Portugal
Antonio Jesús Díaz Honrubia	Polytechnic University of Madrid, Spain
Arkadiusz Kowalski	Wroclaw University of Technology, Poland
Álvaro Michelena Grandío	University of A Coruña, Spain
Borja Sanz Urquijo	University of Deusto, Spain
Bruno Baruque	University of Burgos, Spain
Camelia Serban	Babes-Bolyai University, Romania
Camelia Pintea	Technical University of Cluj-Napoca, Romania
Carlos Carrascosa	Polytechnic University of Valencia, Spain
Carlos Pereira	ISEC, Portugal
Damian Krenczyk	Silesian University of Technology, Poland
David Zamora Arranz	University of Deusto, Spain
David Iclanzan	Sapientia – Hungarian Science University of Transylvania, Romania
David Buján Carballal	University of Deusto, Spain
Diego P. Ruiz	University of Granada, Spain
Dragan Simic	University of Novi Sad, Serbia
Eiji Uchino	Yamaguchi University, Japan

Eneko Osaba	Tecnalia, Spain
Enol García González	University of Oviedo, Spain
Enrique Onieva	University of Deusto, Spain
Enrique De La Cal Marín	University of Oviedo, Spain
Esteban Jove Pérez	University of A Coruña, Spain
Federico Divina	Pablo de Olavide University, Spain
Fermin Segovia	University of Granada, Spain
Fidel Aznar	University of Alicante, Spain
Francisco Martínez-Álvarez	University Pablo de Olavide, Spain
Francisco Javier Martínez de Pisón	University of La Rioja, Spain
Francisco Zayas Gato	University of A Coruña, Spain
George Papakostas	EMT Institute of Technology, Greece
Georgios Dounias	University of the Aegean, Greece
Giorgio Fumera	University of Cagliari, Italy
Giuseppe Psaila	University of Bergamo, Italy
Gloria Cerasela Crisan	University of Bacau, Romania
Gonzalo A. Aranda-Corral	University of Huelva, Spain
Henrietta Toman	University of Debrecen, Hungary
Ignacio J. Turias Domínguez	University of Cádiz, Spain
Ignacio Angulo	University of Deusto, Spain
Igor Santos Grueiro	Mondragon University, Spain
Iñigo López Gazpio	University of Deusto, Spain
Ioana Zelina	Technical University of Cluj-Napoca, Romania
Ioannis Hatzilygeroudis	University of Patras, Greece
Javier Del Ser Lorente	Tecnalia, Spain
Javier De Lope	Polytechnic University of Madrid, Spain
Jon Ander Garamendi Arroyo	University of Deusto, Spain
Jorge García-Gutiérrez	University of Seville, Spain
José Dorronsoro	Autonomous University of Madrid, Spain
José García-Rodriguez	University of Alicante, Spain
José Alfredo Ferreira	Federal University, Brazil
José Luis Calvo-Rolle	University of A Coruña, Spain
José Luis Casteleiro-Roca	University of A Coruña, Spain
José Luis Verdegay	University of Granada, Spain
José M. Molina	University Carlos III of Madrid, Spain
José Manuel Lopez-Guede	University of the Basque Country, Spain
José María Armingol	University Carlos III of Madrid, Spain
José Ramón Villar	University of Oviedo, Spain
Juan Pavón	Complutense University of Madrid, Spain
Juan Humberto Sossa Azuela	CIC-IPN, Mexico
Juan José Gude Prego	University of Deusto, Spain
Julio Ponce	Autonomous University of Aguascalientes, México

Lidia Sánchez González	University of Leon, Spain
Luis Alfonso Fernández Serantes	FH Joanneum University of Applied Sciences, Austria
Mohammed Chadli	University of Paris-Saclay, France
Manuel Castejón-Limas	University of Leon, Spain
Manuel Graña	University of the Basque Country, Spain
Nashwa El-Bendary	Arab Academy for Science, Technology and Maritime Transport, Egypt
Noelia Rico	University of Oviedo, Spain
Oscar Llorente-Vázquez	University of Deusto, Spain
Pau Figuera Vinué	University of Deusto, Spain
Paula M. Castro	University of A Coruña, Spain
Peter Rockett	University of Sheffield, UK
Petrica Claudi Pop	Technical University of Cluj-Napoca, Romania
Qing Tan	Athabasca University, Canada
Robert Burduk	Wroclaw University of Technology, Poland
Rubén Fuentes-Fernández	Complutense University of Madrid, Spain
Sean Holden	University of Cambridge, UK
Sebastián Ventura	University of Córdoba, Spain
Theodore Pachidis	International Hellenic University, Greece
Urko de la Torre	Azterlan Metallurgy Research Centre, Spain
Urszula Stanczyk	Silesian University of Technology, Poland

Organizing Committee

David Buján Carballal	University of Deusto, Spain
José Gaviria De la Puerta	University of Deusto, Spain
Juan José Gude Prego	University of Deusto, Spain
Iker Pastor López	University of Deusto, Spain
Hugo Sanjurjo González	University of Deusto, Spain
Borja Sanz Urquijo	University of Deusto, Spain
Alberto Tellaeche Iglesias	University of Deusto, Spain
Pablo García Bringas	University of Deusto, Spain
Héctor Quintian	University of A Coruña, Spain
Emilio Corchado	University of Salamanca, Spain

Contents

Learning Algorithms

Visual Analysis and Advanced Data Processing Techniques

Machine Learning Applications

Hybrid Intelligent Applications

Deep Learning Applications

Optimization Problem Applications

Data Mining, Knowledge Discovery and Big Data

Document Similarity by Word Clustering with Semantic Distance

Toshinori Deguchi[1][(✉)] and Naohiro Ishii[2]

[1] National Institute of Technology (KOSEN), Gifu College, Gifu, Japan
`deguchi@gifu-nct.ac.jp`
[2] Advanced Institute of Industrial Technology, Shinagawa, Tokyo, Japan
`nishii@acm.org`

Abstract. In information retrieval, Latent Semantic Analysis (LSA) is a method to handle large and sparse document vectors. LSA reduces the dimension of document vectors by producing a set of topics related to the documents and terms statistically. Therefore, it needs a certain number of words and takes no account of semantic relations of words.

In this paper, by clustering the words using semantic distances of words, the dimension of document vectors is reduced to the number of word-clusters. Word distance is able to be calculated by using WordNet or Word2Vec. This method is free from the amount of words and documents. For especially small documents, we use word's definition in a dictionary and calculate the similarities between documents. For demonstration in standard cases, we use the problem of classification of BBC dataset and evaluate their accuracies, producing document clusters by LSA, word-clustering with WordNet, and word-clustering with Word2Vec.

Keywords: Document similarity · Semantic distance · WordNet · Word2Vec

1 Introduction

In information retrieval, a document is represented as a vector of index terms, which is called a document vector, and a set of documents is represented as a term-document matrix arranging document vectors in columns.

Usually, term-document matrices become large and sparse. Therefore, it is popular to use Latent Semantic Analysis (LSA) [1], Probabilistic Latent Semantic Analysis (pLSA), or Latent Dirichlet Allocation (LDA) to reduce the dimension of document vectors to find the latent topics by using statistical analysis. These methods are based on a large corpus of text, and the words are treated statistically but not semantically. Because the words are treated literally, for example, 'dog' and 'canine' are treated as totally different words, unless they co-occur in several documents.

Morave et al. [2] proposed a method to use WordNet [3] ontology. They used ℓ top level concepts in WordNet hierarchy as the topics, which means the topics

© Springer Nature Switzerland AG 2021
H. Sanjurjo González et al. (Eds.): HAIS 2021, LNAI 12886, pp. 3–14, 2021.
https://doi.org/10.1007/978-3-030-86271-8_1

are produced semantically. For given ℓ, the concepts that is used for the topics are determined independently of the target documents.

In this paper, we propose a semantic method to reduce the dimension of term-document matrices. Instead of latent topics by statistics or ℓ top level concepts in WordNet, this method uses word-clusters generated by hierarchical clustering. For hierarchical clustering, we use a semantic distance between words, which is able to be calculated by using WordNet or Word2Vec. The dimension of document vectors is reduced to the number of word-clusters.

When the number of documents or words are not large, statistical method is not suitable, whereas the word-clustering method uses semantic distances of words and is applicable even to that case.

For demonstration in small sets, we take especially small documents which are word's definitions in a dictionary, and show the results with LSA, and word-clustering methods.

For demonstration in standard sets, we use BBC news dataset [4] for document classification, and show the results with LSA and word-clustering methods.

2   Vector Space Model and Dimension Reduction

2.1   Vector Space Model

In vector space model, a document is represented as a vector of terms (or words). For the values of a document vectors, term frequency (tf) or term frequency-inverse document frequency (tf-idf) is used to show the importance of the word in the document in the collection.

Although there are several definitions for tf and idf, in this paper, we use

$$\text{tf-idf}(t,d) = \text{tf}(t,d) \times \text{idf}(t), \tag{1}$$

$$\text{tf}(t,d) = n_{t,d}, \tag{2}$$

$$\text{idf}(t) = \log \frac{|D|}{df_t}, \tag{3}$$

where $n_{t,d}$ is the raw count of the word w_t in the document doc_d, D is the set of document, and df_t is the number of documents where the word w_t appears.

A term-document matrix (or document matrix) X is a matrix that is constructed by lining up document vectors as column vectors.

$$X = (\text{tf-idf}(i,j)) \tag{4}$$

2.2   Distance by WordNet

WordNet [3] is a lexical database. In WordNet, nouns and verbs are arranged in concept (synset in WordNet) hierarchies by hypernyms. For example, a part of hierarchies including the concepts of the word "whale" and the word "horse" is shown in Table 1.

Table 1. An example of hierarchies in WordNet.

Depth	Concept
1	entity
2	physical_entity
3	physical_object
4	whole
5	living_thing
6	being
7	animate_being
8	chordate
9	vertebrate
10	mammalian
11	placental
12	aquatic_mammal hoofed_mammal
13	blower odd-toed_ungulate
14	whale equine
15	horse

There are several definitions to measure semantic relatedness using WordNet. Among them, we adopt Wu and Palmer's method [5] because the similarity value of two words is from 0 to 1 and the distance is calculated as subtracting the similarity from 1.

According to Budanitsky and Hirst [6], the relatedness $\mathrm{rel}(w_1, w_2)$ between two words w_1 and w_2 can be calculated as

$$\mathrm{rel}(w_1, w_2) = \max_{c_1 \in s(w_1), c_2 \in s(w_2)} [\mathrm{rel}(c_1, c_2)], \tag{5}$$

where $s(w_i)$ is the set of concepts in WordNet that are senses of word w_i. We consider the relatedness as the similarity of two words as:

$$\mathrm{sim}(w_1, w_2) = \max_{c_1 \in s(w_1), c_2 \in s(w_2)} [\mathrm{sim}(c_1, c_2)] \tag{6}$$

In Wu and Palmer's method, the similarity of two concepts is defined as follows:

$$\text{sim}(c_i, c_j) = \frac{2\,\text{depth}(\text{lso}(c_i, c_j))}{\text{len}(c_i, \text{lso}(c_i, c_j)) + \text{len}(c_j, \text{lso}(c_i, c_j)) + 2\,\text{depth}(\text{lso}(c_i, c_j))} \quad (7)$$

where $\text{lso}(c_1, c_2)$ is the lowest super-ordinate (or most specific common subsumer) of c_1 and c_2.

For example, from Table 1, letting c_1 = whale and c_2 = horse,

$$\text{lso}(\text{whale}, \text{horse}) = \text{placental}, \quad (8)$$

$$\text{depth}(\text{placental}) = 11, \quad (9)$$

$$\text{len}(\text{whale}, \text{placental}) = 3, \quad (10)$$

$$\text{len}(\text{horse}, \text{placental}) = 4, \quad (11)$$

$$\text{sim}(\text{whale}, \text{horse}) = \frac{2 \times 11}{3 + 4 + 2 \times 11} \simeq 0.759. \quad (12)$$

In computation of $\text{sim}(c_i, c_j)$, there are cases that more than one synset for $\text{lso}(c_i, c_j)$ exists, For example, if you take water and woodland, the hierarchy of these concepts is obtained as shown in Table 2. Because $\text{depth}(\text{physical\_entity}) = \text{depth}(\text{abstract\_entity}) = 2$, $\text{lso}(\text{water}, \text{woodland})$ can be both physical_entity and abstract_entity.

If you use physical_entity for $\text{lso}(\text{water}, \text{woodland})$, you will get

$$\text{sim}(\text{water}, \text{woodland}) = \frac{2 \times 2}{4 + 3 + 2 \times 2} \simeq 0.364. \quad (13)$$

Otherwise,

$$\text{sim}(\text{water}, \text{woodland}) = \frac{2 \times 2}{6 + 4 + 2 \times 2} \simeq 0.286. \quad (14)$$

In these cases, we take the maximum as follows:

$$\text{sim}(c_i, c_j) = \max_{\text{lso}(c_i, c_j)} \left[\frac{2\,\text{depth}(\text{lso}(c_i, c_j))}{\text{len}(c_i, \text{lso}(c_i, c_j)) + \text{len}(c_j, \text{lso}(c_i, c_j)) + 2\,\text{depth}(\text{lso}(c_i, c_j))} \right]. \quad (15)$$

Equation (15) is determined uniquely even when there are multi-paths from c_i or c_j to $\text{lso}(c_i, c_j)$, since we defined $\text{len}(c_i, c_j)$ as the length of the shortest path from c_i to c_j.

From above, we also get the definition of the distance of two words as follows:

$$d(w_i, w_j) = 1 - \text{sim}(w_i, w_j) \quad (16)$$

2.3 Distance by Word2Vec

Word2Vec [7,8] is a method to produce a vector space from a large corpus of text. Using trained Word2Vec model, we can get the vector from a word and calculate the distance and similarity of 2 words using these vectors.

Table 2. Hierarchy on water and woodland.

Depth	Concept

We use Python library "gensim" to calculate the distance by Word2Vec. In this library, the distance of two words is the cosine distance as in (16) and the similarity of two words is as follows:

$$\text{sim}(w_i, w_j) = \frac{\text{vec}(w_i) \cdot \text{vec}(w_j)}{\|\text{vec}(w_i)\| \|\text{vec}(w_j)\|} \tag{17}$$

where $\text{vec}(w)$ is a vector representation of word w.

2.4 Word Clustering

By dividing the words in the documents to clusters using the semantic distance of words which is calculated by using WordNet or Word2Vec, the term-document matrix is mapped to cluster-document matrix. For clustering, we use the hierarchical clustering with semantic distance of words.

First, distance matrix $D = (d_{ij})$ for all words in term-document matrix is calculated as:

$$d_{ij} = d(w_i, w_j) \tag{18}$$

Then, using the matrix D, the hierarchical clustering divides the words into clusters by cutting the dendrogram to form c flat clusters. We use Ward's method as the linkage criteria. For the following calculations, the cluster membership matrix $M = (m_{ij})$ is formed as:

$$m_{ij} = \begin{cases} 1, \text{ for } w_j \in C_i, \\ 0, \text{ for } w_j \notin C_i \end{cases} \tag{19}$$

2.5   Dimension Reduction by Semantic Distance

Using the clusters as topics in LSA, we can reduce the dimension of the term-document matrix and turn it into the cluster-document matrix, joining the words in the same cluster as in Fig. 1.

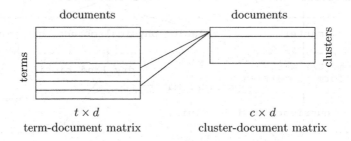

Fig. 1. Dimension reduction by word clustering.

For example, we take the case that there are 13 words in right side of Table 3 in the documents. And the words are divided into 4 clusters as in Table 3. Then, term vectors of "claw", "leg", and "tail" are gathered into cluster 1's vector, and so on. The original term-document matrix which has 13 rows is reduced to 4 rows of clusters.

Table 3. Word-cluster example.

Cluster no.	Words
1	Claw, leg, tail
2	Animal, fur, mouse, pet
3	Catch, guard, hunt, pull
4	Sea, water

In order to convert the matrix, we need the mapping functions from the words to the clusters. In the same way as in LSA, let us assume that each function is represented by a linear combination of the words and that the mapping is represented by a cluster-word relation matrix $R = (r_{i,j})$. Then, we can calculate the cluster-document matrix Y as follows:

$$Y = RX \qquad (20)$$

2.6   Cluster-Word Relation Matrix

To calculate the mapping function from the words to the clusters, we define the grade of membership of the word to the cluster. For word w_j, the grade

$g(C_i, w_j)$ of membership to cluster C_i is considered to have a relation to the distances between w_j and the words in cluster C_i.

Then, we define $R = (r_{ij})$ as the terms used in producing a coefficient are restricted in the cluster as shown in (21).

$$r_{i,j} = \begin{cases} g(C_i, w_j), & \text{for } w_j \in C_i, \\ 0, & \text{for } w_j \notin C_i. \end{cases} \tag{21}$$

Grade for WordNet. Although the inverse proportion to the distance is popular, we will have the multi-synsets problem in which two different word w_j and w_k could be $d(w_j, w_k) = 0$ and the grade becomes infinity. To modify the inverse proportion is one plan, but we chose to use similarity of words.

We define the grade $g(C_i, w_j)$ as the mean similarity of the words as follows:

$$g(C_i, w_j) = \frac{1}{|C_i|} \sum_{w \in C_i} \text{sim}(w, w_j) \tag{22}$$

Grade for Word2Vec. Since Word2Vec maps each word to the numerical vector, we treat the mean vector of the words in a cluster as the exemplar vector for the cluster. The grade $g(C_i, w_j)$ for Word2Vec becomes as follows:

$$g(C_i, w_j) = \text{sim}(\text{vec}(C_i), \text{vec}(w_j)) \tag{23}$$

where

$$\text{vec}(C_i) = \frac{1}{|C_i|} \sum_{w \in C_i} \text{vec}(w) \tag{24}$$

3 Experimental Results

3.1 Conditions of Experiments

For the small documents, we use the first definition of the 6 words *dog*, *cat*, *horse*, *car*, *bus*, and *ship* on online "Cambridge Dictionary" [9]. Each definition sentence is a document. We describe the document defining *dog* as 'dog', and so on.

We use the environment in Table 4. The distances between words with Word-Net are calculated by our tool according to (6), (15), and (16).

WordNet has hypernym hierarchies only for nouns and verbs. Therefore, the input texts are divided into stem words which are POS-tagged, then the document matrices are generated separately for nouns and verbs. For stemming and POS-tagging, we use TreeTagger because it performs the best for tagging nouns and verbs [13].

The two matrices are concatenated to one matrix when they are used in LSA and WordNet. Because Word2Vec has no POS information, for the words in both nouns and verbs, the tfidf's are added. Note that not all the words are stored in WordNet or Word2Vec, so only words in WordNet or Word2Vec are used in those methods.

Table 4. Execution environment.

Language	Python 3.7.5
Stemming	TreeTagger [10]
POS tagging	TreeTagger
LSA	TruncatedSVD (in scikit-learn 0.21.3)
WordNet	Japanese WordNet 1.1 [11]
Word2Vec	GloVe pre-trained vector (42B tokens and 300d vectors) [12]
Hierarchical clustering	hierarchy (in scipy 1.3.2)
linkage method	Ward's method

Note: Japanese WordNet 1.1 includes Princeton WordNet 3.0.

3.2   Similarity of Documents

We get Y reducing the dimension of the document matrix by LSA and word-clustering methods.

For similarity, we use the cosine similarity. For the calculation of the document similarity, let $\boldsymbol{d}_j = (y_{1j}, \ldots, y_{dj})^T$ for $Y = (y_{ij})$. Then, the document similarity between document doc_i and document doc_j is calculated as follows:

$$similarity(i,j) = \frac{\boldsymbol{d}_i \cdot \boldsymbol{d}_j}{\|\boldsymbol{d}_i\| \, \|\boldsymbol{d}_j\|} \tag{25}$$

3.3   Result of Each Method

In this case that the number of words is small, using LSA is questionable. However, we used LSA for a comparison.

The result using LSA is shown in Table 5. The number of topics was decided to 5, as the cumulative contribution ratio was over 80%. Only similarity ('dog', 'cat') and similarity ('dog', 'horse') are high, because there are 4 common terms in 'dog' and 'cat', and 4 terms in 'dog' and 'horse'. As described above, this is not the case to use statistical analyses.

Table 5. Similarities by LSA (5 topics).

	'dog'	'cat'	'horse'	'car'	'bus'
'cat'	0.722				
'horse'	0.693	0.011			
'car'	0.020	0.000	0.041		
'bus'	−0.042	0.009	0.063	0.049	
'ship'	−0.000	−0.000	0.000	0.000	−0.000

The result using word-clustering with WordNet is shown in Table 6. The number of cluster is decided with several pre-experiments checking cluster members. All the terms are divided into 7 clusters.

The results are quite good, in our point of view, because the similarities between the animals are high and the ones between the transporters including 'horse' except similarity ('horse', 'car').

Table 6. Similarities by word-clustering with WordNet (7 clusters).

	'dog'	'cat'	'horse'	'car'	'bus'
'cat'	0.797				
'horse'	0.811	0.523			
'car'	0.344	0.402	0.403		
'bus'	0.491	0.264	0.829	0.553	
'ship'	0.301	0.323	0.550	0.858	0.817

The result using word-clustering with Word2Vec is shown in Table 7. The number of cluster is 7 as same as WordNet case.

The result shows almost the same as WordNet. In this case, the similarity ('cat', 'horse') and similarity ('horse', 'car') are low, and 'ship' has little similarity with others.

Table 7. Similarities by word-clustering with Word2Vec (7 clusters).

	'dog'	'cat'	'horse'	'car'	'bus'
'cat'	0.659				
'horse'	0.789	0.393			
'car'	0.183	0.053	0.466		
'bus'	0.532	0.155	0.935	0.523	
'ship'	0.000	0.000	0.197	0.166	0.234

4 BBC Dataset

For more practical case, we take BBC dataset [4]. This dataset have 2225 documents from the BBC news in 5 categories: business, entertainment, politics, sport and tech.

To categorize the documents to 5 categories, we calculate the document similarities, turn them to the document distances by subtracting from 1, then make 5 clusters by the hierarchical clustering with Ward's method.

To evaluate the results, we adopt the accuracy defined as follows:

$$\text{accuracy}(y, \hat{y}) = \frac{1}{n} \sum_{i=0}^{n-1} 1(\hat{y}_i = y_i) \tag{26}$$

where y_i is a true label of category for the i-th document, $\hat{y}_i$ is an estimated label, n is the number of documents, and $1(x)$ is 1 for $x = $ true and 0 for false.

Because the hierarchical clustering shows no labels of categories, we use the maximum accuracy for all assignments of the clusters to the categories.

The accuracies are calculated using 2 topics to 1200 topics by 10 topics. For word-clustering, the number of clusters follows the number of topics.

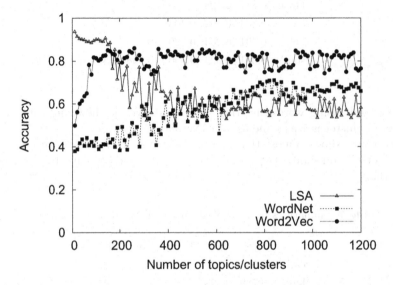

Fig. 2. Accuracy by LSA, word-clustering with WordNet, and word-clustering with Word2Vec.

Figure 2 shows the results of accuracies by LSA, word-clustering with Word-Net, and word-clustering with Word2Vec. LSA produces high accuracies in small topics and the maximum is 0.937 at 10 topics. Word-clustering methods need some number of clusters to obtain high accuracy. The maximum accuracy of WordNet is 0.710 at 840 clusters, and that of Word2Vec is 0.854 at 340 clusters. In this case, there is no advantage for using word-cluster methods.

For the case of less documents, we take first 5 documents in each category, so the number of documents is 25. In this case, we calculate accuracies using 2 topics to 25. The results are shown in Fig. 3.

For shortage of information, accuracies are lower than in Fig. 2. The maximum accuracy of LSA is 0.6 at 20 topics, that of WordNet is 0.52 at 11 and 12 clusters, and that of Word2Vec is 0.64 at 7, 9, 20 and 22–25 clusters. In this

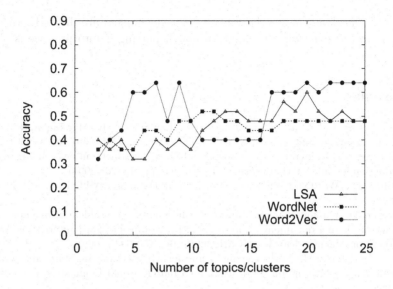

Fig. 3. Accuracy on 25 documents.

case, the accuracy of word-cluster method with Word2Vec is slightly better than that of LSA.

5   Conclusion

Using word-clustering by semantic distance derived from WordNet or Word2Vec, we transformed term-document matrix to cluster-document matrix. This transformation is a dimension reduction of document vectors.

First, we demonstrated this method to calculate the similarities of the small documents. Since the distance of the words is not statistical but semantic, the matrix was able to be reduced even though some words appear only in one document.

Second, we took BBC dataset for larger sets. From the point of accuracy, it turned out that word-clustering methods did not produce higher accuracy than LSA with 2225 documents, whereas word-clustering with Word2Vec produces the highest accuracy with 25 documents. This means that, in the case that there are a large number of documents, the statistical method shows its advantage, but, in the case of less words and documents, this semantic method reveals the its potential.

Unfortunately, we have no other quantitative evaluations, but each result of the experiments suggests that this method is advantageous for a small set of documents. The evaluation for other data sets is a future work.

Beyond the accuracy, the clusters from our method can be regarded as topics which are defined from the words in each cluster and are connected with the meaning. Thus, we can see the way that the documents are divided into categories from the semantic viewpoint, although putting it into practice is also a future work.

References

1. Deerwester, S., Dumais, S.T., Furnas, G.W., Landauer, T.K., Harshman, R.: Indexing by latent semantic analysis. J. Am. Soc. Inf. Sci. **41**(6), 391–407 (1990)
2. Moravec, P., Kolovrat, M., Snášel, V.: LSI vs. wordnet ontology in dimension reduction for information retrieval. In: Dateso, pp. 18–26 (2004)
3. Miller, G.: WordNet: An Electronic Lexical Database. MIT Press, Cambridge (1998)
4. Greene, D., Cunningham, P.: Practical solutions to the problem of diagonal dominance in kernel document clustering. In: Proceedings of 23rd International Conference on Machine learning (ICML 2006), pp. 377–384 (2006)
5. Wu, Z., Palmer, M.: Verbs semantics and lexical selection. In: Proceedings of the 32nd Annual Meeting on Association for Computational Linguistics, pp. 113–138 (1994)
6. Budanitsky, A., Hirst, G.: Evaluating WordNet-based measures of lexical semantic relatedness. Comput. Linguist. **32**(1), 13–47 (2006)
7. Mikolov, T., Chen, K., Corrado, G., Dean, J.: Efficient estimation of word representations in vector space. arXiv preprint arXiv:1301.3781 (2013)
8. Mikolov, T., Sutskever, I., Chen, K., Corrado, G., Dean, J.: Distributed representations of words and phrases and their compositionality. Adv. Neural Inf. Process. Syst. **26**, 3111–3119 (2013)
9. https://dictionary.cambridge.org/
10. Schmid, H.: Probabilistic part-of-speech tagging using decision trees. In: New Methods in Language Processing, pp. 154–164 (2013). http://www.cis.uni-muenchen.de/%7Eschmid/tools/TreeTagger/
11. Bond, F., Baldwin, T., Fothergill, R., Uchimoto, K.: Japanese SemCor: a sense-tagged corpus of Japanese. In: Proceedings of the 6th Global WordNet Conference (GWC 2012), pp. 56–63 (2012). http://compling.hss.ntu.edu.sg/wnja/index.en.html
12. Pennington, J., Socher, R., Manning, C.D.: GloVe: global vectors for word representation. In: Empirical Methods in Natural Language Processing (EMNLP), pp. 1532–1543 (2014). https://nlp.stanford.edu/projects/glove/
13. Tian, Y., Lo, D.: A comparative study on the effectiveness of part-of-speech tagging techniques on bug reports. In: 2015 IEEE 22nd International Conference on Software Analysis, Evolution and Reengineering (SANER), pp. 570–574 (2015)

PSO-PARSIMONY: A New Methodology for Searching for Accurate and Parsimonious Models with Particle Swarm Optimization. Application for Predicting the Force-Displacement Curve in T-stub Steel Connections

Julio Fernandez Ceniceros[1], Andres Sanz-Garcia[2] (ORCID),
Alpha Pernia-Espinoza[1] (ORCID), and Francisco Javier Martinez-de-Pison[1(✉)] (ORCID)

[1] EDMANS Group, Department of Mechanical Engineering, University of La Rioja,
Logroño, Spain
edmans@dim.unirioja.es, fjmartin@unirioja.es
[2] Department of Mechanical Engineering, University of Salamanca, Bejar, Spain
ansanz@usal.es

Abstract. We present PSO-PARSIMONY, a new methodology to search for parsimonious and highly accurate models by means of particle swarm optimization. PSO-PARSIMONY uses automatic hyperparameter optimization and feature selection to search for accurate models with low complexity. To evaluate the new proposal, a comparative study with Multilayer Perceptron algorithm was performed by applying it to predict three important parameters of the force-displacement curve in T-stub steel connections: initial stiffness, maximum strength, and displacement at failure. Models optimized with PSO-PARSIMONY showed an excellent trade-off between goodness-of-fit and parsimony. Then, the new proposal was compared with GA-PARSIMONY, our previously published methodology that uses genetic algorithms in the optimization process. The new method needed more iterations and obtained slightly more complex individuals, but it performed better in the search for accurate models.

Keywords: PSO-PARSIMONY · T-stub connections · Parsimonious modeling · Auto machine learning · GA-PARSIMONY

1 Introduction

Nowadays, there is a growing demand for auto machine learning (AutoML) tools to automatize tedious tasks such as hyperparameter optimization (HO), model selection (MS), feature selection (FS) and feature generation (FG).

In this article, we describe a new methodology, named PSO-PARSIMONY, which uses an adapted particle swarm optimization (PSO) to search for parsimonious and accurate models by means of hyperparameter optimizacion (HO),

© Springer Nature Switzerland AG 2021
H. Sanjurjo González et al. (Eds.): HAIS 2021, LNAI 12886, pp. 15–26, 2021.
https://doi.org/10.1007/978-3-030-86271-8_2

feature selection (FS), and promotion of the best solutions according to two criteria: low complexity and high accuracy.

The present paper describes the new methodology and its application to predict the force-displacement curves in T-stub steel connections. The method was tested with a multilayer perceptron algorithm (MLP) to predict three parameters that define the T-stub curve: initial stiffness, maximum strength, and displacement at failure.

Lastly, we compare the new proposal vs GA-PARSIMONY, our previously published methodology based on GA [12,13,17], that has been successfully applied in a variety of contexts such as steel industrial processes, hotel room-booking forecasting, mechanical design, hospital energy demand, and solar radiation forecasting.

2   Related Works

In recent years, there has been a trend toward providing methods to make ML more accessible for people without expertise in machine learning. The overarching aim is to reduce the human effort necessary in tedious and time-consuming tasks.

Companies like *DataRobot, Strong Analytics, Mighty AI, Akkio, CloudZero* or *Unity Technologies*, among others, are currently providing services to automate a multitude of tasks in machine learning and artificial intelligence. In addition, new AutoML suites have emerged such as *Google Cloud AutoML, Microsoft Azure ML, Alteryx Intelligence Suite,* or *H2O AutoML*. Free software is also available, such as Auto-WEKA, Auto-Sklearn 2.0, Hyperopt, TPOT, or, for Deep Learning, Auto-Keras and Auto-PyTorch. Many companies and tools use hybrid methods based on high computational resources fused with advanced optimization techniques.

However, many data science applications need to construct models with small databases. In this kind of situation, a reliable strategy for finding models that are robust against perturbations or noise, is to select low-complexity models from the most accurate solutions. These will also be easier to maintain and understand [7]. Therefore, such methods can help many scientists and engineers who, though they are not be experts in machine learning, still need to obtain accurate and robust models.

Some studies have focused on the context of automatically seeking parsimonious and accurate models. For example, Ma and Xia [8] used a tribe competition with GA to optimize FS in pattern classification. Wei et al. [22] applied binary particle swarm optimization (BPSO) in HO and FS to obtain an accurate SVM with a reduced number of features. Similarly, Vieira et al. [18] optimized a wrapper SVM with BPSO to predict the survival or death of patients with septic shock. O. Wan et al. [20] combined GA with an ant colony optimization algorithm named MBACO to select an optimal subset of features with minimum redundancy and maximum discriminating ability. Ahila et al. [1] improved accuracy and generalization performance of Extreme Learning Machines (ELM) with PSO for power system disturbances. [21] reported a comparative of chaotic moth-flame algorithm against other methods in an HO and FS optimization strategy of medical diagnoses. This method showed significant improvement in classification performance and obtained a smaller feature subset.

In the field of Deep Learning, there is a great deal of research focused on optimizing the learning process and the structure of neuronal networks. One of the most interesting work is `Auto-Pytorch` [24] for tabular data. Auto-PyTorch automatically tunes the full DL of PyTorch pipeline: data preprocessing, neural architecture, training techniques, and regularization methods.

Similar to our previous works, in previous works we also used GA-PARSIMONY [14], a method to search for parsimonious solutions with GA by optimizing HO, FS, and parsimonious model selection.

3 PSO-PARSIMONY Methodology

Tuning model setting parameters and, at the same time, selecting a subset of the most relevant inputs require efficient heuristic methods to manage such a combinatorial problem. In this article, the search for the best model was based on an optimization technique developed by Kennedy and Eberhart: the particle swarm optimization (PSO) [6]. Its popularity has undoubtedly increased due to its straightforward implementation and demonstrated high convergence ratio.

The PSO algorithm mimics the social behavior of birds flocking and fish schooling to guide the particles towards globally optimal solutions. The movement of a particle is influenced by its own experience (its best position achieved so far) and also by the experience of other particles (the best position within a neighborhood). The formulation of canonical PSO contains expressions for calculating velocities and positions of particles. Considering a D-dimensional search space, $\boldsymbol{X}_i = [x_{i1},\ x_{i2}, \ldots, x_{iD}]$ and $\boldsymbol{V}_i = [v_{i1},\ v_{i2}, \ldots,\ v_{iD}]$ describe the position and velocity of the i-th particle, respectively. The personal best is represented by $\boldsymbol{P}_{best,i} = [p_{best,i1},\ p_{best,i2}, \ldots,\ p_{best,iD}]$ whereas the local best in each neighborhood is described by $\boldsymbol{L}_{best,i} = [l_{best,i1},\ l_{best,i2}, \ldots,\ l_{best,iD}]$. The minimum value among the local bests represents the global best (*gbest*) and, consequently, the optimal solution. In this context, the velocity and position of the next iteration are calculated as follows:

$$\boldsymbol{V}_i^{t+1} = \omega\boldsymbol{V}_i^t + r_1\varphi_1 \times \left(\boldsymbol{P}_{best,i}^t - \boldsymbol{X}_i^t\right) + r_2\varphi_2 \times \left(\boldsymbol{L}_{best,i}^t - \boldsymbol{X}_i^t\right) \qquad (1)$$

$$\boldsymbol{X}_i^{t+1} = \boldsymbol{X}_i^t + \boldsymbol{V}_i^{t+1} \qquad (2)$$

where t represents the current iteration and ω is the inertia weight which depreciates the contribution of the current velocity. The purpose of this parameter is to prevent an uncontrolled increase in particle displacement. The coefficients φ_1 and φ_2 represent the cognitive and social learning rates, respectively. They control the trade-off between global exploration and local exploitation, i.e. how the local and personal bests affect the calculation of the next position. In this study, both of them were set to $\frac{1}{2} + ln\,(2)$, as suggested by Clerc [2]. Finally, r_1

and r_2 are independent and uniformly distributed random vectors in the range $[0, 1]$ whose purpose is to maintain the diversity of the swarm.

In the framework of FS and model optimization, it is desirable to simultaneously minimize the fitness function (J) and the model complexity (M_c) to guarantee both accuracy and generalization capacity. Thus, the modified version of PSO proposed herein includes strategies for updating the personal best (*pbest*) and the global best (*gbest*) considering not only the goodness-of-fit, but also the principle of parsimony.

The process starts with a random initial swarm of models

$$\mathbf{X^0} : \left\{ \boldsymbol{X}_1^0, \boldsymbol{X}_2^0, \ldots, \boldsymbol{X}_s^0 \right\} \tag{3}$$

generated by $\mathbf{X^0} = random_{LHS}\,(s,\ D)$ where $random_{LHS}$ creates a random and uniformly distributed Latin hypercube within the ranges of feasible values for each input parameter. Thus, s and D represent the number of particles in the swarm and the dimensions of the design space, respectively. In this study, the swarm size was set to 40 particles, as empirically suggested in [23].

Each particle $\boldsymbol{X}_i^0$, which represents a model configuration, is characterized by a subset of input variables and the values of setting parameters to be optimized. Similarly, initial velocities $\mathbf{V^0} : \left\{ \boldsymbol{V}_1^0, \boldsymbol{V}_2^0, \ldots, \boldsymbol{V}_s^0 \right\}$ are randomly generated according to the following expression:

$$\mathbf{V^0} = \frac{random_{LHS}\,(s,\ D) - \mathbf{X^0}}{2} \tag{4}$$

Following this process, the training and validation of the models is conducted by 10-fold cross validation (CV). This method is repeated five times to assure the robustness of the model against different subsamples of training data. Model complexity is also calculated. Model complexity depends on the model structure and is related to its robustness to perturbations and noise. In this case, complexity has been defined by $M_c = 10^6 N_{FS} + Int_c$, where N_{FS} is the number of selected features and Int_c is an internal complexity measurement which depends on the ML algorithm. For example, the number of support vectors in Support Vector Machines (SVM), the sum of the squared weights in artificial neuronal networks (ANN), the number of leaves in regression trees (RT), etc. Thus, in this formula, N_{FS} dominates the other measure and, then, Int_c only comes into play when the two compared solutions have the same N_{FS}.

Firstly, the goodness-of-fit of models is assessed by J. The root mean square error (RMSE) of the CV process is utilized herein as fitness function $J\left(\boldsymbol{X}_i^t\right) = RMSE_{CV}$, where $RMSE_{CV}$ denotes the mean of the CV RMSE. The RMSE penalizes large deviations from the target values. This feature is particularly advantageous in the context of steel connections because models should predict the connection response with a similar degree of accuracy throughout the entire design space. Secondly, models are re-evaluated according to their complexity. The principle of parsimony states that in the case of two models with similar accuracy the simplest is the best. In accordance with this principle, we propose strategies for updating the *pbest* and the *gbest*. For each particle of the swarm, the *pbest* is updated to the new position if the fitness value of this new position $J\left(\boldsymbol{X}_i^t\right)$ is clearly lower than the current value. On the other hand, *pbest* is not updated if $J\left(\boldsymbol{X}_i^t\right)$ is clearly higher than the current value. For intermediate cases, where $J\left(\boldsymbol{X}_i^t\right)$ is within a tolerance in regards to the *pbest*, the complexity criterion is applied. Then, the *pbest* is updated to the new particle position only if its complexity $M_c\left(\boldsymbol{X}_i^t\right)$ is lower than the current value. This strategy, therefore, requires a user-defined tolerance (*tol*) to establish the limits wherein the complexity criterion is applicable. Similarly, the update of *gbest* is conducted by comparing those particles that are within a tolerance *tol* in regards to the lowest fitness value. Among those particles, the particle exhibiting the lowest complexity is chosen as *gbest*. A value of *tol* equal to three values [0.01, 0.005, 0.001] was proven to be appropriate for both *pbest* and *gbest* in this particular case.

Once *pbest* and *gbest* have been updated, the next step in the PSO consists of updating velocities and positions of particles in order to evolve towards better solutions. For this particular case where FS is included in the PSO, the binary status of features deserves special treatment. In our proposal, a continuous PSO is applied for both FS and model parameters optimization. The real number x_d corresponding to the feature d is compared with a threshold value, α. Then, feature d is included in the input subset if $x_d \geq \alpha$. Otherwise, the feature is discarded. This approach avoids the premature convergence that characterizes the binary version of particle swarm optimization (BPSO) [19]. In this study, the threshold α was set to 0.6.

The inertia weight is defined as linearly decreasing with the iterations, according to Shi and Eberhart [15], $\omega = \omega_{max} - (\omega_{max} - \omega_{min})\frac{t}{T}$ where t represents the current iteration and T is a predefined maximum number of iterations. ω_{max} and ω_{min} are set to 0.9 and 0.4, respectively [3,16]. Thus, high values of ω in the first iterations have the ability to explore new areas. On the contrary, lower values of ω at the end of the process promote a refined search in local regions.

Concerning the topology that describes the interconnection among particles, we adopted the adaptive random topology initially proposed by Clerc [6]. In this topology, each particle of the swarm informs K particles randomly (the same particle can be chosen several times) and itself, where K is usually set to 3. As a result, each particle is informed by a number of particles (neighborhood) that can vary from 1 to s. If an iteration does not present improvement in the fitness value of *gbest*, new neighborhoods are randomly generated following the same process. This topology provides more diversity in the swarm than using a single global best and is less susceptible to being trapped in local minima. However, the convergence rate is generally slower than in a global best topology.

Finally, this modified version of the PSO also incorporates a mutation operator for the FS similar to that used in GA. The operator tries to prevent the premature convergence of the swarm. The mutation rate m is set to $1/D$, where D represents the dimensionality of the problem. This value guarantees that at least one feature will change its status in each iteration (from 0 to 1 or vice versa).

After updating velocities and positions, the particles are confined in order to avoid out-of-range positions. In this study, the absorbing wall approach is utilized for this purpose [23]. Thus, when a particle is out of the feasible range, the position is set to the boundary and velocity is set to zero.

Optimization ends when the maximum number of iterations T is reached although an early stopping can be used if J does not improve within a predefined number of iterations. The process explained in this section was applied to each ML algorithm technique and each size of training dataset.

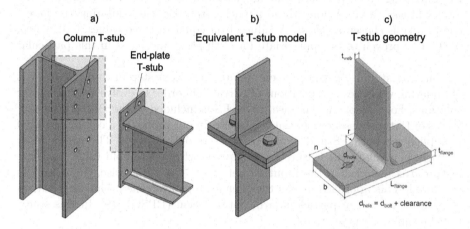

Fig. 1. Tension zone on steel connections. (a) End-plate beam-to-column connection, (b) equivalent T-stub model, and (c) T-stub geometry.

4   Case Study

The case study presented herein focuses on the bolted T-stub component (Fig. 1a), which corresponds to the tension zone in beam-to-column connections. The T-stub component comprises two t-shape profiles tied by their flanges with one or more rows of bolt (Fig. 1b). The tensile load applied to the web is transferred by the flange in bending and by the bolts in tension. During this process, the contact between the flanges produces a prying action that increases the forces developed in the bolts. The contact area, as well as the pressure magnitude, evolves during the loading process, complicating an adequate evaluation of the force-displacement response. Non-linear material laws, large deformations, and the existence of different failure patterns also present further challenges to calculating the T-stub component.

Numerical approaches such as the FE method constitute a reliable tool for assessing steel connections. Figure 2 shows the results of one simulation with an advanced FE model of the T-stub component [4]. The FE model includes

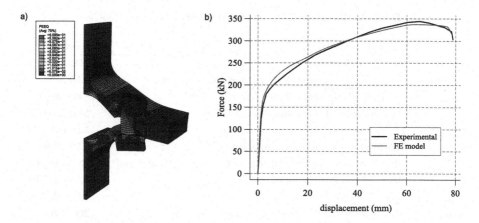

Fig. 2. Advanced FE model of the T-stub bolted component. a) FE simulation: equivalent plastic strain (PEEQ); b) Force-displacement response: FE model vs. Experimental test.

complete stress-strain nonlinear material relationships and a refined characterization of the bolt, including threaded length, nut, and washers. Additionally, the main novelty of the numerical model is the implementation of a continuum damage mechanics model to simulate the failure of the bolted connection. Thus, the force-displacement response of the T-stub can be fully characterized, from the initial stiffness up to the fracture point (Fig. 2b). Interested readers can refer to [4] for an in-depth description of the FE model.

Finally, in this study, the objective was to improve on previous experiments [5,11] by creating more precise and parsimonious models to predict three key parameters of the T-stub force-displacement curve response: initial stiffness (k_i), maximum strength (F_u) and displacement at failure (d_f).

5   Experiment and Results Discussion

The advanced FE model constituted an excellent tool for the generation of a training and testing dataset. Thereby, training data was created with 820 FE simulations from a Design of Computer Experiments (DoCE). DoCE accounts for the deterministic nature of computer experiments, assuming that numerical noise is negligible. For these cases, space-filling sampling techniques are appropriate because they uniformly distribute the points over the entire design space. One of the most widely used space-filling designs is the Latin hypercube sampling (LHS), introduced by McKay et al. in 1979 [10]. LHS divides each input into n equally probable intervals and selects a random value in each interval. The principal advantage of this method is that each input variable is represented in every division of its range. Additionally, a test dataset consisting of 76 samples was generated separately to check the accuracy and generalization capacity of models to predict unseen data.

Table 1. Ranges of the input features included in the DoCE

Variable	Description [units]	Range
$dbolt$	Nominal bolt diameter [-]	M12–M27
$clearance$	Difference between bolt hole and bolt diameter [mm]	0.50–3.50
t_{flange}	Flange thickness of the T-shape profile [mm]	8.00–30.00
t_{web}	Web thickness of the T-shape profile [mm]	5.00–20.00
L_{flange}	Flange length of the T-shape profile [mm]	52.00–180.00
r	Flange-to-web connection radius [mm]	9.75–43.00
n	Dist. from center of the bolt hole to free edge of flange [mm]	15.75–106.00
b	Width of the T-shape profile [mm]	42.00–187.00
L_{thread}	Thread length of the bolt [mm]	2.50–60.25
σy	Yield strength of the structural steel [MPa]	200–400
σ_u	Stress at the maximum tensile load of the structural steel [MPa]	300–800
E_h	Strain-hardening coefficient of the structural steel [MPa]	1000–3000
σ_{yb}	Yield strength of the bolt steel [MPa]	640–1098
σ_{ub}	Stress at the maximum tensile load of the bolt steel [MPa]	800–1200
ϵ_{ub}	Strain at the maximum tensile load of the bolt steel [-]	0.07–0.14

LHS method was used to define the input values of the subsequent FE simulations with different conditions. Table 1 describes the feasible ranges of T-stub geometrical parameters (Fig. 1c) and the mechanical properties of the hot-rolled profiles and bolts used herein. For each combination of input values, a FE simulation was conducted to characterize the response of the T-stub component. Regarding the outputs, the performance of models was evaluated for their prediction of three key parameters of the force-displacement curve: initial stiffness (k_i), maximum strength (F_u) and displacement at failure (d_f).

PSO-PARSIMONY was implemented by the authors in R language. The caret package with nnet method was selected to train and validate multilayer perceptron models (MLP) with a 5-repeated 10-fold cross validation. Every model was optimized by means of the modified PSO for each output variable. In addition, the GAparsimony package [9] was used to obtain results with GA-PARSIMONY method.

Table 2 shows a summary of the PSO-PARSIMONY settings. A description of these parameters and their default values have been described in Sect. 3. The GA-PARSIMONY settings were similar to previous experiments in [5], but with a population size of $P = 40$, a maximum number of generations of $G = 200$, and an early stopping of 35: the same as in the PSO-PARSIMONY experiments.

Experiments were implemented in two separately dual 40-core servers from the University of La Rioja.

Table 3 shows a comparative of PSO-PARSIMONY versus GA-PARSIMONY using MLPs and with three values of *tol*: 0.01, 0.005, 0.001. *tol* corresponds to the margin between the absolute difference of two $J = RMSE_{val}^{norm}$ to be considered similar. In this experiments, $RMSE_{val}^{norm}$ was the validation error of

Table 2. Setting parameters for PSO-PARSIMONY and MLP searching space

Algorithm	Parameter	Values
φ_1	Cognitive learning rate	$0.5 + ln\,(2)$
φ_2	Social learning rate	$0.5 + ln\,(2)$
D	Number of input features	15
s	Number of swarm particles	40
$iter$	Number of iterations in optimization process	200
$early_s topping$	Num of iterations for early stopping	35
tol	Percentage of error to consider J similar	5%
α	Lower feature's probability of being considered in the input subset	0.6
$\omega_{max}, \omega_{min}$	Parameters to calculate inertia weights	$0.9, 0.4$
K	Number of particles to be informed for each one	3
m	Mutation rate	$1/15$
$k - fold$	Number of folds in Repeated-CV	10
$runs$	Number of run in Repeated-CV	5
MLP	Number of hidden neurons (search space)	$[1, 2, ..., 25]$
	Weight decay (search space)	$10^{[-6.99, 6.99]}$

Table 3. Performance of the best MLP model obtained from 10 runs of PSO-PARSIMONY and GA-PARSIMONY with different *tol* values and three different targets: initial stiffness (k_i), maximum strength (F_u), and displacement at failure (d_f) targets.

Target	Method-*tol*	$RMSE_{val}^{norm}$	$RMSE_{tst}$	N_{Iters}	N_{FS}	NH_{net}	$\sum w^2$	Time[min]
k_i	PSO-0.01	**0.0518**	**5.9120**	173	9	14	124.88	181
	GA-0.01	0.0601	8.6535	**37**	**8**	**11**	**104.35**	**45**
	PSO-0.005	**0.0532**	**6.4780**	84	9	17	109.74	114
	GA-0.005	0.0539	8.1935	**41**	9	**14**	**105.42**	**58**
	PSO-0.001	0.0514	8.5613	104	12	**15**	121.04	177
	GA-0.001	**0.0504**	**8.5108**	**75**	**11**	16	**119.83**	**124**
F_u	PSO-0.01	**0.0646**	**10.7591**	129	10	17	**66.30**	187
	GA-0.01	0.0689	11.4850	**40**	**9**	**12**	73.35	**51**
	PSO-0.005	**0.0626**	**10.5941**	172	11	**12**	117.02	251
	GA-0.005	0.0636	11.8633	**43**	**10**	13	**73.50**	**60**
	PSO-0.001	0.0628	**11.1864**	152	13	**13**	**67.32**	217
	GA-0.001	**0.0585**	11.3539	**53**	**12**	14	99.52	**88**
d_f	PSO-0.01	**0.2560**	6.1678	98	12	14	161.73	102
	GA-0.01	0.2637	**4.6624**	**41**	**11**	**12**	**159.46**	**52**
	PSO-0.005	0.2573	6.2690	200	13	**11**	147.73	243
	GA-0.005	**0.2600**	**5.0806**	**69**	**12**	**11**	**111.86**	**82**
	PSO-0.001	0.2627	**4.7767**	85	15	**11**	**115.45**	117
	GA-0.001	**0.2540**	5.3040	**50**	**12**	12	164.05	**70**

the standardized $log(target)$. Logarithmic was used to transform the output to be close to a normal distribution. Therefore, results improved upon the previous experiments in [5]. Additionally, in Table 3, $RMSE_{tst}$ shows the non-normalized testing error, N_{FS} the number of features, N_{Iters} the number of PSO and GA iterations, NH_{net} the number of neurons in the hidden layer, $\sum w^2$ the sum of the MLP squared weights, and $Time[min]$ (in minutes) needed to obtain the best solution. The results shows the best model obtained from 10 different runs of PSO-PARSIMONY and GA-PARSIMONY, respectively. Bold numbers indicate the best value for each tol value and target variable considering the results of both methodologies.

For k_i with $tol = 0.01$, and Fu with $tol = 0.005$, PSO-PARSIMONY obtained better solutions with lower $RMSE_{val}^{norm}$ and $RMSE_{tst}$, respectively, but the computational effort increased and PSO added a feature to N_{FS} in some cases. This can be explained by the fact that the PSO search for the best model used MLPs with a higher number of neurons. Therefore, the PSO search converged later but was more diverse and precise. With $tol = 0.001$, both methodologies obtained better $RMSE_{val}^{norm}$ but with more complex models that produced higher $RMSE_{tst}$. However, the results with displacement at failure (d_f) were different. This measurement is only slightly deterministic and difficult to predict. This can be observed in the low correspondence between $RMSE_{val}^{norm}$ and $RMSE_{tst}$ (lower validation error in PSO is lower test error in GA, or vice versa). Therefore, GA-PARSIMONY obtained a lower $RMSE_{tst}$ using $tol = 0.01$ with more parsimonious and robust models, although its $RMSE_{val}^{norm}$ was worse than in PSO-PARSIMONY.

6   Conclusions

This study has demonstrated that this novel PSO methodology is a promising option for searching for accurate and parsimony machine learning solutions. The present proposal includes a modified version of the PSO algorithm to simultaneously tune model parameters and select the most influential input features. The foremost innovation of this method is the strategy for updating both *pbest* and *gbest* according to a complexity criterion. This enabled us to construct models that were not only accurate, but parsimonious as well.

Our experiments have demonstrated that PSO-PARSIMONY obtain more accurate models than our previous GA-PARSIMONY methodology in deterministic variables, but there is a significant increase in computational effort and a reduction in model parsimony. Nevertheless, more experiments with PSO-PARSIMONY and hybrid strategies are necessary to reduce the computational effort and improve performance in the search for accurate and parsimonious solutions.

Acknowledgement. We are greatly indebted to *Banco Santander* for the REGI2020/41 fellowship. This study used the Beronia cluster (Universidad de La Rioja), which is supported by FEDER-MINECO grant number UNLR-094E-2C-225.

References

1. Ahila, R., Sadasivam, V., Manimala, K.: An integrated PSO for parameter determination and feature selection of ELM and its application in classification of power system disturbances. Appl. Soft Comput. **32**, 23–37 (2015)
2. Clerc, M.: Stagnation Analysis in Particle Swarm Optimisation or What Happens When Nothing Happens, p. 17, December 2006
3. Eberhart, R., Shi, Y.: Comparing inertia weights and constriction factors in particle swarm optimization. In: Proceedings of the 2000 Congress on Evolutionary Computation. CEC00 (Cat. No.00TH8512), vol. 1, pp. 84–88 (2000). https://doi.org/10.1109/CEC.2000.870279
4. Fernandez-Ceniceros, J., Sanz-Garcia, A., Antoñanzas-Torres, F., Martinez-de Pison, F.J.: A numerical-informational approach for characterising the ductile behaviour of the t-stub component. Part 1: Refined finite element model and test validation. Eng. Struct. **82**, 236–248 (2015). https://doi.org/10.1016/j.engstruct.2014.06.048
5. Fernandez-Ceniceros, J., Sanz-Garcia, A., Antoñanzas-Torres, F., Martinez-de Pison, F.J.: A numerical-informational approach for characterising the ductile behaviour of the t-stub component. Part 2: Parsimonious soft-computing-based metamodel. Eng. Struct. **82**, 249–260 (2015). https://doi.org/10.1016/j.engstruct.2014.06.047
6. Kennedy, J., Eberhart, R.: Particle swarm optimization. In: Proceedings of ICNN 1995 - International Conference on Neural Networks, vol. 4, pp. 1942–1948 (1995). https://doi.org/10.1109/ICNN.1995.488968
7. Li, H., Shu, D., Zhang, Y., Yi, G.Y.: Simultaneous variable selection and estimation for multivariate multilevel longitudinal data with both continuous and binary responses. Comput. Stat. Data Anal. **118**, 126–137 (2018). https://doi.org/10.1016/j.csda.2017.09.004
8. Ma, B., Xia, Y.: A tribe competition-based genetic algorithm for feature selection in pattern classification. Appl. Soft Comput. **58**, 328–338 (2017)
9. Martinez-de-Pison, F.J.: GAparsimony: Searching Parsimony Models with Genetic Algorithms (2019). https://CRAN.R-project.org/package=GAparsimony. R package version 0.9.4
10. McKay, M.D., Beckman, R.J., Conover, W.J.: Comparison of three methods for selecting values of input variables in the analysis of output from a computer code. Technometrics **21**(2), 239–245 (1979). https://doi.org/10.1080/00401706.1979.10489755
11. Pernía-Espinoza, A., Fernandez-Ceniceros, J., Antonanzas, J., Urraca, R., Martinez-de Pison, F.J.: Stacking ensemble with parsimonious base models to improve generalization capability in the characterization of steel bolted components. Appl. Soft Comput. **70**, 737–750 (2018). https://doi.org/10.1016/j.asoc.2018.06.005
12. Martinez-de Pison, F.J., Ferreiro, J., Fraile, E., Pernia-Espinoza, A.: A comparative study of six model complexity metrics to search for parsimonious models with GAparsimony R package. Neurocomputing (2020). https://doi.org/10.1016/j.neucom.2020.02.135

13. Martinez-de Pison, F.J., Gonzalez-Sendino, R., Aldama, A., Ferreiro-Cabello, J., Fraile-Garcia, E.: Hybrid methodology based on Bayesian optimization and GA-parsimony to search for parsimony models by combining hyperparameter optimization and feature selection. Neurocomputing **354**, 20–26 (2019). https://doi.org/10.1016/j.neucom.2018.05.136. Recent Advancements in Hybrid Artificial Intelligence Systems

14. Sanz-Garcia, A., Fernandez-Ceniceros, J., Antonanzas-Torres, F., Pernia-Espinoza, A., Martinez-de Pison, F.J.: GA-parsimony: a GA-SVR approach with feature selection and parameter optimization to obtain parsimonious solutions for predicting temperature settings in a continuous annealing furnace. Appl. Soft Comput. **35**, 13–28 (2015). https://doi.org/10.1016/j.asoc.2015.06.012

15. Shi, Y., Eberhart, R.: A modified particle swarm optimizer. In: 1998 IEEE International Conference on Evolutionary Computation Proceedings. IEEE World Congress on Computational Intelligence (Cat. No.98TH8360), pp. 69–73 (1998). https://doi.org/10.1109/ICEC.1998.699146

16. Shi, Y., Eberhart, R.: Empirical study of particle swarm optimization. In: Proceedings of the 1999 Congress on Evolutionary Computation-CEC99 (Cat. No. 99TH8406), vol. 3, pp. 1945–1950 (1999). https://doi.org/10.1109/CEC.1999.785511

17. Urraca, R., Sodupe-Ortega, E., Antonanzas, J., Antonanzas-Torres, F., Martinez-de Pison, F.J.: Evaluation of a novel GA-based methodology for model structure selection: the GA-PARSIMONY. Neurocomputing **271**, 9–17 (2018). https://doi.org/10.1016/j.neucom.2016.08.154

18. Vieira, S.M., Mendonza, L.F., Farinha, G.J., Sousa, J.M.: Modified binary PSO for feature selection using SVM applied to mortality prediction of septic patients. Appl. Soft Comput. **13**(8), 3494–3504 (2013)

19. Vieira, S.M., Mendonça, L.F., Farinha, G.J., Sousa, J.M.: Modified binary PSO for feature selection using SVM applied to mortality prediction of septic patients. Appl. Soft Comput. **13**(8), 3494–3504 (2013). https://doi.org/10.1016/j.asoc.2013.03.021

20. Wan, Y., Wang, M., Ye, Z., Lai, X.: A feature selection method based on modified binary coded ant colony optimization algorithm. Appl. Soft Comput. **49**, 248–258 (2016)

21. Wang, M., et al.: Toward an optimal kernel extreme learning machine using a chaotic moth-flame optimization strategy with applications in medical diagnoses. Neurocomputing **267**, 69–84 (2017). https://doi.org/10.1016/j.neucom.2017.04.060

22. Wei, J., et al.: A BPSO-SVM algorithm based on memory renewal and enhanced mutation mechanisms for feature selection. Appl. Soft Comput. **58**, 176–192 (2017)

23. Zambrano-Bigiarini, M., Clerc, M., Rojas, R.: Standard particle swarm optimisation 2011 at CEC-2013: a baseline for future PSO improvements. In: 2013 IEEE Congress on Evolutionary Computation, pp. 2337–2344 (2013). https://doi.org/10.1109/CEC.2013.6557848

24. Zimmer, L., Lindauer, M., Hutter, F.: Auto-pytorch tabular: multi-fidelity metalearning for efficient and robust AutoDL. IEEE Trans. Pattern Anal. Mach. Intell. 1–12 (2021). Arxiv, IEEE Early Access, to appear

Content-Based Authorship Identification for Short Texts in Social Media Networks

José Gaviria de la Puerta[1]([✉]) [iD], Iker Pastor-López[1] [iD],
Javier Salcedo Hernández[1], Alberto Tellaeche[1] [iD], Borja Sanz[1] [iD],
Hugo Sanjurjo-González[1] [iD], Alfredo Cuzzocrea[2] [iD], and Pablo G. Bringas[1] [iD]

[1] Faculty of Engineering, University of Deusto,
Avda Universidades 24, 48007 Bilbao, Spain
{jgaviria,iker.pastor,javiersalcedo,alberto.tellaeche,borja.sanz,
hugo.sanjurjo,pablo.garcia.bringas}@deusto.es
[2] University of Calabria, Rende, Italy
alfredo.cuzzocrea@unical.it

Abstract. Today social networks contain a high number of false profiles that can carry out malicious actions on other users, such as radicalization or defamation. This makes it necessary to be able to identify the same false profile and its behaviour on different social networks in order to take action against it. To this end, this article presents a new approach based on behavior analysis for the identification of text authorship in social networks.

Keywords: Authorship identification · Word2Vec · Online social networks · Content-based approach

1 Introduction

Today, online social networks (OSNs) are an important factor in people's lives. In recent years, the appearance of these networks has made them the main means of communication in personal relationships, making them present in every aspect of our society, from interpersonal relationships, as in labor relations with companies.

According to Ellison et al. [1], a social network must allow us to do three main things: i) create a profile that defines and represents a person, ii) present a list of different profiles with which to interact, and iii) be able to observe what actions other profiles have performed. In short, a social network is a channel for communication between people who are represented by virtual profiles.

But as in any technology, a series of problems can arise, for example, in the case that a profile does not really represent the real person. A clear example of this is when a false profile could commit a crime of impersonation, defamation,

The work presented in this paper was supported by the European Commission under contract H2020-700367 DANTE.

H. Sanjurjo González et al. (Eds.): HAIS 2021, LNAI 12886, pp. 27–37, 2021.
https://doi.org/10.1007/978-3-030-86271-8_3

cyberbullying or even radicalization. In order to combat these crimes and bring those responsible to justice, false profiles must be linked to their true owners by studying their behaviour. More specifically, the common approach is to find, in the same OSN, the real profile that matches the behaviour of the false profile. This means that it must be assumed that the person behind the false profile has his real profile in the same OSN.

To define this behavior, different approaches are valid. A classic approach is the study of the devices used by the investigated profile, as well as the time period in which the crime was committed. Although it is effective when the offender does not bother to hide, often this information is not enough to find the real profile. This necessitates the inclusion of published texts in the analysis.

In this way, the identification of the perpetrator has begun to be a priority in the processing of modern natural language. This area of research in computer science and artificial intelligence is responsible of the manipulation of large amounts of natural language data in an automated manner. More specifically, as part of natural language processing, authorship identification is the task of identifying the author of a particular text (or set of texts) from a list of possible candidates. From a technical point of view, it can be considered a classification task in which the class or label of a given text (observation) is the author who wrote it.

The main challenge of this task is to extract the characteristics of the texts themselves in order to capture the writing style or tendencies of a given author. Efforts have been made to determine the best approach for extracting those characteristics. Commonly, the characteristics extracted to accomplish this task are stylistic characteristics. Content-based approaches (semantic characteristics) are not common in the identification of authors. In fact, such approaches are generally used for subject classification tasks [2] or sentiment analysis [3]. In addition, it is remarkable that most experiments use large texts [4,5], in which it is easier to extract stylistic traits. Short texts present difficulties in this task of classification with classical approaches. As some social networks restrict the number of characters used in the same message, such as Twitter, it is necessary to find new techniques to overcome these difficulties. Likewise, new approaches can be complementary to those considered avant-garde. They can reduce the size of candidate lists to achieve greater accuracy with other algorithms, as well as to eliminate unnecessary workload.

This article proposes a content-based identification of authorship for short texts. In other words, rather than relying on a particular author's writing style, the bet is made by assuming his or her tendency to write on the same topics and using the same or similar vocabulary.

The rest of this document is organized as follows. The Sect. 2 describes the approach designed for this document. The Sect. 3 shows the steps we took during the experimentation, and also presents the data set and how it is composed and the different methods we use for the process of identifying authorship. The Sect. 4 presents the results obtained from this approach, and discusses them. Finally, the Sect. 5, presents the conclusions that we obtained with this experimentation.

2 Proposed Approach

The main idea underlying the proposed approach is that a user profile (or author) can be defined by the content he or she has written about. Users tend to write about the same topics. People tend to be recurrent. In social networks such as Twitter, it occurs even more frequently than in other media. This approach relies on this tendency in order to discard candidates that do not match with the topics extracted from the text of the subject under investigation or, in some occasions where the data is relevant enough, to identify the author behind the fake profile. In the same way, group users by content is useful to find relations between them. This approach aims to generate these relations too, so the people working on a certain list of candidates can receive useful information. Definitely, the methodology reflected in this paper proposes to define a user by semantic features, to then carry out a process of authorship identification.

3 Experiment

This section shows the steps that we did during the experimentation, and also, present the dataset and how it is composed and the different methods that we used for the authorship identification process.

3.1 Data Gathering

We wanted to develop a solution for constrained environments, such as the one presented in [6], where the number of users under investigation is small. For simulating such environment, we chose 40 public user profiles of the OSN Twitter. They all had a great number of published tweets and likes, as well as a huge number of follows and followers (profiles to interact with).

It is remarkable that our goal was not only to identify an author from a list of candidates, but to relate authors by the content and vocabulary used (semantic features) so the list of candidates can be reduced. That is why these users form a heterogeneous group from two different points of view: the topics they write about and the language they use. Specifically, these users write in three different languages: spanish, catalan and basque. The topics they write about are more numerous: politics, sports and humor, among others. These topics can easily be divided in subtopics (e.g. politics, right wing politics). Regarding this, it is remarkable that each user has been labeled with its main topic, subtopic and language. Although all the selected user profiles were public accounts, they will be anonymised for privacy purposes. We will refer to each user in this paper as User[Number] (e.g. User32). In addition, topics, subtopics and languages will be also anonymised.

All the tweets from the selected user profiles were downloaded by a web crawler system developed by DeustoTech Industry unit for the DANTE[1] project. This project, in which DeustoTech Industry is actively involved, has received funding from the European Union's Horizon 2020 Research and Innovation program. The mentioned web crawler system will be part of a set of tools provided by the project to support law enforcement authorities in counter terrorism activities.

The total amount of downloaded tweets was 142000. We did not keep proportions between users, although all the downloaded tweets were published in the same period of time (2015–2018). From these tweets, 127800 (90%) were assigned to the training set of the model. The rest 14200 tweets (10%) were assigned to the testing set of the model. Again, we did not keep proportions between users, what ensures the robustness of the results. The dataset described in this section is summarized in Table 1.

3.2   Data Pre-processing

We applied a tokenizer to all the tweets downloaded by the system. A tokenizer divides texts into tokens by separator characters (e.g. space). This is often required in natural language processing tasks [7]. Apart from the tokenisation process, stop words[2], user mentions, URLs and other meaningless strings were removed from the downloaded tweets. For instance, the tweet "I love the film I watched yesterday at the cinema with @Markus87:)" would transform to an array of strings such as $\{love, film, watched, cinema\}$. This helps not only the training process of the algorithms by decreasing the workload and fitting its input rules, but it also helps the classification task by reducing the number of meaningless words.

3.3   Word Embeddings

Word embedding is a technique in natural language processing (NLP) where words are assigned to vectors of real numbers. **Word2vec** has been our choice to produce these word embeddings.

Word2vec is a set of related models which produce word embeddings by training two-layer neural networks that aim to reconstruct linguistic contexts of words [8]. This means that, it allows to find not only syntactic relationships between words, but also semantic relationships between them. Semantic features

---

[1] The DANTE (Detecting and analysing terrorist-related online contents and financing activities) project aims to deliver more effective, efficient, automated data mining and analytics solutions and an integrated system to detect, retrieve, collect and analyse huge amount of heterogeneous and complex multimedia and multi-language terrorist-related contents, from both the Surface and the Deep Web, including Dark nets. More information at http://www.h2020-dante.eu/.

[2] Stop words are the most common words in a language, which are normally filtered out during the pre-processing step of a natural language processing experiment.

Table 1. Dataset.

User	Topic	Subtopic	Language	Published tweets	Likes	Follows	Followers	Start date	Crawl date
User1	T0	S01	L0	29,900	87700	1,000	258,000	08/2,012	15/10/18
User3	T0	S01	L0	11,200	37	192	368,000	12/2,009	15/10/18
User2	T0	S02	L0	6,416	2,654	2,222	1,130,000	03/2,011	15/10/18
User9	T0	S02	L0	4,049	1,822	178	211,000	08/2,013	15/10/18
User23	T0	S03	L0	7,294	529	2,976	38,600	02/2,015	16/10/18
User4	T1	S11	L0	3,677	40	722	15,500,000	06/2,010	15/10/18
User11	T1	S12	L0	2,992	532	324	802,000	06/2,010	15/10/18
User12	T1	S12	L0	2,148	489	150	17,100	05/2,011	15/10/18
User13	T1	S12	L0	699	1,669	136	63,500	01/2,013	15/10/18
User5	T1	S12	L0	2,840	2,074	656	5,733	10/2,011	15/10/18
User6	T1	S12	L0	3,298	3,678	171	62,000	04/2,011	15/10/18
User8	T1	S12	L0	4,237	1,909	422	44,000	07/2,011	15/10/18
User7	T1	S13	L0	1,844	10	89	23,700,000	11/2,009	15/10/18
User10	T1	S14	L0	1,959	453	193	11,700,000	03/2,011	15/10/18
User14	T2	S21	L0, L2	58,900	29,300	2,080	56,800	04/2,008	15/10/18
User15	T2	S21	L0, L2	2,965	1,567	286	317,000	02/2,014	15/10/18
User16	T2	S21	L0, L2	12,700	2,299	33,700	128,000	08/2,011	15/10/18
User17	T2	S21	L0, L2	11,100	6,753	1,805	73,500	09/2,010	15/10/18
User18	T2	S22	L0	89,700	18,500	9,906	545,000	05/2,008	15/10/18
User19	T2	S22	L0	24,100	4,041	6,133	921,000	08/2,009	15/10/18
User20	T2	S22	L0	18,700	13,200	2,161	811,000	02/2,012	15/10/18
User24	T2	S22	L0	30,500	7,255	10,400	30,500	11/2,008	16/10/18
User25	T2	S22	L0	19,600	6,668	2,699	2,260,000	06/2,010	16/10/18
User27	T2	S22	L0	94,400	1,464	1,517	1,340,000	01/2,014	16/10/18
User28	T2	S22	L0	88,000	5,075	137,000	634,000	06/2,009	15/10/18
User29	T2	S22	L0	56,100	413	1,175	1,030,000	01/2,008	16/10/18
User32	T2	S22	L0	14,700	29	1,621	18,500	10/2,011	16/10/18
User33	T2	S22	L0	1,789	55	775	172,000	10/2,012	16/10/18
User34	T2	S22	L0	6,520	1,647	1,958	431,000	03/2,009	16/10/18
User40	T2	S22	L0, L2	45,700	698	3,790	91,200	09/2,009	16/10/18
User21	T2	S23	L0	8,885	296	1,139	206,000	11/2,010	15/10/18
User22	T2	S23	L0	53,700	11,300	2,536	1,070,000	01/2,010	15/10/18
User26	T2	S23	L0	2,471	556	1,135	454,000	04/2,012	16/10/18
User30	T2	S23	L0	76,300	33,200	4,432	68,300	02/2,009	16/10/18
User31	T2	S23	L0	42,800	2	377	300,000	12/2,008	16/10/18
User35	T2	S23	L0	120,000	5,180	93,900	484,000	01/2,009	16/10/18
User36	T2	S23	L0	8,684	610	845	50,300	09/2,011	16/10/18
User37	T2	S24	L0, L1	6,430	2,944	918	23,100	06/2,010	16/10/18
User38	T2	S24	L0, L1	28,300	181	64	27,200	10/2,009	16/10/18
User39	T2	S24	L0, L1	5,141	10	29	30,400	06/2,011	16/10/18

can hardly be captured by classical word embedding related techniques as tf-idf [9] or n-grams [10], what makes the use of Word2vec necessary.

Word2vec's implementation can utilize two model architectures to produce those embeddings: continuous bag-of-words (CBOW) or continuous skip-gram. The CBOW architecture predicts the current word based on the context, and the skip-gram predicts surrounding words given the current word. **Continuous**

skip-gram has been our choice. In models using large corpora and a high number of dimensions, the skip-gram model yields the highest overall accuracy, and consistently produces the highest accuracy on semantic relationships, as well as yielding the highest syntactic accuracy in most cases [8]. Since our approach is purely content based, we rely on the meaning of the words to find synonyms (semantic features), the skip-gram model was the best option to choose.

As said before, the skip-gram model predicts the context from a given word. Formally, it means that the skip-gram model's objective is to maximize the average log probability [11], which is shown in formula 1, where $\omega_1, \omega_2, .., \omega_n$ are the sequence of training words (e.g. a sentence) and c is the context window size.

$$\frac{1}{N} \sum_{n=1}^{N} \sum_{-c<i<c, i\neq 0} \log \rho(\omega_{n+i} \mid \omega_n) \tag{1}$$

The basic definition of $\rho(\omega_{n+i} \mid \omega_n)$ by means of a softmax function is shown in formula 2, where $v\prime_\omega$ and v_ω are the input and output vectors of ω and W is the total number of words in the vocabulary [11].

$$\rho(\omega_O \mid \omega_I) = \frac{exp(v\prime_{\omega_O}{}^\top v_{\omega_I})}{\sum_{w=1}^{W} exp(v\prime_\omega{}^\top v_{\omega_I})} \tag{2}$$

Nevertheless, this formulation of $\rho(\omega_{n+i} \mid \omega_n)$ is not efficient when the vocabulary is large enough. For solving this, it is remarkable that we used **hierarchical softmax** as proposed by Mikolov et al. [11] and introduced for the first time in academia by Morin and Bengio [12]. In hierarchical softmax the output layer is represented as a binary tree with W words in the vocabulary as its leaf nodes and the intermediate nodes as internal values. Each node shows the relative probabilities of its child nodes. There are W-1 intermediate nodes. To estimate the probability of a word represented in a leaf node, there exists a distinct path from root node. This approach reduces the complexity to obtain probability distribution from O(N) to O(log (N)). As given by Mikolov et al. in [11], hierarchical softmax defines $\rho(\omega_{n+i} \mid \omega_n)$ by formula 3 where $\sigma(x) = 1/(1 + exp(-x))$. In addition, $ch(n)$ is the left child of n and $[x]$ is a special function defined in formula 4.

$$\rho(\omega \mid \omega_I)$$
$$= \prod_{j=1}^{L(\omega)-1} \sigma([n(w, j+1) = ch(n(w, j))] \cdot v\prime_{n(w,j)}{}^\top v_{\omega_I}) \tag{3}$$

$$|x| = \begin{cases} 1 & \text{if x is true;} \\ -1 & \text{otherwise.} \end{cases} \tag{4}$$

The results of this model are vectors of real numbers $w = (v_1, v_2, \ldots, v_{n-1}, v_n)$ that act as useful word representations for predicting the surrounding words in a sentence.

Apart from the selection of the model, Word2vec has a list of parameters that must be configured to achieve good results:

- **Vector size.** The vector size determines the dimensions of the generated vectors. The higher the vector size, the better the quality of the vectors.
- **Context window size.** The size of the context window sets how many words before and after of a given word would be included in the prediction. Both architectures (CBOW and skip-gram) use this parameter.
- **Minimum count.** The minimum frequency of a word in the training set to compute its vector.
- **Learning rate.** The value which determines how much adjust the weights of the underlying neural networks with respect the loss gradient.

Except the vector size that was set to 400, due to the recommendations in [8], the rest of the parameters where the default ones in Apache Spark's Machine Learning library[3] (version 2.1.1) implementation of the described model. This implementation was used to perform the experiment.

3.4 Tweet Mean Vectors and User Mean Vectors

Word2vec's output can be considered a dictionary where each word that the model has received for training is related to a vector of real numbers. This dictionary allows to find synonyms (context related words) by using the cosine similarity measure[4]. So the word *car* should be synonym of the word *automobile*, as they share the same or similar contexts.

Our idea was to define text context, rather than word context. For achieving this, we considered a tweet the mean vector of all its words, which can be defined as $T = \frac{1}{n}(w_1 + w_2 + \cdots + w_{n-1} + w_n)$ where n is the number of words of a given tweet T that exist in a previously trained model M. Words that have been filtered out by the model are discarded. If the model has not received a certain word (there is no vector for it in the dictionary), it is also discarded. The mean is computed by the rest of existing words in the tweet. In case of discarding all the words, the tweet is also discarded.

This technique is a good approach for texts with few words (e.g. tweets), being able to find synonyms between texts by using the cosine similarity measure. For instance, the text "I love my red car" would be synonym of the text "I bought a red automobile", as they have a word in common (red) and a synonym (car, automobile). Additionally, the rest of the words are normally erased during the pre-processing step (which helps the search of the synonym) as they are considered stop words. Figure 1 illustrates our approach. Similar work has been done before by averaging word vectors to define a text in the context of sentiment analysis [13]. In [14], Foltz et al. also describe the process of averaging word vectors to find similarities between texts, ensuring that they were able to do discourse segmentation (detect topic shifts).

---

[3] More information at https://spark.apache.org/mllib/.

[4] The cosine similarity is a measure that calculates the cosine of the angle between two vectors (orientation). It is commonly used for measuring the similarity between two documents represented in a normalized vector space model.

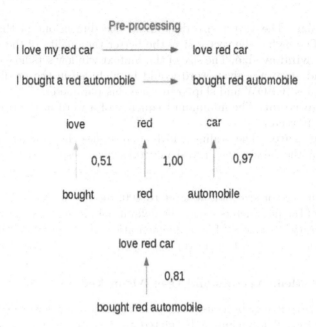

Fig. 1. Definition of text context by averaging word vectors. Use of cosine similarity. (Color figure online)

Equivalently, we considered a user the mean vector of all its tweets, which can be defined as $U = \frac{1}{n}(T_1 + T_2 + \cdots + T_{n-1} + T_n)$ where n is the number of tweets of a given user U. This technique works well to find user profiles clusters (unsupervised learning), but not that well for finding synonyms (supervised learning).

3.5 Supervised Learning

Once the tweet mean vectors are generated, the authorship identification begins. For predicting an author, the **k-nearest neighbors algorithm** has been used. This algorithm is a classic method used for classification and regression in machine learning. As this part of the experimentation consists on assigning a discrete value (author) to a set of tweets, we are in front of a classification problem.

This algorithm commonly uses the euclidean distance for finding the *k-nearest points* to the target that has to be labeled. In our case, we used the cosine similarity measure to find those *k-nearest points*. The value used to label the target is the class belonging to the majority of the *k-nearest points*. The points are the tweet mean vectors in the training set. The targets that have to be labeled are the tweet mean vectors from the testing set of the author to predict. The class is the author who wrote the tweet.

After labeling all the tweet mean vectors from the testing set of the author to predict, a list with the proposed candidates is generated. Each candidate in the list has a probability percentage assigned. This percentage is calculated by dividing the number of tweets assigned to the candidate by the total number of tweets. In other words, the sum of all the candidates probability percentages in the list is equal to 100. The first candidate in the list is the predicted author, although a second or third position is not a bad result.

As a reminder, the aim of the experimentation is not only to predict an author, but to relate authors by semantic features. This is why a candidate list is generated. It is remarkable as well that the value chosen for k was 1. With this value of k, this algorithm can be also considered the **nearest neighbor algorithm**.

The validation of the algorithm has been carried out using the classical method of cross-validation. For this purpose, the dataset was randomly divided into 10 folds, trained with 9 folds and validated with one. This process has been performed 10 times and the average results are shown in the result tables.

After making the predictions for the 40 users, we obtained a 87.5% of accuracy (correct user is first candidate) and a 97.5% of optimistic accuracy (correct user is first, second or third candidate). The results not only show that the approach works for this concrete case, but that the hypothesis of the possibility of relating users by semantic features is correct. It is remarkable as well that the candidates lists show that the mainly proposed authors for each user share topics, what demonstrates again that the prediction is based on semantic features (content and vocabulary).

4   Discussion

As can be shown in the Fig. 2, the obtained results have several considerations. First, the higher number post obtained, the better results. Using the reference value (comparing the value with the same user), we can compare the similarity between users. In all cases, the obtained value is the highest. But even with a small number of publications, the obtained data is relevant. For example, User26 has a very low number of publications (compared with other users), but the results are coherent with our known of this account, even the value obtained is low (only 5.26).

The related accounts (similar topics) show the highest values compared with the others, so the proposed system can identify similarity not only in the related topic, but also in the way to communicate. For example, the User1 and the User3 are very close in real life, and User3 has a high similarity value with User1.

Although the described results are positive, more sophisticated techniques can also be used to define text context. As shown in [13], word vector averaging is outperformed by Paragraph Vector in Stanford Sentiment Treebank Dataset. Other approaches such as Recursive Neural Networks or Matrix Vector-RNN [15] do also outperform word vector averaging. This means that even better results could be achieved in the author identification process described in this paper.

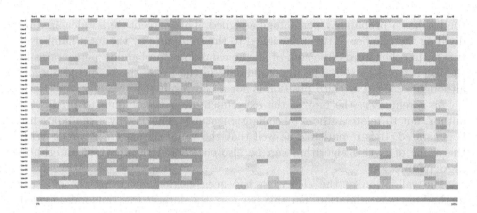

Fig. 2. Heat map of the users and the predictions.

5 Conclusions

This paper describes a new approach than can be complementary to state-of-the-art techniques in author identification processes, showing success in a real use case built from scratch. Until now, no approaches have explored the analysis of semantic features (topics) to identify authors, what means that this new approach opens a new horizon to explore. Likewise, it is remarkable that this approach also aims to generate topic based relations between candidates, what can be useful for different purposes (i.e. recommendation systems). We would like to explore this methodology for such purposes in following papers.

References

1. Boyd, D.M., Ellison, N.B.: Social network sites: definition, history, and scholarship. J. Comput.-Mediat. Commun. **13**(1), 210–230 (2007)
2. Lilleberg, J., Zhu, Y., Zhang, Y.: Support vector machines and word2vec for text classification with semantic features. In: IEEE 14th International Conference on Cognitive Informatics & Cognitive Computing (ICCI* CC), pp. 136–140. IEEE, Beijing (2015)
3. Agarwal, A., and Xie, B., Vovsha, I., Rambow, O., Passonneau, R.: Sentiment analysis of Twitter data. In: Proceedings of the Workshop on Languages in Social Media (LSM 2011), pp. 30–38. Association for Computational Linguistics, Portland (2011)
4. Khonji, M., Iraqi, Y., Jones, A.: Mitigation of spear phishing attacks: a content-based Authorship Identification framework. In: 2011 International Conference for Internet Technology and Secured Transactions, pp. 416–421. IEEE, Abu Dabi (2010)
5. Chunxia, Z., Xindong, W., Zhendong, N., Wei, D.: Authorship identification from unstructured texts. Knowl.-Based Syst. **66**, 99–111 (2014)
6. Galán-García, P., Puerta, J.G.D.L., Gómez, C.L., Santos, I., Bringas, P.G.: Supervised machine learning for the detection of troll profiles in twitter social network: application to a real case of cyberbullying. Log. J. IGPL **24**(1), 42–53 (2016)

7. Webster, J.J., Kit, C.: Tokenization as the initial phase in NLP. In: Proceedings of the 14th Conference on Computational Linguistics, pp. 1106–1110. Association for Computational Linguistics (1992)
8. Mikolov, T., Chen, K., Corrado, G.S., Dean, J.: Efficient estimation of word representations in vector space. arXiv preprint arXiv:1301.3781 (2013)
9. Salton, G., McGill, M.J.: Book Title. McGraw-Hill, Inc. (1986)
10. Cavnar, W.B., Trenkle, J.M.: N-gram-based text categorization. In: Proceedings of SDAIR-94, 3rd Annual Symposium on Document Analysis and Information Retrieval, vol. 161175 (1994)
11. Mikolov, T., Sutskever, I., Chen, K., Corrado, G.S., Dean, J.: Distributed representations of words and phrases and their compositionality. Adv. Neural. Inf. Process. Syst. **26**, 3111–3119 (2010)
12. Morin, F., Bengio, Y.: Hierarchical probabilistic neural network language model. Aistats **5**, 246–252 (2005)
13. Le, Q., Mikolov, T.: Distributed representations of sentences and documents. In: 31st International Conference on Machine Learning, Beijing, pp. 1188–1196 (2014)
14. Foltz, P.W., Kintsch, W., Landauer, T.K.: The measurement of textual coherence with latent semantic analysis. Discour. Process. **25**(2,4), 285–307 (1998)
15. Socher, R., et al.: Recursive deep models for semantic compositionality over a sentiment treebank. In: Proceedings of the 2013 Conference on Empirical Methods in Natural Language Processing, pp. 1631–1642. Association for Computational Linguistics, Seattle (2013)

A Novel Pre-processing Method for Enhancing Classification Over Sensor Data Streams Using Subspace Probability Detection

Yan Zhong[1] , Tengyue Li[1] , Simon Fong[1] , Xuqi Li[2],
Antonio J. Tallón-Ballesteros[3](✉) , and Sabah Mohammed[4]

[1] University of Macau, Taipa, Macau SAR, China
[2] University of Edinburgh, Edinburgh, Scotland
[3] University of Huelva, Huelva, Spain
antonio.tallon@diesia.uhu.es
[4] Lakehead University, Thunder Bay, Canada

Abstract. The rapid development of the Internet of Things has led to the widespread use of sensors in everyday life. Large amounts of data through sensing devices are collected. The data quantity is massive, but most of the data are repetitive and noisy. When traditional classification algorithms are used for classifying sensor data, the performance of the model is often poor because the classification granularity is too small. In order to better data mine the knowledge from the Internet of Things data which is a kind of big data, a new classification model based on subspace probability detection is proposed. This model can be well integrated with traditional data mining algorithms, and the performance on sensor data mining is greatly improved.

Keywords: Internet-of-things · Data pre-processing · Sensor data streams · Big data analytics

1 Introduction

The rapid development of the Internet of Things (IoT) at present, the affordability of sensor equipment and the maturing connectivity technologies, allow us to collect a lot of useful data. The sensing data hence become ingredients to applications that are designed to provide better quality for everyday life. However, this kind of data has unique characteristics. First, the data collected by the sensor is usually numerical data. Second, when the sensor collects the data, the collection frequency is relatively fast, and data can be collected sporadically, within a few seconds at each time depending on the sampling rate. During each period of time, a large amount of data is gathered. But they may be about the same across successive periods because the changes in the environment or the outdoor activities are slow (compared to body activity recognition). Furthermore, the data collected by the sensors over a certain period of time is repetitive but uninteresting. For example, crowd-sensing and security-oriented sensor applications collect a huge

H. Sanjurjo González et al. (Eds.): HAIS 2021, LNAI 12886, pp. 38–49, 2021.
https://doi.org/10.1007/978-3-030-86271-8_4

amount of normal data, in the hope of detecting something that deviates from normal. Therefore, the data collected in a certain period of time may often contain irrelevant data that come from uninteresting and repetitive activities. In short, the data collected by these sensors are characterized by time series, large quantity over a period of time, easy data repetition, and certain noise data. This leads to the consequence that the classifiers that are trained by such data cannot effectively classify tasks. When we conducted data analysis, we found that the reason for the deterioration of the classification effect is often due to that the data was divided too much at high resolution. In daily life, we should pay more attention to the results and phenomena of some abstracted time periods. One extreme example of abstract time period is morning, afternoon and evening. In our new model, a coarse-grained level is adopted for data partitioning to reveal prominent data features while maintaining the data in effective structures. For solving the problem of "diluted data" due to IoT operational nature, a new model is proposed. In this paper, the proposed classification model is empowered by a new probability evaluation classification method treating the input data as data sequence. The advantages of proposed mechanisms can improve the robustness of the model, reduce the sensitivity to noisy training data in the data stream that come from the sensors of the Internet of Things. Therefore, the machine learning model will become better in alignment with real-life prediction objectives. The classification model that is induced using the proposed learning method will be more useful than the direct use of the classification algorithms alone.

2 Related Work

With the advancement of hardware technology, sensor devices are increasing. The sensor-centric Internet of Things has also experienced rapid development. According to statistics [1] until 2017, the total value of the IoT reached 29 billion U.S. dollars. Such a huge market has attracted the attention of industry personnel and academic staff. Over the years, people have been investigating and building smart systems such as smart homes, smart transportation, and smart security [2–4]. The massive increase in IoT devices helps people obtain large amounts of sensory data. How to tap valuable information from this vast amount of data and form knowledge to serve life more effectively is an important issue. Some researchers have tried to use the data mining technology in the development of the IoT to make it more intelligent [5–8]. Clustering is commonly used in data mining of the IoT and the most common clustering method is K-means [9] which is very mature in traditional data mining. The distribution of Internet of Things data in some cases is a clustering problem [10, 11], but the classification results presented by clustering are only similar data and cannot be judged. If people are unfamiliar or unclear with the collected data, they cannot rely on the clustering results to dig out effective knowledge. In supervised learning, people often use decision tree algorithms for data mining in the IoT [12, 13]. In addition, probabilistic models are also widely used, such as the Naïve Bayesian model [14, 15]. In machine learning, there is also a simple and efficient classification method that is SVM [16], combining with the kernel function can linearly separate data in high-dimensional space. However, the traditional classification method has unstable performance on the actual sensor data because of the special nature of IoT data. Through analysis and observation, all collected sensor data have highly repetitive characteristics and often contain noise data whose source may be due to sensor detection errors. This leads to too many samples of negative instances in the classification process, and the accuracy of the trained model decreases. In real life, people pay more attention to the results of a period of time, which is inconsistent with the frequency with which sensors collect data. Therefore, in order to solve this problem, this paper proposes a new type of data mining model, with a pre-processing which consists of constructing the subspace from the initial data set, and finally using the traditional classification method for classification. This model can greatly improve the accuracy of classification. Moreover, this model is robust, and it can be combined with various classification methods.

3 Our Proposed Model

In this section, the paper is going to introduce the model overview and example, formulating the high-level mathematics model for this new pre-processing process (PP). Compared with other classical classifier algorithms and pre-processing process, this pre-processing method combining with other classical classifier algorithms are applied for predicting the major label among a small group of continuous sequential data which are ordered by time. This new method will change the unit of information from a singular time point to a period of time. Each dataset tested in this paper is the data that come from a typical wearable sensor. The data have several labels such as "walking", "running" and so on in the prediction class. In our daily life, be it walking or running, there is a well-known observation that the sensed data would have similar adjacent instances along the data stream. The data instances carry the same label except when the instances are generated located at the boundary of two different actions and the noisy data. When people expect to classify which actions the subject under monitoring is doing through the data, it would be more effective for the machine learning model to be trained with a group of continuous data instances that are grouped with a common target label, than singular data instances with precise but similar data values individually. Our proposed pre-processing method is designed to generate new datasets for training and testing from the original train dataset and test dataset. The meaning of the instance in the new dataset is not merely the information converted from a singular instance which is in the original dataset, but the information from a continuous period of instances. How to convert the relevant information effectively is the most important part of the design, which is reported in this paper. To begin with, the feasibility of this algorithm in overall is defined, and this paper is going to illustrate how this algorithm works in detail in the next section, followed by the experiment result. First of all, when the train dataset is coming, we would like to get some information from them especially from the data which has the same label. Then we collect some sample data instances from each group of instances which have the same label (such as we collect sample instances from a set that each instance inside are labelled by "walking" and then do the same process in the "running" set). And we assume that these sample instances could represent and basically conclude most of traits from class labels. Then these sample subsets are called "standard sets" for each class label which means every class label has one sample set. If there are n labels in the dataset, then there will be n sample dataset. This is just the beginning of the transformation. In the next step, in order to stimulate a normal period of instances in the dataset, we use the sampling method to collect some data from each class to form n label dataset. Theoretically, the size of these label sets is smaller than standard sets because we want to find some similarity index between standard sets and label sets using isolation forest (iForest) detection algorithm [17] which is an algorithm to detect the non-isolation rate between two datasets.

The workflow is divided into three main steps: a) the original sequence training dataset (T0) will be transformed into a new training dataset (T1) while the attributes and the length of T1 are changed and optimized by PP, b) the original testing dataset (D0) will be transformed into a new test dataset (D1) by PP. It needs to be noticed that the core of PP method is based on Isolation Forest algorithm and c) the user could apply these new training dataset and testing dataset for prediction coupled with some known classification algorithms. Abstract flow charts are shown in Figs. 1 and 2, indicating how overall this concept works for reconstructing the training and testing datasets respectively using subspace division.

A set of formula are developed which are used to explain each step in aforementioned figures by explaining the operation pertaining to how the data is processed and converted between successive steps. Suppose $S = \{x_1, ..., x_t, x_{t+1}, ...\}$ to be the original dataset, $x_t \in S$ where $t = 1, 2, ...,$ and the length of dataset S is fixed. Here SP is denoted as a collection of all the subsets of S. For every data $x_t \in S$, there are m attributes characterizing them. And then the attributes space A would be defined as $A = \{(a_1, a_2, ..., a_i..., a_m) | a_i$ is the value of the ith attribute, for $i = 1, ..., m\}$ and the attributes' types are mixed by numeric and nominal data. All the data were labeled, here it is called class and the collection of classes is $C = \{C_1, ..., C_n\}$. Then the data $x_t=(x't, c)$ where $x't = (x1, t', ..., xi, t', ..., xm, t'), x't \in A$ and $c \in C$. In this experiment, cross-validation was applied for splitting the original dataset S into training dataset and testing dataset. We denote one of the training datasets as T0 and testing dataset as D0 whose instances have no labels. Before pre-processing the training data, we need to define some functions and notations to make it more accessible.

Formulae #1
Define a function Class to print out the class c_h of instance h where $h \in S$ and $c_h \in C$.

$$\text{Class} : S \rightarrow C, \text{Class}(h) = c_h \tag{1}$$

Formulae #2
The function Maj is defined on SP which means to print out the major class c_{maj} of dataset w_x and the operation $|\cdot|$ means to calculate the length of the set.

$$\text{Maj} : SP \rightarrow C, \text{Maj}(w_x) = \arg_{c \in C} \frac{|\{h \in w_x | \text{Class}(h) = c\}|}{|w_x|} = c_{maj} \tag{2}$$

Formulae #3
Define a function Div($\cdot$) to collect the instances whose class is c from T_0 then create a subset T_0^c of T_0.

$$\text{Div} : SP \times C \rightarrow SP, \text{Div}(T_0, c) = \{x \in T_0 | \text{Class}(x) = c\} = T_0^c \tag{3}$$

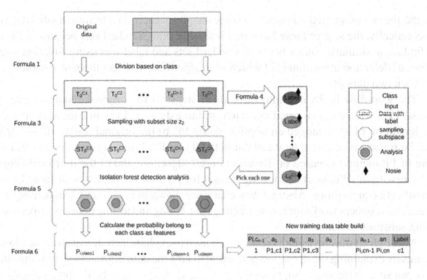

Fig. 1. Block diagram that shows how a new training dataset is reconstructed by our proposed preprocessing method.

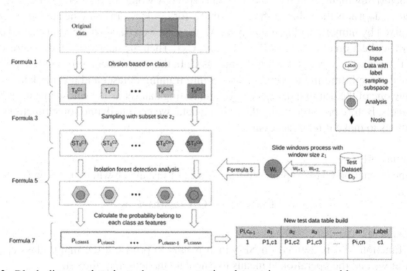

Fig. 2. Block diagram that shows how a new testing dataset is reconstructed by our proposed preprocessing method.

Formulae #4

$Sam_{md}(\cdot)$ is a function that take z samples from T_0^c based on curtain sampling method md, here md could be one of the sample random sampling methods and stratified random sampling. Besides the sample set is named as ST_0^c.

$$Sam_{md} : \ SP \times \ IR \to SP, \ Sam_{md}(T_0, \ z) = ST_0^c \tag{4}$$

Formulae #5

Function $\text{ITR}(\cdot)$ used $\text{ST}_0^{c_j}$ asastandardcase to train an Isolation Forest which is an algorithms created by Prof. Zhihua Zhou [17] for detecting the isolation point and then putting the L_{c_i} sample set into the Isolation Forest model to classify whether there are isolation points or not in the L_{c_i}. Finally computing the rate of data in L^{C_i} that normally obeys the distribution in $\text{T}_0^{sc_j}$. In other words, this function is to compute the non-isolation rate $P_{i,j}$.

$$\text{ITR} : \text{SP} \times \text{SP} \rightarrow [0, 1], \text{ITR}\left(\text{L}^{C_i}, \text{ST}_0^{c_j}\right) = P_{i,j} \tag{5}$$

Step 1: Reconstruct Training Data Table

With the definitions and notations above, the process of this algorithm will be presented below and also be described in the Figs. 1 and 2. When an original dataset S comes, through cross-validation, it could get one of training datasets T0, then divide the T0 into a collection of sub-dataset $\{\text{T}_0^{C_j}\}_{j=1}^n$ where

$$\text{T}_0^{c_j} = \text{Div}(\text{T}_0, c_j) \tag{6}$$

Then the algorithm will do the first time sampling (the reason for why it is first time will be explained at the end of this step) with specific sampling method md to gain the trait from $\{\text{T}_0^{C_j}\}_{j=1}^n$ then it will get a series of sampling dataset $\{\text{ST}_0^{C_j}\}_{j=1}^n$ where

$$\text{ST}_0^{C_j} = \text{Sam}_{md}\left(\text{T}_0^{C_j}, z_2\right) \tag{7}$$

In order to simulate the arbitrary test sliding window w (specific description of w is in the step 2) where $\text{Maj}(w) = C_i$, this pre-processing method will do the first time sampling with specific sampling method md to gain the trait from $\{\text{T}_0^{C_j}\}_{j=1}^n$ then we get a series of Label dataset $\{\text{L}_0^{C_i}\}_{i=1}^n$ where

$$\text{L}_0^{C_i} = \left\{\text{Sam}_{md}\left(\text{T}_0^{C_i}, z_1\right)\right\} \bigcup N_i \tag{8}$$

and N_i is a special noise set to simulate the noise in the arbitrary test sliding windows w whereMaj(w) = C_i. It is clear to find that the length of $\{\text{ST}_0^{C_j}\}_{j=1}^n$ and $\{\text{L}_0^{C_i}\}_{i=1}^n$ are same such as n. Therefore, that is easy to get $n \times n$ combinations such as $\{\{\left(\text{L}_0^{C_i}, \text{ST}_0^{C_j}\right)\}_{i=1}^n\}_{nj}$ = 1. With these $n \times n$ combinations, the preprocess sets $\text{ST}_0^{C_j}$ as the second input element (standardcase) of ITR and $\text{L}_0^{C_i}$ as the first input element (detection case) of ITR. At the end of the whole process, a training table TDT could be generated.

Formulae #6

Construct a Matrix $\text{TDT} \in [[0, 1]^{n \times n} | C^{n \times 1}]$ the element in the ith row and jth column of TDT is computed by function $\text{ITR}\left(\text{L}_0^{C_i}, \text{ST}_0^{C_j}\right)$ and the element in the kth row and n

+ 1 column is C_i:

$$[TDT]_{i,j} = ITR\left(L_0^{C_i}, ST_0^{C_j}\right) \text{for } 1 \leq i \leq n \text{ and} 1 \leq j \leq n$$

$$[TDT]_{k,n+1} = C_i \text{ for } 1 \leq k \leq n \tag{9}$$

Finally, the definition of function PP would be constructed from TDT.
Definition: Function PP is defined to generate a new training dataset T1 from T0 with the setting of z1 (the length of $L_0^{C_i}$) and z2 (the length of$ST_0^{C_j}$).

$$PP1: SP \times IR \rightarrow SP', PP1(T_0, z_1, z_2) = T_1$$

where $T_1 = \{xx_1, \ldots, xx_n\}$ and

$$xx_i = \{[TDT]_{i,j}\}_{j=1}^{n+1} = \left(\{[ITR\left(\{Sam_{md}(Div(T_0, C_i), z_1)\}\bigcup N_i, Sam_{md}(Div(T_0, C_j), z_2)\right)]_{i,j}\}_{j=1}^{n}, C_i\right) \tag{10}$$

Now, we are going to explain why it is written as "the first time". If this algorithm just does sampling for the one time, then the TDT just have n instances which are too few. So, in order to increase the size of the dataset, this algorithm will do the step 1 for many times and the repeating frequency depends on user's choice. Finally, we combine all the training Tables into a new training data set. So, does it work for the test data table. In the following section, the first-time loop is described because the principle is the same.

Step 2: Reconstruct Testing Data Table
This step is to transform the testing dataset D0 to new testing dataset D1 where $D_0 = \{y_1, \ldots, \ldots\}$ and $D_1 = \{yy_1, \ldots, yy_t, \ldots yy_r\}$. Sliding window w will be applied as a instrument to do so and the length of sliding window p can be set by user, e.g. $P = z2$. Let $w_1 = \{x_1, \ldots, x_p\}, w_z = \{x_{1+z}, \ldots, x_{p+z}\}, W = \{w_1, \ldots, w_z, \ldots w_r\}$ where r is determined by the length of slide window and the length of test dataset. Because of the similar technique, the testing dataset is also calculated by ITR with the input. However, the input is no longer the $n \times n$ combinations as TDT but a $r \times n$ combinations such as $\{\{\left(w_t, ST_0^{C_j}\right)\}_{j=1}^{n}\}rt = 1$.

Formulae #7
Construct a Matrix $TDT_1 \in [0, 1]^{r \times n}$:

$$[TDT_1]_{t,j} = ITR(w_t, ST_0^{C_j}) \tag{11}$$

where $ST_0^{C_j} = Sam_{md}\left(T_0^{C_j}, z_2\right)$ for $j = 1 \ldots n$ and $t = 1 \ldots r. w_t$ is the tth sliding window. While the element in the tth row and j^{th} column of TDT_1 is computed by function $ITR\left(w_t, ST_0^{C_j}\right)$.
With the help of TDT1 the final high-level function of step two can be defined.

Formulae #8

Function PP2 is defined for transforming testing dataset D0 into new testing dataset D1 which has same categories of attributes as T1. But part of computing input changes because we no longer utilize T0 to gain sample set $L_0^{C_i}$ whose size is z1 but using sliding windows of D0 while the class of new testing data is replaced by major class of a curtain sliding window.

$$PP2: SP \times IR \to SP'', \quad PP(T_0, D_0, z_2) = D_1$$

where $D_1 = \{yy_1, \ldots, yy_t, \ldots yy_r\}$, and the

$$yy_t = (\{[TDT_1]_{(i,j)}\}_{(j=1)}^n, Maj(w_t)) = (\{ITR(w_t, Sam_{md}(Div(T_0, c_i), z_2))\}_{(j=1)}^n, Maj(w_t)) \quad (12)$$

Step 3: Model Learning

After using the pre-processing method to generate the new training dataset T1 and new testing dataset D1, user could apply different algorithms to make the prediction with the help of T1 and D1. Then a group of high-level equations that represent these processes $\{algoi\}_{i=1}^5$ is defined as follow:

Formulae #9

$\{algo_i\}_{i=1}^5$ is defined for a group of high-level equations that use training data set (such as T_1), testing dataset (such as D_1) as well as a group of parameters for training the model and testing the model, then finally it gets the performance evaluation index pf.

$$algo_i 1 : SP' \times IR \to SP'', \quad Pre \to IR, \quad algo(T_1, D_1, P) = pf \quad (13)$$

where Pre is a collection of all possible specific parameters with respect to the demand of user and $P \in Pre$, pf is a performance evaluation index which is a combination of several statistical parameters of model such as accuracy, recall and F1-score. In this paper, there are five Algorithms involved which are SVM, logistic regression, C4.5, Bayes classifier and KNN. After comparing the performance between before and after per-processing, we found that this new method could improve the accuracy of the prediction. The following section will show the experiment results in detail.

4 Experiment

In this section we evaluate the performance of the proposed model through extensive experiments. We choose a sensor dataset which is a typical data stream in mobile IoT applications and comparing the performance of the algorithms after model optimization to the original algorithms. The Heterogeneity Human Activity Recognition (HHAR) data set, from Smartphones and Smart watches, devised to benchmark human activity recognition algorithms (classification, automatic data segmentation, sensor fusion, feature extraction, etc.) in real-world contexts; specifically, the dataset is gathered with a variety of different device models and use-scenarios, in order to reflect sensing heterogeneities to be expected in real deployments [18, 19]. The data can be obtained free from the public archive at UCI repository.

We choose 5 traditional methods and use three evaluation indicators (Recall, Precision, F1-score). They are: K-Neighbors Classifier, Logistic Regression, Gaussian Naïve Bayes, Decision Tree, and Support Vector Machines. The parameters of the machine learning models are set by their default values. When constructing a new training set, the length of the selected window is 100, and 20 serialized instances are constructed from serialized samples in each category. Then Formula #4 is applied to select the number of T0c. This number should be greater than the number of spaces selected by the window. Therefore, 30 instances are selected as T0c in different categories, and then the iForest algorithm is used to calculate the category probability. The category probability calculations here are using a similar number of percentages. Building a new test set also has the same steps, the length of the selected window is 20, and 20 serialization instances are constructed from random serialized samples in each category of the test set. In the different categories, 30 instances are still selected as T0c, and similar probability is calculated using Eq. 5, so that a new test set is formed. Using a preliminary experiment using the sensor dataset two, it can be seen that the model performed well with large improvement over the classification model that is built without the proposed pre-processing. The comparison results are tabulated in Table 1. The last two columns namely 'Score' and 'Our Model' are the performance indicator values from the classification model which has not been pre-processed and pre-processed, respectively. The running time is only a few minutes for every classifier. In fact, we are more concerned about what has happened over time from the perspective of doing machine learning from time-series. Therefore, the partition of granularity and the length of the window are particularly important. Furthermore, granularity experiments and iterative experiments are designed. If the selected window length is appropriate, the more iterations, the more training instances will be formed. The same parameters in building the training dataset and the testing set are maintained. In the additional experiment W is selected as 20, 30, 40, 50 and ST as 30, 40, 50, 60. The number of training set iterations is set at 100. As it can be seen in Fig. 3, the size of the sliding window which decides how much per pass the instances would enter into the pre-processing and training the classifier, matters. When the window size reaches over 60 approaching 100, almost full score can be obtained in the cross-validation mode of testing of the classifier. This implies sufficient amount of data per pass would help in framing up the subspaces. However, too large the sliding window may lead to a problem of incurring high latency. Having large sliding window is like reverting the incremental learning which is fast as it learns online, to traditional batch learning where the full set of data is used for model induction. The appropriate size of window for balancing between latency and the highest possible accuracy worth in-depth investigation in the future work. Although we can tune the sliding window size to be moderately suitable for accuracy and latency, what if only a limited (small) amount of training samples are available? To test out such extreme situation, another experiment is simulated where only relatively little training and testing data are assumed available and used in the pre-processing. The objective is to test the correlation between accuracy performance and the volume of the training dataset.

From Table 2, when the number of epochs is equal, the more data, the better the effect it shows from our pre-processing approach. Please note that the performance indicator values are averaged over the five classifiers used in our experiments. When the training

Table 1. Comparison results between classifiers built without and with pre-processing

Original algorithm	Performance	Score	Our model
KNeighbours classifiers	Precision	0.23	1.00
	Recall	0.25	0.91
	F1-score	0.35	0.96
Logistic registration	Precision	0.02	1.00
	Recall	0.19	0.92
	F1-score	0.15	0.96
GaussianNB	Precision	0.10	1.00
	Recall	0.25	0.91
	F1-score	0.18	0.95
SVM	Precision	0.10	1.00
	Recall	0.09	0.91
	F1-score	0.14	0.96
Decision tree	Precision	0.10	1.00
	Recall	0.19	1.00
	F1-score	0.12	1.00

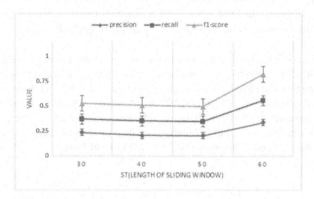

Fig. 3. Averaged performance of classifiers with various W sizes.

size and the testing size gradually increase, although precision, recall, and f1-score may fluctuate, the overall trend is rising. This proves that, to a certain extent, the greater the detection from the probabilistic sample size given sufficient amount of data, the higher the accuracy of detection. From the results, it is found that Pearson coefficients are, 0.620456148, 0.828351883 and 0.728648368 respectively which indicates quite high the correlations between the amount of training data size and the precision, the recall and the balanced F-score.

Table 2. Performance results from various dataset sizes.

Train size	Test size	Train epoch	Precision	Recall	F1 score
30	20	100	0.51	0.29	0.34
40	30	100	0.45	0.31	0.34
50	40	100	0.44	0.31	0.32
60	50	100	0.73	0.48	0.37

5 Conclusions

This paper describes a subspace probabilistic detection pre-processing model based on the subspace-attribute probability calculation. The proposed model is to be used as a pre-processing method that transforms the time diluted dataset to one that can be better characterized by the temporal information from the data, hence better classification model training and prediction results. Five popular classification algorithms are used to test with the pre-processing method by performing classification over sensor data that characterize certain human activities. Such sensor data represent a kind of big data streams that possesses new data mining challenges due to their sheer volumes and sequential nature. This model can effectively solve the problem of repeatability and noise that exist in the sensor data. Through experiments, we can see that this model can effectively improve the performance of traditional machine learning classification algorithms in data mining in the sensor data by large magnitude.

Acknowledgment. The authors are thankful to the financial support from the following research grants: 1) MYRG2016–00069, titled "Nature-Inspired Computing and Metaheuristics Algorithms for Optimizing Data Mining Performance", offered by RDAO/FST, University of Macau and Macau SAR government; 2) FDCT/126/2014/A3, titled "A Scalable Data Stream Mining Methodology: Stream-based Holistic Analytics and Reasoning in Parallel" offered by FDCT of Macau SAR government and 3) TIN2017–88209-C2-R project of the Spanish Inter-Ministerial Commission of Science and Technology (MICYT) and FEDER funds.

References

1. M&M Research Group: Internet of Things (IoT) & M2M communication market - advanced technologies, future cities & adoption trends, roadmaps & worldwide forecasts 2012–2017. Technical report. Electronics.ca Publications (2012)
2. Atzori, L., Iera, A., Morabito, G.: The internet of things: a survey. Comput. Netw. **54**(15), 2787–2805 (2010)
3. Miorandi, D., Sicari, S., De Pellegrini, F., Chlamtac, I.: Internet of things: Vision, applications and research challenges. Ad Hoc Netw. **10**(7), 1497–1516 (2012)
4. Bandyopadhyay, D., Sen, J.: Internet of things: applications and challenges in technology and standardization. Wirel. Pers. Commun. **58**(1), 49–69 (2011)

5. Cantoni, V., Lombardi, L., Lombardi, P.: Challenges for data mining in distributed sensor networks. In: Proceedings of International Conference on Pattern Recognition, vol. 1, pp. 1000–1007 (2006)
6. Keller, T.: Mining the internet of things: Detection of false-positive RFID tag reads using low-level reader data. Ph.D. Dissertation. The University of St. Gallen, Germany (2011)
7. Masciari, E.: A framework for outlier mining in RFID data. In: Proceedings of International Database Engineering and Applications Symposium, pp. 263–267 (2007)
8. Bin, S., Yuan, L., Xiaoyi, W.: Research on data mining models for the internet of things. In: Proceedings of International Conference on Image Analysis and Signal Processing, pp. 127–132 (2010)
9. McQueen, J.B.: Some methods of classification and analysis of multivariate observations. In: Proceedings of Berkeley Symposium on Mathematical Statistics and Probability, pp. 281–297 (1967)
10. Jain, A.K., Murty, M.N., Flynn, P.J.: Data clustering: a review. ACM Comput. Surv. **31**(3) 264–323 (1999). ([43] Xu, R., Wunsch-II, D.C.: Survey of clustering algorithms. IEEE Trans. Neural Netw. 16(3), 645–678 (2005)
11. Xu, R., Wunsch-II, D.C.: Survey of clustering algorithms. IEEE Trans. Neural Netw. **16**(3), 645–678 (2005)
12. Safavian, S., Landgrebe, D.: A survey of decision tree classifier methodology. IEEE Trans. Syst. Man Cybern. **21**(3), 660–674 (1991)
13. Friedl, M., Brodley, C.: Decision tree classification of land cover from remotely sensed data. Remote Sens. Environ. **61**(3), 399–409 (1997)
14. McCallum, A., Nigam, K.: A comparison of event models for Naivebayes text classification. In: Proceesings of National Conference on Artificial Intelligence, pp. 41–48 (1998)
15. Langley, P., Iba, W., Thompson, K.: An analysis of Bayesian classifiers. In: Proceedings of National Conference on Artificial Intelligence, pp. 223–228 (1992)
16. Cristianini, N., Shawe-Taylor, J.: An Introduction to Support Vector Machines and other kernel-based learning methods. Cambridge University Press, Cambridge (2000)
17. Liu, F.T., Ting, K.M., Zhou, Z.H.: Isolation forest. In: Eighth IEEE International Conference on Data Mining. ICDM 2008, pp. 413–422. IEEE (2008)
18. Fong, S., Song, W., Cho, K., Wong, R., Wong, K.K.L.: Training classifiers with shadow features for sensor-based human activity recognition. Sensors **17**(3), 476, 27 (2017)
19. Fong, S., Liu, K., Cho, K., Wong, R., Mohammed, S., Fiaidhi, J.: Improvised methods for tackling big data stream mining challenges: case study of human activity recognition. J. Supercomput. Springer **16**, 1–33 (2016)

Bio-inspired Models and Evolutionary Computation

Remaining Useful Life Estimation Using a Recurrent Variational Autoencoder

Nahuel Costa$^{(\boxtimes)}$ and Luciano Sánchez$^{(\boxtimes)}$

University of Oviedo, Gijón, Asturias, Spain
{costanahuel,luciano}@uniovi.es

Abstract. A new framework for the assessment of Engine Health Monitoring (EHM) data in aircraft is proposed. Traditionally, prognostics and health management systems rely on prior knowledge of the degradation of certain components along with professional expert opinion to predict the Remaining Useful Life (RUL). In order to avoid reliance on this process while still providing an accurate diagnosis, a data-driven approach using a novel recurrent version of a VAE is introduced. The latent space learned by this model, trained with the historical data recorded by the sensors embedded in these engines, is used to visually evaluate the deterioration progress of the engines. High prognostic accuracy in estimating the RUL is achieved by building a simple classifier on top of the learned features of the VAE. The superiority of the proposed method is compared with other popular and state-of-the-art approaches using Rolls Royce Turbofan engine data. The results of this study suggest that the proposed data-driven prognostic and explainable framework offers a new and promising approach.

Keywords: Remaining Useful Life · Prognostics and health management · Recurrent neural networks · Variational autoencoder

1 Introduction

Engineering maintenance and prognostics are a must in modern aircraft engines. Data is routinely collected from the engine to monitor and prevent it from operating in undesirable conditions. The knowledge of the system built into the engine and aircraft is configured to trigger alerts that highlight the need for pilot action, maintenance action or directly shut the engine down if a significant condition is encountered. Over the years, the number of variables and data collected has increased substantially which on the one hand is positive in terms of making more accurate diagnoses, but at the same time increases the difficulty to reach them since traditional strategies such as corrective maintenance of failures and scheduled preventive maintenance are becoming less capable of meeting the

Partially supported by the Ministry of Economy, Industry and Competitiveness ("Ministerio de Economía, Industria y Competitividad") of Spain/FEDER under grants TIN2017-84804-R and PID2020-112726-RB.

© Springer Nature Switzerland AG 2021
H. Sanjurjo González et al. (Eds.): HAIS 2021, LNAI 12886, pp. 53–64, 2021.
https://doi.org/10.1007/978-3-030-86271-8_5

growing industrial demand for efficiency and reliability. On the other hand, smart Prognostics and Health Management (PHM) technologies are showing promising capabilities for application in industries [14]. Remaining useful life (RUL) is a key metric in this regard and can be estimated from the historical data that sensors record on each trip, which is very important to improve maintenance schedules and avoid engineering, safety and reliability failures and, as a consequence, determine engine deterioration, increase engine flight time and reduce maintenance costs.

Accurate diagnosis can be achieved with model-based approaches if the degradation of the complex system is accurately modeled, some examples are Weibull distribution [1] or Eyring model [4]. The main limitation of these approaches is that they require extensive prior knowledge about the physical systems that is usually not available in practice. This is precisely why data-driven approaches have been gaining popularity in recent years, as they are able to model degradation characteristics based solely on historical sensor data from which the underlying correlations and causalities in the collected data can be modeled. In other words, knowledge can be generated from the collected data with little prior prognostic experience to infer valuable system information, such as RUL.

In this paper we propose a new Deep Learning approach for RUL estimation, based on a visual diagnosis capable of assessing the evolution of RUL. To this end, we present a novel recurrent version of a Variational Autoencoder (VAE). Aircraft data is captured over time, comprising a time series, however, VAE research is very much oriented to the field of images, and not so much to that of time series, although some work is beginning to emerge [7]. To the best of our knowledge, this is the first contribution in which a recurrent VAE is used for RUL estimation.

Besides, despite achieving very promising results, most Machine Learning models focus their efforts on predicting a number or a label, leaving aside how they got there [3]. By making use of the internal representation learned by the VAE we can elaborate a strong explanatory component, since a simple prediction may not be informative enough to determine engine deterioration, thus giving insight into the state of an engine with a simple look to a self-explanatory map.

2 RUL Estimation

In the last decade, the relationship between the use of monitored system data and the RUL of engines has gained the attention of data-driven prognostic models. Especially, the use of neural networks has had a great impact given that they have the advantage of learning to model highly nonlinear, complex and multidimensional systems without experience in the physical behavior of the system. In this sense, there are works such as [12], where the authors applied multilayer perceptrons (MLP) for estimating the RUL of laboratory-tested bearings. In addition, some researchers have integrated fuzzy logic to capture more information for EHM [5,9]. It is also worth mentioning works like [11] and [6] where Gradient Boosting trees and Support Vectors are applied for engine RUL prediction.

More recently, in this field, as in many other areas such as image or speech recognition, the application of Deep Learning models has been gaining ground over the years for RUL estimation as the raw data obtained from machine health monitoring share a high dimensionality similar to that of image processing studies. There are clearly two trends: the application of Convolutional Neural Networks (CNN) and Recurrent Neural Networks (RNN). In the first group, works as [8] explore RUL estimation using different configurations of CNNs minimizing prior knowledge about prognostics and signal processing. Regarding RNN, the most common architecture that can be found in the literature are LSTMs [10] and in the last years, Bidirectional LSMTs [13] are beginning to gain importance as their ability to make full use of the sensor date sequence in bi-direction seems to have promising results.

Again, as discussed above even if these models achieve good results, in the end, these systems will be used by people outside Machine Learning and what they will be looking for is a high-quality interpretation of the data. One possible way to provide this is to establish a Representation Learning approach since, unlike others, the performance of models following this approach depends directly on internal representations, which in turn can be leveraged in favor of a better understanding of the problem itself. In this direction, we propose applying a VAE.

VAEs are designed to reconstruct the input data while at the same time learn a compressed representation of it, the so-called latent space. That compression depends on a probability distribution, causing the data to be organized in a continuous space, i.e. two nearby points in the latent space should give similar contents when reconstructed. This also means that similar data are located close together in the latent space, forming different clusters depending on their underlying nature.

The framework we propose relies then on a new recurrent VAE to model the time complexity of the engine data. The fact that it has a recurrent input offers the possibility to feed the model each time a new data sample becomes available, allowing online training. The VAE learns different degradation stages to the point of being able to place in the latent space, which can be understood as a two-dimensional map, aircraft with similar RUL values in the vicinity. This is used to, given data of a new aircraft, place it on the map according to the RUL it has, thus offering a simple and intuitive diagnosis.

3   Model Architecture

Variational autoencoders consist of 3 parts: an encoder network, a decoder network and a latent space. The encoder learns to compress the data to the latent space from which the decoder can generate new samples. The latent space is described by the parameters learned by the encoder that initialize the probability distribution to which the data belong, so that the decoder can not only reconstruct the input data, as conventional autoencoders do, but can also generate new samples from that distribution. The loss function introduces the

Kullback-Leibler divergence, which measures how much one probability distribution diverges from another, to learn the above-mentioned parameters. The reconstruction error between the original input and the output of the decoder is also included. In the end, all together allows the model to produce a latent space in which similar data will be located close to each other and also enables new data to be sampled from points that do not belong to the original data, thus having a generative model.

Given the way the model works, the workflow followed for this problem is quite simple: a VAE is trained with data from Turbofan engines to learn a simplified representation. Thus, the learned encoder acts as a feature extractor by projecting onto the latent space the data according to its properties, which are different stages of deterioration in the engines. This section explains how this extraction, can be exploited to create the diagnostic map we are pursuing. Emphasis is also placed on the recurrent architecture proposed to deal with the time series, as well as how the latent features give rise to perform other tasks such as classification or prediction.

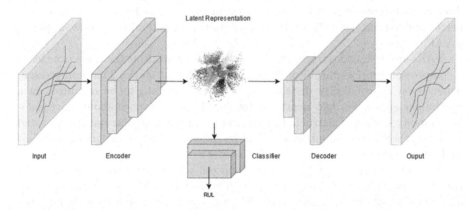

Fig. 1. Network structure of the proposed method. The blue and green blocks are the encoder and decoder respectively and the red blocks refer to the linear classifier. (Color figure online)

Encoder as a Feature Extractor. In a VAE the training process is regularized to avoid overfitting and to ensure that the latent space has good properties that enable the generative process. To obtain these properties, the encoder must be able to map the data in the latent space in such a way that similar data are close. This allows not only that the decoder can reconstruct the data efficiently but also that it can generate new data from a point in the latent space that does not correspond to the encoding of any training sample.

A VAE, given an input, tries to find a latent vector that is capable of describing it and at the same time has the instructions to generate it again. The process can be described as: $p(x) = \int p(x|z)p(z)dz$. Given that the integral of this formula is intractable due to the continuous domain of z, the variational inference is needed via the lower bound of the log likelihood, $\mathcal{L}_{vae}$,

$$\mathcal{L}_{vae} = E_{q_\phi(\mathbf{z}|\mathbf{x})}[\log p_\theta(\mathbf{x}|\mathbf{z})] - D_{\mathrm{KL}}(q_\phi(\mathbf{z}|\mathbf{x})||p_\theta(\mathbf{z})). \tag{1}$$

The first term is the reconstruction of x that tends to make the coding-decoding scheme as efficient as possible by maximizing the log-likelihood $\log p_\theta(\mathbf{x}|\mathbf{z})$ with sampling from $q_\phi(\mathbf{z}|\mathbf{x})$, modeled by a neural network (the encoder) whose output are the parameters of a multivariate Gaussian: a mean and a diagonal covariance matrix (the latent space). That is to say, the main goal of the encoder is to map the input data into a lower-dimensional space, acting as a feature extractor. The second term tends to regularise the organization of the latent space by causing the distributions returned by the encoder to approach a standard normal. It regularises the latent variables (represented by z) by minimizing the KL divergence between the variational approximation and the prior distribution of z.

Based on the representation learned by the encoder, the data, x, is sampled from the conditional probability distribution p(x—z). For generative purposes, this regularization in the latent space is very effective in facilitating random sampling and interpolation for the creation of new data. This is why VAEs are understood to be generative models and precisely it is the most widespread application in the literature. Nevertheless, we place our efforts not on generating new aircraft data but on diagnosing it instead, therefore after training the decoder is not used anymore.

As stated in the introduction, most recent studies make use of recurrent networks to model the time complexity of historical aircraft data. Among the different types of RNNs that can be found, LSTM networks are the most prominent. These networks process data from backward to forward preserving the information of the past through hidden states. However, Bidirectional LSTM networks are in high demand because they provide not only information about the past but also about the future: data is first processed from past to future and then from the future to the past, thus preserving the information from both periods. This is very valuable because the network knows what the data may look like in its future stages, which helps it to understand what kind of information to predict (different stages of engine degradation).

All things considered, we decide to implement the VAE with Bidirectional LSTM networks. In this way, the encoder approximates the Gaussian distribution $p_\theta(\mathbf{z})$ by feeding the output into two linear modules to estimate its mean and covariance. The compression of the input data results in a two-dimensional latent space dominated by the axis represented by the mean and the variance of the approximated distribution. Figure 1 shows the pipeline followed for applying this framework for RUL estimation: the input and output are the same and in the middle it is expected that engines are grouped in different clusters in the latent space according to their features, depicting a simpler representation of their nature. Furthermore, we add a classifier on top of the learned features in order to explicitly report which RUL is the one that best represents each engine that is fed to the model.

Diagnostic Map. The diagnostic tool introduced in this study is a color-coded map that displays the actual state of the engine and the speed of change from healthy to deteriorated. Once the VAE is trained with the engine's data, every input can be compressed into the latent space in terms of the mean and variance of the learned approximated distribution. This information can be projected onto a map as shown in Fig. 2. Every projected point is a sample from the dataset and is colored according to their corresponding RUL: aircraft with low RUL values are painted in red while aircraft with high RUL values are painted in green. It can be seen there is a clear progression in the colors between points since units with no signs of deterioration are located in the upper part of the map (greater values of RUL) while the most deteriorated ones are located in the lower part (lower values of RUL). This representation can be leveraged later on: when new unseen units are used as inputs, the encoder will place them according to their features, giving information about their RUL depending on the proximity to other nearby points that are labeled. That is why it is considered explainable, because the method itself explains the status of each engine. The following section provides further details on the interpretation of this map.

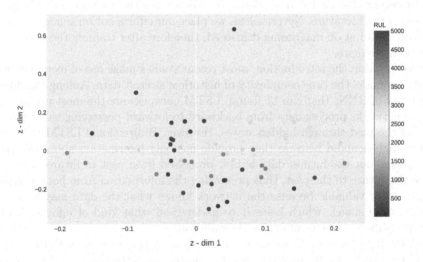

Fig. 2. Latent representation learned by the encoder.

4   Experimental Study

RUL estimation is an actual engineering problem posed by Rolls-Royce therefore, in this paper, the proposed method is evaluated on EHM data from real Rolls-Royce Turbofan engines. This data contains multi-variate temporal information obtained from several built-in sensors. Each time step represents the values of the variables collected for a single flight. Each engine unit may start with different starting conditions all of which, albeit, are considered to be healthy. The expected behavior of an aero-engine is to degrade as time progresses, however, the fact that this degradation is not linear hampers the estimation of the health state, that is the RUL. Anticipating the breakpoint before failure is key to prevent potential problems in the future, thus expanding the lifetime of the engines. Particularly, information is available on the state of the turbine and compressor of the aircraft at the end of the measurement, which allows us to create a dataset that reflects the states through which an aircraft may go through. Based on this, we aim is to generate valuable knowledge that will allow us to estimate the RUL of a new aircraft given its flight history.

4.1 Data Pre-processing

As usual, data capture has its limitations, as a result, in order to prepare a consistent dataset it has to be subjected to a purging and pre-processing process. To begin with, there are quite a few time steps where data is missing, therefore as long as these rows can be dealt with, an imputation is applied based on the column average, i.e. the average of the values captured by the sensor.

Natural factors such as wind, number of passengers or a change of trajectory may cause noticeable peaks in each signal which ultimately leads to noise. As a consequence, it is necessary to apply a smoothing that minimizes the effects of this noise and prioritizes the trend of each signal. For this purpose, an exponential smoothing with an alpha value of 0.4 is applied.

Anomalies, mainly points with spurious values, must be carefully removed because as a final step before passing the data to the network, a normalization is applied to the whole dataset to ease the network to deal with reasonable values and these anomalies can significantly affect the range of the data.

Series length: the useful life of a new aircraft is expected to be at least five thousand cycles. Nonetheless, among the aircraft for which information is available, there are varying lengths due to the fact that there are measurements for different life stages, therefore there are units that may end in eighty cycles, while others may end in seven thousand. It is known that aircraft with less than one thousand cycles had failures linked to different problems other than turbine and compressor, and so it is decided to get rid of them because they are random failures that would add nothing but noise to the dataset. Each aero-engine goes to the workshop every one thousand flights (cycles) and the condition of the turbine and compressor are recorded independently by the mechanics. Precisely, this time window is used to determine the length of the series to be received by the network as input. For this purpose, as one thousand cycles are almost intractable for a neural net, each sequence is transformed by taking each point as the average of the next ten, thus reducing the size of the sequence ten times while maintaining the morphology of the signal.

Finally, it is not so much the values that each signal takes that matter, but rather the increases or decreases that they undergo. As an example, even if two units have different pressures in the turbine, if the pressure has been increasing over the cycles in both units, it means that both have suffered wear in this component, therefore it is more informative to take the derivative of the signal instead of the signal itself. Figure 3 reflects the transformation carried out by this pre-processing for a random aircraft.

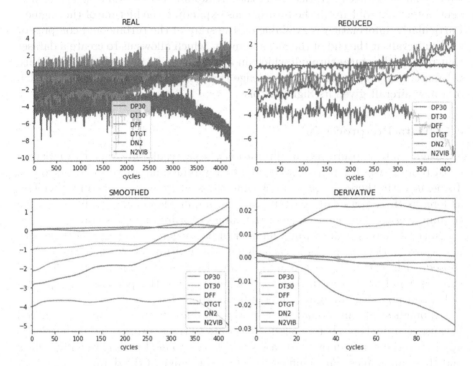

Fig. 3. Every aircraft in the dataset undergoes a transformation that mainly involves the reduction of the number of time steps, noise reduction and calculation of the derivatives of the signals.

4.2   Illustrative Example

To illustrate how the proposed model works, an example that can be understood as a visual fleet diagnosis is presented below. During training, the network learns different degradation patterns which leads the encoder to project the units into the latent space according to their degradation, maintaining coherence in the distances between healthy and compromised engines as described in Fig. 2. This projection is reused as a basis to find out, given undiagnosed units, how their deterioration evolves as the number of trips (cycles) increases. Figure 4 pictures this idea: 8 airplanes have been chosen to project their state into the latent space in four different time steps: t = 0 would correspond to feeding the network with the data corresponding to the cycles from 0...1000, t = 1000 from 1000 to 2000 and so on until t = 3000. Leaving the latent projection obtained in train gives us some insight into the progression of the health status of these units: The latent projection of samples s2, s3, s5 and s8 during all the time steps shown remain over the upper left quadrant, next to other aircraft with similar characteristics, RUL around five thousand and so with no signs of near degradation. On the contrary, there is a clear progression in samples s1, s3, s6 and s7, which move slightly downward and to the right at t = 1000 and t = 2000 to finally at t = 3000 be placed together with units close to their end of life (low values of RUL), thus obtaining an accurate and explainable diagnosis beyond a possible label indicating the predicted health.

In the presented figure only four time steps have been selected to show the update of the engine status according to the data input, however, it is remarkable that once passed the barrier of the first thousand cycles in every posterior trip this update can be done thanks to the fact that we are using recurrent networks and this is where the interest really lies because in the end this can be used as a diagnostic tool: As long as the engine projection remains in the healthy range, its state will be considered positive; on the contrary, if the projection moves towards the red zone, this can be a clear sign of deterioration, information that will be used by the mechanics to make a decision regarding its monitoring, either to make it more exhaustive or to take the aircraft to the workshop for a more complete check-up, just to name a few alternatives. This translates into an extension of the life of these engines by being able to anticipate the break-point where severe deterioration may occur.

4.3 Numerical Results

In this section, we demonstrate that our framework can compete with other modern approaches for time series for the case at hand. It is important to note that RUL estimation is typically a prediction problem but the RUL information available to us are labels that determine the health status of the engine. These labels can be classified into 5 groups from best to worst condition: "Good", "Good To Normal", "Normal", "Normal To High" and "High". It is also understood that each tag corresponds to an approximate number of RUL, being Good ≈5000 cycles, Good to Normal ≈3000 cycles, Normal ≈1000 cycles Normal To High ≈250 cycles and High ≈10 cycles. For this reason, we chose to compare our method with models widely used in the literature: MLP, CNN, LSTM and Bidirectional LSTM, but changing their last layer so that instead of predicting a number they predict a class and thus become a classification problem like the one we have.

Table 1 shows the performance of the different models for each class in terms of accuracy. Each entry in the table is the number of times an engine in a class was recognized by each model for the appropriate class. In addition, to illustrate the performance of each method, the ranking calculated by the Friedman method (ranking by range) for each sub-dataset and the resulting averaged ranking are included.

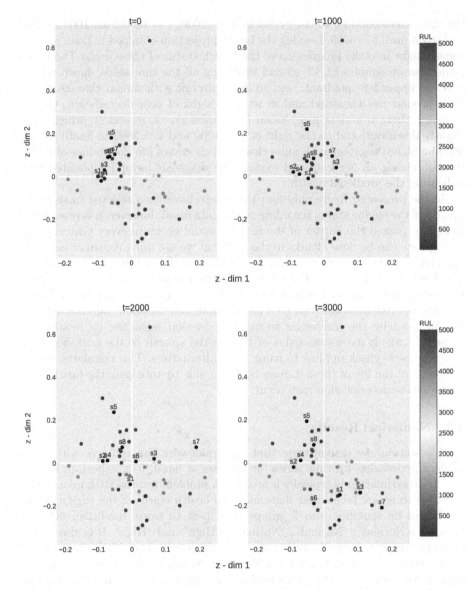

Fig. 4. RUL evolution in a few selected aircraft. As the cycles progress, the fastest degrading aircraft are placed in the zone occupied by aircraft with similar deteriorations.

It can be seen that the best classifier in terms of accuracy is ours, labeled as RVAE. To extend the comparison between the different methods, post-hoc tests were carried out to detect significant differences in pairs between all the classifiers as recommended in [2]. If the significance test yields a p-value lower than a predefined threshold (usually 0.05), then the difference is considered significant, therefore one model is declared superior to another. As a result, our method is the only one that yields a p-value of less than 0.05 (0.0357) for the comparison with MLP. This means that the other methods are not statistically superior to MLP, which makes us value the performance of our method. Additionally, it should be noted that the baseline methods we present do not include any representation of the data, but simply predict the class to which each sample belongs, while interpretability of black-box models, like the one we provide, can present predicted information in a more illustrative way than just a numerical or categorical result.

Table 1. Accuracy of the different classifiers, 5 types of RUL.

	Accuracy				
	MLP	CNN	LSTM	BiLSTM	RVAE
High (50)	0.166 (5)	0.330 (4)	0.500 (2.5)	0.500 (2.5)	0.666 (1)
Normal To High (250)	0.500 (4.5)	0.666 (1.5)	0.666 (1.5)	0.500 (4.5)	0.666 (1.5)
Normal (1000)	0.833 (3)	1.000 (1)	0.750 (4)	0.916 (2)	0.666 (5)
Good To Normal (3000)	0.692 (3.5)	0.615 (5)	0.692 (3.5)	0.846 (1.5)	0.846 (1.5)
Good (5000)	0.692 (5)	0.846 (4)	1.000 (1.5)	0.923 (3)	1.000 (1.5)
Summary results					
Accuracy	0.64	0.74	0.76	0.78	0.80
Average rank	4.2	3.1	2.6	2.7	2.1

5 Concluding Remarks

We have introduced a recurrent VAE architecture based on Bidirectional LSTMs to create a graphical map that describes the condition of engine fleets. The diagnostic tool learns a 2D representation of engine data with different degradation stages to, given a new engine, project its representation near engines with similar deterioration patterns. This allows providing an efficient diagnostic tool on the state of health of the engines without prior knowledge of their physical nature. In addition, the lightness of the model and its recurrent nature would allow incorporating the model as a diagnostic system on any hardware with limited computational capabilities and at the same time updating the learned patterns as more data becomes available, thus coping with the non-stationarity of the data distribution.

References

1. Ali, J.B., Chebel-Morello, B., Saidi, L., Malinowski, S., Fnaiech, F.: Accurate bearing remaining useful life prediction based on Weibull distribution and artificial neural network. Mech. Syst. Signal Process. **56**, 150–172 (2015)
2. Demšar, J.: Statistical comparisons of classifiers over multiple data sets. J. Mach. Learn. Res. **7**(Jan), 1–30 (2006)
3. Doshi-Velez, F., Kim, B.: Towards a rigorous science of interpretable machine learning. arXiv preprint arXiv:1702.08608 (2017)
4. Jouin, M., Gouriveau, R., Hissel, D., Péra, M.C., Zerhouni, N.: Degradations analysis and aging modeling for health assessment and prognostics of PEMFC. Reliabil. Eng. Syst. Saf. **148**, 78–95 (2016)
5. Khawaja, T., Vachtsevanos, G., Wu, B.: Reasoning about uncertainty in prognosis: a confidence prediction neural network approach. In: NAFIPS 2005–2005 Annual Meeting of the North American Fuzzy Information Processing Society, pp. 7–12. IEEE (2005)
6. Khelif, R., Chebel-Morello, B., Malinowski, S., Laajili, E., Fnaiech, F., Zerhouni, N.: Direct remaining useful life estimation based on support vector regression. IEEE Trans. Industr. Electron. **64**(3), 2276–2285 (2016)
7. Li, L., Yan, J., Wang, H., Jin, Y.: Anomaly detection of time series with smoothness-inducing sequential variational auto-encoder. IEEE Trans. Neural Netw. Learn. Syst. **32**, 1177–1191 (2020)
8. Li, X., Ding, Q., Sun, J.Q.: Remaining useful life estimation in prognostics using deep convolution neural networks. Reliabil. Eng. Syst. Saf. **172**, 1–11 (2018)
9. Martínez, A., Sánchez, L., Couso, I.: Engine health monitoring for engine fleets using fuzzy radviz. In: 2013 IEEE International Conference on Fuzzy Systems (FUZZ-IEEE), pp. 1–8. IEEE (2013)
10. Miao, H., Li, B., Sun, C., Liu, J.: Joint learning of degradation assessment and RUL prediction for aeroengines via dual-task deep LSTM networks. IEEE Trans. Industr. Inf. **15**(9), 5023–5032 (2019)
11. Singh, S.K., Kumar, S., Dwivedi, J.: A novel soft computing method for engine RUL prediction. Multimed. Tools Appl. **78**(4), 4065–4087 (2019)
12. Tian, Z.: An artificial neural network method for remaining useful life prediction of equipment subject to condition monitoring. J. Intell. Manuf. **23**(2), 227–237 (2012)
13. Zhang, A., et al.: Transfer learning with deep recurrent neural networks for remaining useful life estimation. Appl. Sci. **8**(12), 2416 (2018)
14. Zhao, Z., Liang, B., Wang, X., Lu, W.: Remaining useful life prediction of aircraft engine based on degradation pattern learning. Reliabil. Eng. Syst. Saf. **164**, 74–83 (2017)

Open-Ended Learning of Reactive Knowledge in Cognitive Robotics Based on Neuroevolution

A. Romero[(⊠)], F. Bellas, and R. J. Duro

Integrated Group for Engineering Research, CITIC Research Center,
Universidade da Coruña, A Coruña, Spain
{alejandro.romero.montero,francisco.bellas,richard}@udc.es

Abstract. Reactive knowledge corresponds to implicit knowledge in the human brain, that is, unconscious knowledge such as reflexes that are executed without "thinking". It is a key aspect in human development, and it is also a key aspect in cognitive architectures for robots, mainly because it avoids inefficient action selection procedures and allows addressing higher-level cognitive processes that make use of it. This paper deals with the acquisition of this type of knowledge in a cognitive architecture for open-ended learning. We propose a method for the learning of policies (reactive knowledge) trough evolution from deliberative models by means of neuroevolution. It is interesting to see in the results presented that this approach of learning reactive knowledge instead of exhaustively selecting the appropriate action every instant of time provides equivalent/better results and more efficient action sequences.

Keywords: Cognitive developmental robotics · Policy learning · Open-ended learning · Neuroevolution

1 Introduction

Dual-process theories [1] indicate that living beings can carry out two different decision processes when it comes to deciding what actions to take at a given moment in time. On the one hand, Type 1 reasoning processes are intuitive, fast and automatic, that is, reactive. Most of the daily activities (like driving, talking, cleaning, etc.) are an example of them. On the other hand, Type 2 reasoning is slower, more cognitively demanding and works deductively to assess a series of possibilities, that is, it is related to deliberation. When solving a difficult math problem or thinking carefully about a philosophical issue, this type of reasoning is used.

If we seek to create cognitive robots capable of operating in open-ended learning environments and learning throughout their lives [2] in such a way that they develop high levels of autonomy, their cognitive architectures must also include these functionalities. We must provide robots with deliberative capabilities (Type 2), which allow them to infer how the world or themselves will behave in a given situation. But we must also give them reactive capacities (Type 1) to be able to do things speedily and "without thinking". Both reactive and deliberative qualities are essential to an optimal action

H. Sanjurjo González et al. (Eds.): HAIS 2021, LNAI 12886, pp. 65–76, 2021.
https://doi.org/10.1007/978-3-030-86271-8_6

selection process. The efficiency of the robot's behavior in the environment will also depend on the correct balance between deliberation and reaction, in order to make proper use of their advantages in different perceptual states.

Deliberative capabilities are necessary when the robot has no experience in the environment or faces a new situation. They should allow it to deliberate and generate new models that allow exploring the environment to gain experience and generate new knowledge. From this perspective, the robot must evaluate what is the best action to achieve its objectives based on its current state. Therefore, it requires the availability of the corresponding world models to perform prospection and of utility models for the evaluation of states.

Reactive or intuitive capabilities are useful when the robot already has enough experience and needs to act fast due to real-time operation requirements, or when it has to deal with dynamic environments in which it is too complex to act based on deliberation. From this point of view, the robot must be able to select a function that directly provides the optimal action to apply based on its current state and the goal to be achieved. These functions are called policies in the reinforcement learning literature. Any policies that the robot uses must have been provided by the designer or learned by the robot itself during its operation. The learning of these models has been carried out for years in traditional robotics in various ways, from Policy gradient [3] and Actor-Critic methods [4] to Model-free policy search [5].

However, in the problems that cognitive robots face, this is not possible, since the robot (nor the designer) does not know the objectives that it will have to face throughout its life. The goals must be discovered. Therefore, the above-mentioned ways of learning are not possible from a developmental point of view. The robot must learn to achieve the objectives as it discovers them while operating in different domains, which will continually change throughout its lifetime. For this reason, one of the solutions to learn this type of reactive behaviors in cognitive robots is to do it from the models that the robot has acquired from its experience with the world. That is, use the learned models to carry out deliberative reasoning processes and reuse and restructure that knowledge to be able to learn reactive policies, which are faster and more useful in certain situations.

This article deals with this problem, that of how it is possible to generate (in a cognitive architecture for robotics) knowledge or reactive policies from the deliberative knowledge previously acquired through the interaction of the robot with its environment. This issue will be studied for the specific case of the e-MDB cognitive architecture [6], but with the aim of providing results that can be generalized to other representations.

The rest of the paper is structured as follows. Section 2 explains in detail how knowledge is acquired and represented within the e-MDB architecture. Section 3 is dedicated to the formal description of the acquisition of reactive knowledge (policies) from declarative knowledge (models) within the e-MDB. Section 4 contains a practical example of application in real robots of this new learning system and its comparison to an exhaustive selection of actions. Finally, in Sect. 5 we draw some relevant conclusions trying to generalize them to other cognitive architectures.

2 Knowledge in the Epistemic Multilevel Darwinist Brain (e-MDB)

The epistemic-Multilevel Darwinist Brain (e-MDB) is a purpose-driven cognitive archi-tecture for lifelong learning in real robotic systems. It has been under development since the early 2000s and it allows artificial agents to learn from their experience in dynamic and unknown domains in order to fulfill their motivations or objectives [7]. The architecture is made up of three main components:

- A motivational system that establishes the robot's motivations and allows it to find new goals and select which ones are active. Its operation is based on domain independent needs and drives [8], since they allow assigning purpose to the robot regardless of the domain in which it works. A detailed explanation of its operation can be found in [7].
- A learning system that allows the creation and learning of utility models [9, 10] to re-achieve goals and learn skills associated with them. It is also devoted with learning world models that represent the domains the robot is in. This system works online.
- And finally, a memory system that allows storing and relating goals, models and all the knowledge that is generated. As this knowledge is domain dependent, the memory is based on a contextual representation [11] to be able to reuse knowledge in the appropriate conditions.

2.1 Knowledge Representation

This subsection provides a description of the different types of knowledge nodes used in the long-term memory of the e-MDB. In addition, for clarity in the explanations, definitions of need and drive are also included. Although they do not represent knowledge acquired by the robot.

- *Need*: A need n_j is an internal state the robot seeks to achieve or maintain.
- *Drive*: A drive, D_j, is associated to a need n_j, and reflects how far the system is from satisfying it.
- *Goal*: It is a perceptual situation (point or area G_r) that when reached within a domain reduces the value of, at least, one drive D_j. These points provide utility.
- *Utility Model (UM):* It is a function associated with a goal G_r, which indicates the probability of reaching it (expected utility, $\hat{u}$) starting from a certain perceptual point S_t.
- *World Model (WM)*: It is a function that represents the behavior of the domain in which the robot is operating. It allows predicting which perceptual state S_{t+1} will result when an action a_i is carried out starting from a given perceptual point S_t.
- *Policy (π)*: A policy π_i is a decision structure associated with a goal G_r that provides the optimal action to be applied in a certain perceptual point S_t.
- *P-node*: The P-nodes or Perceptual classes are discrete entities that group sets of perceptions (continuous). The perceptions have in common that by applying the same action in the same domain, the same perceptual goal is reached.

It is important to note that utility models and world models are prediction structures that conform the deliberative knowledge a robot has. Whereas policies make up the reactive knowledge.

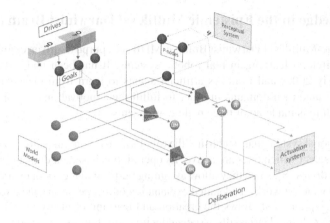

Fig. 1. e-MDB operational structure.

In the e-MDB, the different knowledge nodes are related to each other through the operational structure shown in Fig. 1. In this memory structure, vertices are the knowledge nuggets acquired by the robot, while the links are their activation relationships. This structure of knowledge is based on contexts or C-nodes. C-nodes link P-nodes (initial perceptions), with goals (final perceptions) and the utility models or policies necessary to go from ones to the others. In turn, as the robot works in different domains, they also relate the world models (that represent the domains) to these nodes.

The activation of goals will depend on the drives (that are domain independent). The activation of the P-nodes will depend on the perception of the robot. Whereas the activation of the world models will depend on the identification of the domain by the robot. In this way, the most active context (P-node, goal, and world model) will determine the action or policy to be executed.

2.2 Knowledge Acquisition and Decision Making

As indicated in the previous section, the e-MDB operation (represented in Fig. 1) revolves around the concept of C-node and drives.

The robot designer will provide the robot with a set of operational drives, opD, that define the purpose of the robot and a set of cognitive drives, cgD, that will allow it to explore and learn. Usually, the designer also provides some context independent goals for those cognitive drives and associates them to policies or utility models that can be applied in any domain. This way, the robot can decide on some actions based on cognitive drives. At the beginning of its operation, the robot will start to explore and acquire information related to the domain it is in by executing the actions suggested by its cognitive drives. It will be able to start modeling the domain, that is, learning the corresponding world model.

This stage continues until the robot finds a state S_t that provides utility. This state will lead to the creation of a goal, which will be linked to the drive it satisfied. In turn, it will trigger the creation process of a utility model associated with the goal in that domain. Also, if no P-node was active when the goal was found, a new one will be created with S_t

as the seed. Otherwise, S_t will be added to the active P-node. Finally, a C-node will link the goal, the world model corresponding to the domain in which the robot is located, and the P-node that was active (or that has just been created) with the utility model or the policy that selected the action that led to the goal.

The learning and improvement of models will alternate with the discovery of new goals and the creation of new C-nodes. The activation of the C-nodes will make the robot's actions be directed by reactive or deliberative processes, depending on whether they are connected to policies or utility models.

As a summary, Fig. 2 shows the action decision processes that the robot can take when trying to achieve one goal in a domain.

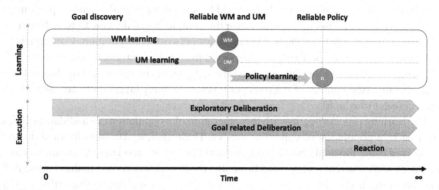

Fig. 2. Action decision processes during the operation time of a robot in a domain when trying to achieve one goal in open-ended fashion.

Of course, deliberation is much slower than reactive action selection and its performance clearly depends on the fidelity of the WMs and UMs. However, it is often the case, especially at the beginning of the interaction of a robot with a new domain, that no policy is available for that context, and the only way an action can be selected is through deliberation over whatever WM and UM the system has available.

3 Policy Learning Process

Given the importance of having policies available, this section is dedicated to the formal description of their learning process from the deliberative knowledge (UMs and WMs) acquired by the robot through its interaction with the world.

As explained above, pre-learning policies directly in open-ended learning problems is not possible, since we do not know the tasks or the domains that the robot will have to face throughout its life. However, this can be done from the deliberative models that the robot has learned through its interaction with the world. These models, once they are properly learnt, contain the necessary information to obtain the optimal action to apply each moment of time when trying to achieve a goal in a domain. They allow carrying out prospection processes (world models) and evaluating possible actions (utility models).

In the specific case of the e-MDB, all the information / knowledge related to how to achieve a certain goal in a given domain is linked through a C-node. Therefore, to generate a reactive knowledge node, it is enough to restructure and reuse this information.

As previously said, the main objective of the e-MDB cognitive architecture is the satisfaction of the motivation of the agent, which may be expressed as the maximization of drive satisfaction, which is represented as expected utility $\hat{u}$ each instant of time. Thus, the policy learning process must obtain the policy that provides the actions with the highest expected utility.

In the current version of the e-MDB, in order to use the most general and powerful prediction structure as possible, both models and policies are represented by means of Artificial Neural Networks (ANN). Thus, policy learning will consist in producing an ANN that implements the policy. This would be something complicated in open-ended problems, since we do not know in advance the complexity of the task to be solved and, therefore, it is difficult to estimate the necessary network configuration. Consequently, model learning has been performed by means of the NEAT neuroevolutionary algorithm [12], since it allows us, not only to adjust the weights of the network, but also to learn its architecture (the number of neurons per layer, how many layers there are, and which layers connect to which).

The learning process begins with the generation of a population of candidate policies (Π), that is, a population of candidate ANNs. These policies are initially random and are evolved using the deliberative models (world and utility models) as "simulators and evaluators of the current reality". Additionally, the P-node connected to the C-node that links the world model and the utility model, indicates for which states that utility model is applicable. Therefore, to evaluate an individual π_i from the policy population Π, the following procedure is applied over a set of representative perceptions/states belonging to the P-node, $P^* \epsilon P_n$:

1. A perception/state belonging to the P-node is taken: p_k.
2. p_k is the input of the policy under evaluation, obtaining an action: a_{ik}:

$$a_{ik} = \pi_i(p_k)$$

3. This action is used as an input in the current world model, WM, together with the perception p_k, providing the predicted perception in $t + 1$:

$$p_{k+1} = WM(p_k, a_{ik})$$

4. From this predicted sensorial information, the current utility model, UM, provides the expected utility for the action a_{ik}:

$$\hat{u}_j = UM(p_{k+1})$$

This process is repeated for all the perceptions in P^*, $p_k \epsilon P^*$. Then, the expected utility is averaged, making up the fitness for policy π_i. The remaining policies in the population are evaluated using the same set of perceptions and the same deliberative models. When evolution finishes, the policy with the highest fitness (highest average expected utility) is selected and linked to the corresponding C-node. Algorithm 1 shows the pseudocode of this approach.

Algorithm 1 Policy evaluation process

p_k: current perceptual point
P^*: set of representative perceptions from the P-node
π_i: individual policy
Π: population of candidate policies
π^*: optimal policy from π_i
WM: world model
UM: utility model
$\hat{u}_k$: expected utility provided by UM for p_k
$\bar{u}_i$: average expected utility for π_i
a_{ik}: action provided by π_i for p_k

1: for $\pi_i \in \Pi$ do
2:　for $p_k \in P^*$ do
3:　　$a_{ik} \leftarrow \pi_i(p_k)$
4:　　$p_{k+1} \leftarrow WM\ (a_{ik}, p_k)$
5:　　$\hat{u}_{k+1} \leftarrow UM\ (p_{k+1})$
6:　$\bar{u}_i \leftarrow getAverage(\hat{u}_{k+1})$
7: $\pi^* \leftarrow \pi_i$ with $max(\bar{u}_i)$

---

4 Real Robot Example

The previous approach for the autonomous learning of reactive knowledge within the e-MDB has been tested in a real robotic experiment. The experimental setup is displayed in Fig. 3 (a), and it includes a Baxter robot, a white table with a target area, a circular box, and a movable item: a yellow cylinder. The final objective is to collect the cylinder and place it in the box whatever the initial positions of the objects on the table. The robot's drive system has been designed for this purpose, and it includes one operational drive that makes the robot try to fulfill the objective, and two cognitive drives that allow the robot to explore the state space (see [7] for more details).

The Baxter robot has two arms with 7 degrees of freedom each. In this experiment we have limited their motion to a set of actions. These actions control the movement of both arms, that is, the change in the direction of movement (from 0 to 360 degrees) of the Baxter effectors and their displacement (from 0 to 5 cm each step in time). Therefore: $A(t) = (\alpha_{left}, v_{left}, \alpha_{right}, v_{right})$. Both arms move at a constant height. Also, some actions are predefined: if the robot reaches the cylinder, it will pick it up automatically. Likewise, if the robot reaches the box while carrying the cylinder, it will drop it inside. Regarding perception, the sensors used in the experiment are a RGB camera placed on the ceiling of the room (to see the scenario) and binary sensors in the grippers indicating whether they are holding something. The camera information was re-described so that the robot can know the x and y position of the objects on the table, as well as the positions of its effectors.

Initially, the robot has no idea where the goal is or how to reach it. Consequently, it has to learn all the knowledge nuggets necessary to solve the problem and associate them with their corresponding contexts to be able to reach the goal regardless of the initial conditions of the scenario. However, to focus on the policy learning process, the explanation of the results starts from the point where the robot has learned the corresponding deliberative models (world model and utility models), as well as the contexts in which to use them. Details of these learning processes can be found in [10, 11].

The memory structure from which the robot must learn the reactive models is like the one shown in Fig. 1. In this case, it contains five utility models that correspond to the different initial situations (represented in the P-nodes) that the robot can encounter. Specifically, it contains one model to approach the cylinder and grasp it with the right arm, a second one to grasp it with the left arm, a third one to approach the box and leave the cylinder inside with the right arm, and the corresponding one with the left arm. In turn, as there are situations in which the cylinder is on the opposite side of the table to the box, there is a utility model to perform an arm exchange, that is, the arms move closer to each other so that the cylinder can be moved from one gripper to the other. This process is shown in Fig. 3 (a).

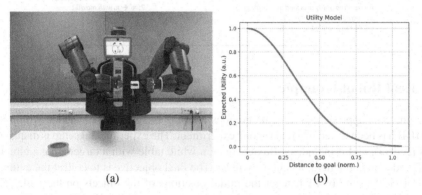

(a)   (b)

Fig. 3. (a) Experimental setup and (b) Utility model used to evaluate the policies.

Following the method explained in Sect. 3, the policies representing the five previous utility models were learned. The learning process has been like the one shown in Fig. 2. As models were considered properly learnt (central part of Fig. 2), the learning process of the different policies was carried out. Of them, we will explain how the process has been in the policies corresponding to putting the cylinder in the box and to the exchange of arms. In what follows, for clarity, we will refer to them as *'Put object in'* and *'Change hands'*. These polices present different complexities, since the former controls only the movement of one arm $(\alpha_{left}, v_{left})$, while the latter controls the movement of both $(\alpha_{left}, v_{left}, \alpha_{right}, v_{right})$.

The parameters of the ANNs that represent the policies were evolved using the MultiNEAT implementation of NEAT, extracted from [13], which was configured as shown in Table 1 (the rest of the algorithm parameters have been kept at their default values). Likewise, to illustrate the evaluation process of the candidate policies, Fig. 3 (b) shows a generic representation (in the form of distance to the goal) of the utility model used to evaluate each policy. In one case (*'Put object in'*), the expected utility is related to the distance between the arm and the box. While in the other, the utility value is related to the distance between arms. In the learning process of the ANNs, the inputs were those corresponding to the perceptions of the robot, while the outputs corresponded to the 4 actions that control the behavior of its arms.

During the knowledge restructuring process, a series of perceptions/states representative of the P-nodes were randomly generated from the P-nodes. For example, in the case of the *'Change hands'* policy, these perceptions corresponded to situations where the robot had the cylinder grasped in the gripper opposite to the position of the box. These points were used, together with the respective world and utility models, to evaluate the utility of the actions proposed by the different candidate policies. Thus, at the end of the evolutionary process, the best of these policies was linked to the corresponding C-node.

Table 1. Parameters of the evolutionary process.

Generations (epochs)	1000
Candidate policies (population)	100
Input/output neurons	4
Number of perceptions / states used for evaluation	1000 points evaluated in batches of 100 Batches changed every 30 generations

To evaluate how optimal the actions provided by the policies were, their behavior has been compared to that of different deliberative action selection systems, which are reference processes that always provide the best possible action from among the candidates. Any deliberative process will choose actions out of a repertoire of possible actions that need to be tested individually (prospection and evaluation needs to be carried out for each one of them). Consequently, the larger the repertoire, the higher the probability of finding just the right action for each situation at the cost of much higher processing times (evaluation of many more possible actions). On the other hand, the smaller the repertoire, lower processing times at the cost of less adapted actions and thus rougher action trajectories. In this paper, we compared the reactive processes produced by the policies to deliberative processes that consider action repertoires that include 1, 5, 50, 500 and 5000 candidate actions. To this end, Fig. 4 shows a representation of 100 random starting points (perceptual situations belonging to the P-nodes) ordered by increasing distance from that point to the goal, and the number of iterations required to reach the goal using the learned policies and the different deliberative models. It should be noted that the candidate actions of the deliberative systems were generated at each instant of time and took random values over the continuous range of possible actions.

Thus, if we look at the results obtained for the *'Put object in'* task (Fig. 4 (a)), it is possible to see how the policy behaves satisfactorily, since it is 2 steps away from solving the task as efficiently as the best of the deliberative systems. On the other hand, if we look at the *'Change hands'* policy, the result obtained is unbeatable. In this case, the optimal number of steps to solve the task (in the maximum distance) was 23, and both the policy and the deliberative system that contemplated 5000 actions solve it in those steps. The difference in the results obtained is because the *'Put object in'* task is more complex to resolve for the ANN that represents the policy. This is due to the fact that the robot needs greater accuracy in the final section of the task because if the displacement of the last step is too large, it can go beyond the position of the box. Thus, to avoid this, the policy chooses to take smaller steps as the robot approaches the box, which makes it less efficient. On the other hand, the fact that in the first case (Fig. 4 (a)) the deliberative system requires fewer candidate actions to solve the task in an optimal way than in the second case (Fig. 4 (b)) is because it only controls one of the arms, which makes the search space for candidate actions smaller.

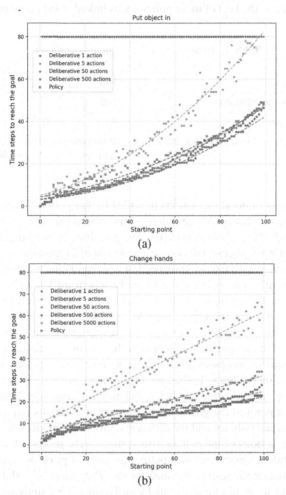

Fig. 4. Comparison of the time steps needed to solve the task with reactive and deliberative decision-making models. (a) Results for the *'Put object in'* task and (b) Results for the *'Change hands'* task. In both cases, the maximum number of time steps allowed to solve the task was 80.

In addition, to compare the decision times using both methods (reactive and deliberative), that is, the time required to obtain the actions, Table 2 shows the orders of magnitude of the average times for deciding on 1 action with each of the methods shown in Fig. 4. This average time has been calculated over 100 actions with the program running and using a processor intel i7-7700HQ CPU @ 2.80GHz. These data show that reactive processes are computationally much less costly to execute than deliberative action selection systems. This is because executing a policy implies evaluating just one

ANN while the deliberation involves evaluating two ANNs (world model and utility model) for each of the candidate actions.

Table 2. Time required to obtain one action with reactive and deliberative decision-making models.

Decision-making system	Average time (seconds)
Reactive model (Policy)	10^{-5}
Deliberative model (1 candidate action)	10^{-5}
Deliberative model (5 candidate actions)	10^{-4}
Deliberative model (50 candidate actions)	10^{-3}
Deliberative model (500 candidate actions)	10^{-2}
Deliberative model (5000 candidate actions)	10^{-1}

Thus, it is easy to see that the policies learned with the proposed method have provided very satisfactory results. In terms of behavior, they are at the level of the best deliberative models and are much better than a deliberative system with small numbers of candidate actions, which would be the closest in response time. In addition, the difference in computing time in favor of the policies, allows the robot to have the type of reflex behaviors that we were looking for.

5 Conclusions

In this article, we have proposed and tested an approach to learning reactive knowledge in open learning environments using the e-MDB cognitive architecture. It is based on the concept of policy, a decision structure that provides the action to be applied based on sensory inputs. These policies are learned using deliberative models (world and utility models in the e-MDB) as simulators of the real world.

The results using the policies obtained with this approach in a real robot experiment are comparable to (or even better than) deliberative action selection procedures, which is very relevant considering the difference in computing time between both methods. The effectiveness of this approach depends on the quality of the perceptual points used to train the ANNs, so a more detailed study of the management of the information present in the P-nodes could help to improve the results obtained. On the other hand, this opens a lot of interesting possibilities since, now, this reactive knowledge can be slightly modified to adapt to new situations, or it can even be composed to achieve complex reactive patterns. In addition, it opens a line of research on the coordination between deliberation and reaction for joint decision-making.

Acknowledgements. This work was partially funded by the Ministerio de Ciencia, Innovación y Universidades of Spain/FEDER (grant RTI2018-101114-B-I00), Xunta de Galicia (grant EDC431C-2021/39) and by the Spanish Ministry of Education, Culture and Sports through the FPU grant of Alejandro Romero. We also acknowledge the support received from "CITIC", funded by Xunta de Galicia and the European Union (European Regional Development FundGalicia 2014–2020 Program), by grant ED431G 2019/01.

References

1. Frankish, K.: Dual-process and dual-system theories of reasoning. Philos. Compass **5**(10), 914–926 (2010). https://doi.org/10.1111/j.1747-9991.2010.00330.x
2. Doncieux, S., et al.: Open-ended learning: a conceptual framework based on representational redescription. Front. Neurorobot. **12**(SEP), 1–6 (2018). https://doi.org/10.3389/fnbot.2018.00059
3. Peters, J., Schaal, S.: Policy gradient methods for robotics. IEEE. Int. Conf. Intell. Robot. Syst. 2219–2225 (2006). https://doi.org/10.1109/IROS.2006.282564
4. Grondman, I., Busoniu, L., Lopes, G.A.D., Babuska, R.: A survey of actor-critic reinforcement learning: standard and natural policy gradients. IEEE Trans. Syst. Man Cybern. Part C (App. Rev.) **42**(6), 1291–1307 (2012). https://doi.org/10.1109/TSMCC.2012.2218595
5. Deisenroth, M.P.: A survey on policy search for robotics. Found. Trends Robot. **2**(1–2), 1–142 (2011). https://doi.org/10.1561/2300000021
6. Bellas, F., Duro, R.J., Faiña, A., Souto, D.: Multilevel Darwinist brain (MDB): artificial evolution in a cognitive architecture for real robots. IEEE Trans. Auton. Ment. Dev. **2**(4), 340–354 (2010). https://doi.org/10.1109/TAMD.2010.2086453
7. Romero, A., Bellas, F., Becerra, J.A., Duro, R.J.: Motivation as a tool for designing lifelong learning robots. Integr. Comput.-Aid. Eng. **27**(4), 353–372 (2020). https://doi.org/10.3233/ICA-200633
8. Hawes, N.: A survey of motivation frameworks for intelligent systems. Artif. Intell. **175**(5–6), 1020–1036 (2011). https://doi.org/10.1016/j.artint.2011.02.002
9. Romero, A., Prieto, A., Bellas, F., Duro, R.J.: Simplifying the creation and management of utility models in continuous domains for cognitive robotics. Neurocomputing **353**, 106–118 (2019). https://doi.org/10.1016/j.neucom.2018.07.093
10. Prieto, A., Romero, A., Bellas, F., Salgado, R., Duro, R.J.: Introducing separable utility regions in a motivational engine for cognitive developmental robotics. Integr. Comput.-Aid. Eng. **26**(1), 3–20 (2018). https://doi.org/10.3233/ICA-180578
11. Duro, R.J., Becerra, J.A., Monroy, J., Bellas, F.: Perceptual generalization and context in a network memory inspired long-term memory for artificial cognition. Int. J. Neural Syst. **29**(6), 1–26 (2019). https://doi.org/10.1142/S0129065718500533
12. Stanley, K.O., Miikkulainen, R.: Evolving neural networks through augmenting topologies (2012). http://direct.mit.edu/evco/article-pdf/10/2/99/1493254/106365602320169811.pdf. Accessed 16 May 2021
13. GitHub - MultiNEAT/MultiNEAT: Portable NeuroEvolution Library http://MultiNEAT.com. https://github.com/MultiNEAT/MultiNEAT. Accessed 16 May 2021

Managing Gene Expression in Evolutionary Algorithms with Gene Regulatory Networks

Michael Cilliers and Duncan A. Coulter[✉]

University of Johannesburg,
Cnr Kingsway and University Rds, Auckland Park, Johannesburg, South Africa
dcoulter@uj.ac.za

Abstract. This paper evaluates the effectiveness of using gene regulatory networks to manage gene expression in evolutionary algorithms for the purpose of balancing exploitation versus exploration. This builds on previous work that has shown that the introduction of non-coding genes can improve the ability of an evolutionary algorithm to adapt to change in the environment. As part of the paper an algorithm is developed and a prototype is implemented. The developed algorithm is compared to the standard genetic algorithm and previously developed methods for managing gene expression. Results show that the developed algorithm can outperform the standard genetic algorithm in dynamic environments. The algorithm is however not able to outperform all the other developed methods of managing gene expression and avenues for future improvement will be explored.

Keywords: Evolutionary algorithm · Simulated epigenetics · Gene regulatory network

1 Introduction

When applying evolutionary algorithms to a dynamic environment, the algorithm's ability to adapt has to be considered. Small changes in the environment can often be adjusted for, by the rate of mutation that is applied to the individuals in the population. This is however not optimal for larger environmental changes which requires greater exploration. To achieve lager exploratory jumps in the search space, reproduction is required. A child produced by the reproduction process will be dissimilar to both parents if they are dissimilar from each other. Thus for the algorithm to exhibit the required exploration a diverse population is required. Stochastic population diversity is not a problem during the initial stages of the algorithm as the generation process produces sufficient diversity. However as the algorithm converges on a solution the diversity diminishes. The algorithm will not be able to effectively adjust to large changes in a dynamic environment due to a lack of diversity.

© Springer Nature Switzerland AG 2021
H. Sanjurjo González et al. (Eds.): HAIS 2021, LNAI 12886, pp. 77–87, 2021.
https://doi.org/10.1007/978-3-030-86271-8_7

Previous work [1] has shown that the introduction of methylated genes could be effective at improving population diversity after convergence. The non-coding genes act as a pool of diversity that can be accessed when the population needs to adapt. The previous work explored different methods of managing the expression of genes. The first explored method directly stored the expression status for each gene in an accompanying array. The Second method uses a bloom filter to manage which of the genes are expressing and which not. This work expands on the previous work by examining the effectiveness of gene regulatory networks as a method for managing the gene expression. It should be noted that there are other evolutionary algorithms that make use of non-coding genes. A good example of this is gene expression programming [12]. Gene Expression programming stores a sequence of genes that is transformed into a tree. All the genes n the initial sequins will not form part of the expressed tree and will thus be non-coding genes [3].

2   Biological Gene Expression

Gene expression refers to the process where DNA is processed into a protein. First the portion DNA is transcribed into RNA. The RNA then gets translated into the protein. There is however not a one to one mapping of genes that make up the entire DNA sequence and the resulting protein that is produced. A number of mechanisms exists that can result in genes not directly contributing to the protein. These genes are referred to as non-coding genes. One of the mechanisms that results in non-coding genes is gene methylation. Gene methylation is when methyl groups form on some of the base pairs that make up the DNA [7]. During the transcription process if a methyl group is present on a gene it will be skipped over and to not be transcribed. The gene will thus be non-coding.

The expression of genes is however a bit more complicated than the simplified process described above. There are a number of interactions that occur between different portions of DNA. These interactions can occur during at the different steps of the expression process. For example the interaction can be with DNA, RNA, the produced protein and in some cases even between different cells. These complex interactions regulating the expression of genes are referred to as gene regulatory networks [2]. When modeling the gene regulatory network the DNA portions will represent the nodes in the network and the interactions between the DNA pieces will represent the edges.

3   Gene Regulatory Networks

To implement an algorithm inspired by gene regulatory networks, a number of node layers will be connected [6]. The different layers will represent the different steps where the interactions can take place in the biological gene regulation network. Each of the layers will consist of a number of nodes. These nodes match the DNA portions during each specific step. The initial unaltered DNA that is involved during the first step of biological gene expression will matched by having a number of inputs that will be used by the first (input) layer. These values will then pass from layer to layer, being adjusted at each layer based on the nodes in that layer [10]. The output for the network will be a number of Boolean values indicating which of the genes should be expressed.

The gene regulatory networks implemented for the developed algorithm uses decimal values for inputs. Each of the nodes in the network will have an associated real valued weight that will be applied to the value reserved as the input by multiplying the values together. If a node has multiple inputs, the weight will be applied to each input after which the results will be added. The final (output) layer will be produce real values that have to be converted to the required Boolean values. This conversion will be done by rounding the output and evaluating the result. Even values will be interpreted as true and odd values as false. This process is demonstrated in Fig. 1.

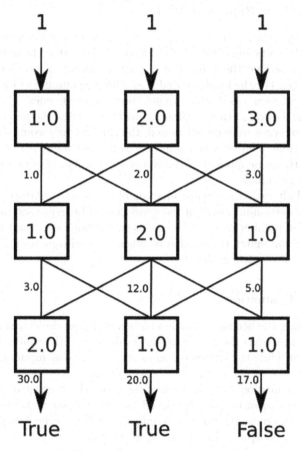

Fig. 1. Gene regulatory network

The number of layers in the network can be adjusted as needed. By decreasing the number of layers it will simplify the network which should make it easier to evolve the network. Increasing the number of layers will make it harder to evolve, but will allow more complex relationships to develop.

4   Proposed Algorithms

This paper proposes the use of a gene regulatory network to manage the expressions of genes in an evolutionary algorithm. The hope is that by using a method that better matches the biological process the algorithm will be better at adapting to changes when required. The standard genetic algorithm [11] will be used as a base and adapted as necessary. This simplifies the development of the algorithm and allows focus to be applied to the relevant portions of the algorithm.

4.1   Chromosome Representation

The proposed algorithm will make two changes to the chromosome representation in the standard genetic algorithm [9]. The first change is that the number of genes will be increased so that the number of genes will exceed the number required for fitness evaluation. For the implemented algorithm the number of inputs required by the fitness function was doubled to get the number of genes. The extra genes that are not used during fitness evaluation will thus act as the non coding genes. If the number of genes were not increased, the chromosome would not be able to provide a sufficient number when some genes are not used. All the individuals will thus have the same number of genes, but the number of expressed genes can vary between individuals.

The second alteration is the introduction of the gene regulatory network that will be used to determine which of the genes should be expressed and which are not coding. The number of layers in the network can vary, but the number of nodes per layer will match the number of genes in the chromosome. The updated chromosome is illustrated in Fig. 2.

4.2   Fitness Evaluation

When performing the fitness evaluation the genes do not directly serve as inputs for the fitness function. Each gene is first evaluated to determine if it is being expressed or not. Only the expressed genes are then used as inputs for the fitness function.

To get the expression state for each of the genes the gene regulatory network is used. The number of outputs produced by the network matches the number of genes in the chromosome. So to determine if a specific gene is being expressed the corresponding output value has to be evaluated.

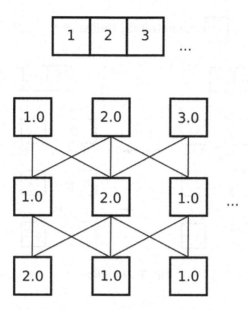

Fig. 2. Chromosome representation

4.3 Mutation

The mutation operation will be divided into two steps. The first step will be to apply mutation to the gene portion of the chromosome. This will remain unaltered from the mutation step used in the standard genetic algorithm [4]. The second step will apply mutation to the gene regulatory network portion of the chromosome. To apply this step a number of the nodes making up the gene regulatory network will be selected. The node selection will use the mutation rate when performing the selection. The weights of the selected nodes will then be altered by a random amount as shown in Fig. 3.

The rate at which the expression of the genes change should be slower than the rate at which the genes themselves change. If the expression updates too rapidly, the algorithm will not be able to converge effectively. To achieve this varying rate of change two separate mutation rates will be used. The mutation rate for the gene regulatory network will be lower than the gene mutation rate.

4.4 Crossover

The crossover operation will also be divided into two steps. The selection of parents will remain unchanged form the standard genetic algorithm, as will the selection of genes making up the children [5]. The added step will control how

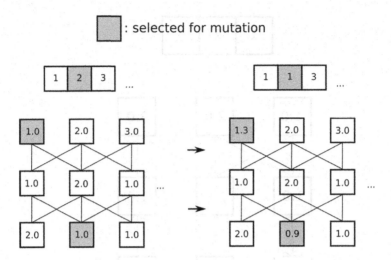

Fig. 3. Mutation

the gene regulatory networks get assigned to children. As mentioned above the change in expression should not occur too rapidly. For this reason the gene expression network will be passed on unaltered. One child will receive the gene regulatory network from one parent and the other child from the other parent as shown in Fig. 4.

5   Evaluation

To evaluate the effectiveness of the developed algorithm it will be applied to find the optima defined by one of the standard fitness landscapes. In this case the Rastrigin function [8] was selected to define the landscape. To simulate a dynamic environment the algorithm will be allowed to converge on a solution. The target will then be transposed to simulate a change in the environment. The optima will thus move to a different point in the search space and the algorithm will have to re-optimize. The algorithm will have to switch focus back to exploration to effectively adapt to the change. The time it takes to converge on the new target can be used to compare the effectiveness of different algorithms.

To eliminate variance in individual runs of the algorithm, the algorithm will be executed multiple times. The fitness at each generation will be averaged out across all the performed runs. The resulting values will then be used for the algorithm comparison.

A parameter study was performed to determine the parameters shared between the standard genetic algorithm and the developed algorithm. A population size of 1000 was used and a mutation rate of 15% for the standard genetic algorithm and the gene portion of the developed algorithm. Each variation was executed 200 times before combining the result.

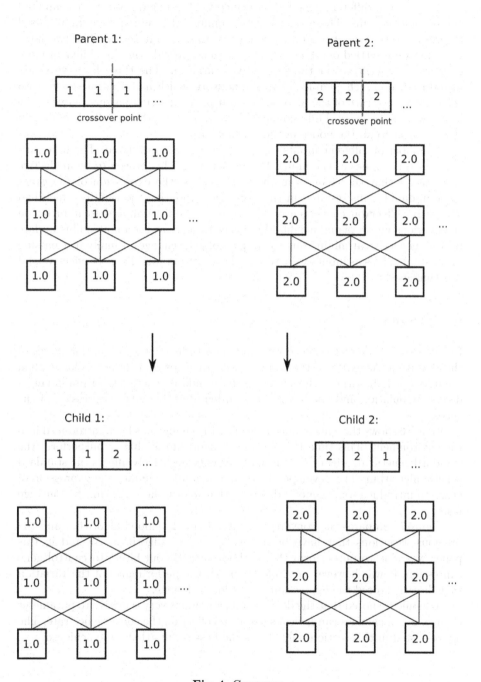

Fig. 4. Crossover

A number of different gene regulatory network configurations were evaluated across multiple runs. These varied in the number of layers and the inputs used for the network. One, two and three layer variation were tested. The two layer variation were settled on. The extra layer would provide the possibility of more complex interactions over the single layer variation. The three layer variation introduced too much complexity in the network which had a negative effect on the convergence rate of the algorithm. Multiple connection patters between the layers were also testes. Fully connected layers were settled on. Each node will thus connect to all the nodes in the previous layer.

A number of different inputs for the network were evaluated. The first variation used the genes as the inputs for the network. This proved to be ineffective as it meant that the changes in the genes caused the expression of the genes to be updated. This resulted in the expression updating too rapidly which had a negative effect on convergence. The next method attempted to improve on the previous by averaging neighboring genes to serve as the inputs. This update proved ineffective at minimizing the impact of gene changes. Finally just passing in constant values as inputs were tested. This method was the most effective and was thus used.

6   Results

Figure 5 shows the convergence rate the algorithms during the initial stages of the algorithm. As expected the standard genetic algorithm converges faster than the developed algorithm. This is due to the standard genetic algorithm having a diverse population and not having the burden of evolving the expression of the genes.

Figure 6 shows the convergence rate after the change has been introduced into the environment. This time the developed algorithm is able to outperform the standard genetic algorithm. The lack of diversity negatively affects the standard genetic algorithm. The developed algorithm is able to update the expression of genes to introduce some genetic diversity that allows the algorithm to converge faster.

Figure 7 compares the convergence rate of the developed algorithm and the previous variations that have been developed. The algorithm covered in this paper was able to outperform the method using Bloom filters to control gene expression. It was however not able to match the performance of the variation that directly stored the expression status for each gene.

It should be noted that the developed algorithm is very sensitive to parameter change. Small changes can have a substantial effect on the convergence rate. The figures listed in this section only shows the best results that were achieved.

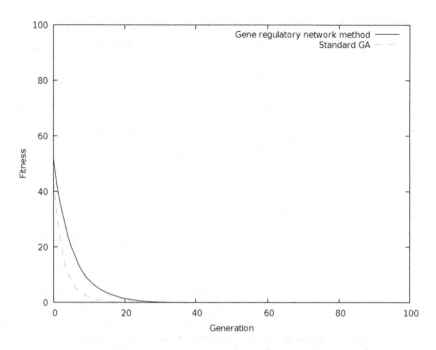

Fig. 5. Convergence rate during start of the algorithm

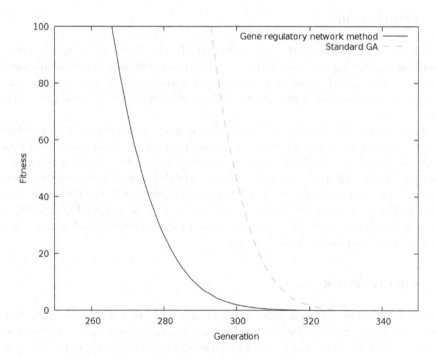

Fig. 6. Convergence rate after change in the environment

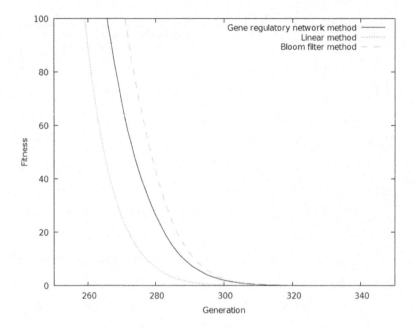

Fig. 7. Convergence rate compared to previous methods

7   Conclusion

Results from the preformed tests show that the developed algorithm is better than the standard genetic algorithm at adjusting to changes in a dynamic environment. This matches previous findings, that the introduction of non coding genes can serve as pool of genetic diversity that improve the adaptability of the algorithm.

Even though the developed algorithm was able to outperform the standard genetic algorithm, it was not able out outperform all the other variation previously developed. It was able to outperform the variation that tracks gene expression with bloom filters, but was not able to match the variation using an array to track the expression of each gene. This is probably due to the increased complexity. The more complex structure requires more evolution to be updated to the required level. So in this case simplicity produces better results than more closely mimicking the biological process.

8   Future Work

Two portions of future work have been identified. The first is to expand the testing of the developed algorithm to include more existing algorithms. An algorithm making use of the island model would be a good candidate for the expanded testing as it also attempts to improve population diversity.

The work presented in this paper can also be expanded by exploring the pre-training of networks. If the best performing networks can be identified, it should be possible to use these networks during future executions of the algorithm. This should cut down on the time required to evolve the gene expression, which cloud in turn cut don on the convergence time of the algorithm.

References

1. Cilliers, M., Coulter, D.A.: Improving population diversity through gene methylation simulation. In: Rutkowski, L., Scherer, R., Korytkowski, M., Pedrycz, W., Tadeusiewicz, R., Zurada, J.M. (eds.) ICAISC 2019. LNCS (LNAI), vol. 11508, pp. 469–480. Springer, Cham (2019). https://doi.org/10.1007/978-3-030-20912-4_43
2. Davidson, E., Levin, M.: Gene regulatory networks. Proc. Natl. Acad. Sci. **102**(14), 4935 (2005)
3. Ferreira, C.: Gene expression programming: a new adaptive algorithm for solving problems. Complex Syst. **13** (2001)
4. Hassanat, A., Almohammadi, K., Alkafaween, E., Abunawas, E., Hammouri, A., Prasath, V.: Choosing mutation and crossover ratios for genetic algorithms–a review with a new dynamic approach. Information **10**(12), 390 (2019)
5. Juneja, S.S., Saraswat, P., Singh, K., Sharma, J., Majumdar, R., Chowdhary, S.: Travelling salesman problem optimization using genetic algorithm. In: 2019 Amity International Conference on Artificial Intelligence (AICAI), pp. 264–268. IEEE (2019)
6. Keedwell, E., Narayanan, A., Savic, D.: Modelling gene regulatory data using artificial neural networks. In: Proceedings of the 2002 International Joint Conference on Neural Networks, IJCNN 2002 (Cat. No. 02CH37290), vol. 1, pp. 183–188. IEEE (2002)
7. Kornienko, A.E., Guenzl, P.M., Barlow, D.P., Pauler, F.M.: Gene regulation by the act of long non-coding RNA transcription. BMC Biol. **11**(1), 59 (2013)
8. Merkuryeva, G., Bolshakovs, V.: Benchmark fitness landscape analysis. Int. J. Simul. Syst. Sci. Technol. **12**(2), 38–45 (2011)
9. Tan, B., Ma, H., Mei, Y.: Novel genetic algorithm with dual chromosome representation for resource allocation in container-based clouds. In: 2019 IEEE 12th International Conference on Cloud Computing (CLOUD), pp. 452–456. IEEE (2019)
10. Turner, A.P., Lones, M.A., Fuente, L.A., Stepney, S., Caves, L.S., Tyrrell, A.M.: The incorporation of epigenetics in artificial gene regulatory networks. Biosystems **112**(2), 56–62 (2013)
11. Whitley, D.: A genetic algorithm tutorial. Stat. Comput. **4**(2), 65–85 (1994)
12. Yang, B., Wang, G., Bao, W., Chen, Y., Jia, L.: CSE: complex-valued system with evolutionary algorithm. IEEE Access **7**, 90268–90276 (2019)

Evolutionary Optimization of Neuro-Symbolic Integration for Phishing URL Detection

Kyoung-Won Park[1]([⊠]), Seok-Jun Bu[2]([⊠]), and Sung-Bae Cho[1,2]([⊠])

[1] Department of Artificial Intelligence, Yonsei University, Seoul 03722, Korea
{pkw408,sbcho}@yonsei.ac.kr
[2] Department of Computer Science, Yonsei University, Seoul 03722, Korea
sjbuhan@yonsei.ac.kr

Abstract. A phishing attack is defined as a type of cybersecurity attack that uses URLs that lead to phishing sites and steals credentials and personal information. Since there is a limitation on traditional deep learning to detect phishing URLs from only the linguistic features of URLs, attempts have been made to detect the misclassified URLs by integrating security expert knowledge with deep learning. In this paper, a genetic algorithm is proposed to find combinatorial optimization of logic programmed constraints and deep learning from given 13 components, which are 12 rule-based symbol components and a neural component. The genetic algorithm explores numerous searching spaces of combinations of 12 rules with deep learning to get an optimal combination of the components. Experiments and 10-fold cross-validation with three different real-world datasets show that the proposed method outperforms the state-of-the-art performance of β-discrepancy integration approach by achieving a 1.47% accuracy and a 2.82% recall improvement. In addition, a post-analysis of the proposed method is performed to justify the feasibility of phishing URL detection via analyzing URLs that are misclassified from either the neural or symbolic networks.

Keywords: Neuro-symbolic integration · Genetic algorithm · Phishing detection

1 Introduction

Phishing is defined as a compound word of private data and fishing. Phishing URL attacks have been on the rise over the past few decades with technology growth, and it referred to as a cybersecurity attack aimed at stealing users' financial or personal information (i.e., password credit card information) by inserting URLs that link to phishing sites within emails or web pages [1–3]. In traditional deep learning-based modeling, phishing URL detection studies were conducted to extract the linguistic features of URLs and achieved approximate 90% accuracy performance.

However, a single undetected URL causes extremely severe damage and cascading impact to thousands of victims. In addition to the linguistic features, existing deep learning methods also had limitations in extracting the intrinsic features of phishing attack patterns and URLs, resulting in degrading recall performance. Besides these linguistic

© Springer Nature Switzerland AG 2021
H. Sanjurjo González et al. (Eds.): HAIS 2021, LNAI 12886, pp. 88–100, 2021.
https://doi.org/10.1007/978-3-030-86271-8_8

features, numerous studies have been conducted to overcome the limitation in recall performance along with phishing URL features based on expert knowledge which is not included in the text itself.

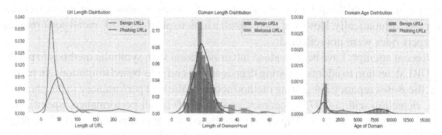

Fig. 1. Different characteristics between benign URL and Phishing URL.

On the other hand, there are distinguishable features between phishing URLs and benign URLs, as shown in Fig. 1. (a) shows that phishing URLs are longer than benign URLs and that they are mainly distributed at lengths less than 50. (b) shows also distinctive features in domain and host length. In addition, (c) shows that phishing URLs are distributed in a period of 0 or 2,500 days. Apart from the features described, precise detection is possible if scripts and domain features of URLs are used as additional features for deep learning.

In this paper, we propose a new approach method to use additional features via the defined 12 rules as well as to find the best features among 12 rules and a neural output in Sect. 3. In Sect. 4, the accuracy and recall performance for each component are compared via using three real-world datasets (ISCX-URL-2016, PhishStorm, and PhishTank), and a 1.47% accuracy and a 2.82% recall improvement are achieved against state-of-the-art integration methods. In addition, a post-analysis of the proposed method is also performed to justify the feasibility of phishing URL detection via analyzing URLs that are misclassified in either the neural or symbol components, respectively.

2 Related Works

The past six years of research in the field of phishing URL detection are summarized as shown in Table 1. Phishing URL detection field can be categorized into four parts: rule-based detection, machine learning-based detection, deep learning-based detection, and integration method of rules and deep learning. Starting with rule-based detection and transition programming of logic programming in 2016, XCS methods to adaptively evolve rules in the training process according to accuracy performance was also studied [4–7]. Furthermore, a method of applying rule-extracted features to machine learning algorithms such as Random Forest and SVM has been studied [8, 9].

However, there was a limit to the recall performance of phishing URL detection as the URL itself could not be reflected in the results. To solve this limitation of recall performance, deep learning is applied to extract the characteristics of characters and words' sequence in latent space [10–14]. By extracting linguistic features, the detection performance of phishing URLs that existing rule-based method could not detect has increased dramatically. However, there is still a limit to phishing URL recall performance as expert rules were not reflected.

Recent attempts have been made to integrate neural with symbolic methods in phishing URL detection to address existing deep learning and rule-based limitations. In particular, the β-discrepancy integrating method achieved the best performance in the phishing URL detection with a 0.9758 accuracy and a 0.9610 recall performance. Specifically, after learning separately the Triplet network-based neural components and 10 numbers of rule-based symbolic components that are defined as the form of first-order logic, the neural components make the main decisions, and the symbolic components support the decisions of the neural components through the loss function defined as $l_{nsi} = l_n + \beta_G \bullet l_s$. However, there are conflicts between the defined rules, and it causes the limitation of phishing URL detection [15, 16]. Thus, this paper proposes a new approach to optimize combinations of rules to remove collision of rules by applying genetic algorithms while not learning two components individually but reflecting all components simultaneously (Fig. 2).

Table 1. A six-year study of phishing URL detection.

Type	Year	Method
Logic programming	2016	Rule-based [4]
	2017	Transition Diagram [5]
	2018	Rule-based [6]
	2019	XCS [7]
Machine learning	2016	Random Forest [8]
	2020	AdaBoost with SVM [9]
Deep learning	2018	SOFM [10]
	2018	Word-level / Char-level CNN [11]
	2019	CNN-LSTM + XGBoost [12]
	2019	Word-level BiLSTM attention [13]
	2020	Char-level/Word-level Texception Network [14]
Neural-symbolic	2021	β-discrepancy loss-based joint learning [15]

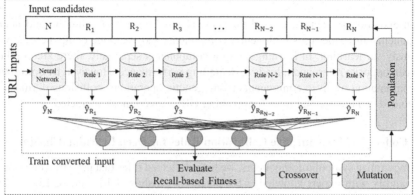

Fig. 2. GA-based neural symbolic integration structure for phishing URL detection.

Table 2. GA-based neural-symbolic integration structure for phishing attack detection.

Type	Rule features	First-Order Logic (FOL) expression
Address - bar	Usage of IP address Digit count Character length Special character count	1. $IPAddress(x_i) \wedge (Digits(x_i) > \mu_d) \wedge (Length(x_i) > \mu_l) \wedge (Special(x_i) > \mu_s) \Rightarrow Phishing\ (x_i)$ 2. $IPAddress(x_i) \wedge \exists c: (Special_\#(c) \wedge Special_\%(c)) \Rightarrow Phishing\ (x_i)$ 3. $Length(x_i) > \tau_l) \wedge (Digits(x_i) > \tau_d) \Rightarrow Phishing\ (x_i)$ 4. $(Dot(x_i) > \tau_d) \Rightarrow Phishing\ (x_i)$
Abnormal request	URL request count SFH	5. $(Request_{src}(x_i) > \mu_r) \Rightarrow Phishing\ (x_i)$ 6. $SFH_{about:blank}(x_i) \vee SFH_{empty}(x_i)) \Rightarrow Phishing\ (x_i)$
Domain	Age of domain Registered top domain Subdomain count Typo-squatted domain Suspicious TLD	7. $HTTPS(x_i) \wedge Typosquatted(x_i) \Rightarrow Phishing\ (x_i)$ 8. $HTTP(x_i) \wedge (Special(x_i) > \mu_s) \Rightarrow Phishing\ (x_i)$ 9. $\neg (Registered(x_i)) \wedge (Subdomain(x_i) > \tau_{sub}) \Rightarrow Phishing\ (x_i)$ 10. $\neg (Age(x_i) > \mu_a) \wedge (Subdomain(x_i) > \tau_{sub}) \Rightarrow Phishing\ (x_i)$ 11. $TLD(x_i) \Rightarrow Phishing\ (x_i)$
Script	Usage of mouseover Usage of pop-up Disabling right click	12. $\neg Rightclick(x_i) \wedge Mouseover(x_i) \wedge Popup(x_i) \Rightarrow Phishing\ (x_i)$

3   The Proposed Method

As an evolutionary method to find the optimal combination of symbolic and neural components, the paper uses both neural components and symbolic components simultaneously for learning. In learning symbol components separately, there are conflicts between opposite rules, and it leads to degrading performance. However, by optimizing the combination of given components using genetic operations, collisions in rules can be avoided and the optimal rules for phishing URL detection are extracted.

3.1   Collect and Structuralize Calibration Rules for Deep Learning Classifier

To structuralize symbolic components, we have collected rules that have proven fairness and validity with expert knowledge for detecting phishing URLs [4, 6, 15, 17, 18]. For example, in domain-based knowledge, a longer URL than 54 characters tends to hide the suspicious part, and it is likely to be classified as a phishing URL [17]. The rules are categorized into four criteria: address-based, abnormal request-based, domain-based, and script-based. We utilize such expert knowledge from former research, and 12 significant rules are evenly selected from each criterion.

A total of 12 rules are collected and organized in the form of first-order logic (FOL) as shown in Table 2. For example, the knowledge that 'the input URL has IP address and contains both '#' and '%' within the URL' is defined as follows: $IPAddress(x_i) \land \exists c : (Special_{\#}(c) \land Special_{\%}(c)) \Rightarrow Phishing(x_i)$.

3.2   Deep Learning Calibration and Optimization to Improve URL Classification Performance

Research has been conducting to find the optimal combination of input values or attributes from the genetic algorithm [19–22]. The genetic algorithm-based method enables broad exploratory search without falling into local minima, and it is possible to optimize the combination of 12 defined rules and a neural network in 8,192 entire combinations.

The genetic algorithm-based flowchart represented in Fig. 3 illustrates the steps of the proposed method. The first step is to initialize a population of chromosomes into a combination that can be generated from 13 components. The chromosomes of the generated combination are denoted by 0 and 1, and 0 is denoted by not selecting the component, while 1 is denoted by selecting the component. For example, if a gene corresponding to a neural component is labeled 1, it can be determined that the usage of the neural component is chosen in the chromosome.

The next step is to convert the denoted value of each gene to the output value of the corresponding component model to compute the recall fitness as the initialized chromosomes. A neural component converts a continuous probability distribution for both classes; phishing URLs and benign URLs, and symbolic components convert the results of each rule to discrete values.

To compute the optimal fitness for phishing URL detection, we define a recall-based fitness function. The function selects chromosomes that return a higher recall value than the highest recall value in the former population for generating a better chromosome in the following populations. The next step is to exchange chromosomes to a certain extent and mutate specific genes, thereby maintaining the diversity of chromosomes and adding flexibility to produce the best chromosomes without being trapped in the local minima.

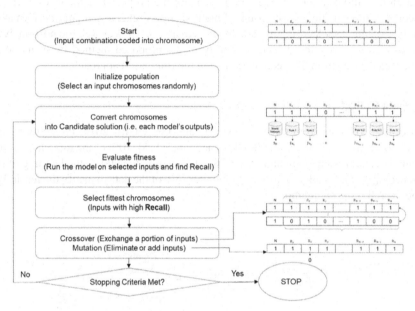

Fig. 3. Genetic algorithm flowchart of proposed method.

3.3 One-Dimensional Convolutional Recurrent Neural Network to Extract Sequence of Character and Word Features

A phishing URL is data that contains both character and word sequence. A shown in Fig. 4, the neural component of this paper is constructed as a one-dimensional convolutional recurrent neural network to extract the character and word sequence features within the URL [14, 15]. We divide into the two separated models to extract features from the input URL by focusing on the character features and word features, respectively, and then add them together sequentially at the end of the models to extract spatiotemporal features. In terms of character sequence, we focus on the spatial features of URLs by creating latent spaces with stacked CNN, which stack multiple convolutional layers. In terms of word sequence, tokenization is carried out, and the bidirectional Long Short-Term Memory (LSTM) model extracts the time-series properties of URLs.

4 Experimental Results

The experiments are completed by three real-world data. In Sect. 4.2, the 10-fold cross-validation is estimated to compare the proposed method's performance with all different integration methods and individual components of the proposed method, respectively. In Sect. 4.3, the genetic algorithm-based method shows relatively higher performance than the existing deep learning method by chi-square test. In Sect. 4.4, the post-analysis is performed to verify the detectable ability of the proposed method by using some misclassified cases from the deep learning and the rule-based approach.

4.1 Data Collection

Datasets are collected from three real-world data: ISCX-URL-2016, PhishStorm, and PhishTank, with a total of 1,048,576 URLs and labels [15]. 70% of the total data is split to train the neural component and 15% of the data is used to validate the learning of the neural component. Finally, we use the remaining 15% of the data to justify the feasibility of the overall proposal method.

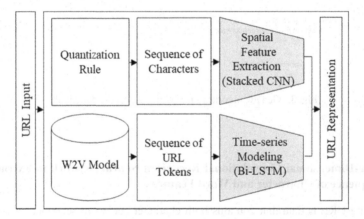

Fig. 4. One-dimensional convolutional neural network for word feature extraction.

4.2 Comparison of Accuracy and Recall Performance

As shown in Fig. 5, 10-fold cross-validation is estimated to compare the proposed method with existing integration methods. The proposed method outperforms state-of-the-art performance and gets improvements of approximately 2.60% accuracy, a 3.90% recall, and especially a 4.50% phishing recall performance compared to other integration methods, such as average, maximum, log average and concatenate mechanism.

As shown in Fig. 6, the proposed method leads to a significant performance improvement with a 9.52% recall and a 5.70% accuracy against a neural component's performance. Moreover, we confirm a 7.00% improvement in phishing recall performance via

10-fold cross-validation. Especially, compared to using only rule-based symbol components, we authenticate a dramatic improvement of a 23.61% recall, a 22.07% accuracy, and a 62.50% phishing recall performance. Since a tiny improvement in accuracy and recall performance is expected to prevent phishing URL attacks, we confirm the feasibility of the combinatorial optimization of logic constraints and deep learning.

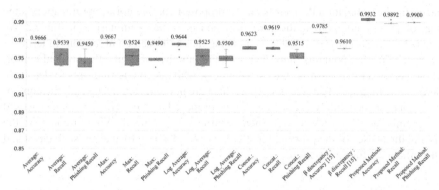

Fig. 5. Comparison accuracy and recall performance by different integration methods.

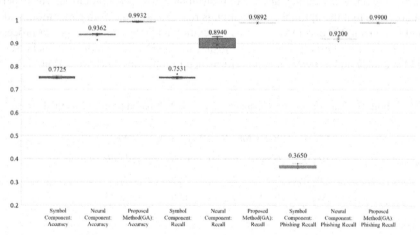

Fig. 6. Comparison of accuracy and recall performance by neural, symbol components and the proposed integration method.

4.3 Classification Performance and Chi-Square Test

We summarize the classification performance report of the proposed method as shown in Table 3 to justify the feasibility of the genetic algorithm-based proposed method, and it reaches an approximate 0.99% recall performance about 15% validation data: 2,289 phishing URLs. In addition, existing convolutional recurrent neural networks are organized in Table 4 by performing chi-square verification. The proposed genetic algorithm-based combinatorial optimization method, which considers existing convolutional recurrent neural networks as the expected frequencies, has verified the validity of its outstanding performance by analyzing quantitatively for phishing URL detection.

4.4 Analysis of Optimized Combination and Misclassified Cases

The optimal combination derived from the proposed combinatorial optimization is the usage of a neural component, the third, fifth, and sixth rules summarized in Table 2. We organize the post-analysis in Table 5 via using the optimized combination. In terms of a sample URL that contains more than a certain number of digits, the optimal combination can detect the phishing URLs accurately, while deep learning fails to detect. In addition, the linguistic features in URLs, which are not reflected from symbolic components alone, are accurately extracted from the proposed method. However, the proposed method misclassifies a few URLs, such as http://k-12.cl/dr/Sm. When it comes to the misclassified cases of the proposed method, we are planning to analyze such cases and reflect them for future research.

4.5 Generations for Genetic Algorithm

Figure 7 summarizes accuracy and recall performance for each generation. The experiment confirms that the optimal combination is reached after the 50th generation, under a condition given 200 chromosomes in a generation and 0.9 rates in crossover process, and 0.1 rates in mutation process. As a result, the optimized combination achieves a 0.9932 accuracy and a 0.9892 recall performance for phishing URL detection.

Table 3. Classification performance report for GA-based proposed method.

Class	Precision	Recall	F1-score	Support
Benign	0.99	1.00	1.00	157,286
Phishing	0.99	0.99	0.99	2,289
Average	0.99	0.99	0.99	159,575

Table 4. The chi-square test of GA-based proposed method and CNN-LSTM.

		Predicted	
		Benign	Phishing
Actual	Benign	0.141139	65
	Phishing	223.1881	7.88880597

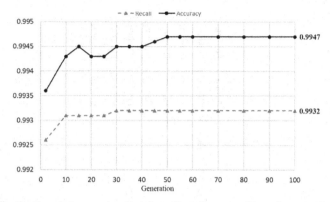

Fig. 7. Comparison accuracy and recall performance for each generation.

5 Conclusion

The paper suggests that the GA-based neuro-symbolic integration method not only considers both neural-based and rule-based learning methods for detecting phishing URLs, but also finds the optimal combination of 13 components to catch new phishing URLs. The genetic algorithm enables to search for the best combination which has the highest recall performance without trapping in the local minima over 50 generations.

Eventually, the best combination of the 13 components is to use neural output, the third, fifth, and sixth rules as summarized in Table 2. The optimal combination accurately classifies the majority of URLs that are confused from either existing deep learning methods or rule-based methods, and the proposed method ultimately justifies the outstanding performance of detection, achieving the highest accuracy (0.9932) and recall (0.9892) performance, compared to the existing state-state-of-art integration method in phishing URL detection.

Table 5. Post-analysis for neural, symbol, and our proposed method.

Class	Sample URL	Actual	Prediction
Neural component	https://1drv.ms/xs/s!Ahtvz T3KrwqMZzLMKnTc8clHnRA?wdFormId=%7BA0F7982D%2D71A4%2D4DE0%2DB4C4%2DC16A	Phishing	Phishing
	http://lido..._cde0f344ac2567a45_j124456733ebe933255beProduct-UserID&emai=anthony.trac@ucalgary.ca	Benign	Phishing
Symbol component	http://market.security***.net	Benign	Benign
	http://protery.com/js/libs/us/delta.com/index.php	Benign	Phishing
Ours	http://lido..._cde0f344ac2567a45_j124456733ebe933255beProduct-UserID&email=anthony.tract@uacalgary.ca	Phishing	Phishing
	http://protery.com/js/libs/us/delta.com/index.php	Phishing	Phishing
	http://k-12.cl/dtr/Sm/	Benign	Phishing

Acknowledgement. This work was supported by an IITP grant funded by the Korean MSIT (No. 2020-0-01361, Artificial Intelligence Graduate School Program (Yonsei University)) and a grant funded by Air Force Research Laboratory, USA.

References

1. Smadi, S., Aslam, N., Zhang, L.: Detection of online phishing email using dynamic evolving neural network based on reinforcement learning. Decis. Support Syst. **107**, 88–102 (2018)
2. Almomani, A., Gupta, B.B., Wan, L.T.C., Altaher, A., Manickam, S.: Phishing dynamic evolving neural fuzzy framework for online detection "zero-day" phishing email. Indian J. Sci. Technol. **6**(1), 1–5 (2013). https://doi.org/10.17485/ijst/2013/v6i1.18
3. Ojugo, A.A., Yoro, R.E.: Forging a deep learning neural network intrusion detection framework to curb the distributed denial of service attack. Int. J. Electr. Comput. Eng. (IJECE) **11**(2), 1498 (2021). https://doi.org/10.11591/ijece.v11i2.pp1498-1509
4. Moghimi, M., Varjani, A.Y.: New rule-based phishing detection method. Expert Syst. Appl. **53**, 231–242 (2016)
5. Liu, W., Zhong, S.: Web malware spread modelling and optimal control strategies. Sci. Rep. **7**, 1–19 (2017)
6. Anand, A., Gorde, K., Moniz, J.R.A., Park, N., Chakraborty, T., Chu, B.T.: Phishing URL detection with oversampling based on text generative adversarial networks. In: 2018 IEEE International Conference on Big Data (Big Data), pp. 1168–1177 (2018)
7. Yadollahi, M.M., Shoeleh, F., Serkani, E., Madani, A., Gharaee, H.: An adaptive machine learning based approach for phishing detection using hybrid features. In: 2019 5th International Conference on Web Research (ICWR), pp. 281–286 (2019)
8. Mamun, M.S.I., Rathore, M.A., Lashkari, A.H., Stakhanova, N., Ghorbani, A.A.: Detecting malicious URLS using lexical analysis. In: International Conference on Network and System Security, pp. 467–482 (2020)
9. Subasi, A., Kremic, E.: Comparison of adaboost with multiboosting for phishing website detection. Procedia Comput. Sci. **168**, 272–278 (2020)
10. Burnap, P., French, R., Turner, F., Jones, K.: Malware classification using self organising feature maps and machine activity data. Comput. Secur. **73**, 399–410 (2018)
11. Le, H., Pham, Q., Sahoo, D., Hoi, S.C.: URLNet: learning a URL representation with deep learning for malicious URL detection (2018). arXiv preprint: arXiv:1802.03162
12. Yang, P., Zhao, G., Zeng, P.: Phishing website detection based on multidimensional features driven by deep learning. IEEE Access **7**, 15196–15209 (2019)
13. Huang, Y., Yang, Q., Qin, J., Wen, W.: Phishing URL detection via CNN and attention-based hierarchical RNN. In: 2019 18th IEEE International Conference On Trust, Security And Privacy In Computing And Communications/13th IEEE International Conference On Big Data Science And Engineering (TrustCom/BigDataSE), pp. 112–119 (2019)
14. Tajaddodianfar, F., Stokes, J.W., Gururajan, A.: Texception: a character/word-level deep learning model for phishing URL detection. In: ICASSP 2020–2020 IEEE International Conference on Acoustics, Speech and Signal Processing (ICASSP), pp. 2857–2861 (2020)
15. Bu, S.J., Cho, S.B.: Integrating deep learning with first-order logic programmed constraints for zero-day phishing attack detection. In: ICASSP 2021–2021 IEEE International Conference on Acoustics, Speech and Signal Processing (ICASSP), pp. 2685–2689 (2021)
16. Wang, W., Pan, S.J.: Integrating deep learning with logic fusion for information extraction. Proc. AAAI Conf. Artif. Intell. **34**, 9225–9232 (2020)

17. Mohammad, R.M., Thabtah, F., McCluskey, L.: An assessment of features related to phishing websites using an automated technique. In: 2012 International Conference for Internet Technology and Secured Transactions, pp. 492–497 (2012)
18. Korkmaz, M., Sahingoz, O.K., Diri, B.: Feature selections for the classification of webpages to detect phishing attacks: a survey. In: 2020 International Congress on Human-Computer Interaction, Optimization and Robotic Applications (HORA), pp. 1–9 (2020)
19. Zhang, Q., Deng, D., Dai, W., Li, J., Jin, X.: Optimization of culture conditions for differentiation of melon based on artificial neural network and genetic algorithm. Sci. Rep. **10**, 1–8 (2020)
20. Afan, H.A., et al.: Input attributes optimization using the feasibility of genetic nature inspired algorithm: Application of river flow forecasting. Sci. Rep. **10**, 1–15 (2020)
21. Cho, S.B., Shimohara, K.: Evolutionary learning of modular neural networks with genetic programming. Appl. Intell. **9**(3), 191–200 (1998)
22. Lee, S.I., Cho, S.B.: Emergent behaviors of a fuzzy sensory-motor controller evolved by genetic algorithm. IEEE Trans. Syst. Man. Cybern. Part B (Cybern.) **31**(6), 919–929 (2001)

Interurban Electric Vehicle Charging Stations Through Genetic Algorithms

Jaume Jordán⬤, Pasqual Martí(✉)⬤, Javier Palanca⬤, Vicente Julian⬤,
and Vicente Botti⬤

Valencian Research Institute for Artificial Intelligence (VRAIN),
Universitat Politècnica de València, Camino de Vera s/n, 46022 Valencia, Spain
{jjordan,jpalanca,vinglada,vbotti}@dsic.upv.es, pasmargi@vrain.upv.es
http://vrain.upv.es/

Abstract. Electric vehicles are one of the strongest ways for society to stop contributing to greenhouse gas emissions. However, for their use to become regular, a good infrastructure of charging stations is needed, allowing a similar convenience to that offered by fossil fuel stations. Our work approaches the location of charging stations to create a nation-wide infrastructure. In this case, we focus on Spain and using genetic algorithms, we search for and evaluate different configurations according to the number of stations desired. Our results show that, with 250 stations, an initial infrastructure that covers most of the territory can be developed.

Keywords: Genetic algorithm · Charging station · Electric vehicle · Interurban

1 Introduction

Electric vehicles (EVs) are currently an essential element in policies oriented towards considerable reductions in gas and noise emissions. The massive introduction of the EV requires the development of new infrastructure to support the EV's charging demands. These needs also have specific requirements, such as the quantity of electric power required at the charging station's location on the energy grid. Every day, more EVs become available to consumers, resulting in significant environmental benefits. Even though the number of EV sales has grown in recent years, according to the European Environment Agency[1], market penetration remains relatively low representing an increase from 2 to 3.5 in 2019 compared to 2018. In the case of Spain, the National Integrated Energy and Climate Plan (PNIEC)[2] foresees a penetration of 5 million EVs by 2030. This plan, in the case of electric mobility, foresees the deployment of fast-charging

---

[1] https://www.eea.europa.eu/data-and-maps/indicators/proportion-of-vehicle-fleet-meeting-5/assessment.

[2] https://www.miteco.gob.es/images/es/pnieccompleto_tcm30-508410.pdf.

© Springer Nature Switzerland AG 2021
H. Sanjurjo González et al. (Eds.): HAIS 2021, LNAI 12886, pp. 101–112, 2021.
https://doi.org/10.1007/978-3-030-86271-8_9

infrastructure corridors, and the widespread deployment of charging infrastructure.

A problem of particular importance is the absence of charging infrastructure to fulfill the projected demand for EVs. This causes a condition called as *range anxiety* [6,8], which is the stress of the people not having enough power to get to the destination. However, because infrastructure development is costly, resources should be concentrated toward the installation of electric charging stations in the best possible location with the greatest possible impact.

According to this, different approaches [3,9] have been used to analyze the distribution of electric charging stations in cities, with an emphasis on the estimated number of charging stations required to increase the vehicle's usefulness. However, much less work has been done on the appropriate location at the intercity level where it is perhaps even more important to have an adequate network of charging stations that provides adequate coverage for the circulation of EVs for intercity travels.

In this sense, the work presented in [5] shows an early interest in the research of EV infrastructure on a countywide level. Authors use activity-based models to estimate the future EV power demand in the Flanders region, in Belgium. Their models flag specific locations of the region which would be available for smart-grid design, although, as authors comment, it lacks accurate data regarding EV energy consumption in real situations. Another example is presented in [10], where a heuristic methodology is used to locate EV fast-charging stations in larger (nationwide) areas. Authors improve previous models by introducing the altitude of the roads as a factor that hardly influences the autonomy of EVs. Their methodology has been applied to define the locations of 34 fast chargers in Costa Rica, as part of the government's plan to regulate the country's charging infrastructure. Their model, however, neglects less populated areas in favor of areas where more mobility data is available, only ensuring fast charging for densely populated areas.

In the case of Spain, authors in [1] propose different configurations for minimum fast-charging infrastructure in Spain. To do so, they define a maximum distance between fast chargers according to highway speed limits, weather conditions, and the average autonomy of the commercial EVs, which ensures mobility with a fully EV from any point of the country to another. Their study, although intensive, it is highly grounded on the technology of the year it was made. Nowadays EVs with higher ranges should be considered. In addition, only an initial minimum infrastructure is considered, as the aim is to reduce the monetary cost.

In [4,7], we presented a genetic algorithm (GA) to measure utility and cost for the optimal location of a set of charging stations in a city considering heterogeneous data sources. Based on the existing solution at the city level, it is possible to extend the idea to an interurban level. Currently, different initiatives trying to improve the use of EVs from the perspective of providing better infrastructures and better information on the current charging point network. Solutions like the provided by Ionity[3], Plugsurfing[4], Plugshare[5], or Electromaps[6] are in that line, but they do not try to optimize the location of new possible charging points to improve the existing network.

In our case, the idea is to propose an efficient location of charging points along a specific geographical area, such as a region, a country, or the European Union. In this way, a possible driver of an EV who wants to move from Cadiz (Spain) to Girona (Spain), which are about 1500 km apart, needs an infrastructure of charging points that ensures adequate battery charging depending on the car's autonomy. The proposed approach must take into account these possible journeys when configuring the best possible solution that guarantees the charging on as many journeys/paths as possible between any two points in the geographical area under consideration.

Therefore, the main contribution of this paper is to provide the optimal location of a network of charging points in an extensive geographical area that ensures the possibility of charging an EV regardless of its route. This optimization proposal must also take into account the minimization of the maximum distance between neighbor charging stations to ensure most of the EVs have enough autonomy to make their journeys. This will also depend on the desired number of charging stations to cover all the considered geographical areas.

Our proposal comes with many advantages. Firstly, it provides the best possible locations for a network of charging points. With this, we achieve a trust improvement in potential EV drivers, which will see them as a more viable option. Finally, the infrastructures presented minimize the necessary investment while offering an adequate service.

According to this proposal, the rest of the paper is structured as follows: Sect. 2 presents the issue and justifies our methodology. Following, Sect. 3 describes our Genetic Algorithm-based solution. Section 4 shows our experimentation and discusses our results. Finally, Sect. 5 assesses our work and comments on future extensions.

2   Problem Description

Finding the best locations within a territory to place charging stations so that they are as optimized as possible in terms of service and cost can be a very costly problem. It would require a search in a very large search space with enormous combinatorics. That is why in this paper we will present a genetic algorithm (GA) that can optimize this search and optimization process.

One of the most important parts of a GA is its fitness function, which allows evaluating how good is a solution generated by the algorithm. To do so, we use a complete set of open data that allows us to score a solution taking into account some variables like the land covered by the layout of the stations, the distance

---

[3] https://ionity.eu/en.

[4] https://www.plugsurfing.com/.

[5] https://www.plugshare.com.

[6] https://www.electromaps.com/.

between stations, the population near each station, or the number of vehicles that potentially benefiting from such stations.

Besides, taking into account each possible coordinate of the land to be covered would result in a huge search problem (and probably useless). To just evaluate those sites that can be possible locations for a station, a set of Points of Interest (PoIs) is provided to the algorithm. This set is big enough to need an algorithm to reduce the search space and look for the best solution that optimizes the station's location but results in a manageable problem with the appropriate algorithm.

Different data sets have been used to compute the fitness function that a GA uses to score a solution. These data sets come from open data repositories and can be used by anyone to reproduce the results.

One of the most important data sets is the initial set of possible points where a station can be placed, that is, the PoIs. In these cases, the chosen points of interest data set have been extracted from the gas station locations in Spain. This is a very interesting initial data set since the infrastructure is already prepared (power lines, access from highways, etc.).

In addition, different data giving indications of the most populated areas, the areas with the most vehicle traffic, and the areas with the most activity have been compiled for consideration. These data are respectively population data of the municipalities and vehicle fleet data by the municipality. Finally, to measure the activity of each area, we have used the number of geolocated tweets from the social network Twitter, which gives us an indication of where people are located throughout the day proportionally.

The following section explains how this data is used by the GA to find the best distribution of EV charging stations in interurban environments.

3   Interurban EV Charging Stations Distribution

The proposed genetic algorithm for the placement of charging stations for interurban electric vehicles receives as main parameters a set of Points of Interest (PoIs) likely to become charging stations, as well as the number of charging stations to be placed. Thus, the number of PoIs as input data must be significantly larger than the stations to be placed. In this way, the GA can better discriminate and find the most suitable distribution for the fitness function to be optimized.

An individual of the GA is an array of bits of the same size as the number of PoIs. A value of 1 at the i-th position of the array represents that a charging station will be placed at the PoI representing that position. A value of 0 implies that a station will not be placed.

3.1   Utility

Since we are dealing with spatial data, the GA takes into account the surrounding area of each PoI to calculate its utility. Thus, the utility of a PoI is calculated as the intersection of this area with the geographic data (population, traffic, and social network activity). The area considered for each of the "active" PoIs, i.e. those being proposed for the placement of a charging station (value 1 in the array), takes into account two geometric figures. On the one hand, the Voronoi diagram between the active PoIs is calculated, thus creating a polygon for each active PoI depending on the neighboring active PoIs. On the other hand, it is also considered a circumference with a particular radius which we define as *influence radius*.

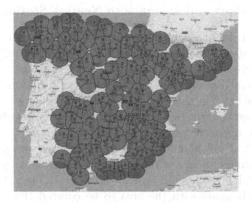

Fig. 1. Example of a set of 150 Points of Interest (PoIs) in which a charging station is placed along with the intersection between the Voronoi polygon and the circumference made with the *influence radius* (50 km) of each of the PoIs.

Thus, to obtain the utility value of each active PoI in an individual, the area of intersection between the corresponding Voronoi polygon and the circumference of radius *influence radius* centered on that PoI is considered. Figure 1 presents an example of active PoIs with their corresponding area of the Voronoi polygon and the intersection with the circumference made with the *influence radius*. In this way, each PoI will have greater or lesser utility depending on the amount of population, traffic, and social network activity it aggregates. In addition, the utility of the PoI in question also depends on whether there is another PoI nearby that already meets the needs in that influence radius. That is, if there are adjacent active PoIs, their Voronoi polygons cut each other, thus reducing their potential area of influence if it would be formed only by the circumference of influence radius. Thus, both PoIs cover less area and their utility is reduced accordingly. With this, we ensure that the final distribution of charging stations has better coverage globally and not all stations are gathered only in the most populated areas (and with more traffic and activity in social networks) that would produce more local utility.

Each of the terms for calculating the utility, i.e., population, traffic, and social network activity, is weighted to give greater or lesser importance to each. In this way, the utility is calculated with the Eq. 1:

$$utility(individual) = \sum_{\forall s_i = 1 \in individual} (p_i \cdot \omega_P + t_i \cdot \omega_T + a_i \cdot \omega_A) \qquad (1)$$

where s indicates whether or not a station is included in the PoI i; p_i, t_i, and a_i are the population, traffic, and social networks activity covered by the i-th PoI; and ω_P, ω_T, and ω_A, are the weights to calibrate the importance of each factor.

3.2   Uncovered Area and Distance to Nearest Station

In addition to the utility of placing charging stations at given PoIs, another value is crucial for calculating an individual's fitness. Since the problem we are dealing with involves the placement of interurban charging stations, it is essential that EV drivers can count on sufficient infrastructure for charging considering the range of their vehicles. Thus, the areas without charging station coverage and the distance between charging stations are important factors to consider.

To calculate the uncovered areas of charging stations of a solution (individual of the GA), the set of polygons resulting from the intersection of each Voronoi polygon with the circumference of radius *influence radius* with the center at the active station or PoI is used again, as shown in Fig. 1. Then, the intersection of all these polygons is made with the area to be covered (for example, the part of the Iberian Peninsula belonging to Spain). With this intersection, we obtain the area or areas that remain without coverage. In this way, the percentage of the area not covered by the areas of influence of the charging stations placed can be obtained. This measure, named *percentage of uncovered area (pua)*, is one of the factors to be minimized in any solution to the problem of placing charging stations in an interurban environment since it ensures that no areas are left uncovered throughout the territory.

The distance to the nearest station for each station, and globally, the highest of those distances, is another measure of the quality of a solution. Therefore, for each station to be placed we calculate the distance to the nearest station. In this way, we will have an array with the minimum distance from each station to one of its neighbors. However, we are interested in a general measure that assesses the suitability of the individual to be evaluated, so we will keep the highest value of this array. This value represents the longest distance that will have to be traveled to get from one station to another, so it becomes a good discriminant for our problem. The smaller this value is, the more suitable the solution represented by the individual since drivers will have more room for maneuver with the autonomy of their EVs.

To calculate this value we assume Euclidean distance between charging stations for simplicity. In this sense, although the Euclidean distance may differ relatively from the actual distance between two points due to non-existing roads, we consider it to be a sufficient approximation for the needs of our GA in this implementation. Furthermore, any values derived from these calculations will be taken into account accordingly in the subsequent analyses.

In order to make this calculation efficient to work in the GA (since many individuals have to be evaluated in each generation), the distances between all the PoIs of the problem are pre-calculated. With this, we have a matrix with all the distances, which avoids costly calculations during the execution of the GA.

For each individual to be evaluated during the evolution of the GA, the Delaunay triangulation[7] with the active PoIs is calculated. With this, we obtain the neighboring PoIs of each PoI, and in this way, we only have to calculate the minimum distance between the PoI in question and its neighbors. Once we have calculated the minimum distance of each PoI to its neighbors, we can obtain the maximum value among all the distances, which is the one we will use as discriminant since it is the worst value in the distance of the individual, i.e., the *longest distance (ld)* to travel between two stations.

3.3   Implementation and Fitness

The DEAP[8] package in Python was used to implement the GA. In addition, since a multi-objective genetic is required, the well-known Non-dominated Sorting Genetic Algorithm II (NSGA-II) [2] has been chosen. This multi-objective algorithm obtains solutions close to the Pareto frontier, thus trying to optimize all the objectives, i.e. the solutions to be obtained are not Pareto dominated. This implies that an objective cannot be improved at the cost of penalizing another objective.

Finally, the multi-objective fitness function consists of the *utility*, the percentage of uncovered area (*pua*), and the longest distance to the nearest station of all stations (*ld*). The objectives of this function is to **maximize the utility** and **minimize** *pua* **and** *ld*.

4   Experimental Results

This section details the experiments carried out to deploy a set of EV charging stations in the peninsular territory of Spain. The Canary and Balearic Islands, as well as Ceuta and Melilla, are omitted since they would be a separate experiment as they are not directly connected to the mainland by road.

The GA described above uses as a basis a set of PoIs which has to be significantly larger than the number of stations to be placed. Thus, given the nature of the interurban problem, all the existing gas stations in the peninsular territory of Spain have been selected. This gives us a set of 9629 PoIs on which to install charging stations. It should be noted that the type of solution we are seeking for the interurban problem is to provide the entire territory with sufficient coverage. Thus, the existing gas stations are a good starting point, since they are located

---

[7] For a set of points, the Delaunay triangulation satisfies that none of those points will be inside the circumference of any of the triangles. This triangulation is closely related to the Voronoi diagram since the circumferences of the Delaunay triangles are the vertices of the Voronoi diagram (i.e., of the polygons that form it).

[8] https://github.com/DEAP/deap.

both within urban centers and especially on highways and roads. In addition, the installation of charging stations would be relatively simple since gas stations already have a connection to the electrical grid (although power would probably have to be upgraded accordingly).

We have run a series of experiments in which the influence radius of the PoIs is 50km. In addition, the weights regulating the importance given to population, traffic, and activity have been set to the same value, i.e. $\omega_P = \omega_T = \omega_A = 0.3333$. On the other hand, the population of individuals considered by the GA is 100, from which individuals are selected for crossover and/or mutation to obtain new individuals and keep the best among the population. The crossover probability is defined at 0.5 and the mutation probability at 0.05. Finally, the evolution process of the algorithm stops after 100 generations.

Table 1. Summary of the results for different numbers of stations to be placed in the peninsular territory of Spain. The values of utility, the longest distance to the nearest station of all stations (*ld*), and the percentage of uncovered area (*pua*) are shown.

Stations	Utility	Distance (km)	%uncovered
50	0.00238	113.7	44.4
100	0.00460	101.1	20.8
150	0.00735	89.5	10.4
200	0.01495	82.9	5.1
250	0.01964	59.2	1.9
300	0.02986	60.3	1.2
400	0.04417	58.9	0.7
500	0.05219	43.8	0.3
750	0.10874	35.0	0.02
1000	0.14614	31.1	0.003

Table 1 summarizes the results obtained from the experiments performed with the GA presented above. The first column shows the number of stations to be placed in the peninsular territory of Spain, while the following columns show the utility of the solution, the longest distance to the nearest station of all stations (*ld*), and the percentage of uncovered area (*pua*), respectively. As might be expected, the utility grows as more charging stations are placed as they can cover more population, traffic, and activity. Similarly, the maximum distance between any station to the nearest station also decreases as the number of stations increases, as does the percentage of uncovered area.

From the results of Table 1 it could be highlighted the significant difference of the solution with 250 stations with respect to the previous ones with fewer stations. In this sense, with 250 stations the maximum distance is significantly reduced, from more than 80km to 59.2km, as well as the uncovered area, which is reduced to 1.9%, while the solutions with fewer stations have a relatively higher uncovered percentage. On the other hand, the improvement of the solutions with more than 250 stations to this solution, although noticeable especially with much larger numbers, does not seem so significant. All this suggests that the solution with 250 stations can be a good initial infrastructure as it provides good coverage without having to install a very large number of stations.

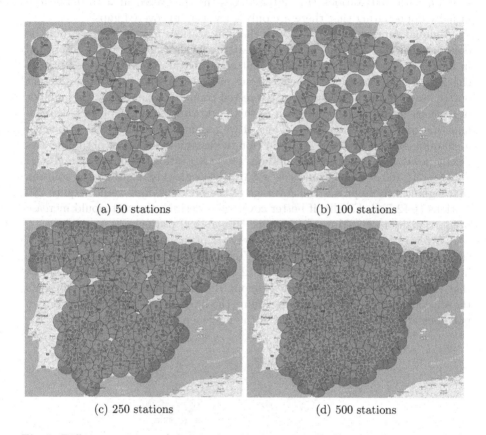

(a) 50 stations   (b) 100 stations

(c) 250 stations   (d) 500 stations

Fig. 2. Different amounts of charging stations covering the Spanish Iberian Peninsula area.

In Fig. 2 we can observe different numbers of stations with the representation of the area covered by each station (with the intersection with the Voronoi polygon). These representations are the result of the experiments summarized in Table 1. In this case, it becomes evident that both the solution with 50 stations (Fig. 2a) and the solution with 100 stations (Fig. 2b) are insufficient to cover the territory since large gaps can be distinguished on the map. In fact, in the case with 50 stations, 44.4% of the area is not covered, while with 100 stations the area not covered is 20.8%.

On the other hand, increasing the number of stations to 250 (Fig. 2c) shows visually that there are hardly any gaps to cover, being these gaps quite small.

For this reason, the percentage of uncovered area is only 1.9% in this case. This shows that considering an influence radius of 50km (and assuming that the calculation has been made with Euclidean distances and not with the real distance by roads), 250 stations are enough to cover the territory in a quite satisfactory way. However, if it is desired to ensure even more this coverage, a solution with 500 stations (Fig. 2d) reduces the uncovered area to 0.3%, also visually appreciating that there are only a few tiny uncovered gaps.

A global interurban installation of charging stations for electric cars could be done both at the government level and at the level of any company interested in having a network of stations, without counting on other stations or private chargers. In any case, the benefit of our proposal is that it obtains different options so that the government or interested companies can decide the number of stations to install depending on both the possible coverage of user demand and the total installation costs. Thus, if we assume an average cost of placing a charging station of about €50,000 (for the associated costs of obtaining sufficient power, construction, installation, etc.), 50 stations would have a cost of €2.5M, and 100 stations would cost €5M. But as already mentioned, a solution that significantly improves the coverage of the previous ones would require 250 stations (€12.5M). Finally, if better coverage is desired, the cost would increase proportionally to €25M or €50M, with 500 or 1000 stations, respectively.

Currently, the Tesla company has a network of charging stations in the peninsular territory of Spain. Specifically, the Tesla network[9] has 38 charging stations in which each one has between 2 and 12 charging points or stalls. This means that there are 284 stalls distributed among the 38 stations. Figure 3 represents the stations of the Tesla network including the intersection between the Voronoi polygons and the circumferences with the influence radius of 50 km.

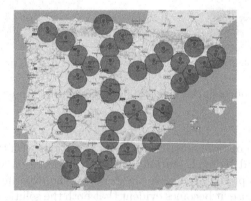

Fig. 3. The current distribution of Tesla chargers on the Spanish mainland along with the intersection between the Voronoi polygon and the circumference made with the *influence radius* (50km) of each station.

---

[9] https://www.tesla.com/es_ES/findus/list/superchargers/Spain.

To compare the solutions proposed in the experiments shown in Table 1 and Fig. 2, we calculated the values of utility, distance, and percentage of uncovered area for the current Tesla installation. In this case, the longest distance to the nearest station of all stations is 210km. This distance is significantly greater than that obtained with our 50-station solution (almost equivalent to Tesla's 38-station solution). Thus, it can be seen that the Tesla station distribution does not follow optimality criteria to reduce these distances. Moreover, if we compare this distance with the solution in which we put 250 stations (of one point), in our case, this 210 km are reduced to 59.2 km.

On the other hand, the percentage of the area not covered by the Tesla solution is 55.6%, which is significantly worse than with our solution of 50 stations with 44.4%, and much higher if we consider the solution of 250 stations, where this percentage is reduced to 1.9%.

Finally, Tesla's solution has a utility of 0.15549 if we consider each station with only one charging point. In case of considering all the charging points of each station, i.e. 284 points in total, the utility of the Tesla solution only goes up to 0.019950, which is practically equivalent to the utility of our 250-station solution, which would have 34 points less (in addition, in our algorithm we have not considered more than one point per station).

All in all, we can state that the solutions of our GA better meet the optimality criteria defined for satisfying the demand compared to the existing Tesla installation. Thus, any government or company interested in the sector should consider these techniques to have a better distribution of EV charging stations.

5   Conclusions

In this paper, we have presented an approach based on a genetic algorithm (GA) to solve the problem of the distribution of a set of electric vehicle (EV) charging stations at an interurban level. Existing gas stations are used as base points for the placement of EV charging stations. The GA considers real data on population, traffic, and social network activity to calculate the utility of stations at different locations. In addition, the longest distance to the nearest station of all stations, as well as the percentage of uncovered area are also considered. These latter factors are attempted to be minimized while trying to maximize the utility.

Experimental results show that a solution with about 250 stations has good coverage of the peninsular territory of Spain with the factors considered. Furthermore, when comparing our solutions with the current installation of Tesla, it is demonstrated that our approach proposes more optimal distributions that better cover the demand with the established optimality criteria.

As future work, we would like to improve the parts referred to below. The use of Euclidean distance constitutes a reasonable approximation, but it would be interesting to use real distances extracted from the routes between stations, as it would better evaluate each solution (individual) of the GA.

We would also like to consider the cost of bringing adequate electric power to each of the PoIs that will allocate a station. However, we need to gather data regarding electric plants and power lines in the area of study.

Finally, we would consider a variable number of charging points per station within the GA. This would cause the utility to be calculated based on the number of charging points placed at each station, and would also increase the search space of the GA. However, such a solution would be more complete by considering more variables to cover the actual demand in each area.

Acknowledgments. This work was partially supported by the MINECO/FEDER RTI2018-095390-B-C31 project of the Spanish government. Pasqual Martí is funded by grant PAID-01-20-4 of Universitat Politècnica de València.

References

1. Colmenar-Santos, A., De Palacio, C., Borge-Diez, D., Monzón-Alejandro, O.: Planning minimum interurban fast charging infrastructure for electric vehicles: methodology and application to Spain. Energies **7**(3), 1207–1229 (2014)
2. Deb, K., Pratap, A., Agarwal, S., Meyarivan, T.: A fast and elitist multiobjective genetic algorithm: NSGA-II. IEEE Trans. Evol. Comput. **6**(2), 182–197 (2002). https://doi.org/10.1109/4235.996017
3. Dong, J., Liu, C., Lin, Z.: Charging infrastructure planning for promoting battery electric vehicles: an activity-based approach using multiday travel data. Transp. Res. Part C: Emerg. Technol. **38**, 44–55 (2014)
4. Jordán, J., Palanca, J., Del Val, E., Julian, V., Botti, V.: A multi-agent system for the dynamic emplacement of electric vehicle charging stations. Appl. Sci. **8**(2), 313 (2018). https://doi.org/10.3390/app8020313
5. Knapen, L., Kochan, B., Bellemans, T., Janssens, D., Wets, G.: Activity based models for countrywide electric vehicle power demand calculation. In: 2011 IEEE First International Workshop on Smart Grid Modeling and Simulation (SGMS), pp. 13–18. IEEE (2011)
6. Neubauer, J., Wood, E.: The impact of range anxiety and home, workplace, and public charging infrastructure on simulated battery electric vehicle lifetime utility. J. Power Sour. **257**, 12–20 (2014)
7. Palanca, J., Jordán, J., Bajo, J., Botti, V.: An energy-aware algorithm for electric vehicle infrastructures in smart cities. Futur. Gener. Comput. Syst. **108**, 454–466 (2020)
8. Pevec, D., Babic, J., Carvalho, A., Ghiassi-Farrokhfal, Y., Ketter, W., Podobnik, V.: A survey-based assessment of how existing and potential electric vehicle owners perceive range anxiety. J. Clean. Prod. **276**, 122779 (2020)
9. Tu, W., Li, Q., Fang, Z., Lung Shaw, S., Zhou, B., Chang, X.: Optimizing the locations of electric taxi charging stations: a spatial-temporal demand coverage approach. Transp. Res. Part C: Emerg. Technol. **65**, 172–189 (2016). https://doi.org/10.1016/j.trc.2015.10.004
10. Victor-Gallardo, L., et al.: Strategic location of EV fast charging stations: the real case of Costa Rica. In: 2019 IEEE PES Innovative Smart Grid Technologies Conference-Latin America (ISGT Latin America), pp. 1–6. IEEE (2019)

An Effective Hybrid Genetic Algorithm for Solving the Generalized Traveling Salesman Problem

Ovidiu Cosma[ID], Petrică C. Pop[✉][ID], and Laura Cosma[ID]

Technical University of Cluj-Napoca, North University Center of Baia Mare,
Dr. V. Babes 62A, 430083 Cluj-Napoca, Romania
petrica.pop@mi.utcluj.ro

Abstract. The Generalized Traveling Salesman Problem (GTSP) looks for an optimal cycle in a clustered graph, such that every cluster is visited exactly once. In this paper, we describe a novel Hybrid Genetic Algorithm (HGA) for solving the GTSP. At the core of the proposed HGA, there is a Chromosome Optimization algoRithm (COR) that is based on Dijkstra's Shortest Path Algorithm (DSPA) and a Traveling Salesman Problem Solver (TSPS). We tested our algorithm on a set of well known instances from the literature and the preliminary computational experiments show promising results. We believe that our HGA can be used as a basis for future developments.

Keywords: Hybrid genetic algorithm · Generalized traveling salesman problem

1 Introduction

The Traveling Salesman Problem (TSP) is one of the most well-known comprehensively investigated and important optimization problems, being studied for a long period of time, since its formulation in 1930.

In the last three decades, there have been considered several variants of the TSP defined on graphs whose nodes are partitioned into a given collection of clusters, especially due to the many interesting real-world applications: the generalized traveling salesman problem [7], the at least generalized traveling salesman problem [16], clustered traveling salesman problem [5], the generalized traveling salesman problem with time windows [30], the clustered generalized traveling salesman problem [2], etc.

The generalized traveling salesman problem belongs to the category of generalized network design problems. This category of problems naturally generalizes the classical combinatorial optimization problem, having the following primary characteristics: the nodes of the underlying graph are divided into a certain number of clusters and, when considering the feasibility constraints of the initial problem, these are expressed in relation to the clusters rather than as individual nodes. For more information on these problems we refer to [6,18,20].

© Springer Nature Switzerland AG 2021
H. Sanjurjo González et al. (Eds.): HAIS 2021, LNAI 12886, pp. 113–123, 2021.
https://doi.org/10.1007/978-3-030-86271-8_10

Due to its challenging theoretical aspects and its practical applications [13], the GTSP has generated a considerable interest, several results concerning complexity aspects, mathematical models, approximation results and heuristic and metaheuristic algorithms have been obtained. Noon and Bean [16] introduced a Lagrangian relaxation approach, while Fischetti et al. [7] described a branch-and-cut algorithm for the GTSP. Pop [23] presented some integer and mixed integer programming formulations of the GTSP. Renaud and Boctor [25] developed an efficient composite heuristic, Gutin and Karapetyan [8] and Bontoux et al. [3] proposed memetic algorithms, Cacchiani et al. [4] described a multistart heuristic, Snyder and Daskin [28] presented a random-key genetic algorithm, Helsgaun [9] and Karapetyan described adaptations of the Lin-Kernighan heuristic algorithm for solving the GTSP. Ant colony approaches have been proposed by Reihaneh and Karapetyan [24] and by Pintea et al. [17], Tasgetiren et al. [29] described a discrete particle swarm optimization algorithm, while Matei and Pop [15] developed an efficient genetic algorithm for solving the problem. Recently, Smith and Imeson [27] proposed an efficient large neighborhood search heuristic and Schmidt and Irnich [26] described new neighborhoods and an iterated local search algorithm for GTSP.

The aim of this paper is to describe a different and novel hybrid heuristic approach which is meant for providing good solutions to the GTSP. Furthermore, we present and discuss the outcomes of preliminary computational experiments on a set of benchmark instances which one can find in the specific literature.

The rest of our paper is organized as follows: in Sect. 2 we define the generalized traveling salesman problem, in Sect. 3 we describe the novel hybrid algorithm which has at his core a chromosome optimization algorithm based on Dijkstra's shortest path algorithm and a TSP solver. In Sect. 3, we present and analyze the achieved preliminary computational results. Finally, the last section contains the conclusions and future research directions.

2   Definition of the Generalized Traveling Salesman Problem

The generalized traveling salesman problem is defined on a graph $G = (V, E)$, where V is a set of vertices and E is a set of edges, having the property that the set of vertices are grouped into a set of n disjoint clusters $C = \{C_i \mid i = 1, 2, ..., n\}$, satisfying the following properties:

1. $\bigcup_{i=1}^{n} C_i = V$;
2. $C_i \cap C_j = \emptyset, \forall i, j \in \{1, ..., n\}$ with $i \neq j$.

Denoting by $[v]$ the cluster to which vertex v belongs to, and by $c_{v_i v_j}$ the cost of the edge $v_i v_j$, then the GTSP looks for a minimum cost cycle with the property that exactly one vertex from each cluster is visited. A cycle having this property will be called a Generalized Hamiltonian Tour (GHT). The cost of a GHT is obtained by summing up all its edges.

Obviously, when all the clusters are singletons, then the GTSP reduces to the classical TSP and therefore it results that the problem is an NP-hard optimization problem.

3 The Proposed Hybrid Algorithm

For solving the GTSP, we propose a Genetic Algorithm (GA) hybridized with Dijkstra's Shortest Path Algorithm (DSPA) and a Traveling Salesman Problem Solver (TSPS).

The chromosomes used in the GA contain a Cluster Numbers List (CNL), in which the clusters visiting order is kept. We developed a Chromosome Optimization algoRithm (COR) based on DSPA and TSPS, that forms the core of our GA.

At the first step of the COR, a subgraph G' of G is associated, by reducing the source cluster C_s to a single vertex (v_s), adding a copy of the source cluster to G' and the CNL, and eliminating all the edges between clusters that are not adjacent in the CNL. Then, we use the DSPA to find the shortest path from v_s in the first cluster, visiting successively all the other clusters, to its copy in the last cluster. Because of the triangle inequality, in the case of euclidean instances, no more than one vertex from each cluster will be visited, and the solution of the DSPA is also a solution of the GTSP.

If the CNL contains the optimal visiting order of the clusters, then, evidently the DSPA will find the optimal solution of the GTSP. Usually, this is not the case, but DSPA finds a promising set of vertices, that are used in the second step of the COR, by the TSPS to adjust the CNL. For running the TSPS, we build a subgraph G'' of G that includes only the vertices in the DSPA solution, and all the edges between them. Then, we run the TSPS that could improve the solution of the GTSP. In the end, the CNL is updated based on the new solution.

The previously described two steps are repeated in the COR as long as they are successful. The COR ends at the first step that fails to improve the solution of the GTSP.

3.1 Chromosome Structure

The chromosome data structure consists of two parts:

- the source vertex v_s
- the CNL which is represented by a circular list $O = \{o_i \mid i = 1, 2, ..., n\}$, where

$$\begin{cases} o_1 = [v_s] \\ o_i \in \{1, 2, ..., n\} \setminus \{[v_s]\}, \ i \in \{2, 3..., n\} \\ o_i \neq o_j, \ \forall i, j \in \{1, 2, ..., n\}, \ i \neq j \end{cases}$$

An illustration of the GSTP problem defined on a graph G and two chromosomes c_1 and c_2 is presented in Fig. 1. The vertices of G are grouped in 5 clusters $C_1, ...C_5$. Because it has the smallest number of vertices, C_3 is selected to be the source cluster. The edges of G are not shown. The chromosomes contain the source vertex v_s and the clusters visiting order O.

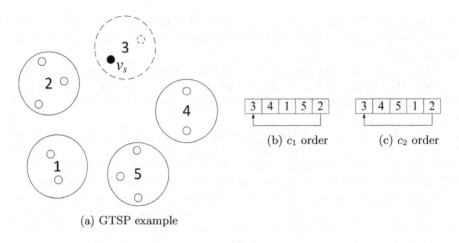

(a) GTSP example

(b) c_1 order

(c) c_2 order

Fig. 1. Example of a GTSP and two chromosomes (c_1 and c_2)

3.2   Chromosome Optimization

Each chromosome created by the genetic algorithm is optimized using Algorithm 1 (COR), as follows:

- Step 3 builds a layered graph $G' = (V', E')$ based on the chromosome c, as follows:

$$V' = (\bigcup_{i=1}^{n} C_i \setminus C_s) \cup \{v_s\}$$

$$E' = \{v_i v_j \mid v_i \in C_{o_k}, \ v_j \in C_{o_{k+1}}, \ k \in \{1, 2, ..., n\}, \ o_{n+1} = o_1\}$$

The resulting graph G' corresponding to the example in Fig. 1 is shown in Fig. 2. After building G', the DSPA algorithm is executed, to find the optimal path from v_s, through the clusters and back to a copy of v_s in the final cluster. The solution is in fact a cycle in G that is also a feasible solution of the GTSP. This solution is shown in Fig. 3 and will be denoted by S. S is the best solution that respects the restrictions specified by chromosome c. If chromosome c would contain the best $\{v_s, O\}$ combination, then the resulting solution S would be optimal. Even if S is not guaranteed to be the optimal GTSP solution, it contains a promising set of vertices that can serve as basis for future improvements.

– If S improved the best-known solution so far, then the optimization algorithm continues to step 9 that tries to improve the order O by running a TSPS. A new graph $G'' = \{S, E''\}$ is build for running the TSPS, where

$$E'' = \{v_i v_j \mid v_i, v_j \in S\}$$

The resulting graph G'' corresponding to the solution presented in Fig. 3 is shown in Fig. 4a. E'' contains all the edges between the vertices in S. By running a TSPS, a new better tour could be found in G''. The result of the TSPS is shown in Fig. 4b, and the updated chromosome c_2 is shown in Fig. 1c. If the solution S cannot be improved in step 9, then the optimization algorithm ends. If the solution S was improved by the TSPS, then it is used to update the chromosome c order list O.

Algorithm 1. Chromosome optimization

---

input: chromosome $c\{v_s, O\}$
1: $z \leftarrow \infty$
2: **do**
3:     Determine solution S by DSPA correction
4:     **if** $c(S) < z$ **then**
5:         $z \leftarrow c(S)$
6:     **else**
7:         STOP
8:     **end if**
9:     Update solution S and order O by TSPS correction
10:    **if** $c(S) < z$ **then**
11:        $z \leftarrow c(S)$
12:    **else**
13:        STOP
14:    **end if**
15: **while** *true*

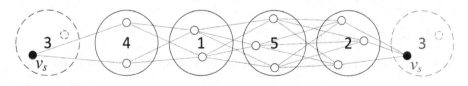

Fig. 2. Layered graph G', corresponding to the example in Fig. 1

Remark. The loop in the chromosome optimization algorithm ends after a finite number of iterations. The TSP optimization performed in step 9 cannot worsen the solution S, because it is run on graph G'' that contains all the vertices

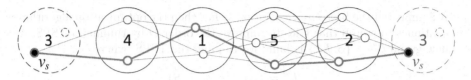

Fig. 3. A feasible solution to the GTSP, found by DSPA executed on the layered graph G' shown in Fig. 2

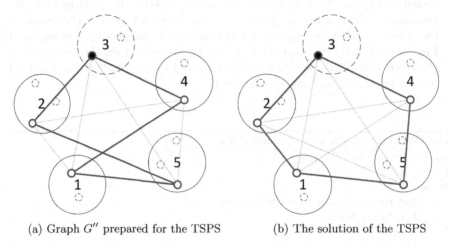

(a) Graph G'' prepared for the TSPS   (b) The solution of the TSPS

Fig. 4. TSP optimization process

from S. In the worst case, the solution remains unchanged. Similarly, the DSPA optimization performed in step 3 improves the solution S, or leaves it unchanged in the worst case. That is because the path defined by S is in fact a subgraph of G', and the DSPA could only improve it, or leave it unchanged. Because each of the two steps improve the solution, the process must end in a finite number of steps. We observed that usually the loop ends after one or two iterations in the cases of the small instances. The largest number of iterations that we observed in the case of the most complicated instances was 5.

3.3 Initial Population

The chromosomes in the initial population are randomly generated as follows:

- The source vertex v_s is picked pseudo-randomly from the source cluster, taking care to generate the same amount of chromosomes for each of the source cluster vertices.
- The clusters visiting order O is initialized randomly.
- Chromosome optimization is performed in the last step, using COR.

3.4   Crossover

The crossover operator creates two offspring chromosomes c_{o1} and c_{o2}, by combining the genes of two parent chromosomes c_{p1} and c_{p2}, as follows:

- The source vertex of c_{o1} is taken from c_{p1}, and the source vertex of c_{o2} is taken from c_{p2}.
- The order list O of each offspring is initialized as follows:
  - Initialize the offspring order list with a cluster taken randomly from one of the two parents order list.
  - Randomly pick a cluster from the other parent order list, that is adjacent with the first or with the last chromosome in the offspring order list, avoiding duplicates. The picked cluster is added at the beginning or at the end of the offspring order list, thus preserving some of the cluster adjacencies in the parents order lists. If there is no solution, the other parent is checked. If no solution is found either, then the next cluster is picked randomly from one of the parents.
  - Repeat the previous step until all the clusters of the instance are added to the order list of the offspring.
  - Finally rotate the offspring order list so that the source cluster reaches the first position.
- The two offspring chromosomes c_{o1} and c_{o2} are optimized with COR.

A crossover operation example is presented in Figs. 5 and 6. Only one of the two resulting offspring is shown.

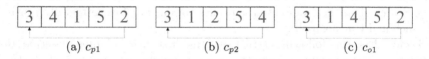

(a) c_{p1} (b) c_{p2} (c) c_{o1}

Fig. 5. Two parents c_{p1}, c_{p2}, and a resulting offspring c_{o1}

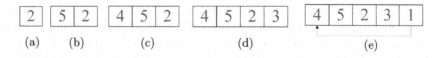

(a)   (b)   (c)   (d)   (e)

Fig. 6. Steps of building the offspring in Fig. 5c

3.5   Mutation

The mutation operator works as follows:

- Create a copy of one of the chromosomes in the current population.
- Randomly pick two different clusters from the order list O of the new chromosome, and switch their position.
- Optimize the new chromosome with COR.

3.6 Genetic Parameters

The following parameters have been optimized experimentally:

- the population dimension constant: $pd = 50$;
- the initial population dimension: $pi = 3 \times pd$;
- the number of crossover operations for building a new generation: $nc = 3 \times pd \times |C_s|$;
- the number of mutation operations: $nm = 0.2 \times pd \times |C_s|$;
- the stopping condition: 10 consecutive generations with no improvement.

If the last generation fails to improve the solution, all the chromosomes from the current population are added to the next generation. Else the size of the next generation is half the size of the current population, without crossing the limit of pd chromosomes.

4 Preliminary Computational Results

Our HGA was implemented in Java language. We used the Concorde TSPS [1] and the DSPA from the Lemon library [14]. We performed 15 independent runs for each instance, on a computer equipped with Intel(R) Core i3-8100 CPU @ 3.6 GHz and 8 GB RAM.

Table 1 displays the results of our hybrid based genetic algorithm (HGA), in comparison with the results obtained by Schmidt et al. [26] using an Iterated Local Search algorithm, and with the results obtained by Fischetti et al. [7] using a Branch and Cut algorithm.

Table 1 has the following structure: the first five columns describe the instances: n is the number of clusters, N is the number of vertices and $BestKS$ is the best known solution. The next two columns contain results obtained with the Branch and Cut (B&C) algorithm and reported in [7]: $BestS1$ is the best found solution, and $Time$ is the running time. The next column contains the best solutions obtained with the Iterated Local Search Algorithm in [26]. The next five columns contain information about our HGA: $BestS3$ is the best found solution, $AvgS$ is the average of the best solutions found in the 15 runs of each instance and $AvgTime$ is the average computational time required to find the best solution. More precisely, $AvgTime$ is the average time spent by our HGA to start the construction of the generation in which the best solution first appears. If $AvgTime = 0$, the best solution appeared each time in the initial population. $AvgTime$ is expressed in seconds. $BestGap$ is the gap between the $BestKS$ and $BestS3$, calculated as follows: $BestGap = (BestS3 - BestKS)/BestKS \times 100$. $AvgGap$ is the gap between the $BestKS$ and $AvgS$, calculated as follows: $AvgGap = (AvgS - BestKS)/BestKS \times 100$. The "=" symbol indicates a value identical to the best known solution for that instance. The "X" symbol indicates that information was not provided in the corresponding paper.

Analyzing the results displayed in Table 1, we observe that our proposed HGA found the best known solutions for 32 of the 34 instances that were tested.

Table 1. Preliminary experimental results

Instance					B&C [7]		ILS [26]	Our HGA				
No.	Name	n	N	BestKS	BestS1	Time	BestS2	BestS3	AvgS	AvgTime	BestGap	AvgGap
1	11eil51	11	51	174	=	2.9	X	=	=	0	0	0
2	14st70	14	70	316	=	7.3	X	=	=	0	0	0
3	16eil76	16	76	209	=	9.4	X	=	=	0	0	0
4	16pr76	16	76	64925	=	12.9	X	=	=	0.99	0	0
5	20kroA100	20	100	9711	=	18.4	X	=	=	0	0	0
6	20kroB100	20	100	10328	=	22.2	X	=	=	0	0	0
7	20kroD100	20	100	9450	=	14.3	X	=	=	0	0	0
8	20kroE100	20	100	9523	=	13	X	=	=	1.41	0	0
9	21eil101	21	101	249	=	25.6	X	=	=	2.91	0	0
10	21lin105	21	105	8213	=	16.4	X	=	=	0	0	0
11	22pr107	22	107	27898	=	7.4	X	=	=	4.47	0	0
12	25pr124	25	124	36605	=	25.9	X	=	=	2.45	0	0
13	26bier127	26	127	72418	=	23.6	X	=	=	0	0	0
14	28pr136	28	136	42570	=	43	X	=	=	25.35	0	0
15	29pr144	29	144	45886	=	8.2	X	=	=	1.57	0	0
16	30kroA150	30	150	11018	=	100.3	X	=	=	16.63	0	0
17	30kroB150	30	150	12196	=	60.6	X	=	12197.27	14.86	0	0.01
18	31pr152	31	152	51576	=	94.8	=	=	=	7.11	0	0
19	32u159	32	159	22664	=	146.4	=	=	=	10.64	0	0
20	39rat195	39	195	854	=	245.9	=	=	=	38.25	0	0
21	40kroA200	40	200	13406	=	187.4	=	=	=	20.08	0	0
22	40kroB200	40	200	13111	=	268.5	=	=	13113.29	38.81	0	0.02
23	45ts225	45	225	68340	=	37875.9	=	=	68365.71	73.38	0	0.04
24	46pr226	46	226	64007	=	106.9	=	=	=	0	0	0
25	53gil262	53	262	1013	=	6624.1	=	=	1016.75	101.77	0	0.37
26	53pr264	53	264	29549	=	337	=	=	=	272.67	0	0
27	56a280	56	280	1079	X	X	=	=	=	83.73	0	0
28	60pr299	60	299	22615	=	812.8	=	=	22620.6	111.55	0	0.02
29	64lin318	64	318	20765	=	1671.9	=	=	=	91.52	0	0
30	88pr439	88	439	60099	=	5422.8	=	=	=	129.16	0	0
31	89pcb442	89	442	21657	=	58770.5	=	=	21728.68	417.28	0	0.33
32	115rat575	115	575	2388	X	X	=	=	2420.5	1035.45	0	1.36
33	157rat783	157	783	3262	X	X	=	3271	3298.18	2163.96	0.28	1.11
34	201pr1002	201	1002	114311	X	X	=	114394	114497.91	1679.38	0.07	0.16

The *BestGap* is very small for the two cases in which the best solution was not found (less than 0.3%). For 25 of the test instances, the best known solution was found in each run. For the remaining 9 instances, the *AvgGap* values are insignificant (less than 1.5%). The running times are significantly smaller than the ones obtained with the B&C algorithm run on a HP 9000/720 CPU and reported in [7], but because the B&C algorithm was tested on a different configuration, the actual performance gain factor of our approach may vary.

5   Conclusions

In this paper, we considered the generalized traveling salesman problem defined on a graph whose vertices are partitioned into a given number of clusters and we are looking for a minimum cost GHT, that has the property that exactly one vertex from each cluster is visited.

In order to solve the investigated optimization problem, we developed an efficient hybrid based genetic algorithm that has at his core a chromosome optimization algorithm based on Dijkstra's shortest path algorithm, and a TSP solver.

The preliminary computational results achieved on a set of 34 benchmark instances from the literature show promising results, proving that our novel HGA can be used as a basis for future developments.

In future work, we plan to increase the efficiency of our HGA, by adding local search operations in the COR, and assess its generality and scalability, by testing it on different types of instances from the literature.

References

1. Applegate, D., Bixby, R., Chvatal, V., Cook, W.: Concorde TSP solver. http://www.tsp.gatech.edu/concorde/index.html. Accessed 1 June 2021
2. Baniasadi, P., Foumani, M., Smith-Miles, K., Ejov, V.: A transformation technique for the clustered generalized traveling salesman problem with applications to logistics. Eur. J. Oper. Res. **285**(2), 444–457 (2020)
3. Bontoux, B., Artigues, C., Feillet, D.: A memetic algorithm with a large neighborhood crossover operator for the generalized traveling salesman problem. Comput. Oper. Res. **37**(11), 1844–1852 (2010)
4. Cacchiani, V., Muritiba, A., Negreiros, M., Toth, P.: A multistart heuristic for the equality generalized traveling salesman problem. Networks **57**, 231–239 (2011)
5. Chisman, J.A.: The clustered traveling salesman problem. Comput. Oper. Res. **2**(2), 115–119 (1975)
6. Cosma, O., Pop, P.C., Zelina, I.: An effective genetic algorithm for solving the clustered shortest-path tree problem. IEEE Access **9**, 15570–15591 (2021)
7. Fischetti, M., Gonzáles, J., Toth, P.: A branch-and-cut algorithm for the symmetric generalized traveling salesman problem. Oper. Res. **45**(3), 378–394 (1997)
8. Gutin, G., Karapetyan, D.: A memetic algorithm for the generalized traveling salesman problem. Nat. Comput. **9**(1), 47–60 (2010)
9. Helsgaun, K.: Solving the equality generalized traveling salesman problem using the Lin-Kernighan-Helsgaun algorithm. Math. Program. Comput. **7**(3), 269–287 (2015). https://doi.org/10.1007/s12532-015-0080-8
10. Karapetyan, D., Gutin, G.: Lin-Kernighan heuristic adaptations for the generalized traveling salesman problem. Eur. J. Oper. Res. **208**(3), 221–232 (2011)
11. Karapetyan, D., Gutin, G.: Efficient local search algorithms for known and new neighborhoods for the generalized traveling salesman problem. Eur. J. Oper. Res. **219**(2), 234–251 (2012)
12. Laporte, G., Semet, F.: Computational evaluation of a transformation procedure for the symmetric generalized traveling salesman problem. INFOR: Inf. Syst. Oper. Res. **37**(2), 114–120 (1999)
13. Laporte, G., Asef-Vaziri, A., Sriskandarajah, C.: Some applications of the generalized travelling salesman problem. J. Oper. Res. Soc. **47**(12), 1461–1467 (1996). https://doi.org/10.1057/jors.1996.190
14. The LEMON library. https://lemon.cs.elte.hu/trac/lemon. Accessed 1 June 2021

15. Matei, O., Pop, P.C.: An efficient genetic algorithm for solving the generalized traveling salesman problem. In: Proceedings of the 2010 IEEE 6th International Conference on Intelligent Computer Communication and Processing, pp. 87–92 (2010)

16. Noon, C.E., Bean, J.C.: A Lagrangian based approach for the asymmetric generalized traveling salesman problem. Oper. Res. **39**(4), 623–632 (1991)

17. Pintea, C.-M., Pop, P.C., Chira, C.: The generalized traveling salesman problem solved with ant algorithms. Complex Adapt. Syst. Model. **5**(1), 1–9 (2017). https://doi.org/10.1186/s40294-017-0048-9

18. Pop, P.C.: The generalized minimum spanning tree problem: an overview of formulations, solution procedures and latest advances. Eur. J. Oper. Res. **283**(1), 1–15 (2020)

19. Pop, P., Oliviu, M., Sabo, C.: A hybrid diploid genetic based algorithm for solving the generalized traveling salesman problem. In: Martínez de Pisón, F.J., Urraca, R., Quintián, H., Corchado, E. (eds.) HAIS 2017. LNCS (LNAI), vol. 10334, pp. 149–160. Springer, Cham (2017). https://doi.org/10.1007/978-3-319-59650-1_13

20. Pop, P.C.: Generalized Network Design Problems. Modeling and Optimization. De Gruyter Series in Discrete Mathematics and Applications, Germany (2012)

21. Pop, P.C., Iordache, S.: A hybrid heuristic approach for solving the generalized traveling salesman problem. In: Proceedings of the 13th Annual Conference on Genetic and Evolutionary Computation, GECCO 2011, pp. 481–488. Association for Computing Machinery (2011)

22. Pop, P.C., Matei, O., Sabo, C.: A new approach for solving the generalized traveling salesman problem. In: Blesa, M.J., Blum, C., Raidl, G., Roli, A., Sampels, M. (eds.) HM 2010. LNCS, vol. 6373, pp. 62–72. Springer, Heidelberg (2010). https://doi.org/10.1007/978-3-642-16054-7_5

23. Pop, P.C.: New integer programming formulations of the generalized traveling salesman problem. Am. J. Appl. Sci. **4**(11), 932–937 (2007)

24. Reihaneh, M., Karapetyan, D.: An efficient hybrid ant colony system for the generalized traveling salesman problem. Algorithmic Oper. Res. **7**, 22–29 (2012)

25. Renaud, J., Boctor, F.F.: An efficient composite heuristic for the symmetric generalized traveling salesman problem. Eur. J. Oper. Res. **108**(3), 571–584 (1998)

26. Schmidt, J., Irnich, S.: New neighborhoods and an iterated local search algorithm for the generalized traveling salesman problem. Gutenberg School of Management and Economics Working Papers, pp. 1–21 (2020)

27. Smith, S.L., Imeson, F.: GLNS: an effective large neighborhood search heuristic for the generalized traveling salesman problem. Comput. Oper. Res. **87**, 1–19 (2017)

28. Snyder, L., Daskin, M.: A random-key genetic algorithm for the generalized traveling salesman problem. Eur. J. Oper. Res. **174**(1), 38–53 (2006)

29. Tasgetiren, M.F., Suganthan, P.N., Pan, Q.Q.: A discrete particle swarm optimization algorithm for the generalized traveling salesman problem. In: Proceedings of the 9th Annual Conference on Genetic and Evolutionary Computation, GECCO 2007, pp. 158–167. Association for Computing Machinery (2007)

30. Yuan, Y., Cattaruzza, D., Ogier, M., Rousselot, C., Semet, F.: Mixed integer programming formulations for the generalized traveling salesman problem with time windows. 4OR (1), 1–22 (2020). https://doi.org/10.1007/s10288-020-00461-y

A Simple Genetic Algorithm
for the Critical Node Detection Problem

Mihai-Alexandru Suciu⬤, Noémi Gaskó⬤, Tamás Képes,
and Rodica Ioana Lung$^{(\boxtimes)}$⬤

Centre for the Study of Complexity, Babeş-Bolyai University, Cluj-Napoca, Romania
{mihai.suciu,noemi.gasko,tamas.kepes,rodica.lung}@ubbcluj.ro
http://csc.centre.ubbcluj.ro/

Abstract. The critical node detection problem describes a class of graph problems that involves identifying sets of nodes that influence a given graph metric. One variant of this problem is to find the nodes that - when removed from the graph - maximize the number of connected components in the remaining graph. This is an example of a practical problem with multiple real-world applications in epidemic control, immunization strategies, social networks, biology, etc. This paper proposes the use of a simple GA to identify the set of the critical nodes of the problem without designing special problem specific variation operators. Problem specific information is used only in the fitness function and the constraint handling technique. We show that this simple approach performs as well as state-of-art methods.

Keywords: Critical node detection · Genetic algorithm · Complex networks

1 Introduction

Nodes in a network can have different importance with respect to different network measures and behavior. Finding these nodes, called critical nodes, is an essential computational task. Critical nodes can be approached also from the general node deletion problem [14], which is a large class of problem composed of several problems, such as the vertex separator problem, the minimum vertex cover problem, the critical node detection problem, etc. Recently, the critical node detection problem (CDNP) gained attention due to its large applicability. A very important class of the critical node detection problem is to identify the set of nodes of a maximal size to remove from the graph in order to maximize the number of connected components. Applications of this problem can be found in epidemic control and immunization strategies, social networks, biology, telecommunications, etc.

This work was supported by a grant of the Romanian National Authority for Scientific Research and Innovation, CNCS - UEFISCDI, project number PN-III-P1-1.1-TE-2019-1633.

H. Sanjurjo González et al. (Eds.): HAIS 2021, LNAI 12886, pp. 124–133, 2021.
https://doi.org/10.1007/978-3-030-86271-8_11

In general, the critical node detection problem consists in finding a set of nodes in a given graph $G = (V, E)$, which deleted maximally degrades the graph according to a given measure σ. CDNP is a central problem in network analysis with applications in several research fields, such as biology [2], network vulnerability [6], social network analysis [3], etc. Regarding the measure σ several studies focus on network centrality measures, such as betweenness centrality, closeness centrality, page rank [11,16].

Although several variants of the CDNP exist, only a few of them deal with computational methods for the variant consisting of removing k nodes in order to maximize the number of remaining components. The main goal of this paper is to approach this problem using a genetic algorithm with minimal problem specific adaptations. The choice of a genetic algorithm came first due to the natural binary encoding of an individual, but this is not the only reason we made it: we believe that it is important to explore different methods and paths and not constrain ourselves to assuming that one method may not work on a certain problem because it has not been tested on it. This is also related to the choice of operators: if there is not need for specific operators that use domain knowledge, we should not use them and keep the approach as general and as flexible as possible.

The rest of the paper is organized as follows: the next section presents the problem and reviews some existing approaches. The third section describes the proposed genetic algorithm. In the fourth section numerical experiments considering synthetic and real world networks are used to compare our results with the existing ones. The articles ends with conclusions and further work.

2   Related Work

Many variants of the critical node detection problem are studied in the literature, among which we mention: minimizing the pairwise connectivity by deleting k nodes (this variant is the most studied in the literature), minimizing the largest component size by deleting k nodes, bound the pairwise connectivity to a given threshold by deleting the minimal set of nodes, etc. A recent survey of the problem can be found in [13].

There are several ways to classify the critical node detection problem (CNDP). In [21] the two types variant is adopted: *CNDP type 1* problems aim to minimize the network connectivity maintaining the number of removed nodes under a given threshold and *CNDP type 2* problems in which the goal is to minimize the number of nodes that are removed such that the network connectivity reaches a given threshold. The type of connectivity measure used depends on the envisaged application, effect or the type of network. Applications are multiple as the CNDP is related to network sustainability and vulnerability [21]. Many practical approaches are devised for wireless sensor networks [7,8,18].

In [21] an exact algorithm for the problem considering the largest connected component is proposed. The k-vertex cut problem, consisting in finding the minimum weight subset whose removal disconnects the graph in at least k components is studied in [9]. Component-Cardinality-Constrained Critical Node Problem (3C-CNP) is approached in [12]. A bi-objective design is presented in [25]. As far as the type of networks, weighted networks are studied for example in [5] and directed graphs in [19].

In [1] the two types of CNDP problems are studied in three versions, among which also *kMaxComp*, the problem of removing a set of maximum k nodes to maximize the number of connected components in the remaining graph. This is one of the less studied CNDP variants, proven to be NP-hard [24]. In [24] a Mixed Integer linear programming approach is presented, [27] present a general integer programming framework. For a special class of graphs (trees and series-parallel graphs) a dynamic programming approach is presented [23]. In [1] a genetic algorithm is designed to solve the problem. The proposed genetic algorithm incorporates in the fitness function a penalization of solutions that are too close to the best solutions, combines a greedy strategy with variation operators and employs a local search mechanism at the end in order to refine solutions.

In this paper we focus on the problem $CDNP_a^3$, denoted here as *kMaxComp*, introduced in [23,24]. The $CDNP_a^3$ is by itself an interesting problem to be studied, with many possible applications. It has received less attention because it does not impose any conditions on the connected components. The problem consists in removing a maximum of k nodes such as the number of remaining components to be maximal. Formally, if S denotes the set of the deleted nodes, and $\mathcal{H}(G[V \setminus S])$ denotes the set of the maximal component of graph G without the set of nodes S, the optimization problem consists in

$$max|\mathcal{H}(G[V \setminus S])|, \text{such that } |S| \leq k, \tag{1}$$

where $|A|$ denotes the cardinality of set A.

3   Maximum Components GA (MaxC-GA)

The goal of this work is to solve the *kMaxComp* problem by using a minimum number of problem specific information during the search. Because we search for a set of nodes from a network out of which some will be included in the critical set S and some not, a binary encoding of an individual of length $N = |V|$ is natural, making a genetic algorithm the first choice in trying to approach this problem. We call this algorithm Maximum Components GA. MaxC-GA is outlined in Algorithm 1. MaxC-GA is a simple approach for the CDNP3a problem, that combines a standard GA with a constraint method based on the marginal contribution of a node to the fitness of an individual, concept borrowed from game theory, where such marginal contributions are used to evaluate the contribution of a player to the value of a coalition when computing the Shapley value [22].

Encoding. An individual has length N equal to the number of nodes in the network. The value 1 on position i indicates that node i is included in S.

Algorithm 1. MaxC-GA outline

Initialize population P of size p_{size} at random.
for a number of generations **do**
    P = Select p_{size} individuals for variation;
    Offsping= variation operators on P;
    Correct and Evaluate Offspring;
    P = offspring;
end for

Variation Operators. Two point crossover and flip-bit mutation are used.

Selection. Tournament selection is used for selection for recombination and mutation.

Fitness Function. The fitness of an individual is computed as the number of connected components the removal of its nodes with value 1 yields. Thus, if individual x encodes the critical set S_x then the fitness $f(x)$ of x is computed as

$$f(x) = |\mathcal{H}(G[V \setminus S_x])|. \tag{2}$$

Constraint Handling. In order to ensure that the size of the corresponding set S does not exceed k, before evaluation each individual is constrained to have only k nodes with value 1 by removing the nodes with the lowest marginal contribution to the fitness of the individuals from S. The marginal contribution of a node to the fitness of the individual is computed as the difference between the fitness of the individual and the fitness of the individual with the node removed from its corresponding set S of critical nodes. For a node i with value 1 in individual x with corresponding critical set S_x the marginal contribution of node i to the fitness of x denoted by $u_i(x)$ is:

$$u_i(x) = f(x) - |\mathcal{H}(G[V \setminus \{S_x \setminus \{i\}\}])|,$$

where $f(x)$ is the fitness defined in Eq. (2).

Parameters. MaxC-GA is a standard GA, and uses typical GA parameters: maximum number of generations, crossover and mutation probabilities, probability to mutate a bit, and tournament size. The effect of these parameters on the search results of a GA has been widely documented [15].

4   Numerical Experiments

The behavior of MaxC-GA is illustrated by using several benchmarks and comparing results with best known found in the literature for this problem.

Table 1. Synthetic benchmark test graphs and basic properties.

Graph	$|V|$	$|E|$	k	$\langle d \rangle$	ρ	l_G
BA1000	1000	999	75	1.998	0.002	6.045
ER466	466	700	80	3.004	0.006	5.973
FF500	500	828	110	3.312	0.007	6.026
WS500	500	1496	125	5.984	0.012	5.304

Benchmarks. A set of synthetic benchmarks[1] was proposed in [26]. The benchmark set contains four different type of graphs: Barabási-Albert (BA), Erdős-Rényi (ER), Forest-fire (FF), Watts–Strogatz (WS) graphs. BA graphs are scale free networks, ER graphs are random networks, FF graphs simulate how fire spreads through a forest, WS graphs are small world graphs with a dense structure.

Table 1 presents some basic measures of the benchmarks used for numerical experiments here: number of nodes ($|V|$), number of edges ($|E|$), average degree ($\langle d \rangle$), density of the graph (ρ), and average path length (l_G). In a similar manner, real networks are described in Table 2 with a reference added for each network.

Table 2. Real-world graphs and basic properties.

Graph	$|V|$	$|E|$	k	$\langle d \rangle$	ρ	l_G	Ref.
Bovine	121	190	3	3.140	0.026	2.861	[20]
Circuit	252	399	25	3.167	0.012	5.806	[17]
EColi	328	456	15	2.780	0.008	4.834	[28]
HumanDis	516	1188	52	4.605	0.008	6.509	[10]
TrainsRome	255	272	26	2.133	0.008	43.496	[4]

Parameter Settings. Several parameter setting are tested: population size set to 25 and 50, maximum number of generations 500, crossover probability 0, 0.5, 0.8, and 1, and mutation rate 0, 0.01, 0.02, 0.03, 0.04, and 0.05.

Results and Discussion. MaxC-GA is compared with three algorithms described in [1]: two greedy algorithms, the first one, G_1 based on node deletion from the candidate critical node set, and the second one, G_2, based on the node addition to the candidate critical node set and a genetic algorithm from an evolutionary algorithm framework using greedy rules (denoted by GA). The genetic algorithm uses a specific fitness function that combines the number of connected components determined by the interval with previous search information, problem specific variation operators and a specific designed local search technique.

---

[1] downloaded from http://individual.utoronto.ca/mventresca/cnd.html, last accessed 05.09.2020.

Table 3. Maximum fitness values for the tested problems. The average over 10 runs is presented for MaxC-GA.

Graph	MaxC-GA	GA	G_1	G_2
ER466	110.0	110	99	105
BA1000	590.0	590	590	590
FF500	215.0	215	214	214
WS500	44.0	44	25	18
Bovine	77.0	77	77	77
Circuit	32.0	31	30	29
EColi	169.0	169	169	168
HumanDis	148.0	148	147	148
TrainsRome	31.0	31	30	31

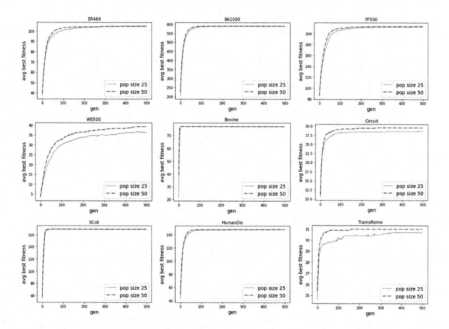

Fig. 1. Search evolution of MaxC-GA for the benchmarks, average best fitness for a population size of 25 and 50 over 10 runs.

Since the problem has been less addressed, we only have one approach based on GAs to compare with, and those results represent only one run. Results presented in the paper are preliminary and promising, supporting the idea that this approach may be extended for larger data sets.

As results presented in [1] include only the maximum number of connected components in one run, therefore statistical comparisons with results reported there are not possible. Table 3 includes these results as well best results reported

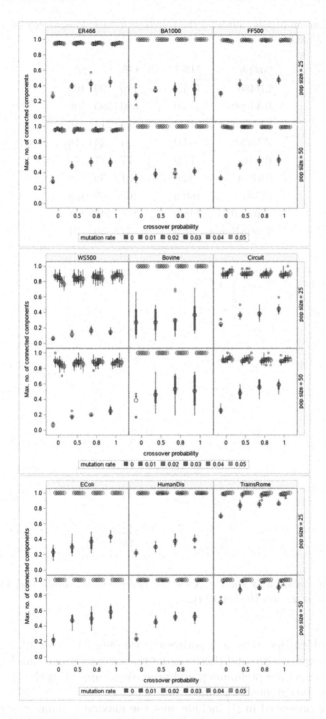

Fig. 2. Box plots presenting results reported by MaxC-GA for the nine benchmarks and different parameter values.

by MaxC-GA. Results reported by MaxC-GA using different parameter settings are illustrated in Fig. 2. Furthermore, Fig. 1 illustrates the evolution of the search of MaxC-GA (average best solutions over 10 runs). We find that the evolution is steady, faster for a larger population size, and that MaxC-GA is capable to find and maintain the optimal solution. Because the behavior of MaxC-GA under different parameter settings is typical for that of a GA, with respect to convergence we have presented only graphs showing that it is capable to detect and maintain the optimal solution during one run. In all other ways it behaves as expected: a larger population size leads to an earlier convergence at a higher computational cost and a small population will eventually converge.

The effect of various parameter settings presented in Fig. 2 as boxplots of the ratio of maximum fitness values reported in 10 runs for each parameter setting and best known result for the benchmark (in order to keep all values between 0 and 1). We find that the algorithm is robust with respect to variation of parameters, with the notable exception that mutation plays an important role in the search, as setting the mutation rate to 0 significantly decreases the performance of the algorithm.

5 Conclusions

The critical node detection problem is approached with MaxC-GA, a simple genetic algorithm that uses a node fitness based on marginal contributions for constraint handling. Numerical results show that this approach is as effective as other, more complex, using more problems specific information.

These results may also be used to advocate for the use of minimal problem specific information in designing new evolutionary algorithms for real-world applications. Overusing specific problem information decreases the adaptability of the presented method, as practitioners will rarely try to adapt an existing algorithm presented in literature to a slightly different problem, mainly because the stochastic nature of these approaches does not guarantee direct portability to a different problem.

References

1. Aringhieri, R., Grosso, A., Hosteins, P., Scatamacchia, R.: A general evolutionary framework for different classes of critical node problems. Eng. Appl. Artif. Intell. **55**, 128–145 (2016)
2. Boginski, V., Commander, C.W.: Identifying critical nodes in protein-protein interaction networks. In: Clustering Challenges in Biological Networks, pp. 153–167. World Scientific (2009)
3. Borgatti, S.P.: Identifying sets of key players in a social network. Comput. Math. Organ. Theory **12**(1), 21–34 (2006). https://doi.org/10.1007/s10588-006-7084-x
4. Cacchiani, V., Caprara, A., Toth, P.: Scheduling extra freight trains on railway networks. Transp. Res. Part B: Methodol. **44**(2), 215–231 (2010)

5. Chen, W., Jiang, M., Jiang, C., Zhang, J.: Critical node detection problem for complex network in undirected weighted networks. Phys. A: Stat. Mech. Appl. **538** (2020). https://doi.org/10.1016/j.physa.2019.122862
6. Cohen, R., Erez, K., Ben-Avraham, D., Havlin, S.: Resilience of the internet to random breakdowns. Phys. Rev. Lett. **85**(21), 4626 (2000)
7. Dagdeviren, O., Akram, V.: An energy-efficient distributed cut vertex detection algorithm for wireless sensor networks. Comput. J. **57**(12), 1852–1869 (2013). https://doi.org/10.1093/comjnl/bxt128
8. Dagdeviren, O., Akram, V., Tavli, B., Yildiz, H., Atilgan, C.: Distributed detection of critical nodes in wireless sensor networks using connected dominating set (2017). https://doi.org/10.1109/ICSENS.2016.7808815
9. Furini, F., Ljubić, I., Malaguti, E., Paronuzzi, P.: On integer and bilevel formulations for the k-vertex cut problem. Math. Program. Comput. **12**(2), 133–164 (2019). https://doi.org/10.1007/s12532-019-00167-1
10. Goh, K.I., Cusick, M.E., Valle, D., Childs, B., Vidal, M., Barabási, A.L.: The human disease network. Proc. Natl. Acad. Sci. **104**(21), 8685–8690 (2007)
11. Iyer, S., Killingback, T., Sundaram, B., Wang, Z.: Attack robustness and centrality of complex networks. PloS One **8**(4), e59613 (2013)
12. Lalou, M., Tahraoui, M., Kheddouci, H.: Component-cardinality-constrained critical node problem in graphs. Discrete Appl. Math. **210**, 150–163 (2016). https://doi.org/10.1016/j.dam.2015.01.043
13. Lalou, M., Tahraoui, M., Kheddouci, H.: The critical node detection problem in networks: a survey. Comput. Sci. Rev. **28**, 92–117 (2018). https://doi.org/10.1016/j.cosrev.2018.02.002
14. Lewis, J.M., Yannakakis, M.: The node-deletion problem for hereditary properties is NP-complete. J. Comput. Syst. Sci. **20**(2), 219–230 (1980)
15. Lobo, F., Lima, C.F., Michalewicz, Z.: Parameter Setting in Evolutionary Algorithms, vol. 54. Springer, Heidelberg (2007). https://doi.org/10.1007/978-3-540-69432-8
16. Lozano, M., García-Martínez, C., Rodriguez, F.J., Trujillo, H.M.: Optimizing network attacks by artificial bee colony. Inf. Sci. **377**, 30–50 (2017)
17. Milo, R., et al.: Superfamilies of evolved and designed networks. Science **303**(5663), 1538–1542 (2004)
18. Min, S., Jiandong, L., Yan, S.: Critical nodes detection in mobile ad hoc network, vol. 2, pp. 336–340 (2006). https://doi.org/10.1109/AINA.2006.136
19. Paudel, N., Georgiadis, L., Italiano, G.: Computing critical nodes in directed graphs. ACM J. Exp. Algorithmics **23** (2018). https://doi.org/10.1145/3228332
20. Reimand, J., Tooming, L., Peterson, H., Adler, P., Vilo, J.: GraphWeb: mining heterogeneous biological networks for gene modules with functional significance. Nucleic Acids Res. **36**, 452–459 (2008)
21. Rezaei, J., Zare-Mirakabad, F., MirHassani, S., Marashi, S.A.: EIA-CNDP: an exact iterative algorithm for critical node detection problem. Comput. Oper. Res. **127** (2021). https://doi.org/10.1016/j.cor.2020.105138
22. Shapley, L.S.: 17. A Value for n-Person Games, pp. 307–317. Princeton University Press (1953). DOIurl10.1515/9781400881970-018
23. Shen, S., Smith, J.C.: Polynomial-time algorithms for solving a class of critical node problems on trees and series-parallel graphs. Networks **60**(2), 103–119 (2012)
24. Shen, S., Smith, J.C., Goli, R.: Exact interdiction models and algorithms for disconnecting networks via node deletions. Discrete Optim. **9**(3), 172–188 (2012)

25. Ventresca, M., Harrison, K., Ombuki-Berman, B.: The bi-objective critical node detection problem. Eur. J. Oper. Res. **265**(3), 895–908 (2018). https://doi.org/10.1016/j.ejor.2017.08.053
26. Ventresca, M.: Global search algorithms using a combinatorial unranking-based problem representation for the critical node detection problem. Comput. Oper. Res. **39**(11), 2763–2775 (2012)
27. Veremyev, A., Prokopyev, O.A., Pasiliao, E.L.: An integer programming framework for critical elements detection in graphs. J. Comb. Optim. **28**(1), 233–273 (2014). https://doi.org/10.1007/s10878-014-9730-4
28. Yang, R., Huang, L., Lai, Y.C.: Selectivity-based spreading dynamics on complex networks. Phys. Rev. e **78**(2), 026111 (2008)

Learning Algorithms

Window Size Optimization for Gaussian Processes in Large Time Series Forecasting

Juan Luis Gómez-González[ID] and Miguel Cárdenas-Montes[(✉)][ID]

Department of Basic Research, CIEMAT, Madrid, Spain
{juanluis.gomez,miguel.cardenas}@ciemat.es
http://www.ciemat.es

Abstract. Many current machine learning applications rely on performance rather than on model interpretability. Robust confidence projection is underrated as well. These qualities are of key importance in experimental sciences where benefits of IA applicability is intended to precisely bound sensible results. A Gaussian Process (GP) constitutes a non-parametric soft computing method which encompasses model interpretability and transparency. Additionally GPs feature rigorous uncertainty estimations by means of convenient kernel specification fitting data stochastic properties. Extensive temporal series of data may efficiently be characterised by GPs. GPs perform selecting the most suitable distribution of solutions conditioning over observations by means of maximizing the logarithm of the marginal-likelihood of data. Selecting an appropriate time-training interval stands out as a requirement and to successfully extract unambiguous structural information from kernel hyperparameters. Nevertheless demanding computational cost in the course of the training stage, scaling as $\mathcal{O}(n^3)$ being n the data set size, sets a trade-off on how much data to include. Our work aims at providing a general procedure to optimize input training selection to be minimum meanwhile retaining unambiguous information to the process modelled by the GP. To address this problem our ideas are evaluated performing forecasting on Madrid air-quality time series.

Keywords: Gaussian processes · Time series · Hyperparameter fitting · Explainable artificial intelligence

1 Introduction

There are efforts pushing towards a **Machine learning** (ML) science accounting for **transparency, interpretability** and **explainability** [10]. From those efforts the field known as **Explainable AI** (XAI) emerges. An ML approach is transparent if algorithm choice and its individual components altogether have a precise motivation and description. A resulting model is interpretable if fitted parameters and hyperparameters have an intuitive meaning concerning input targets. Explainability is the capacity of how a model contributes to a decision based on its predictions.

© Springer Nature Switzerland AG 2021
H. Sanjurjo González et al. (Eds.): HAIS 2021, LNAI 12886, pp. 137–148, 2021.
https://doi.org/10.1007/978-3-030-86271-8_12

A **Gaussian Process** (GP) is a non-parametric soft-computing ML implementation that learns from data interpreting it as a Multivariate Gaussian Distribution [8]. GPs are capable of a broad spectrum of tasks such as regression, classification, inverse engineering, change-point identification, artificial feature sampling, etc. [3,9]. Fitting performance relies on Kernel specification which lets us encode a priori structures that may be masked in data such as change-points, quasi-periodicity, fixed trends, etc. Kernel design is a transparent exercise since kernel components weld precise statistical features. Subsequent resulting models are interpretable because hyperparameters encoding previous features have a clear structural information (i.e. a periodicity in data). A very important contribution GPs bring is being able to provide robust confidence intervals at predictions. If underlying process in data is properly captured by a kernel specification and hyperparameters, predictions follow a Gaussian Distribution with known mean and standard deviation.

Due to the complexity of the evolution of a magnitude over time, the true signal is hindered by a large noisy component. GP naturally captures the probabilistic nature of data, prescribing an average signal within comprehensive confidence interval estimations capturing random variance.

The specific problem to fit GP into Time Series is covered in [11]. But the question how extensive input data size plays a major role remains unanswered. The larger data size is the better stochastic properties are characterized. Nevertheless, the amount of data is a limiting step because of computation costs scaling as $\mathcal{O}(n^3)$ in GP computation. As discussed in [8], GP algorithms admit a strong limit about $10^3 - 10^4$ observations for conservative computational costs due to repeated recurrence in computing large inverted matrices[1]. Even restricting this limit, extraordinary computational resources are required. A GP aimed to fit a time series of about this number of observations, given a minimum complex kernel, can take hours to complete optimization. In scenarios involving more complex kernels is strongly penalized by the extreme computational intensity.

To that extent we have derived a *constant density signature* criteria that helps finding an adequate, minimum valid, window size to perform robust regression on large time series retaining unambiguous information to the process modelled by the GP. The application of this prescription also provides interpretable hyperparameters and subsequent predictions are additionally less computationally demanding since no further optimization or parameter tuning is required. The window size is interpretable itself since provides an optimum interval of time over which meaningful patterns are captured. Different models may yield different optimum sliding windows, helping to discriminate families of models yet with different design ending up yielding a similar signature or not.

These ideas are tested on real data performing short-forecasting on Ground-level urban Ozone pollutant concentrations in Madrid (Spain). Pollution concentration time series forecasting is a state-of-the-art problem due to global concern regarding air quality, and our data-set particularly is of critical importance to

---

[1] There are GP approximations that make use of complex computation methods and sparse matrix techniques [6], but involving approximations.

Europe, addressing a novel case of study with our methodology [2]. Two simple models are proposed distinguished by an important parameter left open to discussion; observational error in pollutant concentrations. Altogether the structure of this work illustrates how a ML problem can follow and intuitive definition, retrieve interpretable results and valuable conclusions not only restricted to statistical metrics using GP.

2   Methodology

2.1   Gaussian Processes

A Gaussian process is a collection of random variables, any finite number of which have a joint Gaussian distribution, hence a GP is completely specified by its mean and covariance functions [8]. Being t the indexes noting the normally distributed observations $\boldsymbol{y}(t)$:

$$y(t) \sim \mathcal{N}(m(t), k(t, t')) \tag{1}$$

In the context of time series indexes $\boldsymbol{t}$ represent time. $k(t, t')$ evaluates the covariance matrix in (Eq. 1) and it is called Kernel function in GP formalism having further dependence on parameters known as hyperparameters. Specification of such Kernel function constitutes a closed approximation for correlations in sampled trajectories $\text{Cov}(y(t), y'(t')) = k(t, t')$. Stochastic processes following previous mean and covariance are suitable to be drawn from a GP, although plausible higher order correlations cannot not be captured. GPs are probability density distributions over functions. Time series are prominent examples of this interpretation, since those can be drawn from a properly characterized GP.

Distribution (Eq. 1) constitutes the prior distribution. Sampled random trajectories, evaluated at desired test points $\boldsymbol{t}_*$ do not include information regarding the collection of data to be modelled. In order to gain insight from a particular process the distribution is conditioned by means of Bayes's rule over a set of given observations at $\boldsymbol{t}$. Since GPs are multivariate Gaussian distributions, conditioning ends up into another Gaussian distribution. Assuming noisy observations with uniform Gaussian random noise $y(t) = f(t) + \epsilon_t$, with $\epsilon_t \sim \mathcal{N}(0, \sigma)$, (Eq. 1) yields the posterior (id. predictive) distribution (Eq. 2) at $\boldsymbol{t}_*$:

$$y(t_*) \sim \mathcal{N}(f(t_*), \text{Cov}(y(t_*), y(t'_*)))$$
$$f(t_*) = m(t_*) + k(t_*, t)\left[k(t, t') + \sigma^2 \mathbb{I}\right]^{-1} y(t')$$
$$\text{Cov}(y(t_*), y(t'_*)) = k(t_*, t'_*) - k(t_*, t)\left[k(t, t') + \sigma^2 \mathbb{I}\right]^{-1} k(t', t'_*) \tag{2}$$

Prediction and statistical inference at $\boldsymbol{t}_*$ is performed. If the conditioned process distribution is well captured, robust statistical uncertainty estimation at $\boldsymbol{t}_*$ is derived from diagonal elements in the predictive covariance matrix (Eq. 2).

Fitting the prior to observations is achieved maximizing the logarithm of the marginal likelihood with respect to hyperparameters. For Gaussian Distributions marginalizing over latent functions is analytically tractable (Eq. 3).

$$\log \rho \left(\boldsymbol{y} | \boldsymbol{t}, \boldsymbol{h}\right) = -\frac{1}{2} \boldsymbol{y}\left(\boldsymbol{t}\right)^{T} K_{h}^{-1}\left(t, t'\right) \boldsymbol{y}\left(\boldsymbol{t'}\right) - \frac{1}{2} \log \left|K_{h}\left(t, t'\right)\right| - \frac{n}{2} \log 2\pi \quad (3)$$

h notes the hyperparameter dependence of the kernel. This is a critical step because h carries the intended inferred information from the data through kernel choice. Once a range of kernels is considered and properly fitted, independent validation on test data conducts to the best model not only performance-wise but also on best confidence interval and favourable structural retrieval.

2.2 Kernel Selection

Kernel selection is not an automatic task. Actual intuition of data behaviour and preliminary analysis over its statistical properties must be carried out before prescribing a proper kernel recipe. Known uncertainty in training observations, characterized with the distribution ϵ_t, has a major role in the predictive distribution too. Kernels grasp global stochastic patterns in a non-parametric nature but its hyperparameters retain interpretable structural information. To validate our ideas we select the **Squared Exponential Kernel plus white noise** function (Eq. 4).

$$k\left(t_{*}, t_{*}'\right) = \alpha \cdot \exp\left(-\frac{|t_{*} - t_{*}'|^{2}}{2\beta^{2}}\right) + \sigma_{wh}^{2} \cdot \delta_{t_{*}=t_{*}'} \quad (4)$$

This kernel of high simplicity states the longer the euclidean distance among input variables, correlation diminishes. It is invariant under input shifts, thus modelling stationary processes. β becomes a characteristic length of the processes drawn from prior. Conditioning to observations, it firstly acknowledges the underlying process is smooth, stationary and single-sourced as only one characteristic length suffices estimation in modulation. Secondly, when tuned to fit the observations, even though the kernel estimates a prior covariance matrix (Eq. 1), data is considered to be sampled from such distribution so β characterizes the length-scale of the process under study as well.

Kernel (Eq. 4) portrays a crude approximation omitting other plausible structures. (i.e. aggregation of several independent processes, periodicity in the signal, inclusion of significant coloured noise contribution, etc.), nevertheless kernel design is versatile as long as it produces semi-definite positive covariance matrices. Addition, multiplication, exponentiation of independent kernels (each accounting for a particular kind of process) is possible. This permits exploring a wide range of arbitrary complex structures.

2.3 General Approach to Large Time Series

In considering GPs as candidates to model time series, there are two possible approaches involving input targets preceding predictions. On one hand defining a concise sliding window being shifted over the whole time domain as long as new predictions are actually observed. On the other hand fitting a GP to the whole data set available. Examples in literature follow the later approach as Mauna

Loa station CO_2 measurements [8], ^{222}Rn radiation level time series at the Canfranc Underground Laboratory [4], solar activity [5] or airline passenger seasonal trends [12]. These have in common the amount of data on the order of 10^2 so that it is feasible to fit a GP to all data and properly optimize hyperparameters in order to grasp maximum pattern retrieval. However the first procedure is necessary when data constitutes a large temporal series because of demanding computational costs in GPs.

Selection of the window size relying on cost-effective considerations would lead to incorrect structural assumptions. For example for O_3 time series one would think 30 days would suffice to capture local correlations (up to a month) and extrapolate predictions successfully, given a good enough kernel based on prediction metrics alone. Nevertheless checking Fig. 1b a clear yearly modulation is present that requires fitting the kernel to a bigger interval of time. There can not be a bounded hyperparameter set due to data behaving on the scale of the short sliding window interval diversely. Not having unique hyperparameters hampers interpretable information from the process as a whole and induces high computational costs since optimization is always required prior to extrapolation.

Because of that an effective approach on modeling time series with GP needs a precise understanding of this sliding window if there is enough data to work with. An appropriate specification also leads to correct assumptions concerning confidence intervals globally, since the posterior covariance matrix depends on hyperparameters as well. This understanding meets a trade off; the selection of a time window both ensures proper capture of time series processes' patterns and to explicitly include just enough amount of previous information required to retain global details meanwhile keeping acceptable computational costs.

Taking the logarithm of the marginal likelihood from Eq. 3 one may be susceptible to inspect its dependence on the size of the sliding window. A GP featuring a kernel that appropriately accounts for the underlying process in data implies already compiled information as input targets can be effectively understood as a sample drawn from the prior distribution. This reasoning implies that hyperparameters are already optimized for that process. Adding new observations will not add extra information (i.e. high variance in hyperparameters) if such optimization is to occur iteratively with no restrictions. In other words, hyperparameters must tend to a stationarity as more data is included in the sliding window frame and there are no process change-points. We propose that the quotient between the log-marginal likelihood over n converges as n tends to infinity or rather, when the Gaussian Process already captures every remarkable pattern in the process (Eq. 5).

$$\frac{\log \rho\left(\boldsymbol{y} | \boldsymbol{t}, \boldsymbol{h}\right)}{n} \xrightarrow[n \longrightarrow \infty]{} constant \tag{5}$$

Previous statement incorporates both the idea that time series statistics over time and hyperparameters converge. Covariance matrices are positive semidefinite so that there exists a change of variable (Eq. 6).

$$\boldsymbol{y}\left(\boldsymbol{t}\right)^T k\left(t, t'\right)^{-1/2} \longrightarrow \boldsymbol{y}_c \tag{6}$$

Simply put being a linear transformation of the input targets. Now log-marginal likelihood reads,

$$\log \rho \left(\boldsymbol{y}_c | \boldsymbol{t}, \boldsymbol{h}\right) = -\frac{1}{2}{\boldsymbol{y}_c}^T \boldsymbol{y}_c - \frac{n}{2}\log 2\pi \tag{7}$$

The main result unfolds when n is sufficiently large

$$\frac{\log \rho \left(\boldsymbol{y}_c | \boldsymbol{t}, \boldsymbol{h}\right)}{n} = -\frac{1}{2}\langle y_c^2 \rangle - \frac{1}{2}\log 2\pi \tag{8}$$

If data points are indefinitely sampled convergence eventually happens. The optimized sized sliding window will be the smallest n such that Eq. 5 maintains. Directly checking that average from Eq. 8 converges is not possible since optimized hyperparameters are unknown beforehand. Computing Eq. 5 continuously optimizing hyperparameters is required. When it is fulfilled, larger intervals of time would not lead to different hyperparameters. Alternatively only checking stationarity on optimized hyperparameters is inappropriate because hyperparameter dimensionality may be high and some variance will always be expected. Hence Eq. 5 remains as a constant density signature to the process under study. Section 3 presents this idea with a practical implementation on actual data.

2.4   Data Set

Regarding Ozone trends restricted to this work, data corresponds to a station located in *Plaza del Carmen* [1]. Data spans from date 1st January, 2010 to 30th June, 2020. This range is well suited since no manipulations have been performed on the monitoring station during that period. In summary there are about 3650 input targets. Few days recordings are missing but over the course of optimizing kernels it does not affect GP extrapolation. Data graphical overview is in Fig. 1.

Ground-level Ozone concentration behavior is characterised in air-quality reports [2]. It showcases a clear seasonal trend reaching maximum concentrations during the summer period when incident solar radiation is more intense. In Fig. 1b, the year-modulation stands out. Worth noting a weekly-modulation arises in the frequency spectrum too (arrow in Fig. 1b). The other main contribution is a short-range correlated noisy pattern (see Fig. 1c for s small). The evolution of this time series has a strong dependence on season, weather, road-traffic density or citizen behaviours. Nevertheless GP is well suited to find confidence bounds. In order to do that, parameter σ is decisive, as it already explicitly appears in Eqs. 2. Rigorously it represents an homogeneous estimation of uncertainty on the input targets. At GP fitting, this parameter plays a major role in loosening/tightening trajectories in the posterior distribution onto actual input targets. Maintenance data repository authorities consulted were not able to provide a precise observational error bounds. Because of that over the course of predictions two values will be taken, $\sigma = 0.01$ representing high precision measurements and $\sigma = 1.0$ for a large uncertainty in order to extend analysis and

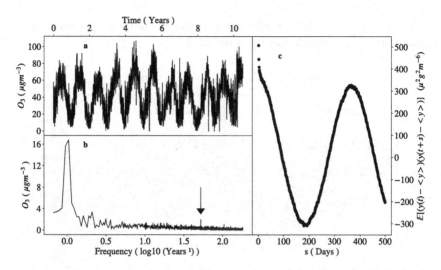

Fig. 1. Ozone data set from *Plaza del Carmen* pollutant monitoring station. Panel **a** shows full temporal series spanning 10 years of daily observation. Panel **b** is the discrete Fourier transform from **a**. It exhibits a yearly modulation (located at 0 in the log-scale). Vertical arrow highlights over the noise background a weekly component contribution, precisely located at $\log_{10} 6.996^{-1}$ days). Panel **c** is the auto-correlation function $\mathbb{E}[(y(t) - \langle y \rangle)(y(t + s) - \langle y \rangle)]$. It portrays a decaying component for short long-range correlations and a yearly modulation that remains over time.

results in plausible ways[2]. As already noted a GP fitted to this data series being extremely accurate at given observations is not a realistic prediction.

3 Results and Analysis

Calculations are performed using Scikit-Learn (V0.24.1) Gaussian Processes implementation running under Python V3.6.8. [7].

3.1 Kernel and Parameter Choice

Due to the lack of an exact observation error magnitude (σ) on recorded measurements, two plausible values are considered to produce two different models. Choice of two distinct σ values show that hyperparameter results are robust within an optimized window yet different explanations arise. This parameter

---

[2] The GP implementation used to compute results performs an automated normalization of the input data, undone when predicting. Nevertheless σ, being an input parameter, has to be provided accordingly standardized prior to fitting. Thus these values are not explicit uncertainty values over observations, rather fractional estimations respect to data variance. Analogously hyperparameters affected by data scaling are retrieved constrained within 0 and 1. Length-scale process β is not altered by this.

becomes critical to validate results on experimental data. A graphical view of GP time-series regression is displayed in Fig. 2.

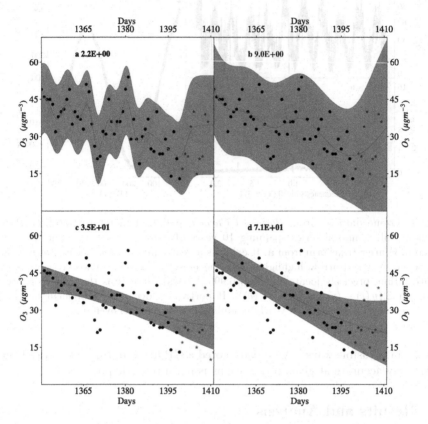

Fig. 2. Ozone prediction examples. GP regression using kernel from Eq. 4. Panel **a,b** consider error magnitude $\sigma_a = \sigma_b = 0.01$, and panels **c, d** $\sigma_c = \sigma_d = 1$. Regarding training observations, panels **a, c** employ 50 days observations to fit the GP, meanwhile panels **b,d** employ the optimized sliding window (379 and 349 days respectively, see Table 1). Same period of time displayed in every panel. Inline texted number refers to the process length-scale magnitude (unit of $days^{-1}$). Results are heavily conditioned to the set of input-parameters. Using an optimized window size retrieves the same set of hyperparameters on average. Regarding results discussion it represents a robust structural retrieval from the GP over data.

Parameter σ is vital at deriving the trajectory distribution that best describes data. As it was mentioned, the wider such uncertainty is the less tight the observations resulting trajectories are. It has profound effects on predictions. In particular it may hinder highly variable (lower length scale) meaningful processes. It also has effects on the confidence intervals. The interest is not strictly capturing the smooth signal but precisely amplitude of variance in detection on the monitoring station.

3.2 Hyperparameter Tuning and Interpretability

Whole data set is involved to fit the optimum hyperparameters and window size. The procedure is as follows: the limit of Eq. 5 versus size of the input time window is explored. Two grids of window sizes are defined, whose observations the GP is probabilistically conditioned, and hyperparameters optimized and extracted. For $\sigma = 0.01$ evaluated sizes in the grid run from 30 to 2200, for $\sigma = 1.0$ window sizes run from 30 to 1010. In each case window sizes increase with a 10 days step. A resulting average value is produced by randomly sampling (50 samples) the training set for each window size. Alternatively this produces the average value of hyperparameters for a particular window size.

Unless ideally aiming at fitting an infinite time series at once, finite size effects are present. Furthermore covering a continuously increasing window size starting from the same initial day adds a bias onto that stream of new pollutant concentrations where the GP is fitted. This is due to the recycling of previous values for new fittings. In order to minimize these difficulties, the Eq. 5 is evaluated by sampling different computations for a given window size.

Results are summarized in Fig. 3. Data points from both columns are connected pair-wise, since both come from a particular window size average. Panels **a1**, **a2** show the same behaviour for the quotient as window size increases. As it can be appreciated in both panels, the magnitude gets stationary as long as the window size surpasses a certain threshold. Stationarity, meanwhile being obvious at a glance, has to be determined automatically. In order to bound the onset for stationarity, including a precise value for the optimized window size, we decided on applying an auxiliary GP onto the full computed trend by means of the kernel from Eq. 4 and $\sigma_{wh} = 0$.

Emerging manifold is smooth which lets us easily compute its numerical derivative. The criteria for stationarity is simply noting where the local derivative first crosses null value (vertical red lines). GP posterior returns a 2-sigma bound interval delimiting the underlying evolution of the quotient presented in Eq. 5. Considering those hyperparameters whose mapping into panels **a1** and **a2** overlap both the stationary manifold and confidence interval (green points in panels **b1** and **b2**), highlights those behaving approximately stationary versus window size as well. We want to stress that this method is a practical approximation which also lets us grasp a confidence interval when selecting the distribution of valid hyperparameters, which may be opted for another, more suitable, approach.

Even though training data set is the same, the value of the limit differs (horizontal red lines in panels **a1** and **a2**) between the low and high observation error models (rows **a1**, **b1** and **a2**, **b2** respectively). Therefore such a limit is a signature of the model, not for the data set, to the extent of our analysis. Since marginal likelihood evaluates the probability of observations fitting the model, this result is expected since the observation error value (σ) has a strong effect on model predictions, as seen in Fig. 2. Conversely, the optimized size for the window sizes are similar for both σ values. For $\sigma = 0.01$ the most suitable window size is 379 days (panel **a1**), whereas for $\sigma = 1$ it is 349 days (panel **a2**).

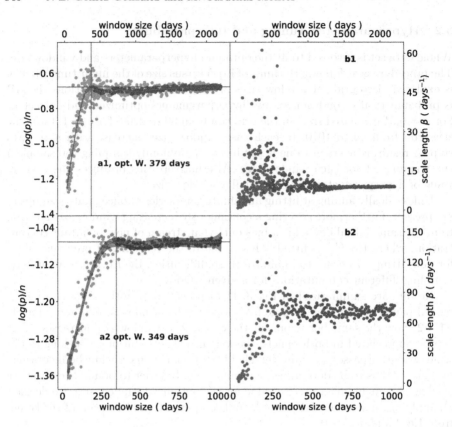

Fig. 3. Optimal interpretable hyperparameter and window size retrieval for low ($\sigma = 0.01$) and high ($\sigma = 1$) observation error models (top and bottom rows respectively). Each point is a 50 averaged GP regression sampled on observation subsets with the same window size. Panels **a1**, **a2** depict the manifold of Eq. 5 limit altogether to auxiliary GP regression and confidence bounds. On panels **b1**, **b2** corresponding hyperparameters (in particular process length-scale) are shown pairwise to **a1**, **a2** respectively. Selected hyperparameters marked as green dots correspond to those points in the left panel column lying on the stationary region. A vertical red line matches the approximate onset on convergence of Eq. 5. (Color figure online)

The optimized window size, not being a kernel hyperparameter, still comprises structural information from the time series, at least from the validating point of view. That is, the minimal unit of meaningful pattern information contained in the time series.

Given the minimum window size that guarantees GP captures all possible distinctive time series features, characteristic processes' hyperparameters are identified as well. As already noted fulfillment of Eq. 5 leads to the fact hyperparameters do not change abruptly when larger window sizes are explored. Stationarity defines a manifold in panels **a1** and **a2** where overlapping values map to optimal hyperparameters in panels **b1** and **b2**.

Results bound to the confidence interval—gray area over the stationary manifold—defined by the fitted auxiliary GP lying beyond the onset of stationarity are valid and compatible hyperparameters to the whole time series (green dots in panels **b1** and **b2** in Fig. 3. Their statistics are used to derive interpretable hyperparameters. This optimized parameter retrieval is presented at Table 1.

Table 1. Model results for optimized window size and interpretable hyperparameters for the large time series. Models with assumed low and high random Gaussian noise error on observations

Model	Opt. window size ($days$)	$\beta\left(days^{-1}\right)$	$\alpha\left(\mu^2 g^2 m^{-6}\right)$	$\sigma_{wh}^2\left(\mu^2 g^2 m^{-6}\right)$
Equation 4, $\sigma = 0.01$	379	9 ± 2	0.60 ± 0.01	0.155 ± 0.006
Equation 4, $\sigma = 1$	349	71 ± 6	0.9 ± 0.1	0

Following this methodology GP modelling of large time series ensures robust optimized hyperparameters retrieval within a particular model. Table 1 displays different hyperparameters for each model, but those are independent for window sizes larger to their respective optimum result (within confidence intervals) and effectively characterize the process under study as a whole. As a result, for each different noisy regime considered, a full and definitive set of hyperparameters is provided. Forecasting O_3 concentrations is immediate using the optimum window size without further training.

4   Conclusions

In this work, a methodology to understand the role of input-targets in large time series regression with Gaussian Processes following transparency and interpretability criteria has been presented. The ratio of log-marginal-likelihood over window size as the latter increases exhibits a stationary manifold that allows characterizing processes' hyperparameter with confidence intervals. The optimum window size gives a compelling work-around when regular GP algorithms are prohibitive computationally demanding as well as proposing the optimum input-data size to capture patterns of interest.

These ideas are illustrated by choosing the Exponential Squared Kernel plus white noise exploring the impact two different observation error values as open parameters have. Our methodology has proven its ability to find the window size that maximizes feature capture while maintaining a minimum size. This assures hyperparameters found for each model characterize the underlying process as a whole. Thus no further optimization is required, improving GP computational efficiency.

Two lines remain open for the future: on the one hand to evaluate the proposed methodology for much more complex kernels, capturing richer patterns from the observations and thus involving a final number of hyperparameters in the order of ten; and on the other hand, to check the performance of these complex kernels with optimized window size versus non-optimized ones.

Acknowledgment. JLGG is co-funded in a 91.89% by the European Social Fund within the Youth Employment Operating Program, as well as the Youth Employment Initiative (YEI), and co-found in a 8,11 by the "Comunidad de Madrid (Regional Government of Madrid)" through the project PEJ-2018-AI/TIC-10290. MCM is funded by the Spanish Ministry of Economy and Competitiveness (MINECO) for funding support through the grant "Unidad de Excelencia María de Maeztu": CIEMAT - FÍSICA DE PARTÍCULAS through the grant MDM-2015-0509.

References

1. Portal de datos abiertos del Ayuntamiento de Madrid. https://datos.madrid.es/portal/site/egob
2. European Environment Agency: Air quality in Europe: 2020 report (2020). https://doi.org/10.2800/786656, https://data.europa.eu/doi/10.2800/786656
3. Camps-Valls, G., Verrelst, J., Munoz-Mari, J., Laparra, V., Mateo-Jimenez, F., Gomez-Dans, J.: A survey on Gaussian processes for earth-observation data analysis: a comprehensive investigation. IEEE Geosci. Remote Sens. Mag. 4(2), 58–78 (2016). https://doi.org/10.1109/MGRS.2015.2510084
4. Cárdenas-Montes, M.: Uncertainty estimation in the forecasting of the ^{222}Rn radiation level time series at the Canfranc Underground Laboratory. Logic J. IGPL (2020). https://doi.org/10.1093/jigpal/jzaa057
5. Duvenaud, D.K.: Automatic model construction with Gaussian processes, pp. 53–55, June 2014
6. Liu, H., Ong, Y.S., Shen, X., Cai, J.: When Gaussian process meets big data: a review of scalable GPs. IEEE Trans. Neural Netw. Learn. Syst. 31(11), 4405–4423 (2020). https://doi.org/10.1109/tnnls.2019.2957109
7. Pedregosa, F., et al.: Scikit-learn: machine learning in Python. J. Mach. Learn. Res. 12, 2825–2830 (2011)
8. Rasmussen, C., Williams, C.: Gaussian Processes for Machine Learning. Adaptive Computation and Machine Learning. MIT Press, Cambridge (2006)
9. Roberts, S., Osborne, M., Ebden, M., Reece, S., Gibson, N., Aigrain, S.: Gaussian processes for time-series modelling. Philos. Trans. R. Soc. A: Math. Phys. Eng. Sci. 371(1984), 20110550 (2013). https://doi.org/10.1098/rsta.2011.0550
10. Roscher, R., Bohn, B., Duarte, M.F., Garcke, J.: Explainable machine learning for scientific insights and discoveries. IEEE Access 8, 42200–42216 (2020). https://doi.org/10.1109/ACCESS.2020.2976199
11. Swastanto, B.A.: Gaussian process regression for long-term time series forecasting. Master's thesis, Electrical Engineering, Mathematics and Computer Science (2016)
12. Wilson, A.G., Adams, R.P.: Gaussian process kernels for pattern discovery and extrapolation. In: Proceedings of the 30th International Conference on Machine Learning, ICML 2013, Atlanta, GA, USA, 16–21 June 2013. JMLR Workshop and Conference Proceedings, vol. 28, pp. 1067–1075. JMLR.org (2013). http://proceedings.mlr.press/v28/wilson13.html

A Binary Classification Model
for Toxicity Prediction in Drug Design

Génesis Varela-Salinas[1], Hugo E. Camacho-Cruz[2]([✉]),
Alfredo Juáŕez Saldivar[4]([✉]), Jose L. Martinez-Rodriguez[1],
Josue Rodriguez-Rodriguez[1], and Carlos Garcia-Perez[3]

[1] Universidad Autónoma de Tamaulipas - UAM Reynosa Rodhe,
Carretera Reynosa - San Fernando, 88779 Reynosa, Tamaulipas, Mexico
[2] Universidad Autónoma de Tamaulipas - FMeISC de Matamoros,
Sendero Nacional Km. 3. H, Matamoros, Tamaulipas, Mexico
[3] Information and Communication Technology Department (ICT),
Helmholtz Zentrum München, Ingolstädter Landstrasse 1, 85764 München, Germany
[4] Laboratorio de Biotecnología Farmacéutica, Centro de Biotecnología Genómica,
Instituto Politécnico Nacional, 88710 Reynosa, Mexico

Abstract. Toxicity in drug design is a very important step prior to
human or animal evaluation phases. Establishing drug toxicity involves
the modification or redesign of the drug into an analog to suppress or
reduce the toxicity. In this work, two different deep neural networks architectures and a proposed model to classify drug toxicity were evaluated.
Three datasets of molecular descriptors were build based on SMILES
from the Tox21 database and the AhR protein to test the accuracy prediction of the models. All models were tested with different sets of hyperparameters. The proposed model showed higher accuracy and lower loss
compared to the other architectures. The number of descriptors played
a key roll in the accuracy of the proposed model along with the Adam
optimizer.

Keywords: Toxicity · Tox21 · Deep learning · Drug design

1 Introduction

Drug development is performed through a series of complex processes. One of
the first steps is defining an enzyme as the drug target. Enzymes are proteins
that act as drug targets for diseases in the drug design. Then small molecules
are identified as active compounds that bind strongly with a protein target. The
active compounds are subjected to various experimental evaluations involving
cell line assays, animal assays, and human clinical trials [17]. In this regard,
during the last decade computational techniques have improved the drug development. Among theses improvements, we can mention the prediction of synergistic drug combinations to avoid drug resistance or increase treatment efficiency
and thus, reduce the drug dose to avoid toxicity [22]. Moreover, the use of large

© Springer Nature Switzerland AG 2021
H. Sanjurjo González et al. (Eds.): HAIS 2021, LNAI 12886, pp. 149–157, 2021.
https://doi.org/10.1007/978-3-030-86271-8_13

volume datasets have led drug research to apply complex calculations, where graphical processing units (GPUs) are used for data processing. Therefore, modern drug development has entered the era of big data [18] and new techniques are required. Nowadays, Deep Learning (DL) is a highly demanded technique to promote drug development in the area of artificial intelligence [23].

Deep Learning is of great interest in the process of drug design, in particular for toxicity prediction. Toxicity (or toxic action) is understood as the ability of a substance to cause a harmful response or severe damage to the body functions at cellular or molecular level, and in some cases death [14]. However, some active compounds can present toxicity in high doses but be harmless and even beneficial in small quantities, thus, failing in the latest phases of the development, even if they have obtained satisfactory results in vivo assays [17,21].

In drug design, toxicity evaluation plays a key role for further phases or the approval for human consumption. Nevertheless, the methods used to determine toxicity are slow, tedious, and expensive, not to mention that some of them raise ethical concerns due to the testing of the active compounds in animals [1,5]. For these reasons, predicting toxicity through computational techniques is convenient to accelerate the development of drugs and thus avoid the use of animals in the process.

Encouraged by these reasons, we decided to apply a DL model to toxicity prediction and contribute it to solve this type of problem. Our proposal uses a binary classification model and a dataset with molecular descriptors as feature elements of the AhR molecule from the Tox21 project [19]. The rest of this paper is organized as follows: Sect. 2 includes the data description and the step methods. Section 3 presents and analyzes the results, and finally, in Sect. 4 we present the discussion and conclusions.

2   Data and Methods

Diverse approaches have been proposed for addressing the toxicity prediction problem through machine learning strategies. For example, multiple heterogeneous neural network types and data representations of chemical compounds as SMILE strings [8] have been introduced. Other approaches are shallow networks via 2D features using PADEL descriptors [7] or Deep Neural Networks (DNN) with static and dynamic features [12].

Deep Learning models can be train to learn and recognize molecular descriptors that are active or toxic to a given type of chemical structures. Therefore, to evaluate the toxicity of drugs and reduce tests, a binary classification model was considered to predict if a drug is toxic or not. Figure 1 shows the proposed pipeline. These steps are described in the following subsections.

2.1   Dataset Creation

Tox21 [4,9,20] is a collaboration program of the NIH's NCATS and the National Toxicology Program at the National Institute of Environmental Health Sciences,

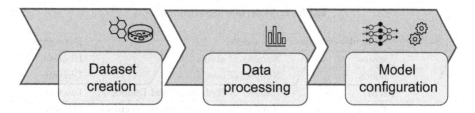

Fig. 1. Proposed method for toxicity prediction

the Environmental Protection Agency, and the Food and Drug Administration. In 2014, the Tox21 members set a Machine Learning challenge to predict the toxicity over 10,000 chemical compounds. Tox21 is divided in 12 assays giving priority to the toxicological evaluation of drugs. We select one assay to predict the toxicity through a Deep Learning model. In the following we will describe the steps to prepare the dataset to train the model. We selected the Aryl hydrocarbon Receptor (AhR) as the target and proceeded to download the list of drugs from the assay in SMILES format (https://tripod.nih.gov/tox21/assays/). The simplified molecular input line entry specification or SMILES is a specification in form of a line notation for describing the structure of a small chemical molecule. It was introduced by Arthur Weininger and David Weininger in the late 1980s. The list contains 8,170 drugs in total, and is divided into active (toxic) or non active (non toxic) regarding the AhR target. Next, we calculate molecular descriptors associated to the AhR target for each drug as shown in Table 1. The molecular descriptors were calculated with Pybel [15].

2.2   Data Processing

Due to the difference in range values between molecular descriptors, we preprocessed the data to ensure a better learning of the features. Data normalization is a recurring technique in Machine Learning for preprocessing data. This type of technique normalizes the data in a range between 0 and 1 for each column. The standard deviation help to avoid differences in values or information loss. For this work, we employed the Normalizer function from the Scikit-learn library [16].

2.3   Proposed Model Architecture

We tried three hyperparameter configurations and architectures designs, as shown in Table 2. We also set a different number of molecular descriptors for the input data (i.e., 4, 8 and 15). All models were run for 64 epochs and with a batch size of 128. We used the Adam [10] and SGD [3] optimizers for the experiments. And we run the training with 10%, 20% and 50% dropout in different models, as is also shown in Table 2.

Having the results of the experiments, we decided to set the proposed model as follows: the input layer as fully connected, 10 nodes in the first hidden layer with sigmoid activation function, and with a dropout at 10%; the second hidden

Table 1. Molecular descriptors.

Id descriptor	Molecular descriptor	Description	Data type
1	atoms	Number of atoms	Discrete
2	bonds	Number of bonds	Discrete
3	HBD	Number of Hydrogen Bond Donors	Discrete
4	HBA1	Number of Hydrogen Bond Acceptors 1	Discrete
5	HBA2	Number of Hydrogen Bond Acceptors 2	Discrete
6	nF	Number of Fluorine Atoms	Discrete
7	logP	Octanol/Water Partition Coefficient	Continuous
8	MW	Molecular Weight Filter	Continuous
9	tbonds	Number of triple bonds	Discrete
10	MR	Molar Refractivity	Continuous
11	abonds	Number of aromatic bonds	Discrete
12	sbonds	Number of single bonds	Discrete
13	dbonds	Number of double bonds	Discrete
14	rotors	Rotatable bonds filter	Discrete
15	MP	Melting point	Continuous

layer with 10 nodes and the RELU activation function, because this avoids gradient fading and saturation. RELU is a rectified function which means that a node will be only activated if the input is above a threshold. Therefore, it rectifies the input values between 0 and 1 regardless of whether they are positive or negative values. The output layer was set to one node to make a binary prediction with sigmoid activation function. We use the Adam [10] optimizer and the Binary Cross-Entropy as loss function. The proposed model is shown in Fig. 2.

3   Experiments and Results

This section presents the experiments used to evaluate the performance of the proposed model. We performed a comparison between the models shown in Table 2 by running the toxicity classification models in a local machine with the following characteristics: 1 node with Intel Core i7 processors at 4.3 Ghz, 8 GB of DDR4 memory, SATA III SSD at 1 TB at 6 GB/, a GPU GEFORCE GTX 1660Ti at 1770 Mhz and 6 GB DDR6. The operating system was LinuxMint version 19.7. Additionally, we applied several libraries such as Pybel [15] for molecular descriptors, Scikit-learn [16] and Pandas [13] for data preprocessing and for creating the input data set. First, the metrics used in the experiments are explained in the following Sect. 3.1. Next, in Sect. 3.2 we will show the scenarios of the experiments. Finally, in Sect. 3.3 we will present the results obtained.

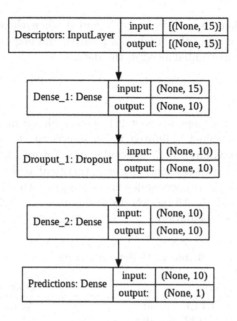

Fig. 2. Binary model architecture

3.1 Metrics

Having two classes denoted as positive and negative causes a binary classification problem. To measure the performance of the trained models, we used the Recall, Precision, and F-1 scores [2,6]. Then,

- Precision. It is the number of correctly classified positive examples divided by the total number that are classified positive. That is,

$$P = \frac{TP}{TP + FP},$$

where TP is the number of true positives, FP the number of false positives, and P precision.
- Recall. Measures the number of how many of the actual positives (true positives and false negatives) were predicted correctly as positives (true positives),

$$R = \frac{TP}{TP + FN},$$

where FN is the number of false negatives and R is the recall.
- F-1 is an harmonic measure that combines precision (P) and recall (R) as shown in

$$\text{F-1} = 2\frac{P \cdot R}{P + R}$$

F-1 is 1 when there is no FP, and FN and 0 when there is no TP. F-1 is particularly useful when the number of positive and negative classes are substantially different or imbalanced in the data.

3.2 Scenarios

As we mentioned in the previous Sect. 2.1, we calculated molecular descriptors to train the models. Pybel is limited to 15 molecular descriptors. Therefore, we calculated the maximum of 15 molecular descriptors, then we tried with a different set of molecular descriptors. After a statistical analysis of the molecular descriptors (not shown here), we made three datasets with the following number of descriptors: 4, 8 and 15. All models use standard deviation as normalization, and the 10 k-fold cross-validation.

Table 2. Model configuration.

Model	Layers	Optimization
M1	1st: RELU - input 2nd: RELU 6 nodes 3rd: Sigmoid– output	Adam
M2	1st: input 2nd: Simoid 10 nodes 3rd: RELU 10 nodes 4th: Sigmoid– output	
	Dropout: 10% 1st hidden layer	
M3	1st: RELU – input 2nd: RELU 16 nodes 3rd: RELU 6 nodes 4th: RELU 64 nodes 5th: sigmoidal– output	SGD
	Dropout: 20% input layer 50% 4th layer	

For each of the three datasets we used three different models (Table 2). We must highlight that the datasets were heavily unbalanced where the toxic samples were the lower class with 950 samples against 7,219 non toxic samples for the AhR receptor. To overcome this, we applied the under-sampling technique provided by the imblearn library [11].

We train the three models with the balanced training set in order to observe the performance according the settings of the hyperparameters. To validate and see if the models generalize well, we wanted to see if any of the models were able to adapt properly to unseen data and classify samples correctly. We used the k-fold cross-validation technique with 10 splits with the Scikit-learn library. We filtered the descriptors to keep only the highest correlated ones to improve the performance of the models.

3.3   Results

The metrics from the three models are shown in Table 3. Model 2 (M2) shows a high F-1 score of 0.89 using the 15 molecular descriptors while model 1 (M1) is the second best (F-1 of 0.88) also using the 15 molecular descriptors. Finally, in model 3 (M3) the F-1 score is 0.84. Evidently, when using 15 molecular descriptors the performance is better for all three models. It is also interesting that M2 and M1 showed a very close F1 score with 8 molecular descriptors. Finally, we can say that M3 performs lower than models M2 and M1. Additionally, we run a Support Vector Machine (SVM) and a Gaussian Naive Bayes (GaussianNB) algorithm from the Scikit-learn library in order to compare traditional ML methods against the three proposed models. Table 3 summarizes that both methods reach a F-1 score of 0.86 with 15 molecular descriptors while M2 has a F-1 score 0.86 with 8 molecular descriptors. It is clear that M2 with 15 molecular descriptors overpasses classical ML approaches. Although the idea of keeping only the descriptors with the highest correlation was supposed to improve the performance of the models (4 and 8 molecular descriptors), the results show that using all the descriptors provides better results in the F-1 score.

Table 3. Metrics for the five models.

Model	Desc	Precision	Recall	F-1
M1	4	0.961	0.776	0.859
	8	0.941	0.806	0.867
	15	0.913	0.854	0.881
M2	4	0.958	0.744	0.837
	8	0.948	0.788	0.860
	15	0.947	0.839	**0.890**
M3	4	0.833	0.737	0.781
	8	0.907	0.764	0.828
	15	0.907	0.791	0.844
SVM	4	0.943	0.755	0.838
	8	0.948	0.782	0.856
	15	0.938	0.798	0.862
GaussianNB	4	0.773	0.823	0.796
	8	0.832	0.843	0.837
	15	0.881	0.851	0.865

4   Conclusions

This paper presents a strategy to develop a binary classifier for toxicity prediction in the drug design pipeline. The dataset from the AhR Tox21 assay was used to calculate molecular descriptors, and it was used as input data to train a set of Machine Learning models. On one side, the experiments showed that the proposed model (M2) achieves promising results shown in the F-1 score when using 15 molecular descriptors and multiple hidden layers with the RELU and sigmoid activation functions. On the other side, M2 performs better than classical ML algorithms as shown in Table 3. In conclusion, the results show that more than 15 molecular descriptors could improve the F1-score for the SVM and the Gaussian Naive Bayes algorithm, and therefore the F1-score from M2.

References

1. Atkinson Jr., A.J., Markey, S.P.: Biochemical mechanisms of drug toxicity. In: Principles of Clinical Pharmacology, pp. 249–271. Elsevier (2007)
2. Bania, R.K.: COVID-19 public tweets sentiment analysis using TF-IDF and inductive learning models. INFOCOMP J. Comput. Sci. **19**(2), 23–41 (2020)
3. Bottou, L.: Large-scale machine learning with stochastic gradient descent. In: Lechevallier, Y., Saporta, G. (eds.) Proceedings of COMPSTAT 2010, pp. 177–186. Springer, Heidelberg (2010). https://doi.org/10.1007/978-3-7908-2604-3_16
4. Collins, F.S., Gray, G.M., Bucher, J.R.: Transforming environmental health protection. Science **319**(5865), 906–907 (2008). https://doi.org/10.1126/science.1154619, https://science.sciencemag.org/content/319/5865/906
5. Dearden, J.C.: In silico prediction of drug toxicity. J. Comput. Aided Mol. Des. **17**(2–4), 119–127 (2003). https://doi.org/10.1023/A:1025361621494
6. Dice, L.R.: Measures of the amount of ecologic association between species. Ecology **26**(3), 297–302 (1945). https://doi.org/10.2307/1932409, https://esajournals.onlinelibrary.wiley.com/doi/abs/10.2307/1932409
7. Karim, A., Mishra, A., Newton, M.H., Sattar, A.: Efficient toxicity prediction via simple features using shallow neural networks and decision trees. ACS Omega **4**(1), 1874–1888 (2019)
8. Karim, A., Singh, J., Mishra, A., Dehzangi, A., Newton, M.A.H., Sattar, A.: Toxicity prediction by multimodal deep learning. In: Ohara, K., Bai, Q. (eds.) PKAW 2019. LNCS (LNAI), vol. 11669, pp. 142–152. Springer, Cham (2019). https://doi.org/10.1007/978-3-030-30639-7_12
9. Kavlock, R.J., Austin, C.P., Tice, R.R.: Toxicity testing in the 21st century: implications for human health risk assessment. Risk Anal. **29**(4), 485–487 (2009). https://doi.org/10.1111/j.1539-6924.2008.01168.x, https://onlinelibrary.wiley.com/doi/abs/10.1111/j.1539-6924.2008.01168.x
10. Kingma, D.P., Ba, J.: Adam: a method for stochastic optimization (2017)
11. Lemaître, G., Nogueira, F., Aridas, C.K.: Imbalanced-learn: a python toolbox to tackle the curse of imbalanced datasets in machine learning. J. Mach. Learn. Res. **18**(17), 1–5 (2017)
12. Mayr, A., Klambauer, G., Unterthiner, T., Hochreiter, S.: DeepTox: toxicity prediction using deep learning. Front. Environ. Sci. **3**, 80 (2016)

13. McKinney, W., et al.: Data structures for statistical computing in python. In: Proceedings of the 9th Python in Science Conference, vol. 445, pp. 51–56 (2010)
14. Muster, W., Breidenbach, A., Fischer, H., Kirchner, S., Müller, L., Pähler, A.: Computational toxicology in drug development. Drug Discovery Today **13**(7–8), 303–310 (2008)
15. O'Boyle, N.M., Morley, C., Hutchison, G.R.: Pybel: a Python wrapper for the OpenBabel cheminformatics toolkit. Chem. Cent. J. **2**(1), 1–7 (2008)
16. Pedregosa, F., et al.: Scikit-learn: machine learning in Python. J. Mach. Learn. Res. **12**, 2825–2830 (2011)
17. Saldívar-González, F., Prieto-Martínez, F.D., Medina-Franco, J.L.: Descubrimiento y desarrollo de fármacos: un enfoque computacional. Educación química **28**(1), 51–58 (2017)
18. Sid, K., Batouche, M.C.: Big data analytics techniques in virtual screening for drug discovery. In: Lazaar, M., Tabii, Y., Chrayah, M., Achhab, M.A. (eds.) Proceedings of the 2nd International Conference on Big Data, Cloud and Applications, BDCA 2017, Tetouan, Morocco, 29–30 March 2017, pp. 9:1–9:7. ACM (2017). https://doi.org/10.1145/3090354.3090363
19. Thomas, R.S., et al.: The US Federal Tox21 program: a strategic and operational plan for continued leadership. ALTEX - Altern. Anim. Exp. **35**(2), 163–168 (2018)
20. Tice, R.R., Austin, C.P., Kavlock, R.J., Bucher, J.R.: Improving the human hazard characterization of chemicals: a Tox21 update. Environ. Health Perspect. **121**(7), 756–765 (2013). https://doi.org/10.1289/ehp.1205784, https://ehp.niehs.nih.gov/doi/abs/10.1289/ehp.1205784
21. Verbist, B., et al.: Using transcriptomics to guide lead optimization in drug discovery projects: lessons learned from the QSTAR project. Drug Discovery Today **20**(5), 505–513 (2015)
22. Wang, X., Song, K., Li, L., Chen, L.: Structure-based drug design strategies and challenges. Curr. Top. Med. Chem. **18**(12), 998–1006 (2018)
23. Zhang, L., Tan, J., Han, D., Zhu, H.: From machine learning to deep learning: progress in machine intelligence for rational drug discovery. Drug Discovery Today **22**(11), 1680–1685 (2017)

Structure Learning of High-Order Dynamic Bayesian Networks via Particle Swarm Optimization with Order Invariant Encoding

David Quesada[(✉)] [iD], Concha Bielza[iD], and Pedro Larrañaga[iD]

Artificial Intelligence Department, Universidad Politécnica de Madrid, Madrid, Spain
dquesada@fi.upm.es

Abstract. Dynamic Bayesian networks usually make the assumption that the underlying process they model is first-order Markovian, that is, that the future state is independent of the past given the present. However, there are situations in which this assumption has to be relaxed. When this order increases, the size of the search space grows greatly, not all structure learning algorithms may be suited to learn higher-order networks, and a new appropriate order has to be found. To address the computational issues of huge networks, we propose a structure learning method that uses particle swarm optimization to search in the space of possible structures. To avoid the additional costs of increasing the Markovian order, we provide an order-invariant encoding that represents the networks as vectors of natural numbers whose length remains constant. Due to this encoding, we only need to set a maximum desired order rather than the exact one. Our experimental results show that this method is efficient in high orders and performs better than similar algorithms in both execution time and quality of the obtained networks.

Keywords: Dynamic Bayesian networks · Structure learning · Particle swarm optimization

1 Introduction

In recent years, the use of dynamic Bayesian networks (DBNs) [3] has received more attention in different areas. Their applications range from bioinformatics, where DBNs have been used to model gene regulatory networks [7,19,22], to more industrial fields like damage assessment of steel decks and estimating the remaining useful life of structures [2,12,21].

Traditionally, DBNs have been modelled with the assumption that they are first-order Markovian models, that is, their future state is independent of the past given the present [9]. However, in real world applications we can find processes

This work was partially supported by the Spanish Ministry of Science, Innovation and Universities through the PID2019-109247GB-I00 project.

H. Sanjurjo González et al. (Eds.): HAIS 2021, LNAI 12886, pp. 158–171, 2021.
https://doi.org/10.1007/978-3-030-86271-8_14

where this assumption does not hold and the future state is determined not only by the last instant, as in the works of Lo et al. [11] and Vinh et al. [18].

If we increase this order, the model will require several time slices to represent the state of the system instead of only two, making it increasingly complex by adding new nodes and arcs. Given that the number of possible different Bayesian network (BN) structures is super exponential in the number of nodes [14], increasing the Markovian order also increases the search space greatly. For this reason, simply applying a structure learning algorithm for first-order Markovian DBNs in a higher-order space may not be an optimal solution.

Meta-heuristic optimization algorithms can be applied to structure learning by searching over the space of possible network structures and assessing the fitness of each solution with some score. In the case of particle swarm optimization (PSO) algorithms [8], they offer a powerful solution for big search spaces but require them to be continuous and real numbered, and the space of possible network structures does not fulfill these requirements. To fix this issue, some authors have translated the space of possible BN graphs into a continuous one over which PSO can be applied [10]. Many other authors [4,6,15,20] have opted for translating the operations of the PSO algorithm to perform discrete movements and then be able to apply the framework of PSO to BN structure learning. In particular, the work of Du et al. [4] proposes to encode particles as binary adjacency matrices that are modified as the particle moves to represent additions and deletions of arcs. Santos and Maciel [15] extend this concept by defining particles as lists with the parents of each node, and so a particle moving means adding or deleting parent nodes. However, these approaches become less efficient as the number of nodes and the Markovian order increase, and the correct order is assumed to be known beforehand. We will expand on both of those methods by defining an encoding for particles that is unaffected by the Markovian order and allows for more efficient searches in larger spaces.

The rest of the paper is organized as follows. In Sect. 2 we introduce the concepts of high-order DBNs and PSO. Section 3 contains the details of our order invariant DBN structure encoding into PSO particles and its operators. Section 4 shows the empirical results. Finally, Sect. 5 concludes the paper and gives some final remarks.

2   Background

In the field of BN structure learning, one family of methods is dedicated to applying PSO to move through the space of possible structures to find an optimal solution. Depending on the type of BN that we want to model, encoding the individuals and moving through the solution space can be done in different ways. In our case, we will center our attention in DBN models and translating the PSO operations to the discrete space defined by our encoding of a particle.

2.1 High-Order Dynamic Bayesian Networks

Dynamic Bayesian networks [13] are a type of probabilistic graphical model that extend the BN framework to the case of time series. As in the static scenario, a DBN is comprised of a directed acyclic graph that defines its structure and a set of parameters that define the probabilistic relationships of the variables. In the case of DBNs, time is discretized into time slices that represent consecutive instants. Let $\mathbf{X}^t = \{X_0^t, X_1^t, \ldots, X_n^t\}$ be the set of all variables in a single time slice t. For some horizon t = T, we can define the joint probability distribution of the network as:

$$p(\mathbf{X}^T, \mathbf{X}^{T-1}, \ldots, \mathbf{X}^0) = p(\mathbf{X}^{T:0}) = p(\mathbf{X}^T) \prod_{t=T-1}^{0} p(\mathbf{X}^t | \mathbf{X}^{T:t+1}), \qquad (1)$$

where $p(\mathbf{X}) = \prod_{i=1}^{n} p(x_i | \mathbf{Pa}(x_i))$ represents the probability distribution of a set of nodes $\mathbf{X}$ and $\mathbf{Pa}(x_i)$ represents the set of parent nodes of X_i in the graph. Usually, in the literature the oldest instant is the one defined as $\mathbf{X}^0$, but we will reverse this notation and define $\mathbf{X}^0$ as the most recent instant to differentiate it from the others. In a dynamic scenario, nodes can have parents in previous time slices and so all variables in all instants $\mathbf{X}^{T:0}$ have to be taken into account to calculate the joint probability distribution as in Eq. 1. In this situation, a very common assumption is to suppose the DBN to be first-order Markovian [9]. This assumes that the future state of the system is independent of the past given the present, that is, $p(\mathbf{X}^t | \mathbf{X}^{T:t+1}) = p(\mathbf{X}^t | X^{t+1})$. Although the model is greatly simplified, it cannot represent systems where the future state is not determined only by the state of the variables in the last instant. We call a high-order dynamic Bayesian network (HO-DBN) a DBN model where the first-order Markovian assumption is relaxed and the network is represented with more than two time slices.

To leverage the increase in complexity of this kind of model, we can restrict the arcs in the network so that they can only be directed to nodes in the most recent time slice, in our case $\mathbf{X}^0$. This kind of DBN structures are called transition networks [15] and they avoid by definition any kind of cycles, which simplifies the search. The space of possible structures that transition networks allow is also much smaller than that of regular DBN models. To prove this, let G be a DBN network structure and G' be a transition network, both with the same number of nodes n_0 per time slice, total number of nodes n and Markovian order m. Let $\mathbf{D}$ be the set of all possible inter-slice arcs in G and $\mathbf{T}$ be the set of all possible arcs in G'. We can calculate the number of possible DBN structures g^{DBN} as the different combinations of the elements in $\mathbf{D}$, that is, the different combinations of all the possible arcs in the network:

$$g^{DBN} = \sum_{i=0}^{|\mathbf{D}|} \binom{|\mathbf{D}|}{i} = \sum_{i=0}^{|\mathbf{D}|} \frac{|\mathbf{D}|!}{i!(|\mathbf{D}| - i)!}. \qquad (2)$$

By definition, $|\mathbf{T}| \leq |\mathbf{D}|$ because even though both have the same number of nodes, the arcs in $\mathbf{T}$ restricted in a way that satisfies $\mathbf{T} \subseteq \mathbf{D}$. If we apply

Eq. (2) to calculate the number of possible transition networks g^{TN}, we can see that:

$$|\mathbf{T}| < |\mathbf{D}| \implies g^{TN} << g^{DBN}. \tag{3}$$

If we would also take into account the possible intra-slice arcs in $\mathbf{D}$, the inequality in Eq. (3) would be super-exponentially bigger. Moreover, increasing m by one translates into adding only n_0^2 arcs in the case of the transition network, as we only allow arcs from the new n_0 nodes to the nodes in $\mathbf{X}^0$. This means that increasing m by one increases $|\mathbf{T}|$ by the constant n_0^2. On the other hand, this operation means adding $n_0 * n$ new inter-slice arcs in the case of a regular DBN, which increases $|\mathbf{D}|$ exponentially with each increase in m. This shows that not only $|\mathbf{T}|$ is always smaller than $|\mathbf{D}|$, but it also increases drastically slower when we increase the Markovian order of the network. For both the lack of cycles in transition networks and their reduced, although still vast, space of possible structures we will be using this kind of network in the rest of the paper. An example of a HO-DBN with the restrictions of a transition network can be seen in Fig. 1.

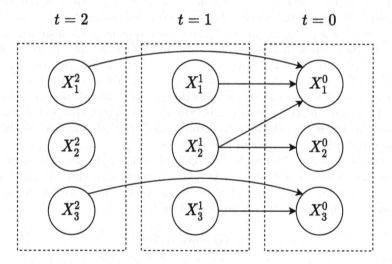

Fig. 1. An example of an order 2 Markovian transition network with tree nodes per time slice. Only arcs directed to t_0 from earlier time slices are allowed.

2.2   Particle Swarm Structure Learning

The PSO algorithm [8] is a meta-heuristic technique that simulates a swarm consisting of n particles moving in a k-dimensional space to find the optimal solution. Particles can be defined as $\mathbf{Pr}_i = \{P_i, V_i, P_l, P_g\}$, where P_i is its current position, V_i is its current velocity, P_l represents the best position found by the particle so far and P_g is the best position found by the whole swarm. A position represents one specific solution of the optimization problem, and the velocities modify these positions, allowing the particles to move in the solution space. To calculate in each iteration t the next position P_i^{t+1} and the next velocity V_i^{t+1} of each particle, the following updating rules are applied:

$$V_i^{t+1} = wV_i^t + c_1r_1(P_l - P_i^t) + c_2r_2(P_g - P_i^t), \qquad (4)$$
$$P_i^{t+1} = P_i^t + V_i^{t+1}, \qquad (5)$$

where $w, c_1, c_2 \in \mathbb{R}$, w is the inertia factor of the last velocity, c_1 is the factor that weighs the importance of the local best position, c_2 weighs the global best position and r_1 and r_2 are two real numbers sampled uniformly from the interval $[0, 1]$. One of the key factors in PSO is the pondered effects that the global and local best positions have on the current velocity of a particle, which can change in each iteration if a particle finds a position that has a better score than P_l or P_g. The inertia factor defines how much importance is given to the random search of each particle, increasing or decreasing the exploratory capabilities of the swarm. A higher inertia factor will mean that the $t+1$ velocity of the particle will be very similar to the one it had in the previous instant t.

Although PSO was originally designed for continuous and real valued spaces, there are adaptations to discrete scenarios. In particular, the approach established by Du et al. [4] defines positions and velocities as the binary adjacency matrix of a BN. In this case, velocities matrices can take any value from the set $\{-1, 0, 1\}$, representing deletions, non modifications or additions of arcs respectively. This same approach is taken by Santos and Maciel [15], but instead of adjacency matrices they define a structure called *causality list* that establishes positions and velocities as sets of parent nodes.

To be able to apply PSO to the problem of learning DBN structures, we need a score that measures how likely it is that a network structure fits some data. Let $\mathcal{D}$ be our training data and let G be the network structure represented by a position. Our objective can now be defined as:

$$\underset{G \in g^{DBN}}{\arg\max} \, score(G, \mathcal{D}). \qquad (6)$$

This score of fitness is necessary to assess which particles in the solution space fit the training data better and guide the exploration towards them. There are many examples of scores in the literature, such as the Bayesian information criterion (BIC) [16] and the Akaike information criterion (AIC) [1] scores for discrete networks, or the Bayesian Gaussian equivalent (BGe) [5] score and the adapted BIC and AIC scores for Gaussian Bayesian networks. Depending on the type of score used, the same algorithms can be used to learn both discrete and Gaussian Bayesian networks.

3   Encoding and Operators

One of the crucial elements of meta-heuristic optimization algorithms is the encoding used. A suboptimal encoding could generate losses in efficiency by having redundant solutions, having to fix invalid individuals each iteration or by not allowing an even exploration of the solution space.

Our proposed encoding maps each possible transition network structure to a vector of natural numbers. This mapping is bijective in both sets: a transition network structure can only be represented with one specific vector, and a vector only represents one specific transition network structure.

3.1   Natural Vector Encoding

In a particle swarm scenario, each particle has a position and a velocity. In our case, positions represent specific transition network structures and velocities represent additions and deletions of arcs.

In a transition network, there are several nodes representing the same variable in different instants of time. We define the concept of a temporal family of nodes $\mathbf{X}_i^f = \{X_i^0, X_i^1, \ldots, X_i^T\}$ as the set of nodes representing a single variable X_i in all existing time slices of the network. Inside a temporal family, we call a *receiving node* the node X_i^0 in the present time slice. This node is the only one in a temporal family that can have arcs pointing to him from any other node in earlier time slices. In our encoding, we will divide a vector in as many sections as receiving nodes X_i^0 there are in the network. Each of these sections is further subdivided into a subsection consisting of a single natural number for each existing temporal family $\mathbf{X}_i^f$ in the network. This number defines with its binary representation the existing arcs from a certain temporal family to a specific receiving node. Each 1-bit encodes an arc from a specific member of the temporal family to the receiving node. By definition, this encoding does not allow invalid individuals because only receiving nodes can have arcs pointing to them and no cycles can appear. Furthermore, the length of the encoded vectors only depends on the number of existing receiving nodes, and not on the Markovian order. Higher orders will only mean bigger natural numbers in the vector. To clarify this explanation, an example of a position and the network it encodes can be seen in Fig. 2.

The velocities follow the same encoding, but represent arc additions or deletions instead of the presence or absence of an arc. Each velocity is composed of two vectors, $\mathbf{V}_p$ and $\mathbf{V}_n$, defining additions and deletions of arcs respectively.

3.2   Position and Velocity Operators

To perform the operations of the PSO, we will adapt them to the discrete space defined by our encoding. In essence, we will need to be able to add positions and velocities, subtract positions and multiply velocities by real numbers. All operators shown are supposed to be bitwise logical operations. Both positions and velocities have the same vector length, so when operated together this bitwise

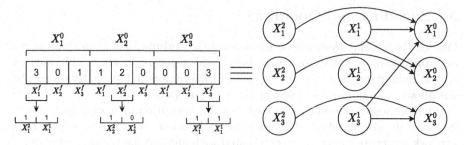

Fig. 2. Representation of a position natural vector on the left and its equivalent transition network on the right. The transition network has three nodes per time slice and a Markovian order 2 for simplicity and clarity. Notice how increasing the order, thus adding many possible nodes and arcs, only implies bigger natural numbers in the vector, but it does not increase its length. For example, increasing the order up to 3 in the figure would only mean that natural numbers up to 7 can now appear in the vector.

operations will be performed throughout both vectors to each pair of natural numbers.

Position Plus Velocity

To add a velocity to a position, first we add all the arcs in $\mathbf{V}_p$ by performing a logical 'or' in the form of $\mathbf{P}' = \mathbf{P} \vee \mathbf{V}_p$. As for the negative part, we define the $\ominus$ operator as:

$$x_1 \ominus x_2 = x_1 \wedge \neg x_2. \tag{7}$$

This operator is equivalent to a 1-bit subtractor without borrow. By performing $\mathbf{P}' \ominus \mathbf{V}_n$, we remove the 1-bits that are present in both the position and negative velocity temporal families and maintain the rest unaffected. The consecutive positive and negative operations will add and remove the marked arcs of the velocity in the original position. An example of this operation can be seen in Fig. 3.

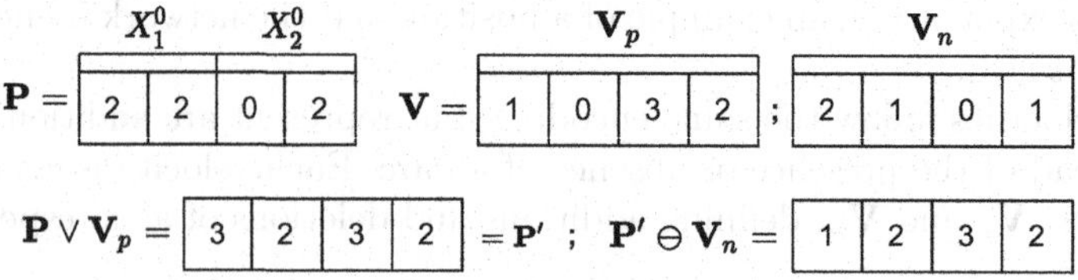

$$\mathbf{P} = \begin{array}{|c|c|c|c|} \hline 2 & 2 & 0 & 2 \\ \hline \end{array} \quad \mathbf{V} = \begin{array}{|c|c|c|c|} \hline 1 & 0 & 3 & 2 \\ \hline \end{array} ; \begin{array}{|c|c|c|c|} \hline 2 & 1 & 0 & 1 \\ \hline \end{array}$$

$$\mathbf{P} \vee \mathbf{V}_p = \begin{array}{|c|c|c|c|} \hline 3 & 2 & 3 & 2 \\ \hline \end{array} = \mathbf{P}' ; \quad \mathbf{P}' \ominus \mathbf{V}_n = \begin{array}{|c|c|c|c|} \hline 1 & 2 & 3 & 2 \\ \hline \end{array}$$

Fig. 3. An example of adding a position and a velocity encoding a network with two receiving variables X_0^0 and X_1^0 and maximum Markovian order 2 for simplicity. All the operations are performed bitwise on the natural numbers of all vectors.

Addition of Velocities

Given two velocities $\mathbf{V}^1$ and $\mathbf{V}^2$, to add them we first need to combine their positive and negative parts. For this, we operate $\mathbf{V}'_p = \mathbf{V}^1_p \vee \mathbf{V}^2_p$ and $\mathbf{V}'_n = \mathbf{V}^1_n \vee \mathbf{V}^2_n$. Afterwards, we perform a bitwise logical 'and' operation to identify redundancies in the form of $\mathbf{Rd} = \mathbf{V}'_p \wedge \mathbf{V}'_n$. Any 1-bit present in $\mathbf{Rd}$ means that an arc is being added and deleted at the same time in the resulting velocity, and it has to be set to 0 in both positive and negative vectors of $\mathbf{V}'$ with the 'xor' operator by performing $\mathbf{V}'_p \oplus \mathbf{Rd}$ and $\mathbf{V}'_n \oplus \mathbf{Rd}$ respectively. An example of this operation is shown in the following lines:

$$\mathbf{V}^1 = [1,0,2,0]; [2,0,0,3], \mathbf{V}^2 = [0,1,2,1]; [0,2,0,2],$$
$$\mathbf{V}^1 \vee \mathbf{V}^2 = [1,1,2,1]; [2,2,0,3] = \mathbf{V}',$$
$$\mathbf{V}'_p \wedge \mathbf{V}'_n = [0,0,0,1] = \mathbf{Rd},$$
$$\mathbf{V}' \oplus \mathbf{Rd} = [1,1,2,0]; [2,2,0,2].$$

Subtraction of Positions

Given two positions $\mathbf{P}_1$ and $\mathbf{P}_2$, the operation $\mathbf{P}_1 - \mathbf{P}_2 = \mathbf{V}'$ returns the velocity $\mathbf{V}'$ such that $\mathbf{P}_1 + \mathbf{V}' = \mathbf{P}_2$. This effect is obtained by using the operator $\ominus$ defined in Eq. (7) to calculate both the positive part $\mathbf{V}'_p = \mathbf{P}_2 \ominus \mathbf{P}_1$ and the negative part $\mathbf{V}'_n = \mathbf{P}_1 \ominus \mathbf{P}_2$ of the velocity. Notice that the $\ominus$ operator is not commutative, and so the inverted positions in each operation give different results. The $\ominus$ operator can be used to get the bits that need to be added to transform a position into another. An example to clarify this operation is shown in the following lines:

$$\mathbf{P}_1 = [1,0,2,1], \mathbf{P}_2 = [1,1,0,3],$$
$$\mathbf{V}'_p = \mathbf{P}_2 \ominus \mathbf{P}_1 = [0,1,0,2],$$
$$\mathbf{V}'_n = \mathbf{P}_1 \ominus \mathbf{P}_2 = [0,0,2,0],$$
$$\mathbf{V}' = [0,1,0,2]; [0,0,2,0].$$

Multiplication of Velocities by Real Numbers

Let $|\mathbf{V}|$ be the total population count in a velocity, that is, the total number of 1-bits in both positive and negative vectors. As proposed by [15], multiplying a velocity by a real number increases or decreases $|\mathbf{V}|$ in the form of $\lfloor \alpha * |\mathbf{V}| \rfloor = |\mathbf{V}|'$. This means that we will randomly add or delete 1-bits in the velocity until the new total number of operations is obtained. We will follow a uniform distribution when sampling a temporal family in the vector and the open bits in the natural numbers. If $\alpha < 0$, $\mathbf{V}_p$ and $\mathbf{V}_n$ will be swapped with each other to invert all additions and deletions of arcs in the velocity and the absolute value of α will be used.

4 Results

In this section we will first discuss the implementation of our PSO structure learning algorithm (natPSOHO) and compare it with two other algorithms: the PSO for HO-DBNs proposed by Santos and Maciel [15] and a variation of the dynamic max-min hill climbing algorithm [17]. This comparison will consist of recovering several synthetic randomly generated networks from sampled datasets, evaluating the execution time and how many of the original arcs are recovered from the data. The Markovian order and the number of receiving nodes will vary, to assess the efficiency and precision of the algorithms as both these factors increase. The number of iterations of the PSO algorithms is set to 50 with populations of 300 particles, and all the datasets will consist of 10.000 instances. These parameters remain constant through all the experiments.

4.1 Implementation

All the algorithms have been implemented in R and C++. The code of the natPSOHO algorithm, the experiments and the generation of the synthetic datasets have been combined into an R package that is publicly available in a *GitHub* repository[1]. All experiments were conducted on an Ubuntu 18 machine with an Intel i7-4790K processor and 16 Gb of RAM.

Due to the translation of the position and velocity operations to our specific encoding, we can use the normal pipeline of the PSO algorithm shown in Fig. 4 without any change. For the evaluation of the positions based on the dataset, we will use the BGe score.

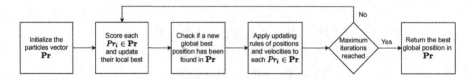

Fig. 4. Pipeline of the PSO algorithm. The particles move applying the updating rules described in Sect. 2.2 and the operators described in Sect. 3.2. The position with the best fitness is translated into its DBN equivalent and returned as the best solution found.

As recommended in other PSO works [10], the inertia value w is set high at the beginning and slowly decreases as the iterations advance to favour exploration at first, and c_1 is set high and decreases over time while c_2 is set low and increases over time, so that the positions close to the global optimum are properly explored prior to finishing the execution.

---

[1] https://github.com/dkesada/natPSOHO.

4.2   Experimental Comparison

The results for Markovian orders 1 and 2 networks can be found in Table 1. We can see that for smaller networks with few nodes and low Markovian order, the DMMHC algorithm is much faster than the particle swarm ones, but its execution time scales rapidly with the number of variables. We will not be testing the DMMHC algorithm on higher orders, given its poor scalability and the similar performance to the other two algorithms in terms of number of real arcs recovered. On the other hand, the natPSOHO algorithm is less efficient in time than the binary PSOHO for low-order networks and many receiving nodes, but it consistently outperforms the other two algorithms in terms of real recovered arcs.

Table 1. Results for low-order networks

[Order, n_0, Arcs]	Algorithm	Rec. arcs	Exec. time
[1, 10, 47]	natPSOHO	47	1.49 m
	PSOHO	42	1.55 m
	DMMHC	36	0.22 s
[1, 15, 111]	natPSOHO	96	3.24 m
	PSOHO	78	2.81 m
	DMMHC	69	2.54 s
[1, 20, 201]	natPSOHO	152	5.81 m
	PSOHO	118	4.22 m
	DMMHC	100	42.21 s
[2, 10, 96]	natPSOHO	86	3.37 m
	PSOHO	68	2.88 m
	DMMHC	71	24.1 s
[2, 15, 234]	natPSOHO	175	7.1 m
	PSOHO	139	12.68 m
	DMMHC	132	40.73 m
[2, 20, 398]	natPSOHO	288	14.24 m
	PSOHO	215	23.15 m
	DMMHC	233	19.1 h

The results for recovering high-order networks from data can be seen in Table 2 and in Fig. 5. We can see how the execution time for the natPSOHO algorithm scales better than the other method as we increase the Markovian order of the networks. We can also see that the number of recovered arcs is consistently higher in the case of the natPSOHO algorithm. In order to evaluate the networks, the BGe score is used to find the fitness of the particles. We have enhanced this score by omitting the scoring of nodes outside of t_0, due to the

Table 2. Results for high-order networks

[Order, n_0, Arcs]	Algorithm	Rec. arcs	Exec. time
[3, 10, 147]	natPSOHO	125	4.93 m
	PSOHO	85	7.56 m
[3, 20, 616]	natPSOHO	398	23.8 m
	PSOHO	308	37.06 m
[4, 10, 208]	natPSOHO	157	6.87 m
	PSOHO	110	11.06 m
[4, 20, 825]	natPSOHO	533	38.02 m
	PSOHO	423	1.04 h
[5, 10, 247]	natPSOHO	181	9.23 m
	PSOHO	148	16.87 m
[5, 20, 982]	natPSOHO	622	57.45 m
	PSOHO	425	1.3 h
[6, 10, 294]	natPSOHO	208	12.43 m
	PSOHO	170	21.6 m
[6, 20, 1171]	natPSOHO	739	1.4 h
	PSOHO	517	2.04 h

fact that in transition networks these nodes never have parents and their structure remains constant. Regardless of this, the score takes longer to compute for networks with a higher number of arcs. Given that the natPSOHO algorithm consistently recovers more arcs from the real networks that generated the synthetic data, the execution time of the score is also consistently higher. This makes it impossible to achieve a constant execution time on the evaluation of the networks, but the constant execution time in the operations of the particles is shown in the overall better performance of the algorithm.

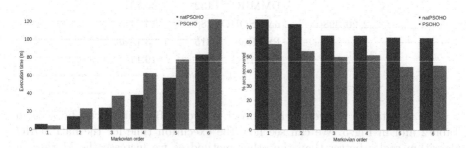

Fig. 5. Execution time in minutes and percentage of recovered arcs of both PSO algorithms with 300 particles, 50 iterations and 20 receiving nodes when learning transition networks as we increase the Markovian order.

The percentage of recovered arcs decreases in both algorithms as we increase the order due to the PSO not being close to convergence. The solutions obtained are the best ones found after 50 iterations, but it is expected that the algorithms did not yet converge to a solution due to the number of iterations being constant in all experiments. As the size of the search space increases, the number of iterations should also increase accordingly.

5   Conclusions

We have presented a new high-order dynamic Bayesian network structure learning algorithm that employs particle swarm optimization to find the network structure that best fits the training data provided. Its order invariant encoding allows a good scalability to bigger networks and high Markovian orders. It also offers the possibility to search up to a maximum desired order rather than having to specify it beforehand.

When learning high-order networks, the search space becomes huge rapidly. Algorithms that rely on independence tests, like the DMMHC algorithm, can become unfeasible in terms of execution time due to the high number of variables and the datasets with thousands of instances. On the other hand, particle swarm algorithms can scale well to finding solutions in bigger search spaces, but suffer slightly in smaller ones.

The execution time of the proposed natPSOHO algorithm scales much better in high orders due to the underlying data structures being constant in size and the operations being performed bitwise. It is shown to be an effective algorithm when dealing with processes with big search spaces generated by high-order networks. In situations where the exact Markovian order is not known beforehand, a higher number of iterations and a maximum desired order can be provided and the algorithm will search for a fitting network in that scenario.

In future work, we would like to refine our encoding and generalize it to be used with any meta-heuristic algorithm, not only with PSO. The main issue that it presents is that even though it relies on vectors of natural numbers, the operators must be bitwise. This would require a transition to a different encoding that is able to take advantage of the natural numbers, like the Gray code, or a generalization of the bitwise treatment before being able to extend it to other frameworks.

References

1. Akaike, H.: Information theory and an extension of the maximum likelihood principle. In: Parzen, E., Tanabe, K., Kitagawa, G. (eds.) Selected Papers of Hirotugu Akaike. SSS, pp. 199–213. Springer, New York (1998). https://doi.org/10.1007/978-1-4612-1694-0_15

2. Cai, B., et al.: Remaining useful life estimation of structure systems under the influence of multiple causes: subsea pipelines as a case study. IEEE Trans. Ind. Electron. 67(7), 5737–5747 (2019)

3. Dean, T., Kanazawa, K.: A model for reasoning about persistence and causation. Comput. Intell. **5**(2), 142–150 (1989)
4. Du, T., Zhang, S.S., Wang, Z.: Efficient learning Bayesian networks using PSO. In: Hao, Y., et al. (eds.) CIS 2005. LNCS (LNAI), vol. 3801, pp. 151–156. Springer, Heidelberg (2005). https://doi.org/10.1007/11596448_22
5. Geiger, D., Heckerman, D.: Learning Gaussian networks. In: Uncertainty in Artificial Intelligence Proceedings 1994, pp. 235–243. Elsevier (1994)
6. Gheisari, S., Meybodi, M.R.: BNC-PSO: structure learning of Bayesian networks by particle swarm optimization. Inf. Sci. **348**, 272–289 (2016)
7. Godsey, B.: Improved inference of gene regulatory networks through integrated Bayesian clustering and dynamic modeling of time-course expression data. PloS One **8**(7) (2013)
8. Kennedy, J., Eberhart, R.: Particle swarm optimization. In: Proceedings of 1995 IEEE International Conference on Neural Networks, vol. 4, pp. 1942–1948. IEEE (1995)
9. Koller, D., Friedman, N.: Probabilistic Graphical Models: Principles and Techniques. The MIT Press, Cambridge (2009)
10. Liu, X., Liu, X.: Structure learning of Bayesian networks by continuous particle swarm optimization algorithms. J. Stat. Comput. Simul. **88**(8), 1528–1556 (2018)
11. Lo, L.Y., Wong, M.L., Lee, K.H., Leung, K.S.: High-order dynamic Bayesian network learning with hidden common causes for causal gene regulatory network. BMC Bioinform. **16**(1), 1–28 (2015)
12. Ma, Y., Wang, L., Zhang, J., Xiang, Y., Liu, Y.: Bridge remaining strength prediction integrated with Bayesian network and in situ load testing. J. Bridge Eng. **19**(10) (2014)
13. Murphy, K.P.: Dynamic Bayesian Networks: Representation, Inference and Learning (2002)
14. Robinson, R.W.: Counting unlabeled acyclic digraphs. In: Little, C.H.C. (ed.) Combinatorial Mathematics V. LNM, vol. 622, pp. 28–43. Springer, Heidelberg (1977). https://doi.org/10.1007/BFb0069178
15. Santos, F.P., Maciel, C.D.: A PSO approach for learning transition structures of higher-order dynamic Bayesian networks. In: 5th ISSNIP-IEEE Biosignals and Biorobotics Conference: Biosignals and Robotics for Better and Safer Living, pp. 1–6. IEEE (2014)
16. Schwarz, G.: Estimating the dimension of a model. Ann. Stat. **6**(2), 461–464 (1978)
17. Trabelsi, G., Leray, P., Ben Ayed, M., Alimi, A.M.: Dynamic MMHC: a local search algorithm for dynamic Bayesian network structure learning. In: Tucker, A., Höppner, F., Siebes, A., Swift, S. (eds.) IDA 2013. LNCS, vol. 8207, pp. 392–403. Springer, Heidelberg (2013). https://doi.org/10.1007/978-3-642-41398-8_34
18. Vinh, N.X., Chetty, M., Coppel, R., Wangikar, P.P.: Gene regulatory network modeling via global optimization of high-order dynamic Bayesian network. BMC Bioinform. **13**(1), 1–16 (2012)
19. Wang, Y., Berceli, S.A., Garbey, M., Wu, R.: Inference of gene regulatory network through adaptive dynamic Bayesian network modeling. In: Zhang, L., Chen, D.-G.D., Jiang, H., Li, G., Quan, H. (eds.) Contemporary Biostatistics with Biopharmaceutical Applications. IBSS, pp. 91–113. Springer, Cham (2019). https://doi.org/10.1007/978-3-030-15310-6_5
20. Xing-Chen, H., Zheng, Q., Lei, T., Shao, L.P.: Research on structure learning of dynamic Bayesian networks by particle swarm optimization. In: 2007 IEEE Symposium on Artificial Life, pp. 85–91 (2007)

21. Zhu, J., Zhang, W., Li, X.: Fatigue damage assessment of orthotropic steel deck using dynamic Bayesian networks. Int. J. Fatigue **118**, 44–53 (2019)
22. Zou, M., Conzen, S.D.: A new dynamic Bayesian network (DBN) approach for identifying gene regulatory networks from time course microarray data. Bioinformatics **21**(1), 71–79 (2005)

Prototype Generation for Multi-label Nearest Neighbours Classification

Stefanos Ougiaroglou[1]($\boxtimes$), Panagiotis Filippakis[2], and Georgios Evangelidis[3]

[1] Department of Digital Systems, School of Economics and Technology, University of the Peloponnese, 23100 Kladas, Sparta, Greece
stoug@uop.gr
[2] Department of Information and Electronic Engineering, School of Engineering, International Hellenic University, 57400 Sindos, Thessaloniki, Greece
filipp1977@gmail.com
[3] Department of Applied Informatics, School of Information Sciences, University of Macedonia, 54636 Thessaloniki, Greece
gevan@uom.gr

Abstract. Numerous Prototype Selection and Generation algorithms for instance based classifiers and single label classification problems have been proposed in the past and are available in the literature. They build a small set of prototypes that represents as best as possible the initial training data. This set is called the condensing set and has the benefit of low computational cost while preserving accuracy. However, the proposed Prototype Selection and Generation algorithms are not applicable to multi-label problems where an instance may belong to more than one classes. The popular Binary Relevance transformation method is also inadequate to be combined with a Prototype Selection or Generation algorithm because of the multiple binary condensing sets it builds. Reduction through Homogeneous Clustering (RHC) is a simple, fast, parameter-free single label Prototype Generation algorithm that is based on k-means clustering. This paper proposes a RHC variation for multi-label training datasets. The proposed method, called Multi-label RHC (MRHC), inherits all the aforementioned desirable properties of RHC and generates multi-label prototypes. The experimental study based on nine multi-label datasets shows that MRHC achieves high reduction rates without negatively affecting accuracy.

Keywords: Multi-label classification · Data reduction · Prototype generation · k-NN classification · Binary relevance · RHC · BRkNN

1 Introduction

Multi-label classification [13] is a challenging problem that has attracted the interest of the machine learning and data mining research communities. Contrary to traditional single-label classification problems, where an instance may belong to only one class, in multi-label problems an instance may belong to more than one classes (or labels). Nowadays, multi-label classification is needed

H. Sanjurjo González et al. (Eds.): HAIS 2021, LNAI 12886, pp. 172–183, 2021.
https://doi.org/10.1007/978-3-030-86271-8_15

in numerous real-life applications of different problem domains. Typical examples of multi-label classification problems are image, text, protein, music and video/movies classification. For instance, a movie can simultaneously be "Crime" and "Drama". A music track may belong to more than one genre. An image may depict "mountain", "sea" and "beach".

The k-Nearest Neighbours (k-NN) classifier [5] is a typical instance-based (or lazy) classification algorithm. For each unclassified instance x, it searches the available training data and retrieves the k nearest to x instances. These instances are called nearest neighbours. Then, x is classified by a majority vote. In effect, x is classified to the most common class among the classes of the retrieved k nearest neighbours. The k-NN classifier is simple, easy to implement and has good performance. Moreover, it can be easily adapted for multi-label datasets. However, the k-NN classifier is CPU and memory intensive since all distances between a new unclassified instance and all training instances must be computed, and all training instances need to always be available in memory. Therefore, it cannot deal with large volumes of training data. Thus, in single-label classification, k-NN is usually applied in conjunction with a Prototype Selection (PS) [7] or Prototype Generation (PG) [12] algorithm.

PS and PG algorithms replace the initial training dataset by a small representative set. This set is called the condensing set and is used by k-NN to achieve comparable accuracy as when using the initial training dataset but at a much lower computational cost. PS algorithms select instances (or prototypes) that represent as much as possible the initial training set. The algorithms of this category are also called condensing algorithms. In contrast, PG algorithms generate prototypes by summarizing similar training instances of the same class. The prototypes generated by a PG algorithm are artificial instances that represent a specific data area in the metric space. Most of PS condensing and PG algorithms are based on the following simple idea: only the close to the class decision boundaries training instances are needed for classification tasks. The training instances that lie to the "internal" area of a class (far from decision boundaries) burden computational cost of the classifier, do not contribute to accurate classification and can be safely removed without loss of accuracy. Therefore, PS condensing and PG algorithms try to select or to generate a sufficient number of prototypes that lie close to class decision boundary data areas.

A sub-category of PS algorithms are called editing algorithms. Editing aims to improve accuracy rather than reduce data. To achieve this, editing algorithms try to improve the quality of the training data by removing noise and by smoothing the decision boundaries between the distinct classes. Of course, in case of non-separable classes, editing may fail to achieve these goals. Thus, contrary to PS condensing algorithms that condense the data by keeping only the instances that lie close to class decision boundaries, PS editing algorithms "clear" the class decision boundary areas by removing the close to the decision boundary instances. Although editing has a completely different goal, it can be used to improve the performance of PS condensing and PG algorithms. More specifically, the size of the condensing set built by PS condensing and PG algorithms depends on the level of noise in the training data. High levels of noise in the training set prevent many PS condensing or PG algorithms from achieving high reduction rates on the training data. In other words, the condensing set does not become small enough with respect to the original training set. Therefore, effective application of such algorithms requires removal of noise from the data, i.e., application of an editing algorithm beforehand.

PS and PG algorithms are appropriate for single-label classification problems. The Label Powerset (LP) transformation technique [13] could be an intuitive approach in order to use a PS or PG algorithm in a multi-label problem. LP transforms a multi-label dataset into singe-label datasets by considering each label combination (labelset) as a distinct class. However, LP can be applied only if the number of labels and the possible labelsets are small and there is a sufficient number of instances for each labelset. Otherwise, the total number of different combinations may increase exponentially, the reduction rate will be low and some combinations may be poorly represented.

Binary Relevance (BR) is the most widely-used problem transformation technique for multi-label problems. It transforms the multi-label problem into multiple binary problems, one for each label. A binary problem is a single-label problem with two labels. More specifically, for each label $l \in L$, BR builds a classifier in order to predict whether an instance belongs to l or not. In effect, BR copies the original training set $|L|$ times. Each copy concerns a different label l and contains all instances of the original training set, labelled as "1" if the original instance is labeled by l and as "0" otherwise.

The k-NN classifier in conjunction with BR is called BRkNN [6] and seems to be an ideal combination because k-NN classifier is a lazy classifier and does not build any classification model. BRkNN does not make $|L|$ copies of the training set. When an instance x needs to be classified, BRkNN searches for the k nearest to x neighbours once, like the single-label k-NN does. Then, the nearest neighbours voting procedure is repeated $|L|$ times, once for each label, and BRkNN makes $|L|$ label predictions for x. Therefore, x obtains the predicted labels of each voting procedure. Since each voting procedure concerns a binary classification problem, k should be odd to prevent ties.

The k-NN classifier stops being lazy when it is used in conjunction with a PS or PG algorithm. In effect, the condensing set is a classification model. In multi-label classification, if BR is used, one condensing set must be constructed for each label. Therefore, the goal of data reduction is not achieved and the k-NN classifier must search for nearest neighbours in each condensing set in order to make each individual label prediction. Hence, the computational cost remains high. This observation makes clear that the PS and PG algorithms must be adapted so that they can be used for multi-label datasets and this is the motive behind the present work. This paper proposes a PG algorithm for multi-label datasets. It constitutes a variation of Reduction through Homogeneous Clustering (RHC) [10], which is a fast and parameter-free PG algorithm for single label classification problems. The proposed algorithm is called Multilabel RHC (MRHC). It inherits the aforementioned desirable properties of RHC and builds a multi-label condensing set that is then used by BRkNN. The experimental study conducted shows that MRHC achieves noteworthy reduction rates while classification accuracy is not affected.

The rest of this paper is organized as follows: Sect. 2 briefly presents the related work, while Sect. 3 reviews RHC. Section 4 presents the proposed MRHC algorithm. Section 5 presents the experimental study, and, Sect. 6 concludes the paper and gives directions for future work.

2   Related Work

While most of research works in multi-label classification focus on proposing accurate classification algorithms, the issue of the computational cost on large multi-label training sets is somehow marginalized. There are only few research works that focus on speeding-up lazy classifiers in large multi-label training sets and even fewer that concern condensing algorithms for that type of dataset. As far as we know, there are no PG algorithms in the literature for multi-label training sets. In this section, we review the limited related works for fast multi-label classification.

The algorithm presented in [4] can be considered as the first PS algorithm for multi-label datasets. However, it focuses on editing of imbalanced datasets. In effect, the authors presented an under-sampling method for imbalanced training sets inspired by the Edited Nearest Neighbour rule (ENN-rule) [15]. Kanj et al. in [8] proposed a PS editing algorithm also based on ENN-rule. The proposed algorithm uses Hamming loss to determine the noisy instances. The idea behind the algorithm is simple: the instances with high Hamming loss probably lie in close decision boundaries and for this reason they should be removed, like in the case of ENN-rule. Both aforementioned works concern data editing. Thus, they are not considered further in the present work.

To the best of our knowledge, the work presented in [1] is a first attempt that adapts existing PS algorithms that edit and condense multi-label data. In this work, the previous proposed PS LSBo and LSSm algorithms [9] are adapted. LSBo condenses the data while LSSm edits it. The proposed algorithms as well as their predecessors are based on the concept of local sets [3] and on the LP transformation technique. In single-label problems, the local set of an instance x is the largest set of instances centered on x, so that all instances are of the same class. In multi-label datasets, the authors define that there is no need for a local set to contain the exact same labelset. The labelset of the instances in a local set may slightly differ. The authors use the Hamming loss calculated over labelsets to measure the difference in the labelsets. If the Hamming loss between the labelsets of two instances is greater than a pre-specifed threshold, the instances are considered to be of different "classes". The proposed PS algorithms are not parameter-free and their performance depends on the pre-specified threshold. Therefore, they are not considered further in this paper.

The paper in [2] uses BR, LP and other transformation methods in conjunction with single label PS algorithms. In case of BR and its variants, the proposed strategy copies the original training set $|L|$ times. Each copy concerns a different label l and contains all instances of the original training set, labelled as "1" if the original instance labeled by l and as "0" otherwise. Then, the strategy utilizes a PS algorithm on each copy and builds $|L|$ condensing sets for each label. Every time an instance is selected, it receives a vote that is accumulated in a vector with the votes for all instances. The strategy builds a complete condensing set by selecting all instances with a number of votes that exceeds a pre-specified threshold. As we mentioned before, we are interested in parameter-free approaches. Hence, these strategies are not considered further in the present work. Furthermore, the same paper [2] points out the LP drawbacks, mentioned in Sect. 1, when it is used in conjunction with a PS algorithm.

The work presented in [11] proposes a scalable lazy classifier for large multi-label datasets. The authors proposed an implementation that takes advantage of the GPU architecture. In effect, the work implements MLkNN [16] on GPUs. The proposed method implements the execution stages of MLkNN in parallel on GPUs, offering computational speedup without loss of accuracy.

3   The Reduction Through Homogeneous Clustering (RHC) Algorithm

Reduction through Homogeneous Clustering (RHC) [10] is a PG algorithm that is based on the well known k-means clustering algorithm. RHC is parameter-free, hence the size of the condensing set is determined automatically. Also, it is a fast algorithm since it avoids costly and time-consuming pre-processing tasks on the training set, which may be prohibitive for large datasets.

Initially, RHC computes the class representatives by averaging the instances of each class. If a dataset has ten classes, RHC will compute ten class representatives. These class representatives are used as initial means for k-means clustering. Then, k-means discovers as many clusters as the number of classes in the training set. If a discovered cluster has instances of only one class (i.e., it is homogeneous), the cluster centroid is placed in the condensing set as a prototype. Otherwise (i.e., the cluster is non-homogeneous), the aforementioned procedure is recursively repeated on that cluster. RHC terminates when all the discovered clusters are homogeneous.

The centroid/representaive m of each cluster/class C is computed by averaging the n attribute values of instances x_i, $i = 1, 2 \ldots |C|$ that belong to C. More formally, the n attributes $m.d_j$ of m is estimated as follows:

$$m.d_j = \frac{1}{|C|} \sum_{x_i \in C} x_i.d_j, j = 1, 2, \ldots, n$$

Obviously, RHC generates more prototypes for the close to the decision boundary areas and fewer prototypes for the "internal" class areas. By using the class representatives as initial means for k-means clustering, RHC increases the probability of quickly finding large homogeneous clusters and achieving a high reduction rate without costly k-means iterations (the larger the homogeneous clusters constructed, the higher the reduction rate achieved). RHC can become even faster if k-means clustering without full cluster consolidation is used. Full clusters consolidation means that k means clustering stops only when there are no moves of instances among clusters.

Contrary to many PS and PG algorithms, RHC always builds the same condensing set regardless the order of the data in the training set. Moreover, RHC is simple and quite easy to implement. The experimental study presented in [10] shows that RHC is faster and achieves higher reduction rates and than state-of-the-art PS condensing and PG algorithms without harming classification.

4   The Proposed Multi-label RHC (MRHC) Algorithm

The Multi-label Reduction through Homogeneous Clustering (MRHC) algorithm is a variant of RHC for multi-label training data. It works quite similar to RHC. However, MRHC builds a multi-label condensing set. The key question that should be answered is how to define homogeneity on clusters that contain multi-label data. In the case of MRHC, a cluster is considered homogeneous, when it contains instances that share at least one common label.

This is how the MRHC algorithm works. Initially, it builds a mean (representative) for each label l by averaging the instances that their label-set contains l. Then, it runs k-means clustering and discovers as many clusters as the number of labels in the training set. For each cluster C with instances that do not share a common label (i.e., C is non-homogenous), the aforementioned procedure is repeated considering only the instances that belong to C. For each homogeneous cluster, MRHC stores the cluster centroid in the condensing set as a prototype. The generated prototype is labeled by the common label(s) in C along with each label that appears in the label-set of more than half the instances in C. We call such labels majority labels. Like the case of RHC, MRHC terminates when all clusters become homogeneous.

Suppose a training set has three attributes (a, b, c) and three labels (x, y, z). Moreover, suppose that MRHC discovered a cluster C which contains the instances $inst_1$, $inst_2$ and $inst_3$ with the following BR representation:

– $inst_1$: $\{a, b, c, x, y, z\} = \{2, 1, 1, 0, 1, 1\}$
– $inst_2$: $\{a, b, c, x, y, z\} = \{1, 2, 3, 0, 1, 0\}$
– $inst_3$: $\{a, b, c, x, y, z\} = \{3, 2, 1, 1, 1, 1\}$

Since all instances contain label y, C is homogeneous. The label-set of the generated prototype p will be $\{y, z\}$ because y is the common label that renders C homogeneous and z is a majority label in C. Consequently, in a BR representation, p is $\{a, b, c, x, y, z\} = \{2, 1.67, 1.33, 0, 1, 1\}$.

Figure 1 represents a two dimensional example. Suppose that a dataset contains sixteen instances (Fig. 1(a)). MRHC computes a representative for the squares, a representative for the circles, and, a representative for the stars (Fig. 1(b)). Then, k-means clustering uses the three label representatives as initial means and discovers three clusters (Fig. 1(c)). Two of them are homogeneous

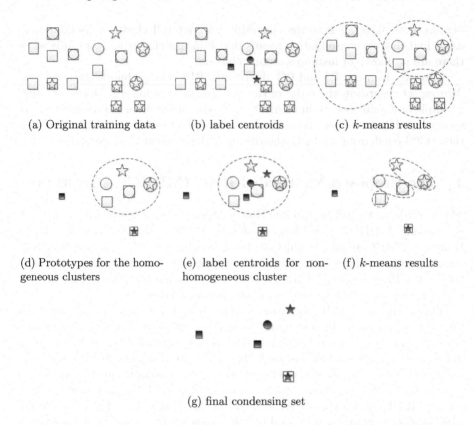

(a) Original training data   (b) label centroids   (c) k-means results

(d) Prototypes for the homo-   (e) label centroids for non-   (f) k-means results
geneous clusters   homogeneous cluster

(g) final condensing set

Fig. 1. MRHC excution example

because their instances have a common class. Thus, the cluster centroids constitute prototypes (Fig. 1(d)). One prototype is labeled only by "square" because
there is no majority label in the cluster. The other prototype is labeled by
"square" and "star", because "star" is the common label and "square" is a
majority label in a that cluster. For the instances of the non homogeneous cluster, MRHC recursively builds three homogeneous clusters (Figs. 1(e, f)). Consequently, three more prototypes are stored in the condensing set. Thus, the final
condensing set contains five prototypes instead of the sixteen instances of the
initial training set (Fig. 1(g)).

Algorithm 1 shows a non-recursive MRHC implementation. It uses a queue
data structure, Q, to hold clusters. Initially, the whole training set is an unprocessed cluster and is placed in q (line 2). At each repeat-until iteration, MRHC
dequeues cluster C from Q (line 5) and checks whether C is homogeneous (has
at least one common label) or not. If it is (line 6), its centroid is placed in
the condensing set (CS) as a prototype (line 15) labeled by the common and
majority labels in C (lines 9–14). Otherwise, MRHC computes a list of label-
representatives (M), one for each of the labels that exist in C (lines 17–21).

Algorithm 1. MRHC

Input: TS
Output: CS

```
 1: Q ← ∅
 2: Enqueue(Q, TS)
 3: CS ← ∅
 4: repeat
 5:     C ← Dequeue(Q)
 6:     if all the instances in C have at least one common label then
 7:         r ← centroid of C
 8:         r_labelset ← ∅
 9:         for each label l in C do
10:             n ← count the instances ∈ C with l in their labelset
11:             if n > |C/2| then
12:                 r_labelset ← r_labelset ∪ l
13:             end if
14:         end for
15:         CS ← CS ∪ {r}
16:     else
17:         M ← ∅ {M is the set of label-centroids}
18:         for each label l ∈ C do
19:             m_l ← centroid of l
20:             M ← M ∪ {m_l}
21:         end for
22:         NewClusters ← K-MEANS(C, M)
23:         for each cluster C ∈ NewClusters do
24:             Enqueue(Q, C)
25:         end for
26:     end if
27: until IsEmpty(Q)
28: return CS = 0
```

Then, MRHC calls k-means clustering, with parameters the non-homogeneous cluster C and the list of the initial label-representatives M to be used as initial means. The result is a new set of unprocessed clusters ($NewClusters$) (line 22) all of which are put into Q (lines 23–25). The repeat-until loop continues until Q becomes empty (line 27), i.e., there are no more clusters to process.

MRHC inherits all the properties of RHC. Therefore, it is a fast and parameter-free PG algorithm. Also, MRHC builds the same condensing set regardless the order of the instances in the training set. MRHC can be ideally combined with BRkNN. For each unclassified instance x, the BRkNN classifier runs over the condensing set (instead of the initial large training set) once and retrieves the k nearest to x prototypes. Then, the label-set of x is predicted by as many voting procedures as the number of labels. For each label l, the same k retrieved nearest prototypes vote in order to predict if x is labeled by l or not.

5   Performance Evaluation

5.1   Datasets

The performance of MRHC was evaluated by conducted experiments on nine multilabel datasets distributed by Mulan datasets repository [14][1]. Table 1 summarizes the key characteristics of the datasets used. The last two columns present the cardinality and the density of the datasets. Cardinality is the mean of the number of labels of the instances. Density is the mean of the number of labels of the instances divided by the number of labels. The second column of Table 1 lists the domain of each dataset.

Table 1. Dataset characteristics

Datasets	Domain	Size	Attributes	Labels	Cardinality	Density
CAL500 (CAL)	Music	502	68	174	26.044	0.150
Emotions (EMT)	Music	593	72	6	1.869	0.311
Water quality (WQ)	Chemistry	1060	16	14	5.073	0.362
Scene (SC)	Image	2407	294	6	1.074	0.179
Yeast (YS)	Biology	2417	103	14	4.237	0.303
Birds (BRD)	Sounds	645	260	19	1.014	0.053
CHD49 (CHD)	Medicine	555	49	6	2.580	0.430
Image (IMG)	Image	2000	294	5	1.236	0.247
Mediamill (MDM)	Video	43907	120	101	4.376	0.043

5.2   Experimental Setup

MRHC was coded in C and it runs only as a pre-processing step to build the condensing set. BRkNN was implemented in Python. We compared the performance of BRkNN running over the condensing set built by MRHC against the performance of BRkNN running over the original training set. The Euclidean distance was used as the distance metric. We measured two metrics: (i) Hamming loss and (ii) Reduction Rate, that were derived via a five-fold-cross-validation schema. Initially, we normalized the datasets in the $[0-1]$ range, and then, we split them into random subsets appropriate for five-fold-cross-validation. Since the computational cost of the BRkNN classifier depends on the size of the training set used, the CPU time needed for the classification is not reported. The Hamming Loss (HL) is the fraction of the wrong predicted labels to the total number of labels. It is computed as follows:

$$HL = \frac{1}{m} \sum_{i=1}^{m} \frac{|Y_i \Delta Z_i|}{|L|}$$

---

[1] http://mulan.sourceforge.net/datasets-mlc.html.

Table 2. Comparison in terms of Hamming Loss (HL (%)) and Reduction Rate (RR (%))

Dataset		BRkNN $k=1$	BRkNN $k=5$	BRkNN $k=9$	MRHC BRkNN $k=1$	MRHC BRkNN $k=5$	MRHC BRkNN $k=9$
CAL	HL	0.19	0.15	0.14	0.17	0.14	0.14
	RR	–	–	–	40.50	40.50	40.50
EMT	HL	0.24	0.20	0.19	0.22	0.20	0.20
	RR	–	–	–	65.73	65.73	65.73
WQ	HL	0.38	0.35	0.33	0.37	0.34	0.32
	RR	–	–	–	40.64	40.64	40.64
SC	HL	0.13	0.11	0.12	0.12	0.12	0.12
	RR	–	–	–	85.13	85.13	85.13
YS	HL	0.24	0.20	0.19	0.23	0.21	0.21
	RR	–	–	–	51.85	51.85	51.85
BRD	HL	0.09	0.08	0.08	0.09	0.08	0.09
	RR	–	–	–	42.70	42.70	42.70
CHD	HL	0.36	0.33	0.31	0.35	0.32	0.30
	RR	–	–	–	65.47	65.47	65.47
IMG	HL	0.29	0.27	0.27	0.28	0.25	0.25
	RR	–	–	–	71.71	71.71	71.71
MDM	HL	0.041	0.033	0.032	0.038	0.032	0.032
	RR	–	–	–	55.86	55.86	55.86

where m is the number of instances in the dataset, $|L|$ is the number of labels, Y_i is the instances' set of real labels and Z_i is the instances' set of predicted labels. Δ is the symmetric difference of two sets and corresponds to the XOR operation. For example, if the labels of an instance are $\{1, 1, 0, 0, 1\}$ and the predicted labels are $\{1, 1, 0, 1, 0\}$, the Hamming loss will be $\frac{2}{5} = 0.4$. Also, we ran experiments using three different k values (1, 5 and 9).

5.3 Experimental Results

Table 2 presents the results obtained by the experimental study. For each dataset, we report Hamming loss and the reduction rate achieved by MRHC. We observe that MRHC achieved reduction rates between 40% and 85%. The average reduction rate is 57.63%. This means that the BRkNN classifier that runs over the condensing set constructed by MRHC is 57.63% faster on average than the BRkNN classifier that runs over the original training set. One could claim that the reduction rates are not very high - certainly, they are lower than the reduction rates achieved by the single-label RHC. We claim that the comparison between the reduction rates achieved on single-label and multi-label datasets does not make sense. Multi-label data is more complex than single-label data. As a result, MRHC cannot identify large homogeneous clusters in multi-label data like RHC does in single-label data.

Moreover, we observe there is no difference in Hamming loss between BRkNN that uses the multi-label condensing set constructed by MRHC and the BRkNN classifier that uses the original training set. BRkNN achieves similar Hamming loss measurements regardless of whether the condensing set or the original training set is used. Therefore, we can safely conclude that MRHC achieves significant gains in reduction rates while accuracy is not negatively affected.

6   Conclusions and Future Work

Prototype Selection and Generation is an essential pre-processing stage in order to avoid the drawbacks of high computational cost and storage requirements in instance based classification. However, the vast majority of existing PS and PG algorithms are not applicable to multi-label classification problems, and also they can not be effectively used in conjunction with a problem transformation method like Binary Relevance or Label Powerset.

This paper initially presented the recent research efforts for speeding-up the k-NN classifier in the context of multi-label classification. Then, it proposed the MRHC algorithm, which is the first PG algorithm for multi-label training data. In effect, MRHC is an adaptation of the fast, single-label RHC algorithm. Like RHC, MRHC is parameter-free and is based on a recursive k-means clustering procedure that discovers homogeneous clusters. In the context of multi-label classification, we considered a cluster to be homogeneous when it contains instances with at least one common label. The centroid of each homogeneous cluster constitutes a prototype labeled by the common labels along with each label that appears in the majority of the cluster instances. Thus, MRHC builds a multi-label condensing set that BRkNN can use to search for nearest neighbours and make multi-label predictions. The experimental study used nine multi-label datasets and demonstrated that there is no difference on the accuracy achieved by BRkNN when using the condensing set built by MRHC or the original training set. On the other hand, the CPU time needed for the classification process when using the condensing set is much lower.

This paper showed that Prototype Selection and Generation for multi-label problems is an open research field in the data mining and machine learning context. We plan to adapt well-known single-label PG algorithms on multi-label problems. Next, we plan to developed new parameter-free PS and PG algorithms as well as scalable classification methods for multi-label training sets.

References

1. Arnaiz-González, Á., Díez-Pastor, J.F., Rodríguez, J.J., García-Osorio, C.: Local sets for multi-label instance selection. Appl. Soft Comput. **68**, 651–666 (2018). https://doi.org/10.1016/j.asoc.2018.04.016

2. Arnaiz-González, Á., Díez-Pastor, J.F., Rodríguez, J.J., García-Osorio, C.: Study of data transformation techniques for adapting single-label prototype selection algorithms to multi-label learning. Expert Syst. Appl. **109**, 114–130 (2018). https://doi.org/10.1016/j.eswa.2018.05.017

3. Brighton, H., Mellish, C.: Advances in instance selection for instance-based learning algorithms. Data Min. Knowl. Discov. **6**(2), 153–172 (2002). https://doi.org/10. 1023/A:1014043630878

4. Charte, F., Rivera, A.J., del Jesus, M.J., Herrera, F.: MLeNN: a first approach to heuristic multilabel undersampling. In: Corchado, E., Lozano, J.A., Quintián, H., Yin, H. (eds.) IDEAL 2014. LNCS, vol. 8669, pp. 1–9. Springer, Cham (2014). https://doi.org/10.1007/978-3-319-10840-7_1

5. Cover, T., Hart, P.: Nearest neighbor pattern classification. IEEE Trans. Inf. Theory **13**(1), 21–27 (1967). https://doi.org/10.1109/TIT.1967.1053964

6. Spyromitros, E., Tsoumakas, G., Vlahavas, I.: An empirical study of lazy multilabel classification algorithms. In: Darzentas, J., Vouros, G.A., Vosinakis, S., Arnellos, A. (eds.) SETN 2008. LNCS (LNAI), vol. 5138, pp. 401–406. Springer, Heidelberg (2008). https://doi.org/10.1007/978-3-540-87881-0_40

7. Garcia, S., Derrac, J., Cano, J., Herrera, F.: Prototype selection for nearest neighbor classification: taxonomy and empirical study. IEEE Trans. Pattern Anal. Mach. Intell. **34**(3), 417–435 (2012). https://doi.org/10.1109/TPAMI.2011.142

8. Kanj, S., Abdallah, F., Denœux, T., Tout, K.: Editing training data for multi-label classification with the k-nearest neighbor rule. Pattern Anal. Appl. **19**(1), 145–161 (2015). https://doi.org/10.1007/s10044-015-0452-8

9. Leyva, E., González, A., Pérez, R.: Three new instance selection methods based on local sets: a comparative study with several approaches from a bi-objective perspective. Pattern Recogn. **48**(4), 1523–1537 (2015). https://doi.org/10.1016/j. patcog.2014.10.001

10. Ougiaroglou, S., Evangelidis, G.: RHC: a non-parametric cluster-based data reduction for efficient k-NN classification. Pattern Anal. Appl. **19**(1), 93–109 (2014). https://doi.org/10.1007/s10044-014-0393-7

11. Skryjomski, P., Krawczyk, B., Cano, A.: Speeding up k-Nearest Neighbors classifier for large-scale multi-label learning on GPUs. Neurocomputing **354**, 10–19 (2019). https://doi.org/10.1016/j.neucom.2018.06.095. Recent Advancements in Hybrid Artificial Intelligence Systems

12. Triguero, I., Derrac, J., Garcia, S., Herrera, F.: A taxonomy and experimental study on prototype generation for nearest neighbor classification. Trans. Sys. Man Cyber Part C **42**(1), 86–100 (2012). https://doi.org/10.1109/TSMCC.2010.2103939

13. Tsoumakas, G., Katakis, I.: Multi-label classification: an overview. Int. J. Data Warehousing Min. **2007**, 1–13 (2007)

14. Tsoumakas, G., Katakis, I., Vlahavas, I.: Mining multi-label data. In: Maimon, O., Rokach, L. (eds.) Data Mining and Knowledge Discovery Handbook, pp. 667–685. Springer, Boston (2009). https://doi.org/10.1007/978-0-387-09823-4_34

15. Wilson, D.L.: Asymptotic properties of nearest neighbor rules using edited data. IEEE Trans. Syst. Man Cybern. **2**(3), 408–421 (1972)

16. Zhang, M.L., Zhou, Z.H.: ML-KNN: a lazy learning approach to multi-label learning. Pattern Recogn. **40**(7), 2038–2048 (2007). https://doi.org/10.1016/j.patcog. 2006.12.019

Learning Bipedal Walking Through Morphological Development

M. Naya-Varela[1]([✉]), A. Faina[2]([✉]), and R. J. Duro[1]([✉])

[1] Integrated Group for Engineering Research, CITIC (Centre for Information and Communications Technology Research), Universidade da Coruña, A Coruña, Spain
martin.naya@udc.es, anfv@itu.dk
[2] Robotics, Evolution and Art Lab (REAL), Computer Science Department, IT University of Copenhagen, Copenhagen, Denmark
richard@udc.es

Abstract. Morphological development has shown its efficiency in improving learning and adaptation to the environment in natural organisms from infancy to adulthood. In the case of robot learning, this is not so clear. The results of a series of experiments that have been carried out in previous work have allowed us to extract, from an analytical perspective, some notions about how and under what conditions morphological development may influence learning. In this paper, we want to adopt an engineering or synthesis perspective and test whether these notions can be used to construct a successful morphological development strategy for a difficult task: learning bipedal locomotion. In particular, we have addressed learning to walk in a 14 degrees of freedom NAO type robot and have designed a morphological development strategy to this end. The results obtained have allowed us to validate the relevance of the assumptions made for the design and implementation of a morphological development strategy.

Keywords: Morphological development · Growing robots · Bipedal walking

1 Introduction

Humans and animals undergo morphological development processes from infancy to adulthood that have been shown to facilitate learning [1, 2]. Some of the developmental principles observed in nature have been applied to different robot morphologies, with the main goal of improving their learning abilities. The implementation of these principles has led to different results, showing that the development of the morphology while learning can be positive [3–6], irrelevant [7, 8] or even detrimental [8, 9] for the learning process. Although the mechanisms through which morphological development may influence learning are still not very well understood, there are several studies that provide some indications on why and how learning may be influenced by morphological development. For example, Bongard and Buckingham [5, 10] relate task complexity to the influence of morphological development, indicating that morphological development does not provide any advantage for a simple problem. In another study [9], Bongard

© Springer Nature Switzerland AG 2021
H. Sanjurjo González et al. (Eds.): HAIS 2021, LNAI 12886, pp. 184–195, 2021.
https://doi.org/10.1007/978-3-030-86271-8_16

also presents instances where morphological development is detrimental for learning, due to the abrupt changes in the controller that occur during the development phase in that particular experiment. Also, Ivanchenko and Jacobs [8] show how morphological development may be beneficial for learning if a suitable development sequence is followed. They show how an inadequate sequence may even produce results that are worse than those of learning without morphological development. The relevance of finding not only a morphological development strategy that is suitable for the particular problem being addressed, but also the necessity of finding its adequate developmental sequence is pointed out by Vujovic et al. [6]. In their article, they find that a suitable developmental sequence may improve learning while an unsuitable one is irrelevant in their case.

With the aim of complementing the insights gained by those authors, Naya-Varela et al. [11, 12] have carried out a series of experiments analyzing the performance of different morphological development strategies over different morphologies. Firstly, on a study that analyzes the performance of morphological development based on growth and on the variation of the Range of Motion (ROM) of the limbs [11], they find that growth-based morphological development improves performance in a quadruped morphology, while ROM is irrelevant. They hypothesize that the success of the growth strategy is motivated by two main reasons: (1) Starting the learning process with an initially smaller morphology lowers the center of gravity, thus increasing the initial stability of the morphology and allowing it to maintain an upright position without falling for more behaviors than in the adult and larger morphology. (2) This increment in the stability increases the exploratory behavior at the beginning of learning, avoiding the stagnation of solutions in local optima as it is observed for the no-development case (learning directly using the adult morphology). Regarding the ROM strategy, they argue that its lack of effectiveness for learning in their case is motivated by the intrinsic characteristics of the strategy and its incorrect alignment with the morphology of the quadruped. Based on these results, they study the influence of the growth strategy in two additional morphologies: a hexapod and an octopod [12]. In this study, they find that the influence of growth decreases with the number of limbs of the morphology, being relevant for the quadruped and irrelevant for the hexapod and octopod. The authors hypothesize that this decrease in the relevance of growth is due to the reduction of the task complexity with the number of limbs: as the number of limbs increase, the stability of the morphology increases, thus the problem becomes easier to learn.

The results and conclusions obtained in all of these studies, as formalized in [13], can be condensed into a series of insights or hypotheses that a problem and a morphology must fulfill to be susceptible of being influenced by morphological development. Section 2 describes them in detail. In fact, these insights can also lead to a series of considerations that should be taken into account and steps that should be followed in order to appropriately design morphological development processes.

The objective of this paper is to provide some experimental results on the application of the insights mentioned in the previous paragraph. To this end, we have addressed a problem that has been classified as quite difficult in the literature: Learning to walk on two legs, with the aim of designing an appropriate morphological development sequence to make learning the task easier and more efficient.

Although there are numerous examples of bipedal legged robots in the literature [14–16], learning to walk in bipedal robots is still a complex task. In fact, most walkers are programmed to walk, they do not learn to walk. There are few examples that study how to learn to walk in bipedal robots from a developmental perspective. In this line, Lungarella and Berthouze [17, 18] analyze the influence of morphological development by freezing and freeing Degrees of Freedom (DOF) of a bipedal robot sustained by a harness attached to its shoulders. More recently, two articles address the problem of learning to walk in bipedal robots using a growth based morphological development strategy [19, 20]. On the one hand, Hardman et al. [19] compared the performance obtained utilizing an annealing optimization algorithm of learning with and without morphological changes. They show how their selected morphological development strategy outperforms learning without it. These results are presented for two different morphological changes, for the case of developing the length of the foot and for the case of increasing the mass and inertia of the body of the robot. Furthermore, they also found that their methodology reduces the number of catastrophic failures, considering as catastrophic failures behaviors in which the robot falls or collides with its own body. On the other hand, Zhu et al. [20] show how a suitable constraint of the morphology in a bipedal robot that learns to walk by a genetic algorithm allows it to improve on the learning performance of the system without any constraint. However, they also pointed out the necessity of a suitable match-up between the task and the selected restrictions, because they also report worse results when different ones are applied.

Summarizing, morphological development has been studied by analyzing different developmental strategies and comparing them to the no-development case. In addition, it has been shown that the morphological development process needs to be carefully chosen as it could lead to completely inadequate results. Based on these results, different authors have extracted some basic knowledge and hypotheses about when it could make sense to apply morphological development. However, to the best of our knowledge, we have not found any work that addresses the opposite problem: Given a morphology and a task, find an appropriate morphological developmental process that makes learning easier.

Thus, this paper presents a first experiment to design or synthesize a morphological developmental process, and we show that it is possible to improve the learning abilities of a selected morphology taking inspiration from the morphological changes that happen in nature. The paper is structured as follows: In Sect. 2, we describe the requirements for designing an application case for morphological development in bipeds. Section 3 is devoted to present the experimental setup we will be using during this experiment. The results of the application case of morphological development, followed by a discussion, are presented in Sect. 4. Finally, we provide some conclusions and future lines of work in Sect. 5.

2 Designing a Morphological Development Strategy

This section is devoted to presenting the design process of a morphological development strategy to improve the learning performance of a bipedal robot [21]. To provide a guide of the aspects that should be taken into account in order to design a morphological development sequence for a particular problem, we have resorted to previous work in the literature. In particular, Naya-Varela et al. [13], after a thorough analysis of the fitness landscapes obtained for several morphological development processes, have suggested a series of general considerations that should be followed. As a summary, they have indicated that:

- As other authors have already mentioned before, the learning problem must be complex enough to justify morphological development. Otherwise, it may not have any impact during the learning phase.
- Learning with the initial morphology must be simpler than learning with the final morphology. This simplifies the problem at the early stages of development, allowing us to gradually increase the complexity of the learning task through the different developmental stages. Thus, the maximum complexity of the problem will be achieved with the final morphology.
- It is necessary to have an adequate synergy between the morphology, the control system, and the selected developmental strategy. Especially, the development of the morphology must be in accordance with the capacity of the controller to adapt to the morphological changes.
- To avoid misleading learning, we consider that optimal solutions must be available from the beginning of the learning process. Reducing the solution search space could imply that an optimal solution would not be available until the final morphology is reached, limiting the capability of the learning algorithm to find optimal solutions and allowing for deceptive paths.

Of course, these are general principles that are mostly related to the evolution of the sequence of fitness landscapes defined by the sequence of morphological changes when viewing the learning strategy as an optimization process. Consequently, they need to be translated into specific features that we want to see in our bipedal robot learning to walk. These features are:

- Problem complexity: In order to have a complex enough problem for learning, we have selected the task of bipedal walking. Learning to walk is a complex task due to the intrinsic difficulties associated with the instability and dynamics [22]. Concretely, we have selected the task of learning to walk in a NAO robot.
- Learning simplification: With the aim to start learning with an initially simpler morphology than the final one, we have selected growth as a morphological development strategy. The reason is that we consider that learning with a lower center of gravity may simplify learning, thanks to an increase in the stability of the morphology.
- Morphology, control system, and development synergy: To maintain a synergy between morphology, control system, and developmental strategy we have selected a progressive and continuous developmental stage, avoiding abrupt changes in the controller or morphology that may distort the relationship between them.
- Availability of solutions: A growth based developmental strategy does not imply any limitation or constraint in the movement of the motor system. Thus, with this developmental strategy, the space of possible solutions is invariant from the initial morphology to the final one.

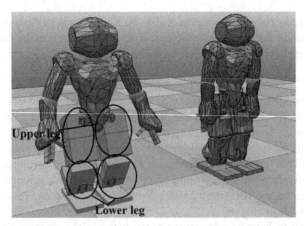

Fig. 1. Different versions of the NAO robot. Right: NAO model without any modification. Left: NAO model with the upper leg and lower leg modified to allow growing.

Thus, to improve NAO's learning performance thanks to growth, a series of different design characteristics have been implemented: (1) To allow the NAO robot to grow, we have modified its legs allowing their extension. Although morphological development could involve more and different parts of the NAO (like the body or arms), as a first approximation, we have decided to apply it only to the legs. We consider that modifications of the length of the legs provide the highest impact in terms of the stability of the robot (Fig. 1); (2) Symmetric growth. We have considered a symmetric growth of the upper leg and the lower leg. This preserves the initial stability as well as the center of gravity as close as possible to the initial position in the xy-plane (bearing in mind that as the morphology grows, it will move upwards along the z-axis), avoiding the possible static and dynamic imbalances that may arise if one part of the leg grows more than the other; (3) Progressive growth. With the aim of avoiding drastic changes in the morphology and control system, a progressive and linear growth sequence has been selected, rather than an abrupt one; (4) Reduction of the maximum ROM available. Finally, we have reduced the maximum available ROM of each joint given by the documentation of the NAO in order to reduce the search space.

3   Experimental Setup

To test the application of the morphological development strategy presented in the previous section, we have created the following experimental setup. As indicated before, we will make use of a NAO platform as the base robotic structure. For convenience, the NAO will be simulated using the CoppeliaSim simulator [23] and the PyRep extension [24]. To apply morphological development based on growth as indicated above, a series of modifications to the legs and feet of the NAO model in CoppeliaSim have been made with the objective of allowing leg growth and increasing stability:

- The upper part of the legs was changed to two links joined by a prismatic joint. Each link is 8 × 8 × 7.2 cm and has a mass of 458.7 g. The prismatic joint has a maximum force of 50 N. The maximum extension of the prismatic joint is 3.5 cm. This group of two links and the prismatic joint, will be considered as a unique link that is able to grow.

- The lower part of the legs was also changed to two links joined by a prismatic joint. The upper link is $8 \times 8 \times 3$ cm and has a mass of 192 g and the lower link is $9 \times 8 \times 3$ cm and a mass of 215.8 g. Both links are equal but present different orientations. The prismatic joint has a maximum force of 50 N. The maximum extension of the prismatic joint is 3.5 cm. Again, this group will be considered as a single link that can grow.
- The size of the feet has also changed from the original NAO foot size, increasing it up to $16 \times 8.5 \times 1.5$ cm in dimension and 204 g in weight.
- The different leg parts, as well as the feet, have been modified to represent, in a simplified manner, the same dimensions as the original NAO. The mass and inertia of the legs and feet are automatically adjusted by the simulator.

The controller of the robot is a neural network whose weights and structure are learnt using NEAT [25], specifically the MultiNEAT implementation [26]. It has 3 inputs plus one bias and 14 outputs, each controlling the actuation of one joint. The inputs are sinusoidal functions of amplitude 2.0 rad and frequency 1.0 rad/s. The phase offsets of the sinusoidal inputs are 0, 3.0 and 5.0 rad respectively.

A series of learning experiments using NEAT have been run over different implementations of the robot and environment using the CoppeliaSim simulator with the ODE physical engine [27] in the CESGA [28] computer cluster. Each NEAT learning run evolves a population of 150 individuals and is trained for 300 generations. A total of 40 independent runs have been carried out for each experiment with the objective of gathering relevant statistical data. Each individual is tested for 5 s with a simulation time step of 50 ms and a physics engine time step of 5 ms.

As the controller is obtained using NEAT, the learning strategy is based on a neuroevolutionary process, where the fitness depends on the distance travelled by the head of the robot in a straight line and whether or not the robot falls during learning. If the NAO does not fall, the fitness value is the distance traveled in a straight line in meters. However, if the NAO falls the simulation is stopped and we consider as the fitness value the distance traveled 16-simulation time steps before the moment the NAO fell. In this sense, we consider that the NAO falls when its head is below 0.3 m. We have selected 16-time steps because 16-time steps before falling, the NAO is still in a stable position.

In order to evaluate the developmental strategy that was designed, we have performed two different types of experiments:

- **Reference Experiment.** This experiment is run with a fixed morphology (the same as the final morphology for the rest of the experiments) from the beginning to the end. The robot starts at generation 0 with the maximum length of the legs and the neuroevolutionary algorithm seeks a neural network-based controller to achieve maximum displacement.
- **Growth Design Experiments.** The robot morphology starts with the shorter version of the legs. That is, at the beginning of learning, the prismatic joints are fully contracted, their extension is 0 cm. The length of the upper legs is 14.334 cm and the length of the

lower leg is 11 cm. The leg length is grown linearly for a number of generations until the upper leg reaches 17.0834 cm of length and the lower length reaches 14.5 cm. This growth takes place in a set number of generations for each experiment. That is, the final morphology is reached at generation 40, 60, 80, 100 and 120, depending on the experiment.

This permits identifying the best growth ratio for the selected morphology and control system and evaluating the relevance of the growth rate with regards to performance.

4   Results and Discussion

The results of the morphological development process designed for the NAO robot can be observed in Fig. 2 and Fig. 3. Figure 2 displays the results obtained after the learning process through neuroevolution in the case of no development and in the design case of growth up to generation 120, as it is the one which results in the highest median. It displays the median of the best fitness obtained for the 40 independent runs at each generation for each configuration. The shaded areas in the graph represent the areas between percentiles 75 and 25 for each case. Figure 3 displays the statistical results at the end of the learning process for the different growth rates and the no development case. Each boxplot represents the median and the 75 and 25 quartiles in the last generation for 40 independent runs of each of the different types of experiments. The whiskers are extended to 1.5 of the interquartile range (IQR). Single points represent values that are out of the IQR. All developmental samples are compared to the no-development case. The statistical analysis has been carried out using the two-tailed Mann-Whitney U test. We want to test whether the performance of the different design cases is similar or not to the reference case. We consider a p-value of 0.05 as the significance value for accepting or rejecting the null hypothesis. All the p-values have been adjusted using the Bonferroni [29] correction. The results show that the design case based on the morphological learning sequence we have designed offers better results than learning without morphological development. Only the growth up to generation 40 case (p-value of 0.37133) does not offer better results than the reference case, while the less representative results, growth up to generation 60 and growth up to generation 100, have a p-value of 0.04128 and 0.02037 respectively, both under the reference value of 0.05. Furthermore, two cases have offered notable improvement concerning the no-development case. These are the growth up to generation 80 (p-value of 0.00189) and growth up to generation 120 (0.0021).

Analyzing Fig. 2 and Fig. 3 it can be observed that:

- In the morphological development case, there is a noisy behavior in the curve representing the median of the fitness value during the developmental period. It is motivated by the adaptation of the controller to the morphological changes that happen during development. This means that the best solution in a specific generation may not be the optimal solution in the next generation due to the variations in the morphology. This is not observed in the curve that represents the median of the no-development experiment. In this case, the fitness value progresses gradually without oscillations as the morphology does not change.

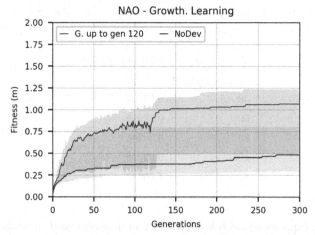

Fig. 2. Results obtained after 40 independent executions for the selected design process, considering different growth speeds and the case without morphological development. For the sake of clarity, we only show the comparative results of the learning process for the case of no development (black) and the design case of growing until generation 120 (blue), which presented the best results.

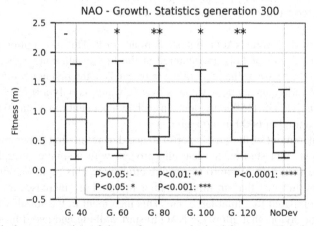

Fig. 3. Statistical representation of the performance obtained from the 40 independent experiments at the end of the neuroevolutionary process. The statistical values of the Man-Whitney test adjusted by the Bonferroni correction have been replaced by asterisks in order to makes the figures clear. "Growth up to generation" is abbreviated by a G.

- The selected morphological development strategy improves the learning ability of the algorithm not only at the end of the optimization process, but from the beginning it already surpasses the performance of the no development case.
- In Fig. 3, it can be observed how the median of the fitness value increases as the growth speed decreases (Table 1). Although this increment is small in absolute values, it clearly shows a tendency. That is, the selected design strategy allows to improve

Table 1. Median fitness for each growth ratio and the no-development case at the end of the experiment

Growth ratio	Median fitness (m)
No-development	0.48292
40	0.85894
60	0.87915
80	0.89766
100	0.92716
120	1.06742

the learning capacity of the NAO, but the control system needs time to adapt to the changes in the morphology. Especially relevant is the comparison between the fitness value of the no-development case and the growth up to generation 120 case. This last one, more than doubles the fitness value of the no-development case.

Analyzing these results in more detail and relating them to our initial design conditions about how to design a specific morphological development strategy based on the insights extracted from the literature, we can say that:

- Learning to walk for the NAO robot is a complex task. This assumption is supported by the high number of falls encountered at the first evaluation of the individuals. Considering the fitness value of all of the individuals in the first generation, we obtain a total of 44 individuals whose value is over 0, which is only a 0.733% of the total number of individuals (150 individuals in each independent run, with 40 independent runs makes a total of 6000 individuals).
- The rapid increment of the fitness performance during the first generations supports our hypothesis that starting the learning process with a smaller morphology than the final one may help to improve learning efficiency. This rapid increment may be motivated by the fact that an initial smaller morphology increases the stability of the NAO, compared to a final large one. This increment in the stability may help to find initial behaviors that allow the NAO to start walking and avoid falls. Behaviors that could be rejected in the adult morphology. Thus, the number of optimal behaviors increases as well as the exploration capacity of the learning algorithm. This hypothesis is supported by Fig. 2, where the fitness value of the growth experiments improves on the fitness achieved by the non-developmental case from the beginning of learning. This shows that robots with shorter legs are able to walk further than robots with longer legs (final morphology).
- Furthermore, we consider that our selected design condition of progressive and gradual development is supported by the results of Fig. 3. On the one hand, the medians with the highest fitness are obtained in those experiments with slower speeds. That is, in those cases with more gradual development. On the other hand, only growth up to

generation 40 has not improved learning, being this growth ratio the one that caused the most abrupt changes in the morphology.

- It seems that the assumption of a symmetrical development in both parts of the legs (in the upper leg and in the lower leg) based on the idea of maximizing stability was also a good choice. However, we do not know if this is the optimal solution and further experimentation analyzing the influence of an asymmetric development strategy would be interesting.

Finally, it is important to mention that during the evaluation time, the morphology of the robot is fixed. What helps to improve learning to walk in the adult morphology is the sequence of developmental stages the morphology follows from the initial morphology to the final one. This is different from cases where the morphology changes during the evaluation time to increase their adaptation to the environmental conditions, without considering initial or final stages of development, such as in Ahmad et al. [30].

5 Conclusion and Future Work

This paper deals with the design of a morphological development strategy with the aim of improving the learning ability of a bipedal robot when learning to walk. The design of the morphological development process was based on design considerations extracted from previous studies and analyses in the field, that can be summarized as: (1) The task must be complex enough to warrant morphological development; (2) Start learning with a morphology that makes the task simpler. In our experiment, this means a more stable initial morphology; (3) Progressive and gradual development. To avoid abrupt changes in the control-morphology relationship we have selected a developmental strategy based on the linear growth of the legs; (4) Finally, we have selected a morphological development strategy that does not omit optimal solutions during the developmental phase, to avoid the learning algorithm getting stuck in suboptimal behaviors while the morphology grows.

The results of applying this morphological development process support the design decisions we have made as morphological development clearly improves the learning performance in the majority of the cases considered. In fact, in the best cases, it doubles the performance of no development. However, much work is needed to provide robust engineering indications about the design considerations we have made. In this sense, further analysis and research about them and their implications should be carried out. For example, it would be interesting to produce a design implementation with slower growth rates than those presented in this paper in order to verify whether the relationship between growth ratio and fitness is consistent or not. Furthermore, it would also be interesting to see whether an asymmetrical growth of the legs could improve the results. Finally, to conclusively state that the selected design considerations are suitable tools for improving morphological development-based learning in general, further research should be carried out using different algorithms and morphologies.

Acknowledgment. This work has been partially funded by the Ministerio de Ciencia, Innovación y Universidades of Spain/FEDER (grant RTI2018–101114-B-I00) and Xunta de Galicia (grant EDC431C-2021/39). We wish to acknowledge the support received from the Centro de Investigación de Galicia "CITIC", funded by Xunta de Galicia and the European Union (European

Regional Development Fund-Galicia 2014–2020 Program), by grant ED431G 2019/01. We also want to thank CESGA (Galician Supercomputing Center) for the use of their resources.

References

1. Piaget, J., Cook, M.: The Origins of Intelligence in Children. International Universities Press, New York (1952)
2. Thelen, E.: Motor development as foundation and future of developmental psychology. Int. J. Behav. Dev. **24**, 385–397 (2000). https://doi.org/10.1080/016502500750037937
3. Kriegman, S., Cheney, N., Bongard, J.: How morphological development can guide evolution. Sci. Rep. **8**, 1–17 (2018). https://doi.org/10.1038/s41598-018-31868-7
4. Baranes, A., Oudeyer, P.-Y.: The interaction of maturational constraints and intrinsic motivations in active motor development. In: IEEE International Conference on Development and Learning (ICDL), pp. 1–8 (2011)
5. Bongard, J.: The utility of evolving simulated robot morphology increases with task complexity for object manipulation. Artif. Life. **16**, 201–223 (2010). https://doi.org/10.1162/artl.2010.Bongard.024
6. Vujovic, V., Rosendo, A., Brodbeck, L., Iida, F.: Evolutionary developmental robotics: improving morphology and control of physical robots. Artif. Life. **23**, 169–185 (2017)
7. Savastano, P., Nolfi, S.: A robotic model of reaching and grasping development. IEEE Trans. Auton. Ment. Dev. **5**, 326–336 (2013)
8. Ivanchenko, V., Jacobs, R.A.: A developmental approach aids motor learning. Neural Comput. **15**, 2051–2065 (2003). https://doi.org/10.1162/089976603322297287
9. Bongard, J.: Morphological change in machines accelerates the evolution of robust behavior. Proc. Natl. Acad. Sci. **108**, 1234–1239 (2011). https://doi.org/10.1073/pnas.1015390108
10. Buckingham, D., Bongard, J.: Physical scaffolding accelerates the evolution of robot behavior. Artif. Life. **23**, 351–373 (2017)
11. Naya-Varela, M., Faina, A., Duro, R.J.: An experiment in morphological development for learning ANN based controllers. In: IEEE World Congress on Computational Intelligence (WCCI). Glasgow (UK) (2020)
12. Naya-Varela, M., Faina, A., Duro, R.J.: Some experiments on the influence of problem hardness in morphological development based learning of neural controllers. In: de la Cal, E.A., Villar Flecha, J.R., Quintián, H., Corchado, E. (eds.) HAIS 2020. LNCS (LNAI), vol. 12344, pp. 362–373. Springer, Cham (2020). https://doi.org/10.1007/978-3-030-61705-9_30
13. Naya-Varela, M.: Study of morphological development as a tool for learning robotic controllers (2021). http://hdl.handle.net/2183/27718
14. Chestnutt, J., Lau, M., Cheung, G., Kuffner, J., Hodgins, J., Kanade, T.: Footstep planning for the Honda Asimo humanoid. In: Proceedings of the 2005 IEEE International Conference on Robotics and Automation, pp. 629–634 (2005)
15. Hartley, R., et al.: Legged robot state-estimation through combined forward kinematic and preintegrated contact factors. In: 2018 IEEE International Conference on Robotics and Automation (ICRA), pp. 1–8 (2018)
16. Feng, S., Xinjilefu, X., Atkeson, C.G., Kim, J.: Optimization based controller design and implementation for the atlas robot in the darpa robotics challenge finals. In: 2015 IEEE-RAS 15th International Conference on Humanoid Robots (Humanoids), pp. 1028–1035 (2015)
17. Lungarella, M., Berthouze, L.: On the interplay between morphological, neural, and environmental dynamics: a robotic case study. Adapt. Behav. **10**, 223–241 (2002). https://doi.org/10.1177/1059712302919993005

18. Berthouze, L., Lungarella, M.: Motor skill acquisition under environmental perturbations: on the necessity of alternate freezing and freeing of degrees of freedom. Adapt. Behav. **12**, 47–64 (2004). https://doi.org/10.1177/105971230401200104

19. Hardman, D., George Thuruthel, T., Iida, F.: Towards growing robots: a piecewise morphology-controller co-adaptation strategy for legged locomotion. In: Mohammad, A., Dong, X., Russo, M. (eds.) TAROS 2020. LNCS (LNAI), vol. 12228, pp. 357–368. Springer, Cham (2020). https://doi.org/10.1007/978-3-030-63486-5_37

20. Zhu, J., Rong, C., Iida, F., Rosendo, A.: Scaffolded Learning of Bipedal Walkers: Bootstrapping Ontogenetic Development. bioRxiv. (2020). https://doi.org/10.1101/2020.10.03.324632

21. Ha, D.: Reinforcement learning for improving agent design. Artif. Life. **25**, 352–365 (2019). https://doi.org/10.1162/artl_a_00301

22. Pratt, J., Dilworth, P., Pratt, G.: Virtual model control of a bipedal walking robot. In: Proceedings of International Conference on Robotics and Automation, pp. 193–198 (1997)

23. Robotics, C.: CoppeliaSim. https://www.coppeliarobotics.com/

24. James, S., Freese, M., Davison, A.J.: Pyrep: Bringing v-rep to deep robot learning. arXiv Prepr. arXiv1906.11176 (2019)

25. Stanley, K.O., Miikkulainen, R.: Evolving neural networks through augmenting topologies. Evol. Comput. **10**, 99–127 (2002)

26. Chervenski, P., Ryan, S.: MultiNEAT, project website (2012). http://www.multineat.com/

27. Smith, R.L.: Open Dynamics Engine. https://www.ode.org/

28. CESGA: CESGA. Centro de Supercomputacion de Galicia. http://www.cesga.es/

29. Abdi, H.: Holm's sequential Bonferroni procedure. Encycl. Res. Des. **1**, 1–8 (2010)

30. Ahmad, H., Nakata, Y., Nakamura, Y., Ishiguro, H.: PedestriANS: a bipedal robot with adaptive morphology. Adapt. Behav. 1059712320905177 (2020). https://doi.org/10.1177/1059712320905177

Ethical Challenges from Artificial Intelligence to Legal Practice

Miguel Ramón Viguri Axpe[✉]

University of Deusto, Avenida de Las Universidades, 24, 48007 Bilbao, Bizkaia, Spain
mrviguri@deusto.es

Abstract. The increasing implementation of software and data analysis tools based on Artificial Intelligence in the field of law, raises technical and ethical questions regarding the legal nature of the actions that take place in an area of eJustice. In the first place, the ethical problems derived from Jurimetrics appear: the increasing capacity to detect hidden patterns between the data of sentences and procedural forms, allow establishing statistical guidelines of reliability in terms of the application of a certain line of defense or another. This endangers the duty of every legal agent (not only of the judges) to ensure justice, turning the orientations and the choices of argumentative strategies into a commercial practice that distorts the ethical nature of the legal profession. On the other hand, the irruption of new communication technologies and data analysis can modify the conditions of establishment and development of both the procedures and the judicial process itself (both civil and criminal). Finally, AI confronts us with a series of ethical problems derived from the predictive function applied to research. This work will try to briefly show these issues and conclude with a series of ethical indications or recommendations to avoid dangers and turn digitization into a tool at the service of the legal operator and at the service of defending the rights and dignity of people.

Keywords: Digitization of justice · Jurimetrics · Ethics of algorithmic transparency

1 Introduction. From Justice to eJustice

The Internet has created a digital space in which it is necessary to rethink our form of social organization and its meanings, also with regard to the legal dimension that everything social entails. Legal actors have radically modified their way of operating in recent years, and this modification continues at an unstoppable pace.

The appearance of expert systems and later software to argue writings, requests or resolutions, is a sign of the change in the legal world, with the incorporation of new information processing technologies. This question is deeply developed by Barona [5].

Legal actors find invaluable help in the digital environment to carry out many of their tasks: preparing lawsuits, complaints, a predictability reliable enough to guide the defense strategy of clients, information on the degree of estimation or dismissal according to the type of matter by judges, courts, etc.

© Springer Nature Switzerland AG 2021
H. Sanjurjo González et al. (Eds.): HAIS 2021, LNAI 12886, pp. 196–206, 2021.
https://doi.org/10.1007/978-3-030-86271-8_17

Lawyers now routinely use all kinds of legal argumentation and analysis software. Machine Learning, Big Data, Data Mining, have created a jurimetry that has substantially changed the way of approaching any matter compared to just 10 years ago.

The unstoppable reality is that eJustice is penetrating all legal areas, including the estimation of the need for new legal norms, as well as the overall meaning of the same legal system. This obviously requires a legal rethinking as a whole, from the point of view of the responsibility enforceable for the use of such systems [18, 22].

All this technological development, as well as its very rapid implementation, raises problems that imply ethical deliberation, since the legal system, within the framework of a State of Social and Democratic Law, is the fundamental system for the protection of personal and social rights and freedoms. Any change that may affect the way of establishing procedures and even legal processes -like the introduction of Smart Justice- may affect legal certainty and, therefore, the safeguarding of rights, integrity and the dignity of individuals and groups [3].

2   Objectives

The objectives of this article are: a) to identify some of the different Artificial Intelligence tools that are commonly used in the legal field; b) analyze the potential danger of said tools for the integrity of the principles of the procedure and the judicial process and c) summarize the ethical issues related to these problems and their possible courses of action, and understand them as scientifically approachable problems by AI.

3   Methodology

At first, we will study the main digital techniques of legal analysis, based on Artificial Intelligence, that have been introduced in law in recent years. We will establish a first legal and ethical impact of these tools.

Second, we will analyze the result of the application of AI to the legal field. We will see that these new forms of eJustice can modify the principles of the procedure and the legal process, in such a way that they can cause defenselessness or undermine the rights of people.

Third, we will show the positive potential that AI has and suggest some research lines of the same in the face of the ethical problems raised.

We will end with some conclusions as a summary of the different ethical issues and proposals or courses of action that have been appearing.

4   Developing. The Application of Algorithms in Justice

We are witnessing a great increase in the use of algorithmic systems, based on AI systems, among legal professionals, which increase the capacity for work, documentation and estimation of possible legal actions. This implies a profound change in day-to-day legal practice, in the way of advising, in the way of writing an opinion, etc. [8].

The factors that have driven increased attention to AI in the legal world can focus: on technological advances in relation to Machine Learning, natural language processing, Big Data and argumentative technology systems. On the other hand, the growing availability of legal data on the internet, with the implementation of the Electronic Legal File, and the improvement of the technological quality of digital platforms are greatly favoring the development of the eJustice system.

Initially, expert systems were introduced, which basically offered interesting advisory information for lawyers [25]. These systems expedited the task of gathering information that could be useful for the presentation of the case.

But intelligent software has created a legal market in which tools are offered that obtain, analyze and screen information and that offer arguments and strategies; that's mean, legal solutions.

Then the legal systems appeared an analysis to apply models derived from legal logic to legal norms [12]. These systems use previous sentences on identical or similar matters, data on judges who handed down the sentences, courts, prosecutors who intervened, action by the administration, etc. Based on these analyzes, they offer courses of action to the legal professional.

But we are already on another level. These products of legal argumentation, until relatively recently, were considered the exclusive prerogative of human reasoning. For this reason, some authors, like Brazier [9], warn about the need for a regulation of eJustice carried out by jurists, who can guide those who design and develop computer systems. Interdisciplinarity, in the field of technologies applied to the field of justice, is the only thing that guarantees that digital tools work in accordance with the legal and ethical principles that are the basis of the Law.

The concern for the explanation of the ethical dimension of these products goes back a few decades. Norbert Wiener [28] himself considered the application of this new technology to Law, but with enormous concern about the irreversibility of the consequences, not only technical, but also ethical and moral, posed by digital development. Therefore, he advocated joint work between jurists and technologists.

4.1 Artificial Legal Intelligence

Artificial Legal Intelligence is defined by Gray, P. M. [16] as the computer simulation of any of the theoretical practical forms of legal reasoning, or the computer simulation of legal services involving the communication of the legal intelligence. Initially, this computational simulation of theoretical-practical legal reasoning was based on computational models, such as the IBM's Debater Program, which could provide answers using existing legal solutions, but could not justify or offer any legal reasoning for the answer given. They were simply programs that gave mechanical information, based on probabilistic calculation.

But the new computational models do explain the proposals with reasons in which the interpretation of the rules and their meaning in certain contexts is essential [6] and could be used for the resolution of disputes [2].

4.2   Jurimetrics. Legal Decision Support System

It is a logical analysis of the Law in order to apply models derived from logic, and the anticipation of future sentences to legal norms [12].

The legal tools carry out a statistical and predictive jurisprudential analysis, offering enormous information about the modus operandi of all the legal actors involved in the case: lawyers, magistrates, courts, public organisms, etc.

Jurimetrics also raises ethical problems. Does the analysis of the judgments of a judge, with the purpose of establishing a personal profile (in terms of its legal principles, its legal argumentation strategy, its scale of moral values that has an influence on the argumentation of the judgments and legal decisions), violate the right to privacy and intimacy?

In France the legislation has limited the use of these tools, precisely linking said limitation to the protection of personal data. Thus, article 33 of the French Justice Reform Law imposes a penalty of up to five years of deprivation of liberty on those who reuse the personal data of judges in order to analyze, evaluate, compare or predict their professional practices.

Another ethical problem that arises is that the loss of critical reasoning is encouraged that, without questioning the issues, only attends to the percentage or the probabilistic calculation that would support a line of action. This, in addition, makes the possibility of injustices derived from possible biases of the algorithms, which could generate inequalities and discrimination, more dangerous [26].

On the other hand, there is the threat of exchanging the own goods of the legal professions for their external goods. This is always a temptation of the legal profession: to turn the lawyer into a machine to win lawsuits, always opting for the course of action that offers the most promising practical results (from the point of view of optimizing profits and minimizing losses). The continued and generalized use of these technologies (based on statistical factors) can generate a mental habit that guides the option for procedural truth (and not always, since, in many cases, the lawyer will opt for a simply pragmatic line of action, which is based more on the presumed personal inclinations of the judge on duty -suggested by the jurimetric software from precedents- than on the argumentation supported by evidence), ignoring the decisive importance of the material truth of the facts and the moral values involved in.

Let us not forget that the Code of Ethics for Spanish Lawyers maintains in its prologue that "the social function of the legal profession requires establishing deontological standards for its exercise". In other words, professional deontology is justified not so much by the interests of the client, but by the ethical-social purpose of the profession. This purpose implies the defense of the fundamental values of the legal profession: justice, equality and human dignity.

Specifically, paragraph 4 of the aforementioned code establishes the primacy of the defense of human dignity as the supreme good, ethically overcoming the professional task reduced to maximizing the interests of the client. To such an extent that the protection of the value of justice is no longer the prerogative of the judge, but also of the lawyer, being obliged to warn his client about the need to also protect the rights of the opposing party.

In paragraph 20 of the Preamble it is said that the effective defense of individual rights, whose recognition and respect is the backbone of the rule of law, is the high function that society has entrusted the lawyer, so that justice and equality prevail in the same.

4.3 Legal Expert Systems

They are something more with respect to legal systems. They are fed with comprehensive data from five areas: law, jurisprudence, legal literature, expert knowledge and legal meta-knowledge. The most difficult to define and formulate is the last, since the way to achieve it is complex, through a combination of general rules and strategic rules.

They were called Expert Systems because they were built with the assistance of human experts in order to act as experts in certain fields, becoming intelligent assistants. They have three characteristics: a) Transparency, because they can generate explanations of the reasoning that lead them to the proposals or conclusions; b) They are heuristic, because they reason from a type of circumstantial knowledge, which is extracted from experience in a certain field or issue and which is not formalizable a priori; c) Flexibility, because these systems allow modifications in their databases and processing styles.

The evolution of these expert systems produces, therefore, a paradigm shift in the operation of legal operators that may have ethical consequences and, therefore, requires collaboration between scientific-technical specialists and specialists in the legal field.

5   Results: First Ethical Questions Derived from the Use of AI in Legal Procedures and Processes

To carry out an adequate ethical evaluation on the impact of digitization in the legal world, it will be necessary to elucidate how the foundations or principles of the procedural and procedural model of the legal system may be essentially or accidentally affected by said digitization [4].

This is especially important. If the development of the eJustice were to alter these principles in any way, the meaning of all the rules and the legal system itself could be affected, which could not only generate legal uncertainty, but also defenselessness, as well as serious moral problems [19].

5.1   Publicity Principle

The general publicity principle is consistent with the digital field in terms of the possibility of openly attending an electronic procedure in progress.

But, on the other hand, the publicity of the processes is linked to the political and ethical question of transparency and democratic control of institutions, which is directly linked to the ethical question of transparency and control of artificial intelligence systems. By virtue of the ethical requirement of transparency, a series of measures must be incorporated that allow to know and examine the structure, data, automation and operation. In other words, the reason for the action of an electronic judicial agent can be publicly known.

5.2 Principle of Contradiction and Right of Defense

The principle of contradiction implies guaranteeing those who intervene as parts in the process the possibility of knowing all the materials that may influence the judicial decision, in such a way that they can oppose the validity of the evidence of the opposing part, supplying new material if necessary, thus guaranteeing their own defense.

The digitization of the process here implies an ethical principle: transparency. Helplessness may occur when computer systems (software) whose data are unknown, or whose algorithmic processing principles, which can be the basis of argumentation and judicial resolution, are unknown by the parts.

5.3 Principle of Procedural Equality or Equality of the Parts

Equality refers to the fact that all litigants have the same opportunities to act within the process, without any one being in a position of inferiority with respect to the others.

But if the litigation occurs in an environment of justice, there are conditions that can influence a citizen inequality with respect to the ability to protect their rights. It may happen that not everyone has the tools that allow them to access digitized legal systems, or that they do not have sufficient training, or technological ability, to guarantee that the use of digital tools will not become a difficulty for the defense and the exercise of their procedural rights [1].

5.4 Influence of AI on the Principles of Criminal Process

According to Gómez Colomer [15], the use of digital tools and intelligent software implies an urgent transformation of the Criminal Procedure Law.

Given the exponential growth of criminal litigation and the arduous nature of its processing, digital technologies are presented as the optimal remedy. Through these, the modus operandi of the protagonists of criminal prosecution, evidence, the ability to decide, legal argumentation, criminal execution, etc., is reconsidered [23].

In relation to the means or instruments to be used in criminal proceedings, national laws have guaranteed the development of electronic or technological investigation systems, giving them a value of reliability.

Due to this guarantee, AI tools are not mere instruments, but rather they become an intelligent source capable of generating a fundamental change: raising some results obtained through these tools to a proof category (which is equivalent to granting value of proof a statistical analysis of data, regardless of the algorithmic system used to process them).

5.5 Preventive and Predictive AI

Derived from the preventive function, another function of AI emerges: the predictive one. A predictability that, at the moment, is offered in all areas of people's lives, powerfully affecting them.

The basic technique is Data Mining. It is basically a search for analogous previous scenarios, in order to get indications of a new possible infraction. This technique allows finding patterns and summarizing large volumes of data, organizing them by different analogy values and structuring categories [17]. But not only that, given the analogous scenarios found in the form of patterns, these intelligent programs suggest possible solutions, according to Sadin [27].

But this does not eliminate the possibility of the appearance of algorithmic biases that condition the selection of the data with which the machine will argue in one sense or another [22]. It is worth, for example, wondering if a computational tool that allows assessing the credibility or not of the people who give testimony is only a tool to help the judge, or if it transforms the very meaning of judicial argumentation and, therefore, the nature and functions of judges and magistrates [5, 10].

6   Discussion. Future Challenges: Understandability, Intepretability, Explicability, Transparency, and Algorithmic Traceability as Ethical Challenges

Far from falling into a simplification of digitization and AI as something only problematic, it is necessary to highlight the positive interaction that has occurred between AI and all its fields of application (as well as - we have already said it above - the concern ethics of the creators and developers of AI from the beginning).

In addition to this, in this section we want to emphasize that ethical questions are not mere philosophical problems, but must be seen as technical challenges that positively influence scientific and human progress. As Alejandro Barredo and other authors say in a magnificent article [6], Artificial Intelligence has spent decades researching strategies and tools to favor the development of comprehensible, explainable, interpretable, transparent and traceable computational models. These scientific developments are key to enhancing cooperation between human experts in the different disciplines where AI is applied (in this case legal experts) and the scientists and technologists who created it. This is fundamental: AI not only generates ethical problems, but new areas of knowledge and reality, necessary to solve these problems, as well as for human development.

6.1   A Relationship Between the Jurimetrics with the Global Framework of the eJustice that Guarantees the Nature of the Latter as an Effective Attorney of Justice

We could say that jurimetrics is the application of artificial intelligence and machine learning to traditional legal and jurisprudential search engines, obtaining new functionalities that give rise to the development of what has also been called predictive justice. Jurimetrics is an innovative advance in statistical and predictive jurisprudential analytics, systematizing and exhaustively extracting the intelligence that resides in judicial decisions from all instances and jurisdictional orders in Spain (from the First Instance to the Supreme Court).

That these tools can emulate and make legal decisions does not mean that they can do so with the guarantees required by the judicial process. It is necessary to delimit and clarify the how and explore what tools and functionalities could support and strengthen the Administration of Justice, in order to support legal operators, although its potential goes far beyond supporting human judicial judgment [11].

In this sense, the scientific challenge is that the legal software is understandable; that is to say, that the learning results of the ML can be represented as an intelligible description for the human being, at least for the experts. It is important to emphasize that such a description does not necessarily have to be easy. Understanding the meaning of certain explanations, depending on the complexity of the model, may require the knowledge of an expert (an expert in legal matters, for example) [6].

6.2 Respect for Procedural Data by Means of Big Data Techniques that Respect the Rights that Are Set Out in Spanish or European Legislation

Data protection regulations establish certain limits. But the key lies in transparency, that is, in the information that is offered in relation to data processing, in determining the bases of legitimation of the treatment and in the establishment of the ways or possibilities of exercising rights against this.

Regarding guarantees, the possibility of monitoring AI systems against the appearance of biases must be guaranteed, and that decisions made by machines must be adjusted to our system of fundamental rights, analyzable and explainable. It cannot be that our system of fundamental rights is the one that adapts to AI, but rather the opposite [13].

The scientific challenge of the analyzability and explicability of the decisions made by AI implies that the model used and its operation is intelligible by itself, without the need for an explanation from an AI expert. In this case, the model is said to be interpretable; that is, its meaning is expressible in natural language and can even be represented graphically to facilitate its understanding.

6.3 Transparency in a Digitized Jurisdictional Framework

By virtue of the principle of transparency, all judicial decisions should be accessible, in such a way that make the reading understandable, and all this with full respect for the protection of personal data.

The use of artificial intelligence tools by public authorities must guarantee transparency, with the ability to explain as far as possible the process followed by the machine to obtain a certain opinion. It should be possible to audit and verify the reliability of algorithms periodically [24].

For a model to be transparent, in addition to being interpretable and understandable, it must be traceable. The challenge of traceability is that the sequence of operations carried out by the machine is significantly traceable and traceable. This implies that the algorithmic procedures are carried out in a number of steps that makes it possible for a human or group of humans to verify them. Algorithmic traceability is one of the least developed areas of AI, but one of the most promising and in which more efforts must be made to ensure transparency and control of possible biases [6].

6.4 Establishment of Responsibilities for the People Who Define the Prediction Algorithms on the Prosperousness of a Claim

With the development of eJusticia, the need for a joint work of lawyers and technologists is seen more and more clearly. A technically and ethically reliable algorithm can only be created by an interdisciplinary team that enables the communication of the technological, legal and ethical knowledge involved.

Algorithms are the fruit and projection of the human intellect and therefore must be criticized and validated by humans. If this algorithm is applicable in the legal field, it is necessary in its design phase that the creative team knows the legal field. Hence the importance and need for multi and interdisciplinary teams. The interdisciplinary control of the human that validates the work of the machine is essential [14].

The challenge of forming interdisciplinary teams, made up of experts in Artificial Intelligence and experts in legal systems and legal logic and argumentation, involves the effort to create intelligent software that is understandable and explainable. Explainable means that the interface created by the software, and that it can act as a proxy, is understandable to humans. Although these humans have to be experts [6].

AI can help, through the creation of explicability techniques (XAI), interdisciplinary groups of experts to understand the selection of data that the models perform to produce their output, estimate the performance limits of the model (for example, under what stimulus the decision of the model changes direction), or to simplify the models to a point understandable by the group of experts (in this case, legal experts).

6.5 Irreplaceability of Lawyers or Other Judicial Agents by AI Legal Systems

These applications cannot replace lawyers or judges. In the Charter of Digital Rights itself, the right not to be the subject of a decision based solely on automated decision processes is explicitly mentioned; as well as the right to challenge automated or algorithmic decisions and to request human supervision and intervention. The digitization of eJustice will have to be another means for the empowerment of the human being, not its replacement [20].

For this human supervision to be real and effective, significant scientific and technical work is required to favor the interpretability of the results and of the procedures of the analysis and argumentation models used. That is, the meaning of a certain computational procedure and its result are understandable in natural language by the human.

6.6 Guarantee that the Digitization of the Judicial Process Does not Destroy the Analysis and Legal Argumentation that Any Matter Requires

The technological modernization of eJustice does not replace, as we have already seen, judgment or legal analysis and rigor, but should be seen as an additional complement and aid to legal operators [21]. In the judicial system, the basic model should be total human control over the machine.

We know that humans are fallible, so all human creations will have, at least, the same flaws and biases as their creators. Therefore, supervision will always have to be from human to technology and not the other way around, and it will always have to take place in an interdisciplinary and intersubjective environment (the only thing that guarantees a greater degree or approach to objectivity).

We are again faced with the challenge of algorithmic traceability. Traceability that must be linked to the transparency, understandability and explicability of the model.

7 Conclusions

7.1 Need for Codes of Ethics and Legal Regulations Regarding the Use of AI Technologies

In the field of law, the deontological codes and the regulations attached to them are, in principle, sufficient to guarantee that legal procedures do not alter the property of the Law and legal practice (the protection of individual and social rights and the development of a harmonious and fair coexistence) exchanging them for their external assets (professional success obtained by betting only on the most statistically favorable scenarios to the detriment of material truth and even procedural truth).

7.2 Protection of Privacy

The privacy and the right to privacy of legal agents (judges, courts, lawyers, prosecutors, etc.) before legal procedures based on Data Mining and Machine Learning, must be also protected by the different regulations, protocols and guidelines proposed in the European and Spanish legal field.

7.3 Algorithmic Ethics and Research in Explanatory AI (XAI)

Finally, from the ethical point of view, the safeguarding of the guarantees of legal security, impartiality and equality, against possible alterations in the legal principles of the procedures and processes (modifications in the lines of argument, selected data, data that are considered as tests), translate into a scientific challenge: the development of AI tools and software that guarantees the transparency and explicability (XAI) of algorithmic models. This implies reinforcing research in areas of AI such as the interpretability, transparency and traceability of the processes carried out by the machine.

In this way, the formation of interdisciplinary teams of AI experts and legal experts capable of interacting with the machine is made possible, guaranteeing its use as a tool for guiding and facilitating legal argumentation, at the service of the rights of individuals, groups and institutions.

References

1. Mekki, S.A.: Garantías frente a eficiencia. Es lo racional siempre razonable?. In: (Collective Work) Justicia: garantías versus eficiencia?, pp. 31–60, Tirant lo Blanch, Valencia (2019)
2. Ashley, A.D.: Artificial Intelligence and Legal Analytics. Cambridge University Press, Cambridge (2017)
3. Vilar, S.B.: Justicia Penal, Globalización y Digitalización. Thomson-Reuters, Santiago de Chile (2018)

4. Barona Vilar, S.: Cuarta revolución industrial (4.0) o ciberindustria en el proceso penal: revolución digital, inteligencia artificial y el camino hacia la robotización de la justicia. Revista jurídica digital UANDES 3 (1), 1–17 (2019)
5. Vilar, S.B.: Algoritmización del derecho y de la justicia. De la Inteligencia Artificial a la Smart Justice. Tirant lo Blanch, Valencia (2021)
6. Barredo, A., et al.: Explainable Artificial Intelligence (XAI): concepts, taxonomies, opportunities and challenges toward responsible AI. Inf. Fusion, **58**, 82–115 (2020)
7. Bench-Capon, T., Atkinson, K.; Wyner, A.Z., et al.: A history of AI and Law in 50 papers: 25 years of the international conference on AI and Law. Artif. Intell. Law **20**(3) (2012)
8. Bex, F., Prakken, H., Van Engers, T., Verheij, B.: Introduction to the special issue on artificial intelligence for justice. Artif. Intell. Law, **25**(3), 1–3 (2017). https://doi.org/10.1007/s10506-017-9198-5
9. Brazier, F., Oskamp, A., Prins, C., Schellekens, M., Wijgaards, N.: Law-abiding and integrity on the internet: a case for agents. Artif. Intell. Law **12**, 5–37 (2004)
10. Mata, F.B.: Robótica y derecho procesal: retos inminentes. In: Mata, F.B. (Ed.) Fodertics 6.0. Los nuevos retos del derecho ante la era digital, pp. 255–264, Comares, Granada (2017)
11. Hueso, L.C.: Ética en el diseño para el desarrollo de una inteligencia artificial, robótica y big data confiables y su utilidad desde el derecho. Rev. Catalana de Dret Públic 58, 29–48 (2019)
12. Fernández Gómez, L.: La revolución informática, nueva frontera del derecho. Temas de Filosofía del Derecho UCAB Publicaciones, Caracas **13**(2), 137–145 (1988)
13. Ferrer, C.: Cómo cumplir el RGPD si manejas datos biométricos?. In: Centro de Prevención de Delitos Informáticos de Canarias (2019)
14. Fischer, J.M., Ravizza, M.: Responsibility and control: a theory of moral responsibility. Cambridge University Press, Cambridge (2000)
15. Colomer, J.L.G.: Los retos que debe afrontar una nueva Ley de Enjuiciamiento Criminal en España. Otrosí.net, Ilustre Colegio de Abogados de Madrid, 4, 1–22 (2020)
16. Gray, P.N.: Artificial Legal Intelligence. Dartmouth Publishing Company, Brookfield (1997)
17. Han, J., Pei, J., Kamber, M.: Data Mining: Concepts and Techniques. Elsevier, EEUU (2011)
18. Delgado, I.M.: Naturaleza, concepto y régimen jurídico de la actuación administrativa automatizada. Rev. de Administración Pública 180, 353–386. Dialnet (2009)
19. Martín Diz, F.: Justicia digital post-covid19: El desafío de las soluciones extrajudiciales electrónicas de litigios y la inteligencia artificial. Revista de Estudios Jurídicos y Criminológicos 2, 41–74. Dialnet (2020)
20. Llinares, F.M.: Inteligencia Artificial y Justicia Penal: más allá de los resultados lesivos causados por robots. Rev. de Derecho penal y Criminología 3(20), 87–130. Dialnet (2018)
21. Mommers, L.: Legitimacy and the virtualization of dispute resolution. Artif. Intell. Law **13**, 207–232 (2005)
22. Neuhäuser, C.: Roboter und moralische verantwortung, In: Hilgendorf, E. Robotik im Kontext von Recht und Moral, pp. 269–287. Nomo Verlag, Baden-Baden (2014)
23. Nieva Fenoll, J.: Inteligencia Artificial y Proceso Judicial. Marcial Pons, Madrid (2018)
24. Pasquale, F.: The Black Box Society. The Secret Algorithms That Control Money and Information. Cambridge University Press, Cambridge (2015)
25. Pietrosanti, E., Graziadio, B.: Advanced techniques for legal document processing and retrieval. Artif. Intell. Law **7**, 341–361 (1999)
26. Rose, W.: Crimes of color: risk, profiling and the contemporary racialization of social control. Int. J. Polit. Cult. Soc. **16**(2), 179–205 (2002)
27. Sadin, E.: La inteligencia artificial: el superyó del siglo XXI. Nueva Sociedad **279**(1), 141–148 (2019)
28. Wiener, N.: Some moral and technical consequences of automation. Science **131**(3410), 1355–1358 (1960)

Visual Analysis and Advanced Data Processing Techniques

Classification of Burrs Using Contour Features of Image in Milling Workpieces

Virginia Riego del Castillo⬤, Lidia Sánchez-González(✉)⬤,
and Claudia Álvarez-Aparicio⬤

Departamento de Ingenierías Mecánica, Informática y Aeroespacial,
Universidad de León, 24071 León, Spain
{vriec,lidia.sanchez,calvaa}@unileon.es

Abstract. Fulfilment of quality standards in manufacturing processes is an essential task and often increases production costs. Specifically, the appropriate edge finishing of machine workpieces is one of the requirements so as to avoid the presence of burrs. In this paper, a vision-based system that employs contour features is proposed to detect and classify images of edge workpieces. In the first stage, we locate the region of the image that contains the edge of the part and in the second one, more precised operations provide detailed information in order to detect the edge type of the machined part. Calculated feature vector feeds supervised classifiers to determine the best approach to this dataset. Random Forest Classifier yields the best results obtaining a 90% of precision, recall and F1-score in the test dataset, which satisfies the experts demand to these processes.

Keywords: Quality estimation · Milling machined parts · Burrs in workpiece · Burr classification · Contour features

1 Introduction

The fourth industrial revolution is a fact that allows us to improve the control of quality standards. It assists operators in decision making in manufacturing processes, which need to achieve high quality standards while offering a competitive price in the market [8].

The ISO 13715 standard [18] defines burr as *"rough remainder of material outside the ideal geometrical shape of an external edge, residue of machining or of a forming process"*. The study of the edge finishing of milling workpieces is an active research field due to the importance that the presence of burrs represents for cutting costs. Following the definition, [7] proposes a 3D scanning of the surface and compares with the ideal one to detect anomalies in the surface of the piece. Other authors study different mechanisms which produce burrs and how to remove them [1]. In [14], different tests were made to reduce burr formation in the process of hole making in drilling aluminium sheets.

© Springer Nature Switzerland AG 2021
H. Sanjurjo González et al. (Eds.): HAIS 2021, LNAI 12886, pp. 209–218, 2021.
https://doi.org/10.1007/978-3-030-86271-8_18

There is evidence that micro-milling processes are affected by variables such as cutting force, vibration, burr formation, and surface quality [2]. For that reason, some researches have proposed how to detect burrs in drilling processes studying the internal signal that belongs to the torque of electro-spindle during the drilling of a hole [9].

Manual inspection presents several drawbacks since human eye involves some limits; they can be minimised with the use of microscopical images and computer vision. In this sense, [12] proposes a burr contour tracking method (BCTM) that employes the maximum histogram valley to determine the most adequate threshold value to binarize the image, followed by morphological operations with small kernels. Similarly, an image processing method to detect edges using an iterative method to calculate the optimum threshold that allows us to distinguish between the background and the workpiece is proposed in [6]. In addition to this, other researches have studied the wear of microdrill tools measuring the flank wear using a machine vision system [16], finding higher wear at the corners of the cutting lips because these are the locations of the highest cutting speed and at the chisel edge as it is the first contact point.

In this paper, we propose a different method to classify burrs in contrast to [5], which uses linear regression and the percentage of white pixels over the height of the image as is detailed in [13]. The method proposed in this paper computes contour features of the ending workpiece that represent the type of burr in order to classify a bigger dataset of images, from 126 images to 1073. The proposed method eliminates operator subjectivity and improves the results of quality processes with lower costs due to the possible automation of the process, and consequently, the reduction of the inspection time.

This paper is structured as follows. Section 2 explains the computer vision method and how the image is processed to obtain the feature vector. Section 3 details the experiments carried out to validate the method and discuss the obtained results. Finally, Sect. 4 presents the conclusions reached from the experimental results.

2   Inspection Method

In [5], a method to detect burrs in workpiece images is proposed; that method uses computer vision to analyse the images obtained by an industrial boroscope linked to a microscope camera.

In this paper the same acquisition system is used to capture the images. Those images are processed in order to detect the region that contains the edge finishing and, subsequently, a feature vector of the section is computed. A flow chart of the entire process is shown in Fig. 1.

As it can be seen in Fig. 1, firstly, the original image is processed as it is explained in Sect. 2.1 so as to remove the noise in the image and normalised its contrast. Then, the edge finishing region of the workpiece is located by following the procedure detailed in Sect. 2.2. In this stage, the goal is to determine where the part is, distinguishing it from the background, and to locate the region that

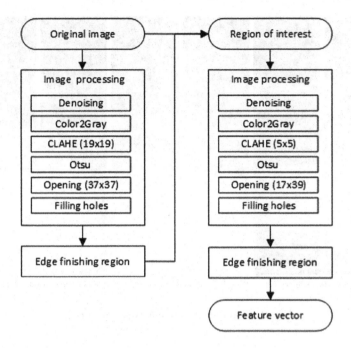

Fig. 1. Scheme of the pipeline to convert each image in a feature vector.

includes its finishing to analyse it deeply. After that, the obtained region of interest is processed with smaller kernels to obtain as much detailed information of the edge as possible in order to be able to compute a feature vector that represents precisely the type of the burr that the part presents, as it is explained in Sect. 2.3.

Regarding the burr types, three categories were considered and each image was assigned to one category depending on the level of imperfections. Therefore, when there is no imperfections the part presents a knife-type burr (K) that means a clear finishing as it can be seen in Fig. 2 (a). However, if the edge presents small splinters the category is named as saw-type burr (S) since it is more irregular, as Fig. 2 (b) shows. Finally, if the edge finishing has suffered a large deformation such as the ones shown in Fig. 2 (c), the part is assigned to the burr-breakage (B) category [5].

2.1   Image Processing

Once the images are acquired, they are transformed into binary images so as to distinguish the piece from the background.

In the first place, the noise of the original image (Fig. 3 (a)) is removed with the Non-local Means Denoising algorithm considering a 7-pixel block size to compute weights and a 21-pixel window size to compute the weighted average [4]. Then, the resultant image (Fig. 3 (b)) is converted to gray scale. After that,

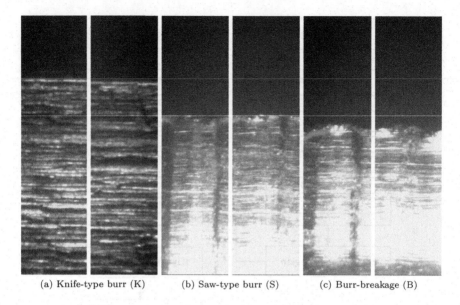

(a) Knife-type burr (K)   (b) Saw-type burr (S)   (c) Burr-breakage (B)

Fig. 2. Example of different types of burrs.

the over-brightness that usually presents metal surfaces is corrected by applying a Contrast Limited Adaptive Histogram Equalization (CLAHE) with a equalization Grid Size of 19×19 (Fig. 3 (c)). Next, the binary image is obtained using Otsu threshold (Fig. 3 (d)). In order to remove small points on the object, opening morphological operations are carried out such as erosion and dilation (opening) with a 37×37 pixel kernel (Fig. 3 (e)). Finally, holes are filled by searching the contours with an area lower than the 5% of the total pixels of the image (Fig. 3 (f)). The resulting image presents two different areas, a white region that represents the background and a black region where the part is. Next steps of the proposed method are focused on the region located at the limit of the two areas that presents a rectangular shape (Fig. 3 (g)).

Once the mask of the part is determined, next stage deals with the region of interest that encloses the edge finishing in order to detect the contour accurately. Although the stages are similar, they now are focused on the contour of the part so some parameters change. In that sense after denoising the original image (Fig. 4 (a) and (b)), the CLAHE equalization is carried out now with a 5×5 equalization Grid Size (Fig. 4 (c)). By applying again the Otsu threshold the binary image is obtained (Fig. 4 (d)) and then small points are removed with opening operation with a kernel of 17×39 pixels (Fig. 4 (e)). Final step consists of filling small holes in the image (Fig. 4 (f)). As a result, the region that contains the edge finishing is located and is used to determine the contour of the part (Fig. 4 (g)) as the next Section explains.

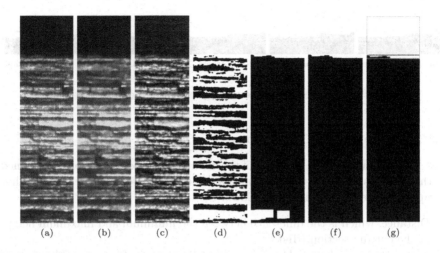

Fig. 3. Process to detect the edge finishing of the workpiece in the original image.

Fig. 4. Process to detect the end of the workpiece in the region of interest.

2.2   Region of Interest Identification

Once we have applied the image processing detailed in the previous section, the edge finishing is located at the image and the next step consists of determining the contour of the part with a high accuracy.

Since the obtained binary image differentiates clearly between the background and the object (Fig. 5 (a)), it is used to calculate the biggest contour of the processed image. Figure 5 (b) shows such contour in green. To obtain the contours, the existing edges in the image are calculated and the continuity of the pixels that are edges is followed [17]. Then the corners of the contour are identified by searching those points which x coordinate is 0 and coincides with the width of the image, Fig. 5 (c). Corners are used to obtain the height of the ending of the workpiece in the image, by calculating the maximum distance from the corners to the boundaries of the image. The calculated height is employed to divide vertically the image and calculate the percentage of white pixels, which is used to distinguish between the section that corresponds to the workpiece and the background.

From the biggest contour, the line that generates the edge finishing of the workpiece is extracted (Fig. 5 (d)), that yields the minimum and maximum height, which is shown in red and corresponds with the region that contains the edge finishing of the image (Fig. 5 (e)).

(a) Binary image   (b) Contour   (c) Corners   (d) Line   (e) Cutting section

Fig. 5. Process to calculate the cutting section in the processed image.

2.3 Feature Vector Calculation

The line generated by the finishing of the workpiece, whose detection is explained in the previous section, is used to calculate the feature vector. The features that compose the vector are detailed below:

- Distance, which is the difference between the minimum and maximum height of the cutting section (**dist**).
- Slope, which is calculated by doing Linear Regression of the line of the contour (**slope**).
- Percentage of white pixels at the limit of the cutting section (**pwp limit**).
- Percentage of white pixels in the middle of the cutting section (**pwp middle**).
- Extent, which is the ratio of the contour area to the bounding rectangle area. The considered contour is the one made with the line joined with the corners of the cutting section (**extent**).

2.4 Classification

After feature extraction, different supervised classification algorithms are compared:

- **Adaptive Boosting** [10], known as AdaBoost, trains one weak learner and the fails of the first model are used to train another weak learner. We can establish the maximum number of learners (n_estimators) that can be generated during the training.
- **Gradient Boosting** [11], follows an iterative procedure by adding a new weak learner to improve on the erroneous data of its predecessor. We can establish the maximum number of learners (n_estimators) and also the maximum depth of nodes of the tree (max_depth).
- **Random Forest** [3], is an ensemble of decision trees by building multiple decision trees that are merged to get a more accurate and stable prediction. We can configure the maximum number of trees in the forest (n_estimators), how to measure the quality of a split (criterion) and the maximum depth of the tree (max_depth).
- **K Neighbours**, optimises the position of K centroids to group the neighbour points of the same class. We can configure the number of centroids (n_neighbors) and the function to measure the weight (weights).

Training is based on Grid Search with ten-fold cross validation and the mean accuracy measures the performance of each classifier.

3   Experimental Results

In order to validate the proposed method, not only do experiments detect the finishing of the part but also the existence of any imperfections along its edge. That is because the existence of these imperfections involves the non-fulfilment of the desired quality requirements.

For experiments, a dataset of 1073 images classified by an specialist are considered. Those images were acquired with the vision system detailed in Sect. 2 that employs a microscope camera and an industrial boroscope. The dataset comprises 429 images of K-type, 400 of S-type and 244 of B-type.

For each image, the feature vector is calculated as is explained in Sect. 2, then a pairs plot is obtained in order to visualise the behaviour of the selected features in burr classification. As Fig. 6 shows, there is an important overlapping area among the distributions.

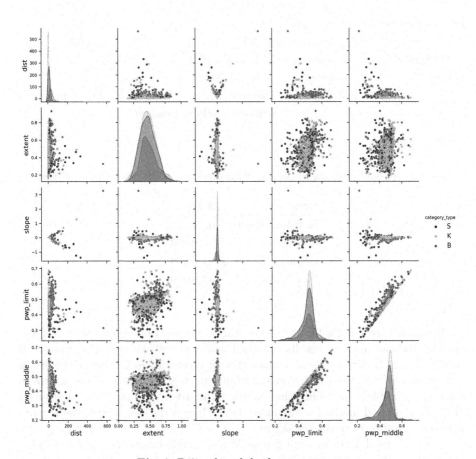

Fig. 6. Pairs plot of the feature vector.

By using a random seed, the dataset is split into two subsets, one formed by the 80% of the samples is used for training and the remaining 20% or the images is considered for testing. Training dataset was trained with different parameters shown in Table 1. Results of the Grid Search are shown in Table 2, which establishes the best results when Random Forest Classifier is used with the following parameters: a maximum depth of the tree of 14, a number of trees in the forest of 95 and entropy as criterion to measure the quality of a split.

Table 1. Hyperparameters used in Grid Search

Model	Parameters	Values	Best value
Ada Boost	n_estimators	50, 100	100
Gradient Boosting	max_depth	5, 7	5
	n_estimators	50, 100	100
Random Forest	max_depth	12, 14, 16, 18	**14**
	n_estimators	85, 95	**95**
K-Neighbors	n_neighbors	3, 4, 5, 6, 7, 8, 9	8
	leaf_size	5, 10, 15, 20, 25	5
	Weights	Distance (inverse of their distance) Uniform (uniform weights)	Distance

Table 2. Ten-fold cross validation mean accuracy results for the training dataset

Model	Accuracy
Random Forest	**0.749480**
K-Neighbors	0.749453
Gradient Boosting	0.749439
Ada Boost	0.719220

Random Forest Classifier, as the best model, was trained with the complete training dataset and the results obtained with the test dataset are gathered in Table 3. Final results show more than 90% of precision, recall, and F1-Score in the average of the three categories. Comparing with [5], results are improved from 76% to 90% which fulfil the quality standards in industry. In addition to this, the proposed method is validated with a bigger dataset, which yields more robustness to the burr classification.

We have used Python with OpenCV and Scikit-Learn. For more detailed information, the code of the complete experiment can be found in [15].

Table 3. Random Forest Classifier results for the test dataset

Category	Precision	Recall	F1-Score	Number of images
Knife-type burr (K)	0.8571	0.9630	0.9070	81
Saw-type burr (S)	0.9315	0.8095	0.8662	84
Burr-breakage (B)	0.9216	0.9400	0.9307	50
Average	0.9034	0.9042	0.9013	215

4 Conclusions

The finishing of the machined parts should present the cleanest possible edges to reduce the cost of removing the generated burrs. Detection of burrs is an important aspect in milling in order to satisfy the industrial quality standards. In this paper, a proposed method to detect and classify the edges of workpieces automatically is proposed. After generating a feature vector for each image, composed by the contour features of the edge of the part, different classification algorithms were tested. Random Forest yields the best results with a 90% of precision, recall and F1-score in the test dataset. These results improve the ones achieved by existing methods. In addition to this, this method provides an automatic quality control tasks what involves not only that the required standards are achieved but also the reduction of personnel costs.

Acknowledgements. We gratefully acknowledge the financial support of Spanish Ministry of Economy, Industry and Competitiveness, through grant PID2019-108277GB-C21. Virginia Riego would like to thank Universidad de León for its funding support for her doctoral studies. Claudia Álvarez would like to thank the regional Government of Castilla y León for its funding support under the grant BDNS (487971).

References

1. Aurich, J., Dornfeld, D., Arrazola, P., Franke, V., Leitz, L., Min, S.: Burrs-analysis, control and removal. CIRP Ann. Manuf. Technol. **58**(2), 519–542 (2009). https://doi.org/10.1016/j.cirp.2009.09.004
2. Balázs, B.Z., Takács, M.: Experimental investigation and optimisation of the micro milling process of hardened hot-work tool steel. Int. J. Adv. Manuf. Technol. 5289–5305 (2020). https://doi.org/10.1007/s00170-020-04991-x
3. Breiman, L.: Random forests. Mach. Learn. **45**(1), 5–32 (2001). https://doi.org/10.1023/A:1010933404324
4. Buades, A., Coll, B., Morel, J.M.: Non-local means denoising. Image Process. On Line **1**, 208–212 (2011)
5. del Castillo, V.R., Sánchez-González, L., Fernández-Robles, L., Castejón-Limas, M.: Burr detection using image processing in milling workpieces. In: Herrero, Á., Cambra, C., Urda, D., Sedano, J., Quintián, H., Corchado, E. (eds.) SOCO 2020. AISC, vol. 1268, pp. 751–759. Springer, Cham (2021). https://doi.org/10.1007/978-3-030-57802-2_72

6. Chen, X., Shi, G., Xi, C., Zhong, L., Wei, X., Zhang, K.: Design of burr detection based on image processing. In: Journal of Physics: Conference Series, vol. 1237, vol. 032075, June 2019. https://doi.org/10.1088/1742-6596/1237/3/032075

7. Claes, K., Koninckx, T., Bruyninckx, H.: Automatic burr detection on surfaces of revolution based on adaptive 3D scanning. In: Fifth International Conference on 3-D Digital Imaging and Modeling (3DIM 2005), pp. 212–219 (2005). https://doi.org/10.1109/3DIM.2005.21

8. Dornfeld, D., Min, S.: A review of burr formation in machining. In: Aurich, J., Dornfeld, D. (eds.) Burrs - Analysis, Control and Removal, pp. 3–11. Springer, Heidelberg (2010). https://doi.org/10.1007/978-3-642-00568-8_1

9. Ferreiro, S., Sierra, B., Irigoien, I., Gorritxategi, E.: Data mining for quality control: burr detection in the drilling process. Comput. Ind. Eng. 60(4), 801–810 (2011). https://doi.org/10.1016/j.cie.2011.01.018

10. Freund, Y., Schapire, R.E.: A decision-theoretic generalization of on-line learning and an application to boosting. J. Comput. Syst. Sci. 55(1), 119–139 (1997)

11. Friedman, J.H.: Greedy function approximation: a gradient boosting machine. Ann. Stat. 29(5), 1189–1232 (2001). https://doi.org/10.1214/aos/1013203451

12. Lee, K.C., Huang, H.P., Lu, S.S.: Burr detection by using vision image. Int. J. Adv Manuf. Technol. 8(5), 275–284 (1993). https://doi.org/10.1007/BF01783611

13. Lin, T.R.: Experimental study of burr formation and tool chipping in the face milling of stainless steel. J. Mater. Process. Technol. 108(1), 12–20 (2000). https://doi.org/10.1016/S0924-0136(00)00573-2

14. Pilný, L., De Chiffre, L., Píška, M., Villumsen, M.F.: Hole quality and burr reduction in drilling aluminium sheets. CIRP J. Manuf. Sci. Technol. 5(2), 102–107 (2012)

15. Riego, V., Sánchez, L., Álvarez, C.: Github - Burr contour features. https://github.com/ULE-Informatica/burr_contour_features

16. Su, J., Huang, C., Tarng, Y.: An automated flank wear measurement of microdrills using machine vision. J. Mater. Process. Technol. 180(1–3), 328–335 (2006). https://doi.org/10.1016/j.jmatprotec.2006.07.001

17. Suzuki, S., Be, K.: Topological structural analysis of digitized binary images by border following. Comput. Visi. Graph. Image Process. 30(1), 32–46 (1985). https://doi.org/10.1016/0734-189X(85)90016-7

18. The International Organization for Standardization: ISO 13715:2017(EN), Technical product documentation—Edges of undefined shape—Indication and dimensioning (2017)

Adaptive Graph Laplacian for Convex Multi-Task Learning SVM

Carlos Ruiz[1](✉), Carlos M. Alaíz[1](✉), and José R. Dorronsoro[1,2](✉)

[1] Department of Computer Engineering, Universidad Autónoma de Madrid,
Madrid, Spain
{carlos.ruizp,carlos.alaiz,jose.dorronsoro}@uam.es
[2] Inst. Ing. Conocimiento, Universidad Autónoma de Madrid, Madrid, Spain

Abstract. Multi-Task Learning (MTL) aims at solving different tasks simultaneously to obtain better models. Some Support Vector Machines (SVMs) formulations for the MTL context involve the combination of common and task-independent models, where we can also use an homogeneous graph over the tasks to impose pairwise connections between the independent models. The graph weight matrix is usually taken as a uniform (i.e., maximum entropy) one that connects equally every pair of tasks. If we take instead the identity matrix (i.e., the one of minimum entropy), then no connections are considered. In MTL situations where the real tasks or the relationships among them are unknown, these two rigid approaches can be detrimental to the learning process. In order to identify the true underlying tasks and their structure, we propose here a Convex Adaptive Graph Laplacian MTL SVM, where the task models and the graph that reflects their relationships are learned in a sequential and collaborative way. We illustrate on four different synthetic scenarios the task structure identification capacities of this approach and how it can lead to better models.

1 Introduction

The standard approach in supervised Machine Learning (ML) is to try to find a good hypothesis function that maps an input space to an output space by minimizing some expected risk. This is generally pursued assuming that a single model is enough to fit the entire sample's input-target pairs. But it may also be the case that this matching can be better done considering the sample as containing a number of different tasks. This is the problem addressed in Multitask Learning (MTL), for which the main ML paradigms have been adapted [3,12,16].

The authors acknowledge financial support from the European Regional Development Fund and the Spanish State Research Agency of the Ministry of Economy, Industry, and Competitiveness under the projects TIN2016-76406-P (AEI/FEDER, UE) and PID2019-106827GB-I00. They also thank the UAM–ADIC Chair for Data Science and Machine Learning and gratefully acknowledge the use of the facilities of Centro de Computación Científica (CCC) at UAM.

H. Sanjurjo González et al. (Eds.): HAIS 2021, LNAI 12886, pp. 219–230, 2021.
https://doi.org/10.1007/978-3-030-86271-8_19

Support Vector Machines (SVMs) have also been applied in MTL, either under the direct SVM paradigm [1,2,4,9,13] or through the addition of extra regularization terms to try to capture the interplay between tasks [5,7,8,17].

However, although tasks are in principle defined according to sensible criteria, it may be the case that better task definitions are possible or, simply, that an MTL approach is not beneficial when compared with either a single common model or just a set of purely independent models. A way to ascertain this is to somehow measure the interplay between tasks by viewing them as the nodes of a task graph and incorporating this graph's Laplacian into the overall regularization term. This term should capture task differences and detect, for instance, similar tasks by minimizing the distance between their model representation, as done in [5] and recast under a convex formulation in [14]. However, these formulations use a fixed graph matrix which assigns the same weight to all task pairs, but it would be certainly much better if this graph matrix could be somehow learned. In this work we present the Convex Adaptive Graph Laplacian MTL SVM, which seeks to identify tasks structure, either as a goal on itself or towards learning better MTL SVM models. More precisely, and as our main contributions, we

- Introduce the Convex Adaptive Graph Laplacian MTL SVM.
- Illustrate its capabilities to identify the task structure underlying several synthetic regression problems.
- Show numerically that this tasks structure learning leads to better model performance.

The paper is organized as follows. In Sect. 2 we review the various MTL SVM formulations. We present our proposal in Sect. 3 and give our experimental results in Sect. 4. Finally, in Sect. 5 we give a brief discussion, present some conclusions and mention possible further work initiatives.

2   Multi-Task Learning SVMs

In Multi-Task Learning (MTL) we have data from different tasks, i.e., the global sample is $\mathcal{D} = \cup_{r=1}^{T}\{(\mathbf{x}_i^r, y_i^r), i = 1, \ldots, n_r\}$, where r represents the task, $\mathbf{x}_i^r \in \mathbb{R}^d$ the vector of attributes, y_i^r the target, and n_r is the number of examples that belong to task r. When we deal with Multi-Task data three different approaches are possible:

- **Common Task Learning (CTL)**, which learns a single common model using the entire sample $\mathcal{D}$ and does not distinguish among tasks.
- **Independent Task Learning (ITL)**, which learns an independent model for each task and does not use any information from other tasks.
- **Multi-Task Learning (MTL)**, which combines all tasks in the learning process to obtain task-specialized models.

Several machine learning approaches are possible in MTL problems. Here we will concentrate on Support Vector Machines (SVMs), which we briefly review first and then discuss their application in MTL problems.

2.1   Convex Multi-Task Learning SVM

Support Vector Machines are applied to regression and classification problems under slightly different formulations. However, both problems also admit a unified formulation [10,11], namely one that defines the following primal minimization problem:

$$\underset{\mathbf{w},b,\boldsymbol{\xi}}{\arg\min} \quad J(\mathbf{w},b,\boldsymbol{\xi}) = C\sum_{n=1}^{N}\xi_i + \frac{1}{2}\|\mathbf{w}\|^2$$

$$\text{s.t.} \qquad y_i(\mathbf{w}\cdot\mathbf{x}_i + b) \ge p_i - \xi_i,\ \xi_i \ge 0,\ i=1,\dots,N, \tag{1}$$

and solves its corresponding dual problem

$$\underset{\boldsymbol{\alpha}}{\arg\min} \quad \Theta(\boldsymbol{\alpha}) = \boldsymbol{\alpha}^\mathsf{T}\mathbf{Q}\boldsymbol{\alpha} - \mathbf{p}^\mathsf{T}\boldsymbol{\alpha} \quad \text{s.t.} \quad 0 \le \alpha_i \le C,\ i=1,\dots,N,\ \sum_{i=1}^{N} y_i\alpha_i = 0, \tag{2}$$

where $\boldsymbol{\alpha}^\mathsf{T} = (\alpha_1,\dots,\alpha_N)$ are the dual multipliers, $\mathbf{Q}$ represents the kernel matrix, and $\mathbf{p}^\mathsf{T} = (p_1,\dots,p_N)$ is a target vector which, when adequately chosen, can be used to derive either a Support Vector Regression problem or a Classification one.

There are several proposals to fit SVMs into the MTL framework. After a first formulation in [2], we proposed in [13] to solve a primal problem which involves a convex combinations between a common model and the task-specific ones, namely

$$\underset{\mathbf{w},\mathbf{v}_1,\dots,\mathbf{v}_T,b,\boldsymbol{\xi}}{\arg\min} \quad \sum_{r=1}^{T} C_r \sum_{i=1}^{n_r}\xi_i^r + \frac{1}{2}\sum_r \|\mathbf{v}_r\|^2 + \frac{1}{2}\|\mathbf{w}\|^2$$

$$\text{s.t.} \qquad y_i^r(\lambda(\mathbf{w}\cdot\mathbf{x}_i^r) + (1-\lambda)(\mathbf{v}_r\cdot\mathbf{x}_i^r) + b_r) \ge p_i^r - \xi_i^r,$$

$$\xi_i^r \ge 0,\ i=1,\dots,n_r,\ r=1,\dots,T, \tag{3}$$

with $\lambda \in [0,1]$. When $\lambda = 1$ we recover the CTL approach in which we fit a single SVM for all tasks and, when $\lambda = 0$, the problem (3) separates into T independent models, equivalent to an ITL approach.

2.2   Convex Graph Laplacian MTL SVM

As mentioned in the introduction, the Graph Laplacian MTL SVM proposed in [4] seeks to exploit the relationships between tasks by using a directed graph $G = (V,E)$ with the tasks as the nodes V and where the edges E represent the relationships between tasks. This is done through a weight matrix $\mathbf{A}$, whose entries A_{rs} contain the weights of the edge between tasks r and s.

In [14] we proposed a convex formulation of the Graph Laplacian MTL SVM which includes a common regularization term and whose primal problem is

$$
\begin{aligned}
\underset{\mathbf{w},\mathbf{v}_1,\ldots,\mathbf{v}_T,b,\boldsymbol{\xi}}{\arg\min} \quad & \sum_{r=1}^{T} C_r \sum_{i=1}^{n_r} \xi_i^r + \frac{\nu}{2} \sum_{r=1}^{T}\sum_{s=1}^{T} A_{rs}\|\mathbf{v}_r - \mathbf{v}_s\|^2 + \frac{1}{2}\sum_r \|\mathbf{v}_r\|^2 + \frac{1}{2}\|\mathbf{w}\|^2 \\
\text{s.t.} \quad & y_i^r(\lambda(\mathbf{w}\cdot\mathbf{x}_i^r) + (1-\lambda)(\mathbf{v}_r\cdot\mathbf{x}_i^r) + b_r) \geq p_i^r - \xi_i^r, \\
& \xi_i^r \geq 0, \ i = 1,\ldots,n_r, \ r = 1,\ldots,T.
\end{aligned}
\tag{4}
$$

Here the Reproducing Kernel Hilbert Spaces of the common and task specific models may be different. Note that by using $\mathbf{A} = \mathbf{I}_T$ we would obtain a problem equivalent to (3). The corresponding Lagrangian of (4) is

$$
\begin{aligned}
& \mathcal{L}(\mathbf{w},\mathbf{v}_r,b_r,\xi_i^r,\boldsymbol{\alpha},\boldsymbol{\beta}) \\
&= \sum_{r=1}^{T} C_r \sum_{i=1}^{n_r} \xi_i^r + \frac{\nu}{2}\sum_{r=1}^{T}\sum_{s=1}^{T} A_{rs}\|\mathbf{v}_r - \mathbf{v}_s\|^2 + \frac{1}{2}\sum_r \|\mathbf{v}_r\|^2 + \frac{1}{2}\|\mathbf{w}\|^2 \\
&\quad - \sum_{r=1}^{T}\sum_{i=1}^{n_r} \alpha_i^r [y_i^r(\lambda(\mathbf{w}\cdot\mathbf{x}_i^r) + (1-\lambda)(\mathbf{v}_r\cdot\mathbf{x}_i^r) + b_r) - p_i^r + \xi_i^r] - \sum_{r=1}^{T}\sum_{i=1}^{n_r} \beta_i^r \xi_i^r.
\end{aligned}
\tag{5}
$$

Taking the primal variable partials and equating them to 0, we obtain $\mathbf{w} = \lambda\boldsymbol{\Phi}\boldsymbol{\alpha}$ and $\mathbf{v} = (1-\lambda)\boldsymbol{\Delta}^{-1}\boldsymbol{\Phi}\boldsymbol{\alpha}$. Here $\boldsymbol{\Delta} = \{(\nu(\mathbf{L}+\mathbf{L}^\mathsf{T}) + \mathbf{I}_T) \otimes \mathbf{I}_d\}$ where $\otimes$ stands for the Kronecker product, $\mathbf{L}$ is the Laplacian matrix and we also use the extended vectors and matrices

$$
\boldsymbol{\Phi} = \underbrace{\begin{bmatrix} \mathbf{X}_1 & \cdots & \mathbf{0} \\ \vdots & \ddots & \vdots \\ \mathbf{0} & \cdots & \mathbf{X}_T \end{bmatrix}}_{Td \times N}, \ \boldsymbol{\alpha}^T = (\boldsymbol{\alpha}_1^\mathsf{T},\ldots,\boldsymbol{\alpha}_T^\mathsf{T})_{1\times N}, \ \boldsymbol{\alpha}_r^\mathsf{T} = (\alpha_1^r,\ldots,\alpha_{n_r}^r)_{1\times n_r}, \ \mathbf{v}^\mathsf{T} = (\mathbf{v}_1^\mathsf{T},\ldots,\mathbf{v}_T^\mathsf{T})_{1\times Td}.
$$

Substituting these equalities back into the Lagrangian we get the dual problem

$$
\begin{aligned}
\underset{\boldsymbol{\alpha}}{\arg\min} \quad & \Theta(\boldsymbol{\alpha}) = \boldsymbol{\alpha}^\mathsf{T}\left\{\lambda^2\mathbf{Q} + (1-\lambda)^2\tilde{\mathbf{Q}}\right\}\boldsymbol{\alpha} - \mathbf{p}^\mathsf{T}\boldsymbol{\alpha} \\
\text{s.t.} \quad & 0 \leq \alpha_i^r \leq C_r, \ i = 1,\ldots,n_r, \ \sum_{i=1}^{n_r} y_i\alpha_i^r = 0, \ r = 1,\ldots,T,
\end{aligned}
\tag{6}
$$

where $\mathbf{Q}$ is the standard kernel matrix and $\tilde{\mathbf{Q}}$ is the Graph Laplacian multi-task kernel matrix defined through the following kernel function

$$
\tilde{k}(\mathbf{x}_i^r,\mathbf{x}_j^s) = \{(\nu(\mathbf{L}+\mathbf{L}^\mathsf{T}) + \mathbf{I}_T) \otimes \mathbf{I}_d\}_{rs}^{-1}\, k(\mathbf{x}_i^r,\mathbf{x}_j^s);
\tag{7}
$$

here $k(\cdot,\cdot)$ is some semi-positive definite kernel. We finally observe that this formulation encompasses all the previous MTL ones. In fact, when $\nu = 0$, problem (4) is equivalent to (3) which, in turn, covers the range between the CTL and ITL approaches when we slide λ from 1 to 0. On the other hand, if $\nu \neq 0$, the transition is made between a CTL approach when $\lambda = 1$, and the pure Graph Laplacian approach of [4] when $\lambda = 0$.

3   Convex Adaptative Graph Laplacian MTL SVM

Graph Laplacian regularization requires to define a graph weight matrix $\mathbf{A}$. One intuitive way to interpret $\mathbf{A}$ is to use matrices where the rows $\mathbf{A}_{r\cdot}$ sum up to one and, hence, can be seen as probabilities; this can be done because the parameter ν can absorb any scaling needed. In order to use an agnostic view of the task relationships, in [14] a constant weight matrix $\mathbf{A}$ is used, where all the tasks are uniformly connected. This corresponds to rows of $\mathbf{A}$ with maximum entropy. Conversely, the Convex MTL SVM primal problem corresponds to the case $\mathbf{A} = \mathbf{I}_T$, that is, no connection between tasks is assumed. This corresponds to rows if $\mathbf{A}$ with minimum entropy. Between these two extreme cases we may find matrices that capture better the relationships among the underlying task. To arrive to such matrix $\mathbf{A}$ we consider the following problem:

$$
\begin{aligned}
\underset{\mathbf{A},\mathbf{v}_1,\dots,\mathbf{v}_T,b,\xi}{\arg\min} \quad & \sum_{r=1}^{T} C_r \sum_{i=1}^{n_r} \xi_i^r + \frac{1}{2}\sum_r \|\mathbf{v}_r\|^2 + \frac{1}{2}\|\mathbf{w}\|^2 \\
& + \frac{\nu}{2}\sum_{r=1}^{T}\sum_{s=1}^{T} A_{rs}\|\mathbf{v}_r - \mathbf{v}_s\|^2 - \frac{\mu}{2}\sum_{r=1}^{T}\mathcal{H}(\mathbf{A}_{r\cdot})
\end{aligned}
$$

$$
\text{s.t.} \quad y_i^r\left(\lambda(\mathbf{w}\cdot\mathbf{x}_i^r) + (1-\lambda)(\mathbf{v}_r\cdot\mathbf{x}_i^r) + b_r\right) \geq p_i^r - \xi_i^r,
$$

$$
\xi_i^r \geq 0,\ i=1,\dots,n_r,\ 0 \leq A_{rs} \leq 1,\ \sum_{s=1}^{T} A_{rs} = 1,\ r,s = 1,\dots,T,
$$

$$
\tag{8}
$$

where $\mathcal{H}(\mathbf{A}_{r\cdot})$ is the entropy of the r-th row of matrix $\mathbf{A}$. Note that our restrictions on the matrix $\mathbf{A}$ result on each row $\mathbf{A}_{r\cdot}$ being a probability distribution among tasks. Using small values of μ should give us the minimum entropy solution $\mathbf{A} = \mathbf{I}_T$, while when μ is large we should get the other extreme of a constant matrix $\mathbf{A}$ whose rows have maximum entropy. That is, problem (8) reduces to problems (3) and (4) when we slide the value of μ from 0 to ∞.

Although problem (8) is not jointly convex in $\{\mathbf{w},\mathbf{v},\mathbf{A}\}$, it is separately convex in $\{\mathbf{w},\mathbf{v}\}$, and also in $\mathbf{A}$. Therefore, starting with some initial matrix $\mathbf{A}^{(0)}$ we can find a solution of (8) by the following two-step optimization:

1. Optimize in $\{\mathbf{w},\mathbf{v}\}$ using a fixed weight matrix $\mathbf{A}^{(\tau-1)}$, as described in Sect. 2.2.
2. Optimize in $\mathbf{A}$, using fixed vectors $\mathbf{w}^{(\tau)}$ and $\mathbf{v}^{(\tau)}$, as described below.

We iterate over these steps for a maximum number of iterations or until reaching certain tolerance value in the objective function, always applying as the last step the optimization in $\{\mathbf{w},\mathbf{v}\}$ with the final matrix $\mathbf{A}$.

To obtain an optimal $\mathbf{A}$ matrix we proceed as follows: after the dual problem (6) of the first step is solved and we have obtained an optimal $\boldsymbol{\alpha}$, it is necessary to compute the distances $\|\mathbf{v}_r - \mathbf{v}_s\|^2 = \mathbf{v}_r \cdot \mathbf{v}_r + \mathbf{v}_s \cdot \mathbf{v}_s - 2\mathbf{v}_r \cdot \mathbf{v}_s$ of the regularization term. For this, notice that $\mathbf{v}_r = \mathbf{U}_r\mathbf{v}$, where

$$\mathbf{U}_r \atop d \times dT = \begin{pmatrix} \overbrace{\mathbf{0}}^{1} & \overbrace{\mathbf{0}}^{2} & \cdots & \overbrace{\mathbf{I}_d}^{r} & \cdots & \overbrace{\mathbf{0}}^{T} \\ d \times d & d \times d & & d \times d & & d \times d \end{pmatrix}.$$

Using this matrix, we can write $\mathbf{v}_r$ as $\mathbf{v}_r = U_r \mathbf{v} = (1 - \lambda)\mathbf{U}_r \boldsymbol{\Delta}^{-1} \boldsymbol{\Phi} \boldsymbol{\alpha}$, and

$$\mathbf{v}_r \cdot \mathbf{v}_s = (1-\lambda)^2 \boldsymbol{\alpha}^{\mathsf{T}} \boldsymbol{\Phi}^{\mathsf{T}} \boldsymbol{\Delta}^{-1} \mathbf{U}_r^{\mathsf{T}} \mathbf{U}_s \boldsymbol{\Delta}^{-1} \boldsymbol{\Phi} \boldsymbol{\alpha}, \tag{9}$$

which, after some calculations and applying the kernel trick, can be expressed as $\mathbf{v}_r \cdot \mathbf{v}_s = (1-\lambda)^2 \boldsymbol{\alpha}^{\mathsf{T}} \widetilde{\mathbf{K}}^{rs} \boldsymbol{\alpha}$, where $\widetilde{\mathbf{K}}^{rs}$ is the kernel defined as:

$$\widetilde{k}^{rs}(\mathbf{x}_i^\rho, \mathbf{y}_j^t) = \boldsymbol{\Delta}_{\rho r}^{-1} \boldsymbol{\Delta}_{st}^{-1} k(\mathbf{x}_i^\rho, \mathbf{y}_j^t). \tag{10}$$

Therefore, to compute each distance we need to perform a matrix multiplication, but only using the dual coefficients values corresponding to support vectors. Therefore, the total cost of computing the distance matrix is $O(T^2 N_{sv}^2)$ where T is the number of tasks and N_{sv} the number of support vectors. Once the distances are obtained, the optimization problem for the matrix $\mathbf{A}$ becomes

$$\arg\min_{\mathbf{A}} \quad J(\mathbf{A}) = \frac{\nu}{2} \sum_{r=1}^{T} \sum_{s=1}^{T} A_{rs} \|\mathbf{v}_r - \mathbf{v}_s\|^2 - \frac{\mu}{2} \sum_{r=1}^{T} \mathcal{H}(\mathbf{A}_{r\cdot})$$

$$\text{s.t.} \quad 0 \le A_{rs} \le 1, \ r, s = 1, \ldots, T, \tag{11}$$

$$\sum_{s=1}^{T} A_{rs} = 1, \ r = 1, \ldots, T.$$

Notice that without the entropy term, we would end with the trivial solution $A_{rs} = \delta_{rs}$, i.e., the minimum-entropy solution $\mathbf{A} = \mathbf{I}_T$. Moreover, problem (11) is separable for each row $\mathbf{A}_r$, and taking derivatives and equating to 0 we obtain:

$$\frac{\partial}{\partial A_{rs}} J(\mathbf{A}) = \frac{1}{2} \left(\nu \|\mathbf{v}_r - \mathbf{v}_s\|^2 + \mu \log A_{rs} + \mu \right) = 0$$

$$\implies A_{rs} \propto \exp -\frac{\nu}{\mu} \|\mathbf{v}_r - \mathbf{v}_s\|^2.$$

As a consequence, since $\sum_s A_{rs} = 1$, the solution is:

$$A_{rs} = \frac{\exp -\frac{\nu}{\mu} \|\mathbf{v}_r - \mathbf{v}_s\|^2}{\sum_t \exp -\frac{\nu}{\mu} \|\mathbf{v}_r - \mathbf{v}_t\|^2}. \tag{12}$$

Since this is a closed solution, the computational complexity of getting an optimal $\mathbf{A}^*$ is that of computing the distances: $O(T^2 N_{sv}^2)$.

4   Synthetic Experimental Results

Dataset Description

We shall illustrate our proposals over four synthetic Multi-Task regression problems. In all of them, the target is defined as $y = f(x) + \epsilon$ where ϵ is Gaussian

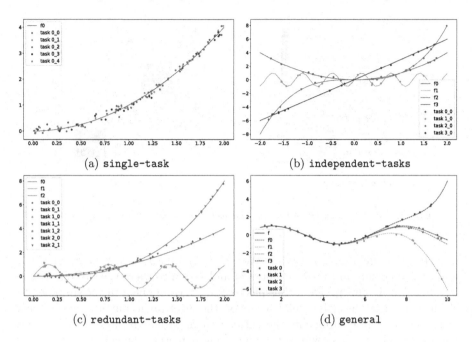

(a) single-task (b) independent-tasks

(c) redundant-tasks (d) general

Fig. 1. Problems illustration.

noise with a standard deviation of 0.1. The first one is the single-task problem, shown in Fig. 1a, where we define 5 tasks by sampling for each one of them 20 points x_j uniformly in the $[0, 2]$ range and taking $f(x_j) = x_j^2$ as their targets. In other words, there is a single underlying task, from which five apparently different tasks are defined by taking different samplings but where for all of them their targets are defined with the same $f(x) = x^2$ function. In problem independent-tasks, shown in Fig. 1b, we consider four different functions: $f_0(x) = x^2$, $f_1(x) = \sin(10x)$, $f_2(x) = x^3$ and $f_3(x) = 3x$, and we sample for each one 15 points uniformly in the interval $[-2, 2]$ and use f_r to them to get their targets. We thus have 60 points under 4 different tasks that do not share any information and, thus, learning them independently is a sensible approach.

In problem redundant-tasks, shown in Fig. 1c, we consider three different functions: $f_0(x) = x^2$, $f_1(x) = \sin(10x)$ and $f_2(x) = x^3$, and generate 7 tasks by performing 7 different uniform samplings of 10 points in the interval $[0, 2]$ and assigning targets to them by applying to the first two samplings the function f_0, to the next three the function f_1, and to the last two samplings the function f_2. We have thus 7 tasks although there are just three independent underlying tasks. Here no common task information is available, but a sensible MTL model should detect three task clusters, i.e., one for each of the underlying tasks. Finally, for problem general, shown in Fig. 1d, we first define four random polynomials, p_1, p_2, p_3 and p_4, with degrees 7, 8, 3 and 9, respectively, and random coefficients with a standard Gaussian distribution. We define then 4 task functions of the

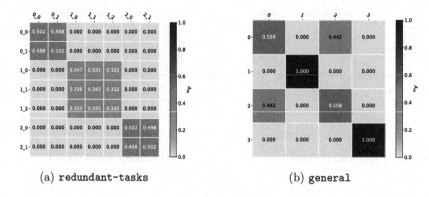

(a) **redundant-tasks**   (b) **general**

Fig. 2. Weights matrices after the last iteration of the training algorithm.

form $f_r(x) = \sin(x) + 10^{-6}p_r(x)$. We finally sample for each function f_r 15 points x_i uniformly in the $[1, 10]$ interval and assign to them the targets $f_r(x_i)$, for $r = 1, 2, 3, 4$. This problem thus has a common component in the sin function which dominates the polynomial terms on the $[0, 5]$ range but afterwards each task's independent component differentiates them, although those of tasks 0 and 2 are quite similar, as shown in the plot. Here detecting the common part may not be enough, and to fully exploit the task structure we need to capture both the common element and the pair-wise connections.

Table 1. Hyperparameters, grids used to select them (when appropriate) and hyperparameter selection method for each model.

	Grid	CTL	ITL	MTL	GL-MTL	AdapGL-MTL
C	$\{10^k : 0 \leq k \leq 4\}$	✓	✓	✓	✓	✓
ϵ	$\{\frac{1}{10^k} : 1 \leq k \leq 3\}$	✓	✓	✓	✓	✓
γ	$\{\frac{10^k}{d} : 0 \leq k \leq 3\}$	✓	✓	✓	✓	✓
λ	$\{0.2k : 0 \leq k \leq 5\}$	–	–	✓	✓	✓
ν	$\{10^k : -2 \leq k \leq 3\}$	–	–	–	✓	✓
μ	$\{10^k : -2 \leq k \leq 3\}$	–	–	–	–	✓

MTL Models Considered

We will apply to each problem the following models:

- CTL: A common SVM using the CTL approach.
- ITL: Independent SVMs hyperparametrized and fitted for each task.
- MTL: A Convex MTL SVM model, as presented in Sect. 2.

Table 2. Optimal values of λ, ν and μ, for the problems considered.

Model		Single-task	Independent-tasks	Redundant-tasks	General
MTL	λ	1	0	0.4	0
GL-MTL	λ	1	0	0.4	0
	ν	10^{-2}	10^{-2}	10^{-1}	10^{-2}
AdapGL-MTL	λ	1	0	0	0.6
	ν	10^{-2}	10^{0}	10^{2}	10^{3}
	μ	10^{-2}	10^{-1}	10^{0}	10^{1}

- GL-MTL: A Convex Graph Laplacian MTL SVM approach, as presented in Sect. 2.
- AdapGL-MTL: The proposed Convex Adaptive Graph Laplacian MTL SVM.

For these models we find optimal hyperparameters via 5-fold Cross Validation (CV) using a grid search. Table 1 shows the grid explored for each hyperparameter. To alleviate computational costs, in all the MTL approaches we will use common values for C, ϵ and γ for all the tasks. As in [4] and [14], we use a constant graph weight matrix for GL-MTL, and the initial weight matrix for AdapGL-MTL is also constant, with maximum entropy rows, i.e., constant $\frac{1}{T}$ entries. The maximum number of iterations for AdapGL-MTL is set to 5, with a tolerance of 10^{-4}. In Table 2 we see the optimal values for λ, μ and ν in each problem, more relevant here than the C, ϵ and γ ones, which we shall omit.

Task Structure Identification

Recall that the main goal of the AdapGL-MTL model proposed here is to detect the real underlying tasks and identify their structure through the optimal graph weight matrix $\mathbf{A}^*$, but notice too that the optimal convex mixing parameter λ^* also affects the possible interplay among tasks, and also between them and a common model, if one does indeed exist.

In problem single-task the optimal λ^* is 1, that is, there is a single relevant common task; as a consequence, the AdapGL-MTL procedure does not identify any specific task. Moreover, the graph matrix $\mathbf{A}^*$ (not shown) is a uniform one, which is to be expected, as the only regularization term left in (11) is the negative entropy, which is minimized with entries $A_{rs}^* = \frac{1}{T}$. Similarly, in problem independent-tasks the optimal λ^* is 0, i.e., there is no common task and only independent ones. But now the final weight matrix $\mathbf{A}^*$ (not shown) is the identity, as the graph regularizer is minimized by not penalizing the difference between tasks, i.e., enforcing total independence among them, which is sensible here given how the tasks were defined. For problem redundant-tasks the optimal λ^* is 0, and Fig. 2a shows the corresponding optimal $\mathbf{A}^*$. As we can see, there are three task clusters corresponding to the three underlying functions which we used to define the 7 tasks. Moreover, in each cluster the weights are evenly distributed among the tasks that conform it, i.e., AdapGL-MTL identifies all

228   C. Ruiz et al.

tasks in each cluster as being the same, which is also the sensible option here. Finally, the optimal λ^* in problem general is 0.6, which implies the presence of a common task besides the four specific ones. Here the optimal $\mathbf{A}^*$ matrix, shown in Fig. 2b, identifies tasks 0 and 2 as similar, as the probabilities in their $\mathbf{A}^*$ rows are almost evenly distributed among both tasks. On the other hand, in the case of tasks 1 and 3, which appear as quite different in Fig. 1d, there are no weights connecting them to other tasks. That is, once AdapGL-MTL has sorted out the common component, it identifies three real underlying tasks, because two of the initial ones can be grouped together.

Table 3. Mean Absolute Errors (MAEs) and the corresponding statistically significant rankings obtained for the considered models in the synthetic problems.

	Single-task	Independent-tasks	Redundant-tasks	General
CTL	0.082 (1)	1.448 (2)	0.946 (4)	0.449 (5)
ITL	0.092 (2)	0.356 (1)	0.165 (2)	0.132 (4)
MTL	0.082 (1)	0.359 (1)	0.207 (3)	0.130 (3)
GL-MTL	0.082 (1)	0.359 (1)	0.209 (3)	0.129 (2)
AdapGL-MTL	0.082 (1)	0.359 (1)	0.108 (1)	0.117 (1)

Model Comparison

While task identification is an important issue on itself, it is also a step toward better models, which we discuss here in terms of their Mean Absolute Errors (MAEs). Table 3 shows them for all the MTL models considered, together with their corresponding statistical significance rankings. To generate them, we order the models using their test MAE, and then we compute a pairwise Wilcoxon test between each pair of consecutive model error distributions. If the null hypothesis is rejected, then the difference between these error distributions is not symmetrical, and we increment the ranking for the model with worse score; otherwise we give both models the same ranking.

We can observe that in the problem single-task, which has a single underlying task, all the MTL models essentially coincide with the single task CTL; recall that here $\lambda^* = 1$, and the ITL model falls behind. The ITL model obtains the best results in the independent-tasks problem but its difference with the MTL models is not significant, and all them are also ranked as first. We point out that there may be multiple strategies for the MTL models to obtain good results in this kind of problems. First of all, they select $\lambda^* = 0$, i.e., there is no common underlying model. Also, for GL-MTL and AdapGL-MTL the Graph Laplacian regularization that imposes similarity among task models should be negligible. In the case of GL-MTL, this is achieved through a small ν value. For AdapGL-MTL, the relation between ν and μ makes less relevant the negative entropy term in (11), allowing to select as $\mathbf{A}^*$ the identity matrix.

Table 4. Number of iterations needed for convergence with AdapGL-MTL.

	Single-task	Independent-tasks	Redundant-tasks	General
n. iterations	1	2	2	3

In the redundant-tasks problem we observe that AdapGL-mtl has a clear advantage over the competition. The capacity of using only the data from related and redundant tasks while ignoring others is more beneficial than having to use a combination of a task-independent model and one that is common to all tasks, which is what MTL and GL-MTL try to do. The ITL model obtains worse results, possibly because it has few training data, and clearly the CTL cannot explain the three underlying task functions using a single model.

Finally, the results for the general problem show that the MTL approaches outperform both CTL and ITL. This is to be expected given the problem structure, more suited for a common part capturing the sine component while individual models capture the different polynomials. However, the best model here is AdapGL-MTL, because tasks 0 and 2 are similar enough to share further information between them, but not with tasks 1 or 3. As mentioned, this is captured in the optimal weight matrix $\mathbf{A}^*$ which enables the model to maximally transfer knowledge among similar tasks.

One possible drawback of AdapGL-MTL is its computational cost, since it is necessary to iterate until convergence, and each iteration implies solving an SVM dual problem. However we can see in Table 4 that it requires few iterations to reach this convergence. Both the performance improvement and task structure identification can justify the greater computational cost.

5 Discussion and Conclusions

The Multi-Task Learning (MTL) paradigm intends to solve different tasks simultaneously to improve the performance of individual models. However, it can also be detrimental in some situations; for instance, if the tasks are arbitrarily chosen, we can introduce an incorrect bias in the model, either by forcing it to transfer knowledge between unrelated tasks, or by splitting a single source of information into several artificial tasks. In this work we propose a method, namely Convex Adaptive Graph Laplacian MTL SVM, that addresses this issue during the learning process by identifying the underlying relations between tasks. More specifically, the structure of the problem is encoded through a graph representing the similarity between tasks, and the graph weight matrix is optimized during the training. We illustrate the characteristics of the proposed method over four synthetic scenarios with different particularities; in all of them their task structure is correctly identified and results on a better model performance in the multi-task situations where it can fully exploit this task structure.

There are still lines for further work. First of all, the preliminary experiments shown here use a *weak* MTL approach, by which we mean that hyperparameters are reused on different models. Therefore, it is interesting to unfold the full capacities of the MTL approaches by performing a full hyperparametrization using more sophisticated methods, such as Bayesian optimization [15] or the covariance matrix adaptation evolution strategy (CMA-ES) [6]. On the other side, MTL model performance should also be considered over classification problems. Finally, the procedure proposed here should also be tested on real-world problems.

References

1. Cai, F., Cherkassky, V.: SVM+ regression and multi-task learning, pp. 418–424. IEEE Computer Society (2009)
2. Cai, F., Cherkassky, V.: Generalized SMO algorithm for SVM-based multitask learning. IEEE Trans. Neural Netw. Learn. Syst. **23**(6), 997–1003 (2012)
3. Caruana, R.: Multitask learning. Mach. Learn. **28**(1), 41–75 (1997)
4. Evgeniou, T., Micchelli, C.A., Pontil, M.: Learning multiple tasks with kernel methods. J. Mach. Learn. Res. **6**, 615–637 (2005)
5. Evgeniou, T., Pontil, M.: Regularized multi-task learning, pp. 109–117. ACM (2004)
6. Hansen, N., Ostermeier, A.: Completely derandomized self-adaptation in evolution strategies. Evol. Comput. **9**(2), 159–195 (2001)
7. Jacob, L., Bach, F.R., Vert, J.: Clustered multi-task learning: a convex formulation, pp. 745–752. Curran Associates, Inc. (2008)
8. Kumar, A., Daume III, H.: Learning task grouping and overlap in multi-task learning. In: International Conference on Machine Learning. ICML.CC/Omnipress (2012)
9. Liang, L., Cai, F., Cherkassky, V.: Predictive learning with structured (grouped) data. Neural Netw. **22**(5–6), 766–773 (2009)
10. Lin, C.: On the convergence of the decomposition method for support vector machines. IEEE Trans. Neural Netw. **12**(6), 1288–1298 (2001)
11. López-Lázaro, J., Dorronsoro, J.R.: Simple proof of convergence of the SMO algorithm for different SVM variants. IEEE Trans. Neural Netw. Learn. Syst. **23**(7), 1142–1147 (2012)
12. Ruder, S.: An overview of multi-task learning in deep neural networks. CoRR abs/1706.05098 (2017)
13. Ruiz, C., Alaíz, C.M., Dorronsoro, J.R.: A convex formulation of SVM-based multi-task learning. In: Pérez García, H., Sánchez González, L., Castejón Limas, M., Quintián Pardo, H., Corchado Rodríguez, E. (eds.) HAIS 2019. LNCS (LNAI), vol. 11734, pp. 404–415. Springer, Cham (2019). https://doi.org/10.1007/978-3-030-29859-3_35
14. Ruiz, C., Alaíz, C.M., Dorronsoro, J.R.: Convex graph Laplacian multi-task learning SVM. In: Farkaš, I., Masulli, P., Wermter, S. (eds.) ICANN 2020. LNCS, vol. 12397, pp. 142–154. Springer, Cham (2020). https://doi.org/10.1007/978-3-030-61616-8_12
15. Shahriari, B., Swersky, K., Wang, Z., Adams, R.P., de Freitas, N.: Taking the human out of the loop: a review of Bayesian optimization. Proc. IEEE **104**(1), 148–175 (2016)
16. Zhang, Y., Yang, Q.: A survey on multi-task learning. IEEE Trans. Knowl. Data Eng. (2021)
17. Zhang, Y., Yeung, D.Y.: A convex formulation for learning task relationships in multi-task learning. arXiv preprint arXiv:1203.3536 (2012)

Fight Detection in Images Using Postural Analysis

Eneko Atxa Landa, José Gaviria de la Puerta$^{(\boxtimes)}$, Inigo Lopez-Gazpio,
Iker Pastor-López, Alberto Tellaeche, and Pablo García Bringas

University of Deusto, Avenida de las Universidades 24, 48007 Bilbao, Spain
eneko.atxa@opendeusto.es, {jgaviria,inigo.lopezgazpio,
iker.pastor,alberto.tellaeche,pablo.garcia.bringas}@deusto.es

Abstract. This paper defines the research process that has been carried out to develop a system for detecting fights in images. The system
takes input frames and evaluates the probability that the frame contains one or more people fighting. Input frames containing images are
initially processed using a well-known neural architecture, called Open-
Pose, which extracts pose information out of images that contain human
postures. Human posture data is then processed by heuristics to extract
both: angles for arms and legs for each person in the image. Angles are
then used to feed an additional neural network that has been trained
to make probability predictions of people being potentially involved in
fights. This paper describes the full pipeline regarding techniques, tools
and assessment required to create a camera based violence detection system.

Keywords: Openpose · Neural networks · Fight detection · Feature
subset selection

1 Introduction

In recent years, in order to increase the security of people in public spaces,
the installation of security cameras has increased considerably [1]. Even so, the
large amount of data generated by each camera, and the very nature of this
data (videos that need to be monitored at all times) make real-time camera
monitoring very costly economically. Considering that the strength of the field of
computer vision, backed by the growing number of new techniques and scientific
articles published in the area of artificial intelligence [2], has been remarkable,
we propose to apply these techniques to make the surveillance problematic more
approachable.

The advances that have been made in the field of computer vision range from
very simple tasks, such as the recognition of handwritten characters, developing
OCR engines (*Optical Character Recognition*) such as Tesseract [3], to complex
vision segmentation tasks, such as the complete processing of a populated road,
to create its full representation, so that a car is able to drive autonomously [4].

H. Sanjurjo González et al. (Eds.): HAIS 2021, LNAI 12886, pp. 231–242, 2021.
https://doi.org/10.1007/978-3-030-86271-8_20

Among these applications, human posture detection and activity detection has become extremely important and has become a major area of research in the field of computer vision. Within activity recognition and in order to improve people's safety efforts have been made to improve the detection of violence in images and videos [5].

The development of the field of computer vision and of artificial intelligence in general has gone hand in hand with the democratisation of tools oriented towards the development of these technologies. General tools for creating applications using computer vision have grown and developed, and are increasingly accessible to researchers. For example, libraries that specialise in computer vision, such as the openCV library , developed by Intel, are free to use, and other high-level libraries such as tensorFlow and Keras allow neural networks to be created much more easily. This has enabled enormous growth in all areas of artificial intelligence, particularly in computer vision, allowing research and development to evolve rapidly. For all these reasons, the aim of this project has been to investigate the different techniques that exist for detecting violence in images and to propose our own pipeline, based on an open source neural network called OpenPose [6], creating a software system that detects violence by recognizing fights between people.

The rest of the article is distributed as follows: Sect. 2 summarises the research carried out in the scientific literature. Section 3 presents an overview of the most important datasets provided in the state-of-the-art. The Sect. 4 presents the methodology developed in this work. The following section resumes the experimentation and discussion of the results obtained to validate the methodology. And the final section collects the conclusions extracted from the research.

2   Related Work

The current section starts by looking at how the field of computer vision pose detection has evolved. Especially, the focus will be placed on 2-D and multi-person detection techniques, as this is the area of OpenPose's focus. Above all, the analysis will be based on a paper by researchers Yucheng Chen, Yingli Tian and Mingyi He [7], which makes a very comprehensive analysis of the state-of-the-art in pose estimation using deep learning, which can be consulted for further information. Posture recognition in humans is one of the most important tasks currently in the field of computer vision, and is applicable to very different tasks, such as: action and activity recognition [8], activity detection [9], human tracking [10], movies and animation, virtual reality, human-computer interaction, video surveillance, medical assistance, automatic driving and posture analysis in sports.

There are studies that extract human activity based on user behaviour, but based on text rather than images[11]. Several techniques using a variety of sensors have been used for this purpose: algorithms using depth sensors [12], infrared sensors [13], radio frequencies [14] and inputs allowing multiple views of the same scene [15]. Still, algorithms using conventional cameras, doing photo and video analysis, have gained momentum amid rapid evolution of computer vision techniques. In the present work, we mainly focus on these last techniques.

Within the techniques that use deep learning, several classifications are made taking into account different aspects. According to [7], the authors make an exhaustive classification, according to several parameters: they use models of human bodies (generative) or not (discriminative), processing level of high abstraction (bottom-up) or at the level of pixel evidence (top-down), dimensions (2 or 3), and so on. One of the most common problems encountered is the visualisation of this data, although studies have already solved this problem [16]. If to this we add graph analysis as a particularity in this type of problem, we find quite a lot of research on this subject, focusing on [17]. Table 1 summarizes the classification given in the cited article, with representative examples for each model category.

Table 1. Classification of posture detection techniques with Deep learning. Dim stands for dimension, SM for Single-member, MM for Multi-member

Dim	# People	Category	Example
2D	SM	Regression	Action recognition in videos [18]
		Detection	Rapid posture estimation [19]
	MM	Top-down	Refined posture estimation [20]
		Bottom-up	PifPafNet [21]
3D	SM	Generative	3D posture detection [22]
		Discriminatory	Exploiting temporal context for 3D pose estimation [23]
	MM	–	3D pose estimation with a single RGB camera [24]

Table 2 is a summary that contains some of the networks for which we have been able to compute the accuracy of the system. The full table is presented in [7]. In each row of the table, the article and its authors are presented, together with the type of neural network they have proposed for the task, and the accuracy they have obtained on the COCO dataset[1]. The table summarizes a variety of papers that have been written on 2-D pose detection for multiple people. It is important to differentiate between top-down and bottom-up techniques: top-down techniques, briefly speaking, are based on first detecting people in an image, and then applying pose detection to each individual; while bottom-up techniques tend to focus more on identifying possible areas of the human body, such as arms or legs, and then merging these areas into different skeletons, corresponding to each person. Top-down techniques tend to give better results

---

[1] COCO is a widely used dataset for object detection, segmentation and labelling tasks. It currently contains around 330000 images, including images of tagged people. It is available at: https://cocodataset.org/.

in terms of accuracy. This can be seen in the third column of the table: techniques in the first subgroup (top-down) present higher accuracies than the ones in the second subgroup (bottom-up). Still, these techniques are computationally more expensive, and produce results more slowly. In the following, we will delve more deeply into the top-down and bottom-up techniques, explaining the development and improvements that have taken place in recent years, and review the literature on each group of techniques.

Table 2. Comparison of different 2D and multi-person posture analysis methods. Acc stands for accuracy and HPE for Human pose estimation. Items marked in bold-face have been evaluated on COCO16 and the remaining ones on COCO17.

Description of method	Type of Network	Acc. (%)
Top-down (pixel-level evidence)		
Keypoint Localization [25]	Faster R-CNN + Inception-v2	**72.2**
Mask R-CNN [26]	Mask R-CNN + ResNet-FPN	**63.1**
HPE and tracking [27]	Faster R-CNN + ResNet	73.7
Cascaded pyramid network for multi HPE [28]	FPN + CPN	73
representation learning for HPE [29]	Faster R-CNN + HRNet	75.5
Bottom-up (more abstraction)		
OpenPose: Realtime multi-person 2d HPE [6]	VGG-19 + CPM	**61.8**
Joint Detection and Grouping [30]	Hourglass	65.5
HPE and instance segmentation [31]	ResNet	68.7
Multi HPE using pose residual network [32]	ResNet-FPN + RetinaNet	69.6
Composite field for HPE [21]	ResNet-50	66.7

3   Existing Datasets for Fighting Detection

This section describes datasets built for the particular purpose of violence detection in images and videos. For this datasets the labelling of images and videos is usually classified into binary labels: violence and non-violence. This makes it easier to train neural networks that need to perform supervised classification, establishing a common evaluation benchmark for all systems. Actually, such datasets have been used to train and evaluate the models involved in distinct phases of the presented research. The current section contains a brief review of some of the datasets that exist for the task of violence detection, based on the article by M. Cheng et al. [33]

According to the cited article, for which we present a summary in Table 3, the datasets used for video violence detection are usually of two types: clipped and unclipped. The former are usually short video clips, tagged at the video level, and the latter are usually longer videos, and periods of violence are usually tagged at the frame level, to determine when the violent behaviour begins and ends.

Table 3. Datasets for violence detection. D stands for video duration, Res stands for resolution, L stands for Label (which can be per Frame (F) or Video (V)), and C for clip. Duration for videos is given in secods.

Dataset	Size	L	Res	L	Type
BEHAVE [34]	4 V (171 C)	0.24–61.92	640 × 480	F	Acted fights
RE-DID [35]	30 V	20–240	1280 × 720	F	Natural
VSD [36]	18 V (1317 C)	55.3–829.4	Variable	F	Films
CCTV-Fights [37]	1000 C	5–720	Variable	F	Natural
Hockey Fight [38]	1000 C	1.6–1.96	360 × 288	V	Hockey matches
Movies Fight [38]	200 C	1.6-2	720 × 480	V	Películas
Crowd Violence [39]	246 C	1.04–6.52	Variable	V	Natural
SBU Kinect [40]	264 C	0.67-3	640 × 480	V	Acted fights
UCF-Crime [41]	1900 C	60–600	Variable	V	Security cameras
RWF-2000 [33]	2000 C	5	Variable	V	Security cameras

4   Methodology

This section presents the methodology used to obtain the results that will be validated in the experimentation and results section.

4.1   Data Gathering

To begin with, the research focuses on obtaining data that can be processed by OpenPose. That is, input images are processed with OpenPose, which produces a JSON file with all the information concerning recognized people. Especially the JSON comprises a list of 25 points coordinate (x, y) tuples of relevant body skeleton parts.

These points, by themselves, are not of much value, as the resolutions of the input images and the positions of the cameras in each scene can vary (because they return absolute pixel values), and therefore the points do not contain useful information. Therefore, we design some heuristics to extract valuable information from the list of skeleton body part coordinates. The proportions and resolutions of the cameras should not affect the data processing process. In other words, first of all, we infer information that is related to the position of people. With valuable information we mean information that can potentially be used to predict whether or not a person is in a fight with significant precision.

4.2   Angles Extraction from Skeleton

As the next step of the investigation we propose to calculate the angles formed by the arms and legs of each person. That is, we propose to make the calculation of angles that can provide valuable information about the posture of each person. Figure 1 shows an example of angle measurement in a skeleton.

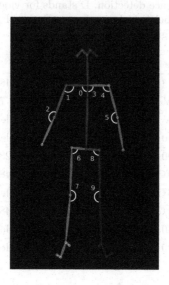

Fig. 1. Angles measured on each person

Using a Python script, skeleton coordinates provided by OpenPose are processed to compute vectors that represent angles for arm, forearms and leg interactions. For this task the following Eq. 1 is employed:

$$v = \frac{x_1 - x_0}{y_1 - y_0} \tag{1}$$

In case the limbs of the body have been identified the cosine between two vectors is calculated as shown in Eq. 2:

$$cos\theta = \frac{u \cdot v}{|u| \cdot |v|} \tag{2}$$

Then, using the inverse function of the cosine, the arccosine, the angle between two vectors is calculated. Thus, at the end of this step, in the best case ten angles will have been calculated for each person, shown above in the Fig. 1. As mentioned above, based on the hypothesis that some specific angles can be representative of specific postures, a priori these ten angles are considered to be valuable information that can be used to infer whether a person is in a fight or not.

4.3   Dataset Generation

The idea is to incorporate a neural network into the project, which learns to classify a photo (or its representation created by the tools) into two classes: fighting and non-fighting. To do this, a network needs to be created and trained with a training dataset.

In order to train the model, the first step was to find the right dataset, containing images or videos of fights and non-fighting situations. After a search for datasets, we used a dataset created by the researchers Enrique Bermejo Nievas et al. [38], which contains scenes obtained from films, some with fights and others without. This dataset contains videos of 2 s each, with 100 videos with fights and 100 without fights. These videos have been separated frame by frame, so that they can be analysed in a whole batch by OpenPose. After frame partitioning, 4790 frames of fights and 5049 frames without fights were obtained.

This dataset has been processed with OpenPose, and 4790 JSON files with encoded information of people fighting and 5049 of people not fighting have been generated. Then, all these JSON files have been processed by a script (the tool mentioned above, which calculates angles), and at the end a single vector information file has been generated, containing 23925 vectors in total.

Each information vector, representing a person, contains 10 calculated angles of the person, ordered as seen in the image 1, and a last bit (1 or 0), which defines whether it belongs to a fight frame (1) or not (0).

5   Experimentation and Results

Several experiments have been carried out on the model itself, in an attempt to improve the results.

5.1   Feature Subset Selection

The first experiment was the *FSS* experiment, which stands for *feature subset selection*. This experiment, very typical in neural network optimisation, consists of the following: starting from the initial set of variables (the 10 angles in this case), try to measure the quality of these variables, measuring how the neural network performs using subsets of these variables.

For example, FSS is used to answer the question: Would leg angles be necessary for fight detection? In this case, we could remove all leg angles from the input dataset, for example, and measure the predictive capacity of the network again. If the capacity does not drop considerably, it would mean that, surely, the leg angles are not really significant in predicting fights, and therefore, we could consider removing them from the input dataset, for ease of training, among other things. However, in case the predictions get worse, it would mean that having the leg angles in the dataset helps the network to predict correctly.

Therefore, this section will test the following subgroups:

– Tests with each angle individually, to analyse them one by one.

- Tests with only the angles of the legs, to see how not having the data of the arms affects them. (angles 7 to 10)
- Tests with only the angles of the arms, to see how it affects not having the data of the legs. (angles 1 to 6)

Table 4 shows the accuracies obtained in each experiment.

Table 4. Results of FSS experiment

Subgroup of variables	Accuracy (%)	Loss (%)
{ang1}	64.70	21.32
{ang2}	65.09	21.23
{ang3}	64.55	20.82
{ang4}	65.14	21.13
{ang5}	65.09	21.51
{ang6}	64.55	20.75
{ang7}	64.65	18.01
{ang8}	74.19	15.51
{ang9}	68.83	17.70
{ang10}	73.99	15.42
{ang1, ang2, ang3, ang4, ang5, ang6} (arms)	71.09	18.38
{ang7, ang8, ang9, ang10} (legs)	77.09	14.42

If we interpret the Table 4, we can see that the network has given much worse results when using a single input variable for training, since the worst accuracies in the table belong to these. It is also noticeable that in the case of angles 8 and 10 the results are much better than with the other angles. The accuracies in these two cases go up to almost 75%, which is quite impressive, considering that these are the angles in each leg. No hypothesis has been made as to why these two angles have a closer correlation than the others for making predictions. Surprisingly, these angles alone have achieved better results than the totality of the arm angles. The set of leg angles, on the other hand, was the best performer.

In this case, we could say that the leg angles are extremely important for the predictions, and that, using only these angles, the network is able to obtain almost the same results as with all 10 angles together. Even so, this test is not conclusive, because tests should be done on more complex networks to see how it would affect the selection of the leg angle subgroup in those networks. In the case of this simple network, although the results are only 3% better in the case of all angles, the training time hardly decreases, and in fact, it is very short in both cases, so, although the improvement obtained is smaller, it is proposed to continue using all angles for training.

5.2   Analysis of Network Learning Curves

After carrying out the FSS experiment, the predictive capability of the network will be analysed according to the size of the input dataset. In other words, in this experiment, the size of the dataset used to train the network will be changed and it will be observed whether the network achieves similar results with different amounts of data. In the case that, with more and more data, the results are better, it would mean that the network is taking advantage of that extra data, to get better and better, and therefore, you could consider increasing even more the size of the input dataset, with more data. If, on the other hand, the results do not improve much even though the amount of data increases, it would mean that the network is reaching its maximum prediction capacity, and one could consider changing its structure to make it more complex and capable of making better predictions.

That said, the results of the learning curve analysis experiment are presented in Table 5. It should be noted that the percentages in the first column correspond to percentages of the training dataset, without taking into account the validation dataset or the test dataset. In other words, the percentage of the 100th percentile corresponds to the accuracy obtained using the entire training dataset. In addition, the sizes of the validation and test datasets have not been changed, and all 10 angles have been taken as input data, as in the original network.

Table 5. Results of the learning curve analysis experiment

Size of the training dataset (% of the original dataset)	Accuracy (%)	Loss (%)
10	79.06	13.95
20	78.52	13.63
30	79.79	13.35
40	80.04	13.28
50	79.79	13.17
60	80.24	13.19
70	80.38	13.01
80	80.24	12.94
90	81.51	13.11
100	80.14	13.02

Analysing the results of the experiment, the result is compelling: the accuracy is almost unchanged even if we reduce the input data. The worst accuracy obtained, using 10% of the original training dataset, is only 1.08% worse than using the entire dataset. Although an overall upward trend is seen, this trend is too small to claim that the results improve much with more data. Therefore, this experiment suggests that, before we attach importance to getting more data, we should change the structure of the original network so that it does not reach its maximum capacity to get results so easily, and that with more complexity we can get better predictions.

The results of the experiment, therefore, were not unexpected, since, as mentioned at the beginning of this section, the investigation started from a very simple initial network, with only a dense intermediate layer. Thus, in the next step, a new architecture will be proposed for the next network proposal.

6   Conclusions

In this article, a new methodology for the detection of fights by analysing the posture of people has been presented. First of all, we have extracted the information about all the points of a skeleton and we have extracted their angles. Secondly, we have checked whether these angles are representative enough to detect for each of them or in groups whether the person's posture correspond to being fighting or not. Finally, we have analyzed the learning curves of the neural network that predicts the probability distribution of plausible fight existence to analyze how the detection generalizes with regards to the training set size.

As we have been able to verify and discuss in the experimentation section of this first approximation, the results have been promising. For future steps we would like to explore new methods for the detection of fights in order to improve the behaviour of the network, based not only on the angles, but also on distances between people.

References

1. SDM: Rise of surveillance camera installed base slows (2016)
2. Voulodimos, A., Doulamis, N., Doulamis, A., Protopapadakis, E.: Deep learning for computer vision: a brief review. Comput. Intell. Neurosci. (2018)
3. Smith, R.: An overview of the tesseract OCR engine. In: Ninth International Conference on Document Analysis and Recognition (ICDAR 2007), vol. 2, pp. 629–633. IEEE (2007)
4. Badue, C., et al.: Self-driving cars: a survey. Expert Syst. Appl. **165**, 113816 (2020)
5. Bermejo Nievas, E., Deniz Suarez, O., Bueno García, G., Sukthankar, R.: Violence detection in video using computer vision techniques. In: Real, P., Diaz-Pernil, D., Molina-Abril, H., Berciano, A., Kropatsch, W. (eds.) CAIP 2011. LNCS, vol. 6855, pp. 332–339. Springer, Heidelberg (2011). https://doi.org/10.1007/978-3-642-23678-5_39
6. Cao, Z., Hidalgo, G., Simon, T., Wei, S.E., Sheikh, Y.: OpenPose: realtime multi-person 2D pose estimation using part affinity fields. IEEE Trans. Pattern Anal. Mach. Intell. **43**, 172–186 (2019)
7. Chen, Y., Tian, Y., He, M.: Monocular human pose estimation: a survey of deep learning-based methods. Comput. Vis. Image Underst. **192**, 102897 (2020)
8. Li, B., Dai, Y., Cheng, X., Chen, H., Lin, Y., He, M.: Skeleton based action recognition using translation-scale invariant image mapping and multi-scale deep CNN. In: 2017 IEEE International Conference on Multimedia & Expo Workshops (ICMEW), pp. 601–604. IEEE (2017)

9. Li, B., Chen, H., Chen, Y., Dai, Y., He, M.: Skeleton boxes: solving skeleton based action detection with a single deep convolutional neural network. In: 2017 IEEE International Conference on Multimedia & Expo Workshops (ICMEW), pp. 613–616. IEEE (2017)

10. Insafutdinov, E., et al.: ArtTrack: articulated multi-person tracking in the wild. In: Proceedings of the IEEE Conference on Computer Vision and Pattern Recognition, pp. 6457–6465 (2017)

11. Cannataro, M., Cuzzocrea, A., Pugliese, A.: Xahm: an adaptive hypermedia model based on xml. In: Proceedings of the 14th International Conference on Software Engineering and Knowledge Engineering, pp. 627–634 (2002)

12. Shotton, J., et al.: Efficient human pose estimation from single depth images. IEEE Trans. Pattern Anal. Mach. Intell. **35**(12), 2821–2840 (2012)

13. Faessler, M., Mueggler, E., Schwabe, K., Scaramuzza, D.: A monocular pose estimation system based on infrared leds. In: 2014 IEEE International Conference on Robotics and Automation (ICRA), pp. 907–913. IEEE (2014)

14. Zhao, M., et al.: Through-wall human pose estimation using radio signals. In: Proceedings of the IEEE Conference on Computer Vision and Pattern Recognition, pp. 7356–7365 (2018)

15. Rhodin, H., et al.: Learning monocular 3D human pose estimation from multi-view images. In: Proceedings of the IEEE Conference on Computer Vision and Pattern Recognition, pp. 8437–8446 (2018)

16. Cuzzocrea, A., Mansmann, S.: OLAP visualization: models, issues, and techniques. In: Encyclopedia of Data Warehousing and Mining, Second Edition, pp. 1439–1446. IGI Global (2009)

17. Cuzzocrea, A., Song, I.Y.: Big graph analytics: the state of the art and future research agenda. In: Proceedings of the 17th International Workshop on Data Warehousing and OLAP, pp. 99–101 (2014)

18. Luvizon, D.C., Picard, D., Tabia, H.: 2D/3D pose estimation and action recognition using multitask deep learning. In: Proceedings of the IEEE Conference on Computer Vision and Pattern Recognition, pp. 5137–5146 (2018)

19. Zhang, F., Zhu, X., Ye, M.: Fast human pose estimation. In: Proceedings of the IEEE/CVF Conference on Computer Vision and Pattern Recognition, pp. 3517–3526 (2019)

20. Moon, G., Chang, J.Y., Lee, K.M.: Posefix: model-agnostic general human pose refinement network. In: Proceedings of the IEEE/CVF Conference on Computer Vision and Pattern Recognition, pp. 7773–7781 (2019)

21. Kreiss, S., Bertoni, L., Alahi, A.: PifPaf: composite fields for human pose estimation. In: Proceedings of the IEEE/CVF Conference on Computer Vision and Pattern Recognition, pp. 11977–11986 (2019)

22. Li, C., Lee, G.H.: Generating multiple hypotheses for 3D human pose estimation with mixture density network. In: Proceedings of the IEEE/CVF Conference on Computer Vision and Pattern Recognition, pp. 9887–9895 (2019)

23. Arnab, A., Doersch, C., Zisserman, A.: Exploiting temporal context for 3D human pose estimation in the wild. In: Proceedings of the IEEE/CVF Conference on Computer Vision and Pattern Recognition, pp. 3395–3404 (2019)

24. Mehta, D., et al.: XNect: real-time multi-person 3D motion capture with a single RGB camera. ACM Trans. Graph. (TOG) **39**(4), 82–1 (2020)

25. Huang, S., Gong, M., Tao, D.: A coarse-fine network for keypoint localization. In: Proceedings of the IEEE International Conference on Computer Vision, pp. 3028–3037 (2017)

26. He, K., Gkioxari, G., Dollár, P., Girshick, R.: Mask R-CNN. In: 2017 IEEE International Conference on Computer Vision (ICCV), pp. 2980–2988 (2017)
27. Xiao, B., Wu, H., Wei, Y.: Simple baselines for human pose estimation and tracking. In: Ferrari, V., Hebert, M., Sminchisescu, C., Weiss, Y. (eds.) ECCV 2018. LNCS, vol. 11210, pp. 472–487. Springer, Cham (2018). https://doi.org/10.1007/978-3-030-01231-1_29
28. Chen, Y., Wang, Z., Peng, Y., Zhang, Z., Yu, G., Sun, J.: Cascaded pyramid network for multi-person pose estimation. In: Proceedings of the IEEE Conference on Computer Vision and Pattern Recognition, pp. 7103–7112 (2018)
29. Sun, K., Xiao, B., Liu, D., Wang, J.: Deep high-resolution representation learning for human pose estimation. In: Proceedings of the IEEE/CVF Conference on Computer Vision and Pattern Recognition, pp. 5693–5703 (2019)
30. Newell, A., Huang, Z., Deng, J.: Associative embedding: end-to-end learning for joint detection and grouping. arXiv preprint arXiv:1611.05424 (2016)
31. Papandreou, G., Zhu, T., Chen, L.-C., Gidaris, S., Tompson, J., Murphy, K.: PersonLab: person pose estimation and instance segmentation with a bottom-up, part-based, geometric embedding model. In: Ferrari, V., Hebert, M., Sminchisescu, C., Weiss, Y. (eds.) Computer Vision – ECCV 2018. LNCS, vol. 11218, pp. 282–299. Springer, Cham (2018). https://doi.org/10.1007/978-3-030-01264-9_17
32. Kocabas, M., Karagoz, S., Akbas, E.: MultiPoseNet: fast multi-person pose estimation using pose residual network. In: Ferrari, V., Hebert, M., Sminchisescu, C., Weiss, Y. (eds.) ECCV 2018. LNCS, vol. 11215, pp. 437–453. Springer, Cham (2018). https://doi.org/10.1007/978-3-030-01252-6_26
33. Cheng, M., Cai, K., Li, M.: Rwf-2000: an open large scale video database for violence detection. arXiv preprint arXiv:1911.05913 (2019)
34. Blunsden, S., Fisher, R.B.: The BEHAVE video dataset: ground truthed video for multi-person behavior classification. Ann. BMVA **4**(1–12), 4 (2010)
35. Rota, P., Conci, N., Sebe, N., Rehg, J.M.: Real-life violent social interaction detection. In: 2015 IEEE International Conference on Image Processing (ICIP), pp. 3456–3460. IEEE (2015)
36. Demarty, C.-H., Penet, C., Soleymani, M., Gravier, G.: VSD, a public dataset for the detection of violent scenes in movies: design, annotation, analysis and evaluation. Multimed. Tools Appl. **74**(17), 7379–7404 (2015). https://doi.org/10.1007/s11042-014-1984-4
37. Perez, M., Kot, A.C., Rocha, A.: Detection of real-world fights in surveillance videos. In: ICASSP 2019–2019 IEEE International Conference on Acoustics, Speech and Signal Processing (ICASSP), pp. 2662–2666. IEEE (2019)
38. Nievas, E.B., Suarez, O.D., Garcia, G.B., Sukthankar, R.: Movies fight detection dataset. In: Computer Analysis of Images and Patterns, pp. 332–339. Springer (2011)
39. Hassner, T., Itcher, Y., Kliper-Gross, O.: Violent flows: real-time detection of violent crowd behavior. In: 2012 IEEE Computer Society Conference on Computer Vision and Pattern Recognition Workshops, pp. 1–6. IEEE (2012)
40. Yun, K., Honorio, J., Chattopadhyay, D., Berg, T.L., Samaras, D.: Two-person interaction detection using body-pose features and multiple instance learning. In: 2012 IEEE Computer Society Conference on Computer Vision and Pattern Recognition Workshops, pp. 28–35. IEEE (2012)
41. Sultani, W., Chen, C., Shah, M.: Real-world anomaly detection in surveillance videos. In: Proceedings of the IEEE Conference on Computer Vision and Pattern Recognition, pp. 6479–6488 (2018)

Applying Vector Symbolic Architecture and Semiotic Approach to Visual Dialog

Alexey K. Kovalev[1,2(✉)], Makhmud Shaban[2], Anfisa A. Chuganskaya[1], and Aleksandr I. Panov[1,3]

[1] Artificial Intelligence Research Institute FRC CSC RAS, Moscow, Russia
[2] HSE University, Moscow, Russia
[3] Moscow Institute of Physics and Technology, Moscow, Russia
panov.ai@mipt.ru

Abstract. The multi-modal tasks have started to play a significant role in the research on Artificial Intelligence. A particular example of that domain is visual-linguistic tasks, such as Visual Question Answering and its extension, Visual Dialog. In this paper, we concentrate on the Visual Dialog task and dataset. The task involves two agents. The first agent does not see an image and asks questions about the image content. The second agent sees this image and answers questions. The symbol grounding problem, or how symbols obtain their meanings, plays a crucial role in such tasks. We approach that problem from the semiotic point of view and propose the Vector Semiotic Architecture for Visual Dialog. The Vector Semiotic Architecture is a combination of the Sign-Based World Model and Vector Symbolic Architecture. The Sign-Based World Model represents agent knowledge on the high level of abstraction and allows uniform representation of different aspects of knowledge, forming a hierarchical representation of that knowledge in the form of a special kind of semantic network. The Vector Symbolic Architecture represents the computational level and allows to operate with symbols as with numerical vectors using simple element-wise operations. That combination enables grounding object representation from any level of abstraction to the sensory agent input.

Keywords: Visual dialog · Vector symbolic architecture · Sign-based world model · Perception

1 Introduction

In recent years, multimodal tasks have attracted increased attention. As an example of such tasks, a Visual Question Answering (VQA) [1] which combines visual and linguistic modalities could be considered. In VQA, a system queried by an image and a question to that image and should output an answer to the given question. This task serves as the starting point for more advanced problems, such as Visual Commonsense Reasoning (VCR) [33] and Visual Dialog [6]. VCR is testing the system's ability to justify its answer by forcing it to choose

© Springer Nature Switzerland AG 2021
H. Sanjurjo González et al. (Eds.): HAIS 2021, LNAI 12886, pp. 243–255, 2021.
https://doi.org/10.1007/978-3-030-86271-8_21

a rationale for it. Visual Dialog presents an example of a possible scenario of interaction between an intellectual assistant and a user. The crucial feature of the data that the dataset has collected during the interaction of two agents (Amazon Mechanical Turk workers) with each other. One agent (answerer) is exposed to an image and its caption and its role to answer questions asked by another agent (questioner) who does not see the image but the caption. Thus the questioner implicitly solves the task of refining the representation of a scene depicted on the image. Explicitly, the collected data is used in a situation where the answerer is exchanged with a computer system and asked to answer the last question about the image in the dialog considering dialog history. This peculiarity of the solving problem determines the types of questions asked. They are very clear-cut, as the questioner tries to refine the scene understanding. Most questions ask about external features (color, shape, size, appearance, etc.) and the existence/counting of objects. The symbol grounding problem [4,10,23] plays a decisive role in the effectiveness of such systems, as the system should relate the information from the image and the textual data with its internal representation of concepts. However, this problem statement limits the number of question types and does not allow for modeling complex cognitive functions as answers are based directly on the system's sensory input.

In this paper, we approach the symbol grounding problem from the semiotic perspective and apply the Sign-Based World Model [22,24] cognitive architecture enriched by hyperdimensional computing [13] (vector symbolic architecture) [17]. In the semiotic approach, the information unit is a sign that differs from a symbol in the sense that it possesses an internal structure and a name. Structurally, a sign consists of four components, namely a meaning, significance, an image, and a name. Sign components are represented by causal matrices. Matrices are represented as vectors of high dimensionality via hyperdimensional computing. Signs are not isolated information units but connected into a special kind of hierarchical semantic network – the semiotic network – and on the lowest level are grounded to the agent's sensory input by the image component. The use of Vector Symbolic Architectures allows for reducing operations on signs components to vector operations.

In psychology, the Visual Dialog task relates to perception and the construction of the image of objects based on sensations. The theory of perception can serve as a model for constructing algorithms in artificial intelligence investigations. The contribution of this paper is twofold: first, psychological groundings for such tasks as Visual Dialog are presented, and the Vector Semiotic Architecture is proposed to address the symbol grounding problem in the context of the Visual Dialog task.

2   Related Works

Perception, one of the leading higher mental functions [29] of a person, makes it possible to form some image of an object and subsequently the image of the world. Perception can be understood in two ways: as an image, the result of sensory systems and categorization work, or as a process that is a structure of actions aimed to obtain such an image. Research in the field of activity and the construction of a system of actions is the most relevant to the development of work algorithms of artificial systems.

The classification of action types and the process of their formation were proposed by Nikolai Bernstein. He identified 5 levels of actions that are characteristic of human activity. Each level was named with a Latin letter: A, B, C, D, E [3]. According to Bernstein, movement levels have their neurophysiological organization. The higher the goal of the action, the higher the level of the corresponding anatomical and physiological organization.

A is the lowest level. This level includes tonic movements, e.g. trembling. They are regulated by simple neurophysiological reactions which are similar both for humans and animals. **B** is the level of coordinated movements. These are coordinated actions without the need for spatial orientation. For example, hand movements while lying on the surface. **C** level needs movement and orientation in space. For example, you need to go around some obstacles. **D** is the subject level that is typical for a person. At this stage, the movements are built according to the logic of the subject. For example, if you need to take a cup without a handle, then a person can find an action consistent with this goal. **E** is the level of the speech muscles movements. This level is carried out when we speak, express our thoughts, or in symbolic movements (dance).

According to the organization of the levels of action, Bernstein proposed the concept of "models of the necessary future". The higher the motive of the activity, the higher the levels connected to its implementation. Levels C and D are significant for the construction of perceptual images.

At the level of perception, actions are associated with the conscious selection of a certain side of a sensory situation and the subsequent categorization of sensory information [32]. Studies of the processes of child perception development show that they are initially included in the external practical actions. The connection of perceptual actions with practical actions (manipulation, movement in space, etc.) is manifested in their expanded motor character, which can be observed externally. In the movements of the hand that feels the object (touch) and in the movements of the eyes that trace the visible contour, there is a continuous comparison of the image with the original, its verification and correction are carried out. In the further development of the activity, there is a reduction in the motor components. This leads to a significant temporary change: the process of perception externally becomes a one-time action. These changes are associated with the formation of a child's system of operations within the framework of perceptual actions and sensory standards. Perceptual actions are implemented using various operations. A similar consideration of the process of constructing a perceptual image is noted in the works of Jean Piaget [25], James Gibson [7], Ulric Neisser [21].

If we consider the process of forming the image of perception in a child of 4–5 years, we can distinguish the following characteristics. The perception is directed for them and carried out in the form of perceptual actions [26,32]: the shape of the object and the ability to group it according to this attribute; the size of items and the ability to group them; dividing the subject into parts and vice versa; the ratio of the integral shape of objects and their parts by size; measurements of objects (length, height, width, etc.); color of objects and their parts; the selection of an object from the surrounding environment based on its spatial position relative to other objects, the placement of objects (including the number of parts), based on the knowledge of their position in the space; understanding the similarity-difference relationship; understanding the general-private relationship; understanding of causal relationships-establishing the cause of a particular phenomenon, action, determining the possible consequence of certain actions and place them in the appropriate order; the necessary amount of knowledge about the objects and phenomena of the surrounding world; mathematical representations of the number, geometric shapes, and magnitudes of objects.

In the paper [26], the four stages of perceptual actions of a preschool child were identified in order to build a holistic image of the subject. At the **first stage**, the subject is perceived as a whole. We can say that at this stage there is a comparison of the general characteristics of the object with sensory standards. There is a primary categorization of the object into a certain class. At the **second stage**, the main parts of the object are isolated and their properties (shape, size, color, etc.) are determined. At the same time, the signs that will relate to the main ones will be updated depending on the perceptual attitude, which updates the field of attention and the appropriate signs. At the **third stage**, the spatial relationships of the parts are distinguished relative to each other (above, below, right, left) and to the entire context. At the **fourth stage**, an examination is carried out by repeated holistic perception of the object. All the data obtained about the object properties is analyzed. The results of the performed perceptual actions are synthesized into a single image.

The cultural significance of an object in the social practice of a person or the biological significance of an object in the life of an animal is described in psychology as the objectivity of perception. Experiments in the field of cognitive psychology [28], as well as data from neuropsychology [5] indicate that the recognition of an object occurs not only based on the geometric features of the configuration, but also within the framework of answers to the following questions: "What is it customary to do with this?"and "How can this be used?". Alexei Leontiev proposed to describe this principle of objectivity through the concept of meaning as the fifth quasi-dimension (existing along with the four dimensions of the space-time continuum), in which the objective world is revealed to a person [19]. The objectivity of perception and mental reflection in general is associated with the use of language for people. Thus, the task that AI specialists set for themselves is completely fundamental to be able to answer questions about the content of the subject scene.

The significant point is the process of analyzing the perceived object and assigning it to a certain class. The question arises about the relationship between the actualization of the perceptual image and semantic information. A number of studies have shown that the selection of semantic features in the pre-adjustment to the process of object perception itself performs a facilitating function if it meets the principle of semantic expediency. In the research of M. Potter was shown a picture of a hammer. The subject was much faster to name the general semantic category "tool"than when showing the word "hammer" [28].

The process of determining the overall value of the image by narrowing down the field of a diversity of response options occurs simultaneously, and sometimes before the selection of geometric features. In the intermediate phases of the microgenesis of perception, the answer is given to the question: "What does it look like?" [28]. In the perceptual processing of complex realistic images, their general semantic content is highlighted. This is done through the analysis and operation of simple filters that work without feedback. Such images are clustered in the coordinates of the semantic space of the scenario character, for example, "apartment", "forest", "sea coast". The overall meaning of the scene is highlighted before the detailed perception of the individual objects that fill it, providing a quick semantic classification [28].

A significant role of semantic clustering and schematization in the process of image perception is also noted in studies of productive processes of memory and reproduction of observed pictures. In experiments by Frederic Bartlett, the respondents were asked to consider the picture [2] or a story. The first person had the opportunity to see the picture and memorize it. In the future, the second and successive respondents were called. The first person tried to clearly describe its contents to the second, and this description was recorded. It was important what the participants forgot and what the final description was left. All colors except one were immediately forgotten. Random details disappeared. There was a progressive forgetting of unimportant details. However, a few units remain dominant.

Bartlett arrived at the conclusion that the scheme was stored in memory. It is understood as a sequential logical-temporal bundle of images and events. Node components or key events are highlighted in the schema. In the case of memorizing images, this is a spatial scheme. When recreating a scheme, additional details usually refer to the following types: details related to circuit nodes; emotionally charged details; details related to personal experience.

There is a replacement of unfamiliar images with similar ones. When recreating an image, the modifications are based on the schema and script. The actions in the scenario are organized according to a given sequence and are directed at the goal.

One of the directions of modern research of visual perception is the identification of those visual parameters that can act in the construction of a complex (semantic) image. For example, when investigating the problem of increasing the number of informative points used by the visual system in comparison with the number that directly falls on the retina of the eye, in [27] is shown that when an image is perceived, a "field picture"is formed, structured not only by the visible (actually indicated in the image itself) but also by the imaginary, or invisible to the eye axes. Such results determine the prospects of research in the field of the extra-sensory basis of human perceptual activity and the need to focus on it when modeling perception by means of artificial intelligence.

3   Vector Semiotic Architecture for Visual Dialog

In this paper, we apply a Vector Semiotic Architecture to the Visual Dialog task. The Vector Semiotic Architecture is a Sign-Based World Model cognitive architecture enriched with Vector Symbolic Architecture on the low level of representation. Such a combination of approaches attempts to address the symbol grounding problem [4,10,23], a fundamental problem of AI that is not solved yet.

The Sign-Based World Model (SBWM) [22,24] is a framework for modeling cognitive tasks. It bases on principles of the cultural-historical approach of Lev Vygotsky and the activity theory of Alexei Leontiev. SBWM relies on the concept of a **sign** representing the agent's knowledge about the environment it operates in, other agents it interacts with, and itself. The signs form a hierarchical semantic network. Conceptually, the sign is a four-component structure. Four components represent different aspects of the agent's knowledge. The meaning component (M) implies the agent's experience. Commonsense knowledge is expressed by the significance component (S). The image component (I) is used to distinguish signs. The name (N) possesses a nominative function. SBWM is successfully applied to different tasks, e.g., planning [8,14], role distribution of a group of agents [15], goal setting [24], and reasoning [16].

The S, M, I components of a sign represented by a special structure called a **causal matrix** (CM). A CM z is defined as a tuple of length t of events e_i. Each event e_i represents the appearance of a particular feature f_j at time step i and is a binary vector of length h. Thus a CM is a binary h by t matrix. The 1 in the position z_{ji} in a CM serves as a link to other matrices. That corresponds to the feature f_j and means that the feature f_j is included in the appropriate component of the sign. Thereby CMs are organized into a hierarchical semantic network where the tuples of CMs are the nodes, and the links are the relations between these tuples. The event index t can serve as discrete time to represent dynamic entities.

Vector Symbolic Architectures (VSA) [13] is an umbrella term for bio-inspired methods of representing and manipulating concepts as vectors of high dimensionality (HD vectors). HD vectors use distributed representations, i.e., the information is distributed across vector positions, and only the whole HD vector can be interpreted as a holistic representation of some entity.

For each concept of interest, an atomic HD vector is generated by sampling vector space. Atomic HD vectors are stored in the item memory (IM). With an extremely high probability, all random HD vectors are quasi-orthogonal to each other, which is an important property of high-dimensional spaces. Hyperdimensional computing defines operations and a similarity measure to manipulate atomic HD vectors. Two key operations for computing with HD vectors are bundling and binding. The nature of a vector space could be different for the different realizations of VSA. In this paper we work with bipolar vectors $\mathcal{S} \in \{-1, +1\}^{[d \times 1]}$.

The **binding** binds two HD vectors together and produces another HD vector that is dissimilar to the bounded HD vectors. The semantic interpretation of this operation is assigning a value to a particular attribute. The Hadamard product is used for the binding operation. The **bundling** is implemented via a position-wise addition. The bundling combines several HD vectors into a single HD vector. The resultant HD vector is similar to all bundled HD vectors. The bundling is used to represent sets.

To map the SBWM structures to VSA operations, we use the approach proposed in [17]. We use **bold font** to denote an HD vector of a corresponding SBWM structure and **H** with an appropriate subscript to show that this HD vector is stored in the IM.

As a causal matrix z could be represented as a set of events e_i, then a suitable VSA operation is a bundling. First, we have to map every event e_i to a corresponding HD vector $\mathbf{H}_{e_i}$ and then apply bundling to the collection of vectors $\mathbf{H}_{e_1}, \mathbf{H}_{e_2}, ..., \mathbf{H}_{e_t}$. To represent a link from a causal matrix z_1 to a causal matrix z_2 we, first, transform z_2 to an HD vector $\mathbf{z}_2$, second, split z_1 into events $e_i^{z_1}$, and map them to HD vectors $\mathbf{H}_{e_i^{z_1}}$, and then bind $\mathbf{z}_2$ with a corresponding HD vector $\mathbf{H}_{e_i^{z_1}}$. From the perspective of VSA, each symbol is seen as a high-dimensional vector. Then we can operate on symbols using vector operations and easily switch between representations using item memory **H**.

The proposed model is in Fig. 1. The image and the question are processed separately. Objects with attributes are extracted from the image. SBWM Module uses this information to construct a hierarchical scene representation based on SBWM via causal matrices. Then, the scene representation is encoded into an HD scene vector $\mathbf{z}_{scene}$. An input question, a caption, and a dialog history are fed into the module, where, if necessary, coreference is resolved. Then, the Seq2seq Parser parses the question into a sequence of VSA procedures (program $\mathcal{P}$). Each procedure is a combination of binding, bundling, and similarity operations. Procedures serve a specific purpose, e.g., find an object with a particular attribute value or the value of an object attribute. After that, the VSemA Reasoner executes the program on the scene representation and outputs the answer.

4 Experiments

In the Visual Dialog task, the asked question relies on the dialog history. Thus to successfully provide an answer, the model has to consider it. In the proposed approach, we use the coreference resolution to process the dialog history and work with the questions independently. Coreference resolution aims at solving the problem of finding all expressions that correspond to the same entity in the text. We replace the pronouns that refer to the objects mentioned in the previous questions with corresponding nouns. It enables question parsing without relying on the history more than using it for coreference resolution.

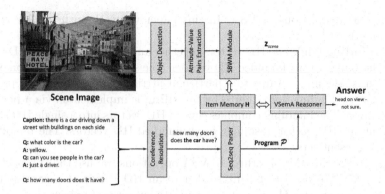

Fig. 1. Vector semiotic architecture for visual dialog.

To replace pronouns, we use a Huggingface coreference resolution model (NeuralCoref[1]), an AllenNLP library[2], and a simple rule-based replacement. AllenNLP model relies on the approach from [18] with the SpanBERT [12] word embeddings. The rule-based replacement is done by using the spaCy[3] parser to extract part-of-speech tags for each word. The rule was to replace every pronoun in a question with an object from the previous question. We present the quality comparison of used coreference resolution methods in Table 1.

Leaving a pronoun in a sentence makes it impossible for the model to answer correctly. Thus it is better to replace more pronouns at the cost of the percentage of wrong replacements. We chose the AllenNLP model as a final coreference resolution model as it copes well with that task.

Table 1. Results of applying coreference resolution to different dialogues.

Questions	GT	Rule-Based	Huggingface	AllenNLP
What color are bikes? Are **they** parked on stock parking?	Bikes	They	**Bikes**	They
Do **they** pose to picture?	People	Helmets	Their	Their
Do you see any buildings? Are **they** single story?	Buildings	**Buildings**	Any buildings	Two
Are there other people in the photo? What color are **They** painting with?	Person	Photo	What color	**Person**

---

[1] https://github.com/huggingface/neuralcoref.
[2] https://docs.allennlp.org/models.
[3] https://spacy.io.

Visual Dialog is a real-world dataset, which means that the questions asked in the dialogues are not standardized (e.g., compared to CLEVR [11]). Thus there is no straightforward way to convert a question to a sequence of template procedures, which will produce the answer if executed. Therefore to demonstrate the proposed approach, in this paper, we narrowed down types of questions to existence (Is there an object?) and counting (How many objects are there?).

The MSCOCO [20] dataset annotation contains 80 classes, which do not cover the entirety of Visual Dialog dataset classes. Therefore we chose a subset of Visual Dialog questions that are about the objects represented in the annotations. There are 20,290 existence questions and 12,472 counting questions. We applied a training pipeline from [31]: the question parser is pretrained on a small subset of question-program pairs in a supervised manner, then REINFORCE [30] is used to fine-tune the parser on the question-answer pairs. There are no program annotations in the Visual Dialog dataset, and we annotated questions manually. It resulted in a total of 39 question-program pairs. Even such a small amount is sufficient to train the model successfully.

We chose the NS-VQA model [31] as a baseline. It achieved an overall accuracy of 50.7%, where 31.7% is the accuracy for counting questions and 71.7% is for existence questions. For Vector Semiotic Architecture, we used HD vectors of size 10000 for scene representation. The proposed model achieved 51.3% accuracy, where the accuracy for counting questions is 31.0%, and the accuracy for existence questions is 73.9%.

5   Discussion

In the experiment, various paintings depicting objects, people, and animals were used. From a psychological point of view, the experiment combines constructing and transmitting a perceptual image. The study design brings it closer, in fact, to the first stage of Bartlett's experiments on productive memory work. The study participant is given a caption as a short description of the image. This semantic framework is the beginning of the construction of a perceptual image. In Example 1 (Fig. 2a), a caption will be "a couple of men riding horses on top of a green field". As we noted in the studies of the microgenesis of perception, this will be the initial focus in a certain scenario – "dressage on the field". In Example 2 (Fig. 2b), we focus on the scenario field "kitchen, dining room" or "catering establishment: restaurant, cafe". Such a process outlines the range of possible objects that can semantically be in such a field of perception. Within the framework of the human psyche, this process acts as an analog of presetting, which significantly reduces the process of identifying the image.

In the further course of the experiment, a dialogue takes place in the form of an answer to questions that will help reconstruct the perceptual image. From a psychological point of view, such activity carries out the process of recognition, successively passing through separate operations of perceptual actions. In the experiment, a model is created at the C level and partially at the D level, according to Bernstein's typology. So, in Example 1, this is the answer to the question about the number of horses. It is necessary to select an object by identifying it by its contour to implement this perceptual action, i.e., primitive counting based on the visual image. This task implements the principle of sensory standards. The system recognizes an image that corresponds to such generalized examples. A mechanism that models the recognition process based on sensory standards is created due to the training of recognition of a particular image. In Example 2, color recognition becomes a perceptual action. There is also a search and correlation with color standards to answer that the napkin and tablecloth are white. The design of the question-answer part of the experiment implies the division of the holistic perceptual action of recreation into separate operations. This scheme is most fully classified in the works on child psychology and is described above. In our experiment, we chose three types of operations, i.e., the detection of the object and its name, color, and quantity. They correspond to simple perceptual operations.

(b) **Caption**: a piece of chocolate dessert on a plate with a napkin and a fork.

Q: what color is the napkin?
A: white.
...
Q: what color is the plate?
A: white.
Q: how many layers are on the cake?
A: 8 very thin.

(a) **Caption**: a couple of men riding horses on top of a greenfield.

Q: how many horses are there?
A: 2 horses.

Fig. 2. Examples from visual dialog [6].

It is clear that the task of creating an artificial psyche, which implements the principle of objectivity of mental reflection, has no solution, and it can hardly be posed by anyone seriously. However, the implementation of this principle in problems with a limited set of "names"and associated "images"(spatial-color configurations), "values"(everyday scenarios where the objects presented in the picture play their roles), and "meanings"(evaluative attitude to these objects) seems quite possible. The organization of "recognition"should be carried out "from top to bottom": from the meaning of the entire scene to the properties description of individual objects. The reasons for such a sequence in the algorithm that models human perception are found in two areas of experimental psychology, i.e., in the study of the microgenesis of visual perception and the research of attribution of motives.

One limitation of a proposed approach, that it extracts objects based on annotated classes, and scanty annotations lead to poor performance. As a solution, an instance segmentation model trained on datasets with many categories [9] could be used.

6   Conclusion

In this paper, we propose the Vector Semiotic Architecture for Visual Dialog. The combination of SBWM and VSA in the Vector Semiotic Architecture allows approaching the symbol grounding problem by connecting high-level representations of concepts with sensory inputs of an agent. The proposed architecture achieved 31.0% accuracy for count questions and 73.9% accuracy for existence questions on a subset of the Visual Dialog dataset. Also, we show the limitations of an existing dataset for the Visual Dialog task.

Acknowledgements. The reported study was supported by RFBR, research Projects No. 19-37-90164 and 18-29-22027.

References

1. Agrawal, A., et al.: VQA: visual question answering. arXiv e-prints arXiv:1505.00468 (2015)
2. Bartlett, F.C.: Remembering: a study in experimental and social psychology. Philosophy **8**(31), 374–376 (1932)
3. Bernstein, A.N.: On dexterity and its development. Publishing House "Physical Culture and Sport", Moscow (1991). (in Russian)
4. Besold, T.R., Kühnberger, K.U.: Towards integrated neural - symbolic systems for human-level AI: two research programs helping to bridge the gaps. Biol. Inspir. Cogn. Architect. **14**, 97–110 (2015)
5. Chomskaya, E.D.: Neuropsychology, 4th edn. Peter (2005). (in Russian)
6. Das, A., et al.: Visual dialog. In: Proceedings of the IEEE Conference on Computer Vision and Pattern Recognition (CVPR) (2017)
7. Gibson, J.: The Perception of the Visual World. Houghton Mifflin, Boston (1950)
8. Gorodetskiy, A., Shlychkova, A., Panov, A.I.: Delta schema network in model-based reinforcement learning. In: Goertzel, B., Panov, A.I., Potapov, A., Yampolskiy, R. (eds.) AGI 2020. LNCS (LNAI), vol. 12177, pp. 172–182. Springer, Cham (2020). https://doi.org/10.1007/978-3-030-52152-3_18
9. Gupta, A., Dollar, P., Girshick, R.: LVIS: a dataset for large vocabulary instance segmentation. In: Proceedings of the IEEE Computer Society Conference on Computer Vision and Pattern Recognition 2019-June, pp. 5351–5359 (2019)
10. Harnad, S.: The symbol grounding problem. Physica D **42**(1), 335–346 (1990)
11. Johnson, J., Hariharan, B., van der Maaten, L., Fei-Fei, L., Zitnick, C.L., Girshick, R.: CLEVR: a diagnostic dataset for compositional language and elementary visual reasoning. In: CVPR (2017)
12. Joshi, M., Chen, D., Liu, Y., Weld, D.S., Zettlemoyer, L., Levy, O.: SpanBERT: improving pre-training by representing and predicting spans. Trans. Assoc. Comput. Linguist. **8**, 64–77 (2019)

13. Kanerva, P.: Hyperdimensional computing: an introduction to computing in distributed representation with high-dimensional random vectors. Cogn. Comput. 1(2), 139–159 (2009). https://doi.org/10.1007/s12559-009-9009-8
14. Kiselev, G., Kovalev, A., Panov, A.I.: Spatial reasoning and planning in sign-based world model. In: Kuznetsov, S.O., Osipov, G.S., Stefanuk, V.L. (eds.) RCAI 2018. CCIS, vol. 934, pp. 1–10. Springer, Cham (2018). https://doi.org/10.1007/978-3-030-00617-4_1
15. Kiselev, G.A., Panov, A.I.: Synthesis of the behavior plan for group of robots with sign based world model. In: Ronzhin, A., Rigoll, G., Meshcheryakov, R. (eds.) ICR 2017. LNCS (LNAI), vol. 10459, pp. 83–94. Springer, Cham (2017). https://doi.org/10.1007/978-3-319-66471-2_10
16. Kovalev, A.K., Panov, A.I.: Mental actions and modelling of reasoning in semiotic approach to AGI. In: Hammer, P., Agrawal, P., Goertzel, B., Iklé, M. (eds.) AGI 2019. LNCS (LNAI), vol. 11654, pp. 121–131. Springer, Cham (2019). https://doi.org/10.1007/978-3-030-27005-6_12
17. Kovalev, A.K., Panov, A.I., Osipov, E.: Hyperdimensional representations in semiotic approach to AGI. In: Goertzel, B., Panov, A.I., Potapov, A., Yampolskiy, R. (eds.) AGI 2020. LNCS (LNAI), vol. 12177, pp. 231–241. Springer, Cham (2020). https://doi.org/10.1007/978-3-030-52152-3_24
18. Lee, K., He, L., Zettlemoyer, L.: Higher-order coreference resolution with coarse-to-fine inference. In: NAACL-HLT (2018)
19. Leontiev, A.N.: Psychology of the image [in russian]. Vestn. Mosk. un-ta. Ser. 14, Psychology, no. 2, pp. 3–13 (1979)
20. Lin, T.-Y.: Microsoft COCO: common objects in context. In: Fleet, D., Pajdla, T., Schiele, B., Tuytelaars, T. (eds.) ECCV 2014. LNCS, vol. 8693, pp. 740–755. Springer, Cham (2014). https://doi.org/10.1007/978-3-319-10602-1_48
21. Neisser, U.: Cognition and Reality: Principles and Implications of Cognitive Psychology. W. H Freeman and Company, New York (1976)
22. Osipov, G.S., Panov, A.I., Chudova, N.V.: Behavior control as a function of consciousness. I. World model and goal setting. J. Comput. Syst. Sci. Int. 53(4), 517–529 (2014)
23. Osipov, G.S.: Signs-based vs. symbolic models. In: Sidorov, G., Galicia-Haro, S.N. (eds.) MICAI 2015. LNCS (LNAI), vol. 9413, pp. 3–11. Springer, Cham (2015). https://doi.org/10.1007/978-3-319-27060-9_1
24. Panov, A.I.: Goal setting and behavior planning for cognitive agents. Sci. Tech. Inf. Process. 46(6), 404–415 (2019)
25. Piaget, J.: Les mécanismes perceptifs. Presses universitaires de France, Paris (1961). (in French)
26. Poddyakov, N.N.: Features of Mental Development of Preschool Children. Professional Education Publishing House, Moscow (1996). [in Russian]
27. Shapoval, A.V.: Description of the image structure in modern art criticism analysis. Izvestiya Samarskogo nauchnogo tsentra Rossiyskoy akademii nauk 13(2), 240–246 (2011). (in Russian)
28. Velichkovsky, B.M.: Cognitive science: fundamentals of the psychology of cognition. In: 2 volumes. Smysl/Akademiya, Moscow (2006). (in Russian)
29. Vygotsky, L.: Collected works in 6 volumes, vol. 3. Pedagogika, Moscow (1983). (in Russian)
30. Williams, R.J.: Simple statistical gradient-following algorithms for connectionist reinforcement learning. Mach. Learn. 8, 229–256 (2004). https://doi.org/10.1007/BF00992696

31. Yi, K., Wu, J., Gan, C., Torralba, A., Kohli, P., Tenenbaum, J.B.: Neural-symbolic VQA: disentangling reasoning from vision and language understanding. arXiv e-prints arXiv:1810.02338 (2018)
32. Zaporozhets, A.V., Lisina, M.I.: Development of Perception in Early and Preschool Childhood. Prosveshchenie Publishing House, Moscow (1966). (in Russian)
33. Zellers, R., Bisk, Y., Farhadi, A., Choi, Y.: From recognition to cognition: visual commonsense reasoning. CoRR abs/1811.10830 (2018)

Transfer Learning Study for Horses Breeds Images Datasets Using Pre-trained ResNet Networks

Enrique de la Cal$^{(\boxtimes)}$, Enol García González, and Jose Ramón Villar

Computer Science Department, University of Oviedo, Oviedo, Spain
{delacal,villarjose}@uniovi.es

Abstract. In previous work, we have carried out an academic study of the automatic classification of horse breed images by pre-trained DL models.

In the present paper, we continue that line of research by extending the former results considering a new dataset including known and unknown breeds. Thus, two main goals are tackled here: i) new experiments of transfer learning considering the known breeds of both former and new datasets, and ii) a study of similarity between the known and unknown breeds. When trying to classify unknown breeds, it is expected that the models obtained in goal i) can be used to analyze the morphological similarity between unknown breeds and known breeds. In order to "evaluate" the results of this analysis, we have relied on the advice of an expert in the field of horses.

From the experts' point of view, the horses' morphology defines some of the typical uses: riding, draught, multi-purpose. Thus, as most of the comparisons agreed with the expert's assessment, the research line into morphological similarities using pre-trained DL models is reliable. Future work will be proposed to carry out similarity studies with other datasets and similarity studies using parts of the horse's body instead of taking full photos.

Keywords: Image classification · Resnet · Deep learning · Classification algorithms · Horse breeds

1 Introduction and Motivation

In [4], preliminary work on the Horse World figures in the light of Data Science was presented. This work concluded two essential issues: i) a quite complete analysis of 108 public datasets on horses available on the web was carried out, stating that there are four popular niches: races, images, herd analysis and horse health, and ii) as an academic challenge, a horse breeds images classification problem was selected from the public datasets analyzed, and a new pre-trained DL model with tuned parameters were obtained outperforming the base-line results [1].

© Springer Nature Switzerland AG 2021
H. Sanjurjo González et al. (Eds.): HAIS 2021, LNAI 12886, pp. 256–264, 2021.
https://doi.org/10.1007/978-3-030-86271-8_22

In this paper, the academic challenge is extended, dealing with the following two objectives: i) to develop several models of horse breed classification using pre-trained DL models, and ii) to study the model's behaviour when classifying unknown breeds. The second one is the main objective of this contribution, and the first one is included to join the preliminary work with this second primary objective.

When trying to classify unknown breeds, the main idea is that the model will analyze the similarity between the breeds. Specifically, it will indicate which known breed is morphological most similar to the unknown breed provided. As this process of similarity is not trivial, the support of an expert in the field will be available to assess the similarity between the different breeds manually.

This work is structured as follows. The following section includes the methodology carried out to deploy the transfer learning study presented, while the experimentation and the discussion of the results are coped in Sect. 3. Finally, conclusions and future work is included in Sect. 4.

2   The Proposal

The experimentation proposal is based on two main parts: a first part in which horse breed classification models are created and evaluated, and a second part in which these classifiers analyze the similarity of horse breeds already known with others that the model does not know.

The first part of the proposal is based on taking a set of Deep Learning architectures for image classification and creating models that can classify images of 6 horse breeds correctly. So, transfer learning techniques will be used to speed up the training stage.

In the second part of the proposal, we will use the models from the first part. It should be remembered that these models were trained to be able to classify images of six specific horse breeds. In this second phase, images of six horse breeds will also be used, but these six breeds will be different from the first. This phase aims to use the models from the first phase to determine how similar two horse breeds are. To cross-check this information, an expert in the horse field will manually analyze the horse breeds included in both phases and evaluate the degree of morphological similarity between the breeds.

Figure 1 summarizes the main ideas of the proposal, the architectures to be used in the experimental phase, and the breeds used in each of the phases.

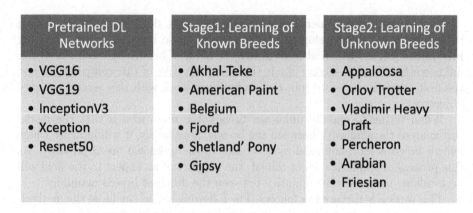

Fig. 1. The general overview of the proposal

3   Numerical Results

3.1   Materials and Methods

Five pre-trained DL architectures have been selected: VGG16 [11], VGG19 [11], Resnet50 [7], InceptionV3 [12], Xception [3]. The runs of the training stage have been carried out with each of these architectures (initialized with pre-trained network weights taken from ImageNet [10]).

For training, the models were set up to be optimized using the stochastic gradient descent method with moment 0.9 and using a learning rate value of $1e^{-1}$. In addition, a Data Augmentation algorithm was used for training in which the images were randomly modified by applying zoom and shear modifications. The shear values vary randomly between 0 and 0.2° and the zoom between 0% and 20%. The number of epochs the training lasts has been set manually for each architecture:

- InceptionV3: 100
- Xception: 150
- Resnet50: 100
- VGG16: 200
- VGG19: 200

As this work proposed using transfer learning, two datasets have been used: i) the source dataset has been taken from Atabay work [1], and the target from Kaggle repository [8]. So let us describe both datasets:

The Atabay dataset, hereafter referred to as D1, consists of 5079 images of six horse breeds, containing approximately the same number of images of each breed. The breeds included in this dataset, along with their abbreviations, are

Akhal-Teke (D1.B1), American Paint (D1.B2), Belgian (D1.B3), Fjord (D1.B4), Shetland' Pony (D1.B5), Gypsy (D1.B6). Figure 2 includes a sample of the images of each of the breeds contained in D1. This dataset will be the one used in the first part of the proposal to train and perform the test of the models.

(a) B1. Akhal-Teke   (b) B2. American Paint   (c) B3. Belgian

(d) B4. Fjord   (e) B5. Shetland's Pony   (f) B6. Gypsy

Fig. 2. Breeds of Dataset 1

The second dataset, referred to as D2, consists of 670 images of 7 horse breeds. As in D1, numerical values will be used for each breed. The breeds included in this dataset, along with their abbreviations, are Akhal-Teke (D2.B1), Appaloosa (D2.B2), Orlov Trotter (D2.B3), Vladimir (D2.B4), Percheron (D2.B5), Arabian (D2.B6), Friesian (D2.B7). Figure 3 includes a sample of the images of each of the breeds contained in D2. Although this dataset will be used in the second part of the proposal, the first breed (Akhal-Teke) is common to D1, which will serve as a control subject.

Table 1 shows the details of each dataset, including the total and deviation of images for each breed.

(a) B1. Akhal-Teke   (b) B2. Appaloosa   (c) B3. Orlov Trotter   (d) B4. Vladimir

(e) B5. Percheron   (f) B6. Arabian   (g) B7. Friesian

Fig. 3. Breeds of Dataset 2

Table 1. Number of images contained in each dataset

	D1.B1	D1.B2	D2.B3	D2.B4	D2.B5	D2.B6		Total	Deviation
D1	876	822	843	810	894	834		5079	32,33

	D2.B1	D2.B2	D2.B3	D2.B4	D2.B5	D2.B6	D2.B7	Total	Deviation
D2	123	105	107	37	56	122	120	670	34,79

3.2   Results

Thus, the two studies referred to in the proposal section are: i) the classification of the known breeds from the datasets D1 as well as the inclusion of the samples D2.B1 in order to check the hardiness of the models, and ii) a similarity study between the breeds from D1 and D2. In addition, a discussion section is included comparing the similarity results obtained with our models for unknown breeds and the corresponding assessment of a human expert.

Comparative Performance Between Our Models and Atabay for Dataset D1. Table 2 shows the accuracy of our proposal for the six breeds of dataset D1 as well as the whole dataset (All B) compared to Atabay results also for the whole dataset (All B). Furthermore, our proposal outperforms the Atabay results for all network architectures (Network) for the whole dataset (All B). Concerning the single breed performance, it can be seen that the worst behaviour is obtained for B3 in all models.

Besides, the results of Table 3 show that the performance of the proposed models is reduced lightly concerning the original models, after including the

Table 2. Accuracy of our models (Our Proposal columns) compared with Atabay results for the test fold of dataset D1

Network	Our proposal							Atabay
	B1	B2	B3	B4	B5	B6	All B	All B
InceptionV3	0.9875	0.9714	0.9463	0.9806	0.9947	0.9755	0.9762	0.8879
Xception	0.9893	0.9390	0.9333	0.9690	0.9610	0.9831	0.9626	0.9300
ResNet50	0.9840	0.9829	0.9759	0.9884	0.9699	0.9755	0.9793	0.9590
VGG16	0.9840	0.9771	0.9556	0.9884	0.9734	0.9718	0.9750	0.9069
VGG19	0.9822	0.9657	0.9574	0.9826	0.9876	0.9755	0.9753	0.9005

images from the D2.B1 dataset into the D1 test dataset, bearing the hardiness of the models out.

Table 3. Accuracy for the test fold of dataset D1 including the D2.B1 samples

	D1.B1	D1.B1+D2.B1	B2	B3	B4	B5	B6	All B
InceptionV3	0.9875	0.9882	0.9714	0.9463	0.9806	0.9947	0.9755	0.9775
Xception	0.9893	0.9882	0.9390	0.9333	0.9690	0.9610	0.9831	0.9651
ResNet50	0.9840	0.9828	0.9829	0.9759	0.9884	0.9699	0.9755	0.9795
VGG16	0.9840	0.9731	0.9771	0.9556	0.9884	0.9734	0.9718	0.9731
VGG19	0.9822	0.9720	0.9657	0.9574	0.9826	0.9876	0.9755	0.9734

Figure 4 also contains a comparison of the two evaluations detailed in the tables.

Study of Similarity Between D1 and D2 Breeds. In this second part, we will perform a similarity study between the breeds of both datasets to determine how similar the D2 breeds are to the D1 breeds. Figure 5 shows the proportion of D2 images classified as each of the classes known to the models. It is assumed that the more significant the proportion of images classified as a class, the greater the similarity between those two races.

Discussion. An expert in the horse world assessed the degree of morphological similarity manually, with values in the range 0%–100%, of each pair of different breeds from both datasets. Figure 6 shows the figures of automatic similarity estimation (System) and expert assessment for each pair of breeds (Expert), where it can be remarked that some agreement between the system and the human expert (in the dotted pattern). The similarity estimated by the system agrees with the expert for the Akhal-Teke/Shetland'pony compared to the D2 breeds barely. Moreover, in general terms, almost the 50% of estimations agree. Moreover, in general terms, the 50% of estimations agree, including the grade of comparison between the system and expert.

D1
D1+D2.B1
VGG19
VGG16
ResNet50
Xception
InceptionV3
Breed 6
Breed 5
Breed 4
Breed 3
Breed 2
Breed 1
0,9000
0,9200
0,9400
0,9600
0,9800
1,0000

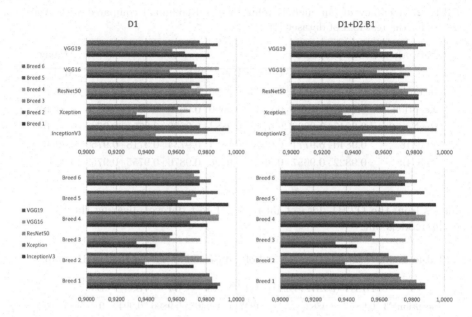

Fig. 4. Accuracy comparative between VGG16, VGG19, ResNet50, Exception and InceptionV3 for D1 and D2 breeds

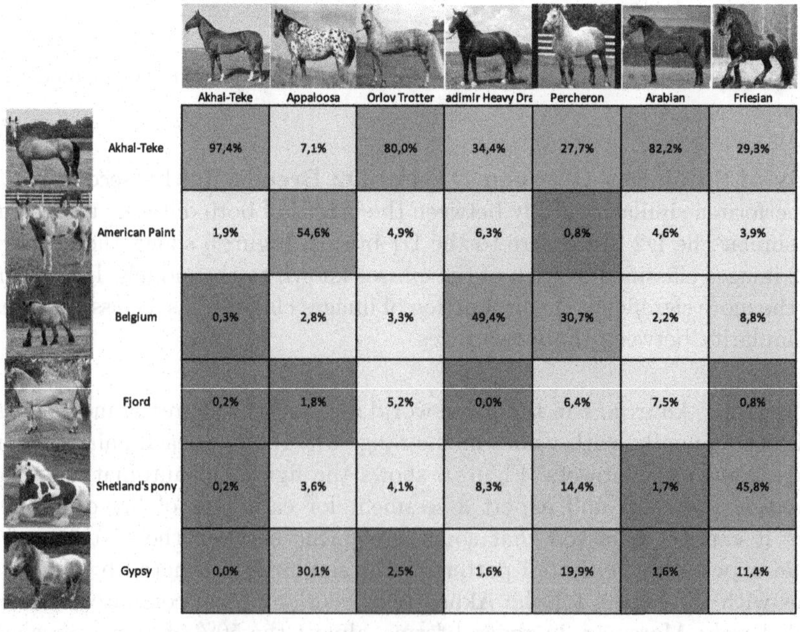

Akhal-Teke
Appaloosa
Orlov Trotter
adimir Heavy Dra
Percheron
Arabian
Friesian
Akhal-Teke 97,4% 7,1% 80,0% 34,4% 27,7% 82,2% 29,3%
American Paint 1,9% 54,6% 4,9% 6,3% 0,8% 4,6% 3,9%
Belgium 0,3% 2,8% 3,3% 49,4% 30,7% 2,2% 8,8%
Fjord 0,2% 1,8% 5,2% 0,0% 6,4% 7,5% 0,8%
Shetland's pony 0,2% 3,6% 4,1% 8,3% 14,4% 1,7% 45,8%
Gypsy 0,0% 30,1% 2,5% 1,6% 19,9% 1,6% 11,4%

Fig. 5. Similarity between known and unknown breeds

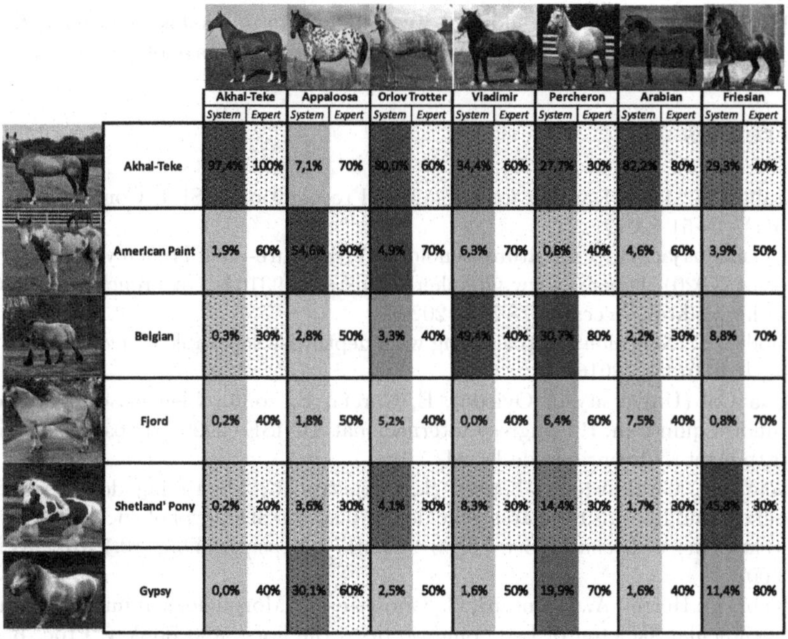

	Akhal-Teke		Appaloosa		Orlov Trotter		Vladimir		Percheron		Arabian		Friesian	
	System	Expert	System	Expert	System	Expert	System	Expert	System	Expert	System	Expert	System	Expert
Akhal-Teke	97,4%	100%	7,1%	70%	80,0%	60%	34,4%	60%	27,7%	30%	82,2%	80%	29,3%	40%
American Paint	1,9%	60%	54,6%	90%	4,9%	70%	6,3%	70%	0,8%	40%	4,6%	60%	3,9%	50%
Belgian	0,3%	30%	2,8%	50%	3,3%	40%	49,4%	40%	30,7%	80%	2,2%	30%	8,8%	70%
Fjord	0,2%	40%	1,8%	50%	5,2%	40%	0,0%	40%	6,4%	60%	7,5%	40%	0,8%	70%
Shetland' Pony	0,2%	20%	3,6%	30%	4,1%	30%	8,3%	30%	14,4%	30%	1,7%	30%	45,8%	30%
Gypsy	0,0%	40%	30,1%	60%	2,5%	50%	1,6%	50%	19,9%	70%	1,6%	40%	11,4%	80%

Fig. 6. Comparative study of the breeds similarity assess by the proposal versus an human expert

4 Conclusion and Future Work

We have developed an exhaustive analysis of five pre-trained models for horse-breeds images classification from two different datasets, D1 and D2. On the one hand, the obtained results for D1 outperformed the base-line algorithm by Atabay [1]; on the other hand, the models were used to compare images from different breeds, obtaining similar results to the ones by a human expert.

For future work, it is proposed to continue with academic studies analyzing the performance of the architectures used for this study with experiments using D2 and new datasets composed by modifications of the images contained in D2.

Concerning the automatic breeds images classification, it can be stated that applications on real-world of these methods can be arranged in three topics: i) morphological research for breeding [2,13,16], ii) identifying archaeological deposits [5,6,9], and iii) visual identification for automatic shepherd [14,15]. So, an exciting extension of the proposal studied in the current work can be training the models from isolated or combined parts of the horse body like the head, legs, body, rump, and tail. This way, it is possible selecting just some desired parts of the studs in the breeding process.

Acknowledgement. This research has been funded partially by the Spanish Ministry of Economy, Industry and Competitiveness (MINECO) under grant TIN2017-84804-

R/PID2020-112726RB-I00. In addition, we would like to thank Jose Sánchez Cebollada for their valuable assistance as the expert on horse breeds morphology.

References

1. Atabay, H.: Deep learning for horse breed recognition. CSI J. Comput. Sci. Eng. **15**(1), 45–51 (2017)
2. de Castelbajac, H.: Viticulture/oenology -Viticulture-: 3 of French winegrowers use horses (2020). https://www.vitisphere.com/news-93104-3-of-French-winegrowers-use-horses.htm. Accessed 18 May 2021
3. Chollet, F.: Xception: deep learning with depthwise separable convolutions. CoRR abs/1610.02357 (2016)
4. de la Cal (University of Oviedo), E, García, E.U.o.O.: Ciencias de Datos en el Mundo Equino. In: II Congreso Internacional AINISE (2020). https://www.ainise.org/ponentes/dr-enrique-de-la-cal/
5. Hanot, P., Guintard, C., Lepetz, S., Cornette, R.: Identifying domestic horses, donkeys and hybrids from archaeological deposits: a 3D morphological investigation on skeletons. J. Archaeol. Sci. **78**, 88–98 (2017). https://doi.org/10.1016/j.jas.2016.12.002
6. Hanot, P., Herrel, A., Guintard, C., Cornette, R.: Morphological integration in the appendicular skeleton of two domestic taxa: the horse and donkey. Proc. R. Soc. B: Biol. Sci. **284**(1864) (2017). https://doi.org/10.1098/rspb.2017.1241
7. He, K., Zhang, X., Ren, S., Sun, J.: Deep residual learning for image recognition. CoRR abs/1512.03385 (2015)
8. Kaggle: Kaggle.com (2020). https://www.kaggle.com
9. Merkies, K., Paraschou, G., McGreevy, P.D.: Morphometric characteristics of the skull in horses and donkeys-a pilot study. Animals **10**(6), 1002 (2020). https://doi.org/10.3390/ani10061002
10. Russakovsky, O., et al.: ImageNet large scale visual recognition challenge. Int. J. Comput. Vis. **115**(3), 211–252 (2015). https://doi.org/10.1007/s11263-015-0816-y
11. Simonyan, K., Zisserman, A.: Very deep convolutional networks for large-scale image recognition. In: Bengio, Y., LeCun, Y. (eds.) 3rd International Conference on Learning Representations, ICLR 2015, San Diego, CA, USA, 7–9 May 2015, Conference Track Proceedings (2015). http://arxiv.org/abs/1409.1556
12. Szegedy, C., Vanhoucke, V., Ioffe, S., Shlens, J., Wojna, Z.: Rethinking the inception architecture for computer vision. CoRR abs/1512.00567 (2015)
13. The Local Journal (France): French vineyards revive horse-drawn ploughs (2016). https://www.thelocal.fr/20160814/picture-postcard-french-vineyards-revive-horse-drawn-ploughs/. Accessed 1 June 2021
14. van Gemert, J.C., Verschoor, C.R., Mettes, P., Epema, K., Koh, L.P., Wich, S.: Nature conservation drones for automatic localization and counting of animals. In: Agapito, L., Bronstein, M.M., Rother, C. (eds.) ECCV 2014. LNCS, vol. 8925, pp. 255–270. Springer, Cham (2015). https://doi.org/10.1007/978-3-319-16178-5_17
15. Vayssade, J.A., Arquet, R., Bonneau, M.: Automatic activity tracking of goats using drone camera. Comput. Electron. Agric. **162**, 767–772 (2019). https://doi.org/10.1016/j.compag.2019.05.021
16. Wineterroirs.com: Wine Tasting, Vineyards, in France: draft Horse in the vineyard (2010). https://www.wineterroirs.com/2010/04/draft_horse.html

Machine Learning Applications

Fraudulent E-Commerce Websites Detection Through Machine Learning

Manuel Sánchez-Paniagua[1,2(✉)], Eduardo Fidalgo[1,2], Enrique Alegre[1,2],
and Francisco Jáñez-Martino[1,2]

[1] Department of Electrical, Systems and Automation,
University of León, León, Spain
{manuel.sanchez,eduardo.fidalgo,enrique.alegre,
francisco.janez}@unileon.es
[2] INCIBE (Spanish National Cybersecurity Institute), León, Spain

Abstract. With the emergence of e-commerce, many users are exposed to fraudulent websites, where attackers sell counterfeit products or goods that never arrive. These websites take money from users, but also they can stole their identity or credit card information. Current applications for user protection are based on blacklists and rules that turn out into a high false-positive rate and need a continuously updating. In this work, we built and make publicly available a suspicious of being fraudulent website dataset based on distinctive features, including seven novel features, to identify these domains based on recently published approaches and current web page properties. Our model obtained up to 75% F1-Score using Random Forest algorithm and 11 hand-crafted features, on a 282 samples dataset.

Keywords: E-commerce · Fraud detection · Machine learning · Cybersecurity

1 Introduction

Over the past few years, retail companies have started the digital transformation to provide their services and products through the Internet [1]. Most physical and new companies are migrating to e-commerce websites to reach customers and display the available goods [25]. Those websites often have the same structure to make them accessible and intuitive to anybody interested in the brand or the company. Fraudsters take advantage of the similarity between websites and create e-commerce online stores to place counterfeit products into the market or to take customer's money in return for no items.

Due to this change in business model, more and more people are entering e-commerce services to obtain products. Statista, a global business data platform, stated that user penetration will be 50.8% in 2021 and is expected to hit 63.1%

Supported by INCIBE.

by 2025. These predictions promote an increase in the online market, which means that more users will be exposed to e-commerce fraud. The Organisation for Economic Co-operation and Development (OECD) found in 2016, that up to 3.3% of the global trade are counterfeit products [19]. Furthermore, the European Commission found that 62% of surveyed customers have suffered a buying scam[1] [6]. These surveys and reports highlighted the impact of these scams on the final user and the targeted companies.

Current protection for users depends on its knowledge and experience to prevent these attacks. There are also online detection tools, like ScamAdviser[2], ScamFoo[3] or Scamner[4], where users can check the confidence of a certain domain. These services use information systems and rules to collect and analyze data about the domain. However, many users may not known these services or may directly trust in the websites. Thereby, an automatic system to speed up the early detection of fraudulent pages is required without depending on human participation.

The main problem of rule-based systems is that new legitimate websites obtain low confidence scores due to their similarities with fraud ones. Registration date and volume of users for a new domain are the main reason for the low confidence score in many of the aforementioned tools. For both reasons, companies and cybersecurity experts look for implementing automatic systems based on Artificial Intelligence to speed up the early detection of fraudulent websites [4,14,27]. Recent works retrieve features from HTML, text, images and 3rd party services and use traditional machine learning algorithms to detect fraudulent websites [27].

In this paper, we propose a pipeline capable of detecting suspicious of being fake online shops using data from the actual page and 3rd party services. Throughout this manuscript, we use *fraudulent websites* term to refer to *suspicious of being fraudulent websites*. We adapt the set of features of Wu et al. [27] and add novel features from Secure Sockets Layer (SSL) certificate and TrustPilot service information, as well as policies pages, e-commerce development technologies and social media links. We look for exploring the challenge of detecting fraudulent websites to set a baseline for this research line with a novel feature vector. First, we collect a set of e-commerce websites to build a tailored dataset to use for our proposal, called Features from Fraudulent Websites 282 (FFW-282). Second, we evaluate the use of a feature vector based on sample analysis to determine the features to be extracted. Finally, we assess five different machine learning algorithms to state the best performance model.

Using machine learning instead of traditional rules provide our work with a holistic point of view. Some rules may categorize a website by its age [27], therefore if the website is one month old, it is directly identified as fraud, gener-

---

[1] a buying scam includes: fake goods, undelivered goods or services, fake invoices and unwanted monthly subscriptions.
[2] https://www.scamadviser.com/.
[3] https://www.scamfoo.com/.
[4] https://www.scamner.com/.

ating a false positive since old domains have their legitimacy proved over time. To address this issue, we count on different features to correctly identify these threats and prevent users from getting their money stolen.

This paper is structured as follows. Section 2 presents a review of the literature and related work. Section 3 explains the methodology, the features proposed and the metrics used. Section 4 present the results of the different experiments, Sect. 5 contains the conclusion along with the limitations and future work.

2   Related Work

2.1   Literature Review

The detection of fraudulent e-commerce websites is an emerging challenge for cybersecurity agencies due to their fast growth and the potential harm to people. Many authors have been developed machine learning systems to deal with fraudulent websites [4,14,16,18,24,25,27]. Due to its continuous evolution and because of being a dynamic environment, recent works focused their proposals on different technologies to retrieve enough information.

On the one hand, some authors focused their research on the features extraction from the websites [14,18,24,27]. Wadleigh et al. [24] based their feature set on URL, web content and WHOIS properties. Mostard et al. [18] extracted features from HTML code which include web-page length, emails, phone number or payment methods. Khoo et al. [14] also retrieved text and images from the pages to detect fraudulent websites, keeping HTML features. Wu et al. [27] widened the number of features and used a feature vector that includes information from URL, social media and email addresses, payment forms or phone number appearance, WHOIS and content structure. However, previous works have not explored features such as policies pages, Trustpilot data, e-commerce technologies and Secure Sockets Layer (SSL) certificate, that may contain information more in line with current frauds.

On the other hand, other works considered natural language processing techniques to find out similarities among web pages, rather than retrieving features [4,16]. Beltzung et al. [4] used similarity between source code from websites by analyzing its HTML, Javascript, CSS, among others. Maktabar et al. [16] applied techniques of text classification introducing a sentiment analysis model on the textual content of the web-page.

2.2   Online Fraudulent Tools

In this subsection, we briefly describe online tools that are available and help users to detect fraudulent websites.

ScamAdviser relies on an automated algorithm that checks different sources to retrieve information about the domain. They rely on the registration and expiration date from WHOIS, the ranking of the website at Alexa, the reviews about the domain, a verification of the SSL certificate, and the e-commerce technology used by the online shop. By applying these rules, the website generates a confidence number from zero to 100 and left the user obtain its conclusions.

ScamFoo also has a set of rules to check website trustworthiness. It recollects information from external services to generate a confidence score. It implements similar services as ScamAdviser and also checks the domain on different blacklists looking for previous reports.

Scamner has a similar structure as previous tools. Additionally, it uses MozRank to provide more information about the ranking based on the number of websites linking to a target domain. It also displays the social media interactions for users to obtain their conclusions.

Finally, **Web Of Trust** is another tool that states to use community ratings, reviews and machine learning algorithms to obtain their rankings but does not provide further information about their method.

3   Methodology

In this study, (1) we propose a set of feature vectors to describe the most relevant information of a website to be considered as a fraudulent one, (2) we build a novel dataset based on the selected features, and (3) we evaluate five traditional machine learning algorithms. We show the entire assessment process in the Fig. 1.

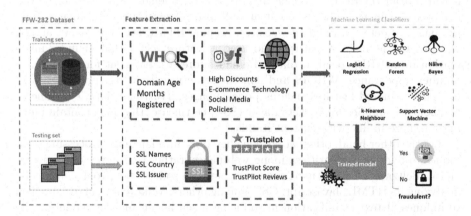

Fig. 1. Graphical Abstract of the evaluation process: (1) we split our dataset in training and testing sets, then (2) we extracted the proposed features from each domain and, finally, (3) we trained and tested five machine learning models to detect fraudulent websites.

3.1   Proposed Features

In this section, we describe the selected features for this task. After reviewing other works [4,24,27], we also consider features used in these works but using them from a novel perspective, such as high discounts, social media, domain age and months registered. To improve the performance of fraudulent detectors against current websites properties, we incorporate seven novel features as are SSL names, country and issuer, Trustpilot score and review, e-commerce technologies and policies. The complete list of features used are the following ones.

High discounts: Users are susceptible of end up tricked when they have the opportunity to obtain a valuable product with a great deal or discount [11]. Fraudsters use large discounts (>70%) to persuade users to buy a bargain. Usually, these offers come with a countdown which is an urgent appeal to increase even more users' susceptibility [26]. Furthermore, some works [24] calculate the average discount for the displayed items and used it as an input feature. For this reason, we look over the HTML code for high discounts, and we include in the feature vector how many discount banners we find.

SSL names: A single Secure Sockets Layer (SSL) certificate can protect multiple domain names. In this way, brands and companies use the same certificate for their online shop on different servers and countries. On the opposite side, attackers do not count on big infrastructures to serve their website in different countries. Therefore, if they have an SSL certificate, it may have only one name registered. We check the number of names registered as a feature for the model.

SSL country: There is a set of banned countries that cannot obtain an SSL certificate due to restrictions from the Certificate Authority (CA) in verification task for organization and domains [5]. We verify if the certificate has any of the banned or risk country codes included in Table 1.

Table 1. Banned and risk countries on SSL certificates.

Country	Country code	Status
Cuba	CU	Banned
Iran	IR	Banned
North Korea	NP	Banned
Sudan	SD	Banned
Syria	SY	Banned
Eritrea	ER	Risk
Guinea	GN	Risk
Iraq	IQ	Risk
Lebanon	LB	Risk
Pakistan	PK	Risk
Rwanda	RW	Risk
Sierra Leona	SL	Risk
Zimbabwe	ZQ	Risk

SSL Issuer: Most legitimate websites use common SSL issuers. After an analysis, we found 16 companies in charge of generating SSL certificates: Let's

Encrypt, Symantec, Geotrust, Comodo, DigiCert, Thawte, Network Solutions, Rapid SSL, Entrust Datacard, SSL.com, Sectigo, Cloudflare, GoDaddy, Google Trust Services, Amazon and CPanel. Let's Encrypt is one of the most common due to their free service [3], which is the main option for fraudsters. However, we should pay attention because it is also used by small companies that are looking for getting into the digital market. Furthermore, CPanel is not a certificate issuer but a service provider. Since there are a great number of websites using CPanel, we also added it to the list.

Trustpilot Score: Trustpilot[5] is a website where users provide opinions about websites, products and services based on their experiences. Reviews go along with a rating score ranging from one to five and they are important for online shops to increase user confidence [21]. Trustpilot calculates the website score with the mean of all rates provided by users. We obtained this score and used it as a legitimate feature.

Trustpilot Reviews: Another important parameter from Trustpilot is the number of reviews. Aged online shops have a great number of reviews but with a mid-range score due to the polarity of customers reviews [22]. By adding the number of reviews, we provide a holistic view of the legitimacy of a web page.

E-commerce Technologies: Most legitimate online shops watch over their design, accessibility and functionality to set an initial trust with the customer [13]. To do that, developers tend to use tools and frameworks with these capabilities to provide the best user experience. Unlike legitimate sites, fraudsters display raw websites, sometimes, implemented from a simple HTML template. We propose to detect the different technologies used by the website since poor and fast designs are commonly related to fraud sites. Wappalyzer allowed us to retrieve the technologies used by the website through the fingerprint exposed in the HTML code. In this feature, we counted the number of e-commerce frameworks or tools were used in the website development, like *Shopfy*, *WooCommerce* or *Zen Cart*.

Social Media: A company developing its digital commerce usually implies consumer brand engagement (CBE) campaigns on social media by creating accounts on different platforms such as Twitter, Instagram or Facebook [12]. We have noticed that some fake websites provide empty social media links of redirections to blank pages. We extracted the social media links from the HTML code and verified two statements: first, if there is a link to any of the three platforms we have mentioned and second, if those links are connected to the actual company profile. We count how many of the three social media accounts are linked to the online shop website.

Policies: Legitimate e-commerce websites clarifies their conditions, terms and policies related to refunds, user's data, shipments and consumer contracts. This information improves the relationship between the user and the online shop [17]. On the other hand, fake shops do not usually provide any of those statements or

---

[5] https://www.trustpilot.com/ Retrieved July 2021.

links to them are blank. Therefore, we check how many links related to policies are responding correctly.

Domain Age: WHOIS information has been used in other fields such as phishing detection to identify short aged servers [2]. Fraud websites also have a short life span since authorities try to take them down once a user reports them. Besides, legitimate websites domains have been registered since their beginning and it is likely that to have a longer life span than fraud web pages. However, this feature can discriminate against recent legitimate websites. We count the number of months from the registration date on WHOIS information for the target domain.

Months Registered: Following the previous feature, attackers do not register their domain for a long time since the attack does not last for so long. We count the number of months from the registration date to the domain expiration date on WHOIS information, i.e., the life period.

3.2   Dataset Creation

The objective of this work is to detect fraudulent websites among e-commerce sites. Furthermore, we need information from the WHOIS service and technology analysis, so we decided to build a dataset that complies with this task. Figure 2 displays the process to collect the dataset used in this work.

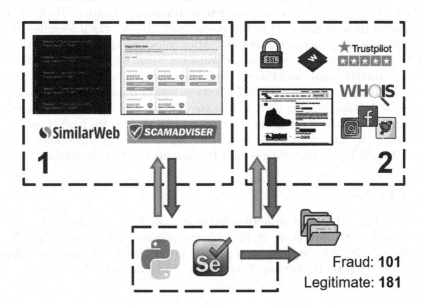

Fig. 2. Dataset recollection process. (1) We obtain the domains from the sources and (2) the script visits the websites and collect the data from it and the 3rd party services

The first step was to identify the sources that provide the target domains. Starting on the fraudulent class, we used user reports on ScamAdviser. These reports contain the domain name submitted by the user and the confidence score calculated using the predefined rules of ScamAdviser. We collected 197 total domains from ScamAdviser from February to June 2020. For those domains, a low confidence score is an indicator of a suspicious website, while higher scores identify legitimate e-commerce websites, which were the most common type among the obtained reports. Since domain names were submitted by users and evaluated with fixed rules, we performed a manual analysis over the collected samples to determine the best threshold to treat a website as suspicious or not. After the examination, we observed that all domains with a score higher than 75% were legitimate, while most of the rest were suspicious. Therefore, legitimate domains with a score under 75% were reviewed and manually labelled to avoid bias in the dataset.

For the legitimate class, we obtained the most visited online stores. We obtained the top 50 worldwide e-commerce domains from SimilarWeb[6]. We also introduced 34 well-known domains that did not appear in the list. The final dataset, Features from Fraudulent Websites 282 (FFW-282), is composed of 181 legitimate e-commerce domains and 101 fraudulent domains and it is publicly available for research purposes[7].

Once we have the domains, we used Selenium Webdriver and Python3 to visit and recollect the features from each domain. First, we obtain the HTML content and use a regex to retrieve if high discounts are in the text. Then, using the SSL module, we call *getpeercert* to retrieve SSL certificate information. To obtain the e-commerce technologies used in the Wappalyzer library, which identifies them by the code fingerprints in the website. After collecting the offline features, we call the external services for further information. First, we introduce the domain name in Trustpilot and collect its score and number of opinions registered on the website. Second, we use the WHOIS library to collect the information related to the actual domain. Finally, we search in the HTML code for social media links, specifically Facebook, Instagram and Twitter. As soon as we have those links, we check if the link corresponds to a valid user or not. All this information is stored in a JSON file that we use later for creating the feature vectors.

3.3   Classifiers

We trained five classifiers to obtain fraudulent prediction models, all of them widely used in the literature [9,10] and very different from each other: Random Forest (RF) [23], Support Vector Machine (SVM) [7], k-Nearest Neighbour (kNN) [20], Logistic Regression (LR) [8], and Naïve Bayes (NB) [15].

---

[6] https://www.similarweb.com/top-websites/category/e-commerce-and-shopping/ Retrieved July 2021.

[7] http://gvis.unileon.es/dataset/features-from-fraudulent-websites-282/ Retrieved July 2021.

4 Experiments and Results

4.1 Experimental Setup

Experiments are executed on an Intel Core i3 8100 at 3.6 Ghz and 16 GB of DDR4 RAM. We used scikit-learn and Python 3. Due to the small dataset size and likely bias, we averaged the output of a 5-Fold Cross Validation.

We have tested different settings to select the best combination of the main parameters for each classifier. For the rest of the parameters, we have used scikit-learn default values since we found no difference in their tuning. In the case of Random Forest, we obtained the best result using *n_estimators = 10* and *max_features = auto*. SVM obtained its best results using *C = 1.0* and Radial Basis Function (RBF) kernel. A high value of C parameter looks for a lower margin of hyperplane separation. Optimal parameters for kNN were 4 neighbours using *manhattan* metric. Logistic Regression was set with *l2* penalty, *C = 1* and *Limited-memory BFGS* solver. Finally, Naïve Bayes obtained its best results with the Bernoulli algorithm.

4.2 Performance Metrics

To assess the performance of the proposed classification models, we employed accuracy, precision, recall and F-score. We denoted the fraud class as the positive class and the legitimate class as the negative one.

We used F1-Score as the main metric for evaluation purposes since our dataset is slightly unbalanced. It can be computed as shown in Eq. 1.

$$F1 = 2 * \frac{Precision * Recall}{Precision + Recall} \tag{1}$$

Precision is also a relevant metric in this field and it is defined as the fraction of correctly classified fraud samples over the number of items classified as fraud, as indicated by Eq. 2.

$$Precision = \frac{TP}{TP + FP} \tag{2}$$

True Positive (TP) indicates the number of fraud samples correctly classified and False Positive (FP) depicts the number of legitimate samples wrongly classified as fraud.

Recall refers to the fraction of correctly classified fraud samples over the total number of fraud instances as indicated by Eq. 3.

$$Recall = \frac{TP}{TP + FN}, \tag{3}$$

True Negative (TN) refers to the number of legitimate samples correctly identified as legitimate and False Negative (FN) denotes the number of fraud samples improperly classified as legitimate.

Finally, the accuracy represents the number of samples that were correctly classified and it is calculated as shown in Eq. 4

$$Accuracy = \frac{TP + TN}{TP + TN + FP + FN} \tag{4}$$

4.3   Evaluation of Machine Learning Algorithms

In this experiment, we compare the proposed algorithms with the best parameters for this task. Results were obtained by calculating the mean value between the 5 folds.

Table 2. Results of the main machine learning algorithms

Algorithm	Precision	Recall	F1-Score	Accuracy
RF	80.00	70.59	**75.00**	**85.96**
kNN	60.87	82.35	70.00	78.95
SVM	61.90	76.47	68.42	78.95
LR	55.56	88.24	68.18	75.44
NB	53.33	94.12	68.09	73.68

Based on the results in Table 2, Random Forest obtained the best results among other classifiers with a 75% F1-Score. Random Forest was the most balanced algorithm and the only one with higher precision than recall, a conservative standpoint where fraud predictions are certain. However, the model misses an important rate of fraud samples (low recall). The opposite happens to the rest of the algorithms. According to the overall precision (80.00%) and recall (70.59%), most of the fraud websites are detected (high recall) but with a higher false-positive rate. Therefore, more legitimate samples are predicted as fraud while they are not. In LR and NB, almost half of the legitimate samples were misclassified as fraud (low precision). Since the increasingly risk of fraudulent websites, like scams or leaked data, for users, we recommend Random Forest.

5   Conclusion and Future Work

In this paper, we have proposed a model for fraudulent e-commerce website detection. We have presented a collection of features, including seven novel ones like SSL properties, Trustpilot metrics, policies analysis and e-commerce technologies, that can help in this task. Those are based on sample inspection and fraudsters techniques that differentiate a legitimate website from a scam one. Using these features, we have created a model to detect these websites with a 75% F1-Score. Results suggest that Random Forest is the algorithm with best performance. Although precision is higher than recall, we recommend its usage for this task. We consider this work as a good baseline to improve the detection of fraudulent websites from different perspectives but also presenting several contributions. First, we introduce and made publicly available our dataset FFW-282, although it contains a small number of samples which may not be enough to train a complex machine learning model. Since ScamAdviser advised that its information should consider as a recommendation rather than a ground truth, we manually inspected the websites to fix a score threshold above 75% labelling its e-commerce websites as legitimate ones, the rest ones were considered as suspicious of fraudulent activities. Therefore, we consider that building a larger dataset could be a high priority task, since there is no data available that can be used for detecting fraudulent e-commerce website. However, building a proper dataset has limitations since the sources of the most visited online shops requires a payment.

Second, the proposed feature vector could be improved to achieve a more descriptive descriptor of each website. We found on fraudulent websites specific identifiers that indicated a high probability of fraud, and they were the core of the proposed features set in this paper. Hence, with the enlargement of the dataset and a profound analysis of the websites and their components, we could look for more valuable features that may improve the performance of the model.

References

1. Adoption rate of emerging technologies in organizations worldwide as of 2020. https://www.statista.com/statistics/661164/worldwide-cio-survey-operational-priorities/. Accessed 25 Feb 2021
2. Adebowale, M., Lwin, K., Sánchez, E., Hossain, M.: Intelligent web-phishing detection and protection scheme using integrated features of images, frames and text. Expert Syst. Appl. **115**, 300–313 (2019). https://doi.org/10.1016/j.eswa.2018.07.067
3. Aertsen, M., Korczyński, M., Moura, G., Tajalizadehkhoob, S., Van Den Berg, J.: No domain left behind: is let's encrypt democratizing encryption? In: ANRW 2017 - Proceedings of the Applied Networking Research Workshop, Part of IETF-99 Meeting, pp. 48–57 (2017). https://doi.org/10.1145/3106328.3106338
4. Beltzung, L., Lindley, A., Dinica, O., Hermann, N., Lindner, R.: Real-time detection of fake-shops through machine learning. In: Proceedings - 2020 IEEE International Conference on Big Data, Big Data 2020, pp. 2254–2263 (2020). https://doi.org/10.1109/BigData50022.2020.9378204
5. Bottarini, J.: List Of Countries Banned & Restricted From Obtaining SSL Certificates (2016). https://www.wiyre.com/list-of-countries-banned-restricted-from-obtaining-ssl-certificates/. Accessed 06 Apr 2021
6. European Commission: Survey on scam and fraud experienced by consumers (2020). https://ec.europa.eu/info/sites/info/files/aid_development_cooperation_fundamental_rights/ensuring_aid_effectiveness/documents/survey_on_scams_and_fraud_experienced_by_consumers_-_final_report.pdf. Accessed 25 Feb 2021
7. Cortes, C., Vapnik, V.: Support-vector networks. Mach. Learn. **20**(3), 273–297 (1995). https://doi.org/10.1007/BF00994018

8. Cox, D.R.: The regression analysis of binary sequences. J. R. Stat. Soc.: Ser. B (Methodol.) **20**(2), 215–232 (1958). https://doi.org/10.1111/j.2517-6161.1958.tb00292.x

9. Das, M., Saraswathi, S., Panda, R., Mishra, A.K., Tripathy, A.K.: Exquisite analysis of popular machine learning-based phishing detection techniques for cyber systems. J. Appl. Secur. Res. 1–25 (2020). https://doi.org/10.1080/19361610.2020.1816440

10. Dou, Z., Khalil, I., Khreishah, A., Al-Fuqaha, A., Guizani, M.: Systematization of Knowledge (SoK): a systematic review of software-based web phishing detection. IEEE Commun. Surv. Tutor. **19**(4), 2797–2819 (2017). https://doi.org/10.1109/COMST.2017.2752087

11. Goel, S., Williams, K., Dincelli, E.: Got phished? Internet security and human vulnerability. J. Associ. Inf. Syst. **18**(1), 22–44 (2017). https://doi.org/10.17705/1jais.00447

12. Hollebeek, L., Glynn, M., Brodie, R.: Consumer brand engagement in social media: conceptualization, scale development and validation. J. Interact. Mark. **28**(2), 149–165 (2014). https://doi.org/10.1016/j.intmar.2013.12.002

13. Karimov, F., Brengman, M., van Hove, L.: The effect of website design dimensions on initial trust: a synthesis of the empirical literature. J. Electron. Commer. Res. **12**(4), 272–301 (2011)

14. Khoo, E., Zainal, A., Ariffin, N., Kassim, M.N., Maarof, M.A., Bakhtiari, M.: Fraudulent e-commerce website detection model using HTML, text and image features. In: Abraham, A., Jabbar, M.A., Tiwari, S., Jesus, I.M.S. (eds.) SoCPaR 2019. AISC, vol. 1182, pp. 177–186. Springer, Cham (2021). https://doi.org/10.1007/978-3-030-49345-5_19

15. Lewis, D.D.: Naive (Bayes) at forty: the independence assumption in information retrieval. In: Nédellec, C., Rouveirol, C. (eds.) ECML 1998. LNCS, vol. 1398, pp. 4–15. Springer, Heidelberg (1998). https://doi.org/10.1007/BFb0026666

16. Maktabar, M., Zainal, A., Maarof, M., Kassim, M.: Content based fraudulent website detection using supervised machine learning techniques. Adv. Intell. Syst. Comput. **734**, 294–304 (2018). https://doi.org/10.1007/978-3-319-76351-4_30

17. McCole, P., Ramsey, E., Williams, J.: Trust considerations on attitudes towards online purchasing: the moderating effect of privacy and security concerns. J. Bus. Res. **63**(9–10), 1018–1024 (2010). https://doi.org/10.1016/j.jbusres.2009.02.025

18. Mostard, W., Zijlema, B., Wiering, M.: Combining visual and contextual information for fraudulent online store classification. In: IEEE/WIC/ACM International Conference on Web Intelligence, pp. 84–90 (2019). https://doi.org/10.1145/3350546.3352504

19. OECD, E.U.I.P. Office: Trends in Trade in Counterfeit and Pirated Goods (2019). https://doi.org/10.1787/g2g9f533-en

20. Peterson, L.E.: K-nearest neighbor. Scholarpedia **4**(2), 1883 (2009). https://doi.org/10.4249/scholarpedia.1883. revision #137311

21. Proserpio, D., Zervas, G.: Online reputation management: estimating the impact of management responses on consumer reviews. Mark. Sci. **36**(5), 645–665 (2017). https://doi.org/10.1287/mksc.2017.1043

22. Schoenmueller, V., Netzer, O., Stahl, F.: The polarity of online reviews: prevalence, drivers and implications. J. Mark. Res. **57**(5), 853–877 (2020). https://doi.org/10.1177/0022243720941832

23. Statistics, L.B., Breiman, L.: Random forests. Mach. Learn. **45**, 5–32 (2001). https://doi.org/10.1023/A:1010933404324

24. Wadleigh, J., Drew, J., Moore, T.: The e-commerce market for "lemons": identification and analysis of websites selling counterfeit goods. In: WWW 2015 - Proceedings of the 24th International Conference on World Wide Web, pp. 1188–1197 (2015). https://doi.org/10.1145/2736277.2741658
25. Weng, H., et al.: Online e-commerce fraud: a large-scale detection and analysis. In: 2018 IEEE 34th International Conference on Data Engineering (ICDE), pp. 1435–1440 (2018). https://doi.org/10.1109/ICDE.2018.00162
26. Williams, E., Hinds, J., Joinson, A.: Exploring susceptibility to phishing in the workplace. Int. J. Hum. Comput. Stud. **120**, 1–13 (2018). https://doi.org/10.1016/j.ijhcs.2018.06.004
27. Wu, K., Chou, S., Chen, S., Tsai, C., Yuan, S.: Application of machine learning to identify Counterfeit Website. In: iiWAS2018: Proceedings of the 20th International Conference on Information Integration and Web-based Applications & Services, pp. 321–324 (2018). https://doi.org/10.1145/3282373.3282407

More Interpretable Decision Trees

Eugene Gilmore[1], Vladimir Estivill-Castro[2]([⊠]) [iD], and René Hexel[1] [iD]

[1] Griffith University, Nathan, QLD 4111, Australia
{eugene.gilmore,r.hexel}@griffithuni.edu.au
[2] Universitat Pompeu Fabra, 08018 Barcelona, Spain
vladimir.estivill@upf.edu

Abstract. We present a new Decision Tree Classifier (DTC) induction algorithm that produces vastly more interpretable trees in many situations. These understandable trees are highly relevant for explainable artificial intelligence, fair automatic classification, and human-in-the-loop learning systems. Our method is an improvement over the Nested Cavities (NC) algorithm. That is, we profit from the parallel-coordinates visualisation of high dimensional datasets. However, we build a hybrid with other decision tree heuristics to generate node-expanding splits. The rules in the DTCs learnt using our algorithm have a straightforward representation and, thus, are readily understood by a human user, even though our algorithm constructs rules whose nodes can involve multiple attributes. We compare our algorithm to the well-known decision tree induction algorithm C4.5, and find that our methods produce similar accuracy with significantly smaller trees. When coupled with a human-in-the-loop-learning (HILL) system, our approach can be highly effective for inferring understandable patterns in datasets.

Keywords: Classification rule mining · Interpretability

1 Introduction

A key reason for decision tree classifiers (DTC) remain a popular choice is their ability to be easily interpreted by humans [13]. The interpretability of classifiers is considered important because applying machine learning models [26,30] involves issues such as fairness and transparency of the model [39], ethical considerations [1], knowledge discovery [10] and accuracy and generalisability to unseen data. Another reason to develop interpretable classifiers is that they are naturally required for systems that allow interactive machine learning (IML) [2] and human in the loop learning (*HILL*) [10,37]. *HILL* is crucial in areas such as medical applications [22] and offers significant advantages such by allowing experts to contribute their domain knowledge [12] as well as to take advantage of the advanced pattern recognition capabilities humans possess, and to tailor a classification model for specific needs (e.g. stressing the importance of accuracy for a specific subset of classes [11]). Using a *HILL* system to exploit humans

© Springer Nature Switzerland AG 2021
H. Sanjurjo González et al. (Eds.): HAIS 2021, LNAI 12886, pp. 280–292, 2021.
https://doi.org/10.1007/978-3-030-86271-8_24

aptitude for pattern recognition also allows for a more natural understanding of classifiers, and new insights into the phenomena source of the datasets.

Decision-tree construction has an ubiquitous presence [19,24,29,31] in machine learning, statistics and data mining. Interestingly, on the issue of what the tests should be in the nodes, the focus is also on how test types impact the accuracy of the tree. However, when selecting test types, it is also important to consider the comprehensibility of the resulting rules. Rules are formed by the conjunction of the tests from the root to a leaf. These are combined into the disjunction of all rules that share the same class label in the leafs. In the case of *HILL*, it is also likely that the comprehensibility will influence the accuracy of the tree that is created. Different types of tests may have different effectiveness, depending on how the tree is learnt. If the tree is learnt using traditional methods with no human intervention, then the best test format may be different than if a human-in-the-loop approach is used. Particularly since more complex formats (for instance, oblique test that involve many attributes) require heuristics [8] to estimate the best possible test when using a machine-only approach.

Similarly, the treatment of the topic of visualisation [32] in knowledge discovery seems to focus on traditional plots (histograms, pie charts, bar charts, box charts commonly found in spread-sheets). What we find remarkable is that the visualisation of the decision-tree construction method and/or its results seems limited to illustrating the tree as an acyclic directed graph [32,34]. More remarkable is that although some textbooks discuss visualisation [33, Chapter 3] and address the topic of parallel coordinates [33, Page 126], the connection between Hunt's algorithm and its potential visualisation with parallel coordinates has remained unnoticed all this time.

We present a new DTC induction algorithm which builds a hybrid of Alfred Inselberg's parallel-coordinates system for high dimensional data visualisation and his accompanying Nested Cavities algorithm with Quinlan's C4.5 [27]. Our new induction algorithm can produce DTCs that, for many datasets, can be easily understood by a human user when using parallel coordinates to visualise the rules of the tree. Section 2 presents Hunts' framework of decision tree induction. We review the Nested Cavities algorithm in Sect. 3 and reformulates it under Hunts' framework. Section 4 describes our new algorithm including how we speed the search for informative splits that grow a decision tree. Section 5 presents our experimental evaluation over more than 45 commonly used datasets, where we show little loss of accuracy but large gains in succinct and human understandable trees. Our conclusions are in Sect. 6.

2 Preliminaries – A Review of Hunt's Algorithm

The fundamental greedy strategy [16] for the construction of decision-trees is 50 years-old and is adequately recognised as Hunt's algorithm [9,33]. The input is a supervised sequence (there may be repeated records) T of vectors of the form $(x_i, y_i)_{i=1,...,n}$, typically we call x the attributes, and we let d be the dimension of x. These encoded cases are considered rows of the training set T from which the method builds a decision-tree for classification among a finite set of categories $\mathcal{C} = \{C_1, C_2, \ldots, C_k\}$; thus $y_i \in \mathcal{C}$, $\forall i = 1, \ldots, n$:

1. If T is such that all y_i meet some homogeneous criteria H, we say T is a "pure" [7] node with a prediction given by a decision procedure D.
2. Alternatively, we select an informative test Q on some of the attributes and horizontally split T using Q into $T_1, \ldots, T_t$ and continue recursively.

Breiman *et al.* [7, Page 22] points out that the entire construction of the tree, revolves around three elements:

1. The criteria Q for the selection of the splits.
2. The decision procedure H to declare a node terminal or to split it.
3. The procedure D for assigning each terminal node to a class.

The diversity of decision-tree induction methods emerges from deciding on the specificities of Hunt's overarching strategy. These details include the following.

1. Often, Q is about one attribute only, and depending on whether the attribute is Boolean, categorical, nominal or numerical the split usually is binary [7].
2. How informative is the question Q is evaluated by how much it reduces the uncertainty about a case that is to be classified or the cost of performing the question Q. Measure such as information gain are the choice of well-known methods like C4.5 [27] but many other criteria (i.e. gini index) exist [7].

2.1   The Question Q or Tests to Split a Node and Expand the Tree

When learning a tree, there are numerous forms the test at the node can take. The simplest method is an unbounded interval, where the test acts on one attribute in the dataset that takes the form of $x > a_1$. It is also possible to have a single-attribute test for a bounded interval, using the form $a_1 > x > a_2$. These two forms are the original type of tests in decision-trees for numerical attributes, resulting in the so-called axis-parallel tests. They are so ubiquitous in decision trees we refer to them as *interval tests*. Intersecting interval tests results in parallelepiped tests; that is test of the form

$$S_i = (l_{i1} < x_{i1} < r_{i1}) \cap (l_{i2} < x_{i2} < r_{i2}) \cap \ldots \cap (l_{it} < x_{it} < r_{it}). \quad (1)$$

Interval tests and parallelepiped tests are equivalent. On the left side of Fig. 1 we can see that an interval just suits left and right tests in sequence in a decision-tree. Figure 1b shows that a parallelepiped can be represented as the corresponding sequence of intervals on the selected set of attributes.

More than one attribute can be used as part of a test for a node. Common examples are linear boundary tests that have the form $a_i x_i + a_j x_j + \ldots + a_t x_t < \theta_t$ (where a_i, a_j, ... a_t and θ_t are constants and x_i, x_j, ..., x_t are attribute values). These tests are called "oblique tests" [8,14,25,35] or "variable combinations" [7, Section 5,2]. The test could be a linear combination on all numerical attributes [7, Section 5.2.2] but for interpretability, a pruning procedure [7,

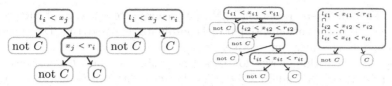

(a) An interval split is representable by two standard decision-tree splits.

(b) A sequence of interval splits represents a parallelepiped.

Fig. 1. Interval and parallelepiped tests have the same the expressive power.

Page 133] is used to involve only a few variables. Oblique splits have been shown to produce trees of comparable accuracy while reducing the size of the trees learnt [14]. We highlight that the simpler the test (for example intervals on one attribute), the smaller the search spaces for picking a particular test that maximises some criteria (such as information gain). The more complex the test, the slower the learning, as far more boundaries (splits) must be evaluated to chose one that maximises an heuristic criteria. Other alternatives are to use Boolean combinations or construct features that group attributes. Further variants explore how to reduce re-computation of the metrics that suggest how informative an attribute is.

Numerous DTC induction algorithms have been developed with the aim of creating smaller trees. Approaches such as MOHEAD-DT [5] use heuristics to attempt to learn a shorter tree while maintaining accuracy however are restricted to single attribute interval tests. Other induction algorithms such as OC1-DE [28] aim to produce smaller DTCs using oblique splits. As the oblique tests used in algorithms such as OC1-DE are variable combinations of attributes, the splits remain uninterpretable to a human.

3   Nested Cavities in the Framework of Hunt's Algorithm

The Nested Cavities [18] (or NC) algorithm uses parallel coordinates [17] to visualise high-dimensional datasets. While NC has been considered nowhere "similar" [20] to the construction of decision-trees, our first contribution is to show that NC is yet another interpretation of Hunt's algorithm, albeit with interesting alternatives for the selection Q of the splits, the tests at a node, and the halting criteria H. The NC is analogous to Ware *et al.*'s [36] USerClassifier. They both are interactive decision-tree building approaches following Hunt's algorithm where, in the divide step, the user selects the split Q. Ware et al. [36] approach (available in WEKA [6]) restricts the split in a node to a 2D-polygon because "it is difficult to construct splits visually on more than two attributes" [36]. Moreover, it seems the predilection for 2D Cartesian visualisations is because "all users are familiar with two-dimensional data representations from drawing programs and graphical packages" [36]. Interestingly, both, NC and USerClassifier [36, Page 282] consider the involvement of the user in configuring the question Q

a "geometric divide-and-conquer" [20], but seem to be unaware of each other. With NC, the visualisation with parallel coordinates bypasses the 2D restrictions when building classifiers interactively [3,4].

The NC algorithm works for two-class problems; thus T's instances are labelled C or $\neg C$. NC splits T into (1) the set S of positive instances (those elements of T labelled with class C), (2) and the negative instances $T \setminus S$ [18,20]. Akin to Hunt's method, initially, NC works with all instances and a root node that is a leaf, and like Hunt's method, NC builds a decision tree iteratively. At each iteration, also as per Hunt's method, the algorithm expands a tree by refining a leaf and building a test Q. The algorithm chooses a set of the attributes to produce a first convex-hull S_1, capturing some elements in positive instances S since T is considered a set in $\Re^d$ (nominal or other attributes are encoded as a real variables in the visualisation). The test at the root of the tree is whether each exemplar is or is not in S_1, and "in many cases using rectangular parallelepipeds for the wrapping suffices [20]". Thus, at depth i in the tree, a test S_i is a parallelepiped test [18, Fig. 3 and Fig. 5] consisting of a set of variables denoting attributes and an interval for each of those selected variables. The split consists of those instances inside the parallelepiped as opposed to those outside. In odd steps, the test wraps all of the positive examples S that arrive at this node ($S \subset C$) while minimising the number of negative examples that are captured by this rule. That is, elements of $T \setminus S \subset \neg C$ may be inside S_i. The role of the classes is reversed in even steps. Now, the test S_i is chosen to include all of the elements of $T \setminus S \subset \neg C$ that passed earlier tests, while hoping that only a few of $S \subset C$ will pass the new test. Thus, NC produces a binary tree with only one node expanded further, and at each level, a decision is made regarding whether the exemplar is C or $\neg C$. This approach produces skinny deep trees. The literature recommends shallow trees since deep tests in the tree are based on significantly fewer instances and thus the learning is subject to over-fitting. Bushy shallow decision trees are believed to generalise better [38].

We note that the NC approach also corresponds to the earlier *Conditional Focusing* [15, Fig. 8.3]; where some of the aim was also to organise knowledge and perhaps understanding the construction of abstractions to gain insight into human thought. The next observation, that seems to have been unnoticed, is that the NC method produces a model/classifier that is representable by standard decision-trees that use interval tests (refer to the discussion of Fig. 1).

The description of the NC algorithm reports astounding accuracy results (see [18, Table 1]) for two datasets (publicly available [21]). For the first dataset (*StatLog*) a remarkable error-rate of only 9% is reported. Algorithms using decision-trees, such as C4.5 and CART, achieve an error-rate of 15% and 13.8%,

Table 1. Typical performance of classifiers on the *StatLog* data set.

Algorithm	Accuracy	Error-rate
RandomForest -1 100 -K 0 -S 1	90.7%	9.3%
NNge -G 5 -I 5	86.5%	13.5%
SimpleCart -S 1 -M 2.0 -N 5 -C 1.0	85.9%	14.1%
J48 -C 0.25 -M 2	85.2%	14.8%
PART -M 2 -C 0.25	84.1%	15.9%
NaiveBayes	79.6%	23.1 %

respectively, on unseen data (the test-set). Similarly, on the vowel-recognition data set the reported accuracy [18, Table 2] for the NC algorithm has only 7.9%

error-rate (using cross-validation). Here, algorithms such as CART and k-NN have reported error-rates of 21.8% and 44.0%, respectively. This indicates a huge impact by involving humans and visualisation in generating the classifier. Using WEKA [6], we re-evaluated the results for the *StatLog* dataset on several algorithms (see Table 1). The classifier is learned from the supplied training set and evaluated in the supplied test set (as indicated by the documentation). These results once more highlight the exceptional results reported for the NC algorithm.

4   Decision Tree Induction with Parallelepipeds

We now detail our algorithm for the induction of DTCs called NCDT. We use the idea that parallelepiped tests are more interpretable than oblique tests, but in contrast to the NC algorithm we can create a DTC for datasets with an arbitrary number of classes. Our parallelepipeds will be the result of the conjunction of one or more interval tests, and we refer to them simply as a box. We can handle datasets with several classes because we evaluate boxes and obtain a best box for each class. The evaluation of a box minimises instances from the other classes. In contrast to NC, rather than constructing a parallelepiped that removes other classes in the next iteration of the algorithm, our iterative step re-evaluates all classes and constructs a box for the class that at that node is the best.

The method of obtaining the convex hull for the original NC algorithm is described only as "an efficient wrapping algorithm" [18]. The description of NC gives no further information on how NC creates this convex hull or even how the set of attributes to use for the convex hull is determined. As mentioned above, Inselberg states that a parallelepiped usually suffices [20]. For situations where a parallelepiped is not sufficient, there is even more mystery behind how the convex hull would be determined. Tests selected in our algorithm take only the form of a parallelepiped. We obtain a candidate parallelepiped for each class as follows. Given a class $C_c \in C$, we find the subset $S_c \subset T$ that contains only the instances labelled with class C_c. For each attribute $x_i \in X$ find the minimum $x_{i,min}$ and maximum $x_{i,max}$ values in S_c. From these minimum and maximum values, we construct a set R of interval tests of the form $x_{i,min} \leq x_i \leq x_{i,max}$. Clearly, the conjunction of the tests in R is a parallelepiped P_R that uses every attribute. Moreover, P_R is a multidimensional box that captures a subset S_R of instances of T. By construction, all instances labelled with class C_c are such that $S_c \subseteq S_R \subset T$. However, there may be instances in T that pass the test of the P_R box and also are not labelled with class S_c. That is, S_c may be a proper subset of S_R ($S_c \subset S_R \subset T$). By computing the number of instances from T captured by the rule R that are not of class C_c, we obtain what we call the impurity level $I_{min} = |S_R \setminus S_c|$. In general, we define an impurity level $I(r, C_c)$ as the number of instances in T which match all intervals in the set r and are not of class C_c. Note that among the sets of intervals defining a parallelepiped which capture all instances in S_c, the set R provides the minimum impurity level I_{min}.

Rather than using every attribute, we want to try to generalise the box P_R as much as possible. To achieve this, rather than using the combination of every test in R, we can explore all possible combination of test $2^R \setminus \{\emptyset\} = P = \{P_1, P_2, \ldots, P_{2^{|R|}-1}\}$, where $P_i \subset R$ is a non-empty subset of tests from R. Our algorithm sorts P on its cardinality, that is, the number of interval tests it uses. Because R uses all the interval tests, it is at the last position in the order, but also with value I_{min} Our algorithm scans P until it finds the first P_i such that $I(P_i) = I_{min}$. In this way, we construct a new test that contains the minimal number of attributes necessary to minimise the impurity with respect to a class. Having a test for each class, the final question Q for the node then simply becomes the test that has the smallest impurity.

4.1 Speeding up the Algorithm

Although the exhaustive approach guarantees finding a box with the minimal number of attributes required to minimise the impurity for a given class, its computational complexity is exponential on the number of attributes in the dataset (we must explore P, which holds all non-empty subsets of attributes). To reduce this computational complexity, we adjust the algorithm and build the box heuristically (using a greedy approach). We build the set $V \subseteq R$ iteratively, adding one interval (and thus one attribute, at a time). Initially, we compute the impurity $I(r)$ resulting from each interval test $r \in R$. That is we compute all the impurity for all singletons in P. We set V initially to the singleton of least impurity. From here, we find the next interval test in $R \setminus V$

Algorithm 1. NCDT Algorithm

1: **procedure** IMPURITY$(T, Rule, C_c)$
2:     $S_R \leftarrow MatchesRule(R, T_i), \forall T_i \in T$
3:     $S_c \leftarrow T_i.class == C_c, \forall T_i \in T$
4:     return $|S_R \setminus S_c|$

5: **procedure** NCDT(T)
6:     $rule \leftarrow \emptyset$
7:     $bestImpurity \leftarrow \infty$
8:     **for** c **do**1—C—
9:         $S_c \leftarrow T_i.class == C_c, \forall T_i \in T$
10:        $R \leftarrow \emptyset$
11:        **for** i **do**1—X—
12:            $x_{i,min} \leftarrow min(v.attribute[i] \forall v \in S_c)$
13:            $x_{i,max} \leftarrow max(v.attribute[i] \forall v \in S_c)$
14:            $R.append(x_{i,min} \leq x_i \leq x_{i,max})$
15:        $I_{min} \leftarrow Impurity(T, R, C_c)$
16:        **if** $I_{min} >= bestImpurity$ **then**
17:            Continue
18:        $V \leftarrow r_n \in R \mid I(R_n) \leq I(r_x) \forall r_x \in R$
19:        **while** true **do**
20:            $V.append(r_n \in R \setminus V \mid I(\{r_n\} \cup V) \leq I(\{r_x\} \cup V) \forall r_x \in R \setminus V)$
21:            **if** $Impurity(V) == I_{min}$ **then**
22:                Break
23:            $rule \leftarrow V$
24:            $bestImpurity \leftarrow I_{min}$
25:    return rule

---

that excludes the most instances of other classes; that is, the one that reduces the impurity the most. That is, if r is a test in $R \setminus V$ such that $I(V \cup \{r\})$ is as small as possible among the tests in $R \setminus V$, then we update $V \leftarrow V \cup \{r\}$. We repeat this step until the impurity $I(V)$ is equal to the best impurity $I_{min} = I(R)$.

Note that this method will always terminate. Although this method does not guarantee that we minimise the number of attributes required for the test, we find that, in practice, the complexity of the rules produced using this method is very similar. We evaluated this claim empirically by growing complete trees using our algorithm for many datasets. At each node, we calculate the rule by calculating the entire set P and by the above greedy method of iteratively adding the next best interval test. Where the interval tests from the two methods are different, we use the interval test obtained from evaluating the entire set P. From the 14 datasets, 223 internal nodes were created. From these 223 internal nodes, both methods produced identical interval tests in 212 cases. In a further three cases, the interval tests were different but contained the same number of attributes. The remaining eight cases used on average 1.125 (just under 13%) extra attributes for boxes generated with the iterative greedy method. Therefore, the greedy method (with complexity $O(d^2)$) approximates exceptionally well the exhaustive search for a box-test with fewest number of attributes.

4.2 Using Bitwise Operations to Compute Impurity

To increase the efficiency of the algorithm, we propose here a further speedup. As described, our greedy method to find a box with few attributes and small impurity could consider each combination of interval tests and recalculate how such combination filters every element in $T \setminus S_c$. To avoid this, we compute a binary representation where each bit at significance i in an integer is set to one if the interval test r_s includes the ith instance. Note that because the path from the root to a node in our algorithm is the conjunction of the box-tests along the way, this does not require recalculation. Under this setting, the iterative step of our greedy algorithm must identify which new interval test has the most positions with a zero (the instance does not pass the interval test, thus is not in the class S_c) where in the representation of V there is a 1.

We take advantage of our implementation utilising fast CPU instructions to compute the number of bits set to one in an integer variable. Naturally, in most cases, the size of most datasets will far exceed the number of available bits in an integer. We, therefore, use an array of integers to represent instances included in the interval test. To find the instances included when combining two interval test in R, we implement a bitwise logical-AND on the corresponding stored arrays and then total the

Table 2. Truth table for $B_V \land \neg B_r$

$B_{V,i}$	$B_{r,i}$	$B_{V,i} \land \neg B_{r,i}$
0	0	0
0	1	0
1	0	1
1	1	0

count using the above mentioned rapid count of bits in an integer. This removes the need to iterate over the entire training set to evaluate every combination of interval-tests. In particular, to quickly find the next interval test to add to the set V, we proceed as follows. For each interval test $r_n \in R \setminus V$ in consideration to be added to V, let B_V the bit representation of impurity of V (1 if the instance is contained in V but not in the class) and let B_r be the bit representation for r_n. The reduction in impurity of adding the test r_n to the current box V is the number of bits set in $B_V \land \neg B_r$. Table 2 shows that this calculation results in only one in cases where V did not exclude the instance($B_{V,i} = 1$) and r_n does exclude the instance($B_{r,i} = 0$).

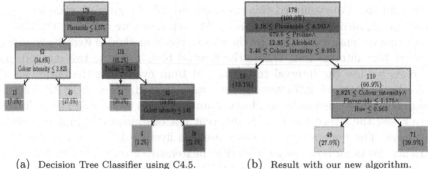

(a) Decision Tree Classifier using C4.5.   (b) Result with our new algorithm.

Fig. 2. Trees for the wine dataset from the UCI repository [21] visualised where each node is coloured according to the distribution of each class that reaches the node.

4.3   The Hybrid Approach with C4.5

While we find that NCDT can often produce very precise and succinct classifiers, it can struggle in its predictive performance for some dataset. Our algorithm aims at interval tests that capture the complete set of instances of at least one class in a node. This clearly is not always possible, and interval test that only covers a large proportion of a class should also be used, and further down the tree, a interval test that covers all instances would be feasible. To introduce other interval tests that do not use the specific interval from minimum value to maximum value of a class in an attribute, we propose a hybrid approach where our newly proposed algorithm shares the splitting with C4.5. In this hybrid approach at each internal node, we calculate two possible questions for the split Q of the node; the box-test that would be produced by our new algorithm and C4.5's suggestion for Q also at this node. We use the same gain ratio metric C4.5 uses to choose between using the test generated by C4.5 or the one from our new algorithm at each internal node. We observe that in the majority of cases, where our new algorithm proposes a sensible test, it has a higher gain ratio than that proposed by C4.5 and as a result, is picked. However, for situations where our newly proposed algorithm would need a test where the interval bounds are along the values of an attribute, adopting the hybrid approach results in the overall accuracy of the classifier improving significantly.

5   Experiments

Figure 2 show a situation where our newly proposed algorithm excels in learning a tree that is easily interpretable by a human user. Our algorithm produces a very small tree with one leaf node for each of the three classes in the dataset. The tree learnt using C4.5 fails to achieve this transparency.

We evaluate the performance of our algorithm by comparing it against our implementation of C4.5, a well known DTC induction algorithm[1]. Our evaluation

---

[1] Implementations available: https://github.com/eugene-gilmore/SwiftDecisionTrees.

measures predictive power as well as the size of the trees learnt on a number of datasets. We use datasets from the literature [23] to compare three different algorithms: our NC based algorithm, our hybrid approach with C4.5 as well as our implementation of the standard C4.5. As in the original implementation of C4.5 the DTCs generated with each algorithm are pruned using pessimistic pruning with a confidence interval of 25% and subtree raising. We performed 5 independent runs of 10-fold stratified cross validation on these datasets. We computed the average F1 Score and the average tree size across each of the five runs for each dataset. We counted the number of internal nodes in the tree (nodes with a split Q) as the size of a DTC. Table 3 shows the resulting F1 scores and tree sizes for each dataset. From this table, we can see that the hybrid approach can substantially increase the performance of the NC-based approach for a number of datasets while still producing smaller trees than C4.5. On average, new sizes are 30% of the original tree size for our method and 50% for the hybrid. Some exceptional data sets (balance-scale and vote) cause a larger tree to achieve comparable precision. The last two columns in Table 3 show the relative difference as a percentage, relative to C4.5 (second last column is C4.5 tree size relative to NCC45 tree size) as well as relative to the NCC45 size (last column).

We used the Wilcoxon signed-ranks test (p-value of 0.05) to check for a significant difference between our hybrid approach and C4.5 for both F1 score and tree size. The minimum sum of ranks for positive and negative difference is 237.5 for F1 score with a critical value of 294 and 68 for a tree size with a critical value of 378. With both of these values being under the exact critical value we can say that both, the reduced accuracy and smaller tree sizes of the hybrid approach are statistically signifi-

Table 3. Average over 5 runs of 10-fold cross validation.

Dataset	F1			Max diff	Tree size			Rel. diff.	
	C45 V1	NC	NCC45		C45 V1	NC	NCC45	as%	as%
arrhythmia	0.38	0.34	0.32	0.06	84	31	31	37%	37%
audiology	0.34	0.41	0.44	0.10	44	35	37	78%	83%
balance-scale	0.55	0.28	0.57	0.29	80	1	103	1%	−29%
breast-cancer	0.63	0.42	0.47	0.21	29	1	7	4%	23%
breast-cancer-Wisconsin	0.94	0.87	0.94	0.07	11	2	9	15%	82%
BreastTissue	0.65	0.67	0.63	0.04	13	8	8	64%	65%
car	0.90	0.65	0.95	0.30	69	12	56	17%	81%
cardiotocography	0.89	0.76	0.85	0.13	106	57	59	53%	56%
chronic	0.96	1.00	1.00	0.04	5	1	1	20%	20%
Climate Sim Crashes	0.74	0.56	0.59	0.18	12	4	4	32%	31%
cmc	0.50	0.24	0.51	0.27	279	8	181	3%	65%
credit	0.84	0.54	0.81	0.30	32	11	22	35%	70%
cryotherapy	0.91	0.92	0.92	0.01	4	3	3	78%	92%
dermatology	0.96	0.92	0.95	0.04	16	16	14	100%	88%
ecoli	0.55	0.48	0.48	0.07	19	19	20	98%	−6%
german	0.65	0.44	0.51	0.21	163	11	74	7%	45%
glass	0.69	0.59	0.61	0.10	25	13	13	52%	52%
haberman	0.57	0.44	0.55	0.13	29	4	13	15%	47%
heart	0.77	0.46	0.70	0.31	19	8	15	39%	80%
hepatitis	0.61	0.67	0.67	0.06	19	12	12	61%	61%
hungarian-14 heart-disease	0.75	0.54	0.78	0.24	36	10	24	27%	67%
ionosphere	0.89	0.91	0.91	0.02	13	3	3	23%	23%
iris	0.94	0.94	0.93	0.01	4	4	4	−16%	−11%
kr-vs-kp	0.99	0.75	0.99	0.24	57	5	35	9%	62%
leaf	0.62	0.55	0.53	0.09	50	30	31	61%	61%
liver	0.65	0.51	0.56	0.14	35	6	15	18%	44%
lymphography	0.17	0.20	0.16	0.04	55	44	56	79%	0%
mfeat-pix	0.52	0.47	0.49	0.05	126	58	60	46%	47%
newthyroid	0.91	0.90	0.91	0.01	15	8	8	54%	54%
optdigits	0.90	0.89	0.91	0.02	407	160	161	39%	40%
page-blocks	0.84	0.77	0.75	0.09	45	44	23	96%	50%
pima-diabetes	0.69	0.53	0.65	0.16	128	15	57	12%	44%
primary-tumor	0.20	0.20	0.20	0.00	86	96	97	−12%	−13%
seeds	0.92	0.90	0.93	0.03	7	4	4	62%	54%
segmentation	0.97	0.96	0.96	0.02	82	38	43	47%	52%
sick	0.93	0.50	0.62	0.43	51	6	7	12%	13%
sonar	0.70	0.58	0.58	0.12	33	7	7	20%	20%
soybean	0.92	0.92	0.90	0.02	87	45	56	52%	65%
tae	0.54	0.48	0.56	0.08	54	31	46	57%	84%
thyroid	0.97	0.98	0.98	0.01	34	13	16	38%	48%
tic-tac-toe	0.91	0.40	0.93	0.53	101	1	86	1%	85%
transfusion	0.63	0.47	0.58	0.16	25	4	13	18%	51%
vehicle	0.72	0.70	0.69	0.03	89	31	31	35%	35%
vote	0.96	0.38	0.95	0.58	10	1	11	10%	−8%
vowel	0.79	0.69	0.72	0.10	97	42	40	43%	41%
wine	0.93	0.93	0.93	0.00	4	2	2	47%	47%
zoo	0.85	0.89	0.89	0.04	16	13	13	83%	83%

cant. Despite the slight loss of accuracy when using the hybrid system, we argue that in most situations this loss is relatively small. Furthermore, our approach

here can be added to our effective *HILL* system [11], and in this case, users
can request notifications of scenarios where accuracy will suffer using this newly
proposed approach. We argue that the significant improvements in the size of
the trees and the ability to easily understand and visually evaluate tests involving numerous attributes make this algorithm a distinctively powerful tool for
assisting a user in a human-in-the-loop approach to learning.

6   Conclusions

We have presented a new decision-tree construction method that is essentially
equivalent in accuracy to previous decision-tree induction algorithms, but produces much shallower trees. Moreover, node tests are simpler, even if they involve
more than one attribute. These tests are parallelepipeds that can be readily visualised with parallel coordinates and are thus well suited to inclusion in human-in-
the-loop learning. We argue that a hybrid, building the trees with a traditional
C4.5 search for the splits gives the best trade-off between classification accuracy
and complexity of the decision-tree.

References

1. Ala-Pietilä, P., et al.: Ethics guidelines for trustworthy AI. Technical report, European Commission – AI HLEG, B-1049 Brussels (2019)
2. Amershi, S., Cakmak, M., Knox, W.B., Kulesza, T.: Power to the people: the role
of humans in interactive machine learning. AI Mag. **35**(4), 105–120 (2014)
3. Ankerst, M., Elsen, C., Ester, M., Kriegel, H.P.: Visual classification: an interactive approach to decision tree construction. In: 5th ACM SIGKDD International
Conference on Knowledge Discovery and Data Mining, KDD '99, NY, USA, pp.
392–396 (1999)
4. Ankerst, M., Ester, M., Kriegel, H.P.: Towards an effective cooperation of the user
and the computer for classification. In: 6th ACM SIGKDD International Conference on Knowledge Discovery and Data Mining, KDD '00, NY, USA, pp. 179–188
(2000)
5. Basgalupp, M.P., Barros, R.C., Podgorelec, V.: Evolving decision-tree induction
algorithms with a multi-objective hyper-heuristic. In: 30th ACM Symposium on
Applied Computing, pp. 110–117. ACM (2015)
6. Bouckaert, R.R., et al.: WEKA Manual V 3-6-2. University of Waikato (2010)
7. Breiman, L., Friedman, J., Stone, C., Olshen, R.: Classification and Regression
Trees. Wadsworth, Monterrey (1984)
8. Cantú-Paz, E., Kamath, C.: Inducing oblique decision trees with evolutionary algorithms. IEEE Trans. Evol. Comput. **7**(1), 54–68 (2003)
9. Cohen, P.R., Feigenbaum, E.A.: The Handbook of Artificial Intelligence, vol. III.
HeurisTech Press, Stanford (1982)
10. Estivill-Castro, V.: Collaborative knowledge acquisition with a genetic algorithm.
In: 9th International Conference on Tools with Artificial Intelligence, ICTAI '97,
pp. 270–277. IEEE Computer Society, Newport Beach (1997)
11. Estivill-Castro, V., Gilmore, E., Hexel, R.: Human-in-the-loop construction of decision tree classifiers with parallel coordinates. In: 2020 IEEE International Conference on Systems, Man, and Cybernetics, SMC, pp. 3852–3859. IEEE (2020)

12. Fails, J.A., Olsen, D.R.: Interactive machine learning. In: 8th International Conference on Intelligent User Interfaces. IUI '03, pp. 39–45. ACM (2003)
13. Freitas, A.A.: Comprehensible classification models: a position paper. SIGKDD Explor. **15**(1), 1–10 (2013)
14. Heath, D. G. et al. : Induction of oblique decision trees. In: 13th International Joint Conference on Artificial Intelligence, pp. 1002–1007. Morgan Kaufmann (1993)
15. Hunt, E.: Concept Learning – An Information Processing Problem, 2nd edn. Wiley, New York (1962)
16. Hunt, E., Martin, J., Stone, P.: Experiments in Induction. Academic Press, New York (1966)
17. Inselberg, A.: Parallel Coordinates: Visual Multidimensional Geometry and its Applications. Springer, New York (2009). https://doi.org/10.1007/978-0-387-68628-8
18. Inselberg, A., Avidan, T.: Classification and visualization for high-dimensional data. In: 6th ACM SIGKDD International Conference on Knowledge Discovery and Data Mining, Boston, MA, pp. 370–374 (2000)
19. Kotsiantis, S.B.: Decision trees: a recent overview. Artif. Intell. Rev. **39**(4), 261–283 (2013). https://doi.org/10.1007/s10462-011-9272-4
20. Lai, P.L., Liang, Y.J., Inselberg, A.: Geometric divide and conquer classification for high-dimensional data. In: DATA International Conference on Data Technologies and Applications, pp. 79–82. SciTePress (2012)
21. Lichman, M.: UCI machine learning repository (2013). http://archive.ics.uci.edu/ml
22. Maadi, M., Akbarzadeh Khorshidi, H., Aickelin, U.: A review on human-AI interaction in machine learning and insights for medical applications. Int. J. Environ. Res. Public Health **18**(4), 2121 (2021)
23. Mantas, C.J., Abellán, J.: Credal decision trees to classify noisy data sets. In: Polycarpou, M., de Carvalho, A.C.P.L.F., Pan, J.-S., Woźniak, M., Quintian, H., Corchado, E. (eds.) HAIS 2014. LNCS (LNAI), vol. 8480, pp. 689–696. Springer, Cham (2014). https://doi.org/10.1007/978-3-319-07617-1_60
24. Murthy, S.K.: Automatic construction of decision trees from data: a multidisciplinary survey. Data Min. Knowl. Discov. **2**(4), 345–389 (1998). https://doi.org/10.1023/A:1009744630224
25. Murthy, S.K., Kasif, S., Salzberg, S.: A system for induction of oblique decision trees. J. Artif. Int. Res. **2**(1), 1–32 (1994)
26. Pedraza, J.A., García-Martínez, C., Cano, A., Ventura, S.: Classification rule mining with iterated greedy. In: Polycarpou, M., de Carvalho, A.C.P.L.F., Pan, J.-S., Woźniak, M., Quintian, H., Corchado, E. (eds.) HAIS 2014. LNCS (LNAI), vol. 8480, pp. 585–596. Springer, Cham (2014). https://doi.org/10.1007/978-3-319-07617-1_51
27. Quinlan, J.: C4.5: Programs for Machine Learning. Morgan Kaufmann, San Mateo (1993)
28. Rivera-Lopez, R., Canul-Reich, J., Gámez, J.A., Puerta, J.M.: OC1-DE: a differential evolution based approach for inducing oblique decision trees. In: Rutkowski, L., Korytkowski, M., Scherer, R., Tadeusiewicz, R., Zadeh, L.A., Zurada, J.M. (eds.) ICAISC 2017. LNCS (LNAI), vol. 10245, pp. 427–438. Springer, Cham (2017). https://doi.org/10.1007/978-3-319-59063-9_38
29. Rokach, L., Maimon, O.: Top-down induction of decision trees classifiers - a survey. Trans. Syst. Man Cyber Part C **35**(4), 476–487 (2005)
30. Rudin, C.: Stop explaining black box machine learning models for high stakes decisions and use interpretable models instead. Nat. Mach. Intell. **1**, 206–215 (2019)

31. Safavian, S.R., Landgrebe, D.A.: A survey of decision tree classifier methodology. IEEE Trans. Syst. Man Cybern. **21**(3), 660–674 (1991)
32. Soukup, T., Davidson, I.: Visual Data Mining: Techniques and Tools for Data Visualization and Mining. Wiley, New York (2002)
33. Tan, P.N., Steinbach, M., Kumar, V.: Introduction to Data Mining. Addison-Wesley, Reading (2006)
34. Nguyen, T.D., Ho, T.B., Shimodaira, H.: Interactive visualization in mining large decision trees. In: Terano, T., Liu, H., Chen, A.L.P. (eds.) PAKDD 2000. LNCS (LNAI), vol. 1805, pp. 345–348. Springer, Heidelberg (2000). https://doi.org/10.1007/3-540-45571-X_40
35. Utgoff, P.E., Brodley, C.E.: An incremental method for finding multivariate splits for decision trees. In: 7th International Conference on Machine Learning, pp. 58–65. Morgan Kaufmann (1990)
36. Ware, M., et al.: Interactive machine learning: letting users build classifiers. Int. J. Hum.-Comput. Stud. **55**(3), 281–292 (2001)
37. Webb, G.I.: Integrating machine learning with knowledge acquisition. In: Expert Systems, vol. 3, pp. 937–959. Academic Press, San Diego (2002)
38. Witten, I.H., Frank, E.: Data Mining: Practical Machine Learning Tools and Techniques with Java Implementations. Morgan Kaufmann, Burlington (1999)
39. Zemel, R., Wu, Y., Swersky, K., Pitassi, T., Dwork, C.: Learning fair representations. In: 30th International Conference on Machine Learning, ICML, vol 28, pp. 325–333 (2013)

Algorithms Air Quality Estimation: A Comparative Study of Stochastic and Heuristic Predictive Models

Nadia N. Sánchez-Pozo[1,2](✉) [iD], Sergi Trilles-Oliver[1,5] [iD], Albert Solé-Ribalta[1] [iD],
Leandro L. Lorente-Leyva[2] [iD], Dagoberto Mayorca-Torres[2,3,4] [iD],
and Diego H. Peluffo-Ordóñez[2,6] [iD]

[1] Universitat Oberta de Catalunya, Barcelona, España
{nsanchezpo,strilles,asolerib}@uoc.edu
[2] SDAS Research Group, Ibarra, Ecuador
{nadia.sanchez,leandro.lorente,diego.peluffo}@sdas-group.com
[3] Facultad de Ingeniería, Universidad Mariana, Pasto (Nariño), Colombia
dmayorca@umariana.edu.co
[4] Programa de Doctorado en Tecnologías de la Información y la Comunicación, Universidad de
Granada, Granada, España
[5] Institute of New Imaging Technologies, Universitat Jaume I, Castelló, Spain
strilles@uji.es
[6] Modeling, Simulation and Data Analysis (MSDA) Research Program, Mohammed VI
Polytechnic University, Ben Guerir, Morocco
peluffo.diego@um6p.ma

Abstract. This paper presents a comparative analysis of predictive models applied to air quality estimation. Currently, among other global issues, there is a high concern about air pollution, for this reason, there are several air quality indicators, with carbon monoxide (CO), sulfur dioxide (SO_2), nitrogen dioxide (NO_2) and ozone (O_3) being the main ones. When the concentration level of an indicator exceeds an established air quality safety threshold, it is considered harmful to human health, therefore, in cities like London, there are monitoring systems for air pollutants. This study aims to compare the efficiency of stochastic and heuristic predictive models for forecasting ozone (O_3) concentration to estimate London's air quality by analyzing an open dataset retrieved from the London Datastore portal. Models based on data analysis have been widely used in air quality forecasting. This paper develops four predictive models (autoregressive integrated moving average - ARIMA, support vector regression - SVR, neural networks (specifically, long-short term memory - LSTM) and Facebook Prophet). Experimentally, ARIMA models and LSTM are proved to reach the highest accuracy in predicting the concentration of air pollutants among the considered models. As a result, the comparative analysis of the loss function (root-mean-square error) reveled that ARIMA and LSTM are the most suitable, accomplishing a low error rate of 0.18 and 0.20, respectively.

Keywords: Air quality · Contamination · Predictive models · Forecasting

© Springer Nature Switzerland AG 2021
H. Sanjurjo González et al. (Eds.): HAIS 2021, LNAI 12886, pp. 293–304, 2021.
https://doi.org/10.1007/978-3-030-86271-8_25

1 Introduction

Today, there are several environmental agencies worldwide that develop their own policies and have established air quality standards and indicators regarding permitted levels
of air pollutants. Immission or air quality can be defined as the amount of pollutant
that reaches a receptor, more or less distant from the emission source [1]. Air quality
is determined especially by the geographic distribution of pollutant emission sources;
when there are few pollutants, air quality is said to be good [2]. In 1993, the London Environmental Monitoring Network was established, this network has pollutant monitoring
systems in 30 of the city's suburban areas [3].

Currently, several air quality indicators show the effects of pollution on people's
health. Among the most important are carbon monoxide (CO), sulfur dioxide (SO2) and
nitrogen dioxide (NO2) [4]. When the concentration level of an indicator exceeds an
established air quality safety threshold, it can affect human health. The measurement
results from the sensor network are equally spaced and ordered observations over time,
resulting in a series of pollutant concentrations [4]. These observations are stored in freely
accessible repositories, which benefits research tasks as such data can be analyzed using
machine learning techniques, obtaining data-driven models.

Data-driven research has benefited the development of efficient algorithms bringing
new knowledge to science and technology [5, 6]. For this reason, this paper compares the
efficiency of stochastic and heuristic predictive models in forecasting the concentration
of air pollutants to estimate London's air quality more accurately. The forecasting methods compared in this paper are classical, an autoregressive integrated moving average
(ARIMA) model, and other more current methods. Among the main machine learning
algorithms used in different areas are: Random Forest, algorithms based on Artificial
Neural Networks (ANNs), Support Vector Machine (SVM), and Support Vector Regression (SVR) [7]. In [8] the authors have used neural networks to obtain satisfactory results.
Currently, Long Short-Term Memory (LSTM) models have been used significantly in
time series prediction [9–11].

This paper presents a comparative study that carries out a comparison between four
predictive methods: ARIMA, SVR, neural networks (LSTM) and Facebook Prophet.
The Root Mean Squared Error (RMSE), the Mean Absolute Error (MAE) and the Mean
Squared Error (MSE) were used as evaluation measures to assess the quality of the
model and evaluate its efficiency. After comparing these metrics, it was concluded that
the best predictive models for the evaluated scenario were the ARIMA models and neural
networks (LSTM).

This paper is structured as follows: Sect. 2 describes the data set employed and the
predictive models considered. Section 3 presents the main results obtained and the loss
function analysis. Finally, Sect. 4 presents the conclusions of the research. A remarkable
contribution of this work is the demonstration of the predictive capacity of stochastic
and heuristic models applied to air quality estimation.

2 Materials and Methods

2.1 Dataset

The dataset analyzed in this paper was retrieved from the open portal London Average Air Quality Levels from the website [23], the output data found in the data directory consists of two csv files: monthly-averages.csv, time-of-day-per-month.csv dating from 01/01/2008 to 31/07/2019.

These data sets consist of the average on-road and background readings for nitrogen dioxide, nitric oxide, nitrogen oxides, ozone, particulate matter (PM10 and PM2.5) and sulfur dioxide.

The dataset consists of 16 variables and 3336 measurements that were recorded every hour. It is worth mentioning that there are no records for some variables such as nitrogen oxides. In addition, data holds outliers. After performing a correlation analysis of variables, it is observed that some variables present multicollinearity. The variable O3 (ozone) is the least correlated with the data.

In this study, the data processed to make the predictions are the hourly measurements of Ozone O3, the data set "air-quality-london-time-of-day.csv", which contains 3336 samples.

2.2 Predictive Models

Autoregressive Integrated Moving Average Model (ARIMA)
ARIMA it is one of the most typical linear methods for time series forecasting [10]. The model uses the existing variations and regressions between the data to determine the intrinsic patterns in the series and, from them, it can generate a forecast [12]. This model is characterized by its low computational cost and its performance depends on few parameter settings. Errors must be estimated period by period once the model is fitted to the data [13].

ARIMA is presented as an ARIMA (p, d, q) model, where p is the number of autoregressive terms, d is the number of non-seasonal differences necessary for stationarity, and q stands for the number of lagged forecasts errors in the prediction equation [14].

Facebook Prophet
Prophet is an open-source software procedure for forecasting time series data based on an additive model where non-linear trends are adjusted for annual, weekly, and daily seasonality. Released by Facebook's Core Data Science team, Prophet is robust to missing data and trend shifts, and it handles outliers well. Prophet is an additive regression model with four components:

Trend: it automatically detects changes in trend by selecting the different trend breaks within the data set and thus assembles the (piecewise-defined) linear trend or logistic growth function (reaching saturation level) [15].

Annual seasonality modeled using Fourier series, and weekly seasonality, modeled with dummy variables.

Important dates, holidays, etc.: the user can define them in advance if they mean a break to be considered by the model [15].

The variables mentioned above are combined in the following equation:

$$y(t) = g(t) + s(t) + h(t) + \epsilon_t \tag{1}$$

g(t): piecewise linear or logistic growth curve for modeling non-periodic changes in time series.

s(t): periodic changes (e.g., weekly/yearly seasonality).

h(t): holiday effects (provided by the user) with irregular schedules.

ϵ_t: the error term accounts for any unusual changes that the model does not take into account.

Prophet uses time as a regressor trying to fit several linear and nonlinear time functions as components.

Modeling seasonality as an additive component is the same approach taken by exponential smoothing in the Holt-Winters technique. Indeed, the forecasting problem is being framed as a curve fitting exercise rather than explicitly looking at the time-based dependence of each observation within a time series [15].

Support Vector Regression (SVR)

Nonlinear SVR models have been widely used to predict time series [16]. SVR allows the use of functions called Kernel, that consist of the non-linear mapping in the feature space. A Kernel must satisfy Mercer's theorem [17]. Some kernel functions are specified in Table 1 [18].

Table 1. Kernel mathematical formulation.

Kernel	Definition
RBF	$\kappa^{RBF}\left(\chi, \chi'\right) = e^{-\sum_{i=1}^{d} \gamma(\chi_i - \chi_i')^\beta}$ $\gamma > 0, \beta \in (0, 2]$
Triangular	$\kappa^{Tri}\left(\chi, \chi'\right) = \begin{cases} \|x - x'\| \le a \to 1 - \frac{\|x - x'\|}{a} \\ \|x - x'\| > a \to 0 \\ a > 0 \end{cases}$
Truncated euclidean	$\kappa^{Tri}\left(\chi, \chi'\right) = \frac{1}{d} \sum_{1}^{j} \max\left(0, \frac{|x_i - x_i'|}{y}\right)$ $\gamma > 0$

Neural Networks

Neural networks are one of the most widely used heuristic techniques to address prediction issues such as air quality forecasting, especially by setting up Multi-Layer Perceptron
(MLP) Radial Basis Functions (RBFN) models [19].

Recurrent neural network (RNN) is a type of neural network used to process sequential data. Long short-term memory (LSTM) compensates for the problems of vanishing
gradient, exploding gradient, and insufficient long-term memory of RNN [20]. It can
make full use of long-distance time sequence information [21]. LSTM is a special type
of RNN consisting of an input layer, an output layer, and a series of recurrently connected hidden layers known as blocks [9]. In neural networks, some parameters must be
declared at the beginning of the training process, such as the number of hidden layers
and nodes, learning rates, and activation function [14].

2.3 Metrics

Root Mean Squared Error (RMSE), Mean Absolute Error (MAE), Mean Squared Error
(MSE) were used as evaluation measures to assess the accuracy of the predictive models.
Mean Absolute Percentage Error (MAPE) was used for the selection of hyperparameters.

For all subsequent formulas, the y_j notation is considered the original series, and the
$\hat{y}_j$ notation, the estimated series.

Root Mean Squared Error (RMSE)

It is one of the most common metrics to evaluate a regression model since it measures the
amount of error between two sets of data. In the case of predictive models, it determines
the difference between the value predicted by the model and the actual value.

$$RMSE = \sqrt{\frac{1}{n} \sum_{j=1}^{n} \left(y_j - \hat{y}_j\right)^2} \tag{2}$$

Mean Absolute Error (MAE)

It is the mean of the absolute errors. The absolute error is the absolute value of the
difference between the predicted value and the actual value. The MAE value indicates
how big, on average, is the error that can be expected from the forecast.

$$MAE = \frac{1}{n} \Sigma \left| y_j - \hat{y}_j \right| \tag{3}$$

Mean Squared Error (MSE)

It calculates the mean square error between the prediction and the actual value.

$$MSE = \frac{1}{n} \sum_{j=1}^{n} (y_j - \hat{y}_j)^2 \tag{4}$$

Mean Absolute Percentage Error (MAPE) [22]

It is the average of the absolute percentage errors of the predictions. The smaller the MAPE, the better the predictions.

$$MAPE = \frac{1}{n} \sum_{j=1}^{n} \frac{|y_j - \hat{y}_j|}{|y_j|} \tag{5}$$

2.4 Data Selection

For the generation of predictive models, it is necessary to have two differentiated data sets. In this case, the train set used to generate the model, and the test set used to validate the quality of the model are defined.

For this study, the data used to generate the model were those corresponding to the years between 2008 and 2018, and the data used to estimate the quality of the model were those corresponding to the last 12 months, from August 2018 to July 2019. The last 12 months correspond to 8% of the total sample, which is enough to evaluate the predictive capacity of the model for the last year. Figure 1 shows the distribution of the data, the train set in red and the test set in blue.

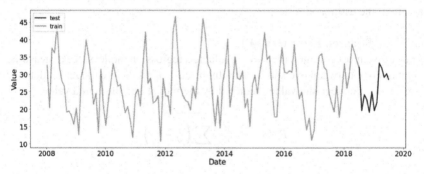

Fig. 1. Ozone concentration train-test (Color figure online)

This study was developed under an experimental methodology consisting of data selection and processing, predictive model selection, model training, and comparative analysis with other models.

In summary, four different models were used for time series prediction: ARIMA, Facebook Prophet, Support Vector Machines, and Neural Networks. Then a comparative evaluation was performed to define the best predictive model. In this project, temporal graphs are used to represent the prediction of the model versus the real series.

3 Results and Discussion

Paramter Settings

This section gathers the information of parameter selection and tuning for all the considered models. As for ARIMA model, the SARIMA extension was found as the best configuration, as it supports the analysis of time series with seasonal or trend components. The best configuration is taken as that obtaining the lowest MSE.

As for the SVR model, three different kernels were used, to set the configuration of the hyperparameters. To do so, a search for the best parameters is performed using the GridSearchCV class, available in SciKit-Learn. For the Facebook Prophet model, the selection of hyperparameters, the GridSearch class is used and the parameters with the lowest MAPE value are selected.

The neural network implementation was carried out using the keras library, which allows for modifying the LSTM network's parameters. The selected parameters are summarized in Table 2. It is worth mentioning that the remaining parameters are used with the by-default settings.

Table 2. Parameter settings

Method	Parameter	Value
SARIMA	p: Trend autoregression order d: Trend difference order q: Trend moving average order	$p = 0$ $d = 0$ $q = 1$
	sp: Seasonal autoregressive order sd: Seasonal difference order sq: Seasonal moving average order s: The number of time steps for a single seasonal period	$sp = 0$ $sd = 1$ $sq = 1$ $s = 12$
Facebook prophet	Seasonality mode	Additive
SVR	RBF KERNEL	$C = 1$ $\gamma = 1$ (gamma)
	Triangular kernel	$C = 1$ $\gamma = 0.01$ (gamma)
	Truncated kernel	$C = 1$ $\gamma = 0.01$ (gamma)
Neural networks	Number of layers	Two LSTM layers with 100 units Two Dropout layers of 0.2, and a dense layer of a single neuron
	Number of training epochs	20
	Loss function	Mean squared_error
	Optimizer	Adam

ARIMA Results

After training and testing the model, the metrics obtained are the following: MAE 0.143, MSE 0.033, RMSE 0.184. Figure 2 shows the model forecast and the test values displayed similar behaviors, although the generated model was not able to reproduce the peaks around first values, as they are outliers. The model was able to reasonably predict the peak in concentration around December.

Facebook Prophet Results

The Facebook Prophet algorithm allows adding holidays to study their impact on the prediction, for this reason, the prediction was performed in two scenarios, with and without holidays. To add holidays, the holidays library, containing London's holidays, was imported. Figure 2 shows the result of the prediction (orange solid line) and contrasts it with the actual measurements (blue solid line). It is important to note that the values are far from reality since it was not possible to predict the peak values.

As a result of the evaluation of the test set, the calculated metrics were the following: MAE 3.606, MSE 21.954, RMSE 4.685.

In the Facebook Prophet model without holidays, there is a slight overlap of the values around 2018–11, but the model failed to predict the actual values, as the predicted ones are far from them. The following values are the result of the algorithm quality evaluation: MAE 3.667, MSE 22.644, RMSE 4.758. The model's failure to capture the peak-to-peak amplitude of weak seasonality explains the high MSE value.

After making the predictions in the two scenarios, the MAE and RMSE values were very similar in both cases. In the case of the model with holidays show in Fig. 2, the MSE value is lower, i.e., by adding holidays to the model, a slight improvement in performance was achieved.

SVR Results

Once the parameter optimization results were obtained, the training of the model was performed, in this case, the model was trained via kernel functions.

The Truncated and Triangular kernels results were identical, as shown in Table 3. The three models have very similar behaviors given the degree of overlap between the three predicted sets.

The predictions from the three models were good since they were able to reproduce the first peak of the real measurements, although the predicted value was far from the real one. Table 3 shows the metrics calculated for each SVR model according to the Kernel function used.

Table 3. SVR model results and Kernel function

Kernel	MAE	MSE	RMSE
Triangular	0.247	0.087	0.295
Truncated	0.247	0.087	0.295
RBF	0.249	0.089	0.298

From Table 3, it should be noted that all the models show similar MAE and RMSE values. The quality metrics show better results for the Triangular and Truncated kernels.

Neural Networks (LSTM) Results

After generating the model, the neural network was successfully trained, since the error values are decreasing in each epoch. Once the model was generated, the test data was evaluated. The following values were obtained as a result: MAE 0.177, MSE 0.057, and RMSE 0.240. Figure 2 shows an overlap between the test values and the prediction, the prediction follows precisely the trend of the original series. Visually, it is a good prediction considering the 0,177 MAE value.

Results of Predictive Models

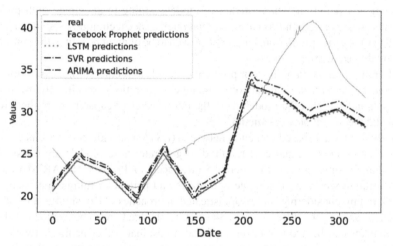

Fig. 2. Results ozone prediction (Color figur online)

The LSTM model detected all the existing peaks in the test set, which indicates a good prediction, the prediction follows precisely the trend of the original series. The predicted values for the first peak are still above the test values, but they are significantly better compared with the Facebook Prophet model ones. The metrics calculated in each predictive model are presented below in Table 4. The ARIMA, SVR, and neural networks models have RMSE values less than 1, therefore, they are the models with better accuracy. Compared with the ARIMA and Facebook Prophet models, the SVR and neural networks models captured the spatio-temporal correlations more effectively, as well as reached better prediction performance.

Table 4. Results of predictive models

Model	MAE	MSE	RMSE
ARIMA	0.143	0.034	0.184
SVR	0.247	0.087	0.295
Facebook Prophet	3.606	21.954	4.685
LSTM	0.166	0.041	0.203

4 Conclusions and Future Work

In this paper, we compared different stochastic and heuristic predictive models applied to air quality estimation since several alternatives are currently being studied to counteract environmental pollution. Thus, predictive models aiming to anticipate high pollution scenarios are generated. In this case, artificial intelligence techniques, specifically those related to time series prediction, have been proven to be of great benefit to the automation of air quality estimation processes.

After the application of the four predictive models to predict the concentration of air pollutants based on historical data of pollutant concentrations, specifically ozone (O3), the results indicate that the models with the best predictive capacity were the ARIMA model and the neural network model.

Predictive models based on neural networks (LSTM) provide good results, however, the time and computational cost required to generate these models are considerably higher than the time and cost to generate SVR, Facebook Prophet, and ARIMA models.

As a future work, more suitable optimizers and parameter tuning approaches will be explored by considering more sophisticated approaches. Also, significant statistical tests for measuring the differences on the results obtained are to be carried out. Future developments can be addressed, on one hand, to evaluate other neural network architectures combining hyperparameters to improve the accuracy of the predictive model. And on the other hand, to perform a comparative study of predictive models for estimating air quality based on kernel functions comparison, using meteorological data (e.g., temperature, humidity, etc.).

Acknowledgment. This work is supported by the SDAS Research Group (www.sdas-group.com). Authors are in debt with the SDAS Group internal editor J. Mejía-Ordóñez for the manuscript reviewing and editing.

Sergio Trilles has been funded by the Juan de la Cierva - Incorporación postdoctoral programme of the Ministry of Science and Innovation - Spanish government (IJC2018–035017-I).

References

1. Rybarczyk, Y., Zalakeviciute, R.: Machine learning approaches for outdoor air quality modelling: a systematic review. Appl. Sci. **8**, 2570 (2018). https://doi.org/10.3390/app812 2570

2. Hood, C., et al.: Air quality simulations for London using a coupled regional-to-local modelling system. Atmos. Chem. Phys. **18**, 11221–11245 (2018). https://doi.org/10.5194/acp-18-11221-2018
3. Gaitán, M., Cancino, J., Eduardo, B.: Análisis del estado de la calidad del aire en Bogotá. Rev. Ing. Unknown, 81–92 (2007). https://doi.org/10.16924/riua.v0i26.299
4. Silva, C., Alvarado, S., Montaño, R., Pérez, P.: Modelamiento de la contaminación atmosférica por particulas: Comparación de cuatro procedimientos predictivos en Santiago, Chile, pp. 113–127 (2003)
5. Gil-Alana, L.A., Yaya, O.S., Carmona-González, N.: Air quality in London: evidence of persistence, seasonality and trends. Theoret. Appl. Climatol. **142**(1–2), 103–115 (2020). https://doi.org/10.1007/s00704-020-03305-1
6. Yadav, M., Jain, S., Seeja, K.R.: Prediction of air quality using time series data mining. In: Bhattacharyya, S., Hassanien, A.E., Gupta, D., Khanna, A., Pan, I. (eds.) International Conference on Innovative Computing and Communications, pp. 13–20. Springer, Singapore (2019)
7. Lorente-Leyva, L.L., Alemany, M.M.E., Peluffo-Ordóñez, D.H., Herrera-Granda, I.D.: A Comparison of machine learning and classical demand forecasting methods: a case study of Ecuadorian textile industry. In: Nicosia, G., et al. (eds.) LOD 2020. LNCS, vol. 12566, pp. 131–142. Springer, Cham (2020). https://doi.org/10.1007/978-3-030-64580-9_11
8. Brownlee, J.: Time Series Prediction with LSTM Recurrent Neural Networks in Python with Keras, https://machinelearningmastery.com/time-series-prediction-lstm-recurrent-neural-networks-python-keras/. Accessed 17 May 2020
9. Li, X., et al.: Long short-term memory neural network for air pollutant concentration predictions: Method development and evaluation. Environ. Pollut. **231**, 997–1004 (2017). https://doi.org/10.1016/j.envpol.2017.08.114
10. Ma, J., Cheng, J.C.P., Lin, C., Tan, Y., Zhang, J.: Improving air quality prediction accuracy at larger temporal resolutions using deep learning and transfer learning techniques. Atmos. Environ. **214**, 116885 (2019). https://doi.org/10.1016/j.atmosenv.2019.116885
11. Siami-Namini, S., Tavakoli, N., Siami Namin, A.: A Comparison of ARIMA and LSTM in forecasting time series. In: Proceedings of 17th IEEE International Conference on Machine Learning Applications, pp. 1394–1401, ICMLA 2018 (2019). https://doi.org/10.1109/ICMLA.2018.00227
12. Riofrío, J., Chang, O., Revelo-Fuelagán, E.J., Peluffo-Ordóñez, D.H.: Forecasting the consumer price index (CPI) of Ecuador: a comparative study of predictive models. Int. J. Adv. Sci. Eng. Inf. Technol. **10**, 1078–1084 (2020). https://doi.org/10.18517/ijaseit.10.3.10813
13. Al-Musaylh, M.S., Deo, R.C., Adamowski, J.F., Li, Y.: Short-term electricity demand forecasting with MARS, SVR and ARIMA models using aggregated demand data in Queensland. Australia. Adv. Eng. Inform. **35**, 1–16 (2018)
14. Rani Patra, S.: Time series forecasting of air pollutant concentration levels using machine learning. Time Ser. Anal. **4**, 280–284 (2017)
15. López, J.: Análisis de Series deTiempo Pronóstico de demanda de uso de aeropuertos en Argentina al 2022, (2018). https://doi.org/10.3726/978-3-0352-0094-2/1.
16. Raimundo, M.S., Okamoto, J.: SVR-wavelet adaptive model for forecasting financial time series. In: 2018 International Conference Information and Computing Technology, pp. 111–114, ICICT 2018 (2018). https://doi.org/10.1109/INFOCT.2018.8356851.
17. Aghelpour, P., Mohammadi, B., Biazar, S.M.: Long-term monthly average temperature forecasting in some climate types of Iran, using the models SARIMA, SVR, and SVR-FA. Theoret. Appl. Climatol. **138**(3–4), 1471–1480 (2019). https://doi.org/10.1007/s00704-019-02905-w
18. Awad, M., Khanna, R.: Support vector regression. In: Awad, M., Khanna, R. (eds.) Efficient Learning Machines: Theories, Concepts, and Applications for Engineers and System

Designers, pp. 67–80. Apress, Berkeley, CA (2015). https://doi.org/10.1007/978-1-4302-599
0-9_4

19. Hermiyanty, H., Wandira Ayu, B., Sinta, D.: Predicción de sistemas caóticos con redes neu-
ronales: un estudio comparativo de los modelos de perceptrón multicapa y funciones de base
radial. J. Chem. Inf. Model. **8**, 1–58 (2017). https://doi.org/10.1017/CBO9781107415324.004

20. Freeman, B.S., Taylor, G., Gharabaghi, B., Thé, J.: Forecasting air quality time series using
deep learning. J. Air Waste Manage. Assoc. **68**, 866–886 (2018)

21. Ying, C.: Voltages prediction algorithm based on LSTM recurrent neural network. 10 (2020).
(pre-proof)

22. Li, C., Hsu, N.C., Tsay, S.-C.: A study on the potential applications of satellite data in air
quality monitoring and forecasting. Atmos. Environ. **45**, 3663–3675 (2011). https://doi.org/
10.1016/j.atmosenv.2011.04.032

23. London, K.C.: London Average Air Quality Levels. https://data.london.gov.uk/dataset/lon
don-average-air-quality-levels

Slicer: Feature Learning for Class Separability with Least-Squares Support Vector Machine Loss and COVID-19 Chest X-Ray Case Study

David Charte[1]([✉]) [iD], Iván Sevillano-García[1] [iD], María Jesús Lucena-González[3], José Luis Martín-Rodríguez[3], Francisco Charte[2] [iD], and Francisco Herrera[1] [iD]

[1] Department of Computer Science and Artificial Intelligence, Andalusian Research Institute in Data Science and Computational Intelligence (DaSCI), University of Granada, Granada, Spain
{fdavidcl,isevillano}@ugr.es,herrera@decsai.ugr.es
[2] Department of Computer Science, Andalusian Research Institute in Data Science and Computational Intelligence (DaSCI), University of Jaén, Jaén, Spain
fcharte@ujaen.es
[3] Hospital Universitario Clínico San Cecilio de Granada, Granada, Spain

Abstract. Datasets from real-world applications usually deal with many variables and present difficulties when modeling them with traditional classifiers. There is a variety of feature selection and extraction tools that may help with the dimensionality problem, but most of them do not focus on the complexity of the classes. In this paper, a new autoencoder-based model for addressing class complexity in data is introduced, aiming to extract features that present classes in a more separable fashion, thus simplifying the classification task. This is possible thanks to a combination of the standard reconstruction error with a least-squares support vector machine loss function. This model is then applied to a practical use case: classification of chest X-rays according to the presence of COVID-19, showing that learning features that increase linear class separability can boost classification performance. For this purpose, a specific convolutional autoencoder architecture has been designed and trained using the recently published COVIDGR dataset. The proposed model is evaluated by means of several traditional classifiers and metrics, in order to establish the improvements caused by the extracted features. The advantages of using a feature learner and traditional classifiers are also discussed.

Keywords: Feature learning · Class separability · Autoencoders

1 Introduction

Data classification [3] is one of the most studied problems in machine learning, and is applicable in many real-world contexts such as medicine, banking,

© Springer Nature Switzerland AG 2021
H. Sanjurjo González et al. (Eds.): HAIS 2021, LNAI 12886, pp. 305–315, 2021.
https://doi.org/10.1007/978-3-030-86271-8_26

robotics, natural sciences and other industries. Class complexity [5,17] is a term which encompasses all the intrinsic traits in data that can hinder the performance of a classifier. There are several categories of metrics that can be used to gauge the complexity of a dataset: feature overlap, linearity, neighborhoods, dimensionality, class balance and network properties. Datasets that present high levels of complexity in some of these categories have been shown to cause poor classifier performance [14].

When it comes to addressing data complexity during a preprocessing [9] phase, several specific methods can be found on the literature, but they usually tackle high dimensionality (feature selection and extraction methods) and class imbalance (resampling methods). Little research has been published on how to address other complexity types during a preprocessing phase and most of it is centered around feature selection [20,23].

In this work, we present Slicer (supervised linear classifier error reduction), an automatic feature extractor designed with linear class separability in mind. It is based on an autoencoder (AE) model [8], using a special loss function inspired by least-squares support vector machines (LSSVM) [18]. The objective of this model is to learn an alternative representation for each instance where classes are more easily distinguishable. Once trained, the model is able to project any new instance onto the learned feature space without knowing its class. This allows to work with compact representations of the samples instead of the original, high-dimensional ones, in a way that facilitates the work of traditional classifiers, which are usually hindered by high dimensionality [4,6] unless they specifically select features internally. Extracting features can also help when it is necessary to combine variables from different sources (e.g. images and clinical data), and working with traditional classifiers makes it easier to understand the decision-making process.

The proposed model is applied in a specific use case, aiming both to analyze the level of performance that could be gained and to open new possibilities for combination of imagery and other data, as well as interpretability of classifiers. The chosen application is recognition of COVID-19 in chest X-ray images, using the COVIDGR dataset [19] for this purpose. The fitness of the set of features learned by Slicer is evaluated by means of classification metrics using several standard classifiers and is compared against using a basic AE and learning from the original, unmodified features. The results show a noticeable advantage of the Slicer-generated variables except when using support vector machines as classifier, even though specific traits of this dataset such as "apparently negative" positive samples might affect the learned representation.

The rest of this document is organized as follows. Section 2 describes the new feature learner named Slicer. Afterwards, Sect. 3 outlines the main aspects of the experimentation, including the dataset and the evaluation strategy. Section 4 discusses the results obtained in the experimentation above and, lastly, Sect. 5 draws some conclusions.

2   Class-Informed Autoencoder for Complexity Reduction

This section is dedicated to introducing a new model designed to learn features with improved class separability. The proposed model, Slicer, is based on the minimization of the error of a linear classifier, at the same time that it attempts to maximize its reconstruction abilities. As a result, the learned features are influenced both by the overall information within the data as well as their relation to the class.

2.1   Autoencoder Fundamentals

An AE is an artificial neural network that is trained to reconstruct the inputs at its output [8]. It includes a certain bottleneck where the representation of the data is somehow restricted: e.g. it is lower dimensional, more sparse or robust against noise. This prevents the AE from simply copying the input instance throughout the network. Instead, an encoder f learns a new representation for the data while a decoder g must be able to recover the original features. This transformation is learned by optimizing the reconstruction error $\mathcal{J}_{\mathrm{RE}}(x, (g \circ f)(x))$ which typically measures a distance between the original samples and their reconstructions, but different penalties and regularizations can allow to influence other aspects of the learned representation.

AEs are usually unsupervised tools, in the sense that they do not receive any information about the labels of the data nor the desired encodings for each instance. Their learning mechanism is, as a result, self-supervised [13]. This means that they can be applied in many contexts where label information is not necessarily available or just partially so: anomaly detection, semantic hashing, data compression, among others [7]. Nonetheless, some AEs do use label information [16], even though the objective is other than learning more separable features. Our objective is to regularize an AE so that it learns from the class labels during training, but does not need them during the prediction phase, and thus facilitates classification tasks.

2.2   Slicer Model: The Loss Function

Slicer is an AE model regularized by a special penalty function which takes class separability into account. In order to do this, an additional component is introduced to simultaneously fit a LSSVM to the encoded samples as well as evaluate the encoding based on the fitness of said classifier.

Figure 1 shows a diagram with the main components of Slicer. f represents the encoder, which is a neural network that transforms the inputs onto encodings. g refers to the decoder, a similar network whose objective is to reconstruct the original data inputs out of the encodings. These encodings are also fed to a single fully connected layer with no activation function, which acts as the support vector machine. Each of the decoder and this last layer are evaluated with their loss functions: the reconstruction error and the LSSVM minimization objective, respectively.

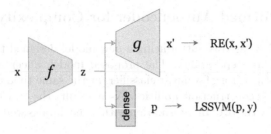

Fig. 1. Schematic illustration of the Slicer model

The reconstruction error chosen for the purposes of modeling input samples is cross entropy: a measure of disagreement between two probability distributions, for the case of Bernoulli distributions. It is usually the better option when all values in each instance are in the $[0, 1]$ interval. Its formulation for n instances with k variables is shown in Eq. 1.

$$
\mathcal{J}_{\mathrm{RE}}(x, \theta) = -\frac{1}{nk} \sum_{i=1}^{n} \sum_{j=1}^{k} x_j^{(i)} \log \left[(g \circ f) \left(x^{(i)} \right)_j \right] \\
+ \left(1 - x_j^{(i)} \right) \log \left[1 - (g \circ f) \left(x^{(i)} \right)_j \right].
$$ (1)

If the variables are not scaled to the $[0, 1]$, other reconstruction errors like the mean squared error could be used.

For its part, the LSSVM objective assumes that labels are in $\{-1, 1\}$ and is simply a sum of quadratic errors e_i^2 subject to the equality constraint $1 - e_i = y^{(i)} - w^T f \left(x^{(i)} \right) + b$, resulting as formulated in Eq. 2.

$$
\mathcal{J}_{\mathrm{LSSVM}}(x, y, \theta) = \frac{\mu}{2} w^T w + \frac{\zeta}{2} \sum_{i=1}^{n} \left(y^{(i)} - w^T f \left(x^{(i)} \right) + b \right)^2.
$$ (2)

In the previous equation, f usually refers to the kernel used in the model but, in this case, it represents the encoder of the neural network. w holds the weights of the SVM, which are associated to a penalty with coefficient μ. The term that compares classes (y) to the LSSVM output is weighted by ζ.

The resulting loss function for the Slicer model is the sum of both the reconstruction error and the LSSVM loss:

$$
\mathcal{J}(x, y, \theta) = \mathcal{J}_{\mathrm{RE}}(x, \theta) + \mathcal{J}_{\mathrm{LSSVM}}(x, y, \theta)
$$ (3)

3   Experimental Framework

The objective of this experimentation is to apply the proposed complexity reduction method in a real world practical case. In particular, we aim to improve the performance of simple classifiers when dealing with a chest X-ray image dataset for COVID-19 classification. This would open several promising research lines, such as the combination of these extracted features with other clinical and laboratory variables or the possibility of using easily interpretable classifiers with the generated features.

3.1   COVIDGR Dataset

The COVIDGR dataset of X-ray chest images was introduced in [19]. These images were collected under a collaboration with expert radiologists of the Hospital Universitario San Cecilio in Granada, Spain. In total, 852 images were annotated under a strict protocol: positive images correspond to patients who have been tested positive for COVID-19 using RT-PCR within a time span of at most 24h between the X-ray image and the test. Every image was taken using the same type of equipment and always with the posterior-anterior view. It is important that all images are consistent since, otherwise, classifiers could find cues to distinguish COVID-positive samples from negative ones different from the intended aspects of the X-ray that characterize the pneumonia associated to the disease [15]. Figure 2 includes one positive example and a negative one.

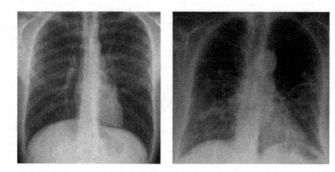

Fig. 2. A negative sample (left) and a positive one (right)

More information about class distribution is provided in Table 1. In the following experiments, the exact same partitions used in [19] are employed, in order to ease comparisons with previous results.

3.2   Evaluation Strategy

Since the final objective of the proposed model is to improve classification performance, the evaluation framework will consist in a variety of simple classifiers that will be trained with either the original features or the encoded ones. Standard classification metrics will be computed using the predictions over test subsets. A 5-fold cross validation scheme will be applied 5 times for a total of 25 train-test

Table 1. Class distribution in COVIDGR dataset. Normal-PCR+ refers to X-rays where COVID-19 was not detected by the experts but the patients tested positive.

Class	#images	Severities	
Negative	426		
Positive	426	Normal-PCR+	76
		Mild	100
		Moderate	171
		Severe	79

runs, so as to prevent errors from statistical chance. Table 2 lists the available feature sets and every classifier and evaluation metric included in the experiment. Each column is independent, in the sense that every feature set has been tested with each one of the classifiers and the performance has always been assessed with all four metrics.

Table 2. Evaluation framework: available feature sets, tested classifiers and evaluation metrics. TP, TN, FP and FN denote true positives, true negatives, false positives and false negatives, respectively.

Feature sets	Classifiers	Evaluation metrics	
Original	Decision tree (DT)	Accuracy	$\dfrac{TP+TN}{TP+TN+FP+FN}$
Basic AE	k nearest neighbors (kNN)	Precision	$\dfrac{TP}{TP+FP}$
Slicer	Support vector machine (SVM)	Recall	$\dfrac{TP}{TP+FN}$
	Gaussian process (GP)	F1-score	$\dfrac{2 \cdot Precision \cdot Recall}{Precision + Recall}$

3.3 Architecture of the Slicer Model Used with the COVIDGR Dataset

The proposed Slicer model has been implemented in the Python language on top of the Tensorflow [1] library. Since the model needs to deal with image data, the specific architecture makes use of convolutional layers for the encoder and deconvolutional (or transposed convolutional) layers for the decoder.

More specifically, most of the AE is composed of residual blocks such as the ones in the ResNet-V2 architectures [11], as can be seen in Fig. 3. The left side of this diagram shows the detailed architecture of the AE-based model, with the encoder ranging from the input to the dense layer with 128 units, and the decoder from there to the last deconvolutional layer. The classification component consists in the fully connected (dense) layer that maps the encoding to one variable and is then connected to the LSSVM loss. The overall loss is simply the sum of both error measures, as explained above.

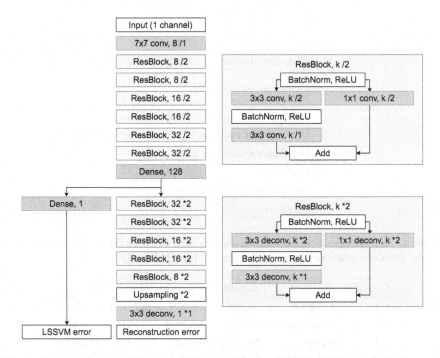

Fig. 3. Structure of the Slicer model used with the COVIDGR dataset. Each convolutional and deconvolutional layer is followed by the number of filters as well as an indication of the stride: /2 indicates a stride of 2 for a convolutional layer (the side of the image is halved), and *2 indicates a stride of 2 for a deconvolutional layer (the side of the image is doubled). Each of the residual blocks unrolls like the corresponding diagram on the right.

Using the previous architecture, a model was trained for each training partition of the total of 25 runs, using the parameters detailed in Table 3. Images were resized to a common resolution of 512×512 pixels (in total, 262144 variables in the range $[0,1]$), and the encoding size was of 128 variables, which gives a reduction ratio of 1:2048, or equivalently, it represents the images using just 0.049% of the original variables. An equivalent basic AE was also trained using the same architecture (except for the LSSVM layer and loss) and partitions.

4   Results and Discussion

In this section, the main results of the experimentation are analyzed. Table 4 contains average metrics for the total of 25 runs that were performed. We can observe drastically different behaviors according to the classifier that was used:

Table 3. Relation of hyperparameters. DA refers to data augmentation techniques (only the techniques shown were applied).

Parameter	Value
μ	0.01
ζ	0.1
Encoding dimension	128
Epochs	50
Batch size	8
Optimizer	Adam
DA: random rotation	$\leq 5°$
DA: horizontal flip	Yes
kNN: k	5
DT: maximum depth	10
Image size	512×512

Table 4. Average classification metrics over 25 runs (5 times 5-fold cross validation). Results from convolutional neural networks are reported in [19].

Classifier	Features	Accuracy	Precision	Recall	F1-score
DT	Original	58.098	58.259	**57.470**	57.665
DT	Autoencoder	58.377	**58.889**	55.919	57.196
DT	Slicer	**58.593**	58.837	57.422	**57.988**
GP	Original	50.024	45.333	1.266	2.452
GP	Autoencoder	50.024	4.000	0.047	0.093
GP	Slicer	**62.656**	**64.339**	**56.673**	**60.032**
kNN	Original	**62.585**	**65.804**	52.780	58.358
kNN	Autoencoder	61.837	65.091	51.469	57.194
kNN	Slicer	62.326	63.624	**57.284**	**60.092**
SVM	Original	**67.329**	66.611	**69.920**	**67.931**
SVM	Autoencoder	67.072	**67.006**	67.622	67.057
SVM	Slicer	65.987	66.235	65.025	65.393

the decision tree had similar performance independently of the set of features that were provided, while the Gaussian process only was competitive when using the features learned by Slicer.

Several deductions can be made out of the results in Table 4. First, the classifiers that take the most advantage from the Slicer-generated features are kNN and GP. In fact, the latter struggles to find an acceptable model of the data using either the original features or the autoencoded ones. For its part, the DT shows little variance with respect to the set of variables it uses, although the

Slicer-generated has a slight lead in F1 score and accuracy. The SVM, however, does not benefit from the more separable features and loses performance with respect to the original and autoencoded ones.

Overall, it is not very surprising that the classifiers that are typically more affected by the quality of features are those which benefit more from the encodings provided by Slicer, whereas classifiers that internally perform their own feature selection or transformations either see small improvements or even decrease their performance.

It is important to note that there exist specific deep learning architectures designed for COVID-19 classification in chest X-rays, such as COVIDNet [21], COVID-CAPS [2] and COVID-SDNet [19]. Comparing the classification performance with these is out of scope for the present work, since we only aim to assess how useful the features extracted by the Slicer model are for traditional classifiers, not to find the best COVID-19 classifier.

As a summary, Table 5 displays the average metrics across all 4 classifiers, for each feature set. The average ranking that each one achieved for each metric is also shown. From this, we can conclude that the Slicer model produces feature sets that are consistently superior to a basic AE. Furthermore, it is able to preserve or even improve the quality of the original features, while simultaneously reducing drastically the dimensionality.

Table 5. Average classification metrics per feature set provided to the classifiers. The average ranking achieved by each feature set in each of the tests is shown in parentheses (lower is better).

Features	Accuracy	Precision	Recall	F1-score
Original	59.509 (2.06)	59.002 (1.97)	45.359 (1.92)	46.602 (1.89)
Autoencoder	59.327 (2.09)	48.747 (2.16)	43.764 (2.38)	45.385 (2.29)
Slicer	62.391 (1.85)	63.259 (1.87)	59.101 (1.71)	60.876 (1.82)

5   Conclusions and Future Work

This work has presented a novel framework for class separability enhancement using an AE-based model with a linear classification component that contributes to the loss function. The model has been implemented as a convolutional AE for the transformation of chest X-ray images onto a more manageable number of variables in order to employ simple classifiers in COVID-19 classification. An exhaustive experimentation has shown that the proposed model improves classification performance over a basic AE with no regularizations and maintains or even improves the performance compared to using the unprocessed data, even though the number of variables is dramatically reduced.

The promising results lead to consider several ways of continuing the work for practical real-world uses:

- Analyze the impact of severity levels on the adequacy of the learned representation. For example, Normal-PCR+ samples appear to look just like negative ones, although they are positive, which could affect the behavior of the model. Removing these images could provide better separation abilities as a consequence.
- Learned features can be combined with other variables that do not come from the chest X-ray, that is, clinical and laboratory data about each patient such as age, gender, comorbidities, etc. Several scores for this kind of data have been proposed but they do not take advantage of the full chest X-ray information [10,12].
- Several ways to provide meaning to the extracted variables, such as feature disentanglement [22], in combination with transparent classifiers like decision trees, would enable more interpretable pipelines for COVID-19 classification, where users could trace predictions back to the original features.
- Combining the proposed loss function with more advanced AE models such as variational or adversarial AEs could add more potential of improving subsequent classification tasks.

Acknowledgments. D. Charte is supported by the Spanish Ministry of Science under the FPU National Program (Ref. FPU17/04069). F. Charte is supported by the Spanish Ministry of Science project PID2019-107793GB-I00/AEI/10.13039/501100011033. F. Herrera is supported by the Spanish Ministry of Science project PID2020-119478GB-I00 and the Andalusian Excellence project P18-FR-4961. This work is supported by the project COVID19RX-Ayudas Fundación BBVA a Equipos de Investigación Científica SARS-CoV-2 y COVID-19 2020.

References

1. Abadi, M., et al.: TensorFlow: large-scale machine learning on heterogeneous systems (2015). https://www.tensorflow.org/
2. Afshar, P., Heidarian, S., Naderkhani, F., Oikonomou, A., Plataniotis, K.N., Mohammadi, A.: COVID-CAPS: a capsule network-based framework for identification of COVID-19 cases from x-ray images. Pattern Recogn. Lett. **138**, 638–643 (2020)
3. Aggarwal, C.C.: Data Classification, pp. 285–344. Springer, Cham (2015). https://doi.org/10.1007/978-3-319-14142-8_10
4. Aggarwal, C.C., Hinneburg, A., Keim, D.A.: On the surprising behavior of distance metrics in high dimensional space. In: Van den Bussche, J., Vianu, V. (eds.) ICDT 2001. LNCS, vol. 1973, pp. 420–434. Springer, Heidelberg (2001). https://doi.org/10.1007/3-540-44503-X_27
5. Basu, M., Ho, T.K.: Data Complexity in Pattern Recognition. Springer, Heidelberg (2006). https://doi.org/10.1007/978-1-84628-172-3
6. Beyer, K., Goldstein, J., Ramakrishnan, R., Shaft, U.: When is "nearest neighbor" meaningful? In: Beeri, C., Buneman, P. (eds.) ICDT 1999. LNCS, vol. 1540, pp. 217–235. Springer, Heidelberg (1999). https://doi.org/10.1007/3-540-49257-7_15
7. Charte, D., Charte, F., del Jesus, M.J., Herrera, F.: An analysis on the use of autoencoders for representation learning: fundamentals, learning task case studies, explainability and challenges. Neurocomputing **404**, 93–107 (2020). https://doi.org/10.1016/j.neucom.2020.04.057

8. Charte, D., Charte, F., García, S., del Jesus, M.J., Herrera, F.: A practical tutorial on autoencoders for nonlinear feature fusion: taxonomy, models, software and guidelines. Inform. Fusion **44**, 78–96 (2018). https://doi.org/10.1016/j.inffus.2017.12.007

9. García, S., Luengo, J., Herrera, F.: Data Preprocessing in Data Mining, vol. 72. Springer, Heidelberg (2015). https://doi.org/10.1007/978-3-319-10247-4

10. Gong, J., et al.: A tool for early prediction of severe coronavirus disease 2019 (COVID-19): a multicenter study using the risk nomogram in Wuhan and Guangdong, China. Clin. Infect. Dis. **71**(15), 833–840 (2020)

11. He, K., Zhang, X., Ren, S., Sun, J.: Identity mappings in deep residual networks. In: Leibe, B., Matas, J., Sebe, N., Welling, M. (eds.) ECCV 2016. LNCS, vol. 9908, pp. 630–645. Springer, Cham (2016). https://doi.org/10.1007/978-3-319-46493-0_38

12. Knight, S.R., et al.: Risk stratification of patients admitted to hospital with COVID-19 using the ISARIC WHO clinical characterisation protocol: development and validation of the 4C mortality score. bmj **370**, 1–13 (2020)

13. Liu, X., et al.: Self-supervised learning: generative or contrastive. arXiv preprint arXiv:2006.08218 **1**(2) (2020)

14. Luengo, J., Fernández, A., García, S., Herrera, F.: Addressing data complexity for imbalanced data sets: analysis of smote-based oversampling and evolutionary undersampling. Soft. Comput. **15**(10), 1909–1936 (2011)

15. Maguolo, G., Nanni, L.: A critic evaluation of methods for COVID-19 automatic detection from x-ray images. Inform. Fusion **76**, 1–7 (2021). https://doi.org/10.1016/j.inffus.2021.04.008

16. Makhzani, A., Shlens, J., Jaitly, N., Goodfellow, I., Frey, B.: Adversarial autoencoders. arXiv preprint arXiv:1511.05644 (2015)

17. Pascual-Triana, J.D., Charte, D., Arroyo, M.A., Fernández, A., Herrera, F.: Revisiting data complexity metrics based on morphology for overlap and imbalance: snapshot, new overlap number of balls metrics and singular problems prospect. Knowl. Inf. Syst. **63**, 1961–1989 (2021)

18. Suykens, J.A., Vandewalle, J.: Least squares support vector machine classifiers. Neural Process. Lett. **9**(3), 293–300 (1999)

19. Tabik, S., Gómez-Ríos, A., Martín-Rodríguez, J.L., Sevillano-García, I., Rey-Area, M., Charte, D., et al.: COVIDGR dataset and COVID-SDNet methodology for predicting COVID-19 based on chest x-ray images. IEEE J. Biomed. Health Inform. **24**(12), 3595–3605 (2020). https://doi.org/10.1109/JBHI.2020.3037127

20. Wang, L.: Feature selection with kernel class separability. IEEE Trans. Pattern Anal. Mach. Intell. **30**(9), 1534–1546 (2008)

21. Wang, L., Lin, Z.Q., Wong, A.: COVID-Net: a tailored deep convolutional neural network design for detection of COVID-19 cases from chest x-ray images. Sci. Rep. **10**(1), 1–12 (2020)

22. Yu, X., Chen, Y., Li, T., Liu, S., Li, G.: Multi-mapping image-to-image translation via learning disentanglement. arXiv preprint arXiv:1909.07877 (2019)

23. Zhang, Y., Li, S., Wang, T., Zhang, Z.: Divergence-based feature selection for separate classes. Neurocomputing **101**, 32–42 (2013). https://doi.org/10.1016/j.neucom.2012.06.036

Hybrid Intelligent Applications

Speech Emotion Recognition by Conventional Machine Learning and Deep Learning

Javier de Lope[1], Enrique Hernández[1], Vanessa Vargas[1], and Manuel Graña[2(✉)]

[1] Computational Cognitive Robotics Group, Department of Artificial Intelligence,
Universidad Politécnica de Madrid (UPM), Madrid, Spain
[2] Computational Intelligence Group, University of the Basque Country (UPV/EHU),
Leioa, Spain
manuel.grana@ehu.es

Abstract. This paper reports experimental results of speech emotion recognition by conventional machine learning methods and deep learning techniques. We use a selection of mel frequency cepstral coefficients (MFCCs) as features for the conventional machine learning classifiers. The convolutional neural network uses as features the mel spectrograms treated as images. We test both approaches over a state of the art free database that provides samples of 8 emotions recorded by 24 professional actors. We report and comment the accuracy achieved by each classifier in cross validation experiments. Results of our proposal are competitive with recent studies.

Keywords: Speech emotion recognition · Deep learning ·
Convolutional neural networks · Mel frequency cepstral coefficients ·
Mel spectrogram

1 Introduction

Speech analysis considers several tasks to solve. Automatic speech recognition has received much attention, reaching quite performing solutions based on statistical models and, recently, deep learning approaches. Speaker identification has been strongly developed in order to help in security environments and to analyze the origin of calls. Speech emotion recognition is a recent line of work that has not received much effort yet. In this paper we contribute experimental results from two competing approaches, the traditional machine learning algorithms versus deep learning techniques.

The rest of the paper is organized as follows. First, we review some of the most popular speech emotion databases, selecting the most interesting one to our objectives that will be exploited to generate the training/validation feature datasets. Then, we describe how we compute the features to be used by our emotion classifiers, which are described in the next sections along with the experimental results. Finally we discuss the results and give some conclusions and future works.

H. Sanjurjo González et al. (Eds.): HAIS 2021, LNAI 12886, pp. 319–330, 2021.
https://doi.org/10.1007/978-3-030-86270-8_27

2 Speech Emotion Recognition Databases

The systems to recognize human emotions from speech use audio databases with samples of several phrases with intonations that correspond to each one of the emotions. These databases are costly to create because they must include a large number of audio samples for the training of the machine learning methods to be applied. Also, the samples must contain a balanced set of voices from male and female subjects.

Sometimes the audios are taken from real world situations such as broadcast television, radio shows, or even call centers. The advantage of these datasets is the samples correspond to phrases in real conversations, in which the emotion is expressed spontaneously by each the subject. However, they can not be freely distributed because of copyright restrictions.

Other databases are composed of recordings specifically played by actors. These approaches are simpler to realize than the previous ones. A special care has to be taken in order to avoid any overacting by the actors in order to get intonations equivalents to the ones produced in real world environments. This type of databases are currently used by many studies found in the literature.

2.1 RAVDESS

RAVDESS (Ryerson Audio-Visual Database of Emotional Speech and Song) [16] is composed of 7356 audio and video files (roughly 25 GB). It provides samples of speech and songs, thus it also allows to use and to analyze the recordings in musical environments. As it is freely available, and one of the most complete and frequently used databases, it is frequently used for research [9–11]. It is the database that we use in this paper for computational experiments. There are 1440 samples of speech audio recordings. They are recorded by 24 professional actors (12 male and 12 female) that read two semantically neutral US English phrases while revealing eight emotions (neutral, calm, happiness, sadness, anger, fear, disgust, surprise). The phrases are "kids are talking by the door" and "dogs are sitting by the door". Two levels of emotional intensity are considered in each phrase (only one intensity in the neutral emotion) and there is just one sample for each recording. Each recordings is provided in uncompressed audio files to avoid the artifacts introduced by lossy compression algorithms. Each recording is about 3 s long.

2.2 Other Databases

Other widely used databases in the last two decades are the following ones. Berlin Emo DB [3] is a German database with audio recordings from 10 actors (5 male and 5 female) whom read 10 texts (5 short phrases and 5 longer texts). It provides seven emotions (anger, neutral, anger, boredom, happiness, sadness, disgust). It is one of the most commonly referred in the technical literature [29]. The works include bioinspired real time speech emotion recognition [17] or conventional classifiers based on features as the used in our work such as MFCC

(Mel Frequency Cepstral Coefficients) and LPCC (Linear Prediction Cepstral Coefficient) [20].

DES (Danish Emotional Speech Database) is an emotional database compiled by the University of Denmark under the VAESS project (Voices, Attitudes and Emotions in Speech Synthesis) [7,8]. The database is composed of 500 audios, which are recorded by 4 actors (2 male and 2 female) in Danish language. There are 13 types of recordings (2 words, 9 short phrases and 2 texts) and it considers five emotions (neutral, surprise, happiness, sadness, anger). It has been one of the most used databases by researchers in this field [4,15,26,31].

3   Feature Extraction

We extract two different kinds of features from the audio files. The first kind of features are *mel frequency cepstral coefficients* (MFCCs), which were proposed for speech representation and recognition [6,19]. The second kind of features are the images corresponding to the waveform spectrogram of the audio files. Specifically we use the mel spectrogram [2,5], which is related with the MFCCs.

3.1   Mel Frequency Cepstral Coefficients

To compute the MFCCs several transformations and operations are applied to the original audio signal in a series of consecutive stages. The first operation on the audio signal is a *Fourier Transform* (FT). It decomposes the signal into its frequencies by means of the well-known expression (1).

$$\hat{g}(f) = \int_{-\infty}^{+\infty} g(f) \, e^{-2\pi i f t} dt \tag{1}$$

Specifically, we apply a *Short-Term Fourier Transform* (STFT) that divides the signal in small segments with equal length. A *Fast Fourier Transform* (FFT) is then applied to each segment in order to know the contents of frequency and phase. The STFT process consists of three stages. Firstly, *framing*, segments the original signal into blocks of N samples that are overlapped by M samples to maintain the signal continuity. Usually the overlapping is about 40–60%. This process is needed because the speech signal is generally non-stationery due to modifications in the pronunciation and random variations in the vocal tract. The signal can be considered as stationery when the small fragments are considered, usually between 10 and 20 ms [12]. Secondly, a function is applied to each new segment (*windowing*). This operation tries to preserve the continuity of the signal as smooth as possible. It is made by flattening and softening both ends of each block and by lifting its center area with a low pass filter. There are many different alternatives but the most frequently used is the *Hamming window*, which is shown in (2) when $a_0 = 25/46 \approx 0.53836$.

$$W(n) = \begin{cases} a_0 - (1 - a_0) \cos(\frac{2\pi n}{N-1}) & 0 \le n \le N - 1 \\ 0 & \text{otherwise} \end{cases} \tag{2}$$

The third stage to finish the STFT process is to apply a transform based on a FT to each segment. In this case a discrete Fourier transform (DFT) is applied to get its frequency magnitude as shown (in 3), which is computed by the means of the FFT algorithm.

$$X(k) = \sum_{n=0}^{N-1} x(n)\, e^{-j2\pi k \frac{n}{N}} \quad ; \quad k = 0, \ldots, N-1 \tag{3}$$

Once applied those basic transforms, a set of 20–40 *mel filters* are applied to the signal for multiplying the magnitude of frequency. The *mel scale* (the term *mel* comes from the word *melody*) is a scale based on the perception of acoustic signals by the human ear. The human ear is able to perceive more clearly differences between two frequencies (for example, 100 Hz) in the low frequency range (for example, between 100 Hz and 200 Hz) than in high frequency range, where they are imperceptible (for example, between 10000 Hz and 10100 Hz). The mel scale adjusts the acoustic signals by reducing the differences according to expression (4).

$$Mel(f) = 2595 \; \log\left(1 + \frac{f}{700}\right) \tag{4}$$

The main concept in MFCCs is the *cepstrum*, that comes from the word *spectrum* by reversing its head. It was introduced by Bogert [2] as a tool to analyze periodic signals. Basically, the resulting spectrogram is obtained from the spectrum computed by applying the logarithm to original signal spectrum as in (5). As this new spectrum does not belong to neither time nor frequency domains, it was defined as *quefrency*.

$$C_p = \left| F\left\{\log\left(|F\{f(t)\}|^2\right)\right\}\right|^2 \tag{5}$$

In this paper, the cepstrums are obtained from the discrete cosine transform (DCT) computed by using expression (6)

$$y_t(k) = \sum_{m=1}^{M} \log\left(Y_t^2(m)\right) \cos\left((m-0.5)\frac{k\pi}{M}\right) \quad ; \quad k = 1, 2, \ldots, J \tag{6}$$

where M is the number of mel filters and J is the number of MFCCs. The computational pipeline of the MFCCs feature extraction is shown in Fig. 1.

3.2 Mel Spectrogram

The mel spectrograms are computed according to expression (5). To be used in some of the classifying techniques proposed in this work (i.e. convolutional neural networks) it is needed to represent them as images. One of the resulting images is shown in Fig. 2.

We use functions provided by the *librosa* software library [18] and other open software tools to generate these images. For this particular case we convert each

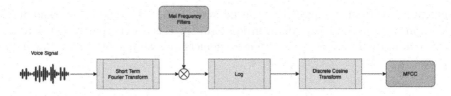

Fig. 1. Obtaining the mel frequency cepstral coefficients (MFCCs).

Fig. 2. Image corresponding to a mel spectrogram of an audio file.

audio file and store them as separate image files. The images dimensions are 128×128 pixels. These numbers have been empirically defined trying to maximize the trade off between the images quality and size, amount of information in them, training time and classifying accuracy.

3.3 Data Augmentation

Although the database has 1440 samples, they are not be enough for training satisfactorily deep learning models. Thus, we have also used some data augmentation techniques in order to avoid overfitting episodes.

We have applied a data augmentation over the mel spectrogram images [23] instead of generating modifications over the audio. The method is initially proposed for automatic speech recognition but it works over the same kind of graphic representation that we use in our approach (i.e. the mel spectrogram). Basically the method applies three masking and warping operations to the images. Firstly, the image is warped by using a *time warping* operation. The operation gets a spectrogram of τ period, a random point along the horizontal axis on the center of the image is warped in the interval $(W, \tau - W)$ by a distance w randomly determined from a uniform distribution $U(0, W)$ (W is a hyperparameter). Secondly, an operation of masking in the frequency domain is applied. The mask is applied over the f channels of consecutive frequencies $[f_0, f_0 + f)$, where f is randomly obtained from a uniform distribution $U(0, F)$, F is a hyperparameter, f_0 is chosen in the range $[0, v - f)$, and v is the number of frequency channels in the spectrogram. Finally, a new masking operation is applied, now in the time domain. The mask is applied in the interval $[t_0, t_0 + t)$, where t is randomly determined from a uniform distribution $U(0, T)$, T is a hyperparameter, and t_0 is chosen in the range $[0, \tau - t)$. Figure 3 depicts an example of the intermediate images.

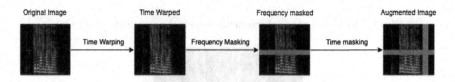

Fig. 3. Intermediate images generated by warping and masking for data augmentation.

Besides, we have used a traditional technique based on adding noise to the original acoustic signal as it is performed in audio processing. Particularly, in a deep learning context, adding noise to the samples contributes to avoiding that the network overfits to the data thanks to the variability in them, and it tends to produce smaller weights between connections, achieving lower error, which improves the generalization. On the other hand, adding noise improves the regularization and allows that more robust networks [1]. Figure 4 shows an example of audio with and without noise.

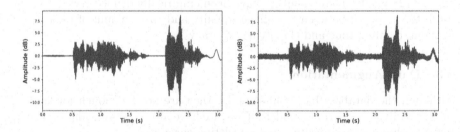

Fig. 4. Original signal (left) and signal with added noise (right).

By applying these two data augmentation techniques we have increased three times the number of samples in the original dataset.

4   Experimental Results

We have employed several methods to classify the audio samples into the corresponding classes. Firstly, we describe and comment the specific management, training and hyperparameters used with the conventional techniques. Then, we briefly describe the convolutional neural network model proposed as well as the training. Finally, we summarize and compare the results with all classifiers.

4.1 Conventional Techniques

After a number of exploratory tests performed with a greater variety of classifiers than reported here, and reviewing the bibliography, the selected classifiers are k-nearest neighbors (kNN), support vector machines (SVM) with both linear and polynomial kernel, random forest (RF) and multi-layer perceptron (MLP). We use the *scikit-learn* library [24].

We get the best results with each classifiers family with the following parameters. We use $k = 1$ with kNN, 20 random states and no bootstrap with RF, and one hidden layer with 200 neurones with MLP. We use the default values for the rest of parameters.

Regarding the audio samples, we also use the default parameters or mandatory values forced by the recordings in the database as for example the sampling rate, which is 22050. We use 20, 13 and 27 coefficients for the experiments, which give acoustic vectors with dimensions $[1536, 2800]$, $[1536, 1820]$ and $[1536, 3780]$, respectively. These vectors are used as dataset to train all the classifiers. We have defined the usual cross-validation approach and we use 80–20 for training and test. Each training is repeated 100 times and we offer averaged values in the results.

Table 1 shows the accuracy achieved by the classifiers with 20, 13 and 27 coefficients over all the 8 emotions in database. The best result is obtained by a SVM classifier with polynomial kernel and 27 acoustic vectors per audio file, about 70%. The larger number of coefficients, the better (only RF makes worse the results). Also the time required as for training as for classifying samples is increased considerably with the number coefficients, about twice in each step. Thus, as the results with 20 acoustic vectors are quite comparable, in order to look for a trading off between time and results, hereinafter we will use this value of MFCCs.

Table 1. Accuracy of tests with 20, 13 and 17 MFCCs.

Classifier	MFCCs 20	MFCCs 13	MFCCs 27
kNN ($k = 1$)	.647	.632	.649
SVM (linear)	.642	.626	.652
SVM (poly)	.679	.651	.681
RF	.587	.581	.572
MLP	.516	.462	.541

Often, the experiments reported in the literature take into account a reduced number of emotions. In order to assess the impact of the number of emotions on the recognition performance we repeat the model training over two reduced sets of emotions. First, we have reduced to six emotions by removing two of the initial ones according to a criterion of similarity between them and the need of complementary information to distinguish the emotions (maybe visual information or the context in which the conversation is carried out). Thus, we have removed *calm* that could be confused with *normal* in some situations, and *disgust* that is easily confused with *anger*. Secondly, we consider a further reduction to four emotions. The criterion is to choose four of the emotions represented in opposite axes in the Plutchik's emotions star [25]. Thus, it would be easier to distinguish these emotions either by a human or a trained automatic system. The four emotions considered in this experiment are *anger*, *sadness*, *hapiness* and *fear*. The sizes of the new reduced datasets are 1152 and 768 samples for six and four emotions, respectively (from 1536 samples in the initial dataset).

Table 2 shows the accuracy achieved by the classifiers over the datasets with reduced number of emotions, and features given by 20 MFCCs. As it was expected the accuracy increases when the number of emotions decreases. Again the best results are achieved with a SVM with polynomial kernel. Note that the MPL does not improve the accuracy performance going from six to four emotions. Table 3 shows an instance of the confusion matrix (eight emotions) of a SVM with polynomial kernel (the best classifier) out from the 100 repetition. We can observe the confusion between *anger* and *disgust*, *happiness* and *surprise*, and *calm* and *sadness*. However, there is no confusion between *calm* and *normal*, the other emotion that was removed.

Table 2. Precision of tests with 20 MFCCs over 8, 6 and 4 emotions.

Classifier	8 emotions	6 emotions	4 emotions
kNN (k = 1)	.641	.673	.702
SVM (linear)	.642	.671	.721
SVM (poly)	.671	.703	.753
RF	.587	.621	.659
MLP	.516	.613	.612

Table 3. Confussion matrix of SVM with 20 MFCCs over 8 emotions.

	Calm	Disgust	Anger	Happiness	Fear	Neutral	Surprise	Sadness
Calm	0.906	0	0	0	0	0	0	0.093
Disgust	0	0.674	0.186	0.023	0.023	0	0.023	0.069
Anger	0	0.075	0.725	0.075	0.05	0	0.05	0.025
Happiness	0.121	0.048	0.122	0.561	0	0.048	0.024	0.073
Fear	0.129	0.064	0.032	0.032	0.580	0.032	0.064	0.064
Neutral	0	0	0	0	0	0.950	0	0.050
Surprise	0	0.051	0.025	0.179	0.128	0.102	0.487	0.025
Sadness	0.230	0.102	0	0.051	0.025	0.102	0	0.487

4.2   Convolutional Model

Instead of applying a transfer learning with some large published network, we prefer to define and train a smaller model based on a LeNet-5 network [13]. Basically, we want to avoid overfitting that troubles those large models when there is a low number of available samples. Also, we need to reduce the training time for lack of computational resources. Thus, we have considered a convolutional module as shown in Fig. 5(a). It is composed of a 2D convolutional layer (the input layer) that applies a convolutional operation with 32 filters, a 3×3 kernel, stride equal to 1, and a relu activation function. Also, we include a max-pooling layer to reduce the input dimensionality with a pool-size of 2×2.

Then, we have added a series of traditional dense layers to learn the patterns identified by the convolutional module. The first of these layers flattens the two dimensional output of the convolutional model that it is followed by one or more intermediate layers and finished with a softmax layer with 8 output neurones as required. After exploratory experiments to determine the number and characteristics of the layers, we define the architecture shown in Fig. 5(b). It is composed of two convolutional layers as described before. The first one receives images with dimension $128 \times 128 \times 3$. The second module uses the same number of filters that the first one but receives images with dimension $64 \times 64 \times 3$ and reduces the output to $32 \times 32 \times 3$. This layer is connected to a dense layer with 512 neurones and relu activation function. As we have previously referred, the last layer uses a softmax activation function with 8 neurones.

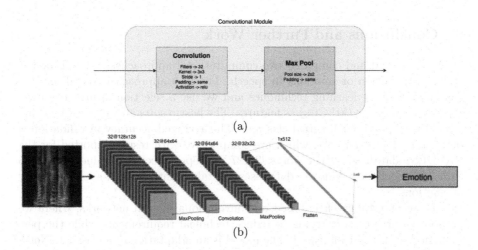

Fig. 5. (a) Convolutional module used in the neural network proposed. (b) Convolutional neural network architecture.

The network is trained 50 epochs with a stochastic gradient descent optimizer, learning rate equal to 0.01, and momentum equal to 0.8. The batch size is 20. We have achieved a training accuracy of 0.901 and a validation accuracy of 0.687 by using the original dataset, which is competitive with the conventional machine learning approaches. In Fig. 6(left) we can observe the overfitting arises during training (mainly by the difference in training and validation results, about 22%) and the network is not able to generalize. The accuracy with the second dataset extended with the data augmentation techniques are 0.332 in training and 0.409 in validation. In Fig. 6(right) the training curves are shown. Here there is not an episode of overfitting but the network is not able to find patterns to differentiate between emotions. The accuracy is much lower than the initial dataset without data augmentation (about 57% in training).

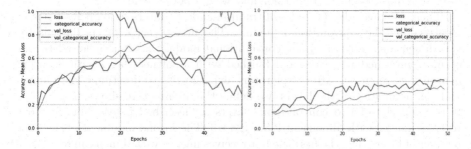

Fig. 6. Training curves with the initial dataset (left) and after the data augmentation (right). The data augmentation prevents the overfitting but reduces the accuracy.

5   Conclusions and Further Work

We have applied and compared two computational approaches to build models for the recognition of emotions in speech. The first approach is based on conventional machine learning techniques and we use a selection of mel frequency cepstral coefficients (MFCCs) as features. The best results are achieved by using a SVM with polynomial kernel classifier. The averaged accuracy in validation is between 67.1% and 75.3%, which are competitive with results reported for the RAVDESS database. Other works [21,27,28] report accuracy values over 90% with equivalent tools, but over databases and taking into account a lower number of emotions.

The second approach is based on convolutional neural networks, where we propose an architecture trying to reduce technical requirements while the performance are kept or increased. The model is an adaptation of a classic LeNet-5 network, which is fed with spectrogram images generated from the speech. Here we also apply data augmentation techniques to control the overfitting. The accuracy in validation improves our own values achieved with conventional methods and it is comparable to the reported ones in the technical literature [14,22,30] with similar techniques and other databases.

The deep learning approach seems quite interesting and promising to study deeply by proposing and testing new and more complex methods. Particularly we have already some initial works with convolutional recurrent neural network models. To use the speech emotion recognition in real time, real world contexts is also a path we are starting to explore.

Acknowledgments. This work has been partially supported by FEDER funds through MINECO project TIN2017-85827-P.

References

1. Bishop, C.: Training with noise is equivalent to Tikhonov regularization. Neural Comput. **7**(1), 108–116 (1995)
2. Bogert, B., Healy, J., Tukey, J.: The quefrency analysis of time series for echoes: cepstrum, pseudo-autocovariance, cross-cepstrum and saphe cracking. In: Proceeddings of the Symposium on Time Series Analysis, pp. 209–243 (1963)
3. Burkhardt, F., Paeschke, A., Rolfes, M., Sendlmeier, W., Weiss, B.: A database of German emotional speech. In: Proceedings of 9th European Conference on Speech Communication and Technology, pp. 1517–1520 (2005)
4. Chavan, V., Gohokar, V.: Speech emotion recognition by using SVM-classifier. Int. J. Eng. Adv. Technol. **1**(5), 11–15 (2012)
5. Chiders, D., Skinner, D., Kemerait, R.: The cepstrum: a guide to processing. Proc. IEEE **65**, 1428–1443 (1977)
6. Davis, S., Mermelstein, P.: Comparison of parametric representations for mono-syllabic word recognition in continuously spoken sentences. IEEE Trans. Acoust. Speech Signal Process. **28**(4), 357–366 (1980)
7. Engberg, I., Hansen, A.: Documentation of the Danish emotional speech database. Technical report, Center for Person Kommunilation, Denmark (1996)
8. Engberg, I., Hansen, A., Andersen, O., Dalsgaard, P.: Design, recording and verification of a Danish emotional speech database. In: Proceedings of EuroSpeech, pp. 1695–1698 (1997)
9. Gao, Y., Li, B., Wang, N., Zhu, T.: Speech emotion recognition using local and global features. In: He, Y., et al. (eds.) BI 2017. LNCS (LNAI), vol. 10654, pp. 3–13. Springer, Cham (2017). https://doi.org/10.1007/978-3-319-70772-3_1
10. Iqbal, A., Barua, K.: A real-time emotion recognition from speech using gradient boosting. In: Proceedings of International Conference on Electrical, Computer and Communication Engineering, pp. 1–5 (2019)
11. Issa, D., Faith-Demirci, M., Yazici, A.: Speech emotion recognition with deep convolutional neural networks. Biomed. Signal Process. Control **59**, 101894 (2020)
12. Kamil, O.: Frame blocking and windowing. Speech Signal **4**, 87–94 (2018)
13. LeCun, Y., Bottou, L., Bengio, Y., Haffner, P.: Gradient-based learning applied to document recognition. Proc. IEEE **86**(11), 2278–2324 (1998)
14. Lim, W., Jang, D., Lee, T.: Speech emotion recognition using convolutional and recurrent neural networks. In: Proceedings of Asia-Pacific Signal and Information Processing Association Annual Summit and Conference, pp. 1–4 (2016)
15. Lin, Y., Wei, G.: Speech emotion recognition based on HMM and SVM. In: IEEE International Conference on Machine Learning and Cybermetics, pp. 4898–4901 (2005)
16. Livingstone, S., Russo, F.: The Ryerson audio-visual database of emotional speech and song (RAVDESS): a dynamic, multimodal set of facieal and vocal expressions in North American English. PLoS One **13**(5), e0196391 (2018)

17. Lotfidereshgi, R., Gournay, P.: Biologically inspired speech emotion recognition. In: Proceedings of IEEE International Conference on Acoustics, Speech and Signal Processing, pp. 5135–5139 (2017)
18. McFee, B., et al.: librosa: Audio and music signal analysis in Python. In: Proceedings of 14th Python in Science Conference, pp. 18–25 (2015)
19. Mermelstein, P.: Distance measures for speech recognition, psychological and instrumental. In: Chen, C. (ed.) Pattern Recognition and Artificial Intelligence, pp. 374–388. Academic Press (1976)
20. Palo, H., Mohanty, M.: Wavelet based feature combination for recognition of emotion. Ain Shams Eng. J. 9(4), 1799–1806 (2018)
21. Pan, Y., Sen, P., Shen, L.: Speech emotion recognition using support vector machines. Int. J. Smart Home 6, 101–108 (2012)
22. Pandey, S., Shekhawat, H., Prasanna, S.: Deep learning techniques for speech emotion recognition: a review. In: Proceedings of 29th IEEE International Conference on Radioelektronika, pp. 1–6 (2019)
23. Park, D., et al.: SpecAugment: a simple data augmentation method for automatic speech recognition. In: Proceedings of Interspeech 2019, pp. 2613–2617 (2019)
24. Pedregosa, F., et al.: Scikit-learn: machine learning in Python. J. Mach. Learn. Res. 12, 2825–2830 (2011)
25. Plutchik, R.: The nature of emotions: human emotions have deep evolutionary roots. Am. Sci. 89(4), 344–350 (2001)
26. Ramakrishnan, S., Emary, I.E.: Speech emotion recognition approaches in human computer interaction. Telecommun. Syst. 52(3), 1467–1478 (2013)
27. Rao, K., Kumar, T., Anusha, K., Leela, B., Bhavana, I., Gowtham, S.: Emotion recognition from speech. Int. J. Comput. Sci. Inf. Technol. 15, 99–117 (2012)
28. Seehapoch, T., Wongthanavasu, S.: Speech emotion recognition using support vector machines. In: International Conference on Knowledge and Smart Technology (2013)
29. Stolar, M., Lech, M., Bolia, R., Skinner, M.: Real time speech emotion recognition using RGB image classification and transfer learning. In: Proceedings of 11th IEEE International Conference on Signal Processing and Communication Systems, pp. 1–8 (2005)
30. Tripathi, S., Kumar, A., Ramesh, A., Singh, C., Yenigalla, P.: Focal loss based residual convolutional neural network for speech emotion recognition. arXiv:1906.05682 (2019)
31. Ververidis, D., Kotropoulos, C.: Automatic speech classification to five emotional states based on gender information. In: 12th IEEE European Signal Processing Conference, pp. 341–344 (2004)

Predictive Maintenance of Vehicle Fleets Using Hierarchical Modified Fuzzy Support Vector Machine for Industrial IoT Datasets

Arindam Chaudhuri[1,2(✉)] and Soumya K. Ghosh[3]

[1] Samsung, R & D Institute Delhi, Noida 201304, India
[2] NMIMS University, Mumbai 400056, India
arindam.chaudhuri@nmims.edu
[3] Department of Computer Science Engineering, Indian Institute of Technology, Kharagpur 701302, India

Abstract. Connected vehicle fleets are deployed worldwide in several industrial internet of things scenarios. With the gradual increase of machines being controlled and managed through networked smart devices, the predictive maintenance potential grows rapidly. Predictive maintenance has the potential of optimizing uptime as well as performance such that time and labor associated with inspections and preventive maintenance are reduced. It provides better cost benefit ratios in terms of business profits. In order to understand the trends of vehicle faults with respect to important vehicle attributes viz mileage, age, vehicle type etc. this problem is addressed through hierarchical modified fuzzy support vector machine which acts as predictive analytics engine for this problem. The proposed method is compared with other commonly used approaches like logistic regression, random forests and support vector machines. This helps better implementation of telematics data to ensure preventative management as part of the desired solution. The superiority of the proposed method is highlighted through several experimental results.

Keywords: Predictive maintenance · Industrial IoT · Support vector machine · Fuzzy sets · Vehicle fleets

1 Introduction

The present business setup has connected vehicles [1] which forms an integral part of operations in every industry today. These vehicles are producing huge amount of data which is projected to grow steadily in coming years. Today organizations are struggling to capture and harness power of internet of things (IoT) information [2] which can be applied for operational insights regarding devices and get ahead of unplanned downtime. Organizations are always looking for faster ways [3] to realize sensor information and transform it into predictive maintenance insights [4].

The cost of vehicle downtime is significant towards customers' demand for higher uptimes and aggressive service level agreements. The service providers look for predictive maintenance techniques using accurate real-time vehicle information. This helps them

© Springer Nature Switzerland AG 2021
H. Sanjurjo González et al. (Eds.): HAIS 2021, LNAI 12886, pp. 331–342, 2021.
https://doi.org/10.1007/978-3-030-86271-8_28

to determine vehicle's condition and its required maintenance. This approach provides savings [5] in terms of cost. The predictive maintenance allows corrective maintenance scheduling and unexpected vehicle failures prevention. The maintenance work can be better planned with prior information. With connected vehicles predictive maintenance solution [6] users achieve timely maintenance towards increased up-times, better plan maintenance in reducing unnecessary field service calls, optimizing repair parts replacement, reducing unplanned stops, improving vehicle performance and service compliance reporting. Vehicles are designed with temperature, infrared, acoustic, vibration, battery-level and sound sensors in order to monitor conditions which can form as initial maintenance indicators as shown in Fig. 1. The predictive maintenance programs are driven by customers which helps them to collect and manage vehicle data alongwith visualization and analytics [7] tools to make better decisions.

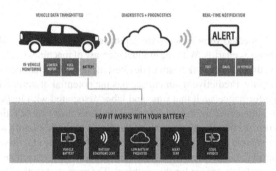

Fig. 1. Predictive maintenance of vehicles

To solve problems posed by predictive maintenance of vehicle fleets in a specified garage for telecom-based company statistical and soft computing techniques [8] are used to develop real time solutions. Here hierarchical version of support vector machine (SVM) viz modified fuzzy support vector machine (HMFSVM) is proposed to achieve predictive analytics task. The predictive analytics problem comprises of classification problem which is performed through HMFSVM, a variant of modified fuzzy support vector machine (MFSVM) [9] and fuzzy support vector machine (FSVM) [10, 11]. In MFSVM classification success lies in selection of fuzzy membership function [12–14]. For this classification problem, vehicle data is paralyzed with population drift incorporating customer behavior. MFSVM is hierarchically extended to HMFSVM which gives better sensitivity to class imbalance problem, reduced number of support vectors, training times and model complexity with better generalization. The experimental results support HMFSVM's superiority over algorithms.

This paper is organized as follows. In Sect. 2 computational method of HMFSVM is highlighted. This is followed by experiments and results in Sect. 3. Finally in Sect. 4 conclusions are given.

2 Computational Method

In this section framework of proposed HMFSVM model is presented. The research problem entails in predictive maintenance for vehicle fleets which in a way involves certain preventive maintenance [15, 16] activities. In order to achieve this, we propose deep learning-based predictor viz HMFSVM to analyse vehicle fleets' maintenance. The prediction task on various strategic aspects of vehicle analytics provides information for several decision-making activities.

2.1 Datasets

The datasets are adopted from telecom-based company [8] garages. The datasets available for predictive analytics include data from 5 garages. The data from first garage spanned over a period of 14 months and 24 months for other garages. The number of data records available from first garage was 3923 of which 2097 was used for prediction analysis. The number of data records available from other garages was 890665 of which 11456 was used for prediction analysis. The reasons behind not using entire available data for prediction include: (a) there were data rows with invalid values which have been considered as outliers (b) vehicle repair types count did not have considerable values for all vehicles (c) fuel economy score and driver behavior score were available only for first garage. There was great degree of skewedness in training data. Before applying data to prediction engine, it was balanced through external imbalance learning method. The dataset was used after majority class under-sampling.

2.2 MFSVM for Predictive Maintenance

In SVM, each sample point is fully assigned to either of two classes. In many applications there are points which may not be exactly assigned to either of two classes. To address this problem, fuzzy membership is assigned to each input point of SVM where input points contribute uniquely for decision surface construction. The corresponding input's membership is reduced with total error decreased. Each input point is treated with higher membership as an input of opposite class. Let us suppose training sample points are considered as $Sample\_Points = \{(S_i, y_i, sp_i); i = 1, \ldots \ldots, P\}$. Here each $S_i \in R^N$ is training point and $y_i \in \{-1, +1\}$ represent class label; $sp_i; i = 1, \ldots \ldots, P$ represents fuzzy membership function with $p_j \leq sp_i \leq p_i; i = 1, \ldots \ldots, P$ where $p_j > 0$ and $p_i < 1$ are sufficiently small constants. $S = \{S_i | (S_i, y_i, sp_i) \in Sample\_Points\}$ containing two classes. One class contains point S_i with $y_i = 1$ as $Class^+$. Other class contains point S_i with $y_i = -1$ as $Class^-$:

$$Class^+ = \{S_i | S_i \in Sample\_Points \wedge y_i = 1\} \tag{1}$$

$$Class^- = \{S_i | S_i \in Sample\_Points \wedge y_i = -1\} \tag{2}$$

Now *Point_Space* = *Class*$^+$ ∪ *Class*$^-$. The quadratic classification proposition is:

$$\min \frac{1}{2}\|\omega\|^2 + C\sum_{i=1}^{P} sp_i\rho_i$$

Subject to

$$y_i(\omega^T \Phi(S_i) + a) \geq 1 - \rho_i, \; i = 1, \ldots \ldots, P, \; \rho_i \geq 0$$

$$(3)$$

In Eq. (3) C is regularization parameter. The fuzzy membership sp_i governs behavior of sample point S_i towards one class and ρ_i measures SVM error. The factor $sp_i\rho_i$ is error measure with different weights. The smaller sp_i reduces ρ_i effect in Eq. (3) where point S_i is less significant [17]. The mean of class *Class*$^+$ and *Class*$^-$ are *mean*$_+$ and *mean*$_-$ respectively and radius of class *Class*$^+$ and *Class*$^-$ ($S_i \in$ *Class*$^+$) are:

$$radius_+ = \max\|mean_+ - S_i\|$$

$$(4)$$

$$radius_- = \max\|mean_- - S_i\|$$

$$(5)$$

The fuzzy membership sp_i [18] is given by:

$$sp_i = \begin{cases} 1 - \frac{\|mean_+ - S_i\|^2}{(radius_+ + \theta)^2} & \text{if } S_i \in Class^+ \\ 1 - \frac{\|mean_- - S_i\|^2}{(radius_- + \theta)^2} & \text{if } S_i \in Class^- \end{cases}$$

$$(6)$$

Now MFSVM is formulated based on FSVM. Consider sample $S_i \in$ *Sample_Space*. Let $\Phi(S_i)$ represent mapping function through input into feature space. The hyperbolic tangent kernel is *Kernel*$(S_i, S_j) = \tanh[\Phi(S_i).\Phi(S_j)]$ [13, 19–22]. As sigmoid kernel has given good results [23], it is used here. Now Φ_+ is defined as center of class *Class*$^+$ as:

$$\Phi_+ = \frac{1}{samples_+}\sum_{S_i \in Class^+} \Phi(S_i) freq_i$$

$$(7)$$

In Eq. (7) *samples*$_+$ is number of samples of class *Class*$^+$ with frequency *freq*$_i$ of ith sample in $\Phi(S_i)$. Again Φ_- is class center of *Class*$^-$ and is defined as:

$$\Phi_- = \frac{1}{samples_-}\sum_{S_i \in Class^-} \Phi(S_i) freq_i$$

$$(8)$$

In Eq. (8) *samples*$_-$ is number of samples of class *Class*$^-$ with frequency *freq*$_i$ in $\Phi(S_i)$. The *Class*$^+$ and *Class*$^-$ radius are:

$$radius_+ = \frac{1}{n}\max\|\Phi_+ - \Phi(S_i)\|$$

$$(9)$$

$$radius_- = \frac{1}{n}\max\|\Phi_- - \Phi(S_i)\|$$

$$(10)$$

With $n = \sum_i freq_i$ we have: $radius_+^2 = \frac{1}{n}\max\|\Phi(s') - \Phi_+\|^2$

$$radius_+^2 = \frac{1}{n}\max[Kernel(S',S') - \frac{2}{samples_+}\sum_{S_i \in Class^+} Kernel(S',S') +$$
$$\frac{1}{samples_+^2}\sum_{S_i \in Class^+}\sum_{S_j \in Class^+} Kernel(S_i,S_j)] \tag{11}$$

In Eq. (11) $S' \in Class^+$ and $samples_+$ is number of training samples in $Class^+$.

$$radius_-^2 = \frac{1}{n}\max[Kernel(S',S) - \frac{2}{samples_-}\sum_{S_i \in Class^-} Kernel(S',S') +$$
$$\frac{1}{samples_-^2}\sum_{S_i \in Class^-}\sum_{S_j \in Class^-} Kernel(S_i,S_j) \tag{12}$$

In Eq. (12) $S' \in Class^-$ and $samples_-$ is number of training samples in $Class^-$. The distance square between sample $S_i \in Class^+$ and its class center is given as:

$$distance_{i+}^2 = Kernel(S_i,S_j) - \frac{2}{samples_+}\sum_{S_j \in Class^+} Kernael(S_i,S_j) +$$
$$\frac{1}{samples_+^2}\sum_{S_j \in Class^+}\sum_{S_k \in Class^+} Kernel(S_j,S_k) \tag{13}$$

Similarly, distance square between sample $S_i \in Class^-$ and its class center is:

$$distance_{i-}^2 = Kernel(S_i,S_j) - \frac{2}{samples_-}\sum_{S_j \in Class^-} Kernel(S_i,S_j) +$$
$$\frac{1}{samples_-^2}\sum_{S_j \in Class^-}\sum_{S_k \in Class^-} Kernel(S_j,S_k) \tag{14}$$

Now $\forall i; i = 1,\ldots\ldots, P$ the fuzzy membership function sp_i is:

$$sp_i = \begin{cases} 1 - \sqrt{\dfrac{\|distance_{i+}^2\|^2 - \|distance_{i+}^2\|\, radius_+^2 + radius_+^2}{\left(\|distance_{i+}^2\|^2 - \|distance_{i+}^2\|\, radius_+^2 + radius_+^2\right) + \epsilon}} & if\ y_i = 1 \\ 1 - \sqrt{\dfrac{\|distance_{i-}^2\|^2 - \|distance_{i-}^2\|\, radius_-^2 + radius_-^2}{\left(\|distance_{i-}^2\|^2 - \|distance_{i-}^2\|\, radius_-^2 + radius_-^2\right) + \epsilon}} & if\ y_i = -1 \end{cases} \tag{15}$$

In Eq. (15) $\epsilon > 0$ such that sp_i is not zero and is function of center and radius of each class in feature space represented through kernel. The training samples are either linear or nonlinear separable. This method accurately represents each sample point contribution towards separating hyperplane construction in feature space [24]. This helps MFSVM to reduce outliers' effect more efficiently with better generalization.

2.3 HMFSVM for Predictive Maintenance

Now hierarchical version of MFSVM is presented. The prediction model was build using HMFSVM. This is a decision-based model that produces predictions as probability values. The computational benefits accrued from MFSVM serve the major prima face. HMFSVM supersedes MFSVM considering similarity-based classification accuracy and running time as data volume increases [25]. The proposed model's architecture is shown in the Fig. 2 with temporal data sequences modeled through MFSVMs. These are

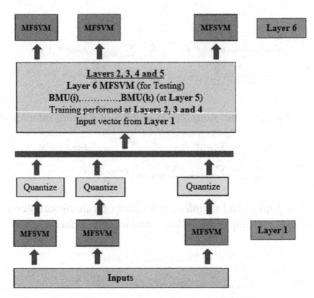

Fig. 2. The architecture of proposed HMFSVM

combined together towards HMFSVM. HMSVM architecture represents data behavior across multiple relevant features.

The model is composed of 6 layers. At 1st and 2nd layers relatively small MFSVMs are utilized. The number of MFSVMs are increased at layers 3, 4 and 5. MFSVMs in last layer are constructed over subset of examples in 5th layer is Best Matching Unit (BMU). MFSVMs at last layer can be larger than used in 1st to 5th layers. It improves resolution and discriminatory capacity of MFSVM. Building HMFSVM requires several data normalization operations. This provides for initial temporal pre-processing and inter-layer quantization between 1st to 5th layers. The pre-processing provides suitable representation for data and supports time-based representation. The 1st layer treats each feature independently. In case of temporal representation standard MFSVM has no capacity to recall histories of patterns directly. A shift-register of length l is employed in which tap is taken at predetermined repeating interval k such that $l \% k = 0$ where $\%$ is modulus operator. The 1st level MFSVMs only receive values from shift register. Thus, as each new connection is encountered (at left), content of each shift register location is transferred one location (to right) with previous item in l^{th} location being lost. In case of n-feature architecture it is necessary to quantize number of neurons between 1st to 5th level MFSVMs. The purpose of 2nd to 5th level MFSVM is to provide an integrated view of input feature specific MFSVMs developed in 1st layer. There is potential in 2nd to 5th layer MFSVM to have an input dimension defined by total neuron count across all 1st layer MFSVMs. This is brute force solution that does not scale computationally.

Given ordering provided by MFSVM neighboring neurons respond to similar stimuli. The structure of each 1st layer SVM is quantized in terms of fixed number of units using potential function classification algorithm [8]. This reduces inputs in 2nd to 5th layers of MFSVM. The units in 4th layer acts as BMU for examples with same class label thus maximizing detection rate and minimizing false positives. The 4th layer SVM acts as BMU for examples from more than one class to partition data. The 5th layer MFSVMs are trained on subsets of original training data. This enables size of 5th layer MFSVMs to increase which improves class specificity with reasonable computational cost. Once training is complete 4th layer BMUs acts to identify which examples are forwarded to corresponding 5th layer MFSVMs on test dataset. A decision rule is required to determine under what conditions classification performance of BMU at 4th layer MFSVM is judged sufficiently for association with 5th layer MFSVM. There are several aspects which require attention such as minimum acceptable misclassification rate of 4th layer BMU relative to number of examples labeled and number of examples represented. The basic implication is that there must be optimal number of connections associated with 4th layer BMU for training of 5th layer MFSVM and misclassification rate over examples associated with 4th layer BMU exceeds threshold.

The input vector to HMFSVM constituted of base and derived input variables. The base input variables included vehicle registration number, registration date, number of garage visits of vehicle, vehicle repair history, odometer reading, repair type of vehicle and vehicle repair types count. The input vector is fed to MFSVM at layer 1 and results are quantized. The quantized results were passed onto layers 2, 3 and 4 where training was performed repeatedly. After successful training BMU was found at layer 5. Finally in layer 6 testing is performed where predicted probability values are achieved which represents chance of vehicle's visit to garage.

Table 1. Classification results of vehicles

Classification category	Number of vehicles
Immediate risk	30
Short term risk	175
Longer term risk	1668
Total number of vehicles	1873

3   Experiments and Results

Here HMFSVM's efficiency is highlighted through experimental results on telecom-based company data [8]. A comparative analysis of HMFSVM is also performed with MFSVM, FSVM and other methods like SVM, logistic regression and random forest algorithms. The predictions are produced as probability values by these models. The algorithm with best prediction accuracy is chosen. HMFSVM is trained on other garages data alongwith first garage data. The computational workflow constitutes processed data comprising of base and derived variables which is fed into HMFSVM. The training process is done iteratively on training data to fine tune learning process. HMFSVM testing is performed on first garage data to create prediction results. The results with probability percentage are translated to time as vehicles come for servicing. The classification is highlighted in Table 1 as (a) Immediate Risk: Probability above 60% (vehicle in bad shape and needs to be in garage immediately) (b) Short Term Risk: Probability between 40% and 60% (vehicle could have problems in near future) and (c) Longer Term Risk: Probability less than 40% (vehicle is in fair condition).

The prediction results obtained are validated with respect to available results. Keeping in view classification results sample vehicles from immediate risk category are considered. Considering an immediate risk vehicle number KE55KZB, total number of jobs is 13, total labor hours is 121.35 and average labor hours is 9.34. These results are shown in Fig. 3. Hence it is required to allocate approximately 9.34 h for this vehicle when scheduling appointment. The typical services required for this vehicle are shown in Fig. 4. As obvious from figure typical services required for this vehicle are vehicle movement, routine service and electrical system repairs. The receiver operating characteristic (ROC) curve or area under curve (AUC) for predicted results from HMFSVM are shown in Fig. 5. The figure presents true positive versus false positive rates resulting from several cut-offs used in prediction engine. It gets better as curve climbs faster. The ROC score obtained is 0.966 and prediction accuracy is about 96%. The predictions being arrived at are in alignment with longer-term objective of using real time sensor data for more accurate predictions.

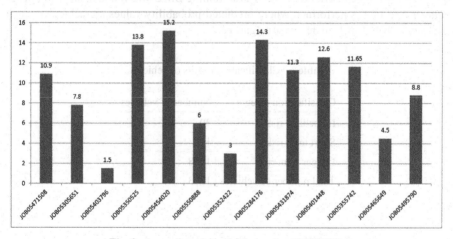

Fig. 3. Immediate risk vehicle number KE55KZB

HMFSVM has proved itself to be promising data classification approach. However, proper parameter selection is required to achieve good outcomes. In this direction cross validation finds best combinations for parameters C and γ which are regularization and kernel parameters respectively. The best parameters are used for training and testing the whole dataset. Here hyperbolic tangent kernel is best suited based on which parameters (C, γ) are chosen. Since best values of (C, γ) for given problem is not known beforehand, model selection or parameter search is performed. This identifies best (C, γ) values where classifier accurately predicts testing data. The dataset is separated into two parts comprising of one unknown component. The unknown component gives prediction accuracy with appreciable performance on classifying an independent dataset. This process is known as v fold cross-validation with training set divided into equal sized v subsets. The classifier trained on remaining $(v - 1)$ subsets with testing being performed on one subset. This results in each instance of whole training set being predicted once. With cross-validation accuracy being correctly classified, overfitting problem is prevented.

Using cross-validation grid search finds values of (C, γ). The feasible values of (C, γ) have range $C = (2^{-5}, 2^{-3}, \ldots \ldots, 2^{18})$ and $\gamma = (2^{-18}, 2^{-17}, \ldots \ldots, 2^4)$. After populating feasible region, a finer grid search results optimal values. A coarse grid is then used to find best (C, γ) as $(2^8, 2^{-4})$ with cross validation rate 89%. Next, finer grid search is conducted on neighborhood of $(2^8, 2^{-4})$ with better cross validation rate of 89.4% at $(2^{8.45}, 2^{-4.45})$ is achieved. After that best (C, γ) is discovered on whole training set to obtain final classifier. The Table 2 shows HMFSVM results as number of support vectors with hyperbolic tangent kernel for different C values.

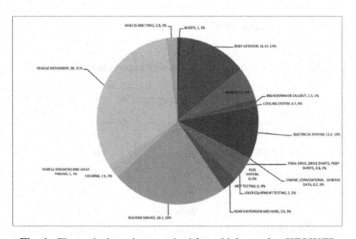

Fig. 4. The typical services required for vehicle number KE55KZB

HMFSVM reduces sensitiveness to class imbalance to a large extent which is present in SVMs [26]. This is achieved through algorithmic modifications to HMFSVM learning algorithm. In lines with [27] a variant cost for classes considered is assigned as misclassification penalty factor. A cost $Cost^+$ is used for positive or minority class while and cost $Cost^-$ is used for negative or majority class. This results in change to fuzzy membership with a factor of $Cost^+Cost^-$. The ratio $Cost^+/Cost^-$ is minority to majority class ratio [28] with penalty for misclassifying minority examples is higher. A grid search was performed to optimise regularization and spread parameters. HMFSVM is an optimization algorithm with reduced support vectors and training times and simplified complexity with improved generalization. Table 3 highlights performance of HMFSVM in terms of sensitivity, specificity, accuracy considering other algorithms. The support vectors in relation to training points $R_{sv/tr}$ is also presented for SVM based algorithms. Figure 6 highlights HMFSVM performance in terms of sensitivity, specificity and accuracy for different classification levels.

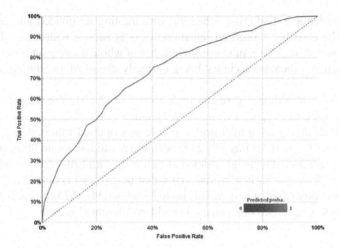

Fig. 5. ROC curve for predicted results

Table 2. The results of HMFSVM with hyperbolic tangent kernel

C	Number of support vectors	Training rate (%)	Testing rate (%)
8	14 (25.5%)	96.9	96.6
16	10 (19.4%)	96.9	96.6
64	10 (19.4%)	96.9	96.6
128	10 (19.4%)	96.9	96.6
256	8 (18.4%)	96.9	96.6
512	8 (18.4%)	96.9	96.6
1024	8 (18.4%)	96.9	96.6

Table 3. The performance of HMFSVM with respect to other algorithms

	Logistic regression	Random forest	SVM	FSVM	MFSVM	HMFSVM
Sensitivity [%]	65.86	69.86	79.86	85.89	89.86	95.66
Specificity [%]	66.90	70.90	80.73	87.00	90.90	96.64
Accuracy [%]	66.86	70.86	80.69	86.96	90.86	96.60
$R_{sv/tr}$			0.909	0.954	0.969	0.986

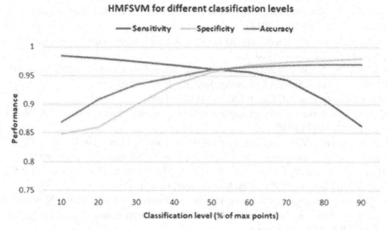

Fig. 6. HMFSVM performance for different classification levels

4   Conclusion

Companies have sent field technicians on fixed schedules to perform routine diagnostic inspections and preventive maintenance on deployed vehicles. But this costly and labor-intensive process. It does not ensure failure won't happen between inspections. Predictive maintenance maximizes vehicle lifespan and ensures optimized productivity. By harnessing vehicle garage data several problems can be anticipated well ahead of failure. With proposed hierarchical SVM-based solution for predictive maintenance organizations can rapidly deploy advanced analytics and gain ability to make decisions in real time. Companies can conduct predictive maintenance and operations management with workflows triggered by analytics output. With these real-time, actionable insights customers can make predictive maintenance a reality based on real-time analytics across their entire field deployment.

References

1. McKinsey & Company: Connected car, automotive value chain unbound, Advanced Industries. McKinsey & Company (2014)
2. Intel: Connected and immersive vehicle systems go from development to production faster. Intel (2016)
3. Intel: Designing next-generation telematics solutions, White Paper In-Vehicle Telematics. Intel (2016)
4. Abel, R.: Uber, Intel and IoT firms join coalition to secure connected cars (2017)
5. Vogt, A.: Industrie 4.0/IoT vendor benchmark 2017: An Analysis by Experton Group AG, An ISG Business, Munich, Germany (2017)
6. Predictive maintenance. https://en.wikipedia.org/wiki/Predictive_maintenance
7. Chowdhury, M., Apon, A., Dey, K. (eds.): Data Analytics for Intelligent Transportation Systems. Elsevier, New York (2017)
8. Chaudhuri, A.: Some investigations in predictive maintenance for industrial IoT solutions. Technical report, Samsung R & D Institute, Delhi, India (2021)

9. Chaudhuri, A.: Modified fuzzy support vector machine for credit approval classification. AI Commun. **27**(2), 189–211 (2014)
10. Chaudhuri, A., De, K.: Fuzzy support vector machine for bankruptcy prediction. Appl. Soft Comput. **11**(2), 2472–2486 (2011)
11. Chaudhuri, A.: Studies in applications of soft computing to some optimization problems, Ph.D. Thesis, Netaji Subhas University, Kolkata, India (2010)
12. Zadeh, L.: Fuzzy sets. Inf. Control **8**(3), 338–353 (1965)
13. Burges, C.: A tutorial on support vector machines for pattern recognition. Data Min. Knowl. Disc. **2**(2), 121–167 (1998)
14. Cortes, C., Vapnik, V.N.: Support vector networks. Mach. Learn. **20**(3), 273–297 (1995)
15. Preventive maintenance. https://en.wikipedia.org/wiki/Preventive_maintenance
16. Lathrop, A.: Preventing Failures with Predictive Maintenance: High-Performance Solutions Using the Microsoft Data Platform. BlueGranite, Portage (2017)
17. Rao, S.S.: Engineering Optimization: Theory and Practice, 4th edn. Wiley, New York (2013)
18. Boixader, D., Jacas, J., Recasens, J.: Upper and lower approximations of fuzzy sets. Int. J. Gen Syst **29**(4), 555–568 (2000)
19. Lin, H.T., Lin, C.J.: A study on sigmoid kernels for SVM and the training of non-PSD kernels by SMO type methods. Technical report, Department of Computer Science and Information Engineering, National Taiwan University (2003)
20. Tang, Y., Guo, W., Gao, J.: Efficient model selection for support vector machine with Gaussian kernel function. In Proceedings of IEEE Symposium on Computational Intelligence and Data Mining, pp. 40–45 (2009)
21. Vapnik, V.N.: The Nature of Statistical Learning Theory. Springer Verlag, New York (1995)
22. Schölkopf, B., Smola, A.J.: Learning with Kernels. MIT Press, Cambridge (2002)
23. Schölkopf, B.: Support vector learning. Ph.D. dissertation, Technische Universität Berlin, Germany (1997)
24. Zimmermann, H.J.: Fuzzy Set Theory and Its Applications. Kluwer Academic Publishers, Boston, MA (2001)
25. Chaudhuri, A., Ghosh, S.K.: Hierarchical rough fuzzy self organizing map for weblog prediction. In: Silhavy, R., Senkerik, R., Kominkova Oplatkova, Z., Prokopova, Z., Silhavy, P. (eds.) CSOC 2017. AISC, vol. 573, pp. 188–199. Springer, Cham (2017). https://doi.org/10.1007/978-3-319-57261-1_19
26. Batuwita, R., Palade, V.: Class imbalance learning methods for support vector machines. In: He, H., Ma, Y. (eds.) Imbalanced learning: foundations, algorithms, and applications, pp. 83–101. John Wiley and Sons Inc., Hoboken, New Jersey (2012)
27. Veropoulos, K., Campbell, C., Cristianini, N.: Controlling the sensitivity of support vector machines, In Proceedings of the International Joint Conference on Artificial Intelligence, pp. 55–60 (1999)
28. Akbani, R., Kwek, S., Japkowicz, N.: Applying support vector machines to imbalanced datasets. In: Boulicaut, J.-F., Esposito, F., Giannotti, F., Pedreschi, D. (eds.) ECML 2004. LNCS (LNAI), vol. 3201, pp. 39–50. Springer, Heidelberg (2004). https://doi.org/10.1007/978-3-540-30115-8_7

Super Local Models for Wind Power Detection

María Barroso$^{(\boxtimes)}$ and Ángela Fernández$^{(\boxtimes)}$ (iD)

Department of Ingeniería Informática, Universidad Autónoma de Madrid,
Madrid, Spain
`maria.barrosoh@estudiante.uam.es`, `a.fernandez@uam.es`

Abstract. Local models can be a useful, necessary tool when dealing with problems with high variance. In particular, wind power forecasting can be benefited from this approach. In this work, we propose a local regression method that defines a particular model for each point of the test set based on its neighborhood. Applying this approach for wind energy prediction, and especially for linear methods, we achieve accurate models that are not dominated by low wind samples, and that implies an improvement also in computational terms. Moreover it will be shown that using linear models allows interpretability, gaining insight on the tackled problem.

Keywords: Local models · Regression · Wind power forecasting

1 Introduction

Local models are a natural option when facing processes with high variance. This is the case when we are dealing with renewable energy power predictions, and in particular with wind energy forecasting. Wind power presents wide, fast changing fluctuations, specially at farm level, making difficult to obtain accurate predictions specially for medium or long term horizons. Nevertheless, nowadays obtaining a good prediction in this context is crucial due to the importance of this kind of energies in the power market, so it is necessary to know in advance how much wind energy will be injected into the power grid, for minimizing operating system costs and maximizing benefits of the electric market agents.

Analyzing more in detail wind power properties, we will check that, as expected, wind speed has a great influence on it. It is well-known that the frequencies of this variable follow a Weibull distribution, that is, a stretched exponential where large frequencies match the lowest wind values, and hence these low wind values are dominating in some sense the distribution. This fact, added to the sigmoid-like structure of wind turbine power curves that clearly shows different regimes at low, medium, and high wind speeds, made local models an attractive option when predicting wind energy [4, 12].

This general situation can be applied to Spain, one of the main producers of wind energy in the World, with 27GW of installed wind-generation capacity

© Springer Nature Switzerland AG 2021
H. Sanjurjo González et al. (Eds.): HAIS 2021, LNAI 12886, pp. 343–354, 2021.
https://doi.org/10.1007/978-3-030-86271-8_29

that covered a 21% of the Spanish electric consumption in 2020 [1]. Moreover, it is expected to be grown up to 50GW extra installed capacity in 2030, covering the 42% of energy consumption by just using renewable energies [3].

In this paper we are going to tackle the wind power forecasting problem for a Spanish wind farm since a local perspective, proposing a new, model-agnostic algorithm. We will see how the proposed local approximation together with a liner model, appears as a good competitor for a Support Vector Regression (SVR) model, state of the art for this particular problem. We will see that, for this particular problem, the proposed Super Local Linear Model outperforms SVRs in computational terms and it has the advantage of been easily interpretable.

This paper is organized as follows: in Sect. 2 we will briefly review the state of the art in local models and wind power prediction, and in Sect. 3 we will present our proposed local algorithm. In Sect. 4 we show the different experiments that we have carried out and, finally, in Sect. 5 we offer some conclusions and point out to lines of further work.

2 Local Models and Wind Power Forecasting

In this section we will briefly review the different methodologies and models applied in the state-of-the-art literature for wind energy prediction. When facing this problem we can find two different basic approximations: using physical or statistical models. Physical models used physics knowledge to define formulas that give us an approximation to the wind energy production at some horizon [10]. By contrast, the statistical approach does not need any knowledge about the system, but it will try to find, in an automatic manner, relations between some features and the predictive target. We will focus our attention on the latter models where, for the problem we are tackling, the features are always numerical weather predictions (NWP) and the target is given by the wind power.

Reviewing the works done in the context of statistical models, we find a wide variety of methods dealing with wind energy forecasting from different points of view. On one side, considering the data as a temporal series, models like ARMA [8], ARIMA [6] or ARX [7] has been defined. This kind of models used to provide good results for short term forecasting, but for horizons longer than 48–72 hours the predictions are weaker [11]. On the other side, traditional machine learning models are also frequent in the state-of-the-art prediction of wind power. In particular, Artificial Neural Networks (ANNs) [15,17] and Support Vector Machines (SVMs) [13] repeatedly appear in the literature, obtaining very good results.

Finally, as previously mentioned, due to its intrinsic properties, wind power forecasting can be correctly modeled following a local approach. For example, in [16] local models based on ANN, Adaptive Neuro-Fuzzy Inference System (ANFIS), or Least Squares Support Vector Machine (LS-SVM) are applied, improving their global versions. Another example can be found in [4] where a specific model for high wind powers is built, defined in terms of the own productions.

Lastly, it should be mentioned that the traditional metric in the energy context is the Normalized Mean Absolute Error (NMAE), defined as:

$$NMAE = \frac{1}{N} \sum_{i=1}^{N} \frac{|y_i - \hat{y}_i|}{pc},$$

where y_i represents each real output (the wind energy), $\hat{y}_i$ its corresponding prediction for a sample with N points and pc represents the installed capacity of the wind farm.

3   Super Local Models

As mentioned before, the idea of building local models is to construct a model for a particular region based on the characteristics of that subset of points, i.e., just using similar points. We will proposed in this section a local algorithm following the idea of local regression models [5]. Local regression (LOESS, or LOWESS) is a well-known method in Statistics, that generalizes the concepts of moving averages and polynomial regression. It consists on defining a linear regression function around a point of interest to model the data, but just using "local" interesting information, i.e. its neighborhood. The LOESS model allows to weight the importance of each neighbor using a kernel function centered at the point in question (LOWESS).

Following this reasoning we define a new local algorithm that can be applied together with any machine learning method. The main idea is to train a different model for each new, unseen pattern, so its training will use just points with similar features to the ones of the point that we want to predict. We will call *Super Local Models* to these individual local models. The neighborhood used for training each model will be defined by the K training nearest neighbors of each test data point, in terms of the Euclidean distance. The most challenging part will be to find the optimal hyperparameters for each model. In this work we will consider a global model built over the whole training dataset, and we will hyperparametrized it using cross validation in this global context. We have considered that the optimal hyperparameter values for the global model will be a good approximation for each super local model hyperparameter.

Validation, training and prediction processes of the proposed method are presented in Algorithm 1. Apart from the input dataset, the number K of nearest neighbors considered is needed to train each particular model. The selection of an optimal K is another challenge that will be addressed in further work.

Looking at the steps in the algorithm, firstly the dataset is divided into training and testing sets to evaluate the performance of the super local model. Second, training and test regressor variables are standardized using the standard deviation of the training set, such that the fitted model does not depend on the scale at which the variables were measured. Next, a validation process is launched over the entire training dataset to obtain the parameters that reduce the overall error model. These parameter values will be used to train each particular model on test data over the training selected neighborhood. Predictions of each model are computed and finally the model is evaluated using some error measure.

Algorithm 1: Super Local Models algorithm

---

input : $\{X, Y\}$ – Dataset
           K - Number of nearest neighbors
output: $\{X_{test}, Y_{pred}\}$ – Prediction dataset

$X_{train}, X_{test}, Y_{train}, Y_{test} \leftarrow \text{split}(X, Y)$;
$\text{standardScaler}(X_{train}, X_{test})$;
$\text{validation}(X_{train}, Y_{train})$;
for (x_{test}, y_{test}) *in* (X_{test}, Y_{test}) **do**
    | $K_x, K_y \leftarrow \text{KNN}(X_{train}, Y_{train}).\text{kneighbors}(x_{test}, K)$;
    | $\text{fit}(K_x, K_y)$;
    | $Y_{pred} \leftarrow \text{predict}(x_{test})$;
end
$\text{Error} \leftarrow \text{evaluation}(Y_{test}, Y_{pred})$;

---

4 Experiments

This section presents the description of the experimental setup, results, and analysis. First, a brief description of used data and their preprocessing is presented. Next, super local regression models based on a linear model, in this case Ridge Regression (RR, [9]), and on a non-linear model, for instance Support Vector Regression (SVR, [14]), are tested in order to prove their efficiency and viability. Results will be compared with those obtained by their corresponding global models. Also, in the linear model, data descriptive statistics of more correlated regressor variables with the target are shown. Finally, a study about computational capability required by each of the models is presented. *scikit-learn* implementations for RR and SVR methods are used.

4.1 Dataset Description

Production data are from Sotavento Wind Farm in Northwestern Spain, collected during 2016, 2017 and 2018, and they consist of hourly mean powers measured in kWh. In order to predict hourly energy production in this farm, NWP variables from European Centre for Medium-Range Weather Forecasts (ECMWF) [2] will be used as predictors. In particular, we will use the next NWP variables:

- Surface pressure (sp).
- 2 m temperature (2t).
- Eastward component of the wind at 10 m (10u).
- Northward component of the wind at 10 m (10v).
- Module of velocity of the wind at 10 m (10Vel).
- Eastward component of the wind at 100 m (100u).
- Northward component of the wind at 100 m (100v).
- Module of velocity of the wind at 100 m (100Vel).

A grid of 0.125° resolution, approximately centered at the farm, is used with southwest and northeast coordinates being (43.25°, −8°) and (43.5°, −7.75°), respectively. This results in a 9 point grid and, using 8 different variables at each point, gives a total of 72 predictive features. We use years 2016 and 2017 for training, and 2018 for testing purposes. In this way, training and test sets dimensions are 17 544 × 72 and 8760 × 72, respectively. Moreover, as previously mentioned, all the regressor variables are standardized using the training set.

4.2 Ridge Regression Models

Our first experiment is focused on evaluating the super local regression model based on Ridge Regression (RR) using K different nearest neighbors. Later, a statistical analysis of the selected neighbors will be done for those variables more correlated with the final power.

The validation process of the proposed model was performed through a 5-fold cross validation procedure in the whole training set using a grid search over $\{10^k : -1 \leq k \leq 3\}$ for optimizing the regularization parameter λ. This validation process resulted in an optimal value of $\lambda = 100$; recall that a high value of λ implies a reduction in variance at the cost of an increase in bias. Now, this λ will be used for training each particular local model based on the nearest neighbors of the test data.

Employing NMAE as metric for evaluating, a global RR model with $\lambda = 100$ on the test set obtains an error equal to 8.44%. This error is far from the state-of-the-art errors for wind power forecasting in Sotavento [13]; thus one of the aims of the experiment is to improve it using the proposed Super Local Models.

As previously said, K-NN is used to select the subset of training data to fit the RR model on. These K values were chosen by rounding different percentages over the 17 544 total training data. Table 1 shows the percentage of testing error made, for each selected K, when predictions of the 2018 data are evaluated using NMAE as indicator.

Table 1. Local RR NMAE for different number of neighbors K.

	0.1%	0.25%	0.5%	1%	5%	10%	25%	50%	75%
K	17	44	87	175	877	1754	4386	8772	13 158
NMAE (%)	6.92	**6.83**	6.88	6.94	7.15	7.26	7.53	7.95	8.22

Looking at NMAE results obtained for each K, certain conclusions can be drawn, such as the fact that a larger number of neighbors implies a larger error in the predictions. Moreover, the bigger the error, the closer to the global model, as expected. This result makes sense since the regularization parameter λ used is fixed. On the other hand, the most accurate local model is obtained by selecting 0.25% of training data, i.e. 44 neighbors, making an error equal to 6.83%.

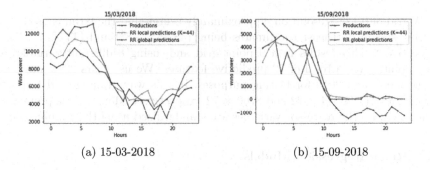

Fig. 1. Real productions and local and global RR predictions.

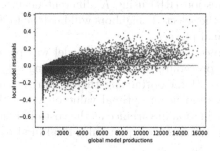

Fig. 2. Local model residual plot.

Therefore, we have got that a linear model as RR is a competitive model in wind power forecasting by making it local.

Next, we analyze predictions obtained by the local Ridge Regression model with 44 neighbors and the global model predictions, and we compare them against real productions for some particular days of 2018 with a high or low production profile.

Figure 1 shows, for two particular days, the comparison between local and global forecasts against the real productions. Local predictions seems to be closer to actual powers than the global ones, specially for high wind energy, and both model predictions maintain the trend of the real curve.

Figure 2 shows the residual plot, where it can be appreciated that high productions are infra-estimated by the proposed method and, by contrast, low productions are overestimated. When the power is null, there are a huge number of big errors that are probably due to hours when the machines stopped production for maintenance work. These errors, added to those derived from meteorological forecasting errors, are difficult to avoid in our model and we should have them in mind when analyzing the results.

In order to understand how the model is selecting the K training patterns to define a particular super local model for each test data point, statistics of these neighbors have been also studied for some regressor variables. This analysis is possible because we are working with a linear regression model that allows us to evaluate the effect of each feature, using the correlation matrix. The possibility of this analysis is one of the main advantages of the proposed method.

Modules of velocity of the wind at 10 and 100 m in the 9 coordinates around Sotavento are the predictor variables that have more correlation with the registered power; temperature and pressure are hardly relevant in the global model. By contrast, correlation matrix of individual models did not show any variable more influential than the others. This is because the value of the regularization parameter is high which implies a reduction of the model coefficients.

We have selected for this purpose two particular models with high and low residual as specified in Table 2.

Table 2. Residuals, local predictions and productions of the selected models.

Datetime model	Residuals	Prediction	Production
09/01/2018 - 07:00:00	1.41×10^{-6}	6563.40	6563.43
13/02/2018 - 20:00:00	0.55	2573.31	12 255.83

Figure 3a shows, using boxplot diagrams, descriptor statistics of some of the most correlated features with the target for the K neighbors selected. The point to be predicted is represented as a blue point, and it can be seen that it is close to the median values of the sample and that the interquartile range is narrow. The corresponding outputs of these neighbors are represented in Fig. 3b, its range goes from 2000 to 12000 kW and it can be seen that the actual production value for this test point is also very near to the median. This model produces, as expected, a good prediction.

In Fig. 4a it can be seen the wind speed modules statistics of the selected neighbors for the second model. In this case, the features of the point to be predicted are, in most of the coordinates, also close to the central values of the sample and the corresponding outputs of these neighbors (see Fig. 4b) have a narrow interquartile range. In this case, the prediction obtained is right on the median but the actual production reaches a value of 12 255.83 kW, high above the expected value according to its neighbors production. This strange result leads us to think that, for this particular point, there should be an error in the weather forecasts used as features.

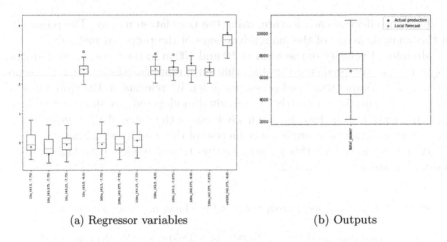

(a) Regressor variables   (b) Outputs

Fig. 3. Neighbors statistics for 09/01/2018 - 07:00:00 model.

4.3 SVR Models

This second experiment is focused on evaluating the performance of an SVR model with an RBF kernel using the proposed super local method.

Again, a 5-fold cross validation procedure over the training set using a grid search over the values specified in Table 3 is launched. In the table, d is the number of dimensions of training data and σ is the standard deviation of the target data.

Table 3. SVR hyperparameters and the grid used to find them.

Hyperparam.	Range
C	$\{10^k : 3 \leq k \leq 5\}$
ϵ	$\{\sigma/2^k : 2 \leq k \leq 6\}$
γ	$\{4^k/d : -2 \leq k \leq 1\}$

After validating the total training set, the smallest error obtained corresponds to the values $C = 10^4$, $\epsilon = \sigma/2^6$ and $\gamma = 4^{-1}/d$. Notice that the low value of γ indicates that the Gaussian shape in the RBF kernel is very flat.

Evaluating the SVR global model over the 2018 test data, using the hyperparameter values described above, we obtain an NMAE equal to **6.35%**. The evaluation of the super local model approach by selecting different K nearest neighbors over the 17 544 total training data points and applying the same percentage method of the previous experiment, is presented in Table 4.

The proposed local model improves as the number of neighbors increases until it stabilizes at $K = 877$, presenting an error similar to the one obtained with the global SVR model. It should be noted that the implicit local character

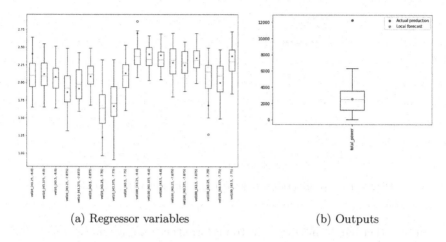

(a) Regressor variables   (b) Outputs

Fig. 4. Neighbors statistics for 13/02/2018 - 20:00:00 model.

Table 4. NMAE for local SVR selecting different number of neighbors K.

	0.1%	0.25%	0.5%	1%	3%	5%	10%	25%	50%
K	17	44	87	175	350	877	1754	4386	8772
NMAE (%)	6.79	6.64	6.60	6.53	6.46	6.37	6.36	6.35	6.34

of the SVR model spotlights the neighboring data of each point through the use of the RBF kernel.

Next, the hourly predictions given by the global SVR model, the local SVR model with 877 neighbors and the real productions will be illustrated, for the same days than in the previous experiment. As shown in Fig. 5, the forecasts of both models hardly differ and they maintain the same trend as the real curve.

4.4   Computation Time

In order to compare the time costs of the aforementioned models, the time taken by each model, either RR or SVR, to predict 8760 hourly powers was calculated. For doing so, we need to measure times to perform the validation of the two models with the 17 544 training data points, as well as to train the models with the optimal hyperparameters.

All the experiments have been executed under the same conditions in the Scientist Computation Center of the Autonomous University of Madrid. Some of the server characteristics used consist in 40 cores, 700GB RAM, 40 TB hard disk drive and 2 GPUs tesla p100.

For RR validation, we only need 0.6 s to find the optimal hyperparameters over the entire training dataset. The SVR validation was very costly and took a total of 12 h to complete. Training and predicting time of RR and SVR global models were negligible.

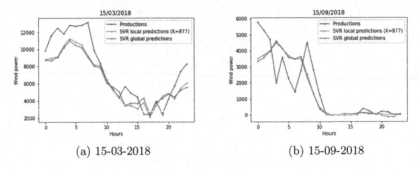

(a) 15-03-2018   (b) 15-09-2018

Fig. 5. Real productions and local and global SVR predictions.

On the other hand, the total time to perform the training and the predictions of the local models corresponding to each test data point has been calculated. The training process of each point to predict has been performed using the optimal parameters of the global model using 44 neighbors for RR and 877 neighbors for SVR.

Training and prediction of each individual linear model costs an average time of 0.17 s. For non-linear super local models, this average time is 0.24 s. Therefore, in a real context of medium-term wind power forecasting, only 5 s are needed to validate, train and predict the power of the next 24 h for the local RR model, compared to more than 12 h for the global SVR model. By contrast, SVR super local model does not represent an improvement in computational times considering that the validation process cannot be lightened.

5   Conclusions

In this work we have proposed a new, model-agnostic local approximation called Super Local Model that we have applied to a real problem, the wind energy forecasting. This new algorithm is based on point-wise models built just over its neighborhood, i.e. its more similar points.

These individual models have appeared as effective methods specially when applied together with linear models for energy forecasting. In the case studied, the Ridge Regression (RR) super local models were able to compete against Support Vector Machines (SVRs), one of the main state-of-the-art models for wind power prediction. Moreover, RR Super Local Models are very efficient in computational terms, outperforming SVRs. Finally, we have analyzed several particular cases detecting possible causes of the biggest errors produced, thanks to the interpretability allowed by linear models.

Nevertheless, some work remains to be done. In particular, we would like to define a proper validation method for detecting the best number of neighbors K for the Super Local Models. Also, other metrics for defining similarity should be studied. Regarding the wind power problem, this prediction could be improved using NWP ensembles, so certain uncertainty will be associated to the energy forecasting in terms of the meteorology reliability. Finally, it will be interesting to try the proposed approximation with other real problems, like for example the solar energy forecasting, and also with other datasets far away from the energy field, so a general good performance of the Super Local Models can be tested.

Acknowledgments. The authors acknowledge financial support from the Spanish Ministry of Science and Innovation, project PID2019-106827GB-I00/ AEI/10.13039/ 501100011033. They also thank the UAM–ADIC Chair for Data Science and Machine Learning and gratefully acknowledge the use of the facilities of Centro de Computación Científica (CCC) at UAM.

References

1. Asociación empresarial eólica. https://www.aeeolica.org/sobre-la-eolica/la-eolica-espana. Accessed 27 Apr 2021
2. Centro europeo de previsiones meteorológicas a medio plazo. https://www.ecmwf.int/. Accessed 10 Mar 2021
3. Instituto para la diversificación y ahorro de la energía. https://www.idae.es/informacion-y-publicaciones/plan-nacional-integrado-de-energia-y-clima-pniec-2021-2030. Accessed 27 Apr 2021
4. Alaíz, C., Barbero, A., Fernández, A., Dorronsoro, J.: High wind and energy specific models for global production forecast. In: Proceedings of the European Wind Energy Conference and Exhibition - EWEC 2009. EWEA, March 2009
5. Cleveland, W.S., Devlin, S.J.: Locally weighted regression: an approach to regression analysis by local fitting. J. Am. Stat. Assoc. **83**(403), 596–610 (1988)
6. De Felice, M., Alessandri, A., Ruti, P.M.: Electricity demand forecasting over Italy: potential benefits using numerical weather prediction models. Electr. Power Syst. Res. **104**, 71–79 (2013). https://doi.org/10.1016/j.epsr.2013.06.004. https://www.sciencedirect.com/science/article/pii/S0378779613001545
7. Duran, M., Cros, D., Santos, J.: Short-term wind power forecast based on ARX models. J. Energy Eng. ASCE **133**, 172–180 (2007). https://doi.org/10.1061/(ASCE)0733-9402(2007)133:3(172)
8. Gallego, C., Pinson, P., Madsen, H., Costa, A., Cuerva, A.: Influence of local wind speed and direction on wind power dynamics – application to offshore very short-term forecasting. Appl. Energy **88**(11), 4087–4096 (2011). https://doi.org/10.1016/j.apenergy.2011.04.051. https://www.sciencedirect.com/science/article/pii/S0306261911002868
9. James, G., Witten, D., Hastie, T., Tibshirani, R.: An Introduction to Statistical Learning: with Applications in R. Springer, Heidelberg (2013). https://doi.org/10.1007/978-1-4614-7138-7. https://faculty.marshall.usc.edu/gareth-james/ISL/
10. Jung, J., Broadwater, R.P.: Current status and future advances for wind speed and power forecasting. Renew. Sustain. Energy Rev. **31**, 762–777 (2014). https://doi.org/10.1016/j.rser.2013.12.054. https://www.sciencedirect.com/science/article/pii/S1364032114000094
11. Jung, S., Kwon, S.D.: Weighted error functions in artificial neural networks for improved wind energy potential estimation. Appl. Energy **111**, 778–790 (2013). https://doi.org/10.1016/j.apenergy.2013.05.060. https://www.sciencedirect.com/science/article/pii/S030626191300473X

12. Pinson, P., Nielsen, H., Madsen, H., Nielsen, T.: Local linear regression with adaptive orthogonal fitting for the wind power application. Stat. Comput. **18**(1), 59–71 (2009)
13. Ruiz, C., Alaíz, C.M., Dorronsoro, J.R.: Multitask support vector regression for solar and wind energy prediction. Energies **13**(23) (2020). https://doi.org/10.3390/en13236308. https://www.mdpi.com/1996-1073/13/23/6308
14. Smola, A., Schölkopf, B.: A tutorial on support vector regression. Stat. Comput. **14**, 199–222 (2004)
15. Wang, J., Yang, W., Du, P., Niu, T.: A novel hybrid forecasting system of wind speed based on a newly developed multi-objective sine cosine algorithm. Energy Convers. Manage. **163**, 134–150 (2018). https://doi.org/10.1016/j.enconman.2018.02.012. https://www.sciencedirect.com/science/article/pii/S0196890418301079
16. Wu, S.F., Lee, S.J.: Employing local modeling in machine learning based methods for time-series prediction. Expert Syst. Appl. **42**(1), 341–354 (2015). https://doi.org/10.1016/j.eswa.2014.07.032. https://www.sciencedirect.com/science/article/pii/S0957417414004394
17. Zhang, J., Yan, J., Infield, D., Liu, Y., Lien, F.: Short-term forecasting and uncertainty analysis of wind turbine power based on long short-term memory network and gaussian mixture model. Appl. Energy **241**, 229–244 (2019). https://doi.org/10.1016/j.apenergy.2019.03.044. https://www.sciencedirect.com/science/article/pii/S0306261919304532

A Comparison of Blink Removal Techniques in EEG Signals

Fernando Moncada[1]($\boxtimes$), Víctor M. González[1]($\boxtimes$) (iD), Beatriz García[2] (iD),
Víctor Álvarez[3] (iD), and José R. Villar[3] (iD)

[1] Electrical Engineering Department, University of Oviedo, Gijón, Spain
{UO245868,vmsuarez}@uniovi.es
[2] Neurophysiology Department, University Hospital of Burgos, Burgos, Spain
bgarcialo@saludcastillayleon.es
[3] Computer Science Department, University of Oviedo, Oviedo, Spain
villarjose@uniovi.es

Abstract. Blink detection and removal is a very challenging task that
needs to be solved in order to perform several EEG signal analyses, especially when an online analysis is required. This study is focused on the
comparison of three different techniques for blink detection and three
more for blink removal; one of the techniques has been enhanced in this
study by determining the dynamic threshold for each participant instead
of having a common value. The experimentation first compares the blink
detection and then, the best method is used for the blink removal comparison. A real data set has been gathered with healthy participants and
a controlled protocol, so the eye blinks can be easily labelled. Results
show that some methods performed surprisingly poor with the real data
set. In terms of blink detection, the participant-tuned dynamic threshold
was found better than the others in terms of Accuracy and Specificity,
while comparable with the Correlation-based method. In terms of blink
removal, the combined CCA+EEMD algorithm removes better the blink
artifacts, but the DWT one is considerably faster than the others.

Keywords: EEG · Blink detection · Blink removal

1 Introduction

Electroencephalography (EEG) is a non-invasive technique to measure the
brain's electrical activity, which has been widely used in the literature for diagnosing and monitoring different neurological disorders. However, the EEG signals
also include the electrical activity due to the muscle movements and other sources

This research has been funded by the Spanish Ministry of Science and Innovation under
project MINECO-TIN2017-84804-R, PID2020-112726RB-I00 and the State Research
Agency (AEI, Spain) under grant agreement No RED2018-102312-T (IA-Biomed).
Additionally, by the Council of Gijón through the University Institute of Industrial
Technology of Asturias grant SV-21-GIJON-1-19.

H. Sanjurjo González et al. (Eds.): HAIS 2021, LNAI 12886, pp. 355–366, 2021.
https://doi.org/10.1007/978-3-030-86271-8_30

(lights, electromagnetic fields, etc.); these undesired components are called arti-facts. This study focuses on the detection and removal of a specific type of arti-facts: those due to the eye blink. Although blink artifact detection and removal is a subject that has been widely studied, new alternatives are still being proposed.

The basic approach for artifact removal while keeping as much EEG infor-mation as possible includes two main approaches. On the one hand, the methods that apply source decomposition belonging to the Blind Source Separation (BSS) family -i.e., using Independent Component Analysis (ICA) [8,15] or Canonical Correlation Analysis (CCA) [14]-. On the other hand, the methods that apply a type of spectral decomposition in any of its variants, i.e., using Wiener filters [1], using the Empirical Mode Decomposition (EMD) or using Ensemble EMD (EEMD), by extracting the basic components called Intrinsic Mode Functions (IMFs) [15] through upper and lower envelopes, or Discrete Wavelet Transform (DWT) [8]. In addition, some authors use Machine Learning to create models to filter the EEG windows [6,11,12].

Focusing on blink artifact removal, a combination of Principal Component Analysis and ICA was proposed in [7], while [4] proposed a combination of CCA and a faster variation of EMD. Besides, [9] proposed ICA to decompose the signals and a classifier to label the ICA channels including an artifact. In [11], the removal is based on an autoencoder that analyses the EEG windows labelled as anomalous that contain a blink artifact using 2 Support Vector Machines models (SVM). A combination of DWT and SVM for blink artifact removal is proposed in [12], while [5] and [13] used extensions of EMD: Multivariate Variational Mode Decomposition (Multivariate VMD) and Variational Mode Extraction (VME), respectively.

In this study, a comparison of blink detection and removal techniques is pre-sented. The detection methods are compared in their performance, and the best blink detection method is then used in combination with the different artifact removal techniques. Due to the difficulty in evaluating these methods, a specific data set of EEG recordings with healthy participants has been gathered and used in the comparison. It is worth mentioning that the majority of the research has been performed with realistic data sets.

The structure of this work is as follows. Firstly, the next Sect. 2 explains in detail the different techniques and compare the different techniques fore auto-matic identification and removal of blink artifacts mentioned in the previous paragraph. Section 3 describes the conditions under which the dataset used was recorded and the experimental setup, while Sect. 4 presents the results and the discussion on each them. Finally, the conclusions are drawn.

2   Blink Detection and Removal Techniques

In this research we consider three basic solutions [4,13], and [3,10,16] for the blink detection and removal problem. The selection of these techniques is based on the ease of their implementation, as they require little coding, and their satisfactory results in the previous studies already mentioned.

This section briefly describes the two stages independently: next section introduces the blink detection part, while Sect. 2.2 focuses on the blink removal components.

2.1 Blink Detection Techniques

Using VME and Peak Detection. VME decomposes the EEG signal into a signal $u(n)$ with a frequency similar to a predefined value ω_d plus the residual $r(n)$ [13]. To do so, VME minimises i) bandwidth around this frequency and ii) the spectral overlap between $u(n)$ and $r(n)$. The bandwidth is minimised using Eq. 1, where $\delta(n)$ is the Dirac distribution and $\left[\left(\delta(n) + \frac{j}{n\pi}\right) * u(n)\right]$ is the Hilbert transform. The spectral overlap is minimised using Eq. 2 and Eq. 3, where (α) is the compactness coefficient which regularises the bandwidth and $\beta(n)$ is the impulse response of the filter β_ω. Finally, the minimisation process is shown in Eq. 4 subject to $u(n) + r(n)$ sums up the original signal.

$$J_1 = \left\| \partial_n \left[\left(\delta(n) + \frac{j}{n\pi} \right) * u(n) \right] e^{-j\omega_d n} \right\|_2^2 \tag{1}$$

$$\beta(\omega) = \frac{1}{\alpha(\omega - \omega_d)^2} \tag{2}$$

$$J_2 = \| \beta(n) * r(n) \|_2^2 \tag{3}$$

$$\min_{\omega_d, u(n), r(n)} \{\alpha J1 + J2\} \tag{4}$$

In [13], $\omega_c = 3$ Hz as long as the eye blinks' frequency is in the interval [0.5, 7.5] Hz and $\alpha = 3000$ after a thoroughly experimentation. Given an EEG-signal window (X_t), VME is applied and the desired mode $u(n)$ is obtained. Assuming the Universal Threshold calculated as $\theta(X_t) = \frac{median(|u(n)|)}{0.6745} * \sqrt{2 * Log(N)}$, with N the number of samples in $u(n)$, those values in X_t higher than $\theta(X_t)$ are considered blinks. For the subsequent removal stage, 500 ms windows are extracted including the samples from 125 ms before the peak to 375 ms after it. If no peak is detected, then X_t keeps unchanged, that is, the window is assumed to include no blink.

Correlation-Based Detection. EMD is a decomposition algorithm that receives a single input signal X_i and extracts a certain number of components called IMFs. For each decomposition level, an IMF is extracted from the initial signal by computing the mean of the upper and lower envelopes using an interpolation algorithm. The difference between the IMF and the initial signal from which it is derived is the residue, which is used as the new initial signal in the next level of decomposition. The Fast-EMD version [4] uses the Akima Spline Interpolation algorithm to calculate the envelopes of the signal instead of the Cubic Spline.

The blink detection method includes two stages: the calibration -where the two first blinking window are found- and the detection stage -that reexamines the data and labels the blinks-. The correlation of the Fp1 and Fp2 EEG channels is computed for every 1.953 s length windows. Whenever a correlation value higher than 0.85 is found a blink is detected the displacement of the amplitude in the Fp1 channel is calculated: the higher the amplitude the higher the difference with the window mean value. The threshold is calculated for each window using Eq. 5, where μ and σ are the mean and the standard deviation of the displacement of the window. The first value within the window higher than the threshold is considered the starting point of a blink, the blinking window is defined from 0.390 s before the starting point and with a length of 1 s.

$$threshold = \mu + 2 * \sigma \tag{5}$$

Each new blinking window is compared with all the previously found through the correlation between them until one pair has a correlation coefficient higher than 0.8. Then, the Fast-EMD variant is applied to these two blink segments, decomposing them up to level 5, and the two blink signals are reconstructed from their respectively third IMF onwards. The blink template is created as the average of both. In the detection phase, a sliding window of the same size as the template is used throughout the raw EEG signal. The correlation between the EEG window and the blink template is computed and the window is classified as contaminated if it is higher than 0.5.

Threshold-Based Detection. The threshold detection method was proposed in [3]; in previous research, an automated threshold tuning algorithm was proposed [10]. Given an EEG window (X_i), its 0–50 Hz the Power Spectrum Density (PSD), p, is computed. Then it is compared using the Bhattacharyya Distance with the PSD (q) from each of the available templates. The distance is computed using Eq. 6 and 7, where d_{BD} is the distance value.

$$s^{BD} = -ln(\sum_{k=1}^{K}(\sqrt{p(k) * q(k)})) \tag{6}$$

$$d_{BD} = \frac{1}{s^{BD}} \tag{7}$$

Firstly, a template of EEG windows with normal activity, i.e. without blink artifact contamination, is created. Let μ and σ be the mean and the standard deviation of the distances between all the normal template signals. These values are calculated once at the beginning and are constant throughout the process. This paper proposes to use the Empirical Rule -a.k.a. 68–9599.7 rule- to determine whether a point represents an outlier.

Once the threshold values are obtained, the current EEG window is compared with each of the normal windows of the template computing the distance between them. The highest distance is extracted: this value represents the lowest similarity between the EEG window and the normal signals. Whenever this maximum distance holds the condition in Eq. 8, the window is labelled as contaminated with a blink artifact.

$$|d_{max}(X_i) - \mu| > 3 * \sigma \tag{8}$$

Two different options will be evaluated: using the same general threshold for all the participants as proposed in [10] and using a specific determined threshold for each participant. This latter threshold will be computed as follows: starting from an empty window, bunches of $Bnch$ samples are introduced to the window and the μ and σ are updated; the process continues until the value of a new sample $d(X_i)$ holds that $|d(X_i) - \mu| > 3 * \sigma$.

2.2 Methods for Blink Removal

DWT-based Blink Removal. DWT is a decomposition algorithm that receives a single input signal and extracts its components by sequentially dividing the continuous spectrum.

In each level of decomposition, the initial signal is passed through two symmetric filters: a low-pass filter and a high-pass filter at a certain frequency value. The former one extracts the low-frequency component, called the approximation component, which will be the new initial signal in the next level until the desired decomposition level is reached; and the latter one extracts the high-frequency component, called the detail component. These filters are designed according to the discrete variant of the mother wavelet function chosen.

Given a blink window, it is processed by applying the DWT algorithm: Daubechies-4 ($db4$) is chosen as the mother wavelet function because its morphology is very similar to the blink one. In terms of the level of decomposition, a skewness-value method is applied: the skewness of the approximation component is calculated in each level of decomposition and compared with the previous one. If the difference between two consecutive skewness values is greater than a threshold, the maximum level of decomposition has been reached. Finally, the approximation component of the last level is removed and the clean EEG window is reconstructed from all the detail components.

CCA Blink Removal. CCA algorithm belongs to the group of BSS techniques, which assume that these signals are formed by the linear combination of their sources (Eq. 9), where X is the measured signals, S is the sources, and A and W are the mixing matrix and its inverse, the unmixing matrix, respectively.

$$X = A * S \rightarrow A = W^{-1} \rightarrow S = W * X \tag{9}$$

All BSS methods try to estimate the unmixing matrix and, therefore, the sources that compose the measured signals, in different ways. Thus, CCA is a decomposition algorithm that receives a set of n input signals and extracts their uncorrelated sources with the autocorrelated values maximized.

The removal process based on the CCA algorithm proposed by [4] is very simple and it is applied as follows: the algorithm is executed to the blink window and the first component computed, which usually corresponds to the artifact, are removed. Then, the EEG window is reconstructed from the remaining components.

Combined CCA-EEMD Blink Removal. Since EMD suffers from mixing and aliasing problems despite performing well, another variant is presented: EEMD is more noise-robust than the basic EMD algorithm. This method adds white noises of different amplitudes to the original signal in a way that each noise signal is applied individually creating different noisy variations of the original. Then, noisy IMFs are extracted from each noisy variation until the chosen level of decomposition is reached. Finally, each ith real IMF of the original signal is calculated as the average of all the i^{ths} noisy IMFs.

The combined CCA+EEMD removal technique was presented in [16]. If a blink is detected in the current window, CCA is applied to separate the uncorrelated source signals and ordered according to their kurtosis value. The artifact component is identified as the source with the highest kurtosis. Then, EEMD is applied only to that component to compute its IMFs and the blink components are identified again by the same method: IMFs are sorted by kurtosis value. The higher ones corresponded to artifacts components. These artifact signals are removed and the rest of the IMFs are preserved to rebuild the CCA component and then all the sources are combined again to reconstruct the EEG signal, now clean and free of artifacts.

3   Materials and Methods

3.1   Data Set Description

One of the main problems in blink analysis is the lack of a specific data set with real blinks; in the development of the different techniques, the authors have been creating a realistic data set combining EEG signals and introducing simulated blinks. The main problem is that this approach generates differences in performance when facing real cases. In this research, we try to compare the different techniques with real EEG signals that contain natural blinks. Therefore, we have designed and recorded an EEG data set with all the blinks labelled.

The recorded EEG data set was recorded with the Ultracortex Mark IV helmet manufactured by OpenBCI [2], including up to 16 electrodes with the positions shown in Fig. 1. Two sessions were planned with each of the 13 healthy participants (5 women, 8 men; ages from 22 to 58); the sessions were video recorded for synchronization and posterior manual labeling. Each session consisted of a 5-min continuous recording, with the participant comfortably seated in front of a screen that provides the instructions. The structure of the session includes: the first 5 s of quickly blinking for synchronization, followed by 5 pairs of 30-s intervals where the participant stayed easy first with the eyes open normally and then with the eyes closed; a beep sound alerted the transitions. All the EEG signals were recorded at a sampling rate 125 Hz and were filtered using a Notch filter with 50 Hz null frequency and a Band-Pass filter 1 Hz 50 Hz cut-off frequencies using the OpenBCI GUI application.

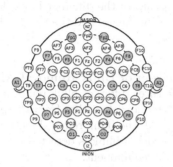

Fig. 1. Position of the 16 recorded channels using the 10–20 system: Fp1, Fp2, F3, F4, F7, F8, T7, T8, C3, C4, P3, P4, P7, P8, O1 and O2.

The signals from the participants were analysed and those recordings with anomalous values (mainly due to ambient noise) were deleted. Each recording was then labeled, marking those timestamps where a blink arose; a neurophysiologist analyzed and monitored this labeling process. In the end, 15 recordings were dropped from the data set; the remaining labeled recordings were used in the comparison. The initial 5 s of quick blinking are erased from the session's data.

3.2   Experimentation Design

The experimentation is split into two stages: the former compares the blink detection techniques, while the latter focuses on the comparison of the blink removal; for this second case, all the removal algorithms run with the outcome of the best blink detection technique.

Experiment I: Comparison of Blink Detection Algorithms. The first experiment aims to compare the three different blink detection techniques described in Sect. 2.1. The VME is an online algorithm, so the EEG signal is windowed -2-s length windows with no overlap- and labeled as normal or blink directly. The Correlation-based is not an online algorithm and requires an initial pass over the signal in order to extract the blink template, so two passes are considered for each participant; for this method, 1-s length sliding windows with 15 samples overlap was used. Finally, the Dynamic Threshold algorithm is an online method that requires threshold tuning.

Therefore, Leave-one participant-out cross-validation is proposed, training the threshold for participant P_i with the training data from the remaining participants (P_j, with $j \neq i$) and evaluating the method with P_i's data. The training data from each participant has been manually extracted by selecting blink-free windows when recording with the eyes opened. In this case, 1-s length sliding windows with 50% of overlap was employed.

To measure the performance of the different blink detection techniques, the Accuracy (Acc), the Sensitivity (Sens), and the Specificity (Spec) measurements will be calculated.

Experiment II: Comparison of Blink Removal Algorithms. This experiment aims to compare the three different blink removal techniques described in Sect. 2.2. The best blink detection algorithm from the previous experiment will be used as the first stage, so all the blink removal methods compete in the same scenario: each window for which a blink is detected is then processed to get the EEG channel rid of the blink component.

To our knowledge, the majority of the previous research made use of data sets consisting of pure and contaminated EEG signals; as a consequence, the performance was measure in terms of the similarity of the outcome of the corresponding blink removal method with the original blink-free EEG signal. However, we can not follow this procedure as long as we do have real blinks and no blink-free signal to compare with. Therefore, in this research, we propose to perform a visual assessment of the removal methods.

4   Results and Discussion

4.1   Comparison of Detection Methods

Table 1 shows the Acc, Sens and Spec values obtained for each of the participants from the gathered data set, and its boxplot representation is presented in Fig. 2. As can be seen, the Dynamic Thresholding with participant tuned threshold is the best method if we consider both the Acc and the Spec; however, it is comparable with the Correlation detection method in terms of Sens. The poor performance of the VME method in terms of Sens is surprising according to the results in the literature: this is only because the Universal Threshold proposed by the authors produces very high threshold values for our recordings; a smaller threshold would allow a huge improvement in detection performance. For the remaining experiments, Dynamic Thresholding with participant-based threshold tuning is used.

4.2   Comparison of Removal Techniques

Figure 3 shows a window of the Fp1 channel and the obtained reconstructions. As can be seen, the three methods can significantly reduce or remove the blink artifacts: the combined method CCA+EEMD is the one that processed better

Table 1. Results of cross-validation using VME (left), correlation (center) and dynamic thresholding either using a threshold value extracted from templates or extracting a value for each participant (right-most columns) as the detection technique. Mn = Mean. Mdn = Median. StD = Standard Deviation.

P_i	VME			Correlation			Template threshold			Participant threshold		
	Acc	Sens	Spec	Acc	Sens	Spec	Acc	Sens	Spec	Acc	Sens	Spec
P_1	0.7006	0.2381	0.8696	0.8032	0.9107	0.7799	0.9079	**1.0000**	0.8880	**0.9429**	**1.0000**	**0.9305**
P_2	0.7089	0.4565	0.8125	0.8644	**1.0000**	0.8346	0.9338	**1.0000**	0.9192	**0.9369**	**1.0000**	**0.9231**
P_3	0.7771	0.4100	0.8371	0.6210	0.9615	0.5903	0.8408	**1.0000**	0.8258	**0.9236**	**1.0000**	**0.9164**
P_4	0.7673	0.3333	0.9386	0.8742	**1.0000**	0.8551	0.8742	**1.0000**	0.8524	**0.9748**	**1.0000**	**0.9705**
P_5	0.7595	0.3077	0.9076	0.6677	**1.0000**	0.6250	0.9367	**1.0000**	0.9273	**0.9525**	0.8049	**0.9745**
P_6	0.7532	0.2571	0.8943	0.6361	**1.0000**	0.5922	**0.9494**	0.9444	**0.9500**	0.9367	**1.0000**	0.9286
P_7	0.8056	0.3462	0.8992	0.7492	**1.0000**	0.7254	0.9035	**1.0000**	0.8944	**0.9711**	**1.0000**	**0.9683**
P_8	0.8608	0.1818	**0.9706**	0.6498	**1.0000**	0.6224	**0.9716**	**1.0000**	0.9693	0.9685	**1.0000**	0.9659
P_9	0.8228	0.6774	0.8583	0.7152	**1.0000**	0.6831	**0.9589**	0.9394	**0.9611**	0.9525	**1.0000**	0.9470
P_{10}	0.7152	0.3125	0.8903	0.8323	**0.9804**	0.8038	0.9430	0.9038	0.9508	**0.9589**	0.8077	**0.9886**
P_{11}	0.8354	0.4412	0.9435	0.8297	**1.0000**	0.8071	0.9495	**1.0000**	0.9424	**0.9621**	0.8205	**0.9820**
Mn	0.7734	0.3602	0.8929	0.7493	**0.9866**	0.7199	0.9245	0.9807	0.9164	**0.9528**	0.9485	**0.9541**
Mdn	0.7673	0.3333	0.8943	0.7492	**1.0000**	0.7254	0.9367	**1.0000**	0.9273	**0.9525**	**1.0000**	**0.9659**
StD	0.0534	0.1350	0.0472	0.0960	**0.0280**	0.1013	0.0393	0.0345	0.0464	**0.0163**	0.0883	**0.0258**

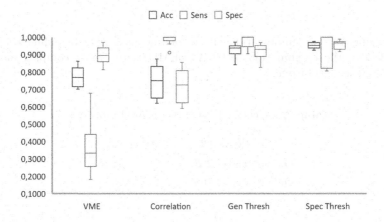

Fig. 2. For each method, the boxplot from the ACC, Sens and Spec are depicted, respectively.

the detected artifacts. Although the performance of the two CCA techniques is reported to improve in those channels contaminated with the blink, in real-life recordings the blinks may not be perfectly diffused, and its removal is penalized when using CCA. DWT algorithm performance is penalized due to the lack of a better-fit skewness threshold.

Concerning the running time of the whole process (detection and removal) -see Table 2 for details-, the DWT method showed the faster both online and offline analysis. The offline analysis considers the complete EEG recording as a whole, while the online analysis evaluates on the fly the EEG signal for artifact detection and removal. As shown, while for offline analysis the DWT seems the

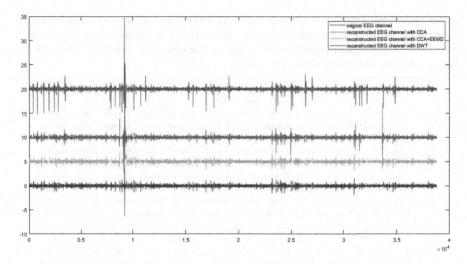

Fig. 3. Visual comparison of blink artifacts removal methods with an Fp1 EEG recording.

most competitive in running time, its advantage decreases in online analysis. Nevertheless, the combination of CCA+EEMD is the most complex method and, thus, the slowest one.

Table 2. Execution time of the removal processes.

	CCA	CCA+EEMD	DWT
Online	0.80 s	2.20 s	0.40 s
Offline	15 s	70 s	0.65 s

5   Conclusions and Future Work

This study compares different techniques for blink artifact detection and removal in EEG raw signals. The VME algorithm, the Correlation-based method and the dynamic threshold technique are compared for artifact detection; while CCA, CCA+EEMD and DWT methods are compared as removal techniques. For this comparison, a specific EEG data set with recordings from healthy participants and alternating periods with the eyes closed and opened to register the blinks has been gathered.

The obtained results show that the presented participant-tuned Dynamic Threshold performs better than the other methods in terms of Acc and Spec, while it is comparable in terms of Sens with the Correlation-based method. Concerning the removal techniques, all the techniques perform well but the DWT showed faster than the others, especially in offline analysis. CCA-EEMD performance is in compromise in those channels where the artifact is not perfectly diffused.

Besides, there is still room for improvement with the VME and the DWT algorithms, especially with the threshold value tuning to make the techniques more competitive with real EEG recordings. Future work includes making modifications to these techniques or introducing Deep Learning techniques for the detection and removal of the blink artifacts, especially in those channels where the blink is hardly noticeable.

References

1. Al-Momani, S., Al-Nashash, H., Mir, H.S.: Comparison between independent component analysis and Wiener-Hopf filter techniques for eye blink removal. In: Middle East Conference on Biomedical Engineering, MECBME (2020). https://doi.org/10.1109/MECBME47393.2020.9265171
2. Open BCI: Open source brain-computer interfaces (2021). http://openbci.com
3. Chen, G., Lu, G., Xie, Z., Shang, W.: Anomaly detection in EEG signals: a case study on similarity measure. Comput. Intell. Neurosci. (2020). https://doi.org/10.1155/2020/6925107
4. Egambaram, A., Badruddin, N., Asirvadam, V.S., Begum, T., Fauvet, E., Stolz, C.: FastEMD-CCA algorithm for unsupervised and fast removal of eyeblink artifacts from electroencephalogram. Biomed. Signal Process. Control (2020). https://doi.org/10.1016/j.bspc.2019.101692
5. Gavas, R., Jaiswal, D., Chatterjee, D., Viraraghavan, V., Ramakrishnan, R.K.: Multivariate variational mode decomposition based approach for blink removal from EEG signal. In: 2020 IEEE International Conference on Pervasive Computing and Communications Workshops, PerCom Workshops (2020). https://doi.org/10.1109/PerComWorkshops48775.2020.9156206
6. Ghosh, R., Sinha, N., Biswas, S.K.: Removal of eye-blink artifact from EEG Using LDA and pre-trained RBF neural network. In: Elçi, A., Sa, P.K., Modi, C.N., Olague, G., Sahoo, M.N., Bakshi, S. (eds.) Smart Computing Paradigms: New Progresses and Challenges. AISC, vol. 766, pp. 217–225. Springer, Singapore (2020). https://doi.org/10.1007/978-981-13-9683-0_23
7. Huang, G., Hu, Z., Zhang, L., Li, L., Liang, Z., Zhang, Z.: Removal of eye-blinking artifacts by ICA in cross-modal long-term EEG recording. In: Proceedings of the Annual International Conference of the IEEE Engineering in Medicine and Biology Society, EMBS (2020). https://doi.org/10.1109/EMBC44109.2020.9176711
8. Issa, M.F., Juhasz, Z.: Improved EOG artifact removal using wavelet enhanced independent component analysis. Brain Sci. 9, 355 (2019). https://doi.org/10.3390/brainsci9120355
9. Markovinović, I., Vlahinić, S., Vrankić, M.: Removal of eye blink artifacts from the EEG signal. Eng. Rev. (2020). https://doi.org/10.30765/er.40.2.11
10. Moncada Martins, F., Gonzalez, V.M., Alvarez, V., García, B., Villar, J.R.: Automatic detection and filtering of artifacts from EEG signals. In: Accepted for Publication on the IEEE CBMEH2021 Conference (2021)

11. Phadikar, S., Sinha, N., Ghosh, R.: Automatic EEG eyeblink artefact identification and removal technique using independent component analysis in combination with support vector machines and denoising autoencoder. IET Signal Process. **14**(6) (2020). https://doi.org/10.1049/iet-spr.2020.0025

12. Phadikar, S., Sinha, N., Ghosh, R.: Automatic eye blink artifact removal from EEG signal using wavelet transform with heuristically optimized threshold. IEEE J. Biomed. Health Inform. **25**, 475–484 (2020). https://doi.org/10.1109/JBHI.2020. 2995235

13. Shahbakhti, M., et al.: VME-DWT: an efficient algorithm for detection and elimination of eye blink from short segments of single EEG channel. IEEE Trans. Neural Syst. Rehabil. Eng. **29**, 408–417 (2021). https://doi.org/10.1109/TNSRE. 2021.3054733

14. Sheoran, P., Saini, J.S.: A new method for automatic electrooculogram and eye blink artifacts correction of EEG signals using CCA and NAPCT. Proc. Comput. Sci. **167**, 1761–1770 (2020). https://doi.org/10.1016/j.procs.2020.03.386

15. Yadav, A., Choudhry, M.S.: A new approach for ocular artifact removal from EEG signal using EEMD and SCICA. Cogent Eng. **7**, 1835146 (2020). https://doi.org/ 10.1080/23311916.2020.1835146

16. Yang, B., Zhang, T., Zhang, Y., Liu, W., Wang, J., Duan, K.: Removal of electrooculogram artifacts from electroencephalogram using canonical correlation analysis with ensemble empirical mode decomposition. Cogn. Comput. **9**(5), 626–633 (2017). https://doi.org/10.1007/s12559-017-9478-0

Hybrid Intelligent Model for Switching Modes Classification in a Half-Bridge Buck Converter

Luis-Alfonso Fernandez-Serantes[1]([⊠]), José-Luis Casteleiro-Roca[1],
Dragan Simić[2], and José Luis Calvo-Rolle[1]

[1] Department of Industrial Engineering, University of A Coruña, CTC, CITIC,
Ferrol, A Coruña, Spain
luis.alfonso.fernandez.serantes@udc.es
[2] Faculty of Technical Sciences, University of Novi Sad, Trg Dositeja Obradovića 6,
21000 Novi Sad, Serbia

Abstract. In this research, a study about the implementation of a hybrid intelligent model for classification applied to power electronics is presented. First of all, an analysis of the chosen power converter, half-bridge buck converter, has been done, differentiating between two operating modes: Hard-Switching and Soft-Switching. A hybrid model combining a clustering method with classification intelligent techniques is implemented. The obtained model differentiate with high accuracy between the two modes, obtaining very good results in the classification.

Keywords: Hard-Switching · Soft-Switching · Half-bridge buck · Power electronics · Classification · Hybrid model

1 Introduction

The efficiency in power electronics has been one of the main topics in the last years, trying to reduce the size and losses of the converters. With the introduction of new materials in this field, the wide band-gap materials such as Silicon Carbide (SiC) and Gallium Nitride (GaN), the silicon power devices were moved out of the focus [1,4]. These new materials are replacing tradicional silicon in state of the art converters and coming more into the market as their price is reducing over the years. Moreover, with these new materials, the use of Soft-Switching techniques become more attractive due to their better characteristics [8,19].

Furthermore, the Artificial Intelligence (AI) starts to take importance in this field as well. The main research is done in the design of components [2,22] and in the control strategies [12,21], with the aim of improving the performance of the converters. The Artificial Intelligence helps the developers to choose the right component for each application, the design of magnetic components or the get the highest performance out of the control strategy.

The classification of the operating mode of a power converter is very important as it indicate us how the efficiency can be improved. When the converter

© Springer Nature Switzerland AG 2021
H. Sanjurjo González et al. (Eds.): HAIS 2021, LNAI 12886, pp. 367–378, 2021.
https://doi.org/10.1007/978-3-030-86271-8_31

is operating in HS-mode, there are losses that can be reduced by just operating the same converter in the SS-mode. The transfer of energy between the input and output, it is increased as the losses are reduced. Therefore, a first step in the efficiency improvement is detecting the operating mode of the converter.

With the aim to continue the research in the introduction of the AI in power electronics, this paper is focused in the detection of Hard- versus Soft-Switching mode detection. The analysis of a half-bridge configuration in a synchronous buck converter is explained in the case study Sect. 2. The model approach describes how the AI can be used to identify the operation mode. Also, in this Sect. 3 the used data-set, the cluster techniques and the hybrid model is described. Then, in the results Sect. 4, the performance of the model is presented and finally, conclusions are drawn, Sect. 5.

2   Case Study

This paper focuses in the analysis of the switching modes o a half-bridge buck converter shown in the Fig. 1. This topology is the base component of many other power converters, like the buck, the boost, the full-bridge, etc. The power components used in this topology are two transistors that operate complementary, when one is close, the other is open. To the switching node, Vsw in Fig. 1, a filter is placed. The filter is made by a inductor and capacitor, absorbing and releasing energy to compensate the pulse voltage of the switching node. Thus, the output voltage and current is constant.

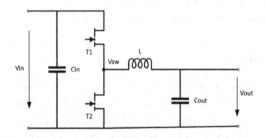

Fig. 1. Synchronous buck converter (half-bridge).

There are mainly two operating modes of this converter regarding the switching losses. Figure 2a shows one of them, Hard-Switching (HS) mode, when during the transition there is voltage and current flowing through the switch. If the transition is analysed: in a first state, the switch is blocking the voltage. When a gate signal is applied to the transistor, the device commutates to on-state: the resistance of the channel of the transistor starts to drop and, thus, the current starts flowing through the device. In a similar way, as the channel starts conducting, the voltage blocked by the device starts decreasing. Once the device saturates, the channel resistance is very low and the current is flowing, causing the on-state losses.

In opposite way, the gate signal is removed, the device starts turning off: the channel of the transistor starts to close, the resistance of the channel increases becoming high impedance, causing the current to decrease while the voltage rises. Thus, the device blocks the applied voltage.

On the other hand, the proposed converter can also operate in other different mode, Soft-Switching (SS) mode. As shown in the Fig. 2b, the turn-on and/or turn-off transitions are done under a certain condition: Zero Current Switching (ZCS) or Zero Voltage Switching (ZVS). The same way as above, when the gate signal is applied, the voltage starts decreasing while the current keeps at zero. Once the voltage has dropped, the current starts rising until its maximum value. In a similar manner, the turn-off transition happens: the current falls to zero before the voltage starts rising. These condition of ZCS or ZVS is mainly achieved by the resonance of components in the circuit that makes the voltage/current to drop or additional components are added to the circuit to achieve this conditions, such as snubber capacitors, resonance LC tanks, etc.

The switching losses during the transitions is highly reduce when the operated mode is forced to be in Soft-Switching condition. Also, in the Fig. 2a and 2b the power losses during the commutation transition are represented. During SS mode, due to the shift of the current or voltage during the commutation the power losses, which are equal to $P(t) = v(t) \cdot i(t)$, can be significantly reduce.

In the converter explained in this section, the resonance components that take place in the proposed circuit are between the filter inductor and the parasitic output capacitance of the transistor (Coss), which can be used as a non-disipative snubber. During this process, the current is reloading the output capacitance while the channel is built, thus making the circuit to operate in Soft-Switching mode [18].

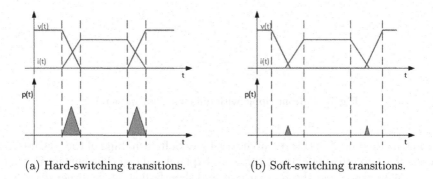

(a) Hard-switching transitions.   (b) Soft-switching transitions.

Fig. 2. Hard- vs Soft-switching transitions.

As the inductor is resonancing in this topology with the output capacitance, its design plays an important role. The converter can operate in different modes depending on the ripple current through the inductance.

Usually, the designs try to keep the ripple current to a low value, to reduce the filtering effort of the output capacitors. As a rule of thumb, the ripple current in a buck converter is kept between 5 % to 20 % of the average current. The design of the inductance is done accordingly to the Eq. 1, where the output is a function of the current and switching frequency.

$$L = \frac{(V_{in} - V_{out}) \cdot D}{f \cdot I_{ripple}} \qquad (1)$$

where L is the inductance value of the inductor, V_{in} is the input voltage to the circuit, V_{out} is the output voltage from the converter, D is the duty cycle, f is the switching frequency and I_{ripple} is the current ripple in the inductor.

Nowadays, thanks to the emergence of new materials for the transistors, the traditional way of designing a filter inductor needs to be reconsidered. In [7,11,19], the Triangular Current Mode (TCM) can be beneficial to the converter allowing the Soft-Switching of the transistors. The selection of the filter inductance as well as the switching frequency in which the converter operates, can determine in which mode the converter operates. As shown in Fig. 3, when the inductance is high or the switching frequency is high, the ripple in the inductor is low, so when commuting the transistors, there is alswalys current flowing (HS mode). On the other hand, if the inductance value is kept low or the switching current is kept low, a high current ripple can be achieve, droping the current down to 0 A, thus allowing to turn-on or turn-off the switching devices in this exact moment. In this way, the switching losses can be highly reduced.

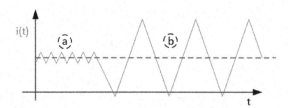

Fig. 3. Current ripple with different filter inductors.

The method used in this research to achieve Soft-Switching of the converter is based on a zero-crossing current ripple, as explained above. The current ripple is always allowed to cross the zero current and slightly flow in the other direction. As explained previously, this operating mode allows a reduction of the switching losses, but as drawback the Root Means Square (RMS) current in the inductor is increased and so in the transistors. This increase of the RMS current causes higher on-state losses. To avoid a extremely high increase of the conduction losses and thanks that the switching losses are highly reduce, the converter would need to operate at higher frequencies. Thus, the increase of the switching frequency allows a reduction of the filter components, main volume parts in the converters, and increase the power densities.

3   Model Approach

The main objective of the research is the development of a hybrid intelligent model used for the detection and classification of the different operating modes, Hard- and Soft-Switching mode. With the aim of classifying the modes, the clustering of the data has been done and then four different methods have been applied to the obtained data-set.

The Fig. 4 shows the structure used in this paper. From the simulation data, first there is a pre-processing process, in which the raw data is evaluated and more representative parameters are calculated. Then, the clustering of this dataset is done for applying later on the four intelligent classification methods.

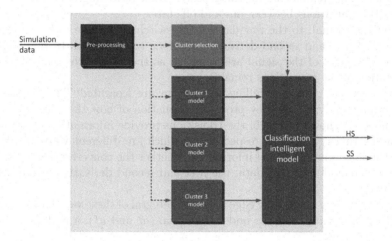

Fig. 4. Model approach.

3.1   Dataset

The used dataset is obtained from the LTSpice simulation tool. Thus, 80 different simulations of the circuit have been done, varying the operating mode, output load and output voltage. The rest of the circuit keeps unchanged to allow reproducibility of the experiment. In total, 80 simulation results have been obtained, a combination of both Hard- and Soft-Switching data, 50% each type. In this circuit, different variables are measured to obtain the dataset:

– Input voltage: the applied input voltage is kept constant at 400 V.

- Output voltage: the output voltage is controlled to be 200 V and has a ripple of maximum 5 %, so varying from 190 V to 210 V.
- Switching node voltage (Vsw node Fig. 1): the voltage at the switching node is a square signal that varies from 0 up to 400 V. The frequency of this signal is also variable, depending on the simulation case, varying from 80 kHz to 2 MHz.
- Inductor current: the current at the inductor has a triangular shape. The current varies accordingly to the load and the switching frequency. In Hard-Switching, the ripple is kept at 20 % of the output current while in Soft-Switching, the ripple is around 2 times the output ripple, to ensure that the current drops to 0 A.
- Output current: the output current depends on the output load, which is a parameter that changes for each of the simulations. Its value is constant with a ripple of 5 %, as the output voltage.

Once the data is obtained, the first step is to analyse it and pre-processed the signals. The main signal for this research is the switching node voltage, as it reflects the transitions between on- and off-states.

From this signal, to the raw data of the switching voltage (Vsw) is used as a base. The first and second derivative are done to the signal. In this way, the on- and off-states of the signal are removed, as their derivative is 0, while the information of the transitions remains.

Moreover, the rising and falling edge of Vsw are separated. These transitions are carefully analyzed, as they provide information of how the commutation is happening. As shown in Fig. 5, the transitions provide information of the rising and falling times (*tr* and *tf*, respectively), which are different when Hard- and Soft-Switching, giving a lot of information whether the converter is operation in one or other mode. To this data, the first and second derivative has also been calculated.

From the rising and falling edge data, the integral of these variables have been calculated, providing the area under the signal (*ar* and *af*), a perfect indicator of the transitions.

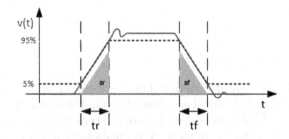

Fig. 5. Rising and falling edge of the switching node voltage, in dashed blue, and the original signal in continuous red.

In total, 8 signals derived from the Vsw are used for each of the simulations: the raw data (red signal in Fig. 5), the first and the second derivatives of the raw data, the rising/falling edge data (dotted blue signal in Fig. 5), the first and second derivatives of rising/falling edge data, the rising edge integral (area at the rising edge, ar, in Fig. 5) and the falling edge integral (area at the falling edge, af, in Fig. 5).

In order to make the data more significant and easier to analyzed, the following statistics have been calculated for each of the 8 variables: average, standard deviation, variance, co-variance, Root Mean Square and Total Harmonic Distortion (THD). Resulting in a matrix of 8×6 for each of the 80 simulations.

3.2   Methods

The classification algorithms used in this research are the Multilayer Perceptron (MLP), the Support Vector Machine (SVM), the Linear Discrimination Analysis (LDA) and the ensemble classifier. These methods are described bellow.

Data Clustering The K-Means Algorithm. An algorithm commonly uses to divide data is the unsupervised technique called K-Means. This clustering data makes groups according to the similarity of the samples [14,20]. The unlabeled samples x are compared with the rest of the samples and make the clusters based on the distance between each sample and the centroids. The clusters are defined by their centroids, that are chosen as the geometrical center of the cluster.

This is an iterative algorithm, and the training phase is the calculation of the centroids. To perform this, the algorithm used to start with randomly centroids, chosen from the dataset. In the following steps, all the samples are assigned to each centroids based on the distance from each samples to all centroids; the sample cluster is the one with the smallest distance. Once all the samples are assigned to a cluster, the centroids are re-calculated as the center of the new group of samples. The procedure finish when the centroids do not change its value between two consecutive iterations.

Multilayer Perceptron. A perceptron is an artificial neural network that has only one hidden layer of neurons. When the same structure is used, but with multiple hidden layer, we refer as a multilayer perceptron. The structure is the following: one input layer, which consists of input features to the algorithm, then the multiple hidden layer which have neurons with an activation function, and one output layer, which number of neurons depends on the desired outputs. All these layer are connected in between by weighted connections. These weights are tuned with the aim of decreasing the error of the output [10,21].

Linear Discrimination Analysis. Another method used for classification is the Linear Discrimination Analysis. This method is based on a dimension reduction, projecting the data from a high dimensional space into a low-dimensional space, where the separation of the classes is done. This method uses a weight vector W, which projects the given set of data vector E in such a way that maximizes the class separation of the data but minimizes the intra-class data [5,23]. The projection is done accordingly to the Eq. 2. The separation is good when the projections of the class involves exposing long distance along the direction of vector W.

$$P_i = W^T \cdot E_i \qquad (2)$$

The LDA provides each sample with its projection and the class label. Two outputs are provided by the analysis, first a gradual decision which is then converted into a binary decision. This method maximizes ratio between the inter-class variance to the intra-class variance, finding the best separation of the classes. The performance of the method increases with the distance between the samples [9,17].

Support Vector Machine. A common used method in classification is the support vector machine, which is a supervised algorithm of machine learning [3,15]. The algorithm tries to find two parallel hyper-planes that maximize the minimum distance between two class of samples [15]. Therefore, the vectors are defined as training instances presented near the hyperplane and a high dimensional feature space is used by a kernel to calculate the projection of the dataset.

Ensemble. The term ensemble is used to define multiple classification methods which are used in combination with the aim of improving the performance over single classifiers [13,16]. They are commonly used for classification tasks. The ensemble does a regularization, process of choosing fewer weak learners in order to increase predictive performance [24].

3.3   Measurement of the Classification Performance

The models are validated with a confusion matrix. The confusion matrix is a commonly used method to asses the quality of a classifier. The entry to this matrix are the true values and the predicted values, where the true values compound the columns while the predicted ones the rows [6].

Usually, the entries to the matrix are two decision classes, positives (P) and negatives (N), and the table entries are called true positives (TP), false positives (FP), true negatives (TN) or false negatives (FN) [9].

Once the confusion matrix have been created, there are 5 indicators that are used to analyzed the performance of the models and compare the results between them. These statistics are the followings: SEnsitivity (SE), SPeCificity (SPC), Positive Prediction Value (PPV), Negative Prediction Value (NPV), and Accuracy (ACC) [9].

3.4   Experiments Description

In this section, the experiments that had been carried out are explained step by step. First of all, the model without clustering has been tested, to see afterwards the comparison with the hybrid model. In all cases, the data is divided into two sub-sets, one used to train the different models and other set used to validate the model. In this case, the data is divided in 75-25%, where 75% of the data is used to train the model and the rest is used to validate it. The division of the dataset is done randomly.

Once the data is grouped in two sets, the different models are trained:

- MLP: the chosen algorithm for the MLP is the Levenberg-Marquardt back-propagation. This algorithm has been trained from 1 to 10 neurons on the hidden layers.
- LDA: the discriminant type is the regularized LDA, in which all classes have the same co-variance matrix
- SVM: The SVM has been trained using the linear kernel function, commonly used for two-class learning. Also, it trains the classifier using the standardized predictors and centers and scales each predictor variable by the corresponding weighted column mean and standard deviation.
- Ensemble: the used ensemble method is an adaptive logistic regression which is used for binary classification. The number of cycles of the ensemble vary from 10 to 100 in steps of 10. The weak-learners used function is the decision tree.

Then, the validation data is used to check if the models have been correctly trained. The predictions obtained from the models are compared with the verified data, using a confusion matrix and the different statistics are calculated.

Once the results without clustering has been obtained, different clusters have been applied to the model, making a hybrid implementation of if. The number of clusters applied vary from 2 up to 10. The used method is K-Means, as explained in previous Sect. 3.2. Then, to each cluster obtained with K-Means, the previous models have been applied and the validation of the model has been also done in the same manner.

4   Results

The obtained results from the hybrid models are shown in this section. As a summary from all the models, the Table 1 shows the performance achieved from each hybrid model configuration, allowing the comparison between them. From a cluster number higher than 4, the model achieves a classification accuracy of 100 %

When the model is run without clustering (one cluster), the best result is achieved by the MLP7 with an accuracy of 0.97059. Then, when the number of clusters is two, 100 % of accuracy has been achieve and the worst case an accuracy of 0.78947. When the number of clusters increase to three, in most

Table 1. Achieved accuracy with each model and cluster configuration.

Clusters	1 cluster	2 clusters		3 clusters			4 clusters			
# of cluster	1	1 of 2	2 of 2	1 of 3	2 of 3	3 of 3	1 of 4	2 of 4	3 of 4	4 of 4
MLP1	0.81538	1	0.94737	0.89474	1	1	1	1	1	1
MLP2	0.73529	1	0.94737	1	1	1	1	1	1	1
MLP3	0.62264	1	0.94737	1	1	1	1	1	1	1
MLP4	0.55882	0.94737	0.94737	0.89474	1	1	1	1	1	1
MLP5	0.79412	0.88889	1	1	1	1	1	1	1	1
MLP6	0.60294	0.94737	0.94737	1	1	1	1	1	1	1
MLP7	0.97059	0.94737	0.94737	0.89474	1	1	1	1	1	1
MLP8	0.73529	0.94737	0.94737	1	1	1	1	1	1	1
MLP9	0.86765	1	0.94737	1	1	1	1	1	1	1
MLP10	0.52941	1	0.94737	1	1	1	1	1	1	1
SVM	0.77941	1	0.94737	1	1	1	1	1	1	1
LDA	0.60294	1	0.94737	1	1	1	1	1	1	1
Ensemble10	0.51471	0.94737	0.78947	1	1	1	0.8	1	1	1
Ensem.20–100	0.51471	1	0.78947	1	1	1	0.8	1	1	1

cases the 100 % of accuracy is achieved. With four clusters and higher, the classification model achieves 100 % accuracy with any model.

The structure of the Table 1 is the following: first a column with just one cluster. Then, the cluster number out of the number of cluster used for the model.

5   Conclusions and Future Works

In this research, a hybrid intelligent model has been applied in power electronics for the detection of the operating mode of a half-bridge buck converter. The model classifies the operating mode between Hard- or Soft-Switching mode. The method implements a hybrid intelligent model based on classification techniques. The data for creating the model is obtained from a simulated circuit of the converter, obtaining the significant variables for the detection. This data is used to differentiate bewteen Hard- or Soft-Switching modes. The input data is are divided into 5 main variables: input voltage, output voltage, switching node voltage, output current and inductor current.

When having just one cluster of data, the best performance of the model is 97 %. On the other hand, when a hybrid model is used with 3 clusters or higher, almost every classification method is able to predict accurately 100 % of the data.

The use of the proposed model in this work can be used as a very useful tool for the detection of the operation modes in power converters and, therefore, helping the design and the increase of efficiency in them.

Future work in this area will be oriented to the development the proposed circuit with the aim of applying this method with real measured data from a circuit.

Acknowledgements. CITIC, as a Research Center of the University System of Galicia, is funded by Consellería de Educación, Universidade e Formación Profesional of the Xunta de Galicia through the European Regional Development Fund (ERDF) and the Secretaría Xeral de Universidades (Ref. ED431G 2019/01).

References

1. Al-bayati, A.M.S., Alharbi, S.S., Alharbi, S.S., Matin, M.: A comparative design and performance study of a non-isolated dc-dc buck converter based on si-mosfet/si-diode, sic-jfet/sic-schottky diode, and gan-transistor/sic-schottky diode power devices. In: 2017 North American Power Symposium (NAPS), pp. 1–6 (2017). https://doi.org/10.1109/NAPS.2017.8107192
2. Aláiz-Moretón, H., Castejón-Limas, M., Casteleiro-Roca, J.L., Jove, E., Fernández Robles, L., Calvo-Rolle, J.L.: A fault detection system for a geothermal heat exchanger sensor based on intelligent techniques. Sensors **19**(12), 2740 (2019)
3. Basurto, N., Arroyo, Á., Vega, R., Quintián, H., Calvo-Rolle, J.L., Herrero, Á.: A hybrid intelligent system to forecast solar energy production. Comput. Electr. Eng. **78**, 373–387 (2019)
4. Casteleiro-Roca, J.L., Javier Barragan, A., Segura, F., Luis Calvo-Rolle, J., Manuel Andujar, J.: Intelligent hybrid system for the prediction of the voltage-current characteristic curve of a hydrogen-based fuel cell. Revis. Iberoam. Autom. Inform. Ind. **16**(4), 492–501 (2019)
5. Crespo-Turrado, C., Casteleiro-Roca, J.L., Sánchez-Lasheras, F., López-Vázquez, J.A., De Cos Juez, F.J., Pérez Castelo, F.J., Calvo-Rolle, J.L., Corchado, E.: Comparative study of imputation algorithms applied to the prediction of student performance. Log. J. IGPL **28**(1), 58–70 (2020)
6. Düntsch, I., Gediga, G.: Indices for rough set approximation and the application to confusion matrices. Int. J. Approx. Reason. **118**, 155–172 (2020). https://doi.org/10.1016/j.ijar.2019.12.008
7. Fernandez-Serantes, L.A., Berger, H., Stocksreiter, W., Weis, G.: Ultra-high frequent switching with gan-hemts using the coss-capacitances as non-dissipative snubbers. In: PCIM Europe 2016; International Exhibition and Conference for Power Electronics, Intelligent Motion, Renewable Energy and Energy Management, pp. 1–8. VDE (2016)
8. Fernández-Serantes, L.A., Estrada Vázquez, R., Casteleiro-Roca, J.L., Calvo-Rolle, J.L., Corchado, E.: Hybrid intelligent model to predict the SOC of a LFP power cell type. In: Polycarpou, M., de Carvalho, A.C.P.L.F., Pan, J.-S., Woźniak, M., Quintian, H., Corchado, E. (eds.) HAIS 2014. LNCS (LNAI), vol. 8480, pp. 561–572. Springer, Cham (2014). https://doi.org/10.1007/978-3-319-07617-1_49
9. Jove, E., et al.: Missing data imputation over academic records of electrical engineering students. Log. J. IGPL **28**(4), 487–501 (2020)
10. Jove, E., Casteleiro-Roca, J., Quintián, H., Méndez-Pérez, J., Calvo-Rolle, J.: Anomaly detection based on intelligent techniques over a bicomponent production plant used on wind generator blades manufacturing. Revis. Iberoam. Autom. Inform. Ind. **17**(1), 84–93 (2020)
11. Jove, E., Casteleiro-Roca, J.L., Quintián, H., Méndez-Pérez, J.A., Calvo-Rolle, J.L.: A fault detection system based on unsupervised techniques for industrial control loops. Expert Syst. **36**(4), e12395 (2019)

12. Jove, E., Casteleiro-Roca, J.L., Quintián, H., Simić, D., Méndez-Pérez, J.A., Luis Calvo-Rolle, J.: Anomaly detection based on one-class intelligent techniques over a control level plant. Log. J. IGPL **28**(4), 502–518 (2020)
13. Jove, E., et al.: Modelling the hypnotic patient response in general anaesthesia using intelligent models. Log. J. IGPL **27**(2), 189–201 (2019)
14. Kaski, S., Sinkkonen, J., Klami, A.: Discriminative clustering. Neurocomputing **69**(1–3), 18–41 (2005)
15. Liu, M.Z., Shao, Y.H., Li, C.N., Chen, W.J.: Smooth pinball loss nonparallel support vector machine for robust classification. Appl. Soft Comput. **98**, 106840 (2020). https://doi.org/10.1016/j.asoc.2020.106840
16. Luis Casteleiro-Roca, J., Quintián, H., Luis Calvo-Rolle, J., Méndez-Pérez, J.A., Javier Perez-Castelo, F., Corchado, E.: Lithium iron phosphate power cell fault detection system based on hybrid intelligent system. Log. J. IGPL **28**(1), 71–82 (2020)
17. Marchesan, G., Muraro, M., Cardoso, G., Mariotto, L., da Silva, C.: Method for distributed generation anti-islanding protection based on singular value decomposition and linear discrimination analysis. Electr. Power Syst. Res. **130**, 124–131 (2016). https://doi.org/10.1016/j.epsr.2015.08.025
18. Mohan, N., Undeland, T.M., Robbins, W.P.: Power Electronics: Converters, Applications, and Design. John Wiley & sons, Hoboken (2003)
19. Neumayr, D., Bortis, D., Kolar, J.W.: The essence of the little box challenge-part a: key design challenges solutions. CPSS Trans. Power Electron. Appl. **5**(2), 158–179 (2020). https://doi.org/10.24295/CPSSTPEA.2020.00014
20. Qin, A.K., Suganthan, P.N.: Enhanced neural gas network for prototype-based clustering. Pattern Recognit. **38**(8), 1275–1288 (2005)
21. Tahiliani, S., Sreeni, S., Moorthy, C.B.: A multilayer perceptron approach to track maximum power in wind power generation systems. In: TENCON 2019–2019 IEEE Region 10 Conference (TENCON), pp. 587–591 (2019). https://doi.org/10.1109/TENCON.2019.8929414
22. Liu, T., Zhang, W., Yu, Z.: Modeling of spiral inductors using artificial neural network. In: Proceedings. 2005 IEEE International Joint Conference on Neural Networks, 2005, vol. 4, pp. 2353–2358 (2005). https://doi.org/10.1109/IJCNN.2005.1556269
23. Thapngam, T., Yu, S., Zhou, W.: DDoS discrimination by linear discriminant analysis (LDA). In: 2012 International Conference on Computing, Networking and Communications (ICNC), pp. 532–536. IEEE (2012)
24. Uysal, I., Gövenir, H.A.: An overview of regression techniques for knowledge discovery. Knowl. Eng. Rev. **14**, 319–340 (1999)

Hybrid Model to Calculate the State of Charge of a Battery

María Teresa García Ordás[1], David Yeregui Marcos del Blanco[1],
José Aveleira-Mata[1], Francisco Zayas-Gato[2], Esteban Jove[2],
José-Luis Casteleiro-Roca[2], Héctor Quintián[2], José Luis Calvo-Rolle[2],
and Héctor Alaiz-Moretón[1](✉)

[1] Department of Electrical and Systems Engineering, University of León, León, Spain
{m.garcia.ordas,jose.aveleira,hector.moreton}@unileon.com,
dmarcb01@estudiantes.unileon.es
[2] CTC, CITIC, Department of Industrial Engineering, University of A Coruña,
Ferrol, A Coruña, Spain
{f.zayas.gato,esteban.jove,jose.luis.casteleiro,hector.quintian,
jlcalvo}@udc.es

Abstract. Batteries are one of the most important component in an
energy storage system; they are used mainly in electric mobility, consumer electronic and some other devices. Nowadays the common battery
type is the liquid electrolyte solution, but it is expected that in some
year, the solid state batteries increase the energy density. Despite the
type of batteries, it is very important that the user knows the energy that
remains inside the battery. The most used ways to calculate the capacity,
or State Of Charge (SOC), is the percentage representation that takes
into account energy that can be stored, and the remained energy. This
research is based on a Lithium Iron Phosphate (LiFePO4) power cell,
because it is commonly used in several applications. This paper develops
a hybrid model that calculate the SOC taking into account the voltage
of the battery and the current to, or from, it. Moreover, there has been
checked two different clustering algorithms to achieve the best accurate
of the model, that finally has a Mean Absolute Error of 0.225.

Keywords: Battery capacity confirmation test · Hybrid model ·
k-Means · Agglomerative clustering · Artificial neural network

1 Introduction

Nowadays, there are a lot of devices that are powered electrically. These devices
cover a wide range of sizes, since small devices like mobiles, until huge industrial
machinery. Although the electricity is present nearly in all places, the energy
generation is changing to renewable sources. The main problem of this type of
energy generation is that the power generation is not linked to the power demand
[22].

The original version of this chapter was revised: an error in the name of one of the
authors has been corrected. The correction to this chapter is available at
https://doi.org/10.1007/978-3-030-86271-8_56

© Springer Nature Switzerland AG 2021, corrected publication 2021
H. Sanjurjo González et al. (Eds.): HAIS 2021, LNAI 12886, pp. 379–390, 2021.
https://doi.org/10.1007/978-3-030-86271-8_32

Energy storage systems are developed to solve, among others, this problem; this storage systems allow to store energy when there is generation and no demand, to continue generation energy when there is not renewable source. In the specific field of the electric mobility, for example, batteries are used as energy storage system, and a lot of researches are focused in analysing the battery behaviour [2,8,19,20]. The main aim is to obtain an accurate model because it allows to improve the efficiency of whole system, and to perform this, it is usual to use artificial intelligent techniques like in [4,5,13,23,27].

One of the possible ways to improve the accurate of the models is training hybrid models. This type of models is used in systems with a high non linearity component, or in systems with a well defined different working points. The main idea under this hybrid models is the division of the whole dataset in several clusters, and train local models for each cluster. Some examples of this hybrid intelligent models can be found in recent researches like [3,6,10,14,17,18], where this approach is develop to predict the solar thermal collector output temperature.

This study creates hybrid intelligent models based on different clustering techniques, comparing the performance of the clustering phase using Silhouette metric, a specific metric to measure the quality of a clustering algorithm. After that, locals models are trained to calculate the State of Charged of a Lithium Iron Phosphate (LiFePO4) power cell. The locals models use an Artificial Neural Network as regression algorithm.

After this introduction, the paper continues with the Case study section where it is described the test used in this research and the dataset obtained from the capacity confirmation test. Then, the Model approach is explained before the Methods section, where the different techniques used in the research are exposed. In the Experiments and results section it is described the performance of the model created. Finally, it is introduced the Conclusions and future work.

2   Case Study

The case study used in this research is based on the capacity confirmation test of a Lithium Iron Phosphate battery (LFP). In this test, that it is shown in Fig. 1, the battery is charged and discharged several times through a constant current power supply. During the test, the voltage is measured to calculate the energy to (and from) the battery [15,26,28]. Moreover, the test bench has two temperature sensors installed in the battery to control its warming and, obviously, the charged and discharge current is registered.

Figure 2 shows the main variables for one cycle test. As it has just been mentioned, the capacity confirmation test consisted in several charged discharged cycles. Each one of this cycle is divide in four different steps, that are indicated in the figure, depending on the state of the test. The limits of the different states are defined by the manufacturer of the battery depending on the maximum and minimum operation voltage. As the real battery used in this research is a $8Ah$ LiFeBATT X-1P [24], its main voltage values can be summarised as:

- Nominal voltage: 3.3 V
- Maximum voltage: 3.65 V
- Minimum voltage: 2.0 V

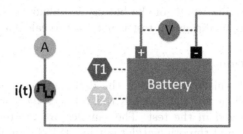

Fig. 1. Representation of the capacity confirmation test

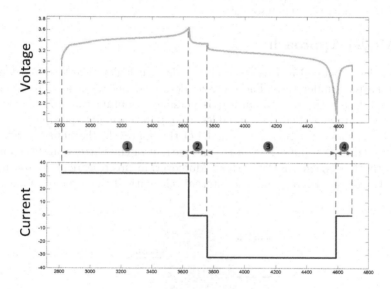

Fig. 2. Evolution of the voltage and current in one cycle test

In the charged state it is applied a positive current to the battery ((1) in Fig. 2), and this state finishes when the battery voltage increase until its maximum (3.65 V). After this charged, it is necessary that the battery recovers its nominal value (3.3 V); in this second state ((2) in Fig. 2), there is no current apply to the battery. Once this nominal value is achieved, the discharged state starts ((3) in Fig. 2) applying a negative current to the battery; this state last until the battery voltage decrease to its minimum value (2.0 V). After the discharge, it is necessary to recover the battery nominal voltage with no current set in the power supply ((4) in Fig. 2).

With the voltage and current evolution thought the time, it is possible to calculate the energy to, and from, the battery in each stage. As the battery has a nominal energy capacity, instead of using the real value of energy inside the battery, it is commonly use the State Of Charge (SOC) variable. This parameter varies from 0 to 100 % according to the energy inside the battery. It must be clarify that, as the initial SOC of the battery is unknown previously, the first cycle in the capacity confirmation test is not used to calculate the energy.

Dataset Description

As it is mentioned, this research uses the data registered from the capacity confirmation test. The raw dataset has the voltage, the current and the two temperatures measured in the test. The samples are recorded with a sample time of one second, and the original dataset has 18130 samples. However, as it is explained, the first cycle of the test is not valid to use in this research because the initial SOC is not known; after the discarding of this cycle, the dataset to create the model has 16269 samples.

3   Model Approach

The model developed in this research is able to calculate the State Of Charge of the battery under test. The dataset, described below, is used to calculate the SOC during the test because it is possible to obtain the energy from the measured voltage and current variables, taking into account the time.

Figure 3 shows the functional model of the approach; the figure clarify that the model uses as inputs the voltage and current of the battery, and the output is directly the capacity of the battery. Moreover, in this research a hybrid model is developed; Fig. 4 represents the internal schematic if this type of models.

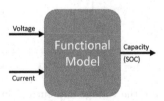

Fig. 3. Functional model

The internal layout of the hybrid models can be divide in two different parts: the cluster selection, and the local models. This hybrid models are based in the division of the model in sub-models (local models), that are used only with a specific group of the dataset. The groups are made with a clustering algorithm, and the local models (in this research) with a regression technique. In the upper left side of the Fig. 4, the cluster selector box is the part that decides what is the correct local model for the specific inputs. Only one local model is calculated, and the output of this local model is route to the output of the hybrid model.

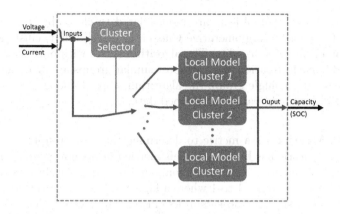

Fig. 4. Internal schematic of a hybrid model

4   Methods

In this section, the clustering and regression techniques used to create the hybrid method presented in this work will be defined.

4.1   Clustering Methods and Metrics

K-Means: K-Means is a well-known clustering algorithm that is responsible for separating samples into groups, minimising inertia [9,12,16,21]. Given an initial set of k centroids $c_1^{(1)}, ..., c_k^{(1)}$, the algorithm performs two steps iteratively. In the assignment step, each piece of data is assigned to the group with the closest mean following Eq. 1.

$$S_i^{(t)} = x_p : ||x_p - c_i^{(t)}|| \leq ||x_p - c_j^{(t)}|| \forall 1 \leq j \leq k \tag{1}$$

where x_p is the data, c are the centroids and each x_p can be assigned only to one $S_i^{(t)}$.

In the update stage, the new centroids are calculated by taking the mean of the group data (see Eq. 2).

$$c_i^{(t+1)} = \frac{1}{|S_i^{(t)}|} \sum_{x_j \in S_i^{(t)}} x_j \tag{2}$$

The algorithm is considered to have converged when the assignments no longer change.

Agglomerative Clustering: Agglomerative clustering is one of the hierarchical grouping strategies. Hierarchical clustering is a well-known cluster analysis method that seeks to build a hierarchy of clusters. Hierarchical clustering is based

on the central idea that objects are more related to close objects than to far-
ther ones. Concretely, Agglomerative Clustering [7,11], builds clusters through
a succession of mergers or splits. The algorithm starts with only one group per
sample and in each iteration, the two most similar groups are merged. When all
samples belong to a single group, the algorithm stops. Dendograms are used as
representation of the hierarchical grouped data.

Silhouette Metric: As a metric to determine the best grouping of the data,
the Silhouette metric has been used. This metric tells us how similar an element
is to the rest of its cluster (cohesion) compared to the other clusters. The value
of the metric ranges from −1 to 1 where a higher value indicates that the object
is similar to the rest of the objects in its cluster. Silhouette is defined in Eq. 3.

$$s(i) = \frac{b(i) - a(i)}{max\{a(i), b(i)\}} \tag{3}$$

if $|C_i|$ is greater than one and $S(i) = 0$ if $|C_i| = 1$. Being $a(i)$ and $b(i)$ defined
as follows (see Eqs. 4 and 5).

$$a(i) = \frac{1}{|C_i| - 1} \sum_{j \in C_i, i \neq j} d(i, j) \tag{4}$$

where $d(i, j)$ is the Squared Euclidean distance between data i and j in the
cluster C_i. The smaller the value of a, the better the assignment.

$$b(i) = min_{k \neq i} \frac{1}{|C_k|} \sum_{j \in C_k} d(i, j) \tag{5}$$

4.2   Regression: MLP

To carry out our hybrid model, a multilayer perceptron is used as a regression
technique. The typical perceptron architecture consists of an input layer, an
output layer, and one or more hidden layers. The most unique feature of MLP is
that each neuron in the next layer is connected to each neuron in the next layer,
so the information is transmitted layer by layer and a complex learning process
takes place.

5   Experiments and Results

In this article, a hybrid method has been proposed to predict the capacity of a
battery based on its voltage and temperature. First, an unsupervised classification is performed to cluster the data. Once we have the class of each data, this
class becomes part of the data features and a regression process is carried out
to predict the capacity. K-Means and Agglomerative Clustering techniques have
been used in the clustering step. The quality of the clustering is measured with
the Silhouette metric in order to determine the optimal number of clusters. In

Table 1. Metric values for k-Means clustering

k	2	3	4	5	6	7	8	9
Silhouette	0.6663	0.6802	0.6106	0.5675	0.5770	0.5895	0.5582	0.5172

Table 1, the metric value for different number of clusters from 2 to 9 carried out using k-Means, is shown.

Performing the grouping with the Agglomerative clustering technique, we obtain the values shown in Table 2.

Table 2. Metric values for Agglomerative clustering

k	2	3	4	5	6	7	8	9
Silhouette	0.6663	0.6744	0.6080	0.5579	0.5688	0.5764	0.5272	0.4799

In view of the results, we obtain the most optimal number of clusters with k = 3 for both clustering methods. Therefore, the data has been grouped into three clusters that will become part of the vector of data characteristics to carry out the regression process.

The clustering of the data with both methods, k-Means and Agglomerative Clustering, can be seen graphically after applying a LDA dimensionality reduction (2 features) in Fig. 5 and it can be seen that they are well differentiated into three groups as indicated by the Silhouette metric.

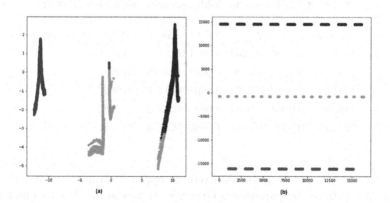

Fig. 5. Graphical representation of data clustering using k-Means (a) and Agglomerative clustering (b) with k = 3.

The second step is oriented to validate the clustering methods and the number of groups established by k-Means and Agglomerative, therefore a MLP has been the regression method chosen for this purpose.

The correct performance of the MLP is based on the implementation of the a Grid Search combined with a Cross-validation in order to find the optimal parameters [1], being the *Mean Squared Error* [25] the error metric chosen for defining the best model in the training stage.

The parameter grid has been defined as follow:

– Number of neurons in the hidden layer: from 2 to 18 neurons.
– Activation function: the hyperbolic tan function ("tanh") or the rectified linear unit function ("relu").
– Solver: optimizer in the family of quasi-Newton methods ("lbfgs"), stochastic gradient descent ("sgd") or stochastic gradient-based optimizer ("adam").
– Nesterov momentum: True for introducing a velocity component for accelerating the convergence.
– Warm Start: if the value is True, use again the solution of the earlier call to fit as initialization, if the value is False, removes the earlier solution.
– Early stopping: it defines a mechanism used for stopping the training process when score used is not improving. It works with solver as "sgd" or "adam".

Results show the two approaches based on the two clustering methods implemented previously. Tables 3 and 4 display the set of error measurements with two different points of view. The first one, a standard average between the three groups based on the hybrid model and the second one, a weighted average based on the hybrid model also, proportional for each error measure to the size of each grouping. Split is configured in the following way: 20% for validating (3273) and the 80% for training-test purposes (12996).

Table 3. MLP error for k-Means clustering with 3 clusters

Cluster	1	2	3	Average	Weighted average
MSE	0.2240	0.1269	0.0001	0.1173	0.1516
MAE	0.3350	0.2732	0.0093	0.2058	0.2617
LMLS	0.0881	0.0574	0.0005	0.0487	0.0627
MASE	0.0104	0.0084	0.002	0.0063	0.0080
SMAPE	0.0093	0.0156	1.0169	0.3473	0.1612

In the tables 5 and 6 can be observed the optimal parameters for each MLP model, chosen from the parameters Grid Search procedure. The two first make reference to hybrid model approach with the optimal parameters for each group while the third one shows best parameters for global model. It should be noted that the parameter "Warm Start" is not applied, due to the optimal solver method is the "lbfgs".

In the Figs. 6 and 7 can be observed a graphical representation of the regression results applying the MLP regressor for each group, the "Y" axis makes reference to Capacity of battery value. Real output is drawn in red while the

Table 4. MLP error for Agglomerative clustering with 3 clusters

Cluster	1	2	3	Average	Weighted average
MSE	0.2367	0.0540	4.70e−5	0.0969	0.13
MAE	0.3346	0.1788	0.0049	0.1728	0.225
LMLS	0.0891	0.0258	2.23e−5	0.0383	0.051
MASE	0.0102	0.0053	0.0001	0.0052	0.0068
SMAPE	0.0137	0.0115	1.1435	0.3895	0.171

Table 5. MLP best parameters for k-Means clustering

Grid Parameter/Cluster	1	2	3
Number of neurons	10	11	3
Activation function	tanh	tanh	tanh
Solver	lbfgs	lbfgs	lbfgs
Nesterov	True	True	True
Warm Start	–	–	–
Early Stopping	True	True	True

Table 6. MLP best parameters for Agglomerative clustering

Grid Parameter/Cluster	1	2	3
Number of neurons	10	12	10
Activation function	tanh	tanh	tanh
Solver	lbfgs	lbfgs	lbfgs
Nesterov	True	True	True
Warm Start	–	–	–
Early Stopping	True	True	True

predicted output displayed in black dashed line. For each figure have been displayed only 100 for cluster in order to clarify the output and how the MLP regressor works.

The results outcomes that three is the optimal number of clusters being the best clustering method the Agglomerative from MAE and MSE point of view. However Agglomerative method requires that the number of neurons for group three was quite bigger than number of neurons required in the hidden layer for Kmeans. Due to is can say that two clustering methods achieve good results choosing Kmeans if it is necessary to work with lighter regression process from computational point of view.

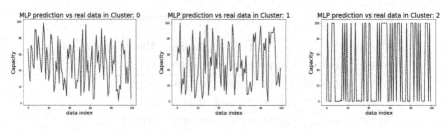

Fig. 6. Real data vs. MLP predictions for k-Means clustering

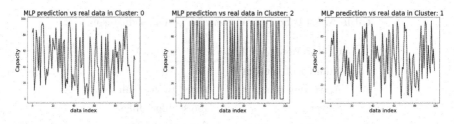

Fig. 7. Real data vs. MLP predictions for Agglomerative clustering

6   Conclusions and Future Works

This research explain the creation of a hybrid intelligent model based on two different clustering algorithms: k-Means and Agglomerative clustering. A specialised metric for cluster quality, Silhouette, is used to define the optimal number of clusters for addressing the problem. The dataset was finally divide in three clusters before train the local models as this division obtain the higher value of the Silhouette metric: 0.6802 and 0.6744 respectively.

The regression phase is performed with several MLPs local models. 6480 model were trained to obtain the better configuration for each clusters. Several error measurements were used to compare the models and select the best ones for the specific case study. After these tests, the MSE and MAE error measurements show that the best results are obtained using Agglomerative cluster: 0.13 and 0.225 against 0.1516 and 0.2617 respectively.

In future researches it will be possible to increase the clustering and regression algorithms in order to allow to increase the accuracy of the prediction of the hybrid intelligent model.

Acknowledgement. CITIC, as a Research Center of the University System of Galicia, is funded by Consellería de Educación, Universidade e Formación Profesional of the Xunta de Galicia through the European Regional Development Fund (ERDF) and the Secretaría Xeral de Universidades (Ref. ED431G 2019/01).

References

1. Aláiz-Moretón, H., Castejón-Limas, M., Casteleiro-Roca, J.L., Jove, E., Fernández Robles, L., Calvo-Rolle, J.L.: A fault detection system for a geothermal heat exchanger sensor based on intelligent techniques. Sensors **19**(12), 2740 (2019). https://doi.org/10.3390/s19122740

2. Casteleiro-Roca, J.L., Barragán, A.J., Segura, F., Calvo-Rolle, J.L., Andújar, J.M.: Fuel cell output current prediction with a hybrid intelligent system. Complexity 2019 (2019)

3. Casteleiro-Roca, J.L., et al.: Solar thermal collector output temperature prediction by hybrid intelligent model for smartgrid and smartbuildings applications and optimization. Appl. Sci. **10**(13), 4644 (2020)

4. Casteleiro-Roca, J.L., Javier Barragan, A., Segura, F., Luis Calvo-Rolle, J., Manuel Andujar, J.: Intelligent hybrid system for the prediction of the voltage-current characteristic curve of a hydrogen-based fuel cell. Rev. Iberoam. Autom. Inform. Ind. **16**(4), 492–501 (2019)

5. Casteleiro-Roca, J.L., Quintián, H., Calvo-Rolle, J.L., Corchado, E., del Carmen Meizoso-López, M., Piñón-Pazos, A.: An intelligent fault detection system for a heat pump installation based on a geothermal heat exchanger. J. Appl. Log. **17**, 36–47 (2016). https://doi.org/10.1016/j.jal.2015.09.007

6. Crespo-Turrado, C., et al.: Comparative study of imputation algorithms applied to the prediction of student performance. Log. J. IGPL **28**(1), 58–70 (2020)

7. Defays, D.: An efficient algorithm for a complete link method. Comput. J. **20**(4), 364–366 (1977). https://doi.org/10.1093/comjnl/20.4.364

8. Fernández-Serantes, L.A., Estrada Vázquez, R., Casteleiro-Roca, J.L., Calvo-Rolle, J.L., Corchado, E.: Hybrid intelligent model to predict the SOC of a LFP power cell type. In: Polycarpou, M., de Carvalho, A.C.P.L.F., Pan, J.-S., Woźniak, M., Quintian, H., Corchado, E. (eds.) HAIS 2014. LNCS (LNAI), vol. 8480, pp. 561–572. Springer, Cham (2014). https://doi.org/10.1007/978-3-319-07617-1_49

9. Forgy, E.W.: Cluster analysis of multivariate data: efficiency versus interpretability of classifications. Biometrics **21**, 768–769 (1965)

10. Gonzalez-Cava, J.M., Reboso, J.A., Casteleiro-Roca, J.L., Calvo-Rolle, J.L., Méndez Pérez, J.A.: A novel fuzzy algorithm to introduce new variables in the drug supply decision-making process in medicine. Complexity 2018 (2018)

11. Jove, E., Casteleiro-Roca, J., Quintián, H., Méndez-Pérez, J., Calvo-Rolle, J.: Anomaly detection based on intelligent techniques over a bicomponent production plant used on wind generator blades manufacturing. Rev. Iberoam. Autom. Inform. Ind. **17**(1), 84–93 (2020)

12. Jove, E., Alaiz-Moretón, H., García-Rodríguez, I., Benavides-Cuellar, C., Casteleiro-Roca, J.L., Calvo-Rolle, J.L.: PID-ITS: an intelligent tutoring system for PID tuning learning process. In: Pérez García, H., Alfonso-Cendón, J., Sánchez González, L., Quintián, H., Corchado, E. (eds.) SOCO/CISIS/ICEUTE -2017. AISC, vol. 649, pp. 726–735. Springer, Cham (2018). https://doi.org/10.1007/978-3-319-67180-2_71

13. Jove, E., et al.: Attempts prediction by missing data imputation in engineering degree. In: Pérez García, H., Alfonso-Cendón, J., Sánchez González, L., Quintián, H., Corchado, E. (eds.) SOCO/CISIS/ICEUTE -2017. AISC, vol. 649, pp. 167–176. Springer, Cham (2018). https://doi.org/10.1007/978-3-319-67180-2_16

14. Jove, E., et al.: Missing data imputation over academic records of electrical engineering students. Log. J. IGPL **28**(4), 487–501 (2020)

15. Jove, E., Casteleiro-Roca, J.L., Quintián, H., Méndez-Pérez, J.A., Calvo-Rolle, J.L.: A fault detection system based on unsupervised techniques for industrial control loops. Expert Syst. **36**(4), e12395 (2019)
16. Jove, E., Casteleiro-Roca, J.L., Quintián, H., Simić, D., Méndez-Pérez, J.A., Luis Calvo-Rolle, J.: Anomaly detection based on one-class intelligent techniques over a control level plant. Log. J. IGPL **28**(4), 502–518 (2020)
17. Jove, E., et al.: Modelling the hypnotic patient response in general anaesthesia using intelligent models. Log. J. IGPL **27**(2), 189–201 (2019)
18. Jove, E., Gonzalez-Cava, J.M., Casteleiro-Roca, J.L., Pérez, J.A.M., Calvo-Rolle, J.L., de Cos Juez, F.J.: An intelligent model to predict ANI in patients undergoing general anesthesia. In: Pérez García, H., Alfonso-Cendón, J., Sánchez González, L., Quintián, H., Corchado, E. (eds.) SOCO/CISIS/ICEUTE -2017. AISC, vol. 649, pp. 492–501. Springer, Cham (2018). https://doi.org/10.1007/978-3-319-67180-2_48
19. Konig, O., Jakubek, S., Prochart, G.: Battery impedance emulation for hybrid and electric powertrain testing. In: Vehicle Power and Propulsion Conference (VPPC), 2012 IEEE, pp. 627–632 (October 2012). https://doi.org/10.1109/VPPC.2012.6422636
20. Luis Casteleiro-Roca, J., Quintián, H., Luis Calvo-Rolle, J., Méndez-Pérez, J.A., Javier Perez-Castelo, F., Corchado, E.: Lithium iron phosphate power cell fault detection system based on hybrid intelligent system. Log. J. IGPL **28**(1), 71–82 (2020)
21. Machón-González, I., López-García, H., Calvo-Rolle, J.L.: A hybrid batch SOM-NG algorithm. In: The 2010 international joint conference on neural networks (IJCNN), pp. 1–5. IEEE (2010)
22. Qian, H., Zhang, J., Lai, J.S.: A grid-tie battery energy storage system. In: Control and Modeling for Power Electronics (COMPEL), 2010 IEEE 12th Workshop on, pp. 1–5 (June 2010). https://doi.org/10.1109/COMPEL.2010.5562425
23. Quintian Pardo, H., Calvo Rolle, J.L., Fontenla Romero, O.: Application of a low cost commercial robot in tasks of tracking of objects. Dyna **79**(175), 24–33 (2012)
24. Technical Specification: LiFeBATT X-1P 8Ah 38123 Cell (March 2011). https://www.lifebatt.co.uk/documents/LiFeBATTX-1P8Ah38123CellMarch2011.pdf
25. Tuchler, M., Singer, A.C., Koetter, R.: Minimum mean squared error equalization using a priori information. IEEE Trans. Signal Process. **50**(3), 673–683 (2002)
26. U.S. Departament of Energy: PNGV Battery Test Manual (February 2001). https://avt.inl.gov/sites/default/files/pdf/battery/pngv_manual_rev3b.pdf
27. Vega Vega, R., Quintián, H., Calvo-Rolle, J.L., Herrero, Á., Corchado, E.: Gaining deep knowledge of android malware families through dimensionality reduction techniques. Log. J. IGPL **27**(2), 160–176 (2019)
28. Wen-Yeau, C.: The state of charge estimating methods for battery: a review. ISRN Appl. Math. 1–7 (2013). https://doi.org/10.1155/2013/953792

Distance Metric Learning with Prototype Selection for Imbalanced Classification

Juan Luis Suárez[✉][iD], Salvador García[iD], and Francisco Herrera[iD]

Department of Computer Science and Artificial Intelligence,
Andalusian Research Institute in Data Science and Computational Intelligence,
DaSCI,University of Granada, 18071 Granada, Spain
{jlsuarezdiaz,salvagl,herrera}@decsai.ugr.es

Abstract. Distance metric learning is a discipline that has recently become popular, due to its ability to significantly improve similarity-based learning methods, such as the nearest neighbors classifier. Most proposals related to this topic focus on standard supervised learning and weak-supervised learning problems. In this paper, we propose a distance metric learning method to handle imbalanced classification via prototype selection. Our method, which we have called *condensed neighborhood components analysis* (CNCA), is an improvement of the classic neighborhood components analysis, to which foundations of the condensed nearest neighbors undersampling method are added. We show how to implement this algorithm, and provide a Python implementation. We have also evaluated its performance over imbalanced classification problems, resulting in very good performance using several imbalanced score metrics.

Keywords: Distance metric learning · Imbalanced classification · Nearest neighbors · Neighborhood components analysis · Undersampling

1 Introduction

One of the most important components in many human cognitive processes is the ability to detect similarities between different objects. The idea of finding patterns in data using similarity notions between them was one of the first methods established in the field of machine learning. The k-nearest neighbors (k-NN) classifier [7] was one of the very first methods used in classification problems. It is still widely used today, due to its simplicity and competitive results.

The k-NN classification algorithm classifies a new sample using the majority label of its k-nearest neighbors in the training set. A similarity measure must be

Our work has been supported by the research project PID2020-119478GB-I00 and by a research scholarship (FPU18/05989), given to the author Juan Luis Suárez by the Spanish Ministry of Science, Innovation and Universities.

H. Sanjurjo González et al. (Eds.): HAIS 2021, LNAI 12886, pp. 391–402, 2021.
https://doi.org/10.1007/978-3-030-86271-8_33

introduced to determine the nearest neighbors. Typically, standard distances, like the Euclidean distance, are used to measure this similarity. However, a standard distance may ignore some important properties available in our dataset, meaning that the learning results might be suboptimal.

Distance metric learning tries to find a distance that brings similar data as close as possible, while moving non-similar data far away from each other. The search for suitable distances for each data set can significantly improve the quality of the similarity-based learning algorithms like k-NN. Although some classic statistical algorithms, like *principal components analysis* (PCA) and *linear discriminant analysis* (LDA) [8], can be considered distance metric learning algorithms, the first steps in the development of this discipline were possibly taken in 2002, with the work of Xing et al. on clustering with side information [26]. Since then, a large number of distance metric learning techniques have been proposed to address different problems related to similarity learning algorithms [20], demonstrating the ability of the former to improve the performance of the latter.

Imbalanced classification [11] is a learning problem with applications in many areas, such as disease diagnosis [17, 23]. In its most common version, one class (the positive class) is highly under-represented with respect to the other class (the negative class). Many of the classical classification algorithms tend to focus on the negative class, providing very poor results for the positive class. Some research about distance metric learning with imbalanced data has been already done [10, 14, 24], but these algorithms rely exclusively on the learned distance to handle the class imbalance. This may not be enough in some situations, since the final classification algorithm would still receive an imbalanced dataset unless an additional balancing algorithm is applied afterwards.

In our paper, we propose a new technique to handle both distance learning and imbalanced classification at the same time. Since most methods that deal with imbalanced classification focus on the data and the algorithms, a distance metric learning approach can also provide a new attribute space in which the most important features of our dataset come to light. Our algorithm takes the objective function of one of the best-known distance metric learning algorithms oriented to nearest neighbors classification: *neighborhood components analysis* (NCA) [15]. We also take the key ideas of the *condensed nearest neighbors* (CNN) rule [16], one of the oldest algorithms for prototype selection. We re-design the NCA objective function by adding the main notions used by CNN to perform its prototype selection, obtaining a new algorithm that can simultaneously learn both a distance metric and a set of prototypes, and whose class distribution is much more balanced than that of the original set. The proposed method, which we have called CNCA (*condensed neighborhood components analysis*), learns a linear transformation that maximizes an expected classification accuracy but, unlike NCA, the expected accuracy is defined with respect to a reduced set of samples. Those samples are selected iteratively during the optimization process itself, allowing the linear transformation and the set of prototypes to be learnt at the same time.

We have tested the learned distance and prototypes over a large collection of imbalanced datasets, showing good performances according to several imbalanced score metrics. The good results have been validated through a set of Bayesian statistical tests.

Our paper is organized as follows. Section 2 introduces the distance metric learning and imbalanced classification problems in more detail, together with the main algorithms that inspired our work. Section 3 describes the method we have developed. Section 4 shows the experimental framework and the results obtained by our algorithm. Finally, Sect. 5 summarizes our contributions on this topic.

2   Background

In this section we will discuss the main problems we have addressed in this paper: distance metric learning and imbalanced classification.

2.1   Distance Metric Learning

Distance metric learning [20] is a field of machine learning that aims to learn a distance using the information available in the training data. A distance over a non-empty set $\mathcal{X}$ is a map $d \colon \mathcal{X} \times \mathcal{X} \to \mathbb{R}$ satisfying the following conditions:

1. Coincidence: $d(x, y) = 0 \iff x = y$, for every $x, y \in \mathcal{X}$.
2. Symmetry: $d(x, y) = d(y, x)$, for every $x, y \in \mathcal{X}$
3. Triangle inequality: $d(x, z) \le d(x, y) + d(y, z)$, for every $x, y, z \in \mathcal{X}$.

The distance metric learning theory is mostly focused on numerical datasets, so we will assume that we have a numerical training set, $X = \{x_1, \ldots, x_N\} \subset \mathbb{R}^d$. In this context, the family of Mahalanobis distances is a set of distances over $\mathbb{R}^d$ that can be easily modeled from a computational perspective. We denote the set of d-dimensional matrices as $\mathcal{M}_d(\mathbb{R})$, and the set of d-dimensional positive semidefinite matrices as $S_d(\mathbb{R})_0^+$. A Mahalanobis distance d_M is parameterized by a positive semidefinite matrix $M \in S_d(\mathbb{R})_0^+$ and is defined by

$$d_M(x, y) = \sqrt{(x - y)^T M (x - y)} \ \forall x, y \in \mathbb{R}^d.$$

Since every $M \in S_d(\mathbb{R})_0^+$ can be decomposed as $M = L^T L$, with $L \in \mathcal{M}_d(\mathbb{R})$ (and L is unique except for an isometry), a Mahalanobis distance can also be understood as the euclidean distance after applying the linear map defined by the matrix L. Indeed, we have

$$d_M(x, y)^2 = (x - y)^T L^T L (x - y) = (L(x - y))^T (L(x - y)) = \|L(x - y)\|_2^2.$$

Thus, distance metric learning comes down to learning a metric matrix $M \in S_d(\mathbb{R})_0^+$, or to learning a linear map matrix $L \in \mathcal{M}_d(\mathbb{R})$. Both approaches have advantages and disadvantages. For instance, learning M usually leads to convex optimization problems, while learning L can be used to force a dimensionality reduction in a simple way.

On the supervised context, distance metric learning algorithms try to optimize objective functions that benefit close samples with the same class and penalize those close samples that are differently labeled. *Neighborhood components analysis* (NCA) [15] aims specifically at improving the performance of the 1-NN classifier. It learns a linear transformation that maximizes the expected leave-one-out accuracy by the 1-NN classifier. To do this, a softmax-based probability that x_i has x_j as its nearest neighbor is first defined, for each x_i, x_j in the training set, and for each linear mapping L, as

$$p_{ij}^L = \frac{\exp\left(-\|L(x_i - x_j)\|^2\right)}{\sum_{k \neq i} \exp\left(-\|L(x_i - x_k)\|^2\right)} \quad (j \neq i), \qquad p_{ii}^L = 0. \tag{1}$$

The sum of the same-class probabilities for each sample in the training set is then optimized using a gradient ascent method. The *large margin nearest neighbors* (LMNN) method [25] minimizes the distance of each sample in the training set to several predefined *target neighbors* while keeping the data of other classes outside a margin defined by the target neighbors themselves.

2.2 Imbalanced Classification

A dataset is said to be imbalanced [11] if there is a significant difference between the number of samples representing each class, that is, one or more classes are underrepresented with respect to the rest of the classes in the dataset. In binary problems the underrepresented class is usually referred to as the positive class, while the majority class is referred to as the negative class. The positive class is usually the class of most interest in the problems that machine learning faces. Standard classifiers, however, are usually biased towards the majority class, because the impact of the minority class is often insignificant in their internal optimization mechanisms. In addition, in these problems the classical acurracy score is not a proper measure, since it is able to reach really high scores that can be as high as the proportion of the majority class, without the need to hit on any positive examples.

There are two non-exclusive main approaches to handling imbalanced classification. The first approach is at *algorithm level*. These procedures adapt base learning methods to bias the learning towards the minority class, by modifying sample weights and other parameters [18]. The second approach is at *data level*. These procedures try to balance the class distribution in the training set by removing training instances from the majority classes (undersampling), by generating new instances of the minority classes (oversampling), or combining oversampling and undersampling. For oversampling, a well-known technique that generates new samples in the line between a minority instance and one of its nearest neighbors is SMOTE [6,12]. For undersampling, the most popular techniques try to remove majority class samples by removing samples that are close to the decision boundary (like Tomek links [21]) or that do not have remarkable effects on that boundary. In the latter, the *condensed nearest neighbors* (CNN) algorithm [16] finds a set of prototypes for which the 1-NN classification of the original dataset remains invariant. This reduces the storage requirements that the nearest neighbors rule needs, and it can also be used to balance the class distribution in a classification problem. The set of prototypes that CNN finds can be minimal by considering the *reduced nearest neighbors* extension of the algorithm [13]. Although CNN performs an instance selection that is oriented to nearest neighbors classification, it is very sensitive to noise at the decision boundary. Several modifications of CNN have been proposed to overcome this issue [5,9]. In our algorithm, we will see that the CNN selection mechanism can be modeled in a probabilistic way to be robust to boundary noise.

3   Condensed Neighborhood Components Analysis

As we have previously mentioned, CNCA learns a linear transformation that maximizes an expected classification accuracy but, unlike NCA, the expected accuracy is defined with respect to a reduced set of samples. Those samples are selected iteratively during the optimization process itself, allowing the linear transformation and the set of prototypes to be learnt simultaneously. The decision to use NCA as the starting point for our algorithm is due to the good results it provided in a previous study [20] in which we compared multiple distance metric learning techniques.

In addition to the possible advantages that the set of prototypes provides as an undersample of the original set, it also determines the learned distance using a smaller (and more relevant) set. Since the linear map is determined by the image of a basis in the data space, by reducing the dataset to prototypes there is a better chance that most of them will belong to a basis, contributing actively to the optimization of the linear map or distance.

Let $X = \{x_1, \ldots, x_N\} \subset \mathbb{R}^d$ be our training set, with corresponding labels $y_1, \ldots, y_N$. We denote the set of (for the moment, arbitrary) prototypes (also called condensed neighbors) with $\mathcal{C}$. Given a sample $x_i \in X$, the classic NCA algorithm p_i^L provides the probability that x_i is correctly classified using the set $X - \{x_i\}$ as

$$p_i^L = \sum_{j \in C_i} p_{ij}^L, \text{ where } C_i = \{j \in \{1, \ldots, N\} \colon y_j = y_i\}. \tag{2}$$

where p_{ij}^L are as defined in 1. We can easily adapt p_i^L to obtain a probability that x_i is correctly classified using the set $\mathcal{C} - \{x_i\}$. We define this new probability as

$$q_i^L = \sum_{y_j = y_i, x_j \in \mathcal{C} - \{x_i\}} q_{ij}^L, \tag{3}$$

where q_{ij}^L are the probabilities that a sample $x_i \in X$ has the sample $x_j \in C$ as its nearest neighbor (in C). It is also redefined from Eq. 1 as the softmax function

$$q_{ij}^L = \frac{\exp\left(-\|L(x_i - x_j)\|^2\right)}{\sum\limits_{x_k \in C - \{x_i\}} \exp\left(-\|L(x_i - x_k)\|^2\right)} \quad (j \neq i), \qquad q_{ii}^L = 0.$$

The functions above are differentiable (in L), therefore the expected classification accuracy using C, given by

$$g(L) = \sum_{i=1}^{N} q_i^L,$$

is also differentiable. To optimize it we can use a stochastic gradient ascent method, in which each sample x_i modifies the linear map L according to the update rule

$$L \leftarrow L + 2L\eta \left(q_i^L \sum_{x_k \in C}^{N} q_{ik}^L O_{ik} - \sum_{y_j = y_i, x_j \in C} q_{ij}^L O_{ij} \right), \tag{4}$$

where η is the learning rate, and $O_{ij} = (x_i - x_j)(x_i - x_j)^T$ is the outer product of the difference of samples x_i and x_j, for each $i, j = 1, \ldots, N$. This update rule will contribute to the maximization of g when C is a constant set.

However, as we have already pointed out, we are interested in finding an appropriate set C, that is, we want to make it adaptive while the gradient optimization is performing. Our proposal consists in determining if x_i is a candidate for condensed neighbor for the current distance L each time a sample x_i updates L during the gradient optimization. If so, it is added to C. Otherwise, it can be removed from C if it was already. Since the modifications of C occur sporadically, the expected result of the gradient optimization is still a local increase of the objective function.

The criterion used to decide whether to add or remove x_i from C is based on the CNN selection rule, using the probabilistic approach that follows throughout the section. If x_i is expected to be misclassified (that is, $q_i^L < \alpha$, where we tipically choose $\alpha = 0.5$), we add it to the set of condensed neighbors. Unlike CNN, the probabilistic approach also allows us to remove redundant samples from the prototype set. If x_i is expected to be correctly classified with high confidence (that is, $q_i^L > \beta$, where $\beta \geq \alpha$), we can remove it from the prototype set. We have observed that the set C tends to stabilize when several gradient ascent iterations have been carried out, resulting in the final condensed neighbors set.

However, this first approach is still vulnerable to the noise in the decision boundary, just as CNN was. To solve this problem, we rely on the probabilistic setup of our model again. We will catalog the sample x_i as noisy if it is expected to be misclassified with high probability, that is, $q_i < \varepsilon$. Noisy samples will never be added to the condensed neighbors set. Now, according to conditional probability laws, a non-noisy sample expected to be misclassified (that is, $\varepsilon < q_i^L < \varepsilon + (1 - \varepsilon)\alpha$) will be added to the prototypes set, and a non-noisy sample expected to be correctly classified with high confidence (that is, $q_i^L > \varepsilon + (1-\varepsilon)\beta$) will be removed from the prototypes set. The adaptive nature of the algorithm allows us to choose a high value for ε (usually 0.5 is a good choice) to get rid of noisy borderline samples.

We have to highlight that the described method performs an undersampling over every class in the dataset, regardless of whether it is majority or minority. When there are highly underrepresented classes, in order to avoid more minority samples to be removed, we can simply force all the samples in those classes to enter and not leave the condensed neighbors set. Then, the undersampling will only act on the majority classes.

We also have observed that, when classes are clearly separable, our method is capable of selecting a single quality prototype per class. This may not be enough for classifiers like k-NN with high values of k. In this case, we complete the instances of single-prototype classes by adding same-class neighbors of those prototypes, until a minimum threshold is reached.

We also highlight that our method works with an arbitrary number of classes, so it can be used beyond binary imbalanced problems. The algorithm is implemented in our Python package pyDML [19], a library containing the most relevant distance metric learning algorithms.

4   Experiments

In this section we describe the experiments we have developed to analyze CNCA performance. The section begins by describing the experimental framework. Then, the results of the experiments are shown. Finally, we analyze the results, with the support of several Bayesian tests conducted to validate the performance of the algorithm.

4.1   Experimental Framework

We have evaluated the distances learned by CNCA using the 1-NN classifier. We have compared it with the 1-NN results with Euclidean distance, the distance learned by NCA and the distance learned by IML [24]. This last algorithm is another metric-learning proposal that applies a local undersampling to deal with imbalanced datasets. It applies LMNN iteratively until a balanced neighborhood of the sample to classify is stable. Then, it performs the classification using this neighborhood. All these algorithms are available in our Python software, pyDML [19].

The classifier with the different distances will be evaluated by a balanced 5-fold cross validation, that is, a cross validation keeping the original class proportions in each fold. We have used 30 binary imbalanced datasets collected from KEEL[1]. All these datasets are numeric, without missing values, and have been

---

[1] KEEL, *knowledge extraction based on evolutionary learning* [22]: http://www.keel.es/.

min-max normalized to the interval $[0, 1]$ prior to the execution of the experiments. The imbalanced ratios (IR, measured as the quotient of the number of negative instances by the number of positive instances) of the datasets vary from 1.86 to 85.88. The datasets and their dimensions, imbalanced ratios and additional descriptions can be found in the KEEL online repository (see footnote 1).

We have chosen some of the most popular metrics for imbalanced classification [3]: the *balanced score* (BSC), which is the geometric mean between the true positive rate and the true negative rate, the *F-score* (F1), given as the harmonic mean between the *precision* and *recall*, and the *area under the ROC curve* (AUC).

All the algorithms have been executed with their default parameters, following the guidelines of the original creators of the different proposals. For CNCA, we have established the default parameters as $\varepsilon = 0.5, \alpha = 0.5$ and $\beta = 0.95$, and the same gradient ascent parameters as those used in NCA. We have also performed undersampling on both classes unless the positive class proportion is lower than 30 %. In this case, we have only undersampled the negative class. When undersampling has produced a very low number of prototypes, we have filled the class with neighbors until 10 instances per class are reached. Since the evaluation in the experimental stage makes use of a high number of datasets, tuning each parameter specifically for each dataset is not feasible. In addition, given that the purpose of this work is to draw a fair comparison between the algorithms and assess their robustness in a common environment with multiple datasets, we have not included a tuning step to maximize any particular performance metric.

4.2   Results

This section describes the results obtained for the different metrics used. Table 1 shows the BSC, F1 and AUC results of the algorithms over the 30 selected datasets. In these tables we also include the average score obtained by the algorithms in each of the metrics, and the average ranking. This ranking has been made by assigning integer values starting from 1 (adding half fractions in case of a tie), according to the position of the algorithms over each dataset.

4.3   Analysis of Results

In the results of the previous section we can see that CNCA usually provides better results in all the chosen metrics. In order to support this observation, we have developed a series of Bayesian statistical tests to assess the extent to which the distance and prototypes learned by CNCA outperforms the other distances. To do this, we have compared CNCA against each of the other algorithms using Bayesian sign tests [1].

The Bayesian sign tests consider the differences between the score metrics (BSC, F1 and AUC) of CNCA and each of the compared methods, assuming that their prior distribution is a Dirichlet Process [2], defined by a prior strength $s = 1$

Table 1. Balanced accuracy, F1 and AUC scores for the algorithms in each dataset.

	BSC				F1				AUC			
	EUC	NCA	IML	CNCA	EUC	NCA	IML	CNCA	EUC	NCA	IML	CNCA
ecoli-0-vs-1	0.961756	0.972691	0.961480	**0.982833**	0.952868	0.966813	0.952321	**0.980146**	0.962865	0.972980	0.962865	**0.983095**
glass-0-1-6_vs_2	0.352886	0.367053	0.251975	**0.585826**	0.263030	0.237143	0.169697	**0.287778**	0.579524	0.582143	0.557381	**0.645238**
glass2	0.344713	0.550550	0.341252	**0.575474**	0.228889	0.373016	0.191111	**0.380000**	0.570000	0.670192	0.570000	**0.690064**
glass4	0.798628	**0.928777**	0.887058	0.883252	0.671429	**0.793333**	0.691746	0.676667	0.820833	**0.937500**	0.896728	0.894167
haberman	0.462212	0.535641	0.518031	**0.594834**	0.291149	0.368080	0.359905	**0.436160**	0.519690	0.558815	0.572794	**0.605907**
iris0	**1.000000**	**1.000000**	**1.000000**	**1.000000**	**1.000000**	**1.000000**	**1.000000**	**1.000000**	**1.000000**	**1.000000**	**1.000000**	**1.000000**
pima	0.657909	0.650308	0.608307	**0.679870**	0.566571	0.555729	0.504360	**0.605322**	0.672737	0.657247	0.630287	**0.708521**
winequality-red-3_vs_5	0.000000	0.000000	0.000000	**0.140900**	0.000000	0.000000	0.000000	**0.100000**	0.494123	0.494853	0.495588	**0.545588**
yeast-2_vs_4	**0.848106**	0.826762	0.798411	0.836520	**0.767807**	0.713651	0.680626	0.717668	**0.861185**	0.839433	0.817615	0.846516
yeast3	0.796700	0.834972	0.804296	**0.848344**	0.664614	**0.702679**	0.694199	0.683596	0.810715	0.843578	0.818291	**0.854004**
shuttle-c0-vs-c4	**0.995959**	**0.995959**	**0.995959**	**0.995959**	**0.995918**	**0.995918**	**0.995918**	**0.995918**	**0.996000**	**0.996000**	**0.996000**	**0.996000**
winequality-red-4	0.266329	**0.350002**	0.118736	0.328329	0.111625	**0.155699**	0.037229	0.114378	0.535368	**0.556274**	0.505245	0.541430
winequality-white-3-9_vs_5	0.178425	0.304262	0.089136	**0.464721**	0.107143	0.187143	0.050000	**0.240904**	0.537253	0.576911	0.515880	**0.629706**
yeast4	0.565627	0.542810	**0.626625**	0.544547	0.352799	0.298249	**0.407306**	0.250611	0.657149	0.642754	**0.694634**	0.636338
yeast5	**0.818932**	0.780894	0.740787	0.816626	**0.676060**	0.596957	0.591462	0.546127	0.835069	0.808333	0.777083	**0.837500**
yeast-1-4-5-8_vs_7	0.273868	0.318154	0.273380	**0.389598**	**0.141503**	0.121616	0.130808	0.114072	**0.553868**	0.544788	0.553845	0.538830
yeast-1-2-8-9_vs_7	0.323914	**0.516359**	0.308881	0.501227	0.167133	**0.313944**	0.168681	0.155981	0.554119	**0.671900**	0.570239	0.607487
ecoli4	0.856683	0.913872	0.858063	**0.915085**	0.782540	0.840635	0.803968	**0.852857**	0.870238	0.918676	0.871825	**0.920288**
ecoli-0-1-3-7_vs_2-6	0.734609	0.734609	**0.736436**	0.734609	0.560000	0.566667	**0.633333**	0.566667	0.842694	0.842694	**0.844512**	0.842694
glass5	0.792544	0.529776	0.736482	**0.932293**	0.714286	0.374286	0.666667	**0.793333**	0.892683	0.735366	0.845122	**0.940244**
glass-0-1-6_vs_5	0.870097	**0.935666**	0.797122	0.797122	0.693333	**0.853333**	0.760000	0.760000	0.885714	**0.944286**	0.894286	0.894286
shuttle-c2-vs-c4	**0.941421**	0.937381	0.937381	0.937381	**0.933333**	0.866667	0.866667	0.866667	**0.950000**	0.946000	0.946000	0.946000
poker-8-9_vs_5	0.089005	0.215297	0.088896	**0.277522**	0.040000	**0.133333**	0.036364	0.108683	0.517317	0.556585	0.517317	**0.587073**
poker-8_vs_6	0.735318	0.624207	**0.794042**	0.619789	0.671111	0.521429	**0.740000**	0.363333	0.782991	0.747945	**0.832991**	0.743151
winequality-white-3_vs_7	0.200000	0.200000	0.199144	**0.408539**	0.160000	0.160000	0.130000	**0.216234**	0.548864	0.548864	0.547727	**0.640341**
winequality-red-8_vs_6-7	0.199400	0.198195	0.198495	**0.335535**	0.133333	0.101587	0.111111	**0.144444**	0.544033	0.541645	0.542233	**0.580892**
winequality-red-8_vs_6	0.339519	0.297240	**0.451087**	0.376620	0.237143	0.164286	**0.243175**	0.202727	0.595300	0.567163	**0.622365**	0.610894
dermatology-6	**1.000000**	**1.000000**	0.973205	**1.000000**	**1.000000**	**1.000000**	0.971429	**1.000000**	**1.000000**	**1.000000**	0.975000	**1.000000**
shuttle-6-vs-2-3	**0.941421**	**0.941421**	**0.941421**	**0.941421**	**0.933333**	**0.933333**	**0.933333**	**0.933333**	**0.950000**	**0.950000**	**0.950000**	**0.950000**
led7digit-0-2-4-5-6-7-8-9_vs_1	0.854219	0.502241	0.501096	**0.884075**	0.596384	0.320000	0.401111	**0.671335**	0.858058	0.624905	0.632694	**0.888318**
AVERAGE SCORE	0.606673	0.616837	0.584573	**0.677628**	0.513778	0.507184	0.497418	**0.525498**	0.739946	0.742594	0.731885	**0.770152**
AVERAGE RANK	2.600000	2.466667	3.133333	**1.800000**	2.566667	2.416667	2.866667	**2.150000**	2.750000	2.533333	2.850000	**1.866667**

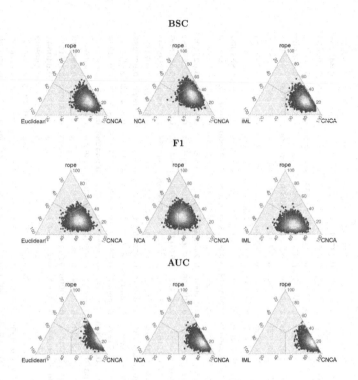

Fig. 1. Bayesian sign test results for the comparison between CNCA against Euclidean distance, NCA and IML using the BSC (first row), F1 (second row) and AUC (third row). Simplex diagrams are shown.

and a prior pseudo-observation $z_0 = 0$. After observing the results, the resulting posterior distribution gives us the probabilities that one algorithm outperforms the other. A *rope* (region of practically equivalent) region is also introduced in the posterior distribution so that the algorithms are considered to have an equivalent performance if their score differences are found in the interval $[-0.01, 0.01]$. Then, the probability that CNCA outperforms the compared algorithm, the probability that the compared algorithm outperforms CNCA and the probability that both algorithms are equivalent can be visualized in a simplex plot of a sample of the posterior distribution, in which a greater tendency of the points towards one of the regions will represent a greater probability.

We have carried out the Bayesian sign tests using the rNPBST library [4]. Figure 1 shows the results of the comparisons using the metrics BSC, F1 and AUC, respectively. For BSC metric, we can see that the samples in all the simplex graphics are clearly concentrated near the right vertex, which determines the CNCA outperforming region. Therefore, CNCA has high probabilities of outperforming each of the other algorithms according to BSC. NCA is the only one that tries to mildly face up to CNCA, slightly biasing the distribution towards the rope. According to AUC, the CNCA outperformance is even more clear, with very high probabilities against all the algorithms tested. The F-Score results present a bit more equality, with higher probabilities for both the rope and the non-CNCA regions, with respect to the other metrics, but all the comparisons are still slightly biased to the CNCA side.

5 Conclusion

In this paper, we have developed a distance metric learning algorithm which shows for the first time how the distance learning process can also be used for prototype selection. The experimental results over a large collection of imbalanced datasets demonstrate that CNCA can outperform NCA, the algorithm on which it is inspired, and it also performs well compared to other distance metric learning proposals for imbalanced classification. The distance-based prototype selection of this approach allows us, beside selecting representative samples according to the similarity to the rest of the dataset, to adapt the distance taking into account those samples, resulting in an effective and efficient (within distance metric learning) preprocessing for imbalanced similarity-based learning problems. Possible future avenues of research involve further experimentation, including hyperparameter analysis, and the possible adaptation of the algorithm to handle very large datasets, taking advantage of the parallelization capability of the method.

Compliance with ethical standards

Declaration of competing interest. The authors declare that there is no conflict of interest.

References

1. Benavoli, A., Corani, G., Demšar, J., Zaffalon, M.: Time for a change: a tutorial for comparing multiple classifiers through bayesian analysis. J. Mach. Learn. Res. **18**(1), 2653–2688 (2017)
2. Benavoli, A., Corani, G., Mangili, F., Zaffalon, M., Ruggeri, F.: A bayesian wilcoxon signed-rank test based on the dirichlet process. In: International Conference on Machine Learning, pp. 1026–1034 (2014)
3. Branco, P., Torgo, L., Ribeiro, R.P.: A survey of predictive modeling on imbalanced domains. ACM Comput. Surv. (CSUR) **49**(2), 1–50 (2016)
4. Carrasco, J., García, S., del Mar Rueda, M., Herrera, F.: rNPBST: an r package covering non-parametric and Bayesian statistical tests. In: Martínez de Pisón, F.J., Urraca, R., Quintián, H., Corchado, E. (eds.) HAIS 2017. LNCS (LNAI), vol. 10334, pp. 281–292. Springer, Cham (2017). https://doi.org/10.1007/978-3-319-59650-1_24
5. Chang, F., Lin, C.C., Lu, C.J.: Adaptive prototype learning algorithms: theoretical and experimental studies. J. Mach. Learn. Res. **7**(10), 2125–2148 (2006)
6. Chawla, N.V., Bowyer, K.W., Hall, L.O., Kegelmeyer, W.P.: Smote: synthetic minority over-sampling technique. J. Artif. Intell. Res. **16**, 321–357 (2002)

7. Cover, T.M., Hart, P.E., et al.: Nearest neighbor pattern classification. IEEE Trans. Inf. Theory **13**(1), 21–27 (1967)
8. Cunningham, J.P., Ghahramani, Z.: Linear dimensionality reduction: survey, insights, and generalizations. J. Mach. Learn. Res. **16**(1), 2859–2900 (2015)
9. Devi, V.S., Murty, M.N.: An incremental prototype set building technique. Pattern Recognit. **35**(2), 505–513 (2002)
10. Feng, L., Wang, H., Jin, B., Li, H., Xue, M., Wang, L.: Learning a distance metric by balancing kl-divergence for imbalanced datasets. IEEE Trans. Syst. Man Cybern. Syst. **99**, 1–12 (2018)
11. Fernández, A., García, S., Galar, M., Prati, R.C., Krawczyk, B., Herrera, F.: Learning from Imbalanced Data Sets. Springer, Heidelberg (2018)
12. Fernández, A., Garcia, S., Herrera, F., Chawla, N.V.: Smote for learning from imbalanced data: progress and challenges, marking the 15-year anniversary. J. Artif. Intell. Res. **61**, 863–905 (2018)
13. Gates, G.: The reduced nearest neighbor rule (corresp.). IEEE Trans. Inf. Theory **18**(3), 431–433 (1972)
14. Gautheron, L., Habrard, A., Morvant, E., Sebban, M.: Metric learning from imbalanced data with generalization guarantees. Pattern Recognit. Lett. **133**, 298–304 (2020)
15. Goldberger, J., Hinton, G.E., Roweis, S., Salakhutdinov, R.R.: Neighbourhood components analysis. Adv. Neural Inf. Process. Syst. **17**, 513–520 (2004)
16. Hart, P.: The condensed nearest neighbor rule (corresp.). IEEE Trans. Inf. Theory **14**(3), 515–516 (1968)
17. Li, Z., Zhang, J., Yao, X., Kou, G.: How to identify early defaults in online lending: a cost-sensitive multi-layer learning framework. Knowl.-Based Syst. **221**, 106963 (2021)
18. Lin, Y., Lee, Y., Wahba, G.: Support vector machines for classification in nonstandard situations. Mach. Learn. **46**(1–3), 191–202 (2002)
19. Suárez, J.L., García, S., Herrera, F.: pyDML: a python library for distance metric learning. J. Mach. Learn. Res. **21**(96), 1–7 (2020)
20. Suárez, J.L., García, S., Herrera, F.: A tutorial on distance metric learning: mathematical foundations, algorithms, experimental analysis, prospects and challenges. Neurocomputing **425**, 300–322 (2021)
21. Tomek, I.: Two modifications of cnn. IEEE Trans. Syst. Man Cybern. **6**, 769–772 (1976)
22. Triguero, I., et al.: Keel 3.0: an open source software for multi-stage analysis in data mining. Int. J. Comput. Intell. Syst. **10**, 1238–1249 (2017)
23. Wang, H., Xu, Y., Chen, Q., Wang, X.: Diagnosis of complications of type 2 diabetes based on weighted multi-label small sphere and large margin machine. Appl. Intell. **51**(1), 223–236 (2020). https://doi.org/10.1007/s10489-020-01824-y
24. Wang, N., Zhao, X., Jiang, Y., Gao, Y.: Iterative metric learning for imbalance data classification. In: IJCAI, pp. 2805–2811 (2018)
25. Weinberger, K.Q., Saul, L.K.: Distance metric learning for large margin nearest neighbor classification. J. Mach. Learn. Res. **10**(2), 207–244 (2009)
26. Xing, E., Jordan, M., Russell, S.J., Ng, A.: Distance metric learning with application to clustering with side-information. Adv. Neural Inf. Process. Syst. **15**, 521–528 (2002)

Enhancing Visual Place Inference via Image Ranking Using Fuzzy Similarity and Agreeability

Piotr Wozniak[2] and Bogdan Kwolek[1(✉)]

[1] AGH University of Science and Technology, 30 Mickiewicza, 30-059 Kraków, Poland
bkw@agh.edu.pl
[2] Rzeszów University of Technology, Al. Powstańców Warszawy 12,
35-959 Rzeszów, Poland
http://home.agh.edu.pl/~bkw/contact.html

Abstract. We present a fuzzy approach to visual place recognition. Our approach consists in designing a hierarchical fuzzy system to leverage matching between query and a given image while taking into account agreeability between different feature extractors, and introducing a fuzzy ranking on the basis of matching and agreeability in order to permit re-ranking of top-ranked images. Fuzzy ranking uses fuzzy similarity and agreeability on features determined by three different CNNs. For each of them the cosine similarity between global features representing an examined image and a query image is calculated. The cosine scores are then fed to a similarity FIS. They are also fed to an agreeability FIS, which operates on linguistic variables representing the agreeability between three cosine scores. The outcomes of Mamdani fuzzy inference systems are fed to a ranking Sugeno-type FIS. The mAP scores achieved by the proposed FIS were compared with mAP scores achieved by the similarity FIS. The algorithm has been evaluated on a large dataset for visual place recognition consisting of both images with severe (unknown) blurs and sharp images with 6-DOF viewpoint variations. Experimental results demonstrate that the mAP scores achieved by the proposed algorithm are superior to results achieved by NetVLAD as well as an algorithm combining crisp outcomes of CNNs that were investigated in this work.

Keywords: Fuzzy similarity and agreeability · Fuzzy ranking · Place recognition

1 Introduction

Visual place recognition (VPR) aims at matching a view of a place with a different view of the same place taken at a different time [1,2]. VPR is becoming an essential component in artificial intelligence, enabling autonomous machines, agents and augmented reality devices a better perceiving and understanding the physical world. Although most of VPR methods have been developed for generating drift-free maps in simultaneous localization and mapping (SLAM), they

© Springer Nature Switzerland AG 2021
H. Sanjurjo González et al. (Eds.): HAIS 2021, LNAI 12886, pp. 403–414, 2021.
https://doi.org/10.1007/978-3-030-86271-8_34

also found applications in several areas, including monitoring of electricity pylons using aerial imagery [3], brain-inspired navigation [4], self-driving cars [5] and image search with an aggregation across single and multiple images [6]. The VPR gets challenging if the visual appearance of places undergoing matching changed in the meantime. The visual differences can be caused by sensor viewpoints and environmental changes, including lighting, shadows as well as changes resulting from different passing the same route by an autonomous robot. The amount of viewpoint variations during scene perception by a humanoid robot is far more complex in comparison to variations experiencing by mobile robots over visual navigation [7].

Visual place recognition is typically formulated as an image retrieval problem. Differences and similarities between VPR and the image retrieval are outlined in a recent survey [8]. Several local and global feature descriptors have been proposed for place recognition [1,9]. Drawing on the advantages of region-based approaches, a system that uses a next convolutional layer as a landmark detector and an earlier one for creating local descriptors in order to match the detected landmarks has been proposed in [10]. The discussed system showed an improved place recognition under severe viewpoint and condition variations. In [11] the performance of data-driven image descriptors in visual place recognition has been studied. A VLAD [12] layer for CNN architecture that could be trained in end-to-end fashion has been proposed in [11]. A comprehensive comparison of ten VPR systems in [11] identified the NetVLAD as the best overall performing approach. A review of VPR in mobile robotics by AI techniques and visual information has recently been published in [2]. In [13] an algorithm for place recognition on blurry images acquired by a humanoid robot has been proposed. A graph-based inference on deep embeddings and blur detections permit visual place recognition and estimating decision confidences.

Generally, fuzzy inference aims at interpreting the values in the input vector and, based on fuzzy set theory and some set of rules, assigning values to the output space. Fuzzy inference systems have successfully been applied in areas such as automatic control, expert systems, data classification, and computer vision [14]. Fuzzy systems usually utilize type-1 fuzzy sets (T1FS), representing uncertainty by numbers in the range [0, 1]. However, the T1FS are not capable of modeling uncertainty, which exists in most real-world applications. Recently, more attention is devoted to fuzzy systems based on type-2 fuzzy sets, which have better ability to accommodate higher levels of uncertainties in the fuzzy system's parameters than type-1 fuzzy sets [15]. In practice, the interval type-2 fuzzy sets are commonly employed for their reduced computational cost. Type-2 fuzzy systems better handle uncertainties than type-1 fuzzy systems [16]. Among others, interval type-2 fuzzy logic systems can handle uncertainties arising due to membership function parameters as well as noisy measurements. The most two common inference methods are Mamdani and Takagi-Sugeno (T-S), which have different consequent of fuzzy rules. The Mamdani inference method has a better interpretation ability, whereas the T-S system has a better approximation accuracy.

One of the strengths of fuzzy inference systems is their interpretability [17]. However, due to the curse of dimensionality, also known as rule explosion, the number of required rules commonly increases exponentially with the number of input variables, which in turn may reduce the interpretability of fuzzy logic systems. In order to cope with rule explosion a hierarchical fuzzy system can be employed in which the fuzzy logic system is decomposed into a number of fuzzy logic subsystems. The rules in such subsystems commonly have antecedents with fewer variables than the rules in the fuzzy logic system with equivalent function, since the number of input variables in each component subsystem is smaller.

In general, image retrieval systems deliver a set of relevant images from large repositories in response to a submitted query image. Algorithms for visual place recognition are based on standard image retrieval pipeline, with a preliminary stage that retrieves the closest candidates to a query image from a previously stored dataset of places, and a subsequent stage in which the retrieved relevant images are re-ranked. The goal of the second stage is to order images by decreasing similarity with the query image. Ranking losses permit learning relative distances between samples. The triplet loss is one of the most commonly utilized ranking losses. Recently, a multi-scale triplet loss for place recognition has been proposed in [18]. As optimizing ranking losses does not guarantee finding the best mAP, a few recent works directly optimize the mAP by leveraging a listwise loss [19, 20].

In this work we propose a hierarchical fuzzy system to get better the mean Average Precision (mAP) of place recognition by using multiple neural networks for feature embedding and then carrying out fuzzy ranking. For the query image and a given image the global feature vectors are extracted by the netVLAD, resNet50 and GoogleNet neural networks. Then, cosine similarities between features representing the query image and given image are calculated. Afterwards, fuzzy similarity between the features representing the query image and given image is determined using a Mamdani fuzzy inference system. At the same time, fuzzy agreeability between the image representations extracted by neural networks is determined by a Mamdani fuzzy inference system. The outputs of the Mamdani systems are fed to a Sugeno-type fuzzy system that is responsible for determining values for re-ranking the images with respect to the similarity with the query image.

2   Input Data for Fuzzy Inference

The place recognition algorithms work by converting images to local or global descriptors and performing a search over previous descriptors for a likely match. Fuzzy inference is the process of expressing the mapping from a given input to an output using fuzzy logic. The proposed fuzzy inference system for place recognition takes the cosine similarities as inputs and computes an ordered list of most relevant places to the query image, see Fig. 1 for an illustrative explanation of the computation of the cosine distance as the visual similarity for the fuzzy inference. The cosine distance is calculated between a global feature vector representing the query image and a global feature vector representing an image for matching. The images to be matched with a given query image are retrieved from the image repository. In [13] far better results have been achieved using information about motion blur detections. In this work, in the same way as in the work mentioned above, only sharp images, i.e. blur-free ones are utilized in the place inference. This means that images with the image label set by the blur detector as blur-free are only utilized in the place recognition.

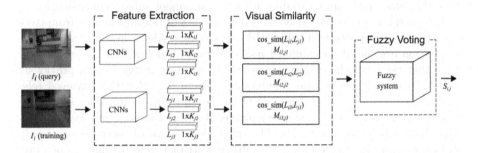

Fig. 1. Schematic diagram illustrating the process of determining the visual similarity for fuzzy place inference.

The global feature vectors have been extracted using deep convolutional neural networks: NetVLAD, ResNet50 and GoogleNet. The NetVLAD features have been extracted using VGG-M network trained on the TokyoTM dataset, GoogleNet trained on Places-365 dataset and ResNet50 trained on the ImageNet dataset, respectively. The NetVLAD features have been determined using the VGG-M network. The size of the feature vector extracted upon pool5-7x7_s1 layer is 1×4096. The size of the GoogleNet-based feature vector is 1×1024 and the features have been extracted from pool5-7x7_s1 layer. The ResNet50-based feature is of size 1×2048 and the features have been extracted from Global-AveragePooling2DLayer, avg_pool layers. The neural networks operate on RGB images of size 640×480 pixels.

3 Proposed Fuzzy Inference Engine

At the beginning of this Section we outline Mamdani as well as Takagi-Sugeno fuzzy models. Afterwards, in Subsect. 3.2 we present our fuzzy engine for place inference.

3.1 Mamdani and Takagi-Sugeno Fuzzy Models

Fuzzy inference system is a rule-based system relying on fuzzy set theory and reasoning that converts the input value into a fuzzy value, applies the rules, and converts the result into a crisp value. It comprises four components: fuzzifier, rule-base, inference engine and defuzzifier. At first, crisp inputs are fuzzified into fuzzy input sets. In this step the fuzzifier maps a crisp number into a fuzzy domain defined by a fuzzy set. The rule-base, which is the heart of fuzzy logic systems is completed by information given by experts or information extracted from data. In the inference engine, fuzzy logic principles are employed in order to combine fuzzy IF-THEN rules from the fuzzy rule-base into a mapping from fuzzy input sets to fuzzy output sets. The IF-part of a rule is its antecedent, whereas the THEN-part of the rule is its consequent. Finally, a defuzzifier determines a crisp output of the fuzzy system on the basis of the fuzzy output set(s) from the inference engine. The aim of defuzzification is to present the output of the reasoning in human understandable form of the fuzzy system. The most popular defuzzification methods are centroid (or centre of gravity), height, and modified height defuzzification. There are two main types of fuzzy inference systems: Mamdani and Takagi-Sugeno. The Mamdani method expects the output membership functions to be fuzzy sets, and thus a defuzzification stage in the inference process in order to convert the fuzzy output into a crisp output is usually executed. The membership functions in the Takagi-Sugeno method are usually singletons. The T-S method employs a combination of linear systems to approximate a nonlinear system.

In this paper we consider only multi-input single-output (MISO) fuzzy models, which are interpreted by rules with multi-antecedent and single-consequent variables. Let us assume that a FLS has M inputs $\{x_m\}_{m=1,2,...,M}$ and a single output y. Let us assume also that an mth input has N_m membership functions (MFs) in its universe of discourse $\mathbb{X}_m$. A membership function (MF) is a two-dimensional function that defines the degree of association of a numeric value under the respective linguistic label using a crisp number assuming values between $[0, 1]$. Let us denote the nth MF in the mth input domain as X_{mn}. A complete rule-base with all possible combinations of the inputs MFs comprises $I = \prod_{m=1}^{M} N_m$ rules, which assume the following form: r^i : IF x_1 is $X_{1,n1i}$ and x_M is $X_{M,nMi}$, then y is y_i, $n_{ki} = 1, 2, ..., N_k$, $i = 1, 2, ..., I$, where y is the rule output. For an input $\mathbf{x} = (x_1, x_2, ..., x_M)$, under assumption that y_i is a constant, which in general assumes different values for different rules, the output of the system can be determined in the following manner: $y(\mathbf{x}) = \sum_{i=1}^{I} f_i y_i / (\sum_{i=1}^{I} f_i)$, where f_i is the firing level of $\mathbf{x}$ for the ith rule, computed as an intersection (t-norm), i.e. $f_i = \mu_{X_{1,n_{1i}}}(x_1) \star \mu_{X_{2,n_{2i}}}(x_2) \star \cdots \star \mu_{X_{M,n_{Mi}}}(x_M)$. The linguistic terms can assume several different shapes such as triangular, trapezoidal or Gaussian.

3.2   Fuzzy Engine for Place Inference

The number of rules increases exponentially with the increase in the number of variables. To cope with this rule-explosion problem, a popular strategy is to hierarchically decompose a fuzzy system into a number of low-dimensional fuzzy systems, i.e., to build a hierarchical fuzzy system. In this work we built a hierarchical fuzzy system for visual place recognition by combining several two-input, one-output fuzzy systems. In the proposed system, input values are incorporated as groups at the lower level, where each input group is fed into a fuzzy inference system. The outputs of the lower level fuzzy systems are aggregated by a higher level fuzzy system. In the first level, two fuzzy systems are employed to infer similarity and agreeability between the image features, respectively, whereas the second level includes a FIS that operates on such outputs and performs re-ranking the images with respect to the similarity with the query image, see Fig. 2.

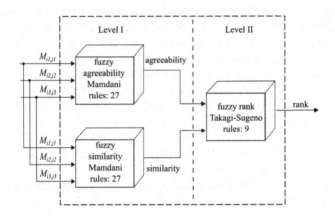

Fig. 2. Schematic diagram of fuzzy system for visual place recognition.

The systems responsible for inference similarity and agreeability operate on three cosine similarities, which are determined on features determined by the NetVLAD, ResNet50 and GoogleNet networks. The system for inference of the similarity has three inputs. In order to describe the similarity between query image and a given image, three linguistic variables corresponding to cosine similarities have been defined: sNetVLAD, sResNet50 and sGoogleNet. They are described by three fuzzy sets: high, medium, and low. Figure 3a depicts the plot of the membership functions for the discussed linguistic variables. The similarity membership functions are described by Gaussian curve, which has the advantage of being smooth and nonzero at all points. The parameters of the membership functions have been tuned manually. Table 1 contains fuzzy rules for modeling the fuzzy similarity.

The fuzzy agreeability is utilized to measure the extent to which an input agrees with the other two inputs. In order to describe the agreeability between the image representations determined by the NetVLAD, ResNet50 and GoogleNet networks, three linguistic variables have been defined: aNetVLAD, aResNet50 and aGoogleNet. They are described by three fuzzy sets: high, medium, and low. Figure 3b depicts the plot of the membership functions for the discussed linguistic variables. The parameters of the triangle membership functions have been tuned manually. Table 1 contains fuzzy rules for modeling fuzzy agreeability. Two Mamdani-type systems have been designed to model the

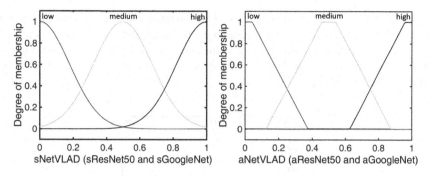

Fig. 3. Membership functions for the input linguistic variables: sNetVLAD, sResNet50 and sGoogleNet, used in modeling similarity (left), and aNetVLAD, aResNet50 and aGoogleNet, used in modeling agreeability (right).

Table 1. Fuzzy rules for modeling similarity and agreeability.

	Similarity				Agreeability			
	sNetVLAD	sResNet50	sGoogleNet	Similar.	aNetVLAD	aResNet50	aGoogleNet	Agreeab.
1	Low	Low	Low	High	Low	Low	Low	High
2	Low	Low	Med.	High	Low	Low	Med.	Med.
3	Low	Low	High	High	Low	Low	High	Low
4	Low	Med.	Low	High	Low	Med.	Low	Med.
5	Low	Med.	Med.	High	Low	Med.	Med.	Med.
6	Low	Med.	High	High	Low	Med.	High	Low
7	Low	High	Low	High	Low	High	Low	Low
8	Low	High	Med.	Med.	Low	High	Med.	Low
9	Low	High	High	Med.	Low	High	High	Low
10	Med.	Low	Low	Med.	Med.	Low	Low	Med.
11	Med.	Low	Med.	Med.	Med.	Low	Med.	Med.
12	Med.	Low	High	Med.	Med.	Low	High	Low
13	Med.	Med.	Low	Med.	Med.	Med.	Low	Med.
14	Med.	Med.	Med.	Med.	Med.	Med.	Med.	High
15	Med.	Med.	High	Med.	Med.	Med.	High	Med.
16	Med.	High	Low	Med.	Med.	High	Low	Low
17	Med.	High	Med.	Med.	Med.	High	Med.	Med.
18	Med.	High	High	Med.	Med.	High	High	Med.
19	High	Low	Low	Med.	High	Low	Low	Low
20	High	Low	Med.	Med.	High	Low	Med.	Low
21	High	Low	High	Low	High	Low	High	Low
22	High	Med.	Low	Low	High	Med.	Low	Low
23	High	Med.	Med.	Low	High	Med.	Med.	Med.
24	High	Med.	High	Low	High	Med.	High	Med.
25	High	High	Low	Low	High	High	Low	Low
26	High	High	Med.	Low	High	High	Med.	Med.
27	High	High	High	Low	High	High	High	High

fuzzy similarity and fuzzy agreeability. The parameters of the Mamdani system are as follows: implication - min, aggregation - max, defuzzification - centroid. The output membership function is Gaussian, see also Fig. 4. In the proposed system, the outputs of the low-level fuzzy systems are utilized as inputs to the high-level fuzzy system.

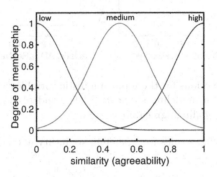

Fig. 4. Output membership functions of Mamdani-type system.

As already mentioned, the outputs of the low-level fuzzy systems are combined using the higher level fuzzy system in order to produce ranking values, which permit ranking the images according to their similarities with the query image. The input membership functions are Gaussian. The system is Sugeno-type and has the following parameters: implication - min, aggregation - max, defuzzification - weighted average. Table 2 contains rules that have been designed to carry out the ranking of images according to their similarities with the query image. They permit retrieving top ranking images.

Table 2. Fuzzy rules for ranking images according to their similarity with the given query image.

	rSimilarity	rAgreeability	Rank
1	Low	Low	Low
2	Low	Med.	Low
3	Low	High	Low
4	Med.	Low	Low
5	Med.	Med.	Med.
6	Med.	High	Med.
7	High	Low	Med.
8	High	Med.	High
9	High	High	High

4   Experimental Results

4.1   The Dataset

The data used in this investigation has been collected using an RGB camera mounted on the head of a humanoid robot. The dataset includes 9000 images of size 640×480 pixels, which have been acquired in nine indoor rooms. Each image has been manually labeled as sharp or blurred or considerably blurred. The training sequence consists of 5287 blurred images and 1913 sharp images. A test image sequence comprises 1366 blurred images and 434 sharp images. It also contains twenty two reference images with corresponding relevant images as well as the irrelevant images.

4.2   Evaluation of Visual Place Recognition

As mentioned above, basic idea of current image-based approaches to visual place recognition is to search a repository of indoor images and return the best match. We analyzed the performance of place recognition on images from Seq. #2 using the NetVLAD, GoogleNet and ResNet50 features. Table 3 presents mean average precision (mAP) scores as well as their average values, which have been achieved in recognition of 22 places in nine rooms. The mAP scores obtained by only look once algorithm [10] (OLN k-NN) and the NetVLAD algorithm [21] (NetVL k-KNN), which usually achieves very good mAP scores in the VPR are collected in the second and third column, respectively. They have been obtained with consideration only sharp images from the test subset, i.e. images that have been indicated by the blur detection algorithm as sharp ones. The next two columns contain mAP scores, which have been obtained on the basis of graph-based decisions on deep embeddings [13] (MST combined desc.). Fourth column contains the mAP score that have been obtained on all images from the test subset. This means that the place inference has been performed without taking into account the blur detection outcomes. The fifth column contains in turn the mAP scores that have been achieved on images, which have been labeled in advance by the blur detection subsystem as sharp. The penultimate column contains mAP scores that have been obtained on the basis of the similarity FIS only. This means that the place inference is performed without taking into account the agreeability.

Comparing results achieved by the discussed algorithm with results obtained by the algorithm [13] that calculates the top ranking images by averaging the indexes from a pool the most similar images determined by the NetVLAD, ResNet50 and GoogleNet networks, we can observe that the fuzzy inference on the basis of the similarity only is not capable of improving the mAP scores. As we can observe, the fuzzy system using both similarity and agreeability and then performing the fuzzy ranking achieves the best result. It is also worth noting that both fuzzy algorithms outperform both the OLN k-NN algorithm and the NetVLAD algorithm. Taking into account the information on blur detections permits considerable improvements of the performance of place recognition.

Table 3. Performance of visual place recognition (bd. - blur detection, mam - Mamdani).

	OLN [10] k-NN bd.	NetVL [21] k-NN bd.	MST [13] combined desc. –	bd.	FS mam similarity bd.	FS mam,sugeno rank bd.
Cor_1	1.0000	1.0000	0.8633	1.0000	0.9931	1.0000
Cor_2	1.0000	1.0000	0.5804	1.0000	1.0000	1.0000
Cor_3	0.7667	0.9750	0.7857	1.0000	0.9750	0.9750
D3A	0.6914	0.6549	0.7852	0.6612	0.5640	0.6097
D7	0.8788	0.8193	0.8041	0.8193	0.9604	0.9604
F102	1.0000	1.0000	0.7833	1.0000	1.0000	1.0000
F104	0.8405	0.8537	0.8632	0.8537	0.8426	0.8537
F105	0.9167	1.0000	0.8851	1.0000	1.0000	1.0000
F107	0.8700	0.8772	0.7224	0.8963	0.8792	0.8792
av. mAP	0.8849	0.9089	0.7859	0.9145	0.9127	**0.9198**

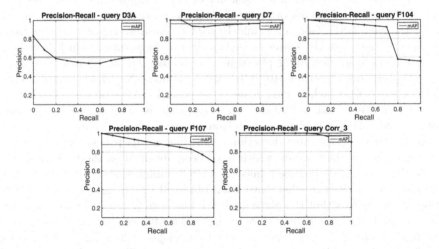

Fig. 5. Precision-recall plots for selected rooms (D3A, D7, F104, F107, Corr_3).

Fig. 5 depicts precision-recall plots for selected rooms. The precision is the fraction/percentage of retrieved images that are relevant. The recall is the fraction/percentage of relevant images that were retrieved. For the rooms: D3A, D7, F104 and F107 the number of landmark points was equal to three, whereas in the room Corr_3 the number of landmark points was equal to two. For the remaining rooms (F102, F105, Corr_1 and Corr_2) the precision-recall curves have been perfect, i.e. the mAP score has been equal to one.

5   Conclusions

In this work we proposed a fuzzy system for visual place recognition. In the first level of the hierarchical system both fuzzy similarity and fuzzy agreeability are inferred and then fed to a FIS in the second level, which is accountable for determining values for re-ranking images with respect to similarity with the query image. We demonstrated experimentally that the proposed algorithm outperforms recent algorithms in terms of mAP scores as well as it is superior to the non-fuzzy algorithm. The algorithm has been evaluated on a challenging dataset for visual place recognition with RGB images acquired by a humanoid robot. The experimental results demonstrated also the usefulness of motion blur detection. Without the blur detection such that the algorithm operates without distinguishing between images with motion blur and without motion blur the mAP scores are considerably worse. One important direction of future work concerns extension of the system about fuzzy type-2 inference. An additional area for future work concerns more advanced use of fuzzy logic for construction of hierarchical FIS for visual place recognition as well as quantitative comparisons of fuzzy type-1 and fuzzy-2 systems for visual place recognition.

Acknowledgment. This work was supported by Polish National Science Center (NCN) under a research grant 2017/27/B/ST6/01743.

References

1. Lowry, S., et al.: Visual place recognition: a survey. IEEE Trans. Robot. **32**, 1–19 (2016)
2. Cebollada, S., Paya, L., Flores, M., Peidro, A., Reinoso, O.: A state-of-the-art review on mobile robotics tasks using artificial intelligence and visual data. Expert Syst. Appl. 114195 (2020)
3. Odo, A., McKenna, S., Flynn, D., Vorstius, J.: Towards the automatic visual monitoring of electricity pylons from aerial images. In: International Conference on Computer Vision Theory and Applications VISAPP (2020)
4. Zhao, J., et al.: Place recognition with deep superpixel features for brain-inspired navigation. Rev. Sci. Instrum. **91**(12), 125110 (2020)
5. Heng, L., et al.: Project AutoVision: localization and 3D scene perception for an autonomous vehicle with a multi-camera system. In: International Conference on Robotics and Automation (ICRA), IEEE (2019)
6. Tolias, G., Avrithis, Y., Jégou, H.: Image search with selective match kernels: aggregation across single and multiple images. Int. J. Comput. Vis. **116**(3), 247–261 (2015)
7. Ovalle-Magallanes, E., Aldana-Murillo, N.G., Avina-Cervantes, J.G., Ruiz-Pinales, J., Cepeda-Negrete, J., Ledesma, S.: Transfer learning for humanoid robot appearance-based localization in a visual map. IEEE Access **9**, 6868–6877 (2021)
8. Zhang, X., Wang, L., Su, Y.: Visual place recognition: a survey from deep learning perspective. Pattern Recognit. **113**, 107760 (2021)
9. Chen, Z., Lam, O., Adam, J., Milford, M.: Convolutional neural network-based place recognition. In: Proceedings of the Australasian Conference on Robotics and Automation, pp. 1–8 (2014)

10. Chen, Z., Maffra, F., Sa, I., Chli, M.: Only look once, mining distinctive landmarks from ConvNet for visual place recognition. In: IEEE/RSJ International Conference on Intelligent Robots and Systems (IROS), pp. 9–16 (2017)
11. Suenderhauf, N., Shirazi, S., Dayoub, F., Upcroft, B., Milford, M.: On the performance of ConvNet features for place recognition. In: IEEE/RSJ International Conference on Intelligent Robots and Systems (IROS), pp. 4297–4304 (2015)
12. Arandjelovic, R., Zisserman, A.: All about VLAD. In: IEEE Conference on Computer Vision and Pattern Recognition, pp. 1578–1585 (2013)
13. Wozniak, P., Kwolek, B.: Place inference via graph-based decisions on deep embeddings and blur detections. In: Paszynski, M., Kranzlmüller, D., Krzhizhanovskaya, V.V., Dongarra, J.J., Sloot, P.M.A. (eds.) ICCS 2021. LNCS, vol. 12746, pp. 178–192. Springer, Cham (2021). https://doi.org/10.1007/978-3-030-77977-1_14
14. Kalibatienė, D., Miliauskaitė, J.: A hybrid systematic review approach on complexity issues in data-driven fuzzy inference systems development. Informatica 32, 85–118 (2021)
15. Mittal, K., Jain, A., Vaisla, K.S., Castillo, O., Kacprzyk, J.: A comprehensive review on type 2 fuzzy logic applications: past, present and future. Eng. Appl. Artif. Intell. 95, 103916 (2020)
16. Wu, D., Mendel, J.M.: Recommendations on designing practical interval type-2 fuzzy systems. Eng. Appl. Artif. Intell. 85, 182–193 (2019)
17. Mikut, R., Jäkel, J., Groell, L.: Interpretability issues in data-based learning of fuzzy systems. Fuzzy Sets Syst. 150(2), 179–197 (2005)
18. Hu, H., Qiao, Z., Cheng, M., Liu, Z., Wang, H.: DASGIL: domain adaptation for semantic and geometric-aware image-based localization. IEEE Trans. Image Process. 30, 1342–1353 (2021)
19. Revaud, J., Almazan, J., Rezende, R., Souza, C.D.: Learning with average precision: training image retrieval with a listwise loss. In: IEEE/CVF International Conference on Computer Vision (ICCV), pp. 5106–5115 (2019)
20. Cakir, F., He, K., Xia, X., Kulis, B., Sclaroff, S.: Deep metric learning to rank. In: IEEE/CVF Conference on Computer Vision and Pattern Recognition (CVPR), pp. 1861–1870 (2019)
21. Arandjelovic, R., Gronat, P., Torii, A., Pajdla, T., Sivic, J.: NetVLAD: CNN architecture for weakly supervised place recognition. IEEE Trans. Pattern Anal. Mach. Intell. 40(6), 1437–1451 (2018)

A Hybrid Intelligent System to Detect Anomalies in Robot Performance

Nuño Basurto$^{(\boxtimes)}$ ⓘ, Ángel Arroyo ⓘ, Carlos Cambra ⓘ, and Álvaro Herrero ⓘ

Grupo de Inteligencia Computacional Aplicada (GICAP),
Departamento de Ingeniería Informática, Escuela Politécnica Superior,
Universidad de Burgos, Av. Cantabria s/n, 09006 Burgos, Spain
{nbasurto,aarroyop,ccbaseca,ahcosio}@ubu.es

Abstract. Although self-diagnosis is required for autonomous robots, little effort has been devoted to detect software anomalies in such systems. The present work contributes to this field by applying a Hybrid Artificial Intelligence System (HAIS) to successfully detect these anomalies. The proposed HAIS mainly consists of imputation techniques (to deal with the MV), data balancing methods (in order to overcome the unbalancing of available data), and a classifier (to detect the anomalies). Imputation and balancing techniques are subsequently applied in for improving the classification performance of a well-know classifier: the Support Vector Machine. The proposed framework is validated with an open and recent dataset containing data collected from a robot interacting in a real environment.

Keywords: Smart robotics · Performance monitoring · Anomaly detection · Missing values · Data unbalance · Hybrid intelligent system

1 Introduction

Smart robots are widely acclaimed as one of the driving forces for technological advance in many application areas. Many and ambitious contributions have been done within this field during last decades. However, there are still open challenges in order to meet reliability demands of smart robots in particular and cyberphysical systems in general. Present requirements of robots demand for an early and accurate detection of anomalies. A wide range of previous works have addressed this topic but it has been pointed out [1] that further research must be conducted on the detection of anomalies associated to robots. It is a complex task whose difficulty increases when facing real-world scenarios where robots interact with people and unexpected circumstances.

One of the very few previous works on the detection of software anomalies is [2]. Authors of this work have released the only publicly-available dataset that contains data from different performance indicators associated to the software of a robot. This dataset [3] (comprehensively described in Sect. 3) comprises detailed an varied information about anomalies induced in a real-world setting.

© Springer Nature Switzerland AG 2021
H. Sanjurjo González et al. (Eds.): HAIS 2021, LNAI 12886, pp. 415–426, 2021.
https://doi.org/10.1007/978-3-030-86271-8_35

As a result, two usual problems of real datasets show up: Missing Values (MV) and data unbalance. In order to overcome such limitations and to improve the successful detection of the anomalies, different alternatives have been independently proposed and validated so far: imputation of MV [4] and application of data balancing methods [5]. The present research goes one step further and combines these solutions in a novel HAIS, validating it with the previously mentioned dataset.

According to the classification of missing data proposed in [6], in this research the problem is missing completely at random, because the probability that a MV is located in one or more attributes of any instance does not depend on particular circumstances. The imputation carried out is a Single Imputation, where the method fills one value for each MV [7]. To optimize the imputed values, it is necessary to do a regression work. In the present research, the applied techniques are: a statistical nonlinear regression [8] and a regression tree [9] called Fine Tree (FT).

To carry out the MV imputation a wide set of techniques have been previously proposed, highlighting those based on Artificial Intelligence (AI) [10]. In robotics context, the research works [11] whose author proposed a probabilistic approach using incomplete data for classification (failure detection). Data samples were classified by calculating, from the data samples that are not missing, the a-priori probability of MV. This proposal was applied to datasets containing only anomalies affecting the hardware and that are outdated (coming from 1999). The present work is the first approach to impute MV in a dataset from a component-based robot in order to improve subsequent classification by using data balancing techniques.

The HAIS, together with the calculated metrics for result comparison, are described in Sect. 2. The real-life problem addressed in the present work is introduced in Sect. 3. The experimental study and the obtained results are compiled in Sect. 4. Finally, the future work proposals are presented, together with the main conclusions, in Sect. 5.

2   Proposed HAIS

The HAIS proposed in present work consists of three main components:

1. MV Imputation. This is the first component and for a comprehensive experimentation, two different methodologies have been defined: (1) Methodology 1 (M1) where the MV imputation is carried out by regressing only from the data samples located sequentially before the set of samples containing MV; Methodology 2 (M2) where the MV imputation is executed by regressing from the whole dataset, (except the data samples whose values are to be imputed). For each of the two methodologies two regression techniques are applied to the original dataset to impute the MV. As a result, there are values for all the attributes of all the data instances. Hence, none of them must be removed in the pre-processing step.

2. Data Balancing. Several balancing techniques are applied to the already imputed dataset (previous component). As a result, a dataset that has a balance between the majority (normal) and the minority (anomalous) classes is obtained.
3. Classification. A standard classifier Support Vector Machine (SVM) is applied in order to assess the effect of the different combinations in previous components. Several classification metrics are calculated in order to compare the obtained results.

The various AI methods that are applied in these components are described in the following subsections. Firstly, the pre-processing methods (for imputation and data balancing) are comprehensively described in Subsects. 2.1 and 2.2 respectively. Then, the applied classifier is introduced in Subsect. 2.3. Finally, the metrics that have are calculated and studied in order to compare the obtained results are explained in Subsect. 2.4.

2.1   Component 1: MV Imputation

The techniques applied in the present work for the MV imputation are described in the following subsections.

Non-Linear Regression. The objective of multiple regressions [12] is about discovering relationships between one or more independent or predictor attributes and the dependent attribute. This relationships can be modelled by a Non-Linear Regression (NLR) which is a model where observational attributes are fixed by a non-linear function. This function depends on (at least) one criterion attribute and combines the input data [13].

The function parameters can take the form of any model of non linear function. To determine this parameter value, an iterative algorithm is applied which is as follows:

$$y = f(X, \beta) + \varepsilon \tag{1}$$

Where X represents the dependent attribute, β is the non linear attribute to be calculated, and ε is the error quantification.

Regression Trees (RT) are often represented as an structure growing from its roots to the top leaves (nodes). An instance runs through the tree across a series of nodes. The tree decides the branch through which the instance should be conducted based on the value of some of the attributes designated as criterion. When a leaf is reached, the value associated to the instance is predicted according to the final node. The values in the end nodes are assigned using the average values of the cases for that particular node [9]. The Fine Tree (FT) a type of RT which works with a small set of sheet nodes to reduce computing time.

2.2 Component 2: Data Balancing

In most binary classification datasets, there are very few samples of one of the classes (minority one) while most samples belong to the other one (majority class). For balancing such datasets, the following techniques can be applied:

1. Oversampling method. These methods generate new instances of the minority class in order to obtain a class balance in the dataset. The chosen one is Random Over-Sampling (ROS), which is in charge of generating new instances of the minority class by performing a random duplication of them.
2. Undersampling method. These techniques eliminate instances of the majority class in order to obtain a more balanced dataset. The Random Under-Sampling (RUS) algorithm is the standard method in this category. In a similar way to that performed by ROS, but in a completely opposite way, it takes on the task of randomly removing the instances of the majority class.
3. Hybrid methods. They combine the use of oversampling and undersampling techniques in order to reduce their impact. For the hybridization of balancing techniques, the previously introduced techniques are applied in unison. ROS + RUS is an hybrid method that eliminates a percentage of the instances of the majority class in a random way and creates new instances of the minority class by replicating them. Both tasks are done randomly.

2.3 Component 3: Classification

The SVM [14,15] is a learning model whose aim is widening the margin of separation between classes. It looks for a hyperplane that universalizes the archetype of data, applying this to new data. In the present experimentation a one-class SVM classifier has been used, as initially selected by the dataset authors [16]. Additionally, the sigmoidal kernel has been used, defined by the following equation:

$$k(x, y) = tanh(ax^T y + c) \tag{2}$$

2.4 Calculated Metrics

To compare the classification results obtained by the different methods, several metrics have been used. Among all these metrics, one of them is the well known Accuracy. The rest of them are more advanced and specifically oriented for the use with unbalanced datasets. They are based on the use of other simpler ones such as False Positive Rate (FPR), Precision, and Recall [17].

1. F_1 Score. Precision and recall metrics are widely used but can not be maximized at the same time. This is why the use of metrics such as this one is introduced as these more advanced ones are capable of obtaining a value that represents in a harmonic way the targets of both precision and recall. The formula of this metric is:

$$F_1 = 2 * \frac{Precision * Recall}{Precision + Recall} \qquad (3)$$

2. ROC Curve and Area Under the Curve. The ROC Curve is based on the ability to visually represent a midpoint of the FPR and Recall indicators. This representation leaves an area at the bottom, whose area is called Area Under the Curve (AUC). The importance of this metric is that it is able to show the capacity of the model to make a distinction between classes. This is why its use is advisable for unbalanced datasets, where other metrics such as Accuracy may take a high value, but in reality there is no class discrimination.
3. g-mean. Following the concordance of the other metrics seen, geometric mean (g-mean) is in charge of finding a balance between Precision and Recall. g-mean can reach high values if both Precision and Recall have high values and are in balance. This is defined as:

$$g - mean = \sqrt{Precision * Recall} \qquad (4)$$

3 Software Performance of Robots

As it has been previously stated, this research focuses on the detection of anomalies in a component-based robot. The analysed dataset [2] was released by researchers from the University of Bielefeld (Germany) and is publicly available at [3]. As it has been previously commented, data were gathered from the software of a component-based robot. It means that the different components that conform the robot can be developed by different manufacturers, but they are all interconnected with each other by means of a middleware. Thus, the robot is decomposed in small parts with their own implementation and they can be independently interchanged.

For the creation of the dataset, the authors carried out 71 experiments in which they replicated the same series of actions that the robot had to carry out, such as moving from one place to another, recognizing people's faces or holding a glass with the mechanical arm. In some of these trials the authors induced anomalies through software. This does not imply that the target action can not be carried out, but its performance is penalized instead. For example, the movement of the arm in the moment of catching the glass; it implies an increase on the number of movements of the arm when it catches the glass, having a direct impact in the performance counters.

Based on the previous work carried out [4], the *armcontrol* component and the *counters* data-source have been selected for the present work. This component is affected by an anomaly called *armServerAlgo*, whose performance impact was mentioned above. The dataset contains a total of 12 attributes of different types, some of which indicate the time spent in core mode, user mode or the amount of information sent by the interface.

In addition to the above, the target class is the state in which the robot is, defined as "normal" (0) or "anomalous" (1). This is unbalanced as there are 20,832 (95%) instances of the "normal" and only 1,055 (5%) of the "anomalous" class. The received_bytes and send_bytes are the only attributes containing MV (15.3% each).

4   Experimental Study

The components and techniques described in Sect. 2 are applied to detect the anomalies affecting the *armcontrol* component of the robot. In the following subsections, the obtained results are shown for each one of the HAIS components.

4.1   Component 1: MV Imputation

The set of techniques detailed in Sect. 2.1 have been applied and the obtained results are presented in this section. To achieve more reliable results, they are validated by the n-fold cross-validation [18] framework. Data partitions (n) has been set to the standard value of 10 for all the experiments.

The regression process was carried out on the attributes containing MV (received_bytes and send_bytes). The attribute on which the regression is performed is set as the predictor attribute and the other 10 attributes are the criterion ones.

The goal of this first component is to impute the existing MV in the send_bytes and received_bytes attributes as accurately as possible. To do this, the regression techniques explained in Sect. 2.1 are applied to the subsets of data with complete samples.

Table 1 shows the mean squared error (MSE) value obtained after applying the imputation techniques to the dataset with M1 Methodology.

Table 1. Obtained MSE values according to the M1 methodology for the component armcontrol.

Attribute/MSE	NLR	FT
received_bytes	**5.72E−18**	1.44E−9
sent_bytes	**1.55E−18**	7.36E−10

The regression techniques that have obtained the lowest MSE values (in bold) is the NLR for both attributes, also the execution times are much lower compared to those of FT.

In a similar way to the Table 1, the results of applying the same two techniques under the frame of the M2 methodology are shown in Table 2.

For the M2 methodology, similar results are obtained for both attributes, achieving the NLR technique obtaining the lowest MSE. It should be noted that

Table 2. Obtained MSE values according to the M2 methodology for the component armcontrol.

Attribute/MSE	NLR	FT
received_bytes	**1.39E–17**	2.85E–9
sent_bytes	**4.27E–18**	1.59E–9

for this M2 methodology the MSE values are slightly higher than those obtained for M1, especially for the received_bytes attribute and the NLR technique. This implies that the values resulting from the M2 imputation are less accurate than the M1 ones.

4.2 Component 2: Data Balancing

After performing the MV calculation, different balancing techniques have been used in order to find a greater equality between the different classes. They have been detailed in Sect. 2.2. The Table 3 shows the different sizes of the data set, after the application of the balancing algorithms.

Table 3. Size and class distribution of the training sets generated by the different balancing methods.

Number of Samples	No Balancing	ROS	RUS	ROS+RUS
Normal Class	15624	15624	792	8185
Anomalous Class	792	15624	792	8231
TOTAL	16416	31248	1584	16416

4.3 Component 3: Classification

As explained in detail in Sect. 2.3, the SVM classifier has been used to perform a prediction on the target class, this class indicates which is the state of the robotic system, either "anomalous" or "normal". For greater reproducibility of the experiment, 75% of the instances have been selected for training and 10% for testing. On the other hand, n-fold CV has been applied with a value of n equal to 10. The parameter values for the SVM classifier are $cost = 10$, $gamma = 0.1$ and $nu = -0.5$.

Results are shown for all the metrics described in Sect. 2.4. For a more complete analysis of the results obtained in this component, executions have also been carried out without imputing MV. That is, the received_bytes and sent_bytes attributes are deleted and hence the classification is carried out on the 10 remaining attributes. This executions are referred as "Deleted MV".

The first experiments have been carried out without using any balancing technique, as shown in Table 4. The results achieved for the accuracy metric stand out. It is mainly because there is such a big unbalance between the two classes that the classifier tend to generate clearly favorable results for the majority class while ignoring the minority one. This evidences the need for other metrics (AUC and g-mean mainly) that take into account the instance distribution among classes. On the other hand, the obtained values for these metrics are clearly lower in comparison with subsequent experiments.

Table 4. Metric values obtained by the SVM classifier on the raw (no balancing method) data.

	FT M1	NLR M1	FT M2	NLR M2	Deleted MV
accuracy	0.9126	0.9209	0.9179	0.9200	0.9149
precision	0.0951	0.0300	0.0684	0.0247	0.0776
recall	0.0943	0.0417	0.0811	0.0325	0.0841
F_1 score	**0.0947**	0.0349	0.0742	0.0279	0.0806
AUC	**0.5245**	0.4979	0.5146	0.4950	0.5174
g-mean	**0.0947**	0.0352	0.0745	0.0282	0.0807

In the execution performed with the oversampling method (ROS), the results (see Table 5) are very different from those obtained in the previous one (no-balancing method). Now the accuracy metric is penalized while the values for the F_1 score, AUC and especially g-mean and precision increase in general terms. These three advances (without considering precision) metrics, always get increased with both methodologies for NLR. On the other hand, FT M1 is worse in all cases. The best value for the F_1 score AUC and g-mean metrics are obtained by FT, in the first and third case with M1 whereas in the second one by M2.

Table 5. Metric values obtained by the SVM classifier on the ROS-balanced data.

	FT M1	NLR M1	FT M2	NLR M2	Deleted MV
accuracy	0.4754	0.4656	0.4776	0.4662	0.4958
precision	0.5247	0.4916	0.4601	0.4787	0.4430
recall	0.0479	0.0443	0.0427	0.0433	0.0427
F_1 score	**0.0877**	0.0813	0.0781	0.0794	0.0779
AUC	0.4988	0.5135	**0.5307**	0.5270	0.5057
g-mean	**0.1585**	0.1476	0.1401	0.1440	0.1375

The results obtained by the undersampling method (RUS) are shown in Table 6. It has achieved fairly good results for most of the imputation techniques. While improving the results achieved by ROS in the vast majority of metrics after the use of the FT imputation technique, the opposite is the case with NLR. As with ROS, it has been the same techniques and methodologies that have stood out with their values over the rest. FT M1 for F_1 score and g-mean and FT M2 for AUC.

Table 6. Metric values obtained by the SVM classifier on the RUS-balanced data.

	FT M1	NLR M1	FT M2	NLR M2	Deleted MV
accuracy	0.4745	0.4460	0.4677	0.4646	0.4958
precision	0.5665	0.5019	0.4905	0.4494	0.4711
recall	0.0512	0.0436	0.0444	0.0407	0.0452
F_1 score	**0.0939**	0.0802	0.0814	0.0746	0.0825
AUC	0.5182	0.5038	**0.5215**	0.5111	0.5148
g-mean	**0.1703**	0.1479	0.1475	0.1352	0.1460

Finally the hybrid data-balancing techniques are also applied. The results obtained by the first hybrid technique (ROS+RUS) are shown in Table 7. Following the current trend, it is to be expected that FT will once again achieve the best values in the different metrics. In this case it is FT M2 that stands out from the rest, but not by a great difference as the different methodologies on which NLR acts obtain similar values. Although in this case the best value for the AUC metric is for the first time the elimination of complete rows. In the other two complex metrics, FT M2 is the best.

Table 7. Metric values obtained by the SVM classifier on the ROS+RUS-balanced data.

	FT M1	NLR M1	FT M2	NLR M2	Deleted MV
accuracy	0.4761	0.4684	0.4942	0.4688	0.4957
precision	0.4943	0.4939	0.4981	0.4916	0.4757
recall	0.0454	0.0448	0.0474	0.0446	0.0457
F_1 score	0.0832	0.0822	**0.0865**	0.0817	0.0834
AUC	0.5152	0.5198	0.5039	0.5039	**0.5291**
g-mean	0.1498	0.1488	**0.1536**	0.1480	0.1474

For a better comparison of all the combinations that have been applied, the average metric values have been calculated for each one of the two methodologies (M1 and M2). Figure 1 shows a radar plot where the average results can be compared per data balancing technique for each one of the key metrics: F_1 score (a), AUC (b) and g-mean (c).

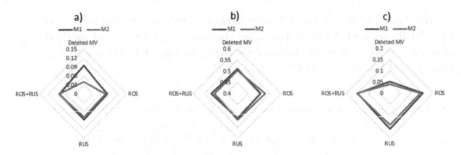

Fig. 1. Radar plot of the metric values per methodology and data balancing technique. a) F_1 Score, b) AUC and c) g-mean

From Fig. 1.a) it can be said that M1 has obtained higher values than M2 for all the applied imputation techniques in the case of the F_1 score. The difference between M1 and M2 is bigger in the case of Deleted MV. When analysing the AUC metric, M2 outperforms M1 for ROS but get similar values in the rest of the balancing techniques. Finally, when considering the g-mean metric, results are very similar to those for the F_1 Score, with smaller differences between M1 and M2, highlighting M1 for RUS. All in all, it is worth mentioning that in general terms, worst results have been always obtained for the Deleted MV experiments less in the case of AUC with the hybrid technique. This supports the proposal of improving the classification results by imputing MV.

5   Conclusions and Future Work

The present work focuses on the detection of the anomalies affecting the arm-control component of a collaborative robot with a novel HAIS. MV are present in two of the data attributes associated to this component. Additionally, the dataset is highly unbalanced (95% vs. 5%). In order to address such problems, a HAIS, including 3 components, has been proposed and validated.

In the first component, the imputation of MV is carried out. Two methodologies (as described at the beginning of Sect. 4) are applied to find the imputation methods that obtain the lowest MSE. The NLR technique obtain the best MSE for both attributes and in both cases (M1 and M2 methodologies). The M1 methodology is more efficient taking into account the computation time and in MSE values.

In the second component, the balancing of classes is carried out. It can be concluded that the resulting datasets obtained by the balancing techniques varies in the total number of instances and the balancing rates, as shown in Table 3.

Finally, in the third component, the SVM classifier is applied in order to detect the anomalous states of the robot by taking advantage of previous components. The main conclusion derived from the obtained results is that the proposed HAIS is useful as better classification results (according to the F_1 score, AUC, and g-mean metrics) are obtained when compared to the MV-deletion strategy.

When considering all the different combinations, the best combinations for each one of the metrics can be identified:

- F_1 score: FT M1 No balancing technique.
- AUC: FT M2 ROS.
- g-mean: FT M1 RUS.

As future work, we consider using new regression and imputation techniques such as those based on neural networks, as well as the use of more advanced balancing techniques to help us improve the quality of the minority class instances.

References

1. Khalastchi, E., Kalech, M.: On fault detection and diagnosis in robotic systems. ACM Comput. Surv. **51**(1), 1–24 (2018). https://doi.org/10.1145/3146389
2. Wienke, J., Meyer zu Borgsen, S., Wrede, S.: A data set for fault detection research on component-based robotic systems. In: Alboul, L., Damian, D., Aitken, J.M.M. (eds.) TAROS 2016. LNCS (LNAI), vol. 9716, pp. 339–350. Springer, Cham (2016). https://doi.org/10.1007/978-3-319-40379-3_35
3. Wienke, J., Wrede, S.: A fault detection data set for performance bugs in component-based robotic systems (2016). https://doi.org/10.4119/unibi/2900911. http://dx.doi.org/10.4119/unibi/2900911
4. Basurto, N., Arroyo, Á., Cambra, C., Herrero, Á.: Imputation of missing values affecting the software performance of component-based robots. Comput. Electr. Eng. **87**, 106766 (2020). https://doi.org/10.1016/j.compeleceng.2020.106766
5. Basurto, N., Cambra, C., Herrero, A.C.: Improving the detection of robot anomalies by handling data irregularities, Neurocomputing In press. https://doi.org/10.1016/j.neucom.2020.05.101. http://www.sciencedirect.com/science/article/pii/S0925231220311954
6. Schafer, J.L.: Multiple imputation: a primer. Stat. Methods Med. Res. **8**(1), 3–15 (1999)
7. Plaia, A., Bondi, A.: Single imputation method of missing values in environmental pollution data sets. Atmos. Environ. **40**(38), 7316–7330 (2006)
8. U. of Yale, Multiple linear regression (2017). http://www.stat.yale.edu/Courses/1997-98/101/linmult.htm
9. Moisen, G.G.: Classification and regression trees (2018)
10. Arroyo, A., Herrero, A., Tricio, V., Corchad, E., Woźniak, M.: Neural models for imputation of missing ozone data in air-quality datasets. Complexity (2018). https://doi.org/10.1155/2018/7238015
11. Twala, B.: Robot execution failure prediction using incomplete data. In: IEEE International Conference on Robotics and Biomimetics (ROBIO), pp. 1518–1523 (2009). https://doi.org/10.1109/ROBIO.2009.5420900
12. Pearson, K., Lee, A.: On the generalised probable error in multiple normal correlation. Biometrika **6**(1), 59–68 (1908). http://www.jstor.org/stable/2331556
13. Neter, J., Kutner, M.H., Nachtsheim, C.J., Wasserman, W.: Applied Linear Statistical Models, vol. 4. Irwin, Chicago (1996)
14. Cortes, C., Vapnik, V.: Support-vector networks. Mach. Learn. **20**(3), 273–297 (1995)

15. Boser, B.E., Guyon, I.M., Vapnik, V.N.: A training algorithm for optimal margin classifiers. In: Proceedings of the Fifth Annual Workshop on Computational Learning Theory, COLT 1992, ACM, New York, NY, USA, 1992, pp. 144–152 (1992)

16. Wienke, J., Wrede, S.: Autonomous fault detection for performance bugs in component-based robotic systems. In: Intelligent Robots and Systems (IROS), 2016 IEEE/RSJ International Conference on IEEE, 2016, pp. 3291–3297 (2016). http://dx.doi.org/0.1109/IROS.2016.7759507

17. Buckland, M., Gey, F.: The relationship between recall and precision. J. Am. Soc. Inf. Sci. **45**(1), 12–19 (1994). https://doi.org/10.1002/(SICI)1097-4571(199401)45:1⟨12::AID-ASI2⟩3.0.CO;2-L

18. Arlot, S., Celisse, A.: A survey of cross-validation procedures for model selection. Stat. Surv. **4**, 40–79 (2010). https://doi.org/10.1214/09-SS054. http://dx.doi.org/10.1214/09-SS054

Counter Intituive COVID-19 Propagation Dynamics in Brazil

Manuel Graña[✉], Goizalde Badiola-Zabala, and Jose Manuel Lopez-Guede

Computational Intelligence Group, University of the Basque Country
UPV/EHU, Leioa, Spain
manuel.grana@ehu.es

Abstract. There are few studies concerning the propagation of COVID-19 pandemic, besides theoretical models that have produced alarming predictions. These models assume wave-like propagation dynamics on an uniform medium. However, it appears that COVID-19 follows unexpected propagation dynamics. In this paper we consider the state-wise data on COVID-19 mortality in Brazil provided by government sources. Conventional propagation models tell us that the pandemic should propagate over the neighboring states from the initial cases in Sao Paulo. We compute several measures of correlation and prediction by random forests finding that the patterns of propagation do not correlate well with the graph distances defined by the spatial neighborhood of the states. We think that this phenomenon deserves further attention in order to understand COVID-19 pandemic.

1 Introduction

Since the early days of 2020, the COVID-19 pandemic has been shocking [5] the world [17,19] with several waves of COVID-19 outbreak hitting differently all countries across the world [2], even showing different incidence inside the administrative partitions of the countries [6]. The main tools that have been proposed [9] to control the damage of a viral infection outbreak are related to either non pharmacological interventions (NPI) impeding the spread of the virus, or pharmacological interventions aiming to treat the associated disease severity. Regarding pharmacological developments, there are many studies trying to assess the benefits of several ancient and new molecules against the SARS-CoV-2 virus [7,8,10,20] while vaccinal approaches are tested at large on the world population [4,13]. Regarding non-pharmacological interventions [11,14,15] many countries or their lower administrative divisions (such as states, regions, or counties) have implemented quarantines and other restrictions of movement, while recommending social distancing, wearing masks and general prophylaxis measures. Concurrently, there is growing concern about the side effects of these measures on the general and at-risk population, specifically children and young adults [1,3,16,18]. For instance, the use of urban green spaces has been greatly affected by the pandemic and it has been valued highly as a resource to overcome the mental burden of the situation [12].

H. Sanjurjo González et al. (Eds.): HAIS 2021, LNAI 12886, pp. 427–435, 2021.
https://doi.org/10.1007/978-3-030-86271-8_36

In this paper we are interested in the propagation dynamics of the pandemic, specifically the mortality. We found that the data from Brazil [6] is rather detailed showing dynamics that in some ways different from European data. Specifically, the sharp peaks of over-mortality that can be observed in the euromomo site (https://www.euromomo.eu/) during the months of march/april 2020 are not so sharp and well defined in the Brazilian data. In a way, Brazilian data appears to have more characteristics of a pure epidemic propagation. Specifically, the research question in this paper is whether the propagation of pandemic follows the spatial neighboring distribution of states. More precisely, does the pandemic propagation follow the spanning tree from the source of the pandemic in the graph of state neighborhoods? In Brazil, the first COVID-19 related deaths were reported in Sao Paulo (SP) and and Rio de Janeiro (RJ), for a single day of difference we choose Sao Paulo as the pandemic origin in Brazil. Figure 1 shows the map of the states and the adjacency matrix of the graph of neighboring states.

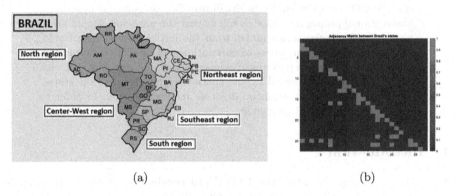

(a)   (b)

Fig. 1. Map of states of Brazil (a) Adjacency matrix of the states following the ordering used in [6].

2   Materials and Methods

The data used for the computational studies are the same used in [6], encompassing data from the beginning of March 2020 until March 2021. Figure 2 shows the unfiltered mortality curves. There are several peaks that appear to be artifacts of the data collection process. Another artifact are the sudden oscillations due to the weekly period of data reporting. The conventional procedure is to apply a 7 day moving average filter that removes most of these artifacts. The processes are non stationary and of quite different magnitudes. As we will be computing correlations, the effect of these magnitude changes should have no effect.

First we have computed the correlations among the entire mortality curves, under the assumption that this map of correlations should correlate well with

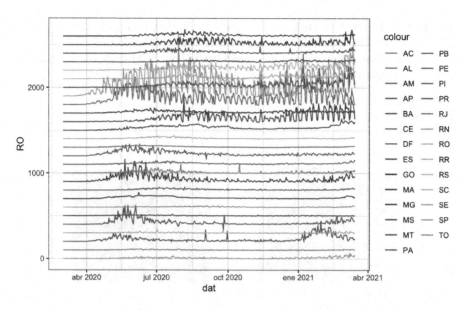

Fig. 2. Plot of the mortality curves of Brazil states

the adjacency graph of the spatial distribution of the states. Secondly, we compute the distances over the spatial adjacency graph between Sao Paulo and each state. We expect that some measures of pandemic evolution relative to the origin will correlate well with this distribution of distances. Thirdly, we compute the time elapsed since the start of the pandemic in Sao Paulo to the start of the pandemic for each state, expecting that it should correlate well with the distribution of distances. Fourthly, we compute the cross-correlation between the first wave of the pandemic of each state relative to Sao Paulo. We focus on the first wave, because the second wave has quite different shapes from the first, and the time span of the data does not cover all second wave which seems to be going down in Brazil at the time of writing. We compute the 7 day moving average filtering in order to remove the weekly artifact from the computation of the cross-correlation. We extract the number of lags corresponding to the maximum correlation. We expect that this number of lags should be always positive, corresponding to the advance of the pandemic from Sao Paulo, and well correlated with the distances over the graph of spatial distribution of the states. Finally, we use random forest regression for the prediction of the mortality at each state alternatively using as regressors the mortality at the remaining states. The training follows 20 times repetition of hold out cross-validation with 60%, 20%, and 20% of data for training, validation, and test, respectively. The training process searches for the optimal number of trees in the forest in each instance of the hold-out cross-validation. We are not interested in the achieved accuracies, but in the role of Sao Paulo in the prediction. We use the variable importance ranking provided by the random forest feature selection, and record the rank of the SP state for each predicted state. We expect that this rank should be well correlated with the graph distance in the graph of spatial distances.

All correlation computations have been carried out in Matlab. For the random forest implementation we use the Jasp software.

3   Results

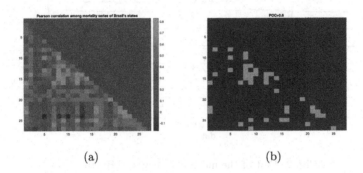

Fig. 3. Correlation among mortality curves (a), and the adjacency after applying a threshold 0.6 (b)

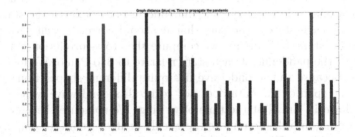

Fig. 4. Distance over the spatial distribution graph of the states versus the time (days) taken by the pandemic to reach the state since its outbreak in Sao Paulo.

Figure 3 shows the Pearson correlation coefficient among all pairs of state mortality curves, and the result of thresholding over 0.6. The expectation is that the correlation induced adjacency graph should be highly similar to the spatial adjacency graph, i.e. neighboring states should have similar mortality curves despite the non-stationarity of the time series. However, correlation among these adjacency matrices is very low ($r < 0.3$) and non significant ($p > 0.05$).

Figure 4 shows the joint plot of the distance over the spatial adjacency graph and the time (days) taken by the pandemic to reach the state. For visualization

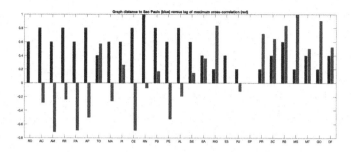

Fig. 5. Distance over the spatial distribution graph of the states versus the lag of the maximal cross-correlation relative to Sao Paulo.

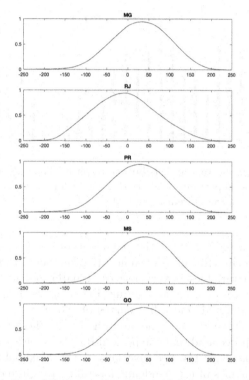

Fig. 6. Cross-correlation of the mortality curves of the states at distance 1 from Sao Paulo with mortality in Sao Paulo (first wave)

the variables have been normalized in the interval $[0,1]$. Correlation between these two variables is very low ($r < 0.1$) and non-significant ($p > 0.05$). It is quite evident from the figure that the propagation of the pandemic arrival does not follow the spatial distribution of states.

Figure 5 shows the joint plot of the distance over the spatial adjacency graph and the number of lags of the maximal cross-correlation between the first wave in each state and the state of Sao Paulo. We notice that there are many negative

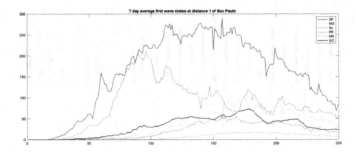

Fig. 7. First wave (filtered) of states at distance 1 from Sao Paulo in the graph of the spatial distribution.

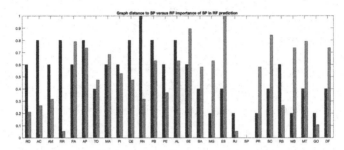

Fig. 8. Distance over the spatial distribution graph of the states versus the ranking of SP in the variable importance ranking of random forest regression predicting the mortality of the state using all remaining states.

lags, indicating a counter flow of the pandemic progression in some way. For instance Rio de Janeiro (RJ) has such a negative number of lags. The correlation between these two variables is very low ($r < 0.01$) and non-significant ($p > 0.5$). In order to have some understanding, Fig. 6 shows the plots of the cross-correlation of the (filtered) mortality curves of the states at graph distance 1 with the mortality curve of Sao Paulo, and Fig. 7 plots these first wave mortality curves. It can be appreciated that the peak in Rio de Janeiro is advanced relative that of Sao Paulo, hence the negative lags of the maximal cross-correlation. Therefore, the dynamics of the pandemic mortality are quite different from one state to another for a multitude of (unexplored) confounding effects. This result questions the view of the pandemic evolution as a progression in an uniform medium, like a wave. This wave model may fit well to explain some influencer propagation in social networks, but it appears that COVID-19 propagation has some complexities that deserve further study.

Figure 8 plots the distance to SP over the spatial graph of states versus the ranking of SP in the random forest regression prediction of mortality in each state. For visualization the variables are normalized in the interval [0,1]. Again, the correlation among these variables is very low ($r < 0.1$) and non-significant ($p > 0.05$). The spatial connectivity between states does not correlate with the predictive influence of the pandemic origin in the prediction of the evolution of the pandemic mortality.

4   Conclusions

The basic assumption of the propagation of a pandemic is that spatially neighboring locations will evolve similarly in the effects of the pandemic. In other words, propagation models assume a wave-like spatial propagation where the infection advances from one site to a neighboring contact site. Modern transport media may allow some spatial jumps, but if we consider the macroscopical level, such as the state spatial distribution in Brazil, then these spatial discontinuities are greatly reduced. We should expect the epidemic wave to propagate from one state to its neighbours, starting from the initial outbourst in Sao Paulo. The beginning of the pandemic also imposed wide strong restrictions on international travel, so that the influence of other countries may be strongly reduced, though microscopic data about travel and contagions should be required to assess these aspects. Therefore, conditions in the actual pandemic are in favor for the wave like propagation assumed by conventional epidemiological models predicting a huge mortality from this COVID-19 pandemic.

We have specifically focused on the mortality, which is the most important outcome and less subject to interpretations. We find that the propagation of the pandemic as reflected by several computed measures does not correlate well with the spatial distribution of the states. Focusing on the pandemic origin in Sao Paulo, we find that propagation time, cross-correlation and importance in the prediction of mortality do not correlate with the spatial distance among states. It can be argued that air travel dissolves these spatial distributions. Since the beginning of the pandemic outbreak, air travel has been extremely impeded, so its influence should be very small in the emergence of the propagation counterintuitive effects that we have noticed. Our conclusions are twofold. First, caution must be taken on the results of predictive models that assume a propagation model that is some kind of wave in an homogeneous medium. Second, the exploration of the propagation data at micro and macro spatial levels, from the small community to the countries, deserves further attention in order to understand the pandemic behavior.

References

1. Aguilar-Farias, N., et al.: Sociodemographic predictors of changes in physical activity, screen time, and sleep among toddlers and preschoolers in Chile during the COVID-19 pandemic. Int. J. Environ. Res. Public Health **18**(1) (2021). https://doi.org/10.3390/ijerph18010176. https://www.mdpi.com/1660-4601/18/1/176
2. Barański, K., Brożek, G., Kowalska, M., Kaleta-Pilarska, A., Zejda, J.E.: Impact of COVID-19 pandemic on total mortality in Poland. Int. J. Environ. Res. Public Health **18**(8) (2021). https://doi.org/10.3390/ijerph18084388. https://www.mdpi.com/1660-4601/18/8/4388

3. Bermejo-Martins, E., et al.: Different responses to stress, health practices, and self-care during COVID-19 lockdown: a stratified analysis. Int. J. Environ. Res. Public Health **18**(5) (2021). https://doi.org/10.3390/ijerph18052253. https://www.mdpi.com/1660-4601/18/5/2253

4. Brown, R.B.: Outcome reporting bias in COVID-19 MRNA vaccine clinical trials. Medicina **57**(3) (2021). https://doi.org/10.3390/medicina57030199. https://www.mdpi.com/1648-9144/57/3/199

5. Cakmakli, C., Demiralp, S., Ozcan, S.K., Yesiltas, S., Yildirim, M.A.: COVID-19 and emerging markets: the case of turkey. Technical report, 2011, Koc University-TUSIAD Economic Research Forum (2020). https://ideas.repec.org/p/koc/wpaper/2011.html

6. Emmerich, F.G.: Comparisons between the neighboring states of amazonas and pará in brazil in the second wave of COVID-19 outbreak and a possible role of early ambulatory treatment. Int. J. Environ. Res. Public Health **18**(7) (2021). https://doi.org/10.3390/ijerph18073371. https://www.mdpi.com/1660-4601/18/7/3371

7. Frediansyah, A., Tiwari, R., Sharun, K., Dhama, K., Harapan, H.: Antivirals for COVID-19: a critical review. Clin. Epidemiol. Global Health **9**, 90–98 (2021)

8. Gentile, I., Maraolo, A.E., Piscitelli, P., Colao, A.: COVID-19: time for post-exposure prophylaxis? Int. J. Environ. Res. Public Health **17**(11) (2020). https://doi.org/10.3390/ijerph17113997. https://www.mdpi.com/1660-4601/17/11/3997

9. Jindal, C., Kumar, S., Sharma, S., Choi, Y.M., Efird, J.T.: The prevention and management of COVID-19: seeking a practical and timely solution. Int. J. Environ. Res. Public Health **17**(11) (2020). https://doi.org/10.3390/ijerph17113986. https://www.mdpi.com/1660-4601/17/11/3986

10. Lagier, J.C., et al.: Outcomes of 3,737 COVID-19 patients treated with hydroxychloroquine/azithromycin and other regimens in Marseille, France: a retrospective analysis. Travel Med. Infect. Dis. **36**, 101791 (2020). https://doi.org/10.1016/j.tmaid.2020.101791. https://www.sciencedirect.com/science/article/pii/S1477893920302817

11. Liu, Y., Morgenstern, C., Kelly, J., Lowe, R., Jit, M.: The impact of non-pharmaceutical interventions on SARS-CoV-2 transmission across 130 countries and territories. BMC Med. **19**(1), 40 (2021). https://doi.org/10.1186/s12916-020-01872-8

12. Mayen Huerta, C., Cafagna, G.: Snapshot of the use of urban green spaces in Mexico City during the COVID-19 pandemic: a qualitative study. Int. J. Environ. Res. Public Health **18**(8) (2021). https://doi.org/10.3390/ijerph18084304. https://www.mdpi.com/1660-4601/18/8/4304

13. Mishra, S.K., Tripathi, T.: One year update on the COVID-19 pandemic: where are we now? Acta Tropica **214**, 105778 (2021). https://doi.org/10.1016/j.actatropica.2020.105778. https://www.sciencedirect.com/science/article/pii/S0001706X20316910

14. Perra, N.: Non-pharmaceutical interventions during the COVID-19 pandemic: a review. Phys. Rep. (2021). https://doi.org/10.1016/j.physrep.2021.02.001

15. Regmi, K., Lwin, C.M.: Factors associated with the implementation of non-pharmaceutical interventions for reducing coronavirus disease 2019 (COVID-19): a systematic review. Int. J. Environ. Res. Public Health **18**(8) (2021). https://doi.org/10.3390/ijerph18084274. https://www.mdpi.com/1660-4601/18/8/4274

16. Salazar-Fernández, C., Palet, D., Haeger, P.A., Román Mella, F.: COVID-19 perceived impact and psychological variables as predictors of unhealthy food and alcohol consumption trajectories: the role of gender and living with children as moderators. Int. J. Environ. Res. Public Health **18**(9) (2021). https://doi.org/10.3390/ijerph18094542. https://www.mdpi.com/1660-4601/18/9/4542

17. Sohrabi, C., et al.: World health organization declares global emergency: a review of the 2019 novel coronavirus (COVID-19). Int. J. Surg. **76**, 71–76 (2020). https://doi.org/10.1016/j.ijsu.2020.02.034. https://www.sciencedirect.com/science/article/pii/S1743919120301977

18. Trabelsi, K., et al.: Sleep quality and physical activity as predictors of mental well-being variance in older adults during COVID-19 lockdown: ECLB COVID-19 international online survey. Int. J. Environ. Res. Public Health **18**(8) (2021). https://doi.org/10.3390/ijerph18084329. https://www.mdpi.com/1660-4601/18/8/4329

19. Yan, Y., et al.: The first 75 days of novel coronavirus (SARS-COV-2) outbreak: recent advances, prevention, and treatment. Int. J. Environ. Res. Public Health **17**(7) (2020). https://doi.org/10.3390/ijerph17072323. https://www.mdpi.com/1660-4601/17/7/2323

20. Zhou, Y., et al.: A network medicine approach to investigation and population-based validation of disease manifestations and drug repurposing for COVID-19. PLOS Biol. **18**(11), 1–43 (2020). https://doi.org/10.1371/journal.pbio.3000970

A Hybrid Bio-Inspired Tabu Search Clustering Approach

Dragan Simić[1]([envelope]) [iD], Zorana Banković[2] [iD], José R. Villar[3] [iD], José Luis Calvo-Rolle[4] [iD], Svetislav D. Simić[1] [iD], and Svetlana Simić[5] [iD]

[1] Faculty of Technical Sciences, University of Novi Sad, Trg Dositeja Obradovića 6, 21000 Novi Sad, Serbia
dsimic@eunet.rs, {dsimic,simicsvetislav}@uns.ac.rs
[2] Frontiers Media SA, Paseo de Castellana 77, Madrid, Spain
[3] University of Oviedo, Campus de Llamaquique, 33005 Oviedo, Spain
villarjose@uniovi.es
[4] Department of Industrial Engineering, University of A Coruña, 15405 Ferrol-A Coruña, Spain
jlcalvo@udc.es
[5] Faculty of Medicine, University of Novi Sad, 21000 Novi Sad, Serbia
svetlana.simic@mf.uns.ac.rs

Abstract. The purpose of a data clustering process is to group a set of objects into multiple classes so that the objects in each class – cluster are similar according to certain rules or criteria, where the definition of similarity can be problem dependent. This paper is focused on a new bio-inspired clustering approach based on a model for combining tabu search algorithm (TS) and firefly algorithm (FF). The proposed hybrid bio-inspired system is tested on two well-known *Iris* and *Wine* data sets. Finally, the experimental results are compared with the parallel tabu search clustering algorithm. The proposed bio-inspired TS-FF clustering system shows a significantly better *accuracy* value for *Iris* data set.

Keywords: Data clustering · Tabu search · Firefly algorithm

1 Introduction

Clustering is a type of classification imposing on a finite set of objects. If the objects are characterised as patterns, or points in a multi-dimensional metrics space, the proximities can be distance of points, such as Euclidian distance, or any other distance metrics. In cluster analysis, the subject relates to the classification of individuals into groups based on the observation on each individual of the values of a number of variables, such as: continuous, discontinuous, dichotomous [1]; or even when applied to several types of data, such as uncertain data, multimedia data, graph data, biological data, stream data, text data, time series data, categorical data and big data.

Clustering is a process in which a group of unlabelled patterns is partitioned into several sets so that similar patterns are assigned to the same cluster, and dissimilar patterns are assigned to different clusters. The purpose of clustering is to identify natural

© Springer Nature Switzerland AG 2021
H. Sanjurjo González et al. (Eds.): HAIS 2021, LNAI 12886, pp. 436–447, 2021.
https://doi.org/10.1007/978-3-030-86271-8_37

groupings of data from a large data set to produce a concise representation of a system's behaviour.

In general, the classification methods, where the objects are labelled, are much more implemented in previous researches than in clustering methods. However, in the real-world setting, the decision-makers do not know the optimal strategy or the best operational decision in advance and therefore it is more important to research and, finally, implement clustering methods than it is to use classification methods. Cluster analysis is one of the most fundamental and essential data analysis tasks with broad applications in: industry [2, 3], logistics [4, 5], medicine [6–8], market research and market segmentation [9, 10].

Among various clustering approaches, the *k-means* clustering algorithm, which is proposed in the research paper [11], is one of the most popular and widely applied. Furthermore, like other heuristics, the *k-means* algorithm suffers from its susceptibility to become trapped in local optima. To address this issue, for a data analytical project, a Tabu Search algorithm is proposed, aiming to produce tighter and more cohesive clusters based on incorporating criteria for these characteristics in the objective function [12]. The algorithm succeeds both in obtaining better global solutions and in producing clusters that are more cohesive, but the computation time is greater than required by a well-implemented *k-means* method.

There are motivations for this research and new challenges for future research to help decision-makers in any fields such as: planning, production, marketing, economics, finances, business, and medicine, in order to make better and more optimal decisions. This paper presents the bio-inspired clustering system in general, combining *Tabu Search Algorithm* (TS) and *Firefly Algorithm* (FF). The proposed hybrid bio-inspired system is tested on data sets with well-known *Iris* and *Wine* data sets from UCI data repository [13].

The rest of the paper is organized in the following way: Sect. 2 provides an overview of the basic idea on clustering, classification and appropriate related works. Modelling the bio-inspired hybrid clustering system combining *Tabu Search Algorithm* and *Firefly Algorithm* is presented in Sect. 3. The preliminary experimental results are presented in Sect. 4. Section 5 provides conclusions and some directions for future work.

2   Clustering and Related Work

Clustering and classification are basic scientific tools used to systematize knowledge and analyse the structure of phenomena. The conventional distinction made between clustering and classification is as follows. Clustering is a process of partitioning a set of items, or grouping individual items, into a set of categories. Classification is a process of assigning a new item or observation to its proper place in an established set of categories. In clustering, little or nothing is known about category structure, and the objective is to discover a structure that fits the observations. The goal of clustering is descriptive, and that of classification is predictive [14].

2.1 Clustering

Clustering groups data instances into subsets in such a manner that similar instances are grouped together, while different instances belong to different groups. The instances are thereby organized into an efficient representation that characterizes the population being sampled.

Formally, the clustering structure is represented as a set of subsets $C = C_1,..., C_k$ of S, such that: $S = \bigcup_{i=1}^{k} C_i$ and $C_i \cap C_j = 0$ for $i \neq j$. Consequently, any instance in S belongs to exactly one and only one subset.

Clustering of objects is as ancient as the human need for describing the salient characteristics of people and objects and identifying them with a type. Therefore, it embraces various scientific disciplines: from mathematics and statistics to biology and genetics, each of which uses different terms to describe the topologies formed using this analysis. From biological "taxonomies" to medical "syndromes" and genetic "genotypes" to manufacturing "group technology" — the problem is identical: forming categories of entities and assigning individuals to the proper groups within it.

Cluster analysis, an important technology in data mining, is an effective method of analysing and discovering useful information from numerous data. Cluster algorithm groups the data into classes or clusters so that objects within a cluster have high similarity in comparison to one another but are very dissimilar to objects in other clusters. General references regarding data clustering and very detail discussion between clustering and classification are presented in textbooks from the beginning of seventies last century [1, 14, 15], and contemporary data mining clustering techniques can be found in the recently textbooks [16–18].

2.2 Related Work in Clustering Data

In data mining and machine learning, the problem of data clustering technique has been extensively studied because of its usefulness in the various applications to segmentation, summarization, target marketing, and bioinformatics [19]. The use of cluster analysis presents a complex challenge because it requires several methodological choices that determine the quality of a cluster solution. The research paper [20] chronicles the application of cluster analysis in strategic management research, where the technique has been used since the late 1970s in order to investigate issues of central importance. Forty-five different strategic studies reveal that the implementation of cluster analysis has been often less than ideal, perhaps detracting from the ability of studies to generate knowledge. Given these findings, suggestions are offered for improving the application of cluster analysis in future inquiry.

A nonconvex problem that has many local minima is considered in research paper [21]. The objects are represented by points in an n-dimensional Euclidean space, and the objective is to classify these m points into c clusters, such that the distance between points within a cluster and its centre is minimized. In that paper, a new algorithm for solving that problem based on a tabu search technique is developed. Preliminary computational experiences developed on the tabu search algorithm are encouraging and compare favourably with both the k-means clustering and the simulated annealing algorithms.

In research paper [22], an algorithm for cluster generation using tabu search approach with simulated annealing is proposed. The main idea of that algorithm is to use the tabu search approach to generate non-local moves for the clusters and apply the simulated annealing technique to select a suitable current best solution in order to speed the cluster generation. Experimental results demonstrate the proposed approach for cluster generation being superior to the tabu search approach with Generalised Lloyd algorithm.

The clustering problem under the criterion of the minimum sum of squares clustering is a nonconvex program which possesses many locally optimal values, resulting that its solution often falls into these traps. In the paper [23], a hybrid tabu search based clustering algorithm called *KT-Clustering* is developed to explore the proper clustering of data sets. Based on the framework of tabu search, *KT-Clustering* gathers the optimization property of tabu search and the local search capability of *k-means* algorithm together, and this is extensively demonstrated for the applied experimental data sets.

3 Modelling the Hybrid Bio-Inspired Clustering System

Clustering algorithms can group the given data set into clusters by different approaches: (i) *hard partitioning methods* – such as – *k*-means, *k*-medoids, *k*-medians, *k*-means ++; and (ii) *fuzzy partitioning methods* – such as – *fuzzy c-means clustering* method, *fuzzy* Gustafson-Kessel clustering method, and *fuzzy* Gath-Geva clustering method.

There are also finer distinctions possible, such as: (1) *strict partitioning clustering*, where each object belongs to exactly one cluster; (2) *strict partitioning clustering with outliers*: objects can also belong to no cluster, and are considered outliers; (3) *overlapping clustering*: objects may belong to more than one cluster; (4) *hierarchical clustering*: objects that belong to a child cluster also belong to the parent cluster; (5) *subspace clustering*: while overlapping clustering, within a uniquely defined subspace, clusters are not expected to overlap.

In this research, a new hybrid bio-inspired approach which combines tabu search algorithm [11] to minimise intra-cluster distance, and firefly algorithm [24] to maximise inter-cluster distance is used. The proposed system is tested with well-known *Iris* and *Wine* data sets from UCI data repository [13].

3.1 Tabu Search Algorithm

Consider a data set of N_p objects, and each data point in the data set has k attributes. A data point x_t will be represented by a vector $x_t = (x_{t1}, x_{t2}, ..., x_{tk})$. The underlying data set can then be represented by.

$$X = \{x_t : t = 1, ..., N_p\} \tag{1}$$

where N_p identifies the total number of data points. Let N_s be the total number of clusters to be built and denote each cluster by C_i, then the resultant cluster set will be:

$$C = \{C_i : i = 1, ..., N_s\} \tag{2}$$

The goal of the clustering problem is to group data points into a pre-defined number of clusters by the criteria previously discussed, minimizing intra-cluster distance and maximizing inter-cluster distance, so that data points lying within the same cluster are as close, *similar*, as possible to the points lying in different clusters which should be as far apart, *dissimilar*, as possible. The similarity is represented by a distance measure, and seeks a collection of clusters that minimizes intra-cluster distance and maximizes inter-cluster distance.

In the research paper [12], for the Tabu Search notation, $Score\ (x_i, x_j)$, representing the similarity of a pair of data points (x_i, x_j), is introduced. $Score\ (x_i, x_j)$ describes the degree of correlation between two data points, so that smaller values indicate a greater desirability for assigning the points to the same cluster and larger values indicate a greater desirability for assigning them to different clusters. Euclidean distance is used to describe the similarity of two data points in a k-dimensional space, but also other distance measures can be used.

$$Score(x_i, x_j) = \sqrt{\sum_{p=1}^{k} (x_{ip} - x_{jp})^2} \tag{3}$$

The algorithm makes use of distances between data points in several different ways depending on the objective, though in general can be helpful in identifying a data point x_i that maximizes or minimizes the sum of the distances from the data point to another data point or to a set of data points. The total distance, from all data points x_j to the centroid x_i, of cluster C_i is obtained by setting $L = \{x_i\}$ and $U = C_i \setminus \{x_i\}$, and it is written as follows:

$$V(i) = \sum_{j=1}^{n} Score(x_i, x_j),\ x_i \in L,\ x_j \in U,\ x_i \neq x_j \tag{4}$$

Under the foregoing settings, the objective function of our clustering problem that we seek to minimize is then denoted by the following equation:

$$objVal = min \sum_{i=1}^{Ns} V(i) \tag{5}$$

TS is a metaheuristic algorithm designed to guide subordinate heuristic search processes to escape the trap of local optimality. TS is distinguished from other metaheuristics by its focus on using adaptive memory and special strategies for exploiting this memory. Memory is often divided into short-term and long-term memory, and the TS strategies for taking advantage of this memory are often classified under the headings of intensification and diversification. A common form of TS short-term memory is a recency-based memory that operates to temporarily prevent recently executed moves from being reversed for a duration (number of iterations) known as the tabu tenure. In this current implementation, a simple version of TS that uses recency-based memory alone is used.

A neighbourhood is the solution subspace of the current solution which defines the available moves for generating a new solution at the next iteration. An inappropriate neighbourhood may miss the opportunity to explore more promising solution spaces or may result in spending too much time examining unnecessary or unpromising spaces. In this research, a neighbourhood definition results from creating a sphere that places a centroid at its centre. For a given cluster C_i with centroid x_i, we define the neighbourhood $N(i)$ of this cluster as follows:

$$N(i) = \{x_j : |x_j - x_i| \leq R_i, x_i, x_j \in C_i, x_i \neq x_j \tag{6}$$

Here R_i is the radius of the neighbourhood sphere, and $|x_j - x_i|$ is defined by the Eq. 3. The sphere radius value plays an important role in our algorithm search strategy and is critical to performance. The following procedure determines the radius value R. Relative to a given cluster C_i with centroid at x_i and the data points, R_i is defined as follows:

$$R_i = \sum_{i=1}^{|C_i|} Score\,(x_j, x_i) \,/\, |C_i|, \; x_i, \, x_j \in C_i, \, x_i \neq x_j \tag{7}$$

Clearly, R_i varies as a function of the individual cluster C_i it is associated with, and in this sense changes dynamically for each cluster as the cluster's composition changes. The overall neighbourhood of the TS algorithm is then:

$$NB = \cup_{i=1}^{N_s} N(i) \tag{8}$$

The *tabu list* (**TL**) holds the information for those solutions that cannot be revisited during the next t iterations. The candidate list **CL** is a subset of the neighbourhood **NB** which is generated to reduce the computational effort of examining the complete neighbourhood.

A variety of more advanced procedures is given in [25] that replaces the maximum distance measure with generalizations of a *MaxMin* distance measure utilizing iterative refinement and adaptive thresholds. In this research *Firefly Algorithm* is used to define maximum distance measure between centroids.

3.2 Firefly Algorithm

Firefly algorithm (FA) imitates the social behaviour of fireflies flying in the tropical summer sky. Fireflies communicate, search for pray and find mates using bioluminescence with varied flashing patterns. By mimicking nature, various metaheuristic algorithms can be designed. Some of the flashing characteristics of fireflies were idealized so as to design a firefly-inspired algorithm [24]. For simplicity's sake, only three rules were followed: (i) All fireflies are unisex so that one firefly will be attracted by other fireflies regardless of their sex; (ii) Attractiveness is proportional to firefly brightness. For any pair of flashing fireflies, the less bright one will move towards the brighter one. Attractiveness is proportional to the brightness which decreases with increasing distance between fireflies; (iii) If there are no brighter fireflies than a particular firefly, this individual will move at random in the space.

Algorithm 1: *The algorithm of firefly algorithm*

Begin

        Step 1:   **Initialization.** *Set the generation counter* **G** *= 1; Initialize the population of n fireflies randomly and each firefly (**centroid**);*
                        define light absorption coefficient γ;
                        set controlling the step size α *and the initial attractiveness* β₀ *at* **r** *= 0.*
        Step 2:   *Evaluate the light intensity* **I** *for each firefly in* **P** *determined by f(x)*
        Step 3:   **While** *the termination criteria are not satisfied* **or** **G** < MaxGeneration **do**
                    for *i=1:n (all n fireflies)* **do**
                        for *j=1:n (n fireflies)* **do**
                            if *($I_j < I_i$),*
                              move firefly i towards j;
                          end if
                          Vary attractiveness with distance r via exp[$-\gamma r^2$];
                          Evaluate new solutions and update light intensity;
                      end for j
                    end for i
                    G = G+1;
        Step 4:   **end while**
        Step 5:   *Post-processing the results and visualization;*

End.

The brightness of a firefly is influenced or determined by the objective function. For a maximization problem, brightness can simply be proportional to the value of the cost function. The basic steps of the FA are summarized by the pseudo code shown in Algorithm 1. The distance between any firefly (centroid) i and firefly (centroid) j (r_{ij}) can be defined as the number of different arcs between them. Hence, the number of different arcs between firefly i and firefly j is > 0. Then, the distance between any two fireflies is calculated using the following equation:

$$r = \frac{A}{N} * 10 \tag{9}$$

where r is the distance between any two centroids, A is the total number of different arcs between two fireflies, and N is the number of centroids. The formula scales r in the interval [0, 10] as r will be used in the attractiveness calculation. The main form of the attractiveness function $\beta(r)$ can be any monotonic decreasing function

$$\beta(r) = \beta_0 e^{-\gamma r^2} \tag{10}$$

where r is the distance between two fireflies (centroids), β_0 is the attractiveness at $r = 0$, and γ is a fixed light absorption coefficient. In this research, γ is in the interval [0.01, 0.15] so that the attractiveness of a firefly is viewed by the others.

3.3 Hybrid Tabu Search and Firefly Based Clustering Algorithm

The method of constructing an initial centroid solution is based on a maximum distance idea expressed in terms of a sum of distances from existing centroids. The first centroid is a randomly selected data point x_j, and the rest of centroids are selected in terms of a maximum sum of distances from existing centroids. Unlike *k-means* algorithm, the centroids in this algorithm are real data points.

Algorithm 2: *The algorithm for* **Hybrid Tabu Search and Firefly clustering algorithm**

Begin

Step 1: *Initialization.* X // the data set; C_i // the ith cluster; N_s // the number of clusters; $V(k)$ // sum of distances for data point N_s determined by equation (4); L // the set of centroids of all clusters;

S // the solution at the current iteration; S' // the locally best solution at the current iteration; $S*$ // the best solution found so far

$Z(.)$ // the objective function value equation (5)

NI // the number of consecutive iterations without improvement of $S*$

$MaxNI$ // a predefined maximum limit on NI

Step 2: *Calling: Create_Initial_Solution() and getting initial solution*

$S = \{Ci : i = 1, ... , Ns\}; NI = 0;$

Setting corresponding values Z(.)

Step 3: **While** *the termination criteria is not satisfied* **do**

Step 4: **for** *all clusters C1, C2 , ... , Cs*

$CL(i) = CreateCandidateList(i)$ *// creating candidate list*

Step 5: **for** *each data point xk* $\in$ *CL(i)*

Use firefly algorithm

Vary attractiveness with distance r via $exp[-\gamma r^2]$;

Evaluate new solutions and update light intensity;

Setting x_k to be the new centroid

end for

Step 6: C = *assigning non-centroid data points to their closest centroids*

if *(S meets tabu condition and Z(S) < Z(S')) or Z(S) < Z(S*)*

$S' = S; Z(S') = Z(S);$

end if

if $Z(S') < Z(S*)$

$S* = S'; NI = 0$

else

$NI = NI + 1$

end if

if $NI = MAXNI$

$S = S'$

Goto **Step 4**

end if

if *S* could not improve*

Terminating the algorithm

else

$NI = 0$

Goto **Step 4**

end if

end for

Step 7: **end while**

Step 8: *Post-processing the results;*

End.

---

4   Experimental Results and Discussion

The proposed bio-inspired clustering system which combines *Tabu Search Algorithm* and *Firefly Algorithm* is tested on data sets collected with well-known *Iris* and *Wine* data sets from UCI data repository [13]. The data types contained in the data sets *Iris* and *Wine*, for the computational experiments, are different so that we may validate the applicability and performance of the proposed system. Furthermore, all data points contained in the data sets have been labelled so that the correctness of the clustering outcomes can be verified relatively easily.

The neighbourhood $N(i)$ described above plays an important role in obtaining satisfactory clustering results. To investigate the impacts of various neighbourhood sizes in more detail, the following three settings for the neighbourhood size are tested: (i) Small: half of that defined by Eq. (7), $R_i/2$ for individual cluster C_i; (ii) Standard: as defined by the Eq. (7), R_i for all clusters C_i; Large: the largest distance among those from the current centroid to all data points.

The experimental results based on the *Iris* data set research with different number of iterations and different radius sizes and the most popular thresholds metrics *Accuracy*, are presented in Table 1. In this research, the focus is on the multi-class problem, and detail description about thresholds metrics on the multi-class problem is presented in [8]. The quality measurements *Precision, Recall* and F_1 *score* for the hybrid bio-inspired clustering system for three classes ("setosa","versicolor","virginica") are presented in Table 2.

The experimental results from this research are compared with experimental results presented in the research paper from 2018, where a tabu search parallel cluster algorithm is implemented [26], and the comparison is presented in Table 3. From Table 3 it can be observed that the bio-inspired TS-FF clustering system, proposed in this research paper, has significantly better *accuracy* value for *Iris* data set for every parameter except for parameters: *number of iterations = 1000* and *Radius size = large*. Also, parallel tabu search algorithm is a significantly better value for all *accuracy* for *Wine* data set.

Table 1. The experimental results for *Iris* data set

No. of iterations	Radius size	Items in cluster 1	Items in cluster 2	Items in cluster 3	Centroids (data points)	Accuracy %
500	Small	50	84	16	14, 79, 119	77.33
500	Standard	50	78	22	23, 91, 119	81.33
500	Large	50	77	23	23, 100, 119	82.00
1000	Small	50	70	30	23, 60, 119	86.66
1000	Standard	50	68	32	14, 81, 118	88.00
1000	Large	50	66	34	23, 80, 119	89.33

Table 2. Quality measurement for the hybrid bio-inspired clustering system for the best experimental results (*Accuracy* = 89.33%)

	Setosa	Versicolor	Virginica	Average
Precision %	100.00	79.31	90.50	89.94
Recall %	100.00	90.48	76.00	88.82
F_1 score %	100.00	84.52	82.62	89.04

Table 3. The experimental results and impacts of algorithm parameters compared with research paper [26]

Data set	No. of iterations	Radius size	Accuracy (%)	Research
Iris	500	Small	61.22	[26]
			77.33	
Iris	500	Standard	69.28	[26]
			81.33	
Iris	500	Large	70.12	[26]
			82.00	
Iris	1000	Small	62.64	[26]
			86.66	
Iris	1000	Standard	**90.52**	[26]
			88.00	
Iris	1000	Large	**92.34**	[26]
			89.33	
Wine	500	Standard	**58.79**	[26]
			54.49	
Wine	1000	Standard	**71.57**	[26]
			55.62	
Wine	1000	Large	**72.15**	[26]
			56.18	

It is not easy to find the main reason why parallel tabu search algorithm has significantly better value for all *accuracy* for *Wine* data set in relation to *accuracy* value for *Iris* data set. Both data sets have similar number of instances: *Iris* − 150, *Wine* − 178; the same number of classes: 3; but very different number of attributes: *Iris* − 3, *Wine* −13. Maybe, the main reason for differences *accuracy* value is the different number of attributes. The second reason could be the usage of Euclidean distance to describe the similarity of two data points in a k-dimensional space.

5 Conclusion and Future Work

The aim of this paper is to propose the new bio-inspired clustering system. The new proposed clustering system is obtained by combining tabu search algorithm and firefly algorithm. The proposed hybrid system is tested on data sets collected with well-known *Iris* and *Wine Data Sets* from UCI data repository. The experimental results of bio-inspired clustering system are presented and then compared with experimental results of parallel tabu search algorithm. The experimental results demonstrate that bio-inspired tabu search and firefly system have significantly better *accuracy* value for *Iris* data set. But, on the other side, parallel tabu search algorithm has better *accuracy* value for *Wine* data set.

This study and preliminary experimental results encourage further research by the authors, since the *accuracy* value for *Iris* data set is between 77–90% for different configurations and parameters. Our future research will focus on creating a hybrid model which combines different Tabu search configurations and different firefly parameters to efficiently solve cluster analysis.

References

1. Kendall, M. G.: The basic problems in cluster analysis. In: Cacoullous, T., (ed.) Discriminamt Analysis and Applications, pp. 179–191. Academic Press, Inc., New York (1973). https://doi.org/10.1016/B978-0-12-154050-0.50016-2
2. Simić, D., Svirčević, V., Sremac, S., Ilin, V., Simić, S.: An efficiency k-means data clustering in cotton textile imports. In: Burduk, R., Jackowski, K., Kurzyński, M., Woźniak, M., Żołnierek, A. (eds.) Proceedings of the 9th International Conference on Computer Recognition Systems CORES 2015. AISC, vol. 403, pp. 255–264. Springer, Cham (2016). https://doi.org/10.1007/978-3-319-26227-7_24
3. Simić, D., Jackowski, K., Jankowski, D., Simić, S.: Comparison of clustering methods in cotton textile industry. In: Jackowski, K., Burduk, R., Walkowiak, K., Woźniak, M., Yin, H. (eds.) IDEAL 2015. LNCS, vol. 9375, pp. 501–508. Springer, Cham (2015). https://doi.org/10.1007/978-3-319-24834-9_58
4. Simić, D., Ilin, V., Tanackov, I., Svirčević, V., Simić, S.: A hybrid analytic hierarchy process for clustering and ranking best location for logistics distribution center. In: Onieva, E., Santos, I., Osaba, E., Quintián, H., Corchado, E. (eds.) HAIS 2015. LNCS (LNAI), vol. 9121, pp. 477–488. Springer, Cham (2015). https://doi.org/10.1007/978-3-319-19644-2_40
5. Simić, D., Ilin, V., Svirčević, V., Simić, S.: A hybrid clustering and ranking method for best positioned logistics distribution Centre in Balkan Peninsula. Logic J. IGPL 25(6), 991–1005 (2017). https://doi.org/10.1093/jigpal/jzx047
6. Krawczyk, B., Simić, D., Simić, S., Woźniak, M.: Automatic diagnosis of primary head- aches by machine learning methods. Open Med. 8(2), 157–165 (2013). https://doi.org/10.2478/s11536-012-0098-5
7. Simić, S., Sakač, S., Banković, Z., Villar, J.R., Simić, S.D., Simić, D.: A hybrid bio-inspired clustering approach for diagnosing children with primary headache disorder. In: de la Cal, E.A., Villar Flecha, J.R., Quintián, H., Corchado, E. (eds.) HAIS 2020. LNCS (LNAI), vol. 12344, pp. 739–750. Springer, Cham (2020). https://doi.org/10.1007/978-3-030-61705-9_62

8. Simić, S., Villar, J.R., Calvo-Rolle, J.L., Sekulić, S.R., Simić, S.D., Simić, D.: An application of a hybrid intelligent system for diagnosing primary headaches. Int. J. Environ. Res. Public Health **18**(4), 1890 (2021). https://doi.org/10.3390/ijerph18041890

9. Herrero, A., Jiménez, A., Alcalde, R.: Advanced feature selection to study the internationalization strategy of enterprises. Peer J. Comput. Sci. **7**, e403 (2021). https://doi.org/10.7717/peerj-cs.403

10. Ližbetinová, L., Štarchon, P., Lorincová, S., Weberová, D., Pruša, P.: Application of cluster analysis in marketing communications in small and medium-sized enterprises: an empirical study in the Slovak Republic. Sustainability **11**, 2302 (2019). https://doi.org/10.3390/su11082302

11. MacQueen, J.B: Some methods for classification and analysis of multivariate observations. In: Proceedings of 5th Berkeley Symposium on Mathematical Statistics and Probability. Berkeley, vol. 5.1, 281–297. University of California Press (1967)

12. Cao, B., Glover, F., Rego, C.: A Tabu search algorithm for cohesive clustering problems. J. Heuristics **21**(4), 457–477 (2015). https://doi.org/10.1007/s10732-015-9285-2

13. UCI Machine Learning Repository: http://archive.ics.uci.edu/ml/datasets.html. Accessed 28 Feb 2018

14. Hartigan, J.: Clustering Algorithms. John Wiley & Sons, New York (1975)

15. Jain, A., Dubes, R.: Algorithms for Clustering Data. Prentice-Hall, NJ (1988)

16. Woźniak, M.: Hybrid Classifiers: Methods of Data, Knowledge, and Classifier Combination. Springer, Berlin (2016)

17. Hennig, C., Meila, M., Murtagh, F., Rocci, R.: Handbook of Cluster Analysis. CRC Press, Chapman & Hall book, Boca Raton (2016)

18. Wierzchoń, S.T., Kłopotek, M.A.: Modern Algorithms of Cluster Analysis. Springer International Publishing AG, Cham, Switzerland (2018)

19. Aggarwal, C.C., Reddy, C.K.: Data Clustering: Algorithms and Applications. CRC Press Inc., Taylor & Francis Group, LLC, Boca Raton (2013)

20. Ketchen, D.J., Jr., Shook, C.L.: The application of cluster analysis in strategic management research: an analysis and critique. Strateg. Manage. J. **17**(6), 441–458 (1996). https://doi.org/10.1002/(SICI)1097-0266(199606)

21. Al-Sultan, K.S.: A Tabu search approach to the clustering problem. Pattern Recogn. **28**(9), 1443–1451 (1995). https://doi.org/10.1016/0031-3203(95)00022-R

22. Chu, S.-C., Roddick, J.F.: A clustering algorithm using the Tabu search approach with simulated annealing. In: Brebbia, C.A., Ebecken N.F.F. (Eds) Data Mining II. WIT Press (2000). ISBN 1–85312–821-X

23. Liu, Y., Liu, Y., Wang, L., Chen, K.: A hybrid Tabu search based clustering algorithm. In: Khosla, R., Howlett, R.J., Jain, L.C. (eds.) KES 2005. LNCS (LNAI), vol. 3682, pp. 186–192. Springer, Heidelberg (2005). https://doi.org/10.1007/11552451_25

24. Yang, X.-S.: Nature-Inspired Metaheuristic Algorithms. Luniver Press, UK (2008)

25. Glover, F.: Pseudo-centroid clustering. Soft. Comput. **21**(22), 6571–6592 (2016). https://doi.org/10.1007/s00500-016-2369-6

26. Lu, Y., Cao, B., Rego, C., Glover, F.: A Tabu search based clustering algorithm and its parallel implementation on Spark. Appl. Soft Comput. **63**, 97–109 (2018). https://doi.org/10.1016/j.asoc.2017.11.038

A Non-parametric Fisher Kernel

Pau Figuera[(✉)] and Pablo García Bringas[(✉)](ID)

University of Deusto - Deustuko Unibertsitatea, D4K Group, University of Deusto,
Unibertsitate Etorb. 24, 48007 Bilbao, Spain
pau.figuera@opendeusto.es

Abstract. In this manuscript, we derive a non-parametric version of the Fisher kernel. We obtain this original result from the Non-negative Matrix Factorization with the Kullback-Leibler divergence. By imposing suitable normalization conditions on the obtained factorization, it can be assimilated to a mixture of densities, with no assumptions on the distribution of the parameters. The equivalence between the Kullback-Leibler divergence and the log-likelihood leads to kernelization by simply taking partial derivatives. The estimates provided by this kernel, retain the consistency of the Fisher kernel.

Keywords: Fisher Kernel · Non-negative matrix factorization · Kullack-Leibler divergence · Non-parametric

1 Introduction

Classification problems appear in many areas in which an assignment based on previous observations is necessary. The problem is solved by introducing a qualitative variable, or label, to entities, observations, or multivalued instances. Then, the labels of a piece of the set are known, and the task consists to assign them to the unknown subset. This problem, central in classical inference, has some problems: asymptotic hypotheses are difficult to justify with finite samples, becoming even worse when the distribution is unknown [11]. Techniques that introduce a Bayesian classifier to the posterior distribution provide good estimates [8]. Equivalent results, without statistical hypotheses, can be obtained when a measure of similarity is used [19]. These techniques and algorithms constitute an important field in Machine Learning. Known as semi-supervised methods, they have gained positions in recent years in Computational Engineering, and a reason is the scalability [26].

The idea behind this kind of methods is to represent the observations as vectors in a space (input space). This structure lets to build a decision surface. The first scheme was to divide into two pieces this space [7], and further, it was generalized to three or more classes [3,18]. The similarity measure of the dot product lets to classify new observations. The structure of the space and the transformations (feature space) to achieve separability, is unique [2]. The core of the dot products space is called the kernel.

© Springer Nature Switzerland AG 2021
H. Sanjurjo González et al. (Eds.): HAIS 2021, LNAI 12886, pp. 448–459, 2021.
https://doi.org/10.1007/978-3-030-86271-8_38

There are many kernels with interesting properties. One of them is the Fisher kernel, attractive for the use of the posterior, connected with the Bayesian Theory, and providing consistent results [27]. The first proposal of this kernel interprets the scalar products as *the basic comparison between the examples, defining what is meant by an "inner product"* [19]. Other derivations of this kernel are obtained in the context of information retrieval [15], within the specific problem of tf-idf (term frequency-inverse document frequency), [12]. The PLSA (Probabilistic Latent Semantic Analysis) in its symmetric formulation [17], leads to a more general derivation [4]. The main limitation in all those derivations is the assumption of a parametric generative model.

The introduction of the NMF (Non-negative Matrix Factorization) algebraic techniques [5] allows for identify matrices with probabilistic mixtures when suitable normalization conditions are imposed [9]. Under these assumptions, kernels can be obtained too. A purely algebraic one, with no statistical connection, is the KNMF (kernel Non-negative Matrix Factorization) [28]. Assuming discrete class distributions, the maximization of the log-likelihood provides a good kernelization too [22]. Moreover, this kernel is relevant since it satisfies statistical and computational requirements, allowing simultaneous application of similarity measurement and visualization techniques, as well as sequential learning when is required [14].

In this work, we present a non-parametric version of the Fisher kernel. In addition, we show that under certain conditions, it lets to classify with an arbitrary (small) error. We have structured the exposition into sections that correspond to the steps we follow to achieve these results.

In Sect. 2 we obtain the NMF formulas, using the KL (Kullback-Leibler) divergence as the objective function. We choice this divergence due to the equivalence with the EM (expectation-maximization) algorithm (used to maximize the likelihood) [21]. With these formulas, in Sect. 3, we derive the Fisher kernel by simply taking the Laplace expansion for the likelihood. Then, identifying the expectation of the second-order term, or Hessian, with the Fisher Information matrix, the kernel is obtained after very simple algebra manipulations. Section 4 is devoted to explaining the conditions for which an arbitrary (small) classification error can be obtained, providing a demonstration. Finally, we illustrate this result with very simple examples (Sect. 5).

2   Non-negative Matrix Factorzation with KL Divergence

A set of m non-negative real valued instances in $\mathbb{R}_+^m$ can be represented as a collection of m vectors $\mathbf{x}_i$. The full set of vectors is a non-negative entries matrix $\mathbf{X}$,[1] and the transformation

$$[\mathbf{Y}]_{ij} = [\mathbf{X}]_{ij}\, \mathbf{D}_X^{-1}\, \mathbf{M}_X^{-1} \qquad (i = 1,\ldots,m \ \text{and}\ j = 1,\ldots,n) \tag{1}$$

with $\mathbf{D}_X$ and $\mathbf{M}_X$ diagonal matrices defined as

$$\mathbf{M}_X = \mathrm{diag}\left(\sum_j x_{ij}\right) \tag{2}$$

$$\mathbf{D}_X = \mathrm{diag}\,(m) \tag{3}$$

ensures that the change of $\mathbf{X} \in \mathbb{R}_+^{m \times n}$ to $\mathbf{Y} \in \mathbb{R}_{[0,1]}^{m \times n}$ is invariant under measurement scale changes.

Taking into account that every non-negative real entries matrix admits a non-negative factorization [5], the approximation

$$[\mathbf{Y}]_{ij} \approx [\mathbf{W}]_{ik}\,[\mathbf{H}]_{kj} \quad \left(\text{with } [\mathbf{W}]_{ik} \geq 0;\ [\mathbf{H}]_{kj} \geq 0 \ \text{and}\ k \in \mathbb{Z}_+\right) \tag{4}$$

holds, and

$$[\widehat{\mathbf{Y}}]_{ij} = [\mathbf{W}]_{ik}[\mathbf{H}]_{kj} \tag{5}$$

approximates the transformed data matrix (as the hat symbol indicates).

To adjust the relation 4 is necessary to minimize some distance or divergence. Formally, a distance d is a map that satisfies, for vectors $\mathbf{a}$, $\mathbf{b}$, and $\mathbf{c}$ the conditions: (i) $d(\mathbf{a}, \mathbf{b}) = d(\mathbf{b}, \mathbf{a})$ (symmetry); (ii) $d(\mathbf{a}, \mathbf{b}) = 0$ iff $\mathbf{a} = \mathbf{b}$ (identity); and (iii) $d(\mathbf{a}, \mathbf{b}) \leq d(\mathbf{a}, \mathbf{c}) + d(\mathbf{c}, \mathbf{b})$ (triangular inequality). A divergence D does not satisfy one of these axioms, usually symmetry.

Roughly speaking, both, distances and divergences induces metrics in a space, and the relative position of the objects (usually distances) is measured by a norm. The L_p type norms are widely used, and denoted as $\| \cdot \|_p$, where the dot represents the object(s). For $p = 2$ we have the euclidean norm. The case of $p = 1$, is the Hilbert norm (which should be not confused with the Hilbert-Schmidt norm, which is another denomination for the L_2 one), and it is the ordinary sum of the involved elements.

---

[1] In the NMF framework it is more convenient to write the matrix $\mathbf{X} = (x_{ij})$ (with the subscripts i and j defined in 1) as

$$[\mathbf{X}]_{ij} = \mathbf{X}$$

This notation is suitable when some index can vary (normally k in 4). If the matrix operations do not depend on the index that varies, eliding the subscripts simplifies the notation (as happens for diagonal matrices). For the rest of the objects, we refer to them as usual, being the matrices indicated as a bold capital letter, and vectors as bold lowercase.

We choose the KL divergence, due to the full equivalence with the log-likelihood. For this divergence, the log-likelihood maximization with the EM algorithm is full equivalent to minimize the KL divergence, as is explained in [21] between many others. For the matrices $\mathbf{Y}$ and $\mathbf{WH}$ is defined as[2]

$$D_{KL}(\mathbf{Y} \| \mathbf{W\,H}) = \sum_{ij} [\mathbf{Y}]_{ij} \odot \log \frac{[\mathbf{Y}]_{ij}}{[\mathbf{W\,H}]_{ij}} \tag{6}$$

$$= \sum_{ij} \left([\mathbf{Y}]_{ij} \odot \log [\mathbf{Y}]_{ij} - [\mathbf{Y}]_{ij} \odot \log [\mathbf{W\,H}]_{ij} \right) \tag{7}$$

Taking the KKT (Karush-Kuhn-Tucker) conditions, with the restriction of nonnegativity entries for all the involved matrices, the solutions are

$$[\mathbf{W}]_{ik} \leftarrow [\mathbf{W}]_{ik} \odot \left(\frac{[\mathbf{Y}]_{ij}}{[\mathbf{W\,H}]_{ij}} [\mathbf{H}]_{kj}^{\top} \right) \tag{8}$$

$$[\mathbf{H}]_{kj} \leftarrow [\mathbf{H}]_{kj} \odot \left([\mathbf{W}]_{ik}^{\top} \frac{[\mathbf{Y}]_{ij}}{[\mathbf{W\,H}]_{ij}} \right) \tag{9}$$

where $\odot$ is the elementwise product (see [29, pp. 72–73] for a definition of this matrix product. For the main properties see [29, chap. 1]).

Selecting a value for k, which we call *model components*, being more extended in literature to refer it as *components* (and they identified too with latent variables), initializing $\mathbf{W}$ and $\mathbf{H}$ (often randomly) and alternating between 8 and 9 until a desired approximation degree in is achieved, the non-negative entries matrix $\widehat{\mathbf{Y}}$ that approximates $\mathbf{Y}$ is obtained. Details for convergence and properties for the choice of k are demonstrated in our work [13].

Then NMF product can be written indistinctly as

$$[\mathbf{Y}]_{ij} \approx [\mathbf{W}]_{ik}[\mathbf{H}]_{kj} \tag{10}$$

$$\approx [\widehat{\mathbf{Y}}]_{ij} \tag{11}$$

The rows or columns normalization in 4 is fully equivalent to conditional probability mixtures [9]. The KL divergence induces the L_1 norm (in fact, is

---

[2] Several authors refers to the KL divergence as

$$D_{KL}(\mathbf{Y} \| \mathbf{W\,H}) = \sum_{ij} \left([\mathbf{Y}]_{ij} \odot \log \frac{[\mathbf{Y}]_{ij}}{[\mathbf{W\,H}]_{ij}} - [\mathbf{Y}]_{ij} + [\mathbf{WH}]_{ij} \right)$$

which we prefer to call it as I-divergence or generalized KL-divergence, according to [6, p. 105], and reserving the name of KL divergence for the mean information (given by the formula 6), following the original denomination of S. Kullback and R.A. Leibler [20].

the sum of the formulas 6 or 7). Imposing this normalization conditions on each column of $\mathbf{Y}$ as

$$\|\mathbf{y}_j\|_1 = \frac{y_{ij}}{\sum_i y_{ij}} \qquad (j = 1, \ldots, n) \tag{12}$$

results the column stochastic (or probabilistic, since every column is a uni-variate density) matrix $\tilde{\mathbf{Y}}$. Proceeding in equal way for the columns of $\mathbf{W}$, and the rows of $\mathbf{H}$ after simplifying, it follows

$$[\widehat{\tilde{\mathbf{Y}}}]_{ij} = [\widetilde{\mathbf{W}}]_{ik}[\widetilde{\mathbf{H}}]_{kj} \tag{13}$$

and

$$[\widehat{\tilde{\mathbf{Y}}}]_{ij} \approx [\widetilde{\mathbf{Y}}]_{ij} \tag{14}$$

indicating the tilde symbol that the matrix is normalized, and being now $\widetilde{\mathbf{W}}$ and $\widetilde{\mathbf{H}}$ a set of k mixtures. In the next sections, we reefer always to Eqs. 13 and 14, and omitting the tilde symbol to simplify the notation.

3   Fisher Kernel in the Non-parametric Framework

The Fisher kernel is defined in [16]

$$K(\mathbf{Y}, \mathbf{Y}^\top) = U_\theta(\mathbf{Y})\mathbf{I}_F^{-1}U_\theta(\mathbf{Y}) \tag{15}$$

being

$$U_\theta(\mathbf{Y}) = -\frac{\partial}{\partial \theta} \log P(\mathbf{Y}|\theta) \tag{16}$$

the Fisher Scores, and

$$\mathbf{I}_F = E_\mathbf{Y}\left[U_\theta U_\theta^\top\right] \tag{17}$$

the Fisher Information Matrix.

In this definition, $\mathbf{Y}$ is a random variable (a mixture is a multivariate random variable), and $f(\mathbf{Y}|\theta)$ is the conditional probability for a known value of the parameter θ. It can be written as $f_\theta(\mathbf{Y}|\theta)$ or simply $f_\theta(\mathbf{Y})$, and $E(\cdot)$ is the expectation. The parametric Fisher kernel, despite requires complicated derivatives (as in [4]), is conceptually simple (particularly in the case of the exponential family). The NMF point of view shares simplicity, without messy differentiation.

Nevertheless, the NMF in the non-parametric case needs some considerations. Now, the parameter is $\theta = \{\mathbf{W}, \mathbf{H}\}$, and taking into account the second term of 7 is the log-likelihood l, we can write the matrix $\mathbf{L}$

$$[\mathbf{L}_{\widehat{\mathbf{Y}}}]_{ij} = [\mathbf{Y}]_{ij} \log[\widehat{\mathbf{Y}}]_{ij} \tag{18}$$

for which, the relation

$$l_{\widehat{\mathbf{Y}}} = \left\| [\mathbf{L}_{\widehat{\mathbf{Y}}}]_{ij} \right\|_1 \tag{19}$$

holds.

The Laplace's method (Taylor expansion of the logarithm around the maximum $\theta^\star$ [25, p. 26]) in 18

$$[\mathbf{L}_{\mathbf{Y}}]_{ij} = [\mathbf{L}_{\widehat{\mathbf{Y}}}]_{ij} - \frac{1}{2}(\theta - \theta^\star)^\top H\left([\mathbf{L}_{\widehat{\mathbf{Y}}}]_{ij}\right)(\theta - \theta^\star) + o(\theta - \theta^\star)^2 \tag{20}$$

being H the Hessian (second partial derivatives matrix; for matrix differentiation techniques see [29, chap. 3], in particular the discussion about what is meant for the matrix gradient [29, p. 133]).

The direct computation of the Hessian provides

$$H\left([\mathbf{L}_{\widehat{\mathbf{Y}}}]_{ij}\right) = \frac{\partial^2}{\partial \widehat{\mathbf{y}} \partial \widehat{\mathbf{y}}^\top}[\mathbf{Y}]_{ij} \log[\widehat{\mathbf{Y}}]_{ij} \quad (\widehat{\mathbf{y}} \in [\widehat{\mathbf{Y}}]_{ij}) \tag{21}$$

$$= -\mathrm{diag}\left(\frac{\mathbf{I}\,[\mathbf{Y}]_{ij}^\top}{[\widehat{\mathbf{Y}}]_{ij}[\widehat{\mathbf{Y}}]_{ij}^\top}\right) \tag{22}$$

being $\mathbf{I}$ the suitable dimension identity matrix.

The relation between the Fisher Information Matrix and the Hessian is [23]

$$\mathbf{I}_F = -E\left(H([\mathbf{L}_{\mathbf{Y}}]_{ij})\right) \tag{23}$$

$$= \mathrm{diag}\left(\frac{[\mathbf{Y}]_{ij}^\top[\widehat{\mathbf{Y}}]_{ij}}{[\widehat{\mathbf{Y}}]^\top[\widehat{\mathbf{Y}}]_{ij}}\right) \tag{24}$$

approximating to the identity when $\widehat{\mathbf{Y}} \approx \mathbf{Y}$.

Writing

$$\Phi(\mathbf{Y}) = [\mathbf{W}]_{ik}[\mathbf{H}_\Phi]_{kj} \tag{25}$$

with $\mathbf{H}_\Phi$ a learned base, and ϕ a map such that

$$\Phi: V \longrightarrow F \tag{26}$$

$$\mathbf{y} \longmapsto \Phi(\mathbf{y}) \tag{27}$$

where V and F are finite, not necessary of the same dimension Hilbert spaces. To avoid the derivatives of 16, we write

$$[\mathbf{W}]_{ik} = \Phi([\mathbf{Y}]_{ij})[\mathbf{H}_\Phi]_{jk}^\dagger \tag{28}$$

with $[\cdot]^\dagger$ the pseudo-inverse.

Algorithm 1: Non-parametric Fisher Kernel.

Procedure (I): Non-negative Factorization.

---

input   : $[\mathbf{X}]_{ij}$ (data matrix or input space vectors)
             k (parameter)
             ϵ (degree of adjustement between $\mathbf{Y}$ and $\mathbf{WH}$)
do   : $[\mathbf{Y}]_{ij} = [\mathbf{X}]_{ij}\,\mathbf{D}_X^{-1}\,\mathbf{M}_X^{-1}$ (transformation formula 1)
             initialize $[\mathbf{W}]_{ik}$ and $[\mathbf{H}]_{kj}$
while $\|\widehat{\mathbf{Y}} - \mathbf{Y}\|_1 \geq \epsilon$ **do**

$$[\mathbf{W}]_{ik} = [\mathbf{W}]_{ik} \odot \left(\frac{[\mathbf{Y}]_{ij}}{[\mathbf{W}\,\mathbf{H}]_{ij}}\, [\mathbf{H}]_{kj}^{t} \right)$$

$$[\mathbf{H}]_{kj} = [\mathbf{H}]_{kj} \odot \left([\mathbf{W}]_{ik}^{t}\, \frac{[\mathbf{Y}]_{ij}}{[\mathbf{W}\,\mathbf{H}]_{ij}} \right)$$

$$[\widehat{\mathbf{Y}}]_{ij} = [\mathbf{W}]_{ik}[\mathbf{H}]_{kj}$$

end
do   : $\widetilde{\mathbf{W}} \leftarrow \mathbf{W}$ (column normalization as indicated in formula 12)
return : $[\mathbf{W}]_{ik}$

Procedure (II): Training.

---

input   : $[\mathbf{W}]_{ik}$
             $q_1, q_2, \ldots, q_q$ (labels)
for $q = 1, 2, \ldots, q$ **do**
   $q_{q_i = -1} = E(q_i|\,[\mathbf{W}]_{ik})$
end
return : $(\mathbf{w}_i, l_q)$

Simple algebra manipulations

$$[\mathbf{W}]_{ik}[\mathbf{W}]_{ik}^{\top} = \left(\varPhi([\mathbf{Y}]_{ij})[\mathbf{H}_{\varPhi}]_{jk}^{\dagger} \right) \left(\varPhi([\mathbf{Y}]_{ij})[\mathbf{H}_{\varPhi}]_{jk}^{\dagger} \right)^{\top} \tag{29}$$

$$= \varPhi([\mathbf{Y}]_{ij})[\mathbf{H}_{\varPhi}]_{jk}^{\dagger}\left([\mathbf{H}_{\varPhi}]_{jk}^{\dagger}\right)^{\top} \varPhi([\mathbf{Y}]_{ij})^{\top} \tag{30}$$

$$= \left\{ \langle \varPhi(\mathbf{y}_{i1}), \varPhi(\mathbf{y}_{i2}) \rangle_{[\mathbf{H}_{\varPhi}]^{\dagger}\mathbf{I}_F^{-1}([\mathbf{H}_{\varPhi}]^{\dagger})^{\top}} \right\}_i \tag{31}$$

with $\mathbf{y}_{i1}$ and $\mathbf{y}_{i2}$ row vectors (or observed entities) of $\widehat{\mathbf{Y}}$.
   Then, the non-parametric Fisher kernel is

$$K(\mathbf{Y}, \mathbf{Y}^{\top}) = \mathbf{W}\mathbf{W}^{\top} \tag{32}$$

and for new observations

$$\left\{ \langle \varPhi(\mathbf{y}_{i1}^{(new)}), \varPhi(\mathbf{y}_{i2}) \rangle_{[\mathbf{H}_{\varPhi}]^{\dagger}\mathbf{I}_F^{-1}([\mathbf{H}_{\varPhi}]^{\dagger})^{\top}} \right\}_{(new)} = <\mathbf{w}^{(new)}, \mathbf{w}> \tag{33}$$

4   Classification Error

From our point of view, the most important property of the Fisher kernel is consistency, and our formulation inherits it. Consistency means reliable estimations at the limit. A consequence is the ability of the kernel to estimate with an arbitrary small classification error. More precisely the kernel given in formula 33, with (i) $k \geq \min(m, n)$, and (ii) $\max(\mathbf{Y} - \mathbf{W}\mathbf{H}_\phi) < 1/m$ performs the task with an arbitrary small error.

We provide a proof of this result, in an equivalent way to the parametric case, in which the relationship between realizability, consistency, and asymptotic error is established [27]. As far as we can, we follow the same scheme, introducing no concepts on realizability nor Information Geometry [1].

Partitioning the known entities with labels $Q = \{q_1, q_2, \ldots, q_q\}$ as

$$Q = \left\{ q_1, \bigcup_{q=q_2}^{q_q} \right\} = \{-1, +1\} \tag{34}$$

the prior

$$P(\mathbf{Y}|Q) = \alpha P(\mathbf{Y}|q = -1) + (1 - \alpha)P(\mathbf{Y}|q = +1) \qquad (0 \leq \alpha \leq 1) \tag{35}$$

can be written, using matrix notation

$$\mathbf{Y}_Q = \cup_i \{\alpha \mathbf{Y}_{-1}, (1 - \alpha)\mathbf{Y}_{+1}\} \tag{36}$$

and estimated, from the data (training phase), as

$$\widehat{\mathbf{Y}} = \cup_i \{\alpha \widehat{\mathbf{Y}}_{-1}, (1 - \alpha)\widehat{\mathbf{Y}}_{+1}\} \tag{37}$$

The ratio

$$\lambda = \log \frac{\mathbf{Y}}{\widehat{\mathbf{Y}}} \tag{38}$$

with expectation

$$E(\lambda) = \sum_{ij} \mathbf{Y} \log \frac{\mathbf{Y}}{\widehat{\mathbf{Y}}} \tag{39}$$

is (again) the KL divergence, which is monotonically decreasing.

For new observations we have

$$E(\lambda) = \sum_{ij} \mathbf{Y} \log \frac{\cup_i \left\{ \alpha \left(\mathbf{Y}_{-1} + \mathbf{Y}_{-1}^{(new)} \right), (1 - \alpha) \left(\mathbf{Y}_{+1} + \mathbf{Y}_{+1}^{(new)} \right) \right\}}{\cup_i \left\{ \alpha' \left(\widehat{\mathbf{Y}}_{-1} + \widehat{\mathbf{Y}}_{-1}^{(new)} \right), (1 - \alpha') \left(\widehat{\mathbf{Y}}_{+1} + \widehat{\mathbf{Y}}_{+1}^{(new)} \right) \right\}} \tag{40}$$

being α and α' the observed proportions labeled with -1.

The conditions to get $\widetilde{\mathbf{Y}} \approx \mathbf{Y}$, for the decomposition given by formulas 8 and 9 are (i) $k \geq \min(m, n)$, and (ii) an approximation error less than $1/m$

for each matrix column. Intuitively, these conditions can be understood when thinking on the empirical distributions of $\widehat{\mathbf{Y}}$ and $\mathbf{Y}$, which are the same for the given approximation. A formal proof of the approximation conditions to get the arbitrary approximation can be found in our work [13].

Then, it approximates no better than condition (ii), being more restrictive when k increases.

Close enough values of $\widetilde{\mathbf{Y}}$ and $\mathbf{Y}$ means that relation 40 approximates to one, and holds only if $\alpha\prime = \alpha$, $\widehat{\mathbf{Y}}_{-1}^{(new)} = \mathbf{Y}_{-1}^{(new)}$ and $\widehat{\mathbf{Y}}_{+1}^{(new)} = \mathbf{Y}_{+1}^{(new)}$

Keeping the negative subscripts $I_{(q=1)} = \{\hat{\mathbf{y}} \in \widehat{\mathbf{Y}}^{(new)}$ s.t. $\text{sign}(\mathbf{w}^\top \phi(\mathbf{y}_i) + b) = -1\}$ (here $\mathbf{w}$ are the here the support vectors, and b a constant), we have a piece of a matrix for the label -1.

Repeating the process for the labels $\{q_2, q_3, \ldots, q_q\}$, the complete predicted values is given by the direct sum

$$\left\{ (\widehat{\mathbf{Y}}^{(new)}, q_i \right\} = \oplus_{Iq} [\widehat{\mathbf{Y}}^{(\mathbf{new})}]_{I_q j} \tag{41}$$

Table 1. Selected UCI repository data sets description.

Data set	Size	Variables	Description	Labels
Seeds	210×7	Non-negative real valued	Geometry of wheat kernels	3
Ecoli	336×8	Non-negative real valued	Microbiological signals	5
Glass	214×10	Non-negative real valued	Chemical composition	6
Bupa	345×23	Non-negative integer/real valued	Blood tests	2

5   Examples

The examples have the sole purpose of illustrating the features of the proposed kernel. In particular, the arbitrary close to zero errors classification capability. The data sets have been selected from the UCI repository [10], with no more criteria to easily handle for our kernel. Strictly non-negativity data are necessary, and the also existence of two or more categories. To simplify, the non-presence of missing values is also desirable. A brief description of the data sets is provided the Table 1.

The content of our work is theoretical, and not to extract information from the data. We only describe how to execute the process. It supposes (i) re-scale the input space vectors of $\mathbf{X}$ to get $\mathbf{Y}$; (ii) the NMF of $\mathbf{Y}$, in a iterative process leading to obtaining the non-negative matrices $\mathbf{W}$ and $\mathbf{H}$; (iii) the column and row normalization of $\mathbf{W}$ and $\mathbf{H}$, respectively. When this is done, is necessary to calculate separated regions for homogeneous labeled data vectors. They are the support vectors, and requires to estimate $E(Q| [\mathbf{W}]_{ik})$ being Q the set of the q labels. All these steps are in the Algorithm 1.

Also, is necessary to fix several parameters. We divide them in two categories: those ones which are necessary for the probabilistic NMF of the data matrix $\mathbf{X}$

(and so, the matrices $\mathbf{Y}$ and $\mathbf{W}$ for a learned basis $\mathbf{H}_\phi$), and the related to the choice of the support vectors. For the first step they are (i) select a value of k; (ii) initialize $\mathbf{W}$ and $\mathbf{H}$ according to k; (iii) execute the iterative process, switching between Eqs. 8 and 9, until a given convergence condition, given by (iv) a desired approximation degree between $\widehat{\mathbf{Y}}$ and $\mathbf{Y}$ is obtained.

We recommend to use $k \geq \min(m, n)$, which lets an arbitrary approximation of $\mathbf{Y}$ to $\widehat{\mathbf{Y}}$ (for a high enough number of iterations). This value is justified in our previous work [13]. We select this approximation condition as $\max(\mathbf{Y} - \mathbf{W}\mathbf{H}_\phi)) < 1/m$ (m is the number of observations, or rows of the matrix $\mathbf{X}$). These two conditions meet the criteria to achieve the result of Sect. 4. The rest of the parameter are related to support vectors. We only set C to 1 (large values of C suppose small margins, and conversely, as required in [24]). All this are in the Algorithm 1.

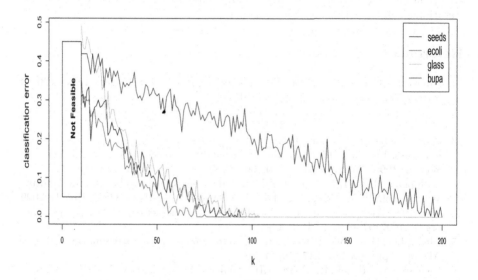

Fig. 1. Misclassification or error rate vs. number of components (value of k in the formulas 7 and 8). The execution parameter for the algorithm is $C = 1$. The zone indicated as not feasible corresponds to the values such that $k < \min(m, n)$, and for which the aproximation given in the formula 10 cannot be arbitrary approximated.

The selection of the support vectors can be a tedious task, and exists extensive literature we do not refer. A practical solution is to write some lines of code as a function and incorporate it within an open code application. Another one, is the extended (but not elegant) trick, which is to write the necessary code for the kernel (Procedure I of the Algorithm 1), and introduce the output as the input for a linear kernel of some reliable package. The linear kernel transforms nothing, calculates the dot products, and selects the support vectors for us. For the experiments, this is the path we take, using the *e1071* R package [24].

To better visualize the potential of the Fisher kernel, we run the algorithm as classification. This procedure is useful to compare results. We run the algorithm is executed for different number of model components ($k = 10, \ldots, 200$), obtaining 190 confusion matrices for each data set. From confusion matrices, the misclassification rate (or the result of subtracting to one the number of valid classifications divided by the total number of observations), is plotted versus k in 1. The sizes of the training-prediction partition are the extended 80%–20%.

6   Conclusion

The main conclusion of this work is that only assuming non-negativity for all the data entries, and with no hypothesis on the distribution of the parameters, a non-parametric Fisher kernel can be obtained. A classification, with an arbitrary small classification error for a high enough number of model components and iterations, can be performed with this kernel.

References

1. Amari, S.: Information Geometry and its Applications. Springer, Tokyo (2016). https://doi.org/10.1007/978-4-431-55978-8
2. Aronszajn, N.: Theory of reproducing kernels. Trans. Am. Math. Soc. **68**(3), 337–404 (1950)
3. Bredensteiner, E.J., Bennett, K.P.: Multicategory classification by support vector machines. In: Pang, J.S. (ed.) Computational Optimization, pp. 53–79. Springer, Boston (1999). https://doi.org/10.1007/978-1-4615-5197-3_5
4. Chappelier, J.-C., Eckard, E.: PLSI: the true fisher kernel and beyond. In: Buntine, W., Grobelnik, M., Mladenić, D., Shawe-Taylor, J. (eds.) ECML PKDD 2009. LNCS (LNAI), vol. 5781, pp. 195–210. Springer, Heidelberg (2009). https://doi.org/10.1007/978-3-642-04180-8_30
5. Chen, J.C.: The nonnegative rank factorizations of nonnegative matrices. Linear Algebra Appl. **62**, 207–217 (1984)
6. Cichocki, A., Zdunek, R., Phan, A.H., Amary, S.I.: Nonnegative Matrix and Tensor Factorizations. Wiley, Hoboken (2009)
7. Cortes, C., Vapnik, V.: Support-vector networks. Mach. Learn. **20**, 273–297 (1995)
8. Ding, C., Li, T., Peng, W.: On the equivalence between non-negative matrix factorization and probabilistic latent semantic indexing. Comput. Statist. Data Anal. **52**(8), 3913–3927 (2008)
9. Ding, C., Xiaofeng, H., Horst D.S.: On the Equivalence of Nonnegative Matrix Factorization and Spectral Clustering, pp. 606–610 (2005). https://doi.org/10.1137/1.9781611972757.70
10. Dua, D., Graff, C.: UCI machine learning repository (2017). School of Information and Computer Sciences, University of California, Irvine. http://archive.ics.uci.edu/ml
11. Durgesh, K.S., Lekha, B.: Data classification using support vector machine. J. Theor. Appl. Inf. Technol. **12**(1), 1–7 (2010)
12. Elkan, C.: Deriving TF-IDF as a fisher kernel. In: Consens, M., Navarro, G. (eds.) SPIRE 2005. LNCS, vol. 3772, pp. 295–300. Springer, Heidelberg (2005). https://doi.org/10.1007/11575832_33

13. Figuera, P., García Bringas, P.: On the probabilistic latent semantic analysis generalization as the singular value decomposition probabilistic image. J. Stat. Theory Appl. **19**, 286–296 (2020). https://doi.org/10.2991/jsta.d.200605.001
14. Franke, B., et al.: Statistical inference, learning and models in big data. Int. Stat. Rev. **84**(3), 371–389 (2016)
15. Hofmann, T.: Learning the similarity of documents: an information-geometric approach to document retrieval and categorization. In: Advances in Neural Information Processing Systems, pp. 914–920 (2000)
16. Hofmann, T., Schölkopf, B., Smola, A.J.: Kernel methods in machine learning. Ann. Stat. **36**, 1171–1220 (2008)
17. Hofmann, T.: Unsupervised learning by probabilistic latent semantic analysis. J. Mach. Learn. Res. **42**(1–2), 177–196 (2000)
18. Hsu, C.W., Lin, C.J.: A comparison of methods for multiclass support vector machines. IEEE Trans. Neural Netw. **13**(2), 415–425 (2002)
19. Jaakkola, T.S., Haussler, D., et al.: Exploiting generative models in discriminative classifiers. In: Advances in Neural Information Processing Systems, pp. 487–493 (1999)
20. Kullback, S., Leibler, R.A.: On information and sufficiency. Ann. Math. Stat. **22**(1), 79–86 (1951)
21. Latecki, L.J., Sobel, M., Lakaemper, R.: New EM derived from Kullback-Leibler divergence. In: KDD06 Proceedings of the 12th ACM SIGKDD International Conference on Knowledge Discovery and Data Minings, pp. 267–276 (2006)
22. Lee, H., Cichocki, A., Choi, S.: Kernel nonnegative matrix factorization for spectral EEG feature extraction. Neurocomputing **72**(13–15), 3182–3190 (2009)
23. Martens, J.: New insights and perspectives on the natural gradient method. J. Mach. Learn. Res. **21**, 1–76 (2020)
24. Meyer, D., Dimitriadou, E., Hornik, K., Weingessel, A., Leisch, F.: e1071: Misc Functions of the Department of Statistics, Probability Theory Group (Formerly: E1071), TU Wien (2021). R package version 1.7-6. https://CRAN.R-project.org/package=e1071
25. Naik, G.R.: Non-negative Matrix Factorization Techniques. Springer, Heidelberg (2016). https://doi.org/10.1007/978-3-662-48331-2
26. Salcedo-Sanz, S., Rojo-Álvarez, J.L., Martínez-Ramón, M., Camps-Valls, G.: Support vector machines in engineering: an overview. Wiley Interdiscip. Rev. Data Mining Knowl. Discov. **4**(3), 234–267 (2014)
27. Tsuda, K., Akaho, S., Kawanabe, M., Müller, K.R.: Asymptotic properties of the fisher kernel. Neural Comput. **16**(1), 115–137 (2004)
28. Zhang, D., Zhou, Z.-H., Chen, S.: Non-negative matrix factorization on kernels. In: Yang, Q., Webb, G. (eds.) PRICAI 2006. LNCS (LNAI), vol. 4099, pp. 404–412. Springer, Heidelberg (2006). https://doi.org/10.1007/978-3-540-36668-3_44
29. Zhang, X.D.: Matrix Analysis and Applications. Cambridge University Press, Cambridge (2017)

Deep Learning Applications

Hybrid Deep Learning Approach for Efficient Outdoor Parking Monitoring in Smart Cities

Alberto Tellaeche Iglesias$^{(\boxtimes)}$, Iker Pastor-López , Borja Sanz Urquijo ,
and Pablo García-Bringas

D4K Group, University of Deusto, Avda. Universidades 24, 48007 Bilbao, Spain
{alberto.tellaeche,iker.pastor,borja.sanz,
pablo.garcia.bringas}@deusto.es

Abstract. The problem of transport is one of the fundamental problems in today's large cities and one of the fundamental pillars on which work is being done under the paradigm of smart cities. Although the use of public transport is encouraged in cities, there are always problems related to private transport parking availability. This work focuses on the development of an algorithm for monitoring parking spaces in open-air car parks, which is effective and simple from the point of view of the costs required for its deployment and robust in detecting the state of the car park. To this end, an algorithm based on Deep Learning has been developed for processing the status images of different car parks, obtaining real-time information on the real status of the car park, without the need for extensive deployment of devices in the car park. The proposed solution has been shown to outperform other image-based solutions for the same problem.

Keywords: Deep learning · Smart cities · Parking monitoring

1 Introduction

The Smart City is an emerging concept that arises from the evolution of what were previously called digital cities. The main concept underlying this concept of the city of the future is the intensive use of information and communication technologies, with the ultimate aim of providing high quality public services to citizens in sectors such as education, security and transport, among others [1].

According to [2], the smart city concept outlined above offers opportunities to mitigate existing problems and inequalities in large cities. According to this work, the main aspects to be addressed to make cities more efficient are environment, economy, safety and quality of life, general city governance and finally, mobility. This last concept, mobility, is also highlighted in [3].

Related to the mobility problem, and, in particular to parking, numerous research studies have been carried out in recent years, both to manage parking spaces in private pay and covered parkings, and to monitor open-air parking spaces in the city's open air environment.

H. Sanjurjo González et al. (Eds.): HAIS 2021, LNAI 12886, pp. 463–474, 2021.
https://doi.org/10.1007/978-3-030-86271-8_39

Alsafery et al. propose an IoT-based information architecture to provide real-time information on private parking spaces in a city [4], using cloud computing. A similar application can be found in detail in [5], where the proposed information architecture uses technologies such as the MQTT protocol and RFID. Works focused on the design of a distributed software architecture to manage this type of information can also be found in [6], where the concept of "parking network" is introduced, or in [7], where an ITS (Intelligent Transportation Systems) architecture is defined, based on hierarchical subsystems that enable the integrated management of different car parks monitored with heterogeneous technologies. More works related to IoT architectures applied to cities and that have application to the problem being defined can be found in [8, 9] or [10].

There are also numerous works related to the intelligent management of parking spaces, mainly in private facilities in which the entire installation is sensorized, in order to have exhaustive control of free or occupied spaces. An example of this type of system can be found in the work published by Alsafery et al. [4].

Similarly, in [7], the integration of smart packing data in smart cities is discussed, but it is assumed that each car park monitors each free space in a different way, assuming one sensor per space. In the work presented by Melnyk et al. [5], they follow a very similar strategy to the previous one, in this case using RFID technologies, as in [11]. This same approach of individually sensing parking spaces in private car parks can be found in many research works such as [12–14] or [15].

For condition monitoring in open-air car parking areas, the solutions described in the previous case are not feasible. In this second case, the vast majority of the solutions that can be found in the literature are based on image processing techniques.

An example of this type of approaches can be found in Baroffio et al. [16], where the obtained image is converted to the HSV color space finally using a SVM classifier with linear kernel to detect the parking spaces that are empty. Another work that evaluates the performance of different machine learning algorithms for this type of applications is presented in [17]. In this case, five types of algorithms (logistic regression, linear-SVMs, rbf-SVMs, decision trees, and random forest) are applied to the monitoring of parking spaces, presenting different results. More examples of the application of these techniques can be found in Tatulea et al. [18] and in Tschentscher et al. [19].

In Lopez et. al [20], however, they use only simple image processing techniques. In this work they use only image histogram operations and edge detection to detect occupied squares. In [21], for example, they take advantage of the benefits of thermal imaging to carry out this type of application regardless of the existing ambient light.

Continuing with image-based techniques, the use of deep learning, and more specifically, the use of convolutional neural networks (CNN), has given rise to new procedures and research work with promising results. A good comparative study of both methods of image processing, deep learning and traditional processing, can be found in [22].

More precisely, in the work developed by Amato et al. [23] a CNN architecture based on AlexNet is developed, which provides promising results in the monitoring of outdoor car parks from different image perspectives. Acharya et al. [24] present a similar solution, in this case using a network based on the VGG architecture. In [25], despite using neural networks, they use another approach, designing an algorithm based on the MobileNet and YOLO architectures.

This research work presents a solution that hybridizes two different approaches based on Deep Learning in such a way that it provides extra robustness in the process of detecting free parking spaces, regardless of the perspective of the acquired image. Special attention has also been paid to the computational capacity required in the proposed solution so that it can run on specific embedded devices. Taking these two pillars into account, a system has been developed that outperforms current state-of-the-art approaches.

This paper is organized as follows. Section 2 proposes the fundamental characteristics that the solution must comply with, while Sect. 3 defines the dataset used for this specific problem, as a fundamental part to reach a proper solution to the problem. Section 4 presents in detail the hybrid deep learning approach proposed to achieve results that overcome current state of the art results and finally Sect. 5 presents numerical results obtained and Sect. 6 the final conclusions of this research work.

2 Problem Formulation

As specified in the previous point, the problem presented focuses on the development of a system that allows monitoring open spaces for car parking with an embedded system that is easy to install, that does not require a large deployment of extra technology in the area to be monitored and that provides reliable data on free and occupied spaces in that area.

Due to the existing problems described in the previous paragraph, a system for monitoring open parking spaces is proposed with the following characteristics:

1. The system must be self-contained and easy to install, i.e. an embedded system that only requires a power supply socket to start operating.
2. It should monitor and classify the occupancy of existing parking spaces in the area being surveyed.
3. It shall provide a system for remote access to real-time status information of the parking spaces. This last point is out of the scope of this research article.

3 Dataset

For the development of this research work, two existing datasets have been used in the literature for this type of applications. The fact that two different datasets have been used is because we have tried to provide the system with a greater capacity for generalization in terms of the perspective of the different parking images used.

The first of these datasets is the so-called PkLot [26]. This dataset contains 12,417 images at a resolution of 1280 * 720 pixel, taken in two different car parks and with different weather conditions (sunny weather, cloudy days, and rainy days). Each image in the dataset has an associated XML file, including the coordinates of each of the parking spaces, and their status, free or occupied. In total, this dataset has 695,900 labelled parking spaces.

This dataset, however, does not include images with a more aerial view, and it is also necessary to train this perspective to correctly detect parking spaces in these situations. Therefore, the interior dataset has been combined with the "aerials cars dataset" [27]. This dataset includes 156 more images of different resolutions together with their associated text files labelling the vehicles present in the images (Fig. 1).

Fig. 1. PkLot and aerials car dataset example images

4   Proposed Solution

The proposed solution for the robust detection of parking spaces in outdoor areas consists of the following distinct processing steps:

1. Acquisition and pre-processing of the image to be processed.
2. Detection of parking spaces by means of a convolutional neural network detector.
3. Processing of each parking space individually by means of a classifier neural network.
4. Hybridization of the results of both neural networks for a reliable final detection of parking spaces.

4.1   Acquisition and Image Pre-processing

The images are acquired using the Raspberry Pi Camera V2, which provides a maximum resolution of 1920 * 1080 pixel, and a pre-processing is applied to them to reduce variability in the images due to different ambient light conditions on different days.

The first of these operations is the application of a Gaussian filter to reduce the possible noise present in the image. The bidimensional Gaussian function can be expressed simply as:

$$G(x, y) = \frac{1}{2\pi\sigma^2} e^{-\frac{x^2+y^2}{2\sigma^2}} \tag{1}$$

In order to eliminate noise while maintaining as much detail as possible in the image, a Gaussian filter of size of 3 * 3 has been selected with the values presented in Fig. 2.

In a second step, the brightness and contrast of the filtered image are automatically modified to maximize its dynamic range. For this purpose, the image in RGB format is

1/16	1/8	1/16
1/8	1/4	1/8
1/16	1/8	1/16

Fig. 2. 3 * 3 Gaussian filter used for image smoothing

converted to grey levels and its histogram is equalized, obtaining the α and β values of the following linear transformation for its equalization:

$$y = \alpha\, y' + \beta \tag{2}$$

Where y is the equalized grey level, and y' is the original grey level of the image to be equalized. The values of α and β are obtained by calculating the cumulative distribution of alpha and beta, according to the Eq. 3

$$cdf_x(i) = \sum_{j=0}^{i} p_x(x = j) \tag{3}$$

Where p_x is the probability of occurrence of a given grey level in the image.

For the calculation of α and β, the cut-off values of the grey levels representing the cumulative distribution of the histogram are obtained at 1% and 99% respectively. With these new cut-off values, the values of α and β are obtained and applied to the different color components of the image. This correction can be observed in Fig. 3.

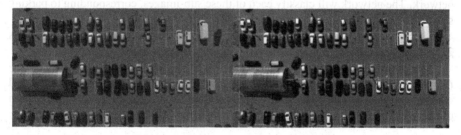

Fig. 3. Example of dataset image enhancement

4.2   Convolutional Network for the Classification of the Status of Parking Spaces.

This problem is, in essence, an object classification problem, in which the generic objects to be classified are the vehicles present in a car park.

Since the aim is to prioritize computational, we have chosen to use the network initially proposed in the work of Amato et al. [23], called mAlexNet (mini AlexNet), as the vehicle classification network. This network is a simplified version of the traditional

AlexNet architecture, reducing the number of trainable parameters from the original 134,268,738 to only 42,260. Some extra modifications were made to this network to try to optimize its performance, modifying its final architecture and obtaining a number of trainable parameters of 140,306.

For this network, we used the stochastic gradient descent optimizer with an α value of 0.01, and the Categorical Crossentropy function as the cost function to optimize, since it is a binary classification problem, that is; only two possible outputs are expected, parking place occupied or free.

The following figure shows the final network architecture used (Fig. 4):

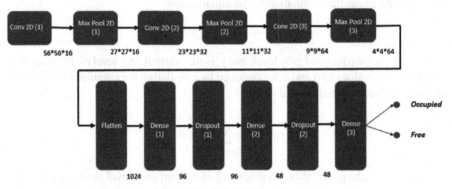

Fig. 4. Modified mAlexNet architecture

Using each of the labelled parking spaces in the dataset, it is possible to obtain a sub-image of each specific parking space, which will be the input image to this classifier network. The network will classify each of these sub-images, corresponding to parking spaces, as free or occupied.

As can be seen in the previous image, the size of the input image for this network is 224 * 224 pixels, so a previous scaling of the images to be analyzed is carried out.

There is an extra problem, the perspective of the image. To avoid this error, a homomorphic transformation of the ROI (Region of Interest) of each square obtained is carried out to obtain an azimuthal view of it. This transformation prior to the image classification can be seen in Fig. 5.

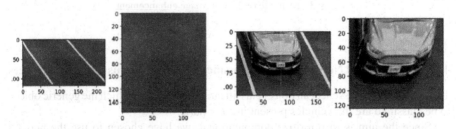

Fig. 5. Homomorphic transformation to avoid perspective problems in classification

Finally, the result obtained for each of the parking spaces is labelled in the original image, where there are cars and those where there are none. An example of this type of result can be seen in Fig. 6.

Fig. 6. Labelled parking lot

4.3 Convolutional Neural Network for Vehicle Detection

For this second object detection neural network, vehicles in our case, we have opted for a model already existing in the literature, which has proved to be very effective in terms of object detection accuracy and processing speed, YOLOv3 [28].

In our problem, therefore, we have used the transfer learning approach from a version of YOLOv3 trained with the COCO dataset [29]. The changes made on the original network have been the following: change in the number of output classes, two in our case, and adaptation of some filter layers in the output layers of the network to adapt them to the new number of classes.

4.4 Hybridization of Results

As explained in previous paragraphs, there are two convolutional neural networks working independently for the detection of occupied parking spaces in a car park. By hybridizing the results of both networks, we obtain a higher capacity for detecting occupied spaces than similar alternatives in the literature for the solution of this problem. Each neural network has certain strengths in detection that need to be identified when deciding on the correct method for hybridizing the results. The conclusions reached after analyzing the results are as follows:

1. The network that works as a vehicle detection network (YOLOv3) does not give false positives (FP), understood as the detection of a vehicle where there is none. The same is not true for false negatives (FN), i.e. the non-detection of an existing vehicle.

2. On the other hand, the classifier neural network (modified mAlexNet) provides more FP due to variable lighting conditions, being more robust against FN.

Therefore, having observed these performance trends empirically, the following hybridization method is proposed for a given image of an outdoor parking lot (Fig. 7):

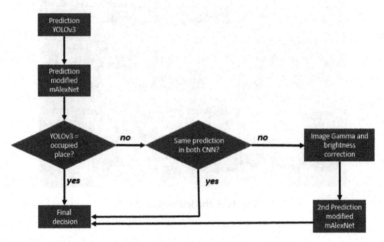

Fig. 7. Hybridation approach for final decision using both CNN solutions

As can be seen, the final decision on the occupancy of a given parking space is made in two stages, thus achieving a higher accuracy in detection than that obtained by each of the independent networks.

5 Obtained Results

5.1 Results for Each CNN Independently

As explained above, the dataset used for this research work has 695900 labelled parking spaces, coming from the PkLot dataset [26], and 4034 instances of aerial views of vehicles, obtained from the "Aerials cars dataset" [27], making a total of 699934 instances of labelled vehicles.

We selected 80% of the available instances (559947) for training, and the remaining 20% for testing, following a 10-fold cross validation approach.

The following table shows the results obtained for each of the networks used for the detection of the two target classes of this work: vehicle or free space. For each fold, the correct classification percentage (CCP) and the average confidence obtained in the detection are presented (Table 1).

As can be seen, both networks perform acceptably on a stand-alone basis, with classification results above 90% in both cases.

Table 1. 10 fold cross validation results for the modified mAlexNet architecture and YOLOv3 using transfer learning.

Fold	CCP (modified mAlexNet)	Mean confidence (modified mAlexNet)	CCP (YOLOv3)	Mean confidence (YOLOv3)
1	96.5%	99.3%	91.5%	90.3%
2	94.6%	94.2%	92.3%	93.6%
3	97.4%	97.5%	90.4%	96.2%
4	92.7%	91%	89.6%	92.5%
5	93.8%	90.8%	90.5%	95.2%
6	98.3%	93.9%	95.1%	90%
7	94.4%	95.1%	94.6%	96.5%
8	96.8%	92.4%	90.3%	97.2%
9	93.9%	96.9%	92.2%	92.3%
10	98.6%	95.4%	90.2%	93.1%

5.2 Final Results After Hybridization and Time Performance

To make the monitoring of parking spaces even more accurate and robust, a method of hybridization of results has been proposed and is explained in detail in Sect. 4.4. Table 2 shows the classification results of the proposed hybrid solution.

Table 2. Final performance results of the proposed hybrid approach

Fold	CCP (hybrid approach)	Mean confidence (hybrid approach)
1	97.4%	93.4%
2	98.5%	90.3%
3	97.8%	95.1%
4	96.9%	98.2%
5	96.7%	89.9%
6	97.1%	90%
7	98.4%	96.7%
8	99.1%	93.9%
9	97.4%	94.1%
10	98.6%	92.7%

After the proposed hybridization method, the average detection accuracy reaches 97.79%, a value higher than the 95.3% provided by the modified mAlexNet network and the 91.67% obtained by YOLOv3.

The processing times required by this system for each frame or captured video image are also affordable. An average computer (Core i5, 8GB RAM takes an average of 2.65 s per image. Embedded systems with limited dedicated hardware for artificial intelligence applications, such as Google Coral or NVidia Jetson cards, this time increases somewhat more, up to an average of 3.72 s per image.

In any case, monitoring a car park every 3 s is more than enough to have an exhaustive control of its occupancy in real time.

6   Conclusions

In this paper, a new approach for monitoring parking spaces in the field of smart cities has been presented. The proposed solution describes a low-cost system that requires practically no infrastructure for its installation in open-air parking lots. Apart from this obvious advantage in its installation, the system described offers an accuracy in the classification of parking spaces of over 97%, monitoring the parking in real time, once every 3 s.

Due to these three characteristics mentioned above, the proposed system presents a substantial advance over the current solutions reviewed in the state of the art for this type of system.

References

1. Santana, E.F.Z., Chaves, A.P., Gerosa, M.A., Kon, F., Milojicic, D.S.: Software platforms for smart cities. ACM Comput. Surv. **50**, 1–37 (2018). https://doi.org/10.1145/3124391
2. Monzon, A.: Smart cities concept and challenges: bases for the assessment of smart city projects. In: Helfert, M., Krempels, K.-H., Klein, C., Donellan, B., Guiskhin, O. (eds.) Smart Cities, Green Technologies, and Intelligent Transport Systems: 4th International Conference, SMARTGREENS 2015, and 1st International Conference VEHITS 2015, Lisbon, Portugal, May 20-22, 2015, Revised Selected Papers, pp. 17–31. Springer International Publishing, Cham (2015). https://doi.org/10.1007/978-3-319-27753-0_2
3. Khatoun, R., Zeadally, S.: Smart cities: concepts, architectures, research opportunities. Commun. ACM. **59**, 46–57 (2016). https://doi.org/10.1145/2858789
4. Alsafery, W., Alturki, B., Reiff-Marganiec, S., Jambi, K.: Smart car parking system solution for the internet of things in smart cities. In: 1st International Conference on Computer Application and Information Security ICCAIS 2018 (2018). https://doi.org/10.1109/CAIS.2018.8442004
5. Melnyk, P., Djahel, S., Nait-Abdesselam, F.: Towards a smart parking management system for smart cities. In: 5th IEEE International Smart Cities Conference, ISC2 2019, pp. 542–546 (2019). https://doi.org/10.1109/ISC246665.2019.9071740
6. Bagula, A., Castelli, L., Zennaro, M.: On the design of smart parking networks in the smart cities: an optimal sensor placement model. Sensors (Switzerland). **15**, 15443–15467 (2015). https://doi.org/10.3390/s150715443
7. Alam, M., et al.: Real-time smart parking systems integration in distributed ITS for smart cities. J. Adv. Transp. **2018**, 1–13 (2018). https://doi.org/10.1155/2018/1485652
8. Ji, Z., Ganchev, I., O'Droma, M., Zhao, L., Zhang, X.: A cloud-based car parking middleware for IoT-based smart cities: design and implementation. Sensors (Switzerland). **14**, 22372–22393 (2014). https://doi.org/10.3390/s141222372

9. Rizvi, S.R., Zehra, S., Olariu, S.: ASPIRE: an agent-oriented smart parking recommendation system for smart cities. IEEE Intell. Transp. Syst. Mag. **11**, 48–61 (2019). https://doi.org/10. 1109/MITS.2018.2876569
10. Koumetio Tekouabou, S.C., Abdellaoui Alaoui, E.A., Cherif, W., Silkan, H.: Improving parking availability prediction in smart cities with IoT and ensemble-based model. J. King Saud Univ. - Comput. Inf. Sci. 1–11 (2020). https://doi.org/10.1016/j.jksuci.2020.01.008
11. Kumar Gandhi, B.M., Kameswara Rao, M.: A prototype for IoT based car parking management system for smart cities. Indian. J. Sci. Technol. **9**, 17 (2016). https://doi.org/10.17485/ ijst/2016/v9i17/92973
12. Wang, T., Yao, Y., Chen, Y., Zhang, M., Tao, F., Snoussi, H.: Auto-sorting system toward smart factory based on deep learning for image segmentation. IEEE Sens. J. **18**, 8493–8501 (2018). https://doi.org/10.1109/JSEN.2018.2866943
13. Khanna, A., Anand, R.: IoT based smart parking system. In: 2016 International Conference on Internet Things Applications, IOTA 2016, pp. 266–270 (2016). https://doi.org/10.1109/ IOTA.2016.7562735
14. Grodi, R., Rawat, D.B., Rios-Gutierrez, F.: Smart parking: parking occupancy monitoring and visualization system for smart cities. In: Conference Proceedings of - IEEE SOUTHEASTCON, 2016-July, pp. 1–5 (2016). https://doi.org/10.1109/SECON.2016.750 6721
15. Sadhukhan, P.: An IoT-based E-parking system for smart cities. 2017 Int. Conf. Adv. Comput. Commun. Informatics. 1062–1066 (2017). https://doi.org/10.1109/ICACCI.2017.8125982.
16. Baroffio, L., Bondi, L., Cesana, M., Redondi, A.E., Tagliasacchi, M.: A visual sensor network for parking lot occupancy detection in smart cities. In: Proceedings of IEEE World Forum Internet Things, WF-IoT 2015, pp. 745–750 (2015). https://doi.org/10.1109/WF-IoT.2015. 7389147
17. Zacepins, A., Komasilovs, V., Kviesis, A.: Implementation of smart parking solution by image analysis. In: VEHITS 2018 – Proceedings of 4th Vehicle Technology and Intelligent Transport Systems, 2018-March, pp. 666–669 (2018). https://doi.org/10.5220/0006629706660669
18. Tatulea, P., Calin, F., Brad, R., Brâncovean, L., Greavu, M.: An image feature-based method for parking lot occupancy. Futur. Internet. **11**, 169 (2019). https://doi.org/10.3390/fi11080169
19. Tschentscher, M., Koch, C., König, M., Salmen, J., Schlipsing, M.: Scalable real-time parking lot classification: An evaluation of image features and supervised learning algorithms. Proceedings of International Joint Conference on Neural Networks, 2015-September (2015). https://doi.org/10.1109/IJCNN.2015.7280319.
20. Lopez, M., Griffin, T., Ellis, K., Enem, A., Duhan, C.: Parking lot occupancy tracking through image processing. In: Proceedings of 34th International Conference on Computers and Their Applications CATA 2019, vol. 58, pp. 265–270 (2019). https://doi.org/10.29007/69m7
21. Paidi, V., Fleyeh, H.: Parking occupancy detection using thermal camera. In: VEHITS 2019 – Proceedings of 5th International Conference on Vehicle Technology and Intelligent Transport System, pp. 483–490 (2019). https://doi.org/10.5220/0007726804830490.
22. Farag, M.S., Mohie El Din, M.M., El Shenbary, H.A.: Deep learning versus traditional methods for parking lots occupancy classification. Indones. J. Electr. Eng. Comput. Sci. **19**, 964–973 (2020). https://doi.org/10.11591/ijeecs.v19i2.pp964-973
23. Amato, G., Carrara, F., Falchi, F., Gennaro, C., Meghini, C., Vairo, C.: Deep learning for decentralized parking lot occupancy detection. Expert Syst. Appl. **72**, 327–334 (2017). https:// doi.org/10.1016/j.eswa.2016.10.055
24. Acharya, D., Yan, W., Khoshelham, K.: Real-time image-based parking occupancy detection using deep learning. CEUR Workshop Proc. **2087**, 33–40 (2018)
25. Chen, L.C., Sheu, R.K., Peng, W.Y., Wu, J.H., Tseng, C.H.: Video-based parking occupancy detection for smart control system. Appl. Sci. **10**, 1079 (2020). https://doi.org/10.3390/app 10031079

26. De Almeida, P.R.L., Oliveira, L.S., Britto, A.S., Silva, E.J., Koerich, A.L.: PKLot-a robust dataset for parking lot classification. Expert Syst. Appl. **42**, 4937–4949 (2015). https://doi.org/10.1016/j.eswa.2015.02.009
27. GitHub-jekhor/aerial-cars-dataset: Dataset for car detection on aerial photos applications. https://github.com/jekhor/aerial-cars-dataset. Accessed 07 Apr 2021
28. Redmon, J., Farhadi, A.: YOLOv3: An incremental improvement (2018). arXiv
29. Lin, T.-Y., et al.: Microsoft COCO: common objects in context. In: Fleet, D., Pajdla, T., Schiele, B., Tuytelaars, T. (eds.) ECCV 2014. LNCS, vol. 8693, pp. 740–755. Springer, Cham (2014). https://doi.org/10.1007/978-3-319-10602-1_48

Personal Cognitive Assistant: Personalisation and Action Scenarios Expansion

Elena Chistova[1(✉)], Margarita Suvorova[1(✉)], Gleb Kiselev[1,2(✉)], and Ivan Smirnov[1,2(✉)]

[1] Artificial Intelligence Research Institute FRC CSC RAS, Moscow, Russia
{chistova,suvorova,kiselev,ivs}@isa.ru
[2] Peoples' Friendship University of Russia, Moscow, Russia

Abstract. This paper examines the problem of insufficient flexibility in modern cognitive assistants for choosing cars. We believe that the inaccuracy and lack of content information in the synthesised responses negatively affect consumer awareness and purchasing power. The study's main task is to create a personalised interactive system to respond to the user's preferences. The authors propose a unique method of supplementing the car buying scenario with deep learning and unsupervised learning technologies to solve the issues, analyse the user's utterances, and provide a mechanism to accurately select the desired car based on open-domain dialogue interaction. The paper examines the problem of changing user interest and resolves it using a non-linear calculation of the desired responses. We conducted a series of experiments to measure the assistant's response to temporary changes in user interest. We ensured that the assistant prototype had sufficient flexibility and adjusted its responses to classify user group interests and successfully reclassify them if they changed. We found it logical to interact with the user in open-domain dialogue when there is no certain response to the user's utterance in the dialogue scenario.

Keywords: Cognitive assistant · Deep learning · NLP · Unsupervised learning · Car purchase scenario

1 Introduction

Dialogue systems are one of the most promising areas of artificial intelligence, as live dialogue is a convenient method of human-computer interaction. The traditional approach to conversational systems design includes several components:

1. A natural language understanding component allows the system to parse the user's text.

The reported study was partially funded by the Russian Foundation for Basic Research (project No. 18-29-22027).

H. Sanjurjo González et al. (Eds.): HAIS 2021, LNAI 12886, pp. 475–486, 2021.
https://doi.org/10.1007/978-3-030-86271-8_40

2. The dialogue state tracker stores a history of the system's user interaction and information about the user.
3. The dialogue manager decides the system's next steps: (a) ask the user an additional question or (b) request data from an external database.
4. Natural language generation is launched; the system receives a command and data and then takes the next step to generate a response using a template. The described architecture is mainly used in task-oriented dialogue systems.

As the dialogue systems developed, they evolved into personal assistants. Such assistants should correctly assess the user's intent, collect information about the interaction, personalise the user's experience, and have the capacity to integrate with external services. For example, send queries to external databases, receive information from them, process and extract the information and transfer it to the user.

This paper describes a technical prototype of a personal assistant that helps with selecting and purchasing a car. The scientific novelty of the paper lies in the structure of the assistant. The proposed structure differs from the traditional one described above. It consists of four modules: (1) a graph of a dialogue scenario, (2) a text processing module (basic natural language understanding), (3) an RL agent acting as a dialogue state tracker and dialogue planner, and (4) an open-domain dialogue module to diversify communication.

The task of the proposed assistant (agent) is to help the user decide about the kind of car they required and then assist the user through the process to the point of purchase. The dialogue system synthesises responses to the user's questions, such as drive preference, transmission, and the country of origin. Questions and possible answers are stored in the system as a graph of the dialogue scenario (the first module of the system). The agent generates informative and coherent sentences from the scenario in response to the user's questions. The second module matches the user's answers to the dialogue scenario graph. The third module is designed as an RL agent at one of the vertices of the dialogue graph. The RL agent selects one of several sentences from the graph. The fourth module is open-domain dialogue mode. Its task is to complement the sentences generated in the previous step to make the dialogue within the scenario more vivid and diverse.

Combining Closed and Open Domain Conversations. The significant human-evaluated metrics of an open domain dialogue system are involvement, emotional understanding, consistency, humanness, and the overall quality of the user's interaction with the system [2,6,16]. In recent years, there has been increasing interest in a similar human evaluation of task-oriented dialog systems of various domains (e.g., flight ticket booking or calendar management [9]). There are various strategies to achieve the best possible performance, such as enriching the task-oriented dialogue agent with chit-chat.

In [18] it is proposed to complement the closed-domain assistant's answer with a relevant chit-chat utterance if the user in their question deviated from the closed domain (e.g. "S: *Where are you living from?* U: *How is your day?*

S: *I am doing great. Where are you living from?"*). The chit-chat capability in this case is rather limited, as the assistant does not maintain an open-domain conversation with the user but tries its best to return to the task-oriented scenario. [14] also suggest enhancing the responses of a goal-oriented assistant with open-domain utterances generated by the GPT-2 [8] pre-trained language model. However, they are not trying to preserve the domain of the conversation, but to make the system's responses more life-like and to increase the engagingness, interestingness, knowledge, and humanness of the closed-domain assistant. The method involves simultaneously generating a closed-domain response and an additional chit-chat utterance, such as *"It's a Pop event starting at 6:30 pm. It's a great way to kick off the summer!"*. Experimental results show that this method achieves higher human evaluation scores, although such a system is still not designed for a continuous chit-chat. Switching between skills in the course of a single interaction, considering open-domain dialogue as an additional skill, is also possible for the assistants. [4] introduced a single end-to-end dialogue model where this switching is done in accordance with the attention vector based on a dialogue history. This model can move freely between skills as well as combine them but requires a significant amount of training data.

Training data for the current deep learning models handling both open and closed domain conversations is created by combining existing English dialogue corpora for different tasks. However, the implementation of learning algorithms for creating a multitask conversational agent for Russian is hampered due to the lack of Russian corpora for individual subtasks. In this work, we propose an assistant allowing both closed and open domain conversation, with switching between skills driven by the relative inflexibility of the closed domain scenario.

Personalising the Goal-Oriented Dialogue. We use non-linear methods to generate a response based on the environment data of a pre-trained Q-learning agent [17]. To create the assistant prototype, we chose this particular reinforcement learning algorithm since all actions of the dialogue system are discrete and their number at each step is not significant. Most conversational systems use quite effective linear models, but they depend on sample data. They cannot predict non-linear functions such as human behaviour based on the material read, the time of day, or any personal relationship where the user is associated with the subject area being studied. When using the RL agent, the assistant's behaviour model is configured based on its Q-table. The classification of user interests is obtained using the pre-training of previous iterations (including, on a small batch, those obtained using the models for classifying user actions). The agent update the Q-table with the user's flow during the interaction process, helping to form a flexible system for predicting user actions and adapting the agent's behaviour to its changes. A fully-fledged RL agent was used [10,12,15], rather than the classic contextual bandits used for solving similar tasks [1,11,15], as it is evident that the agent changed the states of the environment to adapt to the user's behaviour and reaction to new data. This data was presented as messages formed on the response graph's predicted depth basis in the described case. The user then read the information generated and reacts to it, clarifying or choosing a position. Our algorithm considered the immediate reward for responses shown to the user and the impact of the action on future rewards. The system calculated the best option for the user to become familiar with the information presented and predicted their reaction to future indicators.

Further, the paper is organized as follows. Section 2 describes the process of creating a dialogue system capable of switching between a goal-oriented car purchase scenario and chit-chat. In Sect. 3, we discuss the mechanism for personalising goal-oriented responses based on a reinforcement learning algorithm. Section 4 presents experiments.

2   Dialogue System

2.1   System Hierarchy

The assistant is structured as follows: The user enters their request in natural language. The request is parsed in the natural language understanding module and then matched against the dialogue scenario graph to find the most similar sentence. The decision block is then activated. If there is a low similarity between the user's phrase and the utterances in the dialogue graph, the open-domain dialogue mode is triggered. Otherwise, the RL agent is launched, selecting the most suitable response. The structure of the system is shown in Fig. 1.

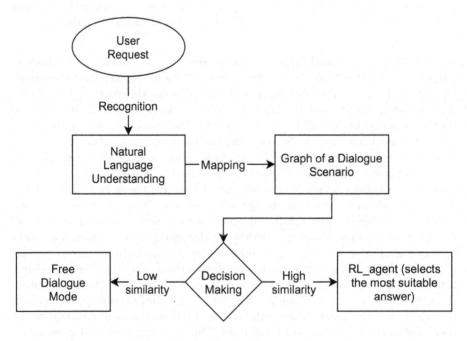

Fig. 1. System structure

Scenarios are a key component of a cognitive assistant's purposeful behaviour: repeated abstract sequences of actions and situations [7], and, in certain circumstances, stereotypical behaviour.

Within the framework of this concept, scenarios are intermediate units between the text and the action plan. The agent extracts scenarios from a text published on the Internet and composes a dialogue scenario graph. In this work, the stage of automatic scenario extraction was omitted. Instead, the dialogue scenario was compiled manually, by an expert, based on a text corpus and assembled as follows: The expert entered search queries, such as "how to choose the first car" and "buying a car step by step". Then the expert looked through the first four pages of results and manually saved appropriate texts. We found that the relevant information is usually published on such kinds of websites as car-related magazines (zr.ru, autoreview.ru), advertisement sites (drom.ru, auto.ru), car-related services (avtocod.ru, pddmaster.ru). Then all texts were manually converted into a list of steps and actions recommended to achieve the goal (e.g. "first, understand why you are buying a car", "then decide on the parameters and budget", and "then look for specific cars"). Where there were different interpretations, the most common option was chosen.

Thus, a scenario was obtained, which is divided into three main stages: selection, inspection, and purchase. There are also branches in the scenario (e.g. a new car or a used one, purchase on credit or purchase with cash). Each branch implies its own sequence of steps. For a more detailed description of the scenario, see [13].

2.2   Open-Domain Dialogue Problem Solving

The switch to an open-domain dialogue is carried out if no correspondence of the user's answer to one of the proposed options is revealed. The correspondence is tested in two steps. First, we calculate the Levenshtein distance to detect typos, and if this value is too high we measure the cosine similarity between the embeddings of expected and received utterances to determine paraphrasing. If the similarity value for the closest match is below the pre-defined threshold, the user's answer is forwarded to the chit-chat module. The utterance similarity confidence threshold is a hyperparameter adjusted between 0.0 and 1.0 based on a small, handcrafted set of utterance pairs the dialogue engine should consider similar or different. This set includes various misspellings and lexical differences between utterances. Examples of user utterances and matches found with the corresponding similarity values are shown in Table 1. In our experiments, we found the similarity threshold to be 0.3. When the cosine similarity value is less than the threshold, the response to the user's utterance is generated by the open-domain dialogue model.

To keep it simple, we use the pre-trained Word2Vec [5] model from RusVectores[1] to extract word embeddings. Then, the word vectors are averaged to obtain an utterance vector. To account for out-of-vocabulary words (including

---

[1] https://rusvectores.org/ru/models/ (ruscorpora_upos_cbow_300_20_2019).

misspellings) and compare them with the vocabulary of the Word2Vec model, the Enchant[2] spell checking library is utilised.

Table 1. Examples of utterances for possible activation of open-domain dialogue.

Utterance	Predicted option	Similarity
для транспотрировки (for transpotration)	для перевозки больших грузов (for carrying large loads)	0.62
на рыбалку кататься (to go fishing)	иногда выбираться на дачу (to get out of town once in a while)	0.40
расскажи про двигатели (tell me about the engines)	для гонок (for racing)	0.25
как дела? (how are you?)	для семьи от 4 человек (for a family of 4 or more)	0.21

The user returns to the main scenario by expressing a desire to return in the cars-related scenario or pressing a special button.

Data. A small dump of pikabu.ru comments is used to train the chit-chat model[3]. From the original set of dialogue exchanges, we excluded the following source-specific frequent utterances (not including additional symbols): *+1, -1, для минусов (for downvotes), что за фильм (what movie), что за игра (what game), пруф (proof), баян (old hat), было (seen), для лл (for the lazy league community)*. User mentions have been removed. The final list consists of 2526143 question-answer pairs. Due to the weak representation of the closed domain in the initial data, we also added 14948 conversational pairs from VKontakte community car enthusiasts' comments.

Method. The GPT-2 pre-trained model for Russian sberbank-ai/ ruGPT3Small[4] is used for chit-chat. A characteristic feature of this model is that it is pre-trained on an extended (from 1024 to 2048 tokens) context. It is the smallest available GPT-2 model for Russian, and it is possible to fine-tune it on an Nvidia Tesla T4 16Gb provided by the Google Colab platform. Of the data, 20% from each source is used for validation. In our experiment, we use batch size 32, number of epochs 8.

There might be not enough information in the user's utterance to generate an answer, or it might continue the previous exchange. Therefore, the chit-chat model generates an answer based on the user's utterance and the context of the previous exchange. As the model is pre-trained on long texts, for practical use

---

[2] https://abiword.github.io/enchant/.

[3] https://github.com/tchewik/datasets.

[4] https://github.com/sberbank-ai/ru-gpts#Pretraining-ruGPT3Small.

in an assistant the length of the GPT-2 model's response length is limited to 30 tokens.

Generative dialogue models tend to predict a common answer (the most frequent one in the training data, or a suitable answer to a wide variety of questions) [3]. Therefore, general answers, except those that may be relevant to the question (*don't know, thank you*), are replaced in the final answer by the "Thinking Face" (U+1F914) emoji placeholder. This can improve the user's general experience with the system because it reduces the possibility of a frequency response that is irrelevant to the user's utterance and informally sends the message that the system cannot provide a relevant answer. To determine common answers, we make the trained model predict answers to 50 open-domain questions, such as (e.g., *"How are you?"*, *"Do you like humans?"*). The similar answers the system gives to the different questions are considered to be common: *тыхорош* (*You're good*), *Что, если я тебе скажу, что это не так?* (*What if I told you it wasn't?*).

3   Q-Learning Agent in Dialogue Refinement

We use the Q-Learning algorithm [17] to personalise responses to user's questions in the context of choosing the desired car model. The Q-learning algorithm is a tabular unsupervised learning method where the agent creates a Q-table that characterizes the choice of the optimal actions in each available state. At each time step, the agent performs an action that corresponds to a policy that describes the probability of choosing all possible actions in the state under consideration. In response to the action, the environment generates a reward and the agent updates the Q-table values based on the received reward. The formula describing the operation of the algorithm is based on the Bellman equation and has the form:

$$Q(s_t, a_t) = Q(s_t, a_t) + \alpha[r_t + \gamma \max_{a_{t+1}} Q(s_{t+1}, a_{t+1}) - Q(s_t, a_t)]$$

Where $\alpha, \alpha \in [0, 1]$ – learning rate, $\gamma, \gamma \in [0, 1]$ – discount factor describing the importance of the current reward r_t compared to future rewards.

A particular "Bandit" environment has been developed[5] because there are no available OpenAI gym environments suitable for solving the current task. The task requires processing user requests to an assistant and choosing an answer suitable for the user; for this, all answer options were conditionally divided into several levels of refinement and saved as a graph. For example, to select a car engine, an answer graph is proposed, where 1 is the most straightforward answer, and 3 is the most detailed:

1. Advantages and disadvantages of gasoline and diesel engines;
2. Advantages and disadvantages of atmospheric, turbocharged, carburettor and injection engines;

---

[5] https://github.com/glebkiselev/contextbandit.

3. Advantages and disadvantages of mechanical and turbocharger turbines, oils for each type of engine, etc.

The environment forms an assessment of the agent's performance and generates a reward based on the reward function:

- -1 – Optimisation of the plan by length and giving the least complete answer;
- -2 – Issuing an answer to the user from the middle level of refinement;
- -3 – Issuing an answer to the user from the highest level of refinement;
- -5 – Prohibiting the agent from performing actions that are impossible in the situation and pressing the refinement button (giving a less accurate answer that the user expects);
- $+1$ – Issuing the expected response to the user based on the correct classification;
- $+10$ – Issuing the expected response to the user in the presence of the user's exact characteristics and the actions he performed in the dataset;
- $+100$ – The user's transition to the next level of communication with the agent and the successful selection of all interest options.

The environment classifies users groups based on a sample of 100 random users and the actions they performed. The classification results provides the ability to predict the most likely action for a user with certain characteristics. Each of the users is characterised by a vector $(s_t, gender, groups, k - words, history)$, where $s_t, s_t \in [0, 1]$ – state number, $gender, gender \in [0, 1]$ – user's gender, $groups, groups \in [0, 1]$ – the ratio of the number of groups about cars in a user's social networks and the total number of groups, $k - words, k - words \in [0, 1]$ – the ratio of keywords (carburettor, injector, turbine, ...) in the request (if any) to the length of the request $history, history \in [0, 1]$ – characterisation of the user's previous requests. The batch size was chosen based on practical experiments and is optimal for the task described in this paper. The agent uses the user's vector to receive a relevant action from the environment in the current state. It receives not only the state, reward, and a flag for reaching the target state, which is classic for the Q-learning algorithm, and a description of the group to which the user belongs. Furthermore, the agent replenishes its Q-table, but for each of the states, it describes not only rewards for all possible actions, but rewards for performing all possible actions by all possible users groups. The table enhancement allows the agent to synthesise policy for any user in any group. After training, the agent provides answers to questions asked by users and saves in a classic way for unsupervised learning: $\langle s_t, a_t, s_{t+1}, r_t \rangle$, gathering a new sample of data. When 100 users are accumulated, the agent is retrained using data randomly selected from the existing dataset and its Q-table is updated.

4   Experiments

Experiments were carried out in which the assistant offered answers to the questions of synthetic users within the framework of the scenario graph for a significant time. We use synthetic users because the assistant is a prototype requiring all bugs to be constantly corrected. Users are generated based on the authors' subjective perception of the intended behaviour of people with different characteristics. Figure 2 displays changes in the users' interest of various groups to the proposed material. In this context, "user interest" is defined as the depth of the graph refinements for each vehicle parameter. On the ordinate, the accuracy scale shows the accuracy of the agent's prediction of the user's desires by sending a response that needs no refinement to the user's request. The abscissa shows the batch scale of user requests. The graph displays information about the prediction adaptation by the agent to the required level of the graph of the response to users in case of a sharp change in their interest. Previous denotes the period preceding a sharp change in interest and characterises the previous stage of the agent's predictions. One of the possible conditions for a sharp change in user interest is the request's processing time, such as increasing interest in the morning or evening hours that would not have been possible to predict linearly without including additional criteria in the vector of user characteristics. The considered approach based on unsupervised learning showed fairly rapid growth in the adequacy of prediction based on a small number of user actions. The experiments were carried out on a laptop with the following characteristics: Intel Core i7-8565U CPU 1.8 GHz 199 GHz RAM 16 Gb.

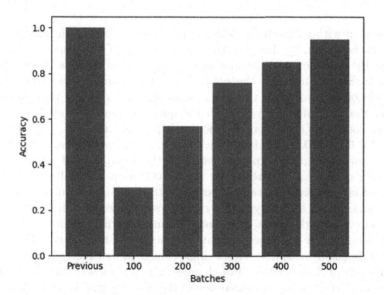

Fig. 2. Accuracy in predicting user actions based on changes in their behaviour

In the experiments, we used accuracy as an estimate of the choice of the correct answer to the user based on their classification by the assistant, as the success rate of the duration of the dialogue. We believe that the user will likely interrupt the dialogue if they receive an unexpected or inaccurate response from

Table 2. User conversing with the GPT-2 model.

U	Какой двигатель выбрать? (Which engine should I choose?)
S	Смотря какой. (It varies what kind.)
U	А какие бывают? (What kinds are there?)
S	В основном, это двигатели внутреннего сгорания и турбины внешнего сгорания. (Mostly they are internal combustion engines and external combustion turbines.)
U	Турбина мне зачем? (What do I use the turbine for?)
S	Чтобы не было проблем с охлаждением двигателя в процессе работы. (So that there are no problems with the engine cooling while running.)
U	А ее не надо охлаждать? (And doesn't it have to be cooled?)
S	Нет, она просто нагревается и работает как турбинный двигатель) (No, it just heats up and runs as a turbine engine))
U	А если очень сильно нагреется? (But what if it gets really hot?)
S	Тогда можно будет использовать как подставку для зажигалки. (Then you can use it as a lighter stand.)
U	понятно (I see)
S	А что, в этом есть логика?) (Why, is there logic in that?))
U	какая-то извращенная логика в этом есть (there's some twisted logic to it)
S	Это не извращение. Это просто шутка юмора такая была) (It's not twisted. It was just a jokey joke))

the assistant without reaching their desired answer. The test included 3000 synthetic users, gradually changing the desired depth of the response graph but leaving the parameter vector unchanged. The vector of parameters, as mentioned in Sect. 3, included the action, the state number, the user's gender, the ratio of car groups in the user's social networks to the total number of groups, the ratio of the keywords in the query to the total number of words, and the coefficient characterising the user's previous requests. Example users: $[4, 0.1, 1.0, 0.1, 0.9, 0.2], [3, 0.1, 0.0, 0.1, 0.2, 0.1], \ldots$ User actions were changed randomly, moving from the expected based on the description of actions towards increasing or decreasing interest for every 1000 users. Modelling showed that the assistant prototype successfully copes with changing user desires. However, when moving from a prototype to a running application, it will be necessary to use a neural network version of the reinforcement learning algorithm to expand the response graph and more accurate diagnostics of user desires. In addition, the experiments reveals that with a single change in user requirements (without the presence of a general trend of changes in requirements), the assistant provided an incorrect result. An expansion of the user description vector is required to solve this problem, achievable by increasing user interactions and gathering more detailed information from the user's social network.

In addition, experiments were carried out with the mechanism of open-domain dialogue that confirmed the adequacy (according to the subjective opinion of experts) of the responses generated by the considered model. An example of interaction between the user (U) and the system (S) in chit-chat mode is presented in Table 2. It shows that the system can maintain a consistent dialog if the user attempts to access unavailable information.

5   Conclusion

This paper presents a model of a cognitive assistant with a flexible structure for buying a car. In addition to the traditional approach based on a branched scenario, the assistant is supplemented with the ability to conduct an open-domain dialogue with the user and provide information on topics of interest. We have developed a mechanism for choosing an answer for users based on their brief characteristics and tested it for resistance to changes in user opinions based on non-linear factors. Experiments show that the mechanism we used to select the appropriate answer for users is suitable for solving the problem, but for processing each user, an extended user description and a neural network version of the reinforcement learning algorithm is required. In further work, the assistant will be improved by expanding the scenario component and deeper personalisation, based on expanding the range of available personal actions for each user and complicating the structure of synthesised constructive cues in the open-domain dialogue. In addition, reinforcement learning algorithms will be applied to reduce the number of illogical synthesised cues.

References

1. Auer, P., Fischer, P., Kivinen, J.: Finite-time analysis of the multiarmed bandit problem. Mach. Learn. **47**(2), 235–256 (2002)
2. Finch, S.E., Choi, J.D.: Towards unified dialogue system evaluation: a comprehensive analysis of current evaluation protocols. In: Proceedings of the 21th Annual Meeting of the Special Interest Group on Discourse and Dialogue, pp. 236–245 (2020)
3. Li, J., Galley, M., Brockett, C., Gao, J., Dolan, W.B.: A diversity-promoting objective function for neural conversation models. In: Proceedings of the 2016 Conference of the North American Chapter of the Association for Computational Linguistics: Human Language Technologies, pp. 110–119 (2016)
4. Madotto, A., Lin, Z., Wu, C.S., Shin, J., Fung, P.: Attention over parameters for dialogue systems (2019)
5. Mikolov, T., Sutskever, I., Chen, K., Corrado, G., Dean, J.: Distributed representations of words and phrases and their compositionality. arXiv preprint arXiv:1310.4546 (2013)
6. Moghe, N., Arora, S., Banerjee, S., Khapra, M.M.: Towards exploiting background knowledge for building conversation systems. In: Proceedings of the 2018 Conference on Empirical Methods in Natural Language Processing, pp. 2322–2332 (2018)
7. Osipov, G., Panov, A.: Rational behaviour planning of cognitive semiotic agent in dynamic environment. Sci. Tech. Inf. Process. **4**, 80–100 (2020). https://doi.org/10.14357/20718594200408
8. Radford, A., Wu, J., Child, R., Luan, D., Amodei, D., Sutskever, I.: Language models are unsupervised multitask learners. OpenAI blog **1**(8), 9 (2019)

9. Rastogi, A., Zang, X., Sunkara, S., Gupta, R., Khaitan, P.: Towards scalable multi-domain conversational agents: the schema-guided dialogue dataset. In: Proceedings of the AAAI Conference on Artificial Intelligence, vol. 34, pp. 8689–8696 (2020)
10. Schrittwieser, J., et al.: Mastering atari, go, chess and shogi by planning with a learned model. Nature **588**(7839), 604–609 (2020)
11. Sezener, E., Hutter, M., Budden, D., Wang, J., Veness, J.: Online learning in contextual bandits using gated linear networks (2020)
12. Silver, D., et al.: Mastering the game of go with deep neural networks and tree search. Nature **529**, 484–489 (2016). https://doi.org/10.1038/nature16961
13. Smirnov, I., et al.: Personal cognitive assistant: scenario-based behavior planning. Inf. Appl. (2021, in press)
14. Sun, K., et al.: Adding chit-chat to enhance task-oriented dialogues. In: Proceedings of the 2021 Conference of the North American Chapter of the Association for Computational Linguistics: Human Language Technologies, pp. 1570–1583 (2021)
15. Sutton, R.S., Barto, A.G.: Reinforcement learning i: Introduction (1998)
16. Tian, Z., Bi, W., Li, X., Zhang, N.L.: Learning to abstract for memory-augmented conversational response generation. In: Proceedings of the 57th Annual Meeting of the Association for Computational Linguistics, pp. 3816–3825 (2019)
17. Watkins, C.J.C.H., Dayan, P.: Q-learning. Mach. Learn. **8**(3–4), 279–292 (1992)
18. Zhao, T., Lu, A., Lee, K., Eskenazi, M.: Generative encoder-decoder models for task-oriented spoken dialog systems with chatting capability. In: Proceedings of the 18th Annual SIGdial Meeting on Discourse and Dialogue, pp. 27–36 (2017)

Nowcasting for Improving Radon-222 Forecasting at Canfranc Underground Laboratory

Tomás Sánchez-Pastor[iD] and Miguel Cárdenas-Montes[✉][iD]

Department of Fundamental Research, Centro de Investigaciones Energéticas
Medioambientales y Tecnológicas, Madrid, Spain
{tomas.sanchez,miguel.cardenas}@ciemat.es

Abstract. Physics' rare event investigation, like the dark matter direct
detection or the neutrinoless double beta decay research, is typically
carried-out in low background facilities like the underground laboratories. Radon-222 (^{222}Rn) is a radionuclide that can be emitted by the
uranium decay in the rock, thus the monitoring and the prediction of Rn
contamination in the air of the laboratory is a key aspect to minimize the
impact of this source of background. In the past, deep learning algorithms
have been used to forecast the radon level, however, due to the noisy
behavior of the ^{222}Rn data, it is very difficult to generate high-quality
predictions of this time series. In this work, the meteorological information concurrent to the radon time series from four distant places has been
considered—nowcasting technique—in order to improve the forecasting
of ^{222}Rn in the Canfranc Underground Laboratory (Spain). With this
work, we demonstrated and quantified the improvement in the prediction
capability of a deep learning algorithm using nowcasting techniques.

Keywords: Forecasting · Deep learning · ^{222}Rn measurements,
Canfranc Underground Laboratory

1 Introduction

Underground laboratories are playing a major role in modern Physics, from the
detection of elusive particles, for example the dark matter WIMPs to the search
of rare phenomena like the neutrinoless double beta decay. A low-background
environment, free from the cosmic rays, is mandatory for this kind of investigation.

Radon-222 (^{222}Rn) is a radionuclide produced by the ^{238}U and ^{232}Th decay
chains, and it is a major contributor to the radiation level in underground laboratories. Being a gas at room temperature, it can be emanated by the rocks
and concrete of the laboratory, diffusing in air of the experimental halls. This
contamination is a potential source of background for any rare event research
experiment, both directly and through the long life radioactive daughters produced in the Rn decay chain, which can stick to the surface of the detector. The

© Springer Nature Switzerland AG 2021
H. Sanjurjo González et al. (Eds.): HAIS 2021, LNAI 12886, pp. 487–499, 2021.
https://doi.org/10.1007/978-3-030-86271-8_41

^{222}Rn contamination in air can be reduced by orders of magnitude only in limited areas, flushing pure N_2 or radon-free air produced by dedicated machines.

For this research, it is critical to obtain accurate prediction of the ^{222}Rn level in order to make an efficient decision-making process. Since the radon can diffuse through the materials composing the detectors and deposit as radionuclides on a surface exposed to air without protection, the relevance of scheduling the maintenance operations during low-activity periods becomes critical. This motivates a long-term effort to produce high-quality predictions of the low Rn-concentration periods at Canfranc Underground Laboratory (LSC).

In the previous work [7], an evident correlation between the radon and the humidity level of the experimental underground area has been found. This humidity comes from the snow melting and the water percolating from the surface, and therefore, it can be correlated with external meteorological variables.

Meteorological variables from distant locations can act as a proxy variables for the radon level. By enriching with these variables the input of deep learning architectures aimed at forecasting the radon values, an improvement in the prediction quality can be expected. Since meteorological variables are concurrent with the radon level, this technique is named **nowcasting**.

The term nowcasting has been used in some disciplines—health and weather—for very short-range forecasting. But, the term nowcasting has also been used in weather forecasting for the prediction of some variables that are happening at the time as the data used to predict [10]. This involves the spatial displacement of these variables.

In the current work, the correlation between meteorological variables of four locations around the LSC with the Radon in the laboratory is studied. The most suitable variables are used for enriching the input of three deep learning architectures, Multilayer Perceptron (MLP) [1], Convolutional Neural Networks (CNN) [5], and Recurrent Neural Networks (RNN) with LSTM cells [6], which are applied to predict the radon level in ranges from 2 to 8 weeks ahead. As a result of the application of nowcasting to the prediction of radon at LSC, an improvement in the forecasting quality is achieved.

1.1   Previous Efforts

Regarding the previous efforts to predict the radon time series at LSC, it can be mentioned the initial work based on non-stochastic algorithms: Holt-Winters, AutoRegressive Integrated Moving Averages, and Seasonal and Trend decomposition using Loess (STL decomposition) [7], where the annual modulation of radon and this correlation with the humidity of the experimental hall have been evidenced. Also in [7] and [8], the forecasting capability of MLP, CNN, and RNN is evaluated for this kind of problem. Furthermore, in [4], improvements are reported when implementing an Ensemble Deep Learning approach composed of multiple deep architectures.

As part of a long-term effort to improve the prediction capability, in [8], the first implementation of the STL decomposition and CNN to improve the forecasting capability is presented. This choice shows promising results but it is penalized by the need of a wider set of parameters and hyperparameters to be optimized: those of STL decomposition and those of the CNN. In this context, additional efforts with evolutionary algorithms have been undertaken. In [11], Population-Based Incremental Learning algorithm (PBIL) is employed to optimize this wide set of parameters.

The decision-making process from time series requires accurate point estimations and of confidence intervals. In [3], the result concerning the propagation of the uncertainty with deep learning architectures is presented in the case of the radon time series at LSC.

Finally, the lessons learned from the application of nowcasting technique to the radon time series can be exported to other time series. Among others, efforts in air-quality forecasting in large urban areas are being undertaken [2].

The rest of the paper is organized as follows: Results are presented and analysed in Sect. 2. This section includes descriptions about data acquisition and preprocessing, the evaluation of proxy variables, and the structure of the data for feeding the deep architectures. Finally, Sect. 3 contains the conclusions of this work, as well as the future work.

2   Results and Analysis

In this section, firstly the capability of the meteorological variables for predicting the radon time series is evaluated. This evaluation is undertaken by studying their correlation with the radon and by evaluating the importance of features with Random Forests when using for predicting purpose. Later, the most relevant meteorological variables are incorporated in the input and the improvement in the prediction when using deep architectures is evaluated. For claiming robust conclusions, the significance of the improvements are checked through non-parametric tests.

2.1   Data Acquisition and Preprocessing

Radon concentration at Hall A of LSC varies from tens to hundreds of $\frac{Bq}{m^3}$ and has been measured every 10 min since July 2013 with an Alphaguard P30. The AlphaGuard is the centerpiece of a compact portable measuring system for the continuous determination of radon and radon progeny concentration in air, as well as some environmental parameters: temperature, pressure and humidity. In standard operation mode, the measuring gas gets by diffusion through a large-scale glass fiber filter into the ionization chamber, i.e. through the glass filter only ^{222}Rn may pass, while the radon daughter nuclei are prevented from entering the ionization chamber. The ingoing ^{222}Rn interacts with the inert gas creating ion-electron pairs. Then, the current strong electric field drives the ions along the chamber with a constant acceleration. At some point, the electrons reach enough energy to ionize the gas, producing a Townsend discharge. Finally the avalanche of electrons arrives at the cathode and the detector counts the resulting electric intensity as one hit of ^{222}Rn.

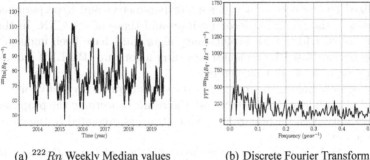

(a) ^{222}Rn Weekly Median values   (b) Discrete Fourier Transform

Fig. 1. ^{222}Rn weekly median concentrations from July 2013 to July 2019 at Hall A of the LSC (Fig. 1(a)), and its frequency domain obtained after applying a Discrete Fourier Transform (FFT) (Fig. 1(b)). The most important period corresponds to 52.7 weeks.

^{222}Rn time series presents a high level of noise. Measurements in the range from 50 to 120 $Bq \cdot m^{-3}$ are observed in any period of time. In [4,7,11], a weak annual-modulation of the monthly and weekly medians was depicted. For this reason, the weekly medians from July 2013 to July 2019 are selected for forecasting (Fig. 1(a)). Lower-period forecasting is infeasible due to the high level of variability of the observation and the absence of any modulation. In Fig. 1(b) the Discrete Fourier Transform of weekly median values is depicted. As it can be observed, there is a dominant frequency corresponding to 52.7 weeks.

Regarding the meteorological data, it has been downloaded from AEMET (Agencia Estatal de Meteorología, Spain). Four cities with airport around the LSC are selected: Pamplona (PMP), Barcelona (BCN), Huesca (HSC) and Zaragoza (ZGZ) (Fig. 2). The four selected stations are located at the airports of each city to avoid most of the missing data. The distance to the LSC are 88 km for HSC, 130 km for PMP, 157 km for ZGZ, and 356 km for BCN. Between the available data that can be downloaded, the significant variables to carry out this study are: the average temperature per day ($\bar{T}(C)$), the average pressure per day ($\bar{P}(hPa)$), and the average wind velocity per day ($\bar{V}(m \cdot s^{-1})$). Although meteorological information is available per day, ^{222}Rn time series is used per week; therefore the weekly mean average is calculated for these variables.

Missing values in meteorological variables have been filled using an autoregressive integrated moving average model (ARIMA(p, q, d)). ARIMA model considers a linear combination of past observations (AR), the past forecast errors (MA) and differentiating terms for making stationary the series (I).

2.2 Data Structure

Regarding the data structure that feed the corresponding deep learning architectures, a loop-back of 52 weeks—including the weekly values of ^{222}Rn and temperature—is created for forecasting the next value—53th—or the next range

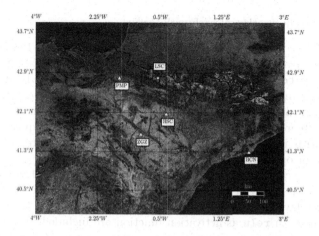

Fig. 2. Map of the area of interest (north-east of Spain) Triangles mark the location of the meteorological stations and the star marks the LSC position. Pamplona is labeled as PMP, Barcelona as BCN, Huesca as HSC and Zaragoza as ZGZ.

of values—53th and 54th for two weeks prediction—of ^{222}Rn. The deep learning architectures are trained with this input for predicting a set of values ahead. Diverse lengths of ahead prediction have been evaluated[1]. Due to this forecasting structure, ahead values are predicting more than once.

The number of predictions undertaken over a certain value is equal to the ahead-length. Therefore, in the prediction graphs a band representing the maximum and minimum values of predictions can be represented.

The loop-back is fixed to one year. In the previous efforts, it was demonstrated that this loop-back size is appropriate for capturing the annual modulation without generating noisy predictions [3,7,8,11].

2.3 Forecasting Evaluation

The dataset has been separated in two different parts: 70% of the total amount for the training set and 30% for the test set. This dataset corresponds to the period from July 2013 to July 2019. Besides 8 weeks observations from August to end September 2019 are put aside for the validation set.

The configuration used in the deep architectures has been selected after a reduced optimization process, a greedy process. One by one the hyperparameters are optimized keeping frozen all the rest hyperparameters. Therefore they can be considered as a high-quality sub-optimal configuration.

---

[1] The number of weeks ahead evaluated ranges from 2 week to 8 weeks (only pair numbers of week). This range allows quick scheduling for short unshielding periods, to intensive maintenance operations taking up to 8 weeks. Consultations with experiments managers indicate that these periods are adequate for planning different types of operations over the experiments.

The final configuration of each architecture is:

- The MLP is configured with two dense layers with 256 and 128 neurons with `relu` as activation function. A dropout layer with a probability of drop of 0.20 is inserted between the two hidden layers. It is trained with 40 epochs when using only radon time series as input, and 60 epochs when adding meteorological variables to the input. This larger number of epochs when adding meteorological variables is required for keeping an equivalent training process with input of higher dimensionality, and therefore, with a higher difficulty for optimizing the weights.
- The CNN is configured with two convolutional blocks of 128 and 64 filters each one with kernel of size of 3, padding `valid` and stride equal to 1. After a flattener layer and before the output layer, two dense layers of 32 and 64 neurons with `relu` as activation function are included. It is trained with 40 epochs when using only radon time series as input, and 80 epochs when adding meteorological variables to the input. As previously, the larger number of epochs when adding meteorological variables is required for keeping an equivalent training process with input of higher dimensionality, and therefore, with a higher difficulty for optimizing the weights.
- The RNN is configured with a hidden layer composed 54 LSTM cells and a dense layer composed of 64 neurons with `relu` as activation function. The network has been trained with 7 epochs.

In all cases, the output layer consists of a number of neurons equal to the number of values to predict ahead with `linear` activation function. Mean absolute error (MAE) has been selected as a figure of merit. In comparison with the mean squared error, MAE produces more reduced impact on the overall error when an observation is poorly forecast. The choice of the number of epochs in the architectures corresponds to the optimal, maximizing the learning and avoiding the overfitting.

In Table 1, the Pearson correlation coefficients between the weekly median ^{222}Rn levels at LSC and the meteorological variables—temperature, pressure and wind velocity—for the selected placements. Only when temperature is considered, a relevant correlation with ^{222}Rn is remarked. This result suggests that temperature could be a proxy of ^{222}Rn level, and consequently reverberates in an improvement of the forecasting capability.

Table 1. Pearson coefficients between the weekly median ^{222}Rn levels at LSC and the meteorological variables—temperature, pressure and wind velocity—for placements: Barcelona (BCN), Pamplona (PMP), Zaragoza (ZGZ) and Huesca (HSC).

City	Rn and T	Rn and P	Rn and V
BCN	0.53	−0.21	0.06
PMP	0.52	−0.19	0.17
ZGZ	0.53	−0.27	0.10
HSC	0.53	−0.16	0.12

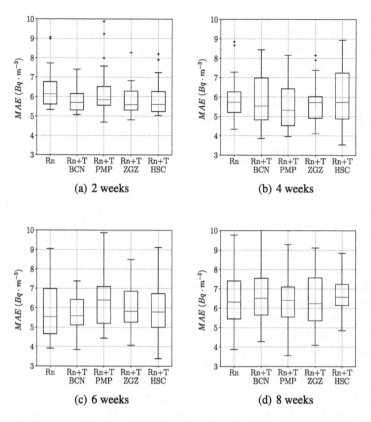

Fig. 3. Boxplots of Mean Absolute Error of testset for 25 independent runs when using MLP architecture for forecasting radon and radon plus temperature time series of locations selected for 2 weeks ahead (Fig. 3(a)), 4 weeks ahead (Fig. 3(b)), 6 weeks ahead (Fig. 3(c)), and 8 weeks ahead (Fig. 3(d)).

MLP Forecasting. In Fig. 3, the boxplots of MAE of MLP architecture of testset for 25 independent runs are shown. Errors when forecasting 2, 4, 6 and 8 weeks ahead by using as input radon time series and radon plus temperature time series in each of the four locations have been calculated. In this second case, only one temperature time series is added to the radon one. Tests including more than one temperature time series, covering all the combinations, have been undertaken without significant improvements. For the sake of the brevity, they are omitted.

For a forecasting of 2 weeks (Fig. 3(a)), improvements are observed when adding in the input the temperature time series, specially for the placements: ZGZ and HSC. Oppositely, shorter errors are not observed when forecasting 4 weeks (Fig. 3(b)), 6 weeks (Fig. 3(c)), and 8 weeks (Fig. 3(d)). In most of the cases, the median errors are higher, or larger variances—dispersion of values—are

Table 2. Mean MAE and standard deviation of the errors of testset for 25 independent runs when using MLP architecture for forecasting radon and radon plus temperature time series of locations selected for diverse weeks ahead).

Weeks	^{222}Rn	BCN	PMP	ZGZ	HSC
2	6.52 ± 1.27	5.87 ± 0.70	6.23 ± 1.23	5.79 ± 0.73	5.93 ± 0.87
4	5.92 ± 1.15	5.91 ± 1.33	5.63 ± 1.53	5.70 ± 1.06	5.90 ± 1.49
6	5.85 ± 1.45	5.98 ± 1.28	6.60 ± 1.75	6.12 ± 1.47	6.12 ± 1.55
8	6.56 ± 1.60	6.72 ± 1.78	6.52 ± 1.50	6.60 ± 1.90	6.88 ± 1.53

obtained. Both aspects indicate a degradation of the performance when forecasting more than 2 weeks ahead.

In Table 2, the mean value and the standard deviation of MAE of testset for 25 independent runs are presented. In this table, it is more clearly appreciated that the mean error and the standard deviation of errors for 2 week forecasting for ZGZ (5.79 ± 0.73), and HSC (5.93 ± 0.87) locations are shorter than when using only the radon time series as input (6.52 ± 1.27).

The application of the Kruskal-Wallis test [9] to the results obtained when forecasting 2 weeks ahead indicates that the differences between the medians of the best result obtained are not significant for a confidence level of 95%. A confidence level of 95% (p-value under 0.05) is used in this analysis. This means that the differences are unlikely to have occurred by chance with a probability of 95%. Furthermore, the application of the Wilcoxon signed-rank test to the values of the MAE between the Rn case and the Rn+T ZGZ case—the best case—points that the differences between the media are not significant for a confidence level of 95% (p-value under 0.05 but with Bonferroni correction p-value under 0.0125), p-value = 0.0229. Moreover, the application of the Wilcoxon signed-rank test to the values of the MAE between the Rn case and the Rn+T HSC and Rn+T BCN cases also point the differences between the media are not significant for a confidence level of 95% (p-value under 0.05 but with Bonferroni correction p-value under 0.0125).

Although a reduction in the MAE is obtained, including lower mean and lower variance after 25 independent runs, this result disallows claiming a significant positive improvement in the prediction when applying the temperature nowcasting to radon forecasting for the locations of ZGZ and HSC.

For the rest of the tests from 4 to 8 weeks ahead forecasting, the application of the Kruskal-Wallis test indicates that the differences between the medians are not significant for a confidence level of 95%. The posterior application of the Wilcoxon signed-rank test with the Bonferroni correction indicates that the differences between the medians are not significant for a confidence level of 95%.

CNN-Based Forecasting. In Fig. 4, the boxplots of MAE of CNN architecture of testset for 25 independent runs when using CNN architecture are shown. For comparison purposed, the same limits are kept along this kind of plots. As it

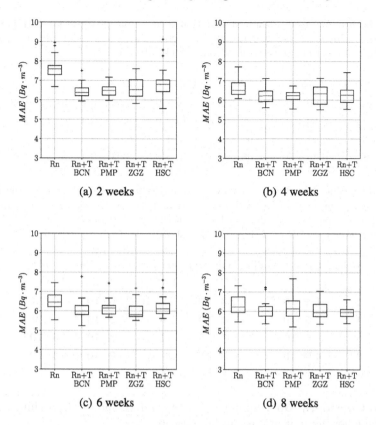

Fig. 4. Boxplots of Mean Absolute Error of testset for 25 independent runs when using CNN architecture for forecasting radon and radon plus temperature time series of locations selected for 2 weeks ahead (Fig. 4(a)), 4 weeks ahead (Fig. 4(b)), 6 weeks ahead (Fig. 4(c)), and 8 weeks ahead (Fig. 4(d)).

can be seen this figure, the addition of the temperature time series as input to the radon time series produces an appreciable reduction of the MAE values in multiple tests. The reduction appears for the four locations selected as well as the number of weeks forecast.

The reduction of the MAE values is confirmed with the mean and standard deviation (Table 3). In all the tests, the CNN architecture with nowcasting—when temperature time series is added as input to the randon time series—outperforms this architecture when only the radon time series is used as input.

The application of the Kruskal-Wallis test to the MAE indicates that the differences between the medians are significant for a confidence level of 95% (p-value under 0.05) for all the test (Table 4). Furthermore, the application of the Wilcoxon signed-rank test with the Bonferroni correction for the pair of test: radon time series versus nowcasting with a placement, points multiples cases where the differences are significant for a confidence level of 95% (p-value under 0.0125). Bonferroni correction forces to divide the p-value by the number of the pair tests: 4 in this case.

Table 3. Mean MAE and standard deviation of the errors of testset for 25 independent runs when using CNN architecture for forecasting radon and radon plus temperature time series of locations selected for diverse weeks ahead).

Weeks	^{222}Rn	BCN	PMP	ZGZ	HSC
2	7.60 ± 0.58	6.45 ± 0.35	6.66 ± 0.97	6.63 ± 0.51	6.87 ± 0.82
4	6.60 ± 0.40	6.22 ± 0.38	6.19 ± 0.27	6.28 ± 0.47	6.31 ± 0.53
6	6.51 ± 0.46	6.09 ± 0.50	6.13 ± 0.38	6.00 ± 0.40	6.23 ± 0.45
8	6.35 ± 0.51	6.06 ± 0.50	6.15 ± 0.52	6.03 ± 0.41	5.93 ± 0.29

Table 4. p-values of the Kruskal-Wallis test and Wilcoxon signed-rank test with Bonferroni correction for the MAE values of CNN architecture. In bold face the cases with positive significance for a confidence level of 95%: p-value under 0.05 for Kruskal-Wallis test and 0.0125 for Wilcoxon signed-rank test with Bonferroni correction.

Weeks	Kruskal-Wallis	Wilcoxon signed-rank with Bonferroni			
		BCN	PMP	ZGZ	HSC
2	$\mathbf{1.4 \cdot 10^{-8}}$	$\mathbf{2.3 \cdot 10^{-5}}$	$\mathbf{2 \cdot 10^{-4}}$	$\mathbf{5 \cdot 10^{-5}}$	$6 \cdot 10^{-2}$
4	$\mathbf{1.5 \cdot 10^{-2}}$	$\mathbf{4 \cdot 10^{-3}}$	$\mathbf{1.6 \cdot 10^{-3}}$	$2 \cdot 10^{-2}$	$11 \cdot 10^{-2}$
6	$\mathbf{7 \cdot 10^{-4}}$	$\mathbf{3 \cdot 10^{-3}}$	$\mathbf{4 \cdot 10^{-3}}$	$\mathbf{3 \cdot 10^{-4}}$	$\mathbf{3 \cdot 10^{-2}}$
8	$\mathbf{4.6 \cdot 10^{-2}}$	$5 \cdot 10^{-2}$	$14 \cdot 10^{-2}$	$2 \cdot 10^{-2}$	$\mathbf{5 \cdot 10^{-4}}$

Global Considerations. Summarizing the performance of the architecture per test, the following indications can be presented:

- For 2 weeks ahead, the best performance (5.79 ± 0.73) is achieved when using MLP architecture with nowcasting for ZGZ placement.
- For 4 weeks ahead, the best performance (5.63 ± 1.53) is achieved when using MLP architecture with nowcasting for PMP placement.
- For 6 weeks ahead, the best performance (5.85 ± 1.45) is achieved when using MLP architecture without nowcasting.
- For 8 weeks ahead, the best performance (5.93 ± 0.29) is achieved when using CNN architecture with nowcasting for HSC placement.

The previous results indicate that nowcasting produces the best results for 3 of 4 number of weeks forecasting, two of which are obtained with MLP architecture and one with CNN architecture. Furthermore, MAE values of the best case are quite similar, without an appreciable degradation as far as the number forecast head weeks increase. Regarding the locations which improve the performance, they are distributed among three locations: PMP, ZGZ and HSC. Finally, RNN is also evaluated but without any positive results, and for this reason their results are omitted.

Finally, in Fig. 5 the prediction with a single run for the validation set (8 weeks) for the same configuration is depicted. These values have not been seen in any time by the deep architectures. As it can be observed, the prediction

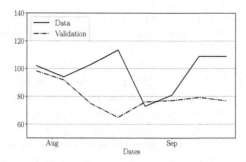

Fig. 5. Forecasting of validation set, 8 weeks (solid line), from August to end September 2019. This prediction (dashed line) has been obtained through a single execution with CNN architecture and nowcasting with HSC temperature.

acceptably reproduces the radon observations for the validation set. This outcome validates the proposed approach providing the necessary efficiency for the decision-making process of scheduling the maintenance operations during the low-radiation periods.

In Table 5, the processing time—in seconds—for the deep learning architectures used in this work with and without nowcasting is presented. Although the activation of nowcasting implies an increment in the processing time, it is enough moderate for not limiting the application of the technique.

Table 5. Mean processing time in seconds and standard deviation of the deep architectures with and without nowcasting.

Architecture	Without nowcasting	With nowcasting
MLP	4.78 ± 0.17	9.19 ± 0.37
CNN	5.54 ± 1.03	11.90 ± 0.33
RNN	6.48 ± 0.47	22.54 ± 0.48

3   Conclusions

In this paper, an implementation of the nowcasting technique to improve the forecasting capability of deep learning architectures has been presented. This technique incorporates information coming from other time series with correlated variables which can act as a proxy of the variables to be predicted. In our study, the nowcasting technique has been applied to the prediction of ^{222}Rn time series from the Canfranc Underground Laboratory (LSC).

Regarding the proxy variables, an analysis of the meteorological variables in four locations around the laboratory (Barcelona, Zaragoza, Huesca and Pamplona) suggests that the temperature time series is acting as a proxy of the radon time series. The nowcasting technique is able to improve the forecasting capability of MLP and CNN, whereas it does not exhibit any capability to improve the performance of RNN.

The improved forecasting capability exhibited by MLP and CNN with nowcasting will help in the decision-making process of scheduling the maintenance operation in the experimental halls of the laboratory, and therefore, avoiding the diffusion of radon in the materials composing the experiments thus reducing this possible source of background for the LSC experiments.

Additional efforts to discover other proxy variables of radon time series at LSC are proposed as future work. For instance, the inclusion of more proxy variables will be linked to techniques of Explainable Artificial Intelligence to ascertain which segments are relevant for the forecasting activity. The inclusion of multiple proxy meteorological variables simultaneously is mandatory in other relevant time series such as the forecasting of air-quality in urban areas, and for this reason is considered as future work.

Acknowledgment. Authors wish to express their thanks for the data support and sharing to Canfranc Underground Laboratory and, particularly to Iulian C. Bandac. TSP is co-funded in a 91.89% by the European Social Fund within the Youth Employment Operating Program, as well as the Youth Employment Initiative (YEI), and co-found in a 8,11 by the "Comunidad de Madrid (Regional Government of Madrid)" through the project PEJ-2018-AI/TIC-10290. MCM is funded by the Spanish Ministry of Economy and Competitiveness (MINECO) for funding support through the grant "Unidad de Excelencia María de Maeztu": CIEMAT - FÍSICA DE PARTÍCULAS through the grant MDM-2015-0509.

References

1. Bishop, C.M.: Neural Networks for Pattern Recognition. Oxford University Press Inc., New York (1995)
2. Cárdenas-Montes, M.: Forecast daily air-pollution time series with deep learning. In: Pérez García, H., Sánchez González, L., Castejón Limas, M., Quintián Pardo, H., Corchado Rodríguez, E. (eds.) HAIS 2019. LNCS (LNAI), vol. 11734, pp. 431–443. Springer, Cham (2019). https://doi.org/10.1007/978-3-030-29859-3_37
3. Cárdenas-Montes, M.: Uncertainty estimation in the forecasting of the ^{222}Rn radiation level time series at the Canfranc Underground Laboratory. Logic J. IGPL (2020). https://doi.org/10.1093/jigpal/jzaa057
4. Cárdenas-Montes, M., Méndez-Jiménez, I.: Ensemble deep learning for forecasting ^{222}Rn radiation level at canfranc underground laboratory. In: Martínez Álvarez, F., Troncoso Lora, A., Sáez Muñoz, J.A., Quintián, H., Corchado, E. (eds.) SOCO 2019. AISC, vol. 950, pp. 157–167. Springer, Cham (2020). https://doi.org/10.1007/978-3-030-20055-8_15
5. Goodfellow, I., Bengio, Y., Courville, A.: Deep Learning. MIT Press, Cambridge (2016)

6. Hochreiter, S., Schmidhuber, J.: Long short-term memory. Neural Comput. 9(8), 1735–1780 (1997). https://doi.org/10.1162/neco.1997.9.8.1735
7. Méndez-Jiménez, I., Cárdenas-Montes, M.: Modelling and forecasting of the ^{222}Rn radiation level time series at the Canfranc Underground Laboratory. In: de Cos Juez, F., et al. (eds.) HAIS 2018. LNCS, vol. 10870, pp. 158–170. Springer, Cham (2018). https://doi.org/10.1007/978-3-319-92639-1_14
8. Méndez-Jiménez, I., Cárdenas-Montes, M.: Time series decomposition for improving the forecasting performance of convolutional neural networks. In: Herrera, F., et al. (eds.) CAEPIA 2018. LNCS (LNAI), vol. 11160, pp. 87–97. Springer, Cham (2018). https://doi.org/10.1007/978-3-030-00374-6_9
9. Sheskin, D.J.: Handbook of Parametric and Nonparametric Statistical Procedures, 4th edn. Chapman & Hall/CRC, London (2007)
10. Tepper, M.: Weather now-casting. In: Bulleting of the American Meteorological Society, vol. 52, p. 1173. AMER Meteorological Society, Boston (1971)
11. Vasco-Carofilis, R.A., Gutiérrez-Naranjo, M.A., Cárdenas-Montes, M.: PBIL for optimizing hyperparameters of convolutional neural networks and STL decomposition. In: de la Cal, E.A., Villar Flecha, J.R., Quintián, H., Corchado, E. (eds.) HAIS 2020. LNCS (LNAI), vol. 12344, pp. 147–159. Springer, Cham (2020). https://doi.org/10.1007/978-3-030-61705-9_13

Analysis of the Seasonality in a Geothermal System Using Projectionist and Clustering Methods

Santiago Porras[1], Esteban Jove[2(✉)], Bruno Baruque[3],
and José Luis Calvo-Rolle[2]

[1] Departamento de Economía Aplicada, University of Burgos,
Plaza Infanta Doña Elena, s/n, 09001 Burgos, Burgos, Spain
sporras@ubu.es

[2] CTC, CITIC, Department of Industrial Engineering, University of A Coruña,
Avda. 19 de febrero s/n, 15495 Ferrol, A Coruña, Spain
{esteban.jove,jlcalvo}@udc.es

[3] Departmento de Ingeniería Informática, University of Burgos, Avd. de Cantabria,
s/n, 09006 Burgos, Burgos, Spain
bbaruque@ubu.es

Abstract. The environmental impact caused by greenhouse gasses emissions derived from fossil fuels, gives rise to the promotion of green policies. In this context, geothermal energy systems has experienced a significant increase in its use. The efficiency of this technology is closely linked with factors such as ground temperature, weather and season. This work develops the analysis of the behaviour of a geothermal system placed in a bioclimatic house during one year, by means of projectionists and clustering methods.

Keywords: Geothermal · LDA · PCA · K-Means · One-class

1 Introduction

The great concern about the climate change has led to the promotion of clean energies that avoid the harmful consequences of fossil fuels use. Furthermore, the strong fossil fuels dependency combined with the rising price of raw materials, have contributed to the acceleration of green policies [25]. In this emergency context, the greenhouse gases emissions reduction plays a significant role and international, national and regional governments have made significant investments that help to develop a sustainable energy generation system, that could be able to mitigate the climate change [7,25].

In spite of the efforts made to increase renewable electric power by many developed countries, it just represented the 15% of the global electric share in 2007, being the hydroelectric power the most used [23]. This value was raised up to a 22% by 2012 [25,29]. Although solar and wind are the most used green

© Springer Nature Switzerland AG 2021
H. Sanjurjo González et al. (Eds.): HAIS 2021, LNAI 12886, pp. 500–510, 2021.
https://doi.org/10.1007/978-3-030-86271-8_42

energies, with their corresponding high level of development, nowadays, there are other fields of renewable energies with significant advances in other disciplines, such as geothermal [18].

Geothermal can be defined as the energy in form of heat under the ground [10]. According to different works, it is calculated that the heat power under the ground reaches 42×10^{12} W [10]. Although this is a valuable amount, its use is limited to specific areas with proper geological conditions [10]. Since the beginning of this century, non-electrical applications of geothermal technologies (heat pumps, bathing, space heating, green houses, aquaculture and industrial processes) represented a more significant share than the electrical ones [10,13, 30].

Different approaches are nowadays considered to improve the geothermal systems, focusing on reducing implementation expenses and optimising the efficiency. They consists of the use of novel materials, the combination of different renewable energies like geothermal and solar, or the application of data science techniques to improve system behaviour [1,26].

Regardless the kind of application, the ground temperature is directly associated with the energy generated, the weather and the season. The work presented in [3] deals with the prediction of geothermal system behavior by means of artificial intelligent techniques. In a similar way, in [6], different time series approaches are proposed to predict the energy generated in a geothermal system located in a bioclimatic house. This work emphasizes the importance of an accurate prediction to optimize the system performance. However, it does not take into consideration the appearance of anomalies or the distribution of system behavior into different months.

In this context, the ground temperature behaviour should be subjected to a deep analysis before during the design stage. This could help to improve a proper system optimisation and helps to estimate the energy produced by the facility. The present work deals with a comparative analysis of the behaviour of a geothermal facility during one year. To do so, correlations between months are analysed by means of different approaches, combining LDA technique and one-class classifiers. This is done using real data registered during one year of normal operation in a geothermal system placed in a bioclimatic house.

The document is structured following this outline: after this section, the bioclimatic house where the geothermal system is placed is described Sect. 2. Then, techniques used to analyse the behaviour of the system under study are detailed in Sect. 3. The experiments and results carried out to analyse the data are presented in Sect. 4. Finally, the last section deals with the conclusions and future works derived from the proposal.

2   Case of Study

2.1   Bioclimatic House of Sotavento

The bioclimatic house of Sotavento is a sustainable facility builded by the Sotavento Galicia Foundation. The main objective of this foundation is to disseminate the use of different renewable energy sources, and promote their positive consequences on the environment. This building is placed between the provinces of A Coruña and Lugo, in Galicia., is located in 43° 21' N, 7° 52' W, at a height of 640 m and is 30 km away from the sea. It is designed to supply electric power and also Hot Domestic Water (HDW) through the following systems:

- Hot Domestic Water:
  - **Solar thermal system.** Eight panels to absorb solar radiation and transfer it to an accumulator.
  - **Biomas system.** This system has a boiler with configurable power, from 7 kW to 20 kW, with a yield of pellets of 90%.
  - **Geothermal system.** A one hundred meters horizontal collector supplies heat from the ground.
- Electricity:
  - **Photovoltaic system.** Twenty two photovoltaic modules with a total power of 2,7 kW.
  - **Wind turbine system.** A low power generator capable of generating 1,5 kW.
  - **Power network.** In charge of supplying electricity when photovoltaic and wind energies do not satisfy the demand.

2.2 Dataset Description

The set of temperatures measured at the geothermal heat exchanger were recorded during one year. The measures consisted of 29 features that were registered using a sample rate of 10 min. To ensure a good interpretation of the dataset, a prior inspection of the dataset was carried out. It is found that three measurements in one day of the year are duplicated. These values are considered not valid, so after discarding them, the 52.645 initial samples were reduced to 52.639.

3 Techniques

3.1 Linear Discriminant Analysis

Linear Discriminant Analysis (LDA) is a supervised classification method for qualitative variables in which two or more groups are known a priori and new observations are classified into one of them based on their characteristics. LDA creates a predictive model for group membership. The model is composed of as many discriminate functions as there are groups less one. Based on linear combinations of the predictor variables that provide the best possible discrimination between the groups. The functions are generated from a sample of cases for which the membership group is known; later, the functions can be applied to new cases that have measurements for the predictor variables but for which the membership group is unknown.

3.2   One-Class Techniques

To evaluate the behaviour of the geothermal installation along one year, two different one-class techniques are evaluated [12].

PCA. The application of dimensional reduction techniques plays a significant role in anomaly detection. In this context, Principal Components Analisys (PCA) has been widely used for one-class classification tasks in many different fields. This technique calculates the eigenvectors through the covariance matrix eigenvalues, which represent the directions with greatest data variability [17,28,28]. These vectors are used to project the original data into a subspace by means of linear combination of the original set. From a geometrical point of view, such operation consist on a rotation of the original axes to a new ones ordered in terms of variance, which can be expressed by Eq. 1.

$$y_i = W_i^T x \tag{1}$$

where:

- x^d in an N-dimensional space onto vectors in an M-dimensional space $(x_1...x_N)$.
- W_i are the N eigenvectors of the covariance matrix.
- y are the projected original data onto the new output M-dimensional subspace $(y_1...y_N)$.

Once the original data is projected onto the new axes, the reconstruction error is computed as the difference between the original dataset and its projection over the new subspace If this value exceeds a threshold, the data remains outside the target set [24].

K-Means. The K-Means is an unsupervised algorithm commonly used in different fields, such as machine learning, image processing or pattern recognition [9,22]. This algorithm aims to divide the dataset into homogeneous groups comprised of data with similar features, called clusters [8,19].

From a given set $X = \{x_1, x_2, ..., x_N\}$, where $x_i \in \mathbb{R}^n$, the K-Means algorithm divides the data into K subsets $G_1, G_2, ..., G_K$, with their corresponding centroids $C = \{c_1, c_2, ..., c_K\}$, where $c_j \in \mathbb{R}^n$, following a clustering error criterion [12,14,15]. This error is usually computed as the sum of all the euclidean distances of each point $x_i \in \mathbb{R}^n$ to its centroid c_j, according to Eq. 2, where $I(A) = 1$ only if A is true.

$$E(c_1, ..., c_K) = \sum_{i=1}^{N} \sum_{j=1}^{K} I(x_i \in G_j) \, ||x_i - c_j|| \tag{2}$$

Once the training set has been divided into clusters and the centroids are calculated, the criteria to determine if a test data is inside the target set is based on the distance of the point to its nearest centroid [5]. If this value exceeds a given threshold, it is considered to be out of the target set [2,16].

4   Experiments and Results

4.1   LDA

The data contains measures of several sensors distributed throughout the circuit. These sensors are allocated one behind another, so it is likely that there is correlation between them. We need to check this hypothesis. A Pearson correlation test [27] has been performed between the input sensor (S28), output sensor (S29) an all intermediate sensors. This test has been run in IBM SPSS. The results can be observed in Table 1 which has been reduced due to space requirements.

Table 1. Pearson correlation test

		S28	S29	S301	S302	S303	S304
S28	Pearson correlation	1	0.993	0.469	0.387	0.461	0.561
	Sig	–	0.000	0.000	0.000	0.000	0.000
S29	Pearson correlation	0.993	1	0.455	0.375	0.448	0.549
	Sig	0.000	–	0.000	0.000	0.000	0.000
S301	Pearson correlation	0.469	0.455	1	0.993	0.994	0.953
	Sig	0.000	0.000	–	0.000	0.000	0.000
S302	Pearson correlation	0.387	0.375	0.993	1	0.991	0.935
	Sig	0.000	0.000	0.000	–	0.000	0.000
S303	Pearson correlation	0.461	0.448	0.994	0.991	1	0.972
	Sig	0.000	0.000	0.000	0.000	–	0.000
S304	Pearson correlation	0.561	0.549	0.953	0.935	0.972	1
	Sig	0.000	0.000	0.000	0.000	0.000	–

All correlations are significant in a bilateral level 0.01. In Table 2 can be noticed that the correlation between sensors s28 and S29 is huge, with a value close to 1. The correlation with other sensors is significant too. The same conclusion can be said of the intermediate sensors: Pearson value among all of them is really close to 1 and significant with S28 and S29. Therefore, the hypothesis is true, all sensors are correlated in the pump heat system, the external control sensor was excluded.

This mean that it can be used a single sensor in the following experiments. In order to design the computational experiments is necessary divide the dataset. The fist logical partition is divide the dataset in months, to preserve the temporal component. In addition, another division in trimesters was made. First, we must check if the division is correct in statistical terms in order to perform the analysis. A parametrical analysis for k independent samples test was performed, where k is the number of months in one test and the number of trimesters in the other.

The test has been run in two cases where the pump heat exchanger is running and for all data. In Table 2 It can be observed the results of the Kruskal-Wallis test [21] using only one variable in this case we have choose S28.

Table 2. Kruskal-Wallis test

	All data		Pump heat running	
	Month	Trimester	Month	Trimester
Square Chi	423.766	183.922	1614.203	1185.164
gl	11	23	11	3
Sig.	5,4724E-84	1,2545E-39	0,0E0	1,2135E-256

As the significance has values lowers than 0.05, it can concluded that exits statistical significant differences between all the months and trimester. So they are different one each other and statically we can perform this division. To reinforce this division, we have performed a discriminant analysis with the same division, this allows to know how similar are the temperatures between the months and trimesters.

Discriminant analysis consists of establishing diagnostic rules that perform classifying a individual in one of two or more well defined groups. The essential goal of discriminant analysis is to use the previously known values of the independent variables to predict the corresponding category of the dependent variable [20].

The Fig. 1 shows the classification in months and trimesters for all data, classification percentage is also provided. In Fig. 2 is show the same analysis but only with pump heat running data. Discriminant analysis for all data: Number of cases: 52645. Percentage of cases classified correctly: By months 95.1%, by trimester 96.4%.

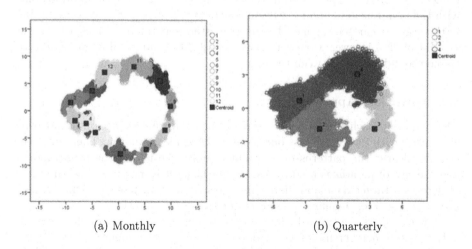

(a) Monthly   (b) Quarterly

Fig. 1. Discriminant analysis for all data

Discriminant analysis for data with pump heat running results are: Number of cases: 6734% of cases classified correctly: By months 96.6%, by trimester 97.6%.

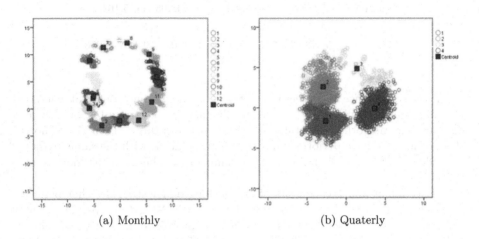

(a) Monthly   (b) Quaterly

Fig. 2. Discriminant analysis heat pump working

According with these results, we can conclude for the analysis of all the whole data that the division will be correct. Only the data for months 3, 4 and 5 corresponding with March, April and May, is confusing, meaning that temperatures in these months are similar. The same conclusion can be done for trimesters 1 and 2. They overlap, because these months. With the data only with the pump heater running, we can conclude the same as the whole data, but in this case we can see the months and trimesters separated more clearly, as the amount of data is lower. However, in the months 3, 4 and 5 will be more difficult to see the differences between them.

4.2   One-Class Approach

The main approach consists of separating the data from each month and train a classifier with data from just one month, called target month. Then, once the classifier learnt the patterns of a specific month, data from the target month and the rest of months is used to test the it. Then, it is measured the capability of detecting target days and discarding days from other months. This process, shown in Fig. 3 for January, is followed for the twelve months.

The classifiers are trained with all temperature sensors (6 variables) and the output is the membership of each sample to the objective month.

The evaluation of the classifier performance has been done though the Area Under the Receiver Operating Characteristic Curve (AUC) [4]. This parameter, in percentage, has offered a really interesting performance, with the particular feature of being insensitive to changes in class distribution, which is a key factor

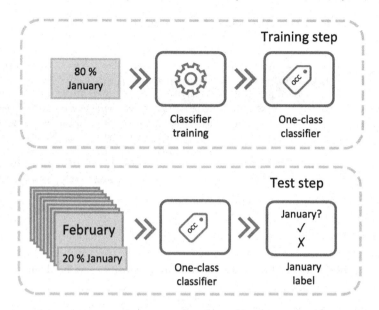

Fig. 3. Example of classifier train and test phases for January

in one-class classification tasks [11]. Furthermore, the time needed to train each classifier is also presented as a measure of the computational cost. The tests were validated following a $k - fold$ cross-validation method, using k = 5.

PCA. Different hyper-parameters configurations are tested to evaluate PCA classifiers. Since it is a dimensional reduction technique, the target subspace size is varied, taking into consideration that it must have, at least, one less dimension than the initial dataset. On the other hand, the reconstruction error threshold considered to determine the membership of a test sample is tested. In the experiments, a reconstruction error greater than 100%, 95%, 90% and 85% of the ones measured during the training stage is evaluated. The highest AUC scores achieved are represented in Table 3.

K-Means. This clustering algorithm has been tested through a wide range of configurations. First, the number of clusters in which the dataset is divided is swept from one unique group to fifteen groups. This ensures that the possibility of different data shapes are considered to achieve good results. Furthermore, the distance level that indicates the non-target nature of a test instance is varied. The percentile distances of the training set are established as criteria, considering 100%, 95%, 90% and 85%. The highest AUC scores achieved are represented in Table 4.

Table 3. Best AUC values for train and test sets using PCA

Target month		Non-target month											
		Jan	Feb	Mar	Apr	May	Jun	Jul	Aug	Sep	Oct	Nov	Dec
	Jan		95.34	99.98	99.99	99.99	100.00	100.00	100.00	99.99	100.00	100.00	98.49
	Feb			96.99	100.00	99.99	100.00	100.00	99.99	100.00	100.00	100.00	99.99
	Mar				98.21	94.78	99.99	99.99	99.99	99.99	99.99	99.99	100.00
	Apr					86.48	99.99	99.99	99.99	100.00	99.99	99.99	99.99
	May						94.10	100.00	100.00	99.99	100.00	99.99	99.99
	Jun							95.45	99.99	99.99	99.99	99.99	100.00
	Jul								99.99	99.99	100.00	100.00	99.99
	Aug									96.02	99.98	99.99	99.99
	Sep										94.97	99.99	99.99
	Oct											97.71	99.99
	Nov												89.57
	Dec												

Table 4. Best AUC values for train and test sets using K-Means

Target month		Non-target month											
		Jan	Feb	Mar	Apr	May	Jun	Jul	Aug	Sep	Oct	Nov	Dec
	Jan		95.87	99.96	97.49	97.95	100.00	100.00	100.00	100.00	100.00	100.00	96.58
	Feb			92.44	97.55	97.33	100.00	100.00	100.00	99.99	100.00	99.99	100.00
	Mar				93.59	95.03	100.00	100.00	100.00	99.99	100.00	99.99	100.00
	Apr					81.18	100.00	100.00	100.00	100.00	100.00	99.99	100.00
	May						94.88	100.00	100.00	100.00	100.00	100.00	100.00
	Jun							97.51	100.00	100.00	100.00	100.00	99.99
	Jul								100.00	100.00	100.00	100.00	100.00
	Aug									96.76	99.99	100.00	100.00
	Sep										97.45	97.69	99.99
	Oct											97.60	99.99
	Nov												95.36
	Dec												

5   Conclusions and Future Work

Once the data has been analyzed, it is seen that the main discrimination problem is the overlaps produced between the end of one month and the beginning of the next. This also occurs with months with similar temperatures. This makes the classification problem complex and challenging since the differences in temperature between consecutive days are minimal. This requires an efficient classifier to detect these anomalies within a small threshold. The one-class methods presented in this work solve this problem efficiently with excellent results. These classifiers are trained with only data from one month and presented good performance when detecting data from other days. This can be presented as a useful tool to detect the appearance of abnormal data in one month. Even in the areas of overlap of the months that are the most difficult to classify, obtaining in the worst case values higher than 80% of correct classification and in general values above 95%.

With all this, as lines of future work, the study and implementation of other classification techniques that are capable of refining the classification in conflicting overlapping areas and improve the detection of anomalies are proposed. Furthermore, a deeper analysis could be carried out considering week classifiers instead of month ones. This could help to identify with even more accuracy the appearance of anomalous measurements.

Acknowledgment. CITIC, as a Research Center of the University System of Galicia, is funded by Consellería de Educación, Universidade e Formación Professional of the Xunta de Galicia through the European Regional Development Fund (ERDF) and the Secretaría Xeral de Universidades (Ref. ED431G 2019/01).

References

1. Aláiz-Moretón, H., Castejón-Limas, M., Casteleiro-Roca, J.L., Jove, E., Fernández Robles, L., Calvo-Rolle, J.L.: A fault detection system for a geothermal heat exchanger sensor based on intelligent techniques. Sensors **19**(12), 2740 (2019)
2. Alaiz-Moretón, H., et al.: Bioinspired hybrid model to predict the hydrogen inlet fuel cell flow change of an energy storage system. Processes **7**(11), 825 (2019)
3. Baruque, B., Porras, S., Jove, E., Calvo-Rolle, J.L.: Geothermal heat exchanger energy prediction based on time series and monitoring sensors optimization. Energy **171**, 49–60 (2019)
4. Bradley, A.P.: The use of the area under the roc curve in the evaluation of machine learning algorithms. Pattern Recogn. **30**(7), 1145–1159 (1997)
5. Casteleiro-Roca, J.L., Barragán, A.J., Segura, F., Calvo-Rolle, J.L., Andújar, J.M.: Fuel cell output current prediction with a hybrid intelligent system. Complexity **2019** (2019)
6. Casteleiro-Roca, J.L., Calvo-Rolle, J.L., Meizoso-López, M.C., Piñón-Pazos, A., Rodríguez-Gómez, B.A.: Bio-inspired model of ground temperature behavior on the horizontal geothermal exchanger of an installation based on a heat pump. Neurocomputing **150**, 90–98 (2015)
7. Casteleiro-Roca, J.L., et al.: Short-term energy demand forecast in hotels using hybrid intelligent modeling. Sensors **19**(11), 2485 (2019)
8. Casteleiro-Roca, J.L., Jove, E., Gonzalez-Cava, J.M., Pérez, J.A.M., Calvo-Rolle, J.L., Alvarez, F.B.: Hybrid model for the ANI index prediction using remifentanil drug and EMG signal. Neural Comput. Appl. **32**(5), 1249–1258 (2020)
9. Crespo-Turrado, C., et al.: Comparative study of imputation algorithms applied to the prediction of student performance. Logic J. IGPL **28**(1), 58–70 (2020)
10. Dickson, M.H., Fanelli, M.: Geothermal Energy: Utilization and Technology. Routledge, Milton Park (2013)
11. Fawcett, T.: An introduction to roc analysis. Pattern Recogn. Lett. **27**(8), 861–874 (2006)
12. Jove, E., Casteleiro-Roca, J., Quintián, H., Méndez-Pérez, J., Calvo-Rolle, J.: Anomaly detection based on intelligent techniques over a bicomponent production plant used on wind generator blades manufacturing. Revista Iberoamericana de Automática e Informática industrial **17**(1), 84–93 (2020)

13. Jove, E., Aláiz-Moretón, H., Casteleiro-Roca, J.L., Corchado, E., Calvo-Rolle, J.L.: Modeling of bicomponent mixing system used in the manufacture of wind generator blades. In: Corchado, E., Lozano, J.A., Quintián, H., Yin, H. (eds.) IDEAL 2014. LNCS, vol. 8669, pp. 275–285. Springer, Cham (2014). https://doi.org/10.1007/978-3-319-10840-7_34

14. Jove, E., Antonio Lopez-Vazquez, J., Isabel Fernandez-Ibanez, M., Casteleiro-Roca, J.L., Luis Calvo-Rolle, J.: Hybrid intelligent system to predict the individual academic performance of engineering students. Int. J. Eng. Educ. **34**(3), 895–904 (2018)

15. Jove, E., et al.: Missing data imputation over academic records of electrical engineering students. Logic J. IGPL **28**(4), 487–501 (2020)

16. Jove, E., Casteleiro-Roca, J.L., Quintián, H., Méndez-Pérez, J.A., Calvo-Rolle, J.L.: Virtual sensor for fault detection, isolation and data recovery for bicomponent mixing machine monitoring. Informatica **30**(4), 671–687 (2019)

17. Jove, E., Casteleiro-Roca, J.L., Quintián, H., Simić, D., Méndez-Pérez, J.A., Luis Calvo-Rolle, J.: Anomaly detection based on one-class intelligent techniques over a control level plant. Logic J. IGPL **28**(4), 502–518 (2020)

18. Kaltschmitt, M., Streicher, W., Wiese, A.: Renewable Energy. Springer, Heidelberg (2007). https://doi.org/10.1007/3-540-70949-5

19. Kaski, S., Sinkkonen, J., Klami, A.: Discriminative clustering. Neurocomputing **69**(1–3), 18–41 (2005)

20. Klecka, W.R., Iversen, G.R., Klecka, W.R.: Discriminant Analysis, vol. 19. Sage, Thousand Oaks (1980)

21. Kruskal, W.H., Wallis, W.A.: Use of ranks in one-criterion variance analysis. J. Am. Stat. Assoc. **47**(260), 583–621 (1952)

22. Likas, A., Vlassis, N., Verbeek, J.J.: The global k-means clustering algorithm. Pattern Recogn. **36**(2), 451–461 (2003)

23. Lund, H.: Renewable energy strategies for sustainable development. Energy **32**(6), 912–919 (2007)

24. Machón-González, I., López-García, H., Calvo-Rolle, J.L.: A hybrid batch SOM-NG algorithm. In: The 2010 International Joint Conference on Neural Networks (IJCNN), pp. 1–5. IEEE (2010)

25. Owusu, P.A., Asumadu-Sarkodie, S.: A review of renewable energy sources, sustainability issues and climate change mitigation. Cogent Eng. **3**(1), 1167990 (2016)

26. Ozgener, O., Ozgener, L.: Modeling of driveway as a solar collector for improving efficiency of solar assisted geothermal heat pump system: a case study. Renew. Sustain. Energy Rev. **46**, 210–217 (2015). http://www.sciencedirect.com/science/article/pii/S1364032115001318

27. Pearson, K.: VII. note on regression and inheritance in the case of two parents. Proc. R. Soc. London **58**(347–352), 240–242 (1895)

28. Quintián, H., Corchado, E.: Beta scale invariant map. Eng. Appl. Artif. Intell. **59**, 218–235 (2017)

29. Vega Vega, R., Quintián, H., Calvo-Rolle, J.L., Herrero, Á., Corchado, E.: Gaining deep knowledge of android malware families through dimensionality reduction techniques. Logic J. IGPL **27**(2), 160–176 (2019)

30. Vega Vega, R., Quintián, H., Cambra, C., Basurto, N., Herrero, Á., Calvo-Rolle, J.L.: Delving into android malware families with a novel neural projection method. Complexity **2019** (2019)

A Model-Based Deep Transfer Learning Algorithm for Phenology Forecasting Using Satellite Imagery

M. Á. Molina, M. J. Jiménez-Navarro, F. Martínez-Álvarez,
and G. Asencio-Cortés[(⊠)]

Data Science and Big Data Lab, Pablo de Olavide University, 41013 Seville, Spain
{mamolcab,mjjimnav}@alu.upo.es, {fmaralv,guaasecor}@upo.es

Abstract. A new transfer learning strategy is proposed for classification in this work, based on fully connected neural networks. The transfer learning process consists in a training phase of the neural network on a source dataset. Then, the last two layers are retrained using a different small target dataset. Clustering techniques are also applied in order to determine the most suitable data to be used as target. A preliminary study has been conducted to train and test the transfer learning proposal on the classification problem of phenology forecasting, by using up to sixteen different parcels located in Spain. The results achieved are quite promising and encourage conducting further research in this field, having led to a 7.65% of improvement with respect to other three different strategies with both transfer and non-transfer learning models.

Keywords: Transfer learning · Deep learning · Classification · Pattern recognition

1 Introduction

The agricultural sector has historically positioned itself as the most important economic sector as it provides the basic livelihood for the population. In recent years, due to a growing world population, increased crop security and climate change, this sector is undergoing strict production controls that are driving up production costs and therefore prices. This is why new technologies are becoming essential to keep track of all this and provide a fundamental aid for the farmer's decision making.

To this end, it is becoming increasingly widespread to take data on crops and to make use of the information provided by these data. However, this data collection is a very recent development and there is still not enough historical information available to allow the application of traditional machine learning techniques.

For this reason, it is necessary to apply new techniques that allow the optimal treatment of this data and to use the existing general information in geographical

© Springer Nature Switzerland AG 2021
H. Sanjurjo González et al. (Eds.): HAIS 2021, LNAI 12886, pp. 511–523, 2021.
https://doi.org/10.1007/978-3-030-86271-8_43

areas where information is still scarce. New deep neural network techniques typically require a very large data history and, consequently, a long time before they can be applied. In these cases, there is a need of widening the knowledge base by extracting information from additional datasets.

The integration of deep learning [1,2] with transfer learning is called deep transfer learning and it makes the most of both paradigms. Thus, deep learning is used to model problems within big data contexts and, afterwards, re-purposed to transfer the knowledge to models with insufficient data [3]. There is a major flaw in transfer learning, which is the lack of interpretability of its models because pretrained models are applied to the new data without any prior information or understanding of the model [4].

A new transfer learning strategy is proposed in this work, based on the application of a convolutional neural network (CNN). In particular, a 8-layer CNN is trained with the source dataset. Then, the last two layers are retrained with a training set from the target dataset. Different training sets, as explained in Sect. 3, are created in order to validate the robustness of the method.

To assess the performance of the proposal, sampling data from phenological stages in different olive grove plots and satellite index data (obtained from images) have been used. However, any other problems could be used to evaluate the goodness of this general-purpose methodology. This dataset is formulated as a classification problem, in four phenological states. Three additional strategies are also evaluated to compare the performance in terms of accuracy. The results achieved are quite promising.

The rest of the paper is structured as follows. Section 2 overviews recent and relevant papers in the field of deep transfer learning and its application to classification datasets. Section 3 describes the proposed methodology Sect. 4 reports and discusses the results achieved. Finally, Sect. 5 summarizes the conclusions.

2   Related Works

Deep transfer learning is becoming one of the research fields in which much effort is being put into [5].

In fact, many applications can be found in the literature currently. However, these techniques are not yet widespread in agriculture sector. Some works have been published introducing these techniques. For example, the one described in [6] where basic algorithms are used to yield predictions, disease detection or crop quality, for example.

It should be noted that this work is a continuation of the experiments started in [7], where the same techniques were applied to a group of images of cells affected or not by malaria.

In [8], deep learning techniques are applied to detect phenological stages of the rice crop through images taken by aerial vehicles to make estimates of production and harvest dates.

Transfer learning techniques have also been applied to the agricultural sector for crop type detection in different regions with limited access [9]. Finally, more advanced techniques such as image-based deep transfer learning have been used for disease detection in crops, as can be read in [10]. Papers explaining how transfer learning techniques work can be found at [11,12].

Many applications relate deep transfer learning with remote sensing problems. In 2016, Xie et al. [13] proposed the use of deep transfer learning techniques on satellite images or another type of sensor with the aim of extracting the most information from the available data on issues of food security or sustainable development.

Also, a transfer learning application for scene classification with different datasets and different deep learning models can be found in [14]. The authors analyzed how the setting up of CNN hyperparameters can influence the transfer learning process. They concluded that, contrary to expectations, transfer learning from models trained on more generic images outperformed transfer learning from those trained directly on smaller datasets.

The inclusion of distance studies in works with deep transfer learning techniques are important in order to see the effects or not of datasets similarity. Thus, in [15] some time series are classified using a distance based approach.

3   Methodology

This section first describes the data preprocessing in Sect. 3.1. Section 3.2 details how source and target datasets have been created. The deep neural network architecture is discussed in Sect. 3.3. The validation schemes are introduced in Sect. 3.4. Finally, Sect. 3.5 is devoted to analyze how the clusters shape may affect the model.

3.1   Data Preprocessing

The first step in the data preprocessing is to link the data of the ten bands obtained from the satellite images and phenology (study of changes in the timing of seasonal events such as budburst, flowering, dormancy, migration and hibernation) data of each parcel (a tract or plot of land). Once we have the dataset with the information of all parcels, the information is transformed in order to relate past events with the phenology of the parcel. In this work, the objective is to predict the phenology of each parcel seven days ahead, so the obtained relations between input and output are oriented in that way.

The second step of the preprocessing is to encode the data labels, in order to have as many outputs of the neural network as data labels. Thus, a predicted probability is returned for each label.

3.2   Creation of Source and Target Subsets

Disjoint source and target subsets of data were extracted from the original dataset. The source subset is the dataset from which the initial model was trained. The target subset is the dataset used both to update such model (transfer process) and to test the updated model.

To extract the source and target subsets, it has been tried that such datasets were as different as possible. Additionally, the source subset is larger than the target one. The idea underlying this strategy is to check if the transfer learning is effective when the source and target subsets contain dissimilar information, comparing when the information is similar, and the target set is a smaller one.

For this purpose, a hierarchical clustering was applied to each different label using such table as input. A dendrogram was generated for each label after applying the hierarchical clustering.

Finally, the first three nodes of the second level of the dendrogram were selected. As will be explained later, these three clusters will be used to build the source subset. Each scheme will be tested for each parcel, being a campaign (season) of each one the target subset.

It can be concluded that source subsets were generated in such a way that they contain dissimilar and similar parcels in order to obtain certain results and the source subset is larger than the target subset.

3.3 Deep Neural Network Architecture

The next step consists in training a neural network and testing it using the subsets described in the previous section. The way these subsets are divided to validate the methodology will be explained in the next subsection.

The deep neural network is composed of six dense and fully-connected layers of the network. The last two layers are the layers used to adapt the neural network in transfer learning schemes as it is indicated in last column in Table 1.

The neural network proposed has 14,069 parameters to be adjusted. The detailed network used is shown in Table 1. To implement the neural network architecture, Keras 2.2.4 over TensorFlow 1.14 was used [16].

Table 1. Deep neural network architecture used for transfer learning.

Layer (type)	Output shape	Parameters	Used for transfer learning
Dense	(None, 115)	2,760	No
Dense	(None, 70)	8,120	No
Dense	(None, 35)	2,485	No
Dense	(None, 15)	540	No
Dense	(None, 8)	128	Yes
Dense	(None, 4)	36	Yes

3.4 Novel Validation Schemes Proposed

Four schemes will be executed for each parcel, so the next explanation is valid for each one. In each experiment, each parcel will be the target subset.

The target subset is divided into two parts: test (last campaign) and training (the remaining campaigns). Fixing the same test part of the target subset provided a fair comparison in each of the four different validation schemes have been proposed.

Scheme 1. The model is generated using the training part (campaigns of each parcel except the last one) of the target subset, and it is tested by evaluating its predictions over the test part (last campaign of each parcel) of the target subset.

Scheme 2. Three schemes are proposed for each parcel in order to prove the effectiveness of the transfer learning techniques.

First, the scheme 2.1. In this scheme, the steps are the following: First, the model is trained using the parcels of the same cluster of the target parcel. Then, such model is updated using the training part (all campaigns of this parcel except the last one) of the target subset. This updating process only optimizes the weights within the two last layers of the neural network, maintaining the rest of its layers without changes. Finally, the updated model is tested by evaluating its predictions over the test part (last campaign) of the target subset.

In the next two schemes, different source subsets are used using the other two clusters in order to show the goodness of transfer learning techniques using dissimilar subsets (belonging to the same branch of the tree or the other one).

In scheme 2.2, the source subset which we train the model is the cluster on the right of the cluster of the target parcel viewing the dendrogram. The steps are the same of the Scheme 2.1.

In scheme 2.3, the source subset which we train the model is the remaining cluster. The steps are the same of the Scheme 2.1.

For each scheme, the methodology has been tested up to 10 times, obtaining very similar but different tests due to the own network characteristics.

3.5 Intra-cluster and Inter-cluster Analysis

Finally, an analysis has been conducted to prove how the effectiveness of the proposed transfer learning methodology varies depending on the distance between parcels classified in the own and another clusters.

For such purpose, the dendrogram obtained with the hierarchical cluster throws the clusters which we are going to use showing the similarity between them depending on the branch to which they belong. The intra-cluster distance is calculated as the average of the distance between the own parcels.

4   Experimentation and Results

This section presents and discusses the results achieved. Thus, Sect. 4.1 describes how the dataset has been generated and interpreted. The metric used to evaluate the performance of the proposal is introduced in Sect. 4.2. The experimental settings are listed and discussed in Sect. 4.3. Finally, the results themselves are reported and commented in Sect. 4.4.

4.1   Dataset

The set of images used to perform the methodology explained in previous section have been taken from two different sources of data.

Table 2. Domains characteristics.

Code	Region	Coordinates	Altitude	Surface	Slope	Soil	Density	Main variety
P01	Jaén	37.71, −2.96	700	6.02	2	Irrigated	178	Picual
P02	Jaén	37.68, −2.94	700	2.86	5	Irrigated	200	Picual, Marteño
P03	Jaén	37.66, −2.93	700	1.09	1	Irrigated	140	Picual, Marteño
P04	Cádiz	36.86, −5.43	360	8.54	19	Dry	58	Lechín, Zorzaleño, Ecijano
P05	Cádiz	36.87, −5.45	770	2.26	19	Dry	138	Lechín, Zorzaleño, Ecijano
P06	Cádiz	36.94, −5.25	350	1.04	5	Dry	134	Picual, Marteño
P07	Cádiz	36.94, −5.30	425	3.10	11	Dry	74	Lechín
P08	Córdoba	37.44, −4.69	280	6.30	1	Dry	194	Picual, Marteño
P09	Córdoba	37.67, −4.33	300	42.53	3	Irrigated	154	Picual, Marteño
P10	Córdoba	37.60, −4.42	460	9.58	9	Dry	76	Picual, Picudo
P11	Córdoba	37.44, −4.71	320	19.52	1	Irrigated	208	Manzanillo
P12	Jaén	37.67, −2.91	700	0.96	3	Irrigated	92	Picual
P13	Sevilla	37.05, −5.01	600	1.93	15	Dry	119	Picual, Marteño
P14	Sevilla	37.10, −5.04	460	31.59	15	Irrigated	156	Manzanillo
P15	Sevilla	37.05, −5.08	520	1.78	20	Dry	120	Hojiblanco
P16	Sevilla	37.07, −5.06	510	2.87	25	Dry	150	Hojiblanco

On the one hand, the information of index and bands (multi-spectral data with 12 bands in the visible, near infrared, and shortwave infrared part of the spectrum) has been collected of the Sentinel satellite images. This satellite captures images from 2015 increasing the sampling rate since then. Now, an image is captured each 4 or 5 days (from 15 days on average in 2015). From these images, different indexes and the values of the colour bands are calculated for each pixel of the image. In this work, the mean value of the pixels of each band will be used for each image to train the model. These bands can be consulted in Table 3.

On the other hand, the olive phenology states of each studied parcel are taken from the open data set, property of *Red de Alerta e Información Fitosanitaria* [17] belonging to *Junta de Andalucía*.

Table 3. Sentinel bands: description.

Band number	Band description	Wavelength range (nm)	Resolution (m)
B1	Coastal aerosol	433–453	60
B2	Blue	458–523	10
B3	Green	543–578	10
B4	Red	650–680	10
B5	Red-edge 1	698–713	20
B6	Red-edge 2	733–748	20
B7	Red-edge	773–793	20
B8	Near infrared (NIR)	785–900	10
B8A	Near infrared narrow (NIRn)	855–875	20
B9	Water vapour	935–955	60
B10	Shortwave infrared/Cirus	1360–1390	60
B11	Shortwave infrared 1 (SWIR1)	1565–1655	20
B12	Shortwave infrared 2 (SWIR 2)	2100–2280	20

Fourteen phenology states are detected in olive. In this work, they have been summarized in four: The state 1 which includes winter bud and moved bud states (2.25% of samples); state 2 with inflorescence, corolla flowering and petal drop states (29% of samples); state 3 which includes set fruit and hardening of the olive stone (65.13% of samples) and finally, state 4 than collected the veraison and matured fruit states (3.62% of samples).

This phenology dataset has been obtained from sixteen parcels for four regions of Andalucía (four for each one) as it can be observed in Fig. 1, with different characteristics between them, like variety, altitude, type of crop (traditional, intensive or super-intensive), etc. The characteristics of these parcels are shown in Table 2, where the altitude is expressed in meters, the surface in hectares the slope in % and the density in plants/hectares.

4.2 Evaluation Metric

In order to quantify the effectiveness of the methodology proposed, Accuracy was computed.

The metric used for Accuracy calculates the mean accuracy rate across all predictions for multi-class classification problems. The formula is:

$$Accuracy = \frac{1}{N} \sum_{k=1}^{|G|} \sum_{x:g(x)=k} I(g(x) = \hat{g}(x)) \tag{1}$$

where I is the indicator function, which returns 1 if the classes match and 0 otherwise.

Fig. 1. Localization of the parcels used for the study of transfer learning (Andalusia, Spain).

4.3 Experimental Settings

In order to set up the neural network for the experimentation carried out, the experimental settings established were the presented in Table 4.

Table 4. Experimental settings.

Parameter	Description
Batch size	With a value of 128, it defines the number of samples that will be propagated through the network
Epochs	One epoch is when an entire dataset is passed forward and backward through the neural network only once. The number of epochs used was 50
Optimizer	The optimizer used was *adam*. This optimizer recommends to leave the parameters at their default values
Hierarchical clustering	The method used to calculate the matrix distance was *ward.D2* with its default values

4.4 Results and Discussion

The results obtained after applying the methodology to the four validation schemes described in the previous section are shown now.

Previously, the parcels have been clustered in the manner explained in the previous section. The dendrogram can be seen in Fig. 2.

Three clusters have been obtained. From now on, Cluster 1 will be made up of parcels P05, P04, P15, P13, P10, P14; Cluster 2 parcels P01, P03, P12 and Cluster 3 will contain parcels P07, P02, P09, P16, P06, P08, P11. Details of the distances between plots can be found at Fig. 3.

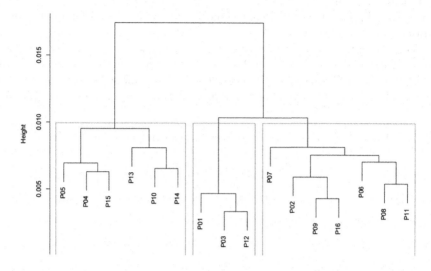

Fig. 2. Cluster dendrogram, with cutoff value leading to three parcel clusters.

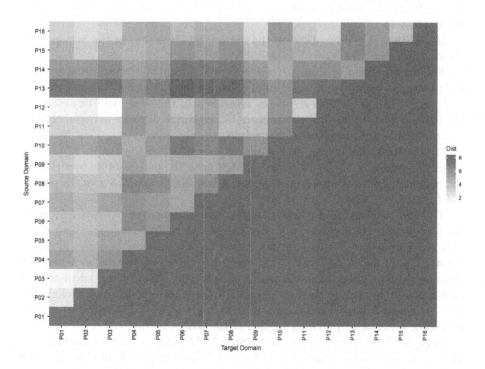

Fig. 3. Geographic distances among the parcels.

In this way, Cluster 2 and Cluster 3 are the most similar because they belong to the same branch. Cluster 1 has the most dissimilar information of the entire dataset.

Calculating the intra-cluster distance for the parcels, it can be observed that Cluster 1 distance is 1.49, Cluster 2 one is 5.42 and in the last one is 4.37. It can be observed that the parcels of Cluster 1 are most similar between them than the parcels of the other two clusters. This evidence is also reflected in Table 5.

Table 5. Average distance for parcels belonging to a same cluster.

Cluster	Intra-cluster distance
1	1.49
2	5.42
3	4.37

In Table 6, the accuracy of the four proposed schemes can be observed. The *Scheme* 1, where we train with the own parcel, throws an average accuracy of 80.67%. The improvement training with the parcels of the same cluster is 8.15%, being the improvements of 5.94% and 7.65% using the parcels of the others clusters.

In addition, it can be observed the improvements assuming the Scheme 1 as the baseline reference.

The improvement of some parcels is very significant using transfer learning techniques, as it can be observed in parcels number P03, P09, P11 and P13.

In order to analyze the results in more depth, the study will be separated by clusters, relating the improvements obtained to the distance between clusters and the number of parcels used for training.

Firstly, the study focuses on the parcels belonging to Cluster 1.

In this case, the *Scheme* 2.1 uses, for training, the parcels of its own cluster, the *Scheme* 2.2 uses the parcels of Cluster 2 and the *Scheme* 2.3 uses the parcels of Cluster 3.

The best average improvement is the one obtained in *Scheme* 2.1, using the parcels of its own cluster to apply transfer learning techniques. Very close, it is the improvement using Cluster 3. It can be explained because, although Cluster 3 is in another branch of the tree, and therefore, the distance between them is bigger, it is composed of many parcels, which provided a lot of general information that can be used to learn with these transfer learning techniques. The *Scheme* 2.2 throws less improvement since Cluster 2 provides less parcels and, also, it belongs to another branch of the tree.

In cases where parcels belong to Cluster 2, the *Scheme* 2.2 uses, for training, the parcels of its own cluster, the *Scheme* 2.2 uses the parcels of Cluster 3 and the *Scheme* 2.3 uses the parcels of Cluster 1.

Table 6. Accuracy for the proposed method for each phenology dataset and validated for Schemes 1, 2.1, 2.2 and 2.3. Cluster assignation for each parcel.

Parcels	Scheme 1	Scheme 2.1	Scheme 2.2	Scheme 2.3	Cluster
P01	86.68%	88.60%	90.90%	90.10%	2
P02	85.42%	90.80%	90.80%	90.80%	3
P03	82.22%	91.10%	91.10%	90.99%	2
P04	82.93%	86.31%	82.59%	87.10%	1
P05	83.84%	87.37%	85.49%	87.20%	1
P06	89.34%	89.63%	88.83%	88.63%	3
P07	90.12%	90.10%	90.80%	90.37%	3
P08	65.29%	71.15%	65.85%	67.13%	3
P09	76.08%	88.60%	85.44%	86.86%	3
P10	79.34%	84.01%	79.72%	88.04%	1
P11	81.49%	92.84%	92.07%	91.89%	3
P12	83.83%	85.37%	84.60%	84.71%	2
P13	27.08%	92.38%	81.35%	87.56%	1
P14	88.41%	91.70%	86.70%	91.70%	1
P15	95.43%	95.60%	94.05%	95.60%	1
P16	93.27%	95.60%	95.60%	94.41%	3
Average	80.67%	88.82%	86.62%	88.32%	–
Improvement	–	8.15%	5.94%	7.65%	–

The best average improvement is the one obtained in *Scheme* 2.2, using the parcels of Cluster 3 to apply transfer learning techniques. It can be explained because these two clusters are in the same branch of the tree and Cluster 3 contains many parcels which contribute with many general and similar information. *Scheme* 2.3, using Cluster 1 (with many parcels), has a big improvement but below the results of *Scheme* 2.2, most probably due to Cluster 1 is another branch of the tree. Finally, *Scheme* 2.1 is the worst of them, even training with the parcels of its own cluster. Although in this case, the distance intra-cluster is very small and the similarity between these parcels is great, the low number of parcels in this cluster affects the final results.

Finally, in cases where parcels belong to Cluster 3, the *Scheme* 2.1 uses, for training, the parcels of its own cluster, the *Scheme* 2.2 uses the parcels of Cluster 1 and the *Scheme* 2.3 uses the parcels of Cluster 2. The best average improvement is the obtained in *Scheme* 2.1, using the parcels of its own cluster to apply transfer -learning techniques. In second position it is the improvement using Cluster 2. It can be explained because, although Cluster 2 has a low number of clusters, these two clusters are in the same branch and the similarity between them is bigger. The *Scheme* 2.2 throws less improvement due to, although cluster

1 has many parcels, it belongs to another branch of the dendrogram and have less similarity among the rest of parcels.

5 Conclusions

In this paper the benefits of transfer learning have been empirically demonstrated using a dataset of phenological states of olive crops. Different experiments have been carried out to learn general knowledge from other image subsets. First, comparing the fourth validation schemes, the use of transfer learning techniques (Scheme 2.3) has provided a 7.65% of improvement with respect to different ways to train non-transfer learning models (Scheme 1, 2.1 and 2.2). Also, transfer learning has provided more robustness, reflected in the smaller standard deviations obtained, bringing more general knowledge of the treated data sets. According to the analysis of improvements, similarities of parcels, class imbalance ratios and a big number of instances, clear improvements have been observed. These works are a starting point to continue exploring the benefits of TL. As future works, the use of different architectures, the application of meta-heuristics to determine optimal values for hyperparameters and the application of statistical tests to determine significance are proposed.

Acknowledgements. The authors would like to thank the Spanish Ministry of Economy and Competitiveness for the support under the project TIN2017-88209-C2-1-R.

References

1. Torres, J.F., Hadjout, D., Sebaa, A., Martínez-Álvarez, F., Troncoso, A.: Deep learning for time series forecasting: a survey. Big Data **9**, 3–21 (2021)
2. Lara-Benítez, P., Carranza-García, M., Riquelme, J.C.: An experimental review on deep learning architectures for time series forecasting. Int. J. Neural Syst. **31**(3), 2130001 (2021)
3. Deng, Z., Lu, J., Wu, D., Choi, K., Sun, S., Nojima, Y.: New advances in deep transfer learning. IEEE Trans. Emerg. Top. Comput. Intell. **3**(5), 357–359 (2019)
4. Kim, D., Lim, W., Hong, M., Kim, H.: The structure of deep neural network for interpretable transfer learning. In: Proceedings of the IEEE International Conference on Big Data and Smart Computing, pp. 1–4 (2019)
5. Tan, C., Sun, F., Kong, T., Zhang, W., Yang, C., Liu, C.: A survey on deep transfer learning. In: Kůrková, V., Manolopoulos, Y., Hammer, B., Iliadis, L., Maglogiannis, I. (eds.) ICANN 2018. LNCS, vol. 11141, pp. 270–279. Springer, Cham (2018). https://doi.org/10.1007/978-3-030-01424-7_27
6. Liakos, K.G., Busato, P., Moshou, D., Pearson, S., Bochtis, D.: Machine learning in agriculture: a review. Sensors **18**(8), 2674 (2018)
7. Molina, M.Á., Asencio-Cortés, G., Riquelme, J.C., Martínez-Álvarez, F.: A preliminary study on deep transfer learning applied to image classification for small datasets. In: Herrero, Á., Cambra, C., Urda, D., Sedano, J., Quintián, H., Corchado, E. (eds.) SOCO 2020. AISC, vol. 1268, pp. 741–750. Springer, Cham (2021). https://doi.org/10.1007/978-3-030-57802-2_71

8. Yang, Q., Shi, L., Han, J., Yu, J., Huang, K.: A near real-time deep learning approach for detecting rice phenology based on UAV images. Agric. For. Meteorol. **287**, 107938 (2020)

9. Hao, P., Di, L., Zhang, C., Guo, L.: Transfer learning for crop classification with cropland data layer data (CDL) as training samples. Sci. Total Environ. **733**, 138869 (2020)

10. Chen, J., Chen, J., Zhang, D., Sun, Y., Nanehkaran, Y.A.: Using deep transfer learning for image-based plant disease identification. Comput. Electron. Agric. **173**, 105393 (2020)

11. Wang, Z., Dai, Z., Póczos, B., Carbonell, J.: Characterizing and avoiding negative transfer. In: Proceedings of the IEEE/CVF Conference on Computer Vision and Pattern Recognition, pp. 11293–11302 (2019)

12. Zhuang, F., et al.: A comprehensive survey on transfer learning. Proc. IEEE **109**(1), 43–76 (2020)

13. Xie, M., Jean, N., Burke, M., Lobell, D., Ermon, S.: Transfer learning from deep features for remote sensing and poverty mapping. In: Proceedings of the AAAI Conference on Artificial Intelligence, pp. 3929–3935 (2016)

14. Pires de Lima, R., Marfurt, K.: Convolutional neural network for remote-sensing scene classification: transfer learning analysis. Remote Sens. **12**(1), 86 (2020)

15. Abanda, A., Mori, U., Lozano, J.A.: A review on distance based time series classification. Data Min. Knowl. Disc. **33**(2), 378–412 (2018). https://doi.org/10.1007/s10618-018-0596-4

16. Chollet, F., et al.: Keras (2015). https://github.com/fchollet/keras

17. Junta de Andalucía: RAIF website of the Consejeria de Agricultura, pesca y desarrollo rural (2020). https://www.juntadeandalucia.es/agriculturapescaydesarrollorural/raif. Accessed 26 Mar 2020

Deep Transfer Learning for Interpretable Chest X-Ray Diagnosis

C. Lago, I. Lopez-Gazpio$^{(\boxtimes)}$, and E. Onieva

Faculty of Engineering, University of Deusto (UD),
Av. Universidades 24, 48007 Bilbao, Spain
carloslago@opendeusto.es, {inigo.lopezgazpio,enrique.onieva}@deusto.es

Abstract. This work presents an application of different deep learning related paradigms to the diagnosis of multiple chest pathologies. Within the article, the application of a well-known deep Convolutional Neural Network (*DenseNet*) is used and fine-tuned for different chest X-Ray medical diagnosis tasks. Different image augmentation methods are applied over the training images to improve the performance of the resulting model as well as the incorporation of an explainability layer to highlight zones of the X-Ray picture supporting the diagnosis. The model is finally deployed in a web server, which can be used to upload X-Ray images and get a real-time analysis.

The proposal demonstrates the possibilities of deep transfer learning and convolutional neural networks in the field of medicine, enabling fast and reliable diagnosis. The code is made publicly available (https://github.com/carloslago/IntelligentXray - for the model training, https://github.com/carloslago/IntelligentXray_Server - for the server demo).

Keywords: X-Ray diagnosis · Deep learning · Convolutional neural networks · Model interpretability · Transfer learning · Image classification

1 Introduction

Artificial Intelligence is poised to play an increasingly prominent role in medicine and healthcare due to advances in computing power, learning algorithms, and the availability of large datasets sourced from medical records and wearable health monitors [2]. In recent literature, deep learning shows promising results in medical specialities such as radiology [15], cancer detection [5], detection of referable diabetic retinopathy, age-related macular degeneration and glaucoma [19], and cardiology, in a wide range of problems involving cardiovascular diseases, performing diagnosis, predictions and helping in interventions [3].

In this article, we investigate the application of deep learning models for multiple chest pathology diagnoses with the objective of designing a fast and reliable method for diagnosing various pathologies by analyzing X-Ray images. For the

© Springer Nature Switzerland AG 2021
H. Sanjurjo González et al. (Eds.): HAIS 2021, LNAI 12886, pp. 524–537, 2021.
https://doi.org/10.1007/978-3-030-86271-8_44

task, we employ, tune, and train a deep learning model using Tensorflow [1], as well as evaluate it on various medical state-of-the-art benchmarks through transfer learning. The used base model, DenseNet [11], is a state-of-the-art deep Convolutional Neural Network (CNN) that employs an innovative strategy to ensure maximum information flow between all the layers of the network. In the model, each layer is connected to all the others within a dense block, as a consequence, all layers can get feature maps from preceding ones. The resulting model is shown to be compact and have a low probability of overfitting [11]. Through this work, DenseNet is adapted to be evaluated in different benchmarks from the medical domain. A post-process training step based on image augmentation is also incorporated in order to increase its accuracy. Our contribution is two-fold. We first adapt DenseNet for chest X-Ray effective diagnosis and then the addition of the explainability layer. The evaluation of the new model is performed on X-Ray classification benchmarks, including (i) pneumonia detection task; (ii) detection of different pathologies which can be evaluated by doctors; (iii) a Covid-ChestXray detection dataset; which consists of an open dataset of chest X-Ray images of patients who are positive (or suspected) of COVID-19 or other viral and bacterial pneumonia.

This article is structured as follows. Section 2 presents the data used for building and evaluating models, in this case, three different medical image databases. Section 3 introduces the original model, as well as the modification proposed. Section 4 presents the inclusion of the explainability layer to the model, in order to make it capable of explaining its diagnoses. Section 5 analyses results obtained. Finally, Sect. 6 presents conclusions and future works.

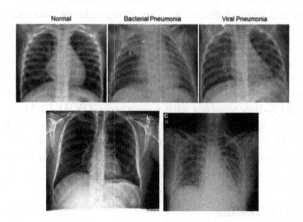

Fig. 1. (Top) Three examples from the pneumonia detection task [14]. (Bottom left) An example image from CheXpert interpretation task [12]. (Bottom right) An example from the Covid-ChestXray benchmark [6].

2   Data Description

This section describes the datasets and benchmarks used to fine-tune, train and evaluate the model. We also describe the preprocessing performed on the original data. The evaluation of the model is carried out in three different fields of the area of X-Ray diagnosis:

(i) *Chest X-Ray Images (Pneumonia)*, an open dataset of chest X-Ray images of patients and their pneumonia diagnosis [14].

(ii) *CheXpert*, a large dataset of chest X-Rays for automated interpretation, featuring uncertainty labels and radiologist-labeled reference standard evaluation sets [12].

(iii) *Covid-ChestXray-Dataset*, an open dataset of chest X-Ray images of patients who are positive (or suspected) of COVID-19 or other viral and bacterial pneumonias [6].

Data for the Pneumonia detection benchmark dataset (i) is publicly available from a Kaggle competition[1], including 5863 X-Ray images associated with two different resulting categories (Pneumonia in 25% of the cases and Normal in the remaining 75%). Automated chest X-Ray sources (ii) have been extracted from CheXpert database[2]. CheXpert is a dataset provided by Stanford ML Group, with over 224,316 samples with both frontal and lateral X-Rays collected from tests performed between October 2002 and July 2017 at Stanford Hospital. Images are labeled to differentiate 14 different diagnoses: no concluding pathology, presence of support devices, and a list of 12 different possible pathologies. The distribution of the classes is outlined in Fig. 2. Chest radiography is one of the most common imaging examinations performed overall, and it is critical for screening, diagnosis, and management of many life-threatening diseases. Automated chest radiography explainability is capable of providing a substantial benefit for radiologists. This research aims to advance the state-of-the-art development and evaluation of medical artificial intelligence applications. Finally, Covid-ChestXray-Dataset sources (iii) can be found at GitHub[3]. It contains around 100 X-Ray samples with suspected COVID-19. Data has been collected from public sources as well as through indirect collection from hospitals and physicians during the previous year. This task aims to differentiate between no pathology, pneumonia, and COVID-19 cases. Examples of X-Ray images used in the experimentation are presented in Fig. 1.

2.1   Data Preprocessing

To prepare the data for training, we first re-scaled all the input image sizes to 224×224 pixels, since not all the images from all the sources have the same resolution. In addition, image augmentation techniques increase the training set

---

[1] https://www.kaggle.com/paultimothymooney/chest-xray-pneumonia.

[2] https://stanfordmlgroup.github.io/competitions/chexpert/.

[3] https://github.com/ieee8023/covid-chestxray-dataset.

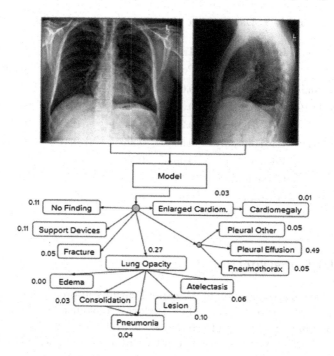

Fig. 2. CheXpert classes distribution [12]

size to avoid overfitting [22]. The CheXpert dataset contains four possible labels, *empty* when the pathology was not considered (they are considered negative), *zero*, denoting the pathology is not present, *one* when the pathology is detected, and a value (−1) denoting that it is unclear if the pathology exists or not. For modeling, all the cases with empty or zero values are considered negative cases. In the case of unclear values, different approaches are applied depending on the pathology, according to the best results showed in [12] when using the U-zeros or U-ones method. Train, validation, and test splits are shown in Table 1 for all the different cases. Both the Pneumonia and the CheXpert cases have a split of 80/10/10, while the COVID-19 case has a split of 60/15/25, as there is less data available.

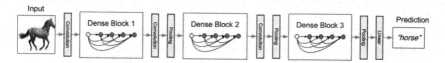

Fig. 3. DenseNet architecture overview. (Figure from [10]).

Table 1. Data balance in training and testing splits

Case	Pathologies	Train and validation		Test	
		Positive	Negative	Positive	Negative
Pneumonia	Pneumonia	0.63	0.37	0.74	0.26
COVID-19	Pneumonia	0.31	0.69	0.34	0.66
	COVID-19	0.29	0.71	0.33	0.67
CheXpert	Cardiomegaly	0.64	0.36	0.47	0.53
	Edema	0.48	0.52	0.29	0.71
	Consolidation	0.21	0.79	0.25	0.75
	Atelectasis	0.63	0.37	0.59	0.41
	Pleural Effusion	0.53	0.47	0.38	0.62

3   Model Construction

CNNs have become the dominant Machine Learning approach for visual object recognition in recent years [10]. They perform convolution instead of matrix multiplication in contrast to fully connected neural networks. As a consequence, the number of weights is decreased resulting in a less complex network that is easier to train [21]. Furthermore, images can be directly imported into the input layer of the network avoiding the feature extraction procedure widely used in more traditional machine learning applications. It should be noted that CNNs are the first truly successful deep learning architectures due to the inherent hierarchical structure of the inner layers [10,17].

Deep CNNs can represent functions with high complexity decision boundaries as the number of neurons in each layer or the number of layers is increased. Given enough labeled training instances and suitable models, Deep CNN approaches can help establish complex mapping functions for operation convenience [17].

This research is based on the well-known DenseNet model [25], which is a popular architecture making use of deep CNNs. The main contribution of DenseNet relies upon that it connects all layers in the network directly with each other, and in that each layer also benefits from feature maps from all previous layers [10]. A visual representation of DenseNet is provided in Fig. 3. In this sense, DenseNet provides shortcut connections through the network that lead to deep implicit supervision, which is denoted in the state-of-the-art as a simple strategy to ensure maximum information flow between layers. This architecture has been used in a wide variety of benchmarks yielding state-of-the-art results as it produces consistent improvements in the accuracy, by increasing the complexity of layers, without showing signs for overfitting [25].

3.1   Fine-Tunning over DenseNet

Originally, DenseNet was trained for object recognition benchmark tasks as CIFAR-10, CIFAR-100, The Street View House Numbers (SVHN) Dataset or

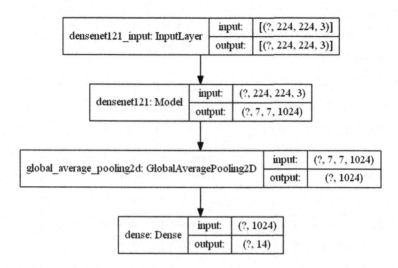

Fig. 4. DenseNet model architecture used for experiments (i) and (ii).

ImageNet. These benchmarks are composed of a big number of training instances for each one of the classes to predict [8,16,18]. The need for alteration of the original model raises when adapting it to environments where the number of training samples is low, as is the case of medical diagnosis based on an image. In order to accommodate the DenseNet network into the medical domain and reuse the latent knowledge condensed in its layers, the final layers of the network are updated to perform classification in the medical domain by retraining only these final layers. Transfer Learning is a great tool for fixing the lack of data problem widely extended in deep learning, which usually employs methods that need more data than traditional machine learning approaches. Transferring the knowledge from the source domain where the network has been trained to another is a common practice in deep learning [23]. Moreover, thanks to the transfer of knowledge, the amount of time it is required to learn a new task is decreased notably, and the final performance that can be achieved is potentially higher than without the transfer of knowledge [24]. All of our DenseNet model variants have been trained with transfer learning for DenseNet121, the only difference between each architecture is the regularization techniques that are implemented. As shown in Fig. 3, DenseNet includes several convolution layers (referred to as transition layers) and dense blocks in an iterative way, connecting all layers in the network directly with each other, with each layer receiving feature maps from all the previous layers [10].

The architectures of the base DenseNet models are showed in Figs. 4 and 5. The final Dense layer from the mentioned figures is the part that we fine-tune for experimentation. It should be noted that L2 regularization is being used on the GlobalAveragePooling2D layer, with a rate of 0.001, which is essential to avoid over-fitting in the network. This difference among architectures was found to produce better results in laboratory experimentation due to the lower sizes

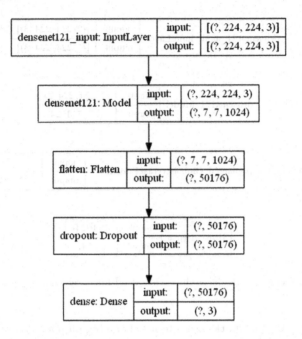

Fig. 5. DenseNet model architecture used for experiment (iii).

of the dataset. For both the first and second evaluation benchmarks, regarding the Pneumonia detection and the CheXpert identification, the same type of regularization techniques have been used. For training both models the following image augmentation techniques have been used: horizontal flip, zoom range both in and out in with a maximum value ($zoom_{range}$) of 0.2, ZCA whitening, to obtain decorrelated features, and rotation within a range (rot_{range}) of -5^a and $5°$. Furthermore, given the lower number of samples in the COVID-19 case, additional data augmentation was included: Height and width shift range, both with a range ($shift_{range}$) of 0.2, moving the image up to that percent vertically or horizontally, shear range ($shear_{range}$) of 0.2, rotation range of $5°$ and, brightness modification, ranging from 0.8 to 1.2. In addition, lateral images are being used for training the model too, which is used in this research as a form of image augmentation, which gives to the model a different viewpoint of a particular sample. Regarding the COVID-19 case, which also contains sample instances from Pneumonia and Covid cases, the network is slightly different as it uses dropout, which was observed useful not to over-train the model as several neuron units are randomly disconnected at training [9]. The specific model for this third experiment is shown in Fig. 5. It should be noted that for this case more image augmentation techniques are being used: horizontal flips, zoom range, rotation range, height and width shift range, shear range, and brightness range.

4   Explainability Layer

As a final step, an explainability layer is added on top of the adapted DenseNet to make the model self-explainable. Developing the explainability of a machine learning model is essential to understand the decision process behind the predictions of a model, analyzing if it makes sense or not, and facilitate human trust in the final decisions. It can also help to gain some insights and confidence in the model, often seen as a black-box tool, by observing clearly how it performs under given circumstances. Adding interpretability to models is a key towards transforming an untrustworthy model into a trustworthy one [20]. Local Interpretable Model-agnostic Explanations (LIME) [20] is used to provide explainability to the final model. LIME is an algorithm able to explain the predictions of a regressor or classifier by approximating its result with an interpretable model. LIME provides local explanations of predictions of a model by fitting an explanation model locally around the data point of which classification is to be explained. LIME supports generating model explanations for text and image classifiers. The layer implements the function in Eq. 1.

$$\xi(x) = g \in G \ argmin_x \ \mathcal{L}(f, g, \pi_x) + \Omega(g), \tag{1}$$

where the fidelity function $\mathcal{L}$ is a measure of how unfaithful an explanation g, an element in the set of possible interpretation models G, in approximating f, the probability of x belonging to a class in the locality defined by the proximity measure π_x. The $\Omega(g)$ term is a measure of complexity of the explanation of $g \in G$. For the explanation to ensure interpretability and local fidelity, it is necessary to minimize $\mathcal{L}(f, g, \pi_x)$ and have a $\Omega(g)$ low enough so it is interpretable by humans. *SP-LIME* is a method that selects a set of instances with explanations that are representative to address the problem of trusting the model. To understand how the classifier works, it is necessary to explain different instances rather than just provide an explanation of a single prediction. This method selects some explanations that are insightful and diverse to represent the whole model.

5   Experimentation and Results

This section presents results of the experimentation for each of the studied benchmark tests. For all cases, the performance of the model is evaluated in terms of Accuracy (Eq. 2), AUC (Eq. 3)[4] and micro and macro F1 scores (Eqs. 4 and 5 respectively).

$$Accuracy = \frac{TrueNegatives + TruePositive}{All \ samples} \tag{2}$$

$$AUC(f) = \frac{\sum_{t_0 \in \mathcal{D}^0} \sum_{t_1 \in \mathcal{D}^1} \mathbf{1}\left[f(t_0) < f(t_1)\right]}{|\mathcal{D}^0| \cdot |\mathcal{D}^1|} \tag{3}$$

---

[4] $\mathbf{1}\left[f(t_0) < f(t_1)\right]$ returns 1 if $f(t_0) < f(t_1)$ and otherwise 0, $\mathcal{D}^1$ is the set of positive examples and $\mathcal{D}^0$ of negatives.

Table 2. Pneumonia benchmark experimentation results.

Case	Accuracy	AUC	Macro-F1	Micro-F1
Training set	94.5%	0.99	0.96	0.96
Test set	93.75%	0.99	0.96	0.96

Table 3. CheXpert benchmark experimentation results.

Case	Accuracy	AUC	Macro-F1	Micro-F1
Training set	88%	0.93	0.88	0.9
Test set	83%	0.88	0.74	0.78

Table 4. Detailed CheXpert benchmark experimentation results.

Case	Accuracy	AUC	Macro-F1	Micro-F1
Cardiomegaly	74.3%	0.74	0.74	0.74
Edema	82.64%	0.78	0.78	0.83
Consolidation	77.78%	0.65	0.66	0.78
Atelectasis	70.83%	0.68	0.68	0.71
Pleural Effusion	82.41%	0.83	0.82	0.82

Table 5. COVID-19 benchmark experimentation results.

Case	Accuracy	AUC	Macro-F1	Micro-F1
Training set	92%	0.98	0.92	0.93
Test set	94%	0.98	0.94	0.94

$$F1_{micro} = 2 \times \frac{recall_{micro} \times precision_{micro}}{recall_{micro} + precision_{micro}} \qquad (4)$$

$$F1_{macro} = \sum_{classes} \frac{F1 \text{ of class}}{number \text{ of classes}} \qquad (5)$$

For the calculation of micro F1, the precision and recall values are obtained with Eqs. 6 and 7, as well as for the calculation of macro F1 (Eq. 5). The calculation per class of the F1 values is done as presented in Eq. 8.

$$precision_{micro} = \frac{\sum_{classes} TP \text{ of class}}{\sum_{classes} TP \text{ of class} + FP \text{ of class}} \qquad (6)$$

$$recall_{micro} = \frac{\sum_{classes} TP \text{ of class}}{\sum_{classes} TP \text{ of class} + FN \text{ of class}} \qquad (7)$$

$$F1 = \frac{2 \times precision \times recall}{precision + recall} \tag{8}$$

Pneumonia detection task results can be seen in Table 2 and the COVID-19 case results are shown in Table 5. We can see that the network has good performance on balanced datasets, proved by the macro and micro F1 results. CheXpert benchmark results can be seen in Table 3, with the results for the most important pathologies in detail in Table 4. Concerning CheXpert, it is much more complex, having 14 different observations and unknown labels. Moreover, only a small subset of the data has been used, as the objective is to prove the efficiency of this type of network but not get the best possible result. The F1 score only takes into account Cardiomegaly, Edema, Consolidation, Atelectasis, and Pleural Effusion, due to the unbalanced samples in the dataset. Model evaluation in benchmarks (i) and (iii) yields an accuracy of over 90%, with AUC scores of 0.98+, and lower results for the benchmark (ii), with accuracies over 83% and AUC scores of 0.9. These results are in line with other neural network approaches applied to chest X-Ray benchmarks. For instance, Çallı et al. in [4] report a total of 9 neural network systems trained and validated in the COVID-19 infection benchmark. Their results indicate that the mean accuracy of these systems is 0.902 with a standard deviation of 0.044. Our experimentation results also confirm that transfer learning can successfully be applied for rapid chest X-Ray diagnosis and help expert radiologists with a system that offers immediate assistance, as supported by [4].

5.1   Evaluating Model Explainability

With the addition of an explainability layer on top of it, the model can explain its behavior, by highlighting the areas of the image that support its diagnosis. The addition of this explanatory layer is the first step towards increasing human confidence in health-related artificial intelligence applications, as the model can self-explain its decisions to gain trust in human-computer interactions. We now show model explanations for each one of the benchmark datasets in which we evaluate our model. First, we show a sample instance from the pneumonia detection dataset (Fig. 6) The figure represents a pneumonia detectable by the airspace consolidation that can be seen on the right lower zone . The model correctly outlines the area affected by pneumonia. For the CheXpert benchmark dataset, the case shown in Fig. 7 has been diagnosed with cardiomegaly . Cardiomegaly is present when the heart is enlarged, as can be easily seen on the frontal X-Ray from Fig. 7. In this scenario, the network successfully performs a correct prediction of the area. Finally, we show an example for the COVID-19 benchmark dataset in Fig. 8 which has been diagnosed with airspace opacities . The model explanations show that the network is focusing on the relevant areas related to these opacities.

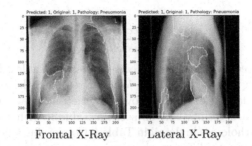

Fig. 6. Pneumonia detection samples including model explanation.

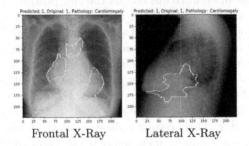

Fig. 7. CheXpert samples diagnosed with cardiomegaly including model explanation.

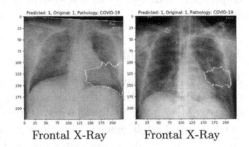

Fig. 8. COVID-19 samples diagnosed with COVID-19 including model explanation.

5.2 Demo

In addition to the development of the models, we build a web prototype where the CheXpert model is deployed. This web prototype enables uploading personal X-Ray images to get a diagnose in less than one second and model explanations in less than ten seconds. Figure 9 shows the web application for performing the predictions and how the results are presented. As shown in the image, it is possible to upload both the frontal and lateral X-Ray or just one for the prediction, the table on the right displays the prediction of the model for each of the pathologies along with its confidence in the prediction.

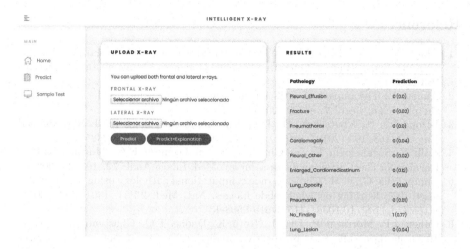

Fig. 9. Web demo

6   Conclusion and Future Work

This research demonstrates how deep learning and CNNs can be useful in the field of medicine, enabling fast and reliable support for diagnosis. For diseases like cancer, where early diagnoses could save millions of lives, they also enable a more accurate and early diagnosis, with models such as [13] achieving an accuracy of more than 90%. Furthermore, the addition of the explainability layer is an important step towards improving trust in model predictions, which is one of the main concerns for public usage. We consider the addition of this explanatory layer as the first step towards increasing human confidence in health-related artificial intelligence applications, as the model can self-explain its decisions.

Although polymerase chain reaction (RT-PCR) is the preferred way to detect COVID-19, the costs and response time involved in the process have resulted in a growth of rapid infection detection techniques, most of them based on chest X-Ray diagnosis [7]. The primary advantage of the automatic analysis of chest X-Rays through deep learning is that the technique is capable of accelerating the time required for the analysis. It should be noted that this study is just an experiment to showcase the capabilities of deep convolutional neural networks in the field of radiology, but shouldn't be considered as a way to replace a radiologist checkup. For future lines, it will be interesting to mix the current datasets used in the project with other existing datasets, enabling the detection of more pathologies and fixing the unbalance problems. More complex architectures could be tested in a bigger sample to improve performance.

Acknowledgments. This work has been supported by The LOGISTAR project, which has received funding from the European Union's Horizon 2020 research and innovation programme under Grant Agreement No. 769142.

References

1. Abadi, M., et al.: TensorFlow: a system for large-scale machine learning. In: 12th USENIX Symposium on Operating Systems Design and Implementation, pp. 265–283 (2016)
2. Ahuja, A.: The impact of artificial intelligence in medicine on the future role of the physician. PeerJ **7**, e7702 (2019)
3. Bizopoulos, P., Koutsouris, D.: Deep learning in cardiology. IEEE Rev. Biomed. Eng. **12**, 168–193 (2019). https://doi.org/10.1109/RBME.2018.2885714
4. Çallı, E., Sogancioglu, E., van Ginneken, B., van Leeuwen, K.G., Murphy, K.: Deep learning for chest x-ray analysis: a survey. Med. Image Anal. **72**, 102125 (2021)
5. Campanella, G., et al.: Clinical-grade computational pathology using weakly supervised deep learning on whole slide images. Nat. Med. **25**(1), 1301–1309 (2019). https://doi.org/10.1038/s41591-019-0508-1
6. Cohen, J.P., Morrison, P., Dao, L., Roth, K., Duong, T.Q., Ghassemi, M.: COVID-19 image data collection: prospective predictions are the future. arXiv preprint arXiv:2006.11988 (2020)
7. Das, N.N., Kumar, N., Kaur, M., Kumar, V., Singh, D.: Automated deep transfer learning-based approach for detection of COVID-19 infection in chest X-rays. IRBM (2020). https://www.sciencedirect.com/science/article/abs/pii/S1959031820301172
8. Deng, J., Dong, W., Socher, R., Li, L.J., Li, K., Fei-Fei, L.: ImageNet: a large-scale hierarchical image database. In: 2009 IEEE Conference on Computer Vision and Pattern Recognition, pp. 248–255 (2009)
9. Gal, Y., Ghahramani, Z.: Bayesian convolutional neural networks with Bernoulli approximate variational inference. arXiv preprint arXiv:1506.02158 (2015)
10. Huang, G., Liu, Z., Van Der Maaten, L., Weinberger, K.Q.: Densely connected convolutional networks. In: Proceedings of the IEEE Conference on Computer Vision and Pattern Recognition, pp. 4700–4708 (2017)
11. Iandola, F., Moskewicz, M., Karayev, S., Girshick, R., Darrell, T., Keutzer, K.: DenseNet: implementing efficient convnet descriptor pyramids. arXiv preprint arXiv:1404.1869 (2014)
12. Irvin, J., et al.: CheXpert: a large chest radiograph dataset with uncertainty labels and expert comparison. In: Proceedings of the AAAI Conference on Artificial Intelligence, vol. 33, pp. 590–597 (2019)
13. Kalra, S., et al.: Pan-cancer diagnostic consensus through searching archival histopathology images using artificial intelligence. NPJ Digit. Med. **3**(1), 1–15 (2020)
14. Kermany, D.S., et al.: Identifying medical diagnoses and treatable diseases by image-based deep learning. Cell **172**(5), 1122–1131 (2018)
15. Kim, J.R., et al.: Computerized bone age estimation using deep learning based program: evaluation of the accuracy and efficiency. AJR Am. J. Roentgenol. **209**(6), 1374–1380 (2017)
16. Krizhevsky, A., Hinton, G.: Learning multiple layers of features from tiny images. Technical report, University of Toronto (2009)
17. Liu, W., Wang, Z., Liu, X., Zeng, N., Liu, Y., Alsaadi, F.E.: A survey of deep neural network architectures and their applications. Neurocomputing **234**, 11–26 (2017). https://doi.org/10.1016/j.neucom.2016.12.038
18. Netzer, Y., Wang, T., Coates, A., Bissacco, A., Wu, B., Ng, A.Y.: Reading digits in natural images with unsupervised feature learning. Technical report, NIPS Workshop (2011)

19. Rahimy, E.: Deep learning applications in ophthalmology. Curr. Opin. Ophthalmol. **29**(3), 254–260 (2018). https://doi.org/10.1097/ICU.0000000000000470
20. Ribeiro, M.T., Singh, S., Guestrin, C.: "Why should i trust you?" Explaining the predictions of any classifier. In: Proceedings of the 22nd ACM SIGKDD International Conference on Knowledge Discovery and Data Mining, San Francisco, CA, USA, 13–17 August 2016, pp. 1135–1144 (2016)
21. Ronao, C.A., Cho, S.-B.: Deep convolutional neural networks for human activity recognition with smartphone sensors. In: Arik, S., Huang, T., Lai, W.K., Liu, Q. (eds.) ICONIP 2015. LNCS, vol. 9492, pp. 46–53. Springer, Cham (2015). https://doi.org/10.1007/978-3-319-26561-2_6
22. Shorten, C., Khoshgoftaar, T.M.: A survey on image data augmentation for deep learning. J. Big Data **6**(1), 1–48 (2019). https://doi.org/10.1186/s40537-019-0197-0
23. Tan, C., Sun, F., Kong, T., Zhang, W., Yang, C., Liu, C.: A survey on deep transfer learning. In: Kůrková, V., Manolopoulos, Y., Hammer, B., Iliadis, L., Maglogiannis, I. (eds.) ICANN 2018. LNCS, vol. 11141, pp. 270–279. Springer, Cham (2018). https://doi.org/10.1007/978-3-030-01424-7_27
24. Torrey, L., Shavlik, J.: Transfer learning. In: Handbook of Research on Machine Learning Applications and Trends: Algorithms, Methods, and Techniques, pp. 242–264. IGI Global (2010)
25. Zhu, Y., Newsam, S.: DenseNet for dense flow. In: 2017 IEEE International Conference on Image Processing (ICIP), pp. 790–794 (2017)

Companion Losses for Deep Neural Networks

David Díaz-Vico[2]([✉]), Angela Fernández[1], and José R. Dorronsoro[1,2]

[1] Department of Computer Engineering, Universidad Autónoma de Madrid,
Madrid, Spain
[2] Inst. Ing. Conocimiento, Universidad Autónoma de Madrid,
Madrid, Spain
david.diaz@iic.uam.es

Abstract. Modern Deep Neuronal Network backends allow a great flexibility to define network architectures. This allows for multiple outputs
with their specific losses which can make them more suitable for particular goals. In this work we shall explore this possibility for classification
networks which will combine the categorical cross-entropy loss, typical of
softmax probabilistic outputs, the categorical hinge loss, which extends
the hinge loss standard on SVMs, and a novel Fisher loss which seeks to
concentrate class members near their centroids while keeping these apart.

1 Introduction

After the seminal work of G. Hinton [8] and J. Bengio [11] and starting
about 2010, Deep Neural Networks (DNNs) have exploded in terms of scientific
advances, technological improvements and great successes on many applications.
There are many reasons for this, but paramount among them is the great flexibility that modern DNN environments such as TensorFlow [1] or PyTorch [9]
allow to define, train and exploit DNN models. Key for this are the modern
tools for automatic differentiation that make possible the definition of very general network architectures and losses. For instance, this has made possible to
incorporate under a DNN framework cost functions such as the hinge and ϵ-
insensitive losses, with models and results that are very competitive with those
of the standard Gaussian SVMs [5].

Once that more general losses are available, a natural next step is to try to
combine some of them in principle independent losses within the same network,
so that they can take advantage of their different goals to jointly improve on their
individual achieved results. For instance, consider for two-class problems the competing cross-entropy loss, customarily used for DNN classification, with the SVM

The authors acknowledge financial support from the European Regional Development
Fund and the Spanish State Research Agency of the Ministry of Economy, Industry,
and Competitiveness under the project PID2019-106827GB-I00. They also thank the
UAM–ADIC Chair for Data Science and Machine Learning and gratefully acknowledge
the use of the facilities of Centro de Computación Científica (CCC) at UAM.

© Springer Nature Switzerland AG 2021
H. Sanjurjo González et al. (Eds.): HAIS 2021, LNAI 12886, pp. 538–549, 2021.
https://doi.org/10.1007/978-3-030-86271-8_45

hinge loss. The goal of the former is to assume a certain posterior probability and estimate a model that maximizes its sample based likelihood, while that of the latter is essentially to find a separating hyperplane with a margin as large as possible given the sample. A similar situation arises in multiclass problems, where the categorical cross-entropy with softmax outputs is used for DNN classifiers while the categorical hinge loss [3] is used for multiclass SVM-like classifiers. In these problems one or several of the losses act as the main one, while the others act as companions in the sense that they accompany the main loss towards a better model. This idea of combining losses has been applied in other areas of knowledge, specially in computer vision [10, 14], although following a different rationale.

Although we will not deal with them here, competing losses for regression problems would be the squared error of regression DNNs and the ϵ-insensitive loss used in support vector regression. Again, the underlying problem is basically the same, but the latter establishes an ϵ-wide tube around the fitted model and only penalizes errors outside the tube. This results on models more robust with respect to outliers but with the drawback of ignoring small errors, which may be important in some settings and that the squared error does not ignore. Given these different but not necessarily competing goals, it is in principle conceivable that both losses could work together towards building a model that improves on those built separately with each loss.

The goal of this work is precisely to explore these possibilities for classification problems. More precisely, we will compare the combination of the cross-entropy and hinge losses for two-class problems, and that of the categorical cross-entropy and hinge losses in multiclass ones. To these we will add a squared loss-based cost function which enforces for its inputs on each class to be concentrated near the class centroids while trying to keep these centroids apart. This approach has been proved [13] to yield linear models whose outputs can theoretically be seen to be equivalent with those provided by the classical Fisher Discriminant Analysis and that has been extended to a DNN setting in [4]. While trying to yield a classifier directly, such a loss can produce a pattern representation on the last hidden layer of a DNN which can make easier the job of a classifier acting on these representations and, hence, result in a better model.

We will work with a substantial number of classification problems and our results point out that this approach can indeed give such results. In fact, and as we will experimentally show, combining the Fisher loss with the cross-entropy or hinge ones improves on the models obtained when only single losses are used. On the other hand, the cross-entropy plus hinge combination ties at best with a single cross-entropy loss. This has to be further studied but a possible reason may be that, at least in our experiments, cross-entropy DNNs yield better results that hinge-base ones (we address reasons for this later in this paper). In summary, our contributions here are:

- The proposal of DNNs with combined losses for two- and multi-classification, which we implement as Keras [2] functional models.
- A substantial experimental work showing positive results that deserve further study.

The rest of the paper is organized as follows. We review the losses used in Sect. 2, discuss how to combine them and give some implementation details. Section 3 contains details on the datasets used, the experimental methodology and results, and a discussion and a final section offers some conclusions as well as pointers to further work. We point out that, throughout the paper, by deep networks we mean artificial neural networks that use modern techniques such as automatic differentiation, Glorot-Bengio initializations [6], ReLU activations or Adam optimizers [7], rather than they having actually deep (i.e. many layered) architectures; in fact, in our experiments we will apply all these techniques but on a single layer network with 100 units.

2 Classification Losses

Throughout this section we will work with DNN architectures that yield models acting on a pattern x with outputs $F(x, \mathcal{W})$, where $\mathcal{W}$ denotes the set of weight matrices and bias vectors associated with the network's architecture. We will denote targets as y, which can be either $\{-1, 1\}$ for two-class problems or one-hot encoded vectors for multiclass ones. We denote the network outputs at the last hidden layer as $z = \Phi(x, \widetilde{\mathcal{W}})$, with $\widetilde{\mathcal{W}}$ the weight and bias set of all layers up to the last one. Such a z is then transformed as $Wz + B$, where W, B are either the transpose of an n_H dimensional vector w, with n_H the number of hidden units in the last hidden layer, and a scalar b, or a $K \times n_H$ matrix and a K dimensional vector in a K-class problem. These $Wz + B$ will be the final network outputs in the case of the deep SVM (or deep Fisher networks, as we describe them below); for the more standard DNN classifiers, the network output is obtained applying to them either a sigmoid or a softmax function.

In more detail, and starting with two-class problems, the usual DNN loss is the binary **cross-entropy**, given by

$$\ell_{bc}(\mathcal{W};\ S) = \ell(w, b, \widetilde{\mathcal{W}}) = -\sum_p y^p \left(w \cdot z^p + b\right) + \sum_p \log(1 + e^{w \cdot z^p + b}) \qquad (1)$$

where S denotes an i.i.d. sample $S = \{(x^p, y^p)\}$ with N patterns, z represent the last hidden layer outputs, w represent the weights which connects the last hidden layer with the network's output and b is the vector bias of the output. We recall that this expression is just the sample's minus log-likelihood associated to the assumption

$$P(y|x) = P(y|x; w, b, \widetilde{\mathcal{W}}) = \frac{1}{1 + e^{-y(w \cdot z + b)}} \qquad (2)$$

for the posterior probability of the y class, and we assume a sigmoid network output. For general multiclass problems, the network output function is the softmax

$$F_j(x; \mathcal{W}) = \frac{e^{w_j \cdot z + b_j}}{\sum_{k=0}^{K-1} e^{w_k \cdot z + b_k}}. \qquad (3)$$

Obviously then $\sum_j F_j(x; \mathcal{W}) = 1$ and we assume $P(j|x) \simeq F_j(x; \mathcal{W})$. For two-class problems this reduces to the previous output if we take $w = w_0 - w_1$. For the loss to be used here, we assume one-hot encoded targets, i.e., the target of the k-th class is $e_k = (0, \ldots, \underbrace{1}_{k}, \ldots, 0)$; then, the probability of getting patterns x^p in class k_p (i.e., $y^p_{k_p} = 1$) within an i.i.d. sample $S = (X, Y)$ is

$$P(Y|X; \mathcal{W}) = \prod_{p=1}^{N} P(k_p|x^p; \mathcal{W}) = \prod_{p=1}^{N} \prod_{m=0}^{K-1} P(m|x^p; \mathcal{W})^{y^p_m} \simeq \prod_{p=1}^{N} \prod_{m=0}^{K-1} F_c(x; \mathcal{W})^{y^p_m},$$

(4)

and we estimate the DNN's weights $\mathcal{W}$ by minimizing the minus log of the approximate sample's likelihood

$$\widetilde{P}(Y|X; \mathcal{W}) = \prod_{p=1}^{N} \prod_{m=0}^{K-1} F_m(x^p; \mathcal{W})^{y^p_m}.$$

(5)

That is, we will minimize the categorical cross-entropy loss

$$\ell_{cce}(\mathcal{W}) = -\log \widetilde{P}(Y|X; \mathcal{W}) = -\sum_{p=1}^{N} \sum_{m=0}^{K-1} y^p_m \log F_m(x^p; \mathcal{W}).$$

(6)

Once an optimal weight set $\mathcal{W}^*$ has been obtained, the decision function on a new x is given by $\arg\max_m F_m(x; \mathcal{W}^*)$, i.e. the class with the maximum posterior probability.

Turning our attention to two-class SVMs, the local loss is now the **hinge loss** $h(x, y) = \max\{0, 1 - yF(x, \mathcal{W})\}$; here the network has linear outputs, i.e., $F(x; \mathcal{W}) = F(x; w, b, \widetilde{\mathcal{W}}) = w \cdot \Phi(x, \widetilde{\mathcal{W}}) + b$. The global loss is now

$$\ell_h(\mathcal{W}; S) = \sum_p \max\{0, 1 - y^p F(x^p, \mathcal{W})\}.$$

(7)

There are several options to extend SVMs for multiclass problems. The usual approach in a kernel setting is to use either a one-vs-one (ovo) or a one-vs-rest (ovr) approach so that just binary kernel classifiers have to be built. Unfortunately, this cannot be directly translated to a DNN setting but there are two other ways to define multiclass local losses. The first one is due to Weston and Watkins [12] which for a pattern x in class k_x (i.e., $y_{k_x} = 1$) propose the local loss

$$\ell(x, y) = \max\left\{0, \sum_{m \neq k_x} (1 + F_m(x) - F_{k_x}(x))\right\}$$

(8)

where we denote as $F_0(x), \ldots F_{K-1}(x)$ the network's lineal outputs. The alternative to this is the local loss proposed by Crammer and Singer [3], namely

$$\ell(x, y) = \max\left\{0, \max_{m \neq y}(1 + F_m(x) - F_{k_x}(x))\right\}.$$

(9)

Notice that both coincide for two-class problems and, moreover, if in this case we require $w_0 = w_1 = \frac{1}{2}w$ and $b_0 = b_1 = \frac{1}{2}b$, they coincide with the local hinge loss. We will use the second one, that results in the categorical hinge global loss

$$\ell_{ch}(\mathcal{W}) = \sum_{p=1}^{N} \max \left\{ 0, 1 - F_{k_{x^p}}(x^p) + \max_{m \neq k_{x^p}} F_m(x) \right\}; \tag{10}$$

here $F_m(x)$ denotes the m-th component of the network's K dimensional linear output, i.e., $F_m(x) = w_m \cdot \Phi(x; \widetilde{\mathcal{W}}) + b_m$. Now, once an optimal weight set $\mathcal{W}^*$ has been obtained, the decision function on a new x is given again by $\arg\max_m F_m(x; \mathcal{W}^*)$, although now no posterior probabilities are involved.

We will finally consider what we call the **Fisher loss**. The goal in standard Fisher Discriminant Analysis is to linearly project patterns so that they concentrate near the projected class means while these are kept apart. To achieve this, one seeks to maximize the trace criterion

$$g(A) = \operatorname{trace}(s_T^{-1} s_B) = \operatorname{trace} \left((A^t S_T A)^{-1} (A^t S_B A) \right), \tag{11}$$

where A is the projection matrix, S_B and S_T denote the between-class and total covariance matrices, respectively, of the sample patterns and s_B and s_T are their counterparts for the projections $z = Ax$. Solving $\nabla_A g = 0$ leads to the system $S_T^{-1} S_B A = A\Lambda$, with Λ the non-zero eigenvalues of $S_T^{-1} S_B$. For such an A we have

$$g(A) = \operatorname{trace}(s_T^{-1} s_W) = \operatorname{trace} \Lambda = \lambda_1 + \ldots + \lambda_q, \tag{12}$$

where s_W represents the within-class matrix. This expression is maximized by sorting the eigenvalues $\{\lambda_1, \lambda_2, \ldots\}$ in Λ in descending order and selecting the $K - 1$ largest ones and some conveniently normalized associated eigenvectors. Since the minimizer of (11) is not uniquely defined, it is usually normalized as $A^t S_T A = I_{K-1}$, being I_N the identity matrix of size N. It turns out that an equivalent solution can be obtained by solving the least squares problem $\min \frac{1}{2} \| Y^f - XW - \mathbf{1}_N B \|^2$, where W is a $d \times K$ matrix, B a $1 \times K$ vector, $\mathbf{1}_N$ is the all ones vector and for a pattern x^p in the p-th row of the data matrix X which is in class m, we have $Y_{pm}^f = \frac{N - N_m}{N \sqrt{N_m}}$ when x^p and as $Y_{pj}^f = -\frac{\sqrt{N_j}}{N}$ for $j \neq m$, with N_j the number of patterns in class j. Then, it can be shown that the optimal W^* is equivalent, up to a rotation, to a solution $\widetilde{V}$ of (11) subject to the normalization $\widetilde{V}^t S_T \widetilde{V} = \Lambda$; see [4,13] for more details. As a consequence, any classifier defined in terms of distances to class means will give the same results with the Fisher's projections using $\widetilde{V}$ than with the least squares ones using W^*.

This can be extended to a DNN setting by solving

$$\min_{W,B,\widetilde{W}} \frac{1}{2} \| Y^f - F(X, \mathcal{W}) \|^2 = \frac{1}{2} \| Y^f - \Phi(X; \widetilde{\mathcal{W}})W - \mathbf{1}_N B) \|^2. \tag{13}$$

As discussed in [4], this could be exploited to define a Fisher-like distance classifier on the network outputs $F(x, \mathcal{W})$; however, those classifiers are in general worse than those based on the categorical entropy or hinge losses. On the other hand, such a loss is likely to enforce the last hidden layer projections z to be concentrated around their class means while keeping these apart and, hence help entropy or hinge based classifiers to perform better.

In this line, the above suggests that instead of using just one of the previous losses, we can try to combine them, something that can be done by defining a

Table 1. Sample sizes, number of features and number of classes.

	Size	Size test	Features	Classes
a4a	4781	27780	123	2
a8a	22696	9865	123	2
australian	690	–	14	2
breast-cancer	569	–	30	2
diabetes	768	–	8	2
digits	1797	–	64	10
dna	2000	–	180	3
german	1000	–	24	2
letter	10500	5000	16	26
pendigits	7494	3498	16	10
protein	14895	6621	357	3
satimage	4435	–	36	6
segment	2310	–	19	7
usps	7291	–	256	10
w7a	24692	25057	300	2
w8a	49749	14951	300	2

DNN with two or even three output sets upon which one of these losses acts. In the most general setting, we may have outputs $\widehat{y}_{ce}, \widehat{y}_{ch}$ and $\widehat{y}_f$, one-hot encoded targets y and the y_f Fisher-like targets just defined for the loss (13), and we define the combined loss

$$\ell(y, y_f, \widehat{y}_{ce}, \widehat{y}_{ch}, \widehat{y}_f) = \ell_{ce}(y, \widehat{y}_{ce}) + \lambda \ell_{ch}(y, \widehat{y}_{ch}) + \mu \ell_f(y_f, \widehat{y}_f), \qquad (14)$$

with $\{\lambda, \mu\}$ appropriately chosen hyperparameters. In our experiments next, we will consider the (ce, fisher), (hinge, fisher), (ce, hinge) and (ce, hinge, fisher) loss combinations; for simplicity we will just take $\lambda = \mu = 1$.

Table 2. Test accuracies of the models considered.

	ce	ce-fisher	hinge	hinge-fisher	ce-hinge	ce-hinge-fisher	max	min
a4a	84.38	84.36	84.39	84.40	84.43	84.45	84.45	84.36
a8a	85.10	85.51	84.96	84.93	85.20	85.09	85.51	84.93
australian	86.52	85.65	85.22	85.07	85.36	85.07	86.52	85.07
breast-cancer	97.72	98.07	96.84	97.19	96.49	96.31	98.07	96.31
diabetes	77.60	77.60	76.82	76.95	77.08	76.95	77.60	76.82
digits	98.22	98.44	98.39	98.27	98.50	98.22	98.50	98.22
dna	95.70	95.85	94.65	94.85	95.55	95.60	95.85	94.65
german	77.90	77.50	75.30	76.40	74.00	74.80	77.90	74.00
letter	95.38	95.28	95.42	95.66	94.98	95.42	95.66	94.98
pendigits	99.52	99.45	99.47	99.52	99.49	99.44	99.52	99.44
protein	69.78	69.76	66.49	66.71	67.62	67.07	69.78	66.49
satimage	91.09	90.55	91.54	91.25	91.75	91.25	91.75	90.55
segment	97.66	97.79	97.01	97.36	97.88	97.79	97.88	97.01
usps	97.79	97.93	97.82	97.53	97.65	97.68	97.93	97.53
w7a	98.79	98.84	98.83	98.81	98.83	98.84	98.84	98.79
w8a	98.97	98.99	98.84	98.88	99.00	99.00	99.00	98.84

3 Experimental Results

In this section we will describe the considered models, describe the datasets we will use, present our experimental methodology and results, and finish the section with a brief discussion.

3.1 Models Considered

We will consider six basic model configurations, involving different network outputs, losses and ways to make predictions, namely

- ce: the model uses softmax outputs and the categorical cross-entropy loss. Class labels are predicted as the index of the output with the largest a posteriori probability.
- hinge: the model uses linear outputs and the categorical hinge loss. Class labels are predicted as the index of the largest output.
- ce_hinge: the model uses two different outputs, one with softmax activations and the other with linear ones; the losses are the categorical cross-entropy and the categorical hinge, respectively. To get predictions, the softmax function is applied to the second output set, so that we can see the entire output vector as made of estimates of posterior probabilities (although this is not true for the second set, as no probability model is assumed for the hinge loss); class labels are predicted as the index of the output with the largest value.
- ce_fisher: the model uses two different output sets, one with softmax activations and the other with linear ones. The categorical cross-entropy is minimized on the first and the Fisher loss introduced in Sect. 2 on the second. Label predictions are computed on the first set, as the one with the largest a posteriori probability.
- hinge_fisher: the model uses two linear outputs, the first one to minimize the categorical hinge loss and the second one the Fisher loss. Class labels are predicted as the index of the largest output of the first set.
- ce_hinge_fisher: the model uses now three different outputs, a first one with softmax activations and the other two with linear outputs; the categorical cross-entropy, categorical hinge and Fisher losses are minimized, respectively, on each output set. Here the softmax function is also applied to the hinge loss linear outputs and class labels are predicted as the index of the first two outputs with the largest value.

3.2 Datasets

We will work with sixteen datasets, namely a4a, a8a, australian, breast_cancer, diabetes, digits, dna, german, letter, pendigits, protein, satimage, segment, usps, w7a and w8a; eight of them are multiclass and the rest binary. All are taken from the LIBSVM data repository, except when pointed out otherwise. Table 1 shows their train and test (when available) sample sizes, dimensions and the number of classes. We give more details about them below.

Table 3. Model rankings for each problem in ascending accuracies.

	ce	ce-fisher	hinge	hinge-fisher	ce-hinge	ce-hinge-fisher
a4a	5	6	4	3	2	1
a8a	3	1	5	6	2	4
australian	1	2	4	5	3	5
breast-cancer	2	1	4	3	5	6
diabetes	1	1	6	4	3	4
digits	5	2	3	4	1	5
dna	2	1	6	5	4	3
german	1	2	4	3	6	5
letter	4	5	2	1	6	2
pendigits	1	5	4	1	3	6
protein	1	2	6	5	3	4
satimage	5	6	2	3	1	3
segment	4	2	6	5	1	2
usps	3	1	2	6	5	4
w7a	6	1	3	5	3	2
w8a	4	3	6	5	1	1
ave	3	2.6	4.2	4	3.1	3.6

- **a4a** and **a8a.** Variations of the `adult` of predicting whether income exceeds $50,000 per year based on census data.
- **australian.** The goal is to decide whether or not an application is credit-worthy.
- **breast_cancer.** The goal is here to predict whether a patient is to be diagnosed with cancer.
- **diabetes.** The objective here is to diagnose the presence of hepatitis on a sample of Pima Indian women.
- **digits.** We want to classify pixel rasters as one of the digits from 0 to 9; the subset is pre-loaded in the *scikit-learn* library.
- **dna.** The goal is to classify splice-junction gene sequences into three different classes.
- **german.** This is another problem where patterns are to be classified as either good or bad credits.
- **letter.** Pixel displays are to be identified as one of the 26 English capital letters.
- **pendigits.** Images of handwritten digits between 0 and 9 are to be classified.
- **satimage.** The goal is to classify the central pixel in a satellite image; we will work only with the 4,435 training subsample.
- **segment.** We want to classify satellite images into one of seven categories.
- **usps.** We want to classify image rasters as a digit between 0 and 9.
- **w7a** and **w8a.** Variants of a classification problem of web pages.

3.3   Experimental Methodology and Results

Recall that all model losses include a L_2 regularization term, which requires the selection of a hyperparameter α so we will proceed first to estimate the optimal

Table 4. Model rankings for each problem in ascending accuracies after putting together models with closer rankings.

	ce	ce-fisher	ce-hinge	ce-hinge-fisher	hinge	hinge-fisher
a4a	2.0	3.0	1.0	1.0	3.0	2.0
a8a	3.0	1.0	2.0	1.0	2.0	3.0
australian	1.0	2.0	3.0	2.0	1.0	2.0
breast-cancer	2.0	1.0	3.0	3.0	2.0	1.0
diabetes	1.0	1.0	3.0	1.0	3.0	1.0
digits	3.0	2.0	1.0	3.0	1.0	2.0
dna	2.0	1.0	3.0	1.0	3.0	2.0
german	1.0	2.0	3.0	3.0	2.0	1.0
letter	1.0	2.0	3.0	2.0	2.0	1.0
pendigits	1.0	3.0	2.0	3.0	2.0	1.0
protein	1.0	2.0	3.0	1.0	3.0	2.0
satimage	2.0	3.0	1.0	2.0	1.0	2.0
segment	3.0	2.0	1.0	1.0	3.0	2.0
usps	2.0	1.0	3.0	2.0	1.0	3.0
w7a	3.0	1.0	2.0	1.0	2.0	3.0
w8a	3.0	2.0	1.0	1.0	3.0	2.0
ave	1.9	1.8	2.2	1.8	2.1	1.9

Table 5. Accuracy spreads across all models as percentages of the difference between the maximum and minimum accuracies over the minimum one.

	a4a	a8a	austr	breast	diab	digits	dna	german
Spread	0.11	0.69	1.7	1.82	1.02	0.28	1.27	5.27

	letter	pendig	protein	satimage	segment	usps	w7a	w8a
Spread	0.72	0.08	4.95	1.32	0.89	0.41	0.05	0.16

α and then evaluate model performance. Eight datasets considered have train-test splits and we will find the optimal α by searching on a one-dimensional logarithmically equally-spaced grid using 5-fold cross validation (CV) on the training set. Then we will evaluate the performance of optimal α^* model by computing its accuracy on the test set. On the other datasets we will apply 5-fold nested cross validation (CV), defining first a 5-fold outer split and applying again 5-fold CV to estimate the optimal α_i^* over the i-th outer train split. Once this α_i^* is obtained, the associated model is trained over the i-th outer train fold and applied on the patterns remaining on the i-th test fold; these class predictions $\widehat{y}$ are then compared with the true target labels y to compute now the accuracy of the model under consideration.

The resulting accuracies are given in Table 2 while Table 3 shows for each problem the model ranking by decreasing accuracies; when two or more give the same accuracy, they receive the same rank. We remark that these rankings are given only for illustration purposes and they do not imply statistically significant differences. This table also shows the mean ranking of each model across all the datasets considered. As it can be seen, the model with the best mean ranking is ce-fisher, followed by ce and ce-hinge. The following one is ce-hinge-fisher while hinge and hinge-fisher perform similarly but behind all others.

These similar performances can also be seen in Table 4, where model rankings are shown after we group together ce-fisher, ce and ce-hinge on a first model group, and ce-hinge-fisher, hinge and hinge-fisher on another. Here ce-fisher still performs best on the first group, while in the second ce-hinge-fisher and hinge-fisher perform similarly and better than hinge. In any case, the test accuracies of all models shown in Table 2 are quite similar. This can also be seen in Table 5, which shows for each problem the difference between the highest accuracy (i.e., the best one) and the smallest one (i.e., the worst) as a percentage of the latter. As it can be seen, and except for the german and protein problems, in all other this percentage is below 2%, and even below 1% in nine problems.

Finally, in Table 6 we give the statistic values returned by the Wilcoxon signed rank test when applied to the columns of the test accuracies in Table 2, and their associated p-values. To obtain them, we have sorted the different losses in increasing order of their rank averages given in the last row of Table 3; this means that in the rows of Table 6 the first model is better ranked than the second one and, hence, expected to perform better. Recall that the test's null hypothesis is that the two paired samples come from the same distribution.

As it can be seen, this null hypothesis can be rejected at the $p = 0.05$ level when comparing the ce-fisher loss with the ce-hinge-fisher, hinge-fisher and hinge, and at the $p = 0.1$ level when doing so with the ce-hinge loss; on the other hand, the p-value when comparing it with the ce is quite high. This suggests that the ce-fisher loss performs similarly to the ce loss, but better than the others. In the same vein, the ce loss appears to perform better than the hinge and hinge-fisher ones; all the other loss pairings give similar performances.

Table 6. Model rankings for each problem in ascending accuracies after putting together models with closer rankings.

	stat	p-val
ce-fisher vs ce	62.0	0.776
ce-fisher vs ce-hinge	32.0	0.065
ce-fisher vs ce-hinge-fisher	27.0	0.036
ce-fisher vs hinge-fisher	23.0	0.018
ce-fisher vs hinge	22.0	0.016
ce vs ce-hinge	43.0	0.211
ce vs ce-hinge-fisher	39.0	0.14
ce vs hinge-fisher	24.0	0.024
ce vs hinge	27.0	0.034
ce-hinge vs ce-hinge-fisher	43.0	0.205
ce-hinge vs hinge-fisher	44.0	0.231
ce-hinge vs hinge	46.0	0.266
ce-hinge-fisher vs hinge-fisher	55.0	0.603
ce-hinge-fisher vs hinge	64.0	0.856
hinge-fisher vs hinge	44.0	0.231

3.4 Discussion

The preceding results indicate that the companion losses proposed here can improve on the accuracies of models based on the single ce and hinge losses. More precisely, adding the Fisher loss results in larger accuracies than those achieved by using just the ce and hinge ones, although this does not extend to the ce-hinge combination (which basically ties with the single ce loss), or when combining the ce-hinge with the Fisher loss, worsens the ce-hinge performance.

It is also to be pointed out that the performance of the hinge-based models is worse than that of the ce-based ones. A possible reason for this is the relatively small number of units in the single hidden layer architecture of all models. In fact, it is well known that for SVMs to achieve good results, the dimension of the projected input space (upon which an SVM acts linearly) must be quite large. For instance, in the case of the commonly used Gaussian kernel, this dimension essentially coincides with the sample's size. Here we are using the last hidden layer activations as a proxy of the projection space but its dimension is 100, much below sample size for all problems. In fact, in [5] deep networks with at least 1,000 units in the last layer were needed to match or improve the performance of a standard Gaussian SVM. Also, better performance should also be possible with more hidden layers, although this will also help ce models. Finally, it is also clear that adding the Fisher loss helps; in fact, it seeks a last hidden layer representation which concentrates each class samples near their centroids while keeping these apart. This should help any classifier acting on that layer while, on the other hand, it doesn't compete directly with ce and hinge.

4   Conclusions and Futher Work

In this paper we have proposed how to combine different classification losses in a single DNN so that each one acts on specific network outputs. The underlying goal is that these competing but, at the same time, complementary objectives result in models with a performance better than that of those built on each individual loss. We give experimental results on sixteen classification problems combining the categorical cross-entropy and hinge losses as well as a least squares one inspired in Fisher's Discriminant Analysis, and they show that, indeed, such combinations yield better accuracies. In fact, using the Fisher based loss as a companion of the cross-entropy or hinge ones improved on the performance of DNN models using individually those losses; the same happens with the combination of the entropy and hinge losses.

These results suggest that further study is warranted. Beside the obvious extension to more complex network architectures than the simple one here, the choice of the hyperparameters used to define the combined loss (14) should be explored. Moreover, the same strategy of adding companion losses to a base one can be applied to regression problems, where natural choices are the mean square, ϵ-insensitive or Huber losses. We are currently pursuing these and related venues.

References

1. Abadi, M., et al.: TensorFlow: a system for large-scale machine learning. In: 12th USENIX Symposium on Operating Systems Design and Implementation (OSDI 2016), pp. 265–283 (2016)
2. Chollet, F.: Keras (2015). https://github.com/fchollet/keras
3. Crammer, K., Singer, Y.: On the algorithmic implementation of multiclass kernel-based vector machines. J. Mach. Learn. Res. **2**, 265–292 (2001)
4. Díaz-Vico, D., Dorronsoro, J.R.: Deep least squares fisher discriminant analysis. IEEE Trans. Neural Netw. Learn. Syst. **31**(8), 2752–2763 (2020)
5. Díaz-Vico, D., Prada, J., Omari, A., Dorronsoro, J.R.: Deep support vector neural networks. Integr. Comput. Aided Eng. **27**(4), 389–402 (2020)
6. Glorot, X., Bengio, Y.: Understanding the difficulty of training deep feedforward neural networks. In: Proceedings of the Thirteenth International Conference on Artificial Intelligence and Statistics, AISTATS 2010. JMLR Proceedings, Chia Laguna Resort, Sardinia, Italy, 13–15 May 2010, vol. 9, pp. 249–256 (2010)
7. Kingma, D.P., Ba, J.: Adam: a method for stochastic optimization. In: 3rd International Conference on Learning Representations, ICLR 2015, Conference Track Proceedings, San Diego, CA, USA, 7–9 May 2015 (2015)
8. Krizhevsky, A., Sutskever, I., Hinton, G.E.: ImageNet classification with deep convolutional neural networks. In: Advances in Neural Information Processing Systems 25, 26th Annual Conference on Neural Information Processing Systems 2012, Proceedings of a Meeting Held 3–6 December 2012, Lake Tahoe, Nevada, United States, pp. 1106–1114 (2012)
9. Paszke, A., et al.: PyTorch: an imperative style, high-performance deep learning library. In: Advances in Neural Information Processing Systems, vol. 32. Curran Associates, Inc. (2019)
10. Tang, M., Perazzi, F., Djelouah, A., Ben Ayed, I., Schroers, C., Boykov, Y.: On regularized losses for weakly-supervised CNN segmentation. In: Proceedings of the European Conference on Computer Vision (ECCV), pp. 507–522 (2018)
11. Vincent, P., Larochelle, H., Lajoie, I., Bengio, Y., Manzagol, P.: Stacked denoising autoencoders: learning useful representations in a deep network with a local denoising criterion. J. Mach. Learn. Res. **11**, 3371–3408 (2010)
12. Weston, J., Watkins, C.: Support vector machines for multi-class pattern recognition. In: ESANN 1999, Proceedings of the 7th European Symposium on Artificial Neural Networks, Bruges, Belgium, 21–23 April 1999, pp. 219–224 (1999)
13. Zhang, Z., Dai, G., Xu, C., Jordan, M.I.: Regularized discriminant analysis, ridge regression and beyond. J. Mach. Learn. Res. **11**, 2199–2228 (2010)
14. Zhao, H., Gallo, O., Frosio, I., Kautz, J.: Loss functions for image restoration with neural networks. IEEE Trans. Comput. Imaging **3**(1), 47–57 (2017)

Financial Forecasting via Deep-Learning and Machine-Learning Tools over Two-Dimensional Objects Transformed from Time Series

Alessandro Baldo[1], Alfredo Cuzzocrea[2,3(✉)], Edoardo Fadda[4,5], and Pablo G. Bringas[6]

[1] ISIRES, Torino, Italy
alessandro.baldo@isires.org
[2] iDEA Lab, University of Calabria, Rende, Italy
alfredo.cuzzocrea@unical.it
[3] LORIA, Nancy, France
[4] Politecnico di Torino, Torino, Italy
edoardo.fadda@polito.it
[5] ISIRES, Torino, Italy
[6] Faculty of Engineering, University of Deusto, Bilbao, Spain
pablo.garcia.bringas@deusto.es

Abstract. In this study, we propose a deeper analysis on the algorithmic treatment of financial time series, with a focus on Forex markets' applications. The relevant aspects of the paper refers to a more beneficial data arrangement, proposed into a two-dimensional objects and to the application of a Temporal Convolutional Neural Network model, representing a more than valid alternative to Recurrent Neural Networks. The results are supported by expanding the comparison to other more consolidated deep learning models, as well as with some of the most performing Machine Learning methods. Finally, a financial framework is proposed to test the real effectiveness of the algorithms.

1 Introduction

The idea of this study was suggested after the analysis of the study-case [17], where the authors presented some advantages in condensing 1D-time series, into 2D data formats enhancing the spatiality.

Starting from that suggestion, this study has the scope to more deeply investigate this paradigm, applying it to a regression domain. The explored environment has been the Forex markets [8], which are widely recognised to be very liquid and volatile markets, and thus more challenging for the algorithmic applications.

A. Cuzzocrea—This research has been made in the context of the Excellence Chair in Computer Engineering – Big Data Management and Analytics at LORIA, Nancy, France.

H. Sanjurjo González et al. (Eds.): HAIS 2021, LNAI 12886, pp. 550–563, 2021.
https://doi.org/10.1007/978-3-030-86271-8_46

After a short explanation on the nature of data, the paper will explore a large variety of techniques. A net distinction will be presented, separating the Deep Learning framework from the Machine Learning counterpart, having here mostly a validation role on the results.

The discussion will then switch into the numerical results, examining both from an algorithmic and a financial points of view the feasibility of these models.

Finally, the last section will propose some suggestions based on the final results, as some key points that could be improved in the future to reach still more consistent results.

Due to the exchanged volumes and its liquidity, Forex markets require an high-precision tracking of the evolution of the trends. The fundamental unit of measure is the *pip*, representing the penultimate significant decimal number. In most cases, that coincides with the 4th–5th decimal number, but it is not so uncommon for it to be the at the 2nd–3rd decimal place. Indeed, pairs where the inflation of the quoted currency is very high are more frequently characterized by a minor precision. The exotic currencies fell in this scenario, having less market power when compared to the occidental ones. Also for this reason, trading over these pairs is often restricted to well-navigated traders who can cope with high volatility scenarios.

However, even though the variations on the price seem to be quite small, large profits and losses in Forex are commonplace, due to its highly noisy and stochastic nature. Furthermore, the quantities required to investors to open a position in the market are often very high and are denoted by the measure called *lot*. A lot is equivalent to 100'000 times the fundamental unit of the quoted currency (e.g., investing one lot over the EUR/USD currency pair means investing 100'000$). The minimum lot size a trader can choose to invest is 1/100 of the lot unit.

2   Description of the Datasets

Data has been retrieved from the open-data directory EaForexAcademy[1], by collecting the historical trends of 11 major currency pairs, under 7 different time-frame intervals (from a 1 min frequency, counting about 200,000 records, to a daily time division, mapping the last 13 years).

These raw data contains the *Open, Close, High, Low* prices and the *Volume* reference, from which the 28 technical indicators were computed for each interval of the series. The list of the indicators comprehends two indexes related to the variations of price and volumes between consecutive time frames, four types of moving averages, mapped into four different periods and statistical oscillators. The considered features are:

– **Yield**

$$Yield = \frac{Close - Open}{Open} \tag{1}$$

---

[1] https://eaforexacademy.com/software/forex-historical-data/.

- **Percentage Volume**

$$PercentageVolume = 10^4 \frac{High - Low}{Volume} \tag{2}$$

- Simple Moving Average (**SMA**) of period n

$$\text{SMA}_i(Close, n) = \frac{1}{n} \sum_{i=0}^{n} Close_{n-i} \tag{3}$$

- Exponential Moving Averages (**EMA**) of period n

$$\text{EMA}_i(Close, n) = \alpha \cdot Close_i + (1 - \alpha) \cdot \text{EMA}_{i-1}(Close, n) \tag{4}$$

- Weighted Moving Averages (**WMA**) of period n

$$\text{WMA}_i(Close, n) = \frac{\sum_{i=0}^{n-1} i \cdot Close_i}{\frac{n(n-1)}{2}} \tag{5}$$

- Hull Moving Averages (**HMA**) of period n

$$\begin{aligned} \text{HMA}_i(Close, n) &= \text{WMA}_i(arg, \sqrt{n}), \\ arg &= (2 \cdot \text{WMA}(Close, \frac{n}{2}) - \text{WMA}(Close, n)) \end{aligned} \tag{6}$$

- Moving Average Convergence/Divergence (**MACD**)

$$\text{MACD} = \text{EMA}(Close, 12) - \text{EMA}(Close, 26) \tag{7}$$

- Commodity Channel Index (**CCI**)

$$\text{CCI} = \frac{TypicalPrice - \text{SMA}(TypicalPrice, 20)}{0.015 \times AvgDev}, \tag{8}$$

where

$$TypicalPrice = \frac{High + Low + Close}{3} \tag{9}$$

- **Stochastic Oscillator**

$$StochasticOscillator = 100 \frac{Close - H14}{H14 - L14} \tag{10}$$

where $H14, L14$ are respectively the Highest and Lowest prices registered in the last 14 time intervals

- Relative Strength Index (**RSI**)

$$\begin{aligned} \text{RSI}(n) &= 100 - \frac{100}{1 + \text{RS}}, \\ \text{RS} &= \frac{U}{D}, \quad n = 14 \end{aligned} \tag{11}$$

where U, D are respectively the average of the differences $Close - Open$ of the last n Bullish/Bearish bars

- Rate of Change (**ROC**)

$$\text{ROC}(n) = 100\frac{Close_i - Close_{i-n}}{Close_{i-n}}, \tag{12}$$
$$n = 14$$

- Percentage Price Oscillator (**PPO**)

$$\text{PPO} = 100\frac{\text{EMA}(Close, 12) - \text{EMA}(Close, 26)}{\text{EMA}(Close, 26)} \tag{13}$$

- Know Sure Thing (**KST**)

$$
\begin{aligned}
\text{KST} &= \text{SMA}(\text{RCMA}_1 + 2\text{RCMA}_2 + 3\text{RCMA}_3 + 4\text{RCMA}_4, 9), \\
\text{RCMA}_1 &= \text{SMA}(\text{ROC}(10), 10), \\
\text{RCMA}_2 &= \text{SMA}(\text{ROC}(15), 10), \\
\text{RCMA}_3 &= \text{SMA}(\text{ROC}(20), 10), \\
\text{RCMA}_4 &= \text{SMA}(\text{ROC}(30), 15)
\end{aligned}
\tag{14}
$$

- Bollinger Bands Middle, Up and Down (**BOLM,BOLU,BOLD**)

$$
\begin{aligned}
\text{BOLM} &= \text{SMA}(TypicalPrice, 20) \\
\text{BOLU} &= \text{BOLM} + 2\sigma_{20}(TypicalPrice) \\
\text{BOLD} &= \text{BOLM} - 2\sigma_{20}(TypicalPrice)
\end{aligned}
\tag{15}
$$

where $\sigma_{20}(TypicalPrice)$ is the standard deviation of the last 20 Typical Prices.

The consequent step for the management of data dealt with the normalization of the features, due to some existing diversities. The generation of the two-dimensional (2D) objects then collected all these records and features, at step of 28 time-frames. The target was set as the *Close* price of the immediately subsequent interval. This process was iteratively repeated, by scaling each 28 intervals window by one step forward. Some data augmentation was performed to improve the model robustness. It represents a form of pre-processing which often grants to improve the overall model's robustness. For the sake of this study, it consisted in randomly selecting a pre-defined percentage of 2D objects (about 30%) and applying on them an "obscuration" (i.e. totally nullifying some 2D object rows) of some of data closer in time to the referred label. In this way, the model was forced to rely more on the data references further in time, avoiding to simply output a prediction based on the last few recorded references, where an high auto-correlation with the label existed.

Other forms of data augmentation can be furtherly thought, by including for example some samples from other correlated currency pairs, but that went out of the scopes of this study.

For the training purposes, the daily time interval was considered as a trade-off between the enlargement of training times and the inclusion of data characterized by higher variances. The latter was particularly beneficial to improve the generalization capabilities of the models on shorter times frames, where the variability resulted to be more distributed in time.

The division among training, validation and test sets was respectively done according to the 72-8-10 proportions for the Deep Learning models and according to the Pareto rule (i.e. 80–20) for the Machine Learning models. As suggested by the choice of the time intervals, the focus mainly dealt with Day Traders and Scalpers scenarios.

2.1   Feature Engineering

Correlation Between Features. Correlations measures at Fig. 1 (left) highlight the two different natures among the features: the Moving Averages family and the Oscillators. The former presents a flat behaviour, comprehending also the Bollinger Bands (which exploit themselves a concept of Moving Average); the latter category is more diversified, underlining not only the similarities between the RSI and the Stochastic Oscillators (which map conceptually the same information), but also evidencing the importance to include an indicator like the Know Sure Thing (KST), apparently characterizing in a different way the evolution of the price.

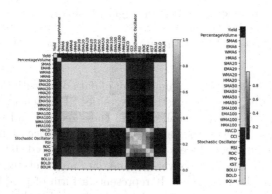

Fig. 1. Correlation matrix among features (left) and between features and target (right)

Correlation Between Features and Labels. The Moving Averages are broadly adopted for their forecasting capabilities and they are often used singularly as predictors in regression tasks. According to this, the correlation measures reported at Fig. 1 (right) are coherent with the nature of those features.

Mutual Information of the Features and Motivations. The Mutual Information represents a standard metric in Machine Learning to perform Feature Engineering. If the correlation measures give a global knowledge on the features, more specific statistics are useful to explain the contribution of each information in the final prediction. Table 1 furthermore confirms the central role of the Moving Averages in the predictions, evidencing how for the intra-daily scenario the fastest moving averages (i.e. the moving averages with a smaller window) are more adequate. However, the support of slower MAs is justified by many financial strategies. Indeed, the financial technical analysis often makes use of combinations of them, by varying the type and the period. A longer period is usually referred to a slow moving average, because the indicator better absorbes variability by averaging it on the last *period* timeframes. On the contrary, a faster moving average is associated to a shorter averaging period, being thus more influenced by short-term variability.

Table 1. Mutual information values of the features

Indicators	Mutual info	Indicators	Mutual info	Indicators	Mutual info	Indicators	Mutual info
HMA6	3,321	HMA50	2,115	EMA50	1,900	KST	0,396
WMA6	2,897	SMA20	2,082	SMA50	1,848	Perc. Vol.	0,184
EMA6	2,814	BOLM	2,070	WMA100	1,805	RSI	0,174
SMA6	2,628	BOLD	2,070	EMA100	1,760	ROC	0,153
HMA20	2,478	BOLU	2,070	SMA100	1,735	CCI	0,133
WMA20	2,277	WMA50	2,002	MACD	0,432	Stoch. Osc.	0,096
EMA20	2,222	HMA100	1,966	PPO	0,417	Yield	0,013

Traders generally look to the crossovers of slow and fast moving averages, since they create some accurate pivot points, useful to determine the entry or exit points for BUY/SELL orders.

The Oscillators, instead, mostly track the different natures of the so called *Momentum*. This has the equivalent meaning of the instantaneous speed in physics, evaluating the price variation speed for a defined period. In addition, indicators like the RSI and the Stochastic Oscillators are frequently used to classify Overbought and Oversold regions in the trend. The main principle is that each trend should have an implicit equilibrium overall, untethered by the bullish or bearish global nature of the market. Therefore, very rapid variations would unbalance the system and these tools are capable to track when it is likely the trend will undergo to a movement in the opposite direction. Conventionally, the period set for these indicators is around 12–14 days.

3   Deep Learning and Machine Learning Models for Financial Forecasting

Neural Networks and deep learning has shown to be a really effective tool in several cincumstances (see [4,6,13]). The model at the core of the study is the

Temporal Convolutional Network (TCN) [2], representing a quite novel structure in the Deep Learning field. Due to its strict connection and derivation with the *Convolutional Neural Networks* (CNN) and the *Recurrent Neural Networks* (RNN), the study also includes an internal analysis with their performances, presenting a standard CNN and a *Long Short-Term Memory Network* (LSTM).

Finally, an external comparison is proposed between the adoption of a Deep Learning and a Machine Learning approach [1,10], including then some of the most performing and traditional ML techniques.

As a baseline, all the deep learning models were implemented through the Keras library and were tuned by using some built-in callbacks, allowing for a dynamic reduction of the learning rate over the epochs, when a plateau was detected and which opportunely stopped the training if some over-fitting was occurring.

3.1   Temporal Convolutional Neural Network

Temporal Convolutional Networks, when firstly introduced in 2018, proved to have great performances on sequence-to-sequence tasks like machine translation or speech synthesis in text-to-speech (TTS) systems.

By merging the benefits of the Convolutional Neural Networks, with the memory-preserving features of the Recurrent-LSTM networks, such structures end to be more versatile, lightweight and faster than the two derivation structures.

It is designed around the two basic principles of *causal convolution*, where no information is leaked from past to future, keeping a memory of the initial "states", and of an architecture mapping dynamically input and output sequences of any length, allowing both the one-to-one, one-to-many and many-to-many paradigms.

Since a simple causal convolution has the disadvantage to look behind at history with size linear in the depth of the network (i.e. the receptive field grows linearly with every additional layer), the architecture employs convolutions with *dilation* (Fig. 2), enabling an exponentially large receptive field. Under a mathematical point of view, this means mapping an input sequence $x \in \mathcal{R}^T$ with a filter $f : \{0, ...k - 1\} \to \mathcal{R}$, using a convolution operator F in this way:

$$F(x) = (x_d f)(x) = \sum_{i=0}^{k-1} x_{s-d \cdot i} \tag{16}$$

where $d = 2^l$ is the dilation factor, l the level of the network and k the kernel size of the filter.

Using larger dilation enables an output at the top level to represent a wider range of inputs, thus effectively expanding the receptive field of a CNN. There are thus two ways to increase the receptive field of a TCN: choosing lager filter sizes k and increasing the dilation factor d, since the effective history of one layer is $(k - 1)d$.

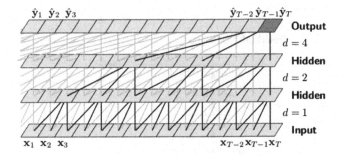

Fig. 2. Dilated convolution

Residual (Temporal) Block. The Residual (Temporal) Block is the fundamental element of a TCN. It is indeed represented as a series of transformations, whose output is directly summed to the input of the block itself. Especially for very deep networks stabilization becomes important, for example, in the case where the prediction depends on a large history size with a high-dimensional input sequence.

Each block is composed of two layers of dilated convolutions and rectified linear units (ReLUs). The block is terminated by a Weight Normalization [16] and a Dropout [18]. Figure 3 shows the structure.

The final TCN structure was conceived by correctly testing different combinations of layers and depth of the convolutional layers. The overall network is reported at Fig. 4 and resulted in a 7-levels network, where each one is represented by a Residual (Temporal) Block. The high number of filters of each convolutional layer combined with the high dilation factor gave the network optimal memory-preserving properties.

The other optimized parameters used during the experiments were: (*i*) Epochs: 200; (*ii*) Learning Rate: 0.01; (*iii*) Batch Size: 32; (*iv*) Kernel Size: 3. To set these parameters, the methodology used was simply finding the best parameters' combination provided by iterative trials.

4   Experimental Results and Analysis

All the results reported were achieved under identical conditions. Each model underwent to a Cross Validation procedure, by computing the final score on the same test dataset constituted by more than 800 days up to the end of October 2020. Incorporating the COVID-19 pandemic financial crisis, this surely made this period the most intriguing and challenging to test the generalization capabilities of the models.

A first batch of results has been reported for an immediate comparison. In Table 2 the eight models are referred to daily Forex data of the 11 currency pairs taken into account. The Mean Squared Error (MSE) measures represented the main benchmark for the goodness of each model. In order to better explore

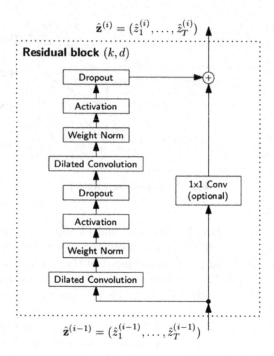

Fig. 3. Residual (temporal) block

the obtained results, we first measure a correlation term which was able to correctly capture the general trend of the true time series. The Pearson Correlation (17) thus represented an as simple as effective mathematical indicator for what concerns the spatial evolution of the financial trend.

$$\rho_{XY} = \frac{cov(X,Y)}{\sigma(x)\sigma(Y)} \tag{17}$$

In the deep learning experiments it was firstly used retrospectively as a method to favour those models which not only minimized the distance with respect to the real trend (i.e. low MSE), but also which were able to maximize the correlation. The Equation (18) models the decision rule adopted at the end of each training procedure:

$$\underset{\theta}{\mathrm{argmin}}\ \mathrm{MSE}_{\hat{y},y} \cdot (1 - \rho_{\hat{y},y}) \tag{18}$$

This simple mathematical combination ensured to have a more global view on the outcomes and to increase the overall optimality.

The next tentative mainly dealt with the plug-in of such decision rule inside the loss function itself, but, due to the consequent increase of complexity and training times of the networks structure, this task was left for future improvements.

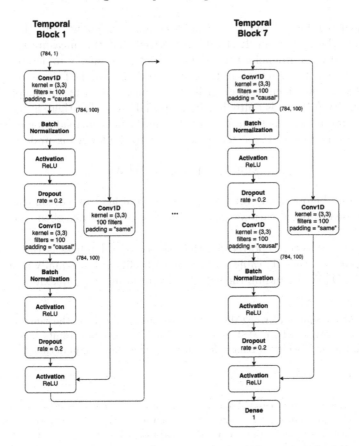

Fig. 4. The final TCN resulting from the hyperparameters' tuning

Given the aforementioned spatial modelling, the intent was then to include in the error measures an indicator which could symmetrically act on the temporal domain. Indeed, when dealing with random walk time series (as the financial case is), predictive models are distinguished by systematically missing the so-called *pivot* points, i.e. where the trend changes its concavity (on the local minima and maxima). This often results in a delay in the prediction of these points, since each model prediction would be obtained based on the high auto-correlation existing with the last known sample.

In literature [9,11,15] there are several studies about the most correct temporal measures/loss to introduce in deep learning models and in this study two of them have been adopted to verify whether an increase of the complexity is justified with respect to their effectiveness.

The first measure is the so-called *Dynamic Time Warping*. Behind the computation of the DTW there is a solution of a Quadratic Optimization Problem making use of Dynamic Programming. The value of the indicator is the value of the shortest path built between two time series, according to a *window* parameter, which regulates a one-to-one unidirectional mapping from the predicted output and the real trend. It should be noticed that this measure is important because of, despite financial time series do not particularly undergo non-linear warps or time shifts, it reveals critical characteristics for the forecasting purpose.

Despite its effectiveness in describing the temporal nature of the predictions, its complexity $(O(n^2))$ is quite limiting, especially whether there is the necessity to plug it into a loss function. As did with the Pearson correlation coefficient, its evaluation was done retrospectively according to the following updated decision rule:

$$\operatorname*{argmin}_{\theta} \mathrm{MSE}_{\hat{y},y} \cdot (1 - \rho_{\hat{y},y}) \cdot \mathrm{DTW}_{\hat{y},y,window} \tag{19}$$

As stated before, some brief experiments were held by customizing the loss function used in deep learning models, but that would have been too time consuming.

The last indicator reported for the sake of completeness is the Fast-DTW (*Fast Dynamic Time Warping*) [12]. It is a simplified, but equally accurate version of the DTW, bounding the calculations to a linear complexity.

Table 2. Mean Squared Error (MSE) statistics of the models on the same test dataset. In **bold**, the results whose MSE is under 5E − 5; in **red** the results below the 10^{-5} threshold; the best achieved results for each currency pair are underlined

	TCN	LSTM	CNN	Linear regression	Bayesian ridge regression	Support vector regression	Random forest regression	KNN regression
AUDCAD	**2,282E-05**	**2,195E-05**	6,050E-05	**1,607E-05**	**1,663E-05**	6,292E-04	**5,946E-06**	**1,050E-05**
AUDUSD	**4,130E-05**	**1,961E-05**	1,132E-04	**1,704E-05**	**1,601E-05**	4,018E-04	**5,477E-06**	3,190E-05
EURAUD	9,708E-05	6,799E-05	8,376E-04	**6,046E-05**	**6,426E-05**	1,123E-03	**2,147E-05**	9,962E-05
EURCAD	5,195E-05	**4,620E-05**	8,827E-05	**3,814E-05**	**3,990E-05**	3,687E-04	**1,540E-05**	2,736E-05
EURCHF	**1,322E-05**	**1,192E-05**	2,327E-04	**9,702E-06**	**8,624E-06**	3,658E-04	**2,730E-06**	1,587E-05
EURGBP	**2,365E-05**	**1,834E-05**	4,288E-05	**1,524E-05**	**1,557E-05**	4,253E-04	**5,225E-06**	1,138E-05
EURJPY	5,406E-01	8,072E+00	2,166E+01	3,209E-01	3,053E-01	2,968E-01	1,004E-01	2,185E-01
EURUSD	**2,403E-05**	**2,081E-05**	1,139E-04	**2,058E-05**	**2,006E-05**	1,360E-03	**7,338E-06**	1,587E-05
GBPUSD	5,015E-05	5,496E-05	3,909E-04	**4,978E-05**	**4,934E-05**	2,341E-04	**1,808E-05**	4,747E-05
USDCAD	**3,763E-05**	**3,633E-05**	1,805E-04	**2,664E-05**	**2,890E-05**	9,294E-04	**1,064E-05**	2,815E-05
USDJPY	2,150E-01	7,112E-01	1,153E+02	1,969E-01	1,996E-01	1,949E-01	6,108E-02	1,387E-01

4.1 Algorithmic Results

Looking at the numerical results of the Machine Learning methods, the Random Forest Regression represented the best performing model under all the adopted metrics (Table 3, Fig. 5).

Conversely, the SVR model did not guarantee to be as accurate as the other models did; however, it represents the evidence of why adopting multiple judgement criteria: it is indeed true that it presents the lowest MSE values, but at the same time, as Table 4 proves, it achieves good correlation scores, catching both the spatiality and temporality of the trend.

Table 3. Results of random forest regression

Currency	MSE	Pearson correlation	DTW	Fast DTW	Number of Estimators
EURGBP	**5,225E-06**	0,993	0,99	0,98	500
USDJPY	6,108E-02	0,994	111,63	110,21	500
AUDUSD	**5,477E-06**	0,998	1,07	1,06	500
AUDCAD	**5,946E-06**	0,997	1,03	1,02	500
EURUSD	**7,338E-06**	0,997	1,18	1,17	500
USDCAD	**1,064E-05**	0,995	1,43	1,41	500
EURCAD	**1,540E-05**	0,995	1,65	1,62	500
GBPUSD	**1,808E-05**	0,996	1,80	1,78	500
EURCHF	**2,730E-06**	0,999	0,78	0,76	500
EURJPY	1,004E-01	0,998	139,53	138,21	500
EURAUD	**2,147E-05**	0,995	1,92	1,91	500

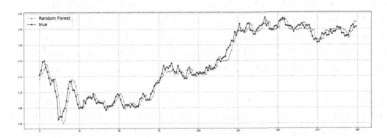

Fig. 5. Random Forest Regression predictions on a 200 days prediction over EUR/USD (Mar–Oct 2020). The Random Forest was composed of up to 500 estimators, minimizing a negative mean squared error measure. Its high capabilities ensured the network to well "follow" the real trend even in a singular event as the COVID-19 pandemic

Table 4. Results of support vector regression

Currency	MSE	Pearson correlation	DTW	Fast DTW	Kernel	Degree	C	Gamma
EURGBP	4,253E-04	0,862	14,61	7,13	Poly	3	10	0,001
USDJPY	1,949E-01	0,981	66,82	65,82	Linear	1	10	0,0001
AUDUSD	4,018E-04	0,968	15,33	5,69	Poly	1	10	0,0001
AUDCAD	6,292E-04	0,956	16,98	11,94	RBF	1	1	0,0001
EURUSD	1,360E-03	0,972	27,50	15,02	Poly	2	10	0,001
USDCAD	9,294E-04	0,919	22,45	10,85	Poly	2	1	0,001
EURCAD	3,687E-04	0,917	11,03	5,52	Poly	2	1	0,01
GBPUSD	2,341E-04	0,943	8,26	4,49	Poly	1	1	0,01
EURCHF	3,658E-04	0,989	14,97	6,78	RBF	1	1	0,0001
EURJPY	2,968E-01	0,993	85,89	84,83	Linear	1	10	0,0001
EURAUD	1,123E-03	0,867	20,41	14,01	Poly	2	10	0,001

It should be considered that these experiments consider simple prediction tasks only, and our best method is, of course, suitable for the targeted tasks.

On the other hand, for more complex tasks (e.g., machine-learning based) there could be better more complex ML/DL models to be exploited.

5   Conclusions and Next Steps

The study explored the potentialities of Deep Learning methods in financial time series, showing internal comparisons between some of the most wide-spread techniques and innovative performing structures, and external comparisons with the Machine Learning domain. The Temporal Convolutional Network proved to be reliable, having similar performances to the Recurrent Neural Long Short-Term Memory Network and presenting advantages in terms of ease of training and scalability. Furthermore, the singular nature of data defines another "hyperparameter" to be considered in the variety of model parameters: some improvements can indeed be retrieved by enhancing spatiality characterizations among features. On the other hand, this assertion has even some limitations. In fact, other powerful models exist, and, as a matter of fact and overall lesson learned from the study, modern Deep Learning methods should not consider a-priori as the best modeling choice to go. A critical review of such limitations is presented in [14].

The discussion then moved to the necessity of introducing more meaningful error metrics and the possibility to plug them into the loss functions. The derived benefits of this operation could ensure better updates (and thus learning), definitely creating a performance gap between Deep Learning and Machine Learning.

Finally, further improvements should then focus particularly on the capabilities of the models to correctly detect pivot points, where the trend changes its direction. This could be done by plugging a classifier on top of the regression models, discriminating between pivot and non-pivot points. In addition, integration with big data features (e.g., [3,5,7]) is foreseen.

References

1. Ahmed, N.K., Atiya, A.F., El Gayar, N., El-Shishiny, H.: An empirical comparison of machine learning models for time series forecasting. Economet. Rev. **29**(5–6), 594–621 (2010)
2. Bai, S., Zico Kolter, J., Koltun, V.: An empirical evaluation of generic convolutional and recurrent networks for sequence modeling. CoRR, abs/1803.01271 (2018)
3. Campan, A., Cuzzocrea, A., Truta, T.M.: Fighting fake news spread in online social networks: actual trends and future research directions. In: 2017 IEEE International Conference on Big Data, BigData 2017, Boston, MA, USA, 11–14 December 2017, pp. 4453–4457. IEEE Computer Society (2017)
4. Castrogiovanni, P., Fadda, E., Perboli, G., Rizzo, A.: Smartphone data classification technique for detecting the usage of public or private transportation modes. IEEE Access **8**, 58377–58391 (2020)

5. Ceci, M., Cuzzocrea, A., Malerba, D.: Effectively and efficiently supporting roll-up and drill-down OLAP operations over continuous dimensions via hierarchical clustering. J. Intell. Inf. Syst. **44**(3), 309–333 (2015). https://doi.org/10.1007/s10844-013-0268-1

6. Cuzzocrea, A., Leung, C.K., Deng, D., Mai, J.J., Jiang, F., Fadda, E.: A combined deep-learning and transfer-learning approach for supporting social influence prediction. Procedia Comput. Sci. **177**, 170–177 (2020)

7. Cuzzocrea, A., Song, I.Y.: Big graph analytics: the state of the art and future research agenda. In: Proceedings of the 17th International Workshop on Data Warehousing and OLAP, DOLAP 2014, Shanghai, China, 3–7 November 2014, pp. 99–101. ACM (2014)

8. Evans, C., Pappas, K., Xhafa, F.: Utilizing artificial neural networks and genetic algorithms to build an algo-trading model for intra-day foreign exchange speculation. Math. Comput. Model. **58**(5), 1249–1266 (2013)

9. Gastón, M., Frías, L., Fernández-Peruchena, C.M., Mallor, F.: The temporal distortion index (TDI). A new procedure to analyze solar radiation forecasts. In: AIP Conference Proceedings, vol. 1850, p. 140009 (2017)

10. Krollner, B., Vanstone, B.J., Finnie, G.R.: Financial time series forecasting with machine learning techniques: a survey. In: ESANN 2010, 18th European Symposium on Artificial Neural Networks, Bruges, Belgium, 28–30 April 2010, Proceedings (2010)

11. Le Guen, V., Thome, N.: Shape and time distortion loss for training deep time series forecasting models. In: Advances in Neural Information Processing Systems, vol. 32. Curran Associates Inc. (2019)

12. Li, H., Yang, L.: Accurate and fast dynamic time warping. In: Motoda, H., Wu, Z., Cao, L., Zaiane, O., Yao, M., Wang, W. (eds.) ADMA 2013. LNCS (LNAI), vol. 8346, pp. 133–144. Springer, Heidelberg (2013). https://doi.org/10.1007/978-3-642-53914-5_12

13. Li, Y., Fadda, E., Manerba, D., Tadei, R., Terzo, O.: Reinforcement learning algorithms for online single-machine scheduling. In: 2020 15th Conference on Computer Science and Information Systems (FedCSIS), pp. 277–283 (2020)

14. Manibardo, E.L., Laña, I., Del Ser, J.: Deep learning for road traffic forecasting: does it make a difference? IEEE Trans. Intell. Transp. Syst. pp. 1–25 (2021)

15. Rivest, F., Kohar, R.: A new timing error cost function for binary time series prediction. IEEE Trans. Neural Netw. Learn. Syst. **31**(1), 174–185 (2020)

16. Salimans, T., Goodfellow, I., Zaremba, W., Cheung, V., Radford, A., Chen, X.: Improved techniques for training GANs. CoRR, abs/1606.03498 (2016)

17. Omer Berat Sezer and Ahmet Murat Ozbayoglu: Algorithmic financial trading with deep convolutional neural networks: time series to image conversion approach. Appl. Soft Comput. **70**, 525–538 (2018)

18. Srivastava, N., Hinton, G., Krizhevsky, A., Sutskever, I., Salakhutdinov, R.: Dropout: a simple way to prevent neural networks from overfitting. J. Mach. Learn. Res. **15**(56), 1929–1958 (2014)

A Preliminary Analysis on Software Frameworks for the Development of Spiking Neural Networks

Ángel M. García-Vico$^{(\boxtimes)}$ ⓘ and Francisco Herrera ⓘ

Andalusian Research Institute in Data Science and Computational Intelligence (DaSCI), University of Granada, Granada, Spain
{agvico,herrera}@decsai.ugr.es

Abstract. Today, the energy resources used by machine learning methods, especially those based on deep neural networks, pose a serious climate problem. To reduce the energy footprint of these systems, the study and development of energy-efficient neural networks is increasing enormously. Among the different existing proposals, spiking neural networks are a promising alternative to achieve this goal. These methods use activation functions based on sparse binary spikes over time that allow for a significant reduction in energy consumption. However, one of the main drawbacks of these networks is that these activation functions are not derivable, which prevents their direct training in traditional neural network development software. Due to this limitation, the community has developed different training methods for these networks, together with different libraries that implement them. In this paper, different libraries for the development and training of these networks are analysed. Their main features are highlighted with the aim of helping researchers and practitioners in the decision making process regarding the development of spiking neural networks according to their needs.

1 Introduction

Nowadays, the size of deep learning-based models that constitute the state of the art is growing exponentially. For example, it is estimated that the cost of training these models from 2012, with the appearance of AlexNet [17], to AlphaGo Zero [30], developed in 2018, has increased by approximately 300,000 times [29]. Such growth needs a huge amount of energy to perform the computation. This is a major problem not only environmentally, but also in contexts where low energy consumption is required such as mobile computing, the Internet of Things or edge computing [36]. This energy consumption is mainly due to the large number of multiply-and-accumulate (MAC) operations that are carried out to calculate the output value of each neuron [32], as shown in Eq. (1)

$$y_i = f\left(\sum_{i=0}^{N_I} w_{ij}x_j + b\right) \tag{1}$$

H. Sanjurjo González et al. (Eds.): HAIS 2021, LNAI 12886, pp. 564–575, 2021.
https://doi.org/10.1007/978-3-030-86271-8_47

where w_{ij} is the connection weight between neuron i and j, x_j is the output value of neuron j, b is the bias and $f(\cdot)$ is a non-linear activation function. Due to this, hardware accelerators for neural networks are mainly based on the optimisation of MAC operations. On the other hand, developers employ techniques such as pruning [18] or quantisation of weights [20] for the reduction of the use and the size of the MAC arrays in order to improve efficiency.

Spiking neural networks [19] (SNNs) follow a different but complementary philosophy to artificial neural networks (ANNs). Strongly inspired by brain biology, SNNs focus on the temporal, distributed, asynchronous processing of sparse binary inputs, known as spikes. Each neuron has an internal state or voltage, which together with the use of binary activation functions emits a spike only when sufficient voltage is accumulated due to the arrival of several spikes from other neurons in a short period of time. Using this technique not only reduces the volume of data transmitted between nodes, but also replaces MAC operations with adders since $x_i \in \{0, 1\}$. Replacing these MAC operations implies an efficiency gain of several orders of magnitude over ANNs [5].

Although the energy efficiency of SNNs is far superior to ANNs, their performance in machine learning tasks is often significantly lower. In addition, they usually require a much longer time to be trained. This is because the inputs need to be presented for a longer period of time, and the activation functions of SNNs are not derivable. Therefore, gradient-descent algorithms such as backpropagation [28], cannot be directly applied. For this reason, throughout the literature, researchers have developed a wide range of learning methods for this type of network. These can be roughly grouped into three broad areas:

1. Local learning methods based on bio-inspired hebbian plasticity.
2. Modification of the spiking activation functions in order to use backpropagation.
3. Conversion methods from trained ANNs to its SNN counterpart.

Recently, a number of tools have emerged to facilitate the development, training and evaluation of such networks for the general public. However, each of these tools have different components and features that can significantly limit the development of SNNs. This paper analyses these libraries and their main characteristics are highlighted. To do so, the main features of SNNs are defined in Sect. 2. Then, the main characteristics of the learning approaches for SNNs described above are detailed in Sect. 3. Next, the main SNN development libraries and their characteristics are presented, described and analysed in Sect. 4. After that, a comparison of these frameworks is carried out on an experimental study presented in Sect. 5. Finally, the conclusions of this work are presented in Sect. 6.

2 Spiking Neural Networks

SNNs are considered the third generation of artificial neural networks [19]. They are more biologically realistic models than ANNs as they explicitly include the concept of time and the processing of their inputs by means of spikes. Therefore,

SNNs can be defined as a dynamic system where neurons and synapses are generally described by ordinary differential equations (ODEs). Among the different kinds of neurons developed, the most employed one in SNNs for machine learning tasks is the Leaky Integrate-and-Fire (LIF) model [10], defined in Eqs. (2) and (3), and graphically presented in Fig. 1.

$$y_i = \begin{cases} 1 & IF\ u_i(t)\ \geq \vartheta \\ 0 & otherwise \end{cases} \tag{2}$$

$$\tau_{RC}\frac{du_i}{dt} = -u_i(t) + \sum_i w_{ij}x_j(t) \tag{3}$$

Equation (2) defines the activation function of the LIF neuron, where $u_i(t)$ is the membrane voltage of neuron i at time t, and ϑ is a threshold value. Equation (3) defines the dynamics of the LIF membrane potential, where w_{ij} is the synaptic weight between neuron i and j, and $x_j(t)$ is the output value of neuron j at time t. Once the neuron reaches a voltage threshold, i.e., $u(t) \geq \vartheta$ it emits a spike. Then its voltage goes to a resting potential u_{rest} and Eq. (3) is not applied during a refractory period τ_{ref}. Other neuron models have been developed which are more biologically realistic such as the Izhikevich [16] or the Hodgkin-Huxley [14], amongst others.

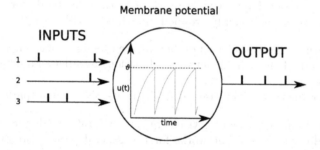

Fig. 1. Graphical representation of a single LIF neuron. Spikes arrive throughout time, where voltage $(u(t))$ is updated according to Eq. (3). Spikes are emitted when reached the threshold ϑ. After that, $u(t)$ goes to its resting potential.

3   Main Learning Approaches

Although the energy efficiency of these models is far superior, the achievement of competitive performance compared to ANNs in real-world problems is still difficult. The main reason is that training multi-layer SNNs is difficult as the spiking activation functions are non-differentiable, so gradient-descent methods are not directly applicable, as opposed to ANNs. In fact, it is hypothesised that the backpropagation algorithm is not biologically plausible [3], so our brain does not use this learning method. Therefore, throughout the literature researchers have developed a large amount of learning methods for SNNs [34]. In a nutshell, researchers try to find learning methods with a trade-off between its performance, its learning speed and its energy efficiency in the training and inference steps. According to this, the learning methods proposed in the literature can be roughly divided into three wide categories: local learning methods based on synaptic plasticity, learning using backpropagation and conversion of a well-trained ANN to it SNN counterpart.

3.1 Local Learning Methods Based on Synaptic Plasticity

These methods are based on biologically-plausible approaches where updates are performed locally using only the pre- and post-synaptic activities of the connected neurons following some kind of Hebbian learning [13]. One of the most employed ones is the spike-time dependent plasticity (STDP) rule [9]. These models are the most interesting ones as they leverage all the characteristics of SNNs: they are energy-efficient in both training and inference, they directly work with spikes and they are highly parallel due to its locality. However they present several drawbacks: they are mainly unsupervised, its learning process is slow, its implementation is non-trivial, they can be not suitable for some application due to its temporal nature, and their performance is still lower than its ANN counterpart. In fact, the development of these learning methods is an active research area nowadays.

3.2 Learning Using Backpropagation

Although SNNs' activation functions are not differentiable, these methods use different elements of the network to produce a differentiable function (also known as surrogate function) in order to employ backpropagation for weights update. For example, SpikeProp [6], one of the former methods proposed in the litera-ture, use the difference of arrival times between the desired spike and the actual one. Others employ the membrane potential $u(t)$ as the differentiable function [35], or they soften the derivative of the activation function [15], amongst oth-ers. These methods, although they are not biologically plausible, they are an interesting alternative for data scientists as the inference step can be efficiently carried out by means of spikes, whereas the training phase can be carried out by backpropagation in order to improve performance of the SNN. Nevertheless, it is non-trivial to develop such approaches, they are not widely available in SNN frameworks and their scalability is much more reduced than local methods.

3.3 Conversion of a Well-Trained ANN to Its SNN Counterpart

These methods tries to map the parameters of a pre-trained ANN into an equiv-alent SNN with the same architecture [7,24,27]. The idea is to minimise the performance error between both networks. Therefore, these methods leverage the high-performing and already trained ANN, and they obtain a highly-efficient SNN for inference with almost the same performance. However, they completely rely on an external training process using an energy-demanding deep learning process.

4   Frameworks for Spiking Neural Networks

The processing of inputs in an SNN is very similar to how the human brain is supposed to work. Due to this fact, together with the lower performance of SNNs in machine learning tasks with respect to ANNs, these networks have been mostly used by neuroscientists in simulation models of the human brain over the years. This is because they can faithfully simulate both neurons and synapses to determine, for example, the effect of a drug on the brain. For this reason, classical SNN libraries are highly oriented towards neuroscience in order to produce accurate simulation models. However, due to the increasing need for efficient machine learning algorithms, the development of SNN libraries oriented towards data science has recently increased. In this section, some of these libraries are analysed and compared, where its main characteristics are highlighted.

For the correct development of SNNs a number of key elements need to be determined, independently of its application. Therefore, in SNN frameworks the presence of several kinds of these elements should be expected. These elements are:

1. The encoding and decoding scheme. Numerical inputs must be translated into a set of spikes distributed over time, known as a spike train, and vice versa (optionally) [25].
2. The model of spiking neuron and its spatial distribution. Experts need to find a trade-off between biological realism and performance when using different neuron models. Moreover, neurons can be distributed in layers, like ANNs, or in groups following another shape, like a 2D mesh or a 3D cube for a more accurate representation of the brain.
3. The type of synapses and the flexibility in the connection configuration between neurons. Synapses can also be dynamically defined by ODEs, e.g., for simulating the effects of a neurotransmitter. In addition, the neurons can be connected one-to-one, one-to-many, many-to-many, and so on.
4. The type of learning algorithm for updating the network weights, if it is required.

In the following subsections, different SNN libraries are briefly analysed. A summary of its main characteristics is shown in Table 1.

4.1   Brian2

Brian2 [31] is a simulator for SNNs mainly oriented towards the simulation of neural processes in neuroscience. In Brian2, neurons and synapse dynamics

Table 1. Main features of the open-source SNN frameworks analysed.

Features	SNN Library						
	BindsNet	Brian2	ANNarchy	Norse	Nengo	NEST	SNN toolbox
Last version	0.2.9	2.4.2	4.6.10	0.0.6	3.1.0	2.20.0	0.6.0
Oriented towards	Data Science	Neuroscience	Neuroscience	Data Science	Data Science	Neuroscience	Data Science
Encoding types	Bernouilli	Poisson	As Brian2	Poisson	NEF	+10 types	Poisson
	Poisson	Custom			Population-based Custom		
	RankOrder						
	Custom						
Neuron Types	LIF-like	Custom	LIF-like	LIF	LIF	+50 types	LIF
	CSRM		Izhikevich	LSNN	Izhikevich		
	Ichikevich		Custom	Izhikevich	Spiking ReLU		
	McCulloch						
Learning Methods	Hebbian	Custom	As Brian2	SuperSpike	Oja	+10 local methods	-
	STDP				BCM		
					Rate-based backprop		
ANN-SNN Conversion methods?	Yes	No	No	No	Yes	No	Yes
Backends	CPU/GPU	CPU/GPU	CPU/GPU/MPI	CPU/GPU	CPU/GPU/MPI	CPU/MPI	CPU
		SpiNNaker			FPGA		SpiNNaker
					SpiNNaker		Loihi
					Loihi		
Programming Language	Python	Python	C++/Python	Python	Python	C++/Python	Python
Documentation	Scarce	Extensive	Extensive	Scarce	Extensive	Extensive	Scarce

can be individually defined by means of its ODEs, instead of writing Python code. Inputs can be also converted to spikes using ODEs, but it has several predefined methods such as Poisson generation. Under the hood, Brian2 solves these ODEs using some method such as Euler. This working scheme allows non-proficient Python users to use and configure Brian2 easily. Moreover, it provides a huge flexibility for researchers, as new neuron or synapse types can be added very quickly. However, it does not provide any predefined neuron, synapse or learning model. This implies that users must define all their models from scratch, which can be a major source of errors for the inexperienced user. Fortunately, it is a mature software with an extensive documentation. Brian2 code can be easily deployed on CPU, GPU or specific neuromorphic hardware, such as the SpiNNaker platform [22].

4.2 BindsNet

BindsNet [12] is a library for an easy development of biologically inspired algorithms for machine learning. Therefore, it implements several learning methods based on STDP for both machine learning and reinforcement learning. It employs PyTorch as backend, so the models can be easily deployed under CPU or GPU. It is based on three main elements for the development of networks: layers, connections and monitors. Layers, as in ANNs, are groups of neurons or input data encoders that can be arranged in a specified shape. Connections link layers between them. Learning methods are applied over connections. Finally, monitors allow to watch the state of the network over time. BindsNet implements a wide variety of neuron models, input encoding methods and connection types that allow data scientists to easily develop models for data mining tasks. Custom neurons, encoders, connections or learning rules can be developed by means of subclassing. This provides enough flexibility in the generation of models. Finally, it also provides a PyTorch ANN to SNN conversion method.

4.3   ANNarchy

ANNarchy [33] is a library for the development of distributed rate-based networks or SNNs. The models can be easily executed on CPU, GPU via CUDA or in a distributed fashion by means of openMPI. To build a network, the dynamics of the network's elements are defined by means of ODEs such as Brian2. After that, the network is compiled in efficient C++ code for a faster execution. As opposite to Brian2, it provides many built-in neuron, synapse and local learning rule types, e.g., Oja [21], BCM [4] or several STDP-based methods, that allow a faster, safer development of networks while keeping the flexibility of Brian2. For input data, it provides the same encoding methods as Brian2. Moreover, it provides a built-in decoding method based on computing the firing rate of the input neuron.

4.4   Nengo

Nengo [1] is a powerful software library that allows the development of SNNs following the principles of the neural engineering framework (NEF) [8]. It provides a series of built-in neuron models such as LIF or Izhikevich, amongst others. These neurons are grouped together in ensembles, where the NEF principles can be applied. Although the framework is oriented towards the NEF, it allows the user to build SNNs without applying the NEF optimisation, where several biologically plausible learning rules such as Oja or BCM are provided. Custom neurons or learning methods are created by means of subclassing. One of the main advantages of Nengo is the wide range of backends where its networks can be deployed. They can run on CPUs, GPUs, FPGAs and specific neuromorphic hardware such as SpiNNaker or Intel Loihi. In addition, Nengo provides a library called NengoDL [26], which provides a backpropagation-based SNN optimisation method for LIF neurons. It is based on smoothing its derivative according to its firing rate [15]. It also provides a conversion method from Keras ANN models to Nengo-based SNNs.

4.5   Norse

Norse [23] is a recent library built on top of PyTorch for the development of SNNs. Therefore, it can be executed on PyTorch supported devices. The development of SNNs in Norse is based on PyTorch's programming style. It provides several LIF-based and Izhikevich neuron models and a spiking version of an LSTM, called LSNN [2]. In addition, it also provides many input encoding methods such as Poisson or population-based coding. However, for maximising the use of the resources of PyTorch, it only provides a single learning algorithm which is a backpropagation-based algorithm called SuperSpike [35]. This method is based on the computation of the derivative with respect to the difference of the membrane potential $u(t)$ and its firing threshold ϑ.

4.6   Nest

Nest [11] is a powerful neural network simulation software written in C++ which provides a Python interface. Nest is highly specialised towards neuroscience brain simulations. The library provides more than 50 neuron models, more than 10 synapse models, i.e., local learning rules, and can only be executed in CPU with possibility of distributed, event-based computation via openMPI. Nest is also highly flexible with respect to the spatial distribution of neurons and how they are connected which allows the user to create highly customised SNNs. These elements can be easily configured by means of Python's dictionaries. Moreover, Nest provides a higher-level API called "topology" that allows an easy, fast creation of interconnected layers of neurons with specific parameters.

4.7   SNN Conversion Toolbox

As implied by the name, SNN conversion toolbox is a software tool that allows the conversion of Keras/PyTorch ANN models to its SNN counterpart. These SNNs can be used on some of the frameworks described previously such as Brian2 or Nest, amongst others, or deployed in neuromorphic hardware such as SpiNNaker or Loihi. The conversion method employed in this software for several ANN layers such as ReLU, Batch Normalization, Softmax, and so on are described in [27]. The converted SNN presents the same topology, i.e., number of connections and neurons, and only LIF neurons are employed. Finally, the software presents an extensive configuration in order to fine-tune the converted SNN model. However, its documentation is quite scarce.

5   Experimental Analysis

This section presents a comparison of the different libraries described in the previous section on a specific problem. In this section, the classic MNIST problem has been used as benchmark. As the learning methods provided by the different frameworks are not the same, this section will focus on how SNNs can be built, simulated and evaluated in order to perform a fair comparison of their characteristics. To do so, a simple ANN-SNN conversion process is carried out consisting of:

1. Build and train an ANN using Keras.
2. Save the ANN weights.
3. Build an SNN on each framework trying to replicate as much as possible the ANN.

4. Set the ANN connection weights as the synaptic weights of the SNN.
5. Perform the inference on SNN using the test data and compute its accuracy.

Specifically, the employed ANN is based on a multilayer perceptron with an input layer, a hidden, densely-connected layer with 128 ReLU units and an output, densely-connected layer with 10 units with the softmax activation that will return the classification value of each MNIST sample data. This ANN is trained by means of the Adam method with learning rate equals to 0.001 for 150 epochs. The loss function is the categorical cross-entropy.

Table 2. ANN-SNN layer correspondence on the MNIST problem.

ANN	SNN
Input (28, 28)	Poisson (784)
Flatten	–
Dense (128 ReLu)	Dense (128 LIF)
Dense (10 Softmax)	Dense (10 LIF)

The correspondence between ANN and SNN layer is shown in Table 2, the SNN counterpart will replicate the ANN architecture, where the ReLU and the softmax activations are replaced by LIF neurons using the default parameters of each framework. The encoding of the input data to spikes is carried out by means of an input layer with 784 Poisson neurons with a firing rate equal to the analog input value. For example, if pixel 100 of an MNIST sample has a value of 180, then the SNN input neuron number 100 will fire Poisson-distributed spikes 180 Hz. This SNN is not trained as the pre-trained ANN weights are used instead. Finally, each input sample is shown to the network for a period of 350ms with a resting time of 100ms whenever it is required. The classification value is the output neuron that fired the maximum number of spikes during the presentation time. All the source code to reproduce this experimental study is available at GitHub: https://github.com/agvico/prelim_snn_frameworks/tree/master.

Table 3. Comparison of accuracy and execution time of the Keras ANN and the SNNs developed in the different frameworks.

	ANN	Annarchy	BindsNet	Brian2	Nengo	Nest	Norse	SNN Toolbox
Accuracy	**0.9775**	0.9715	0.9684	0.9721	0.5491	0.9748	**0.9775**	0.7031
Execution time (s)	**0.4251**	213	1274	9468	1341	28246	**30**	2007

Table 3 presents the accuracy obtained from the SNN developed on the different frameworks. Also, the execution time in seconds for the processing of the 10000 MNIST test samples is shown. From these results, it can be observed that the proposed conversion process obtains almost the same quality as the ANN counterpart. In fact, the same quality has been obtained using Norse. This means that, at least on the MNIST problem, we can obtain huge energy reductions while keeping almost the same performance using this conversion method. However, we need to highlight the significant reduction of performance in SNN Toolbox and Nengo. In the former, it is required to perform additional configuration fine-tuning in order to obtain competitive results. Nevertheless, this is not straightforward as the framework documentation is quite scarce. In Nengo, it is important to remark that the framework is built towards the NEF. Therefore, it is possible that additional computation is performed which degrades the performance. In both cases, additional fine-tuning and knowledge about the internals of each framework is required in order to obtain competitive results.

Finally it is highlighted the wide difference of execution times between the different frameworks. It is important to note that, although SNNs are event-based, i.e., a neuron reacts only when a spike arrives, a computer is governed by a clock which updates its states at regular time intervals. Therefore, these frameworks work by a clock-based approach where the ODEs of all elements are solved for a specific timestep (1ms in this study) and updated at the same time. According to the results obtained, Norse is the fastest method due to all its operations are performed using PyTorch's tensors with an additional time dimension. Thus, it can leverage GPU processing for faster inference at the expense of high energy consumption. On the other hand, Nest and Brian2 are the slowest methods as they are highly-precise simulation frameworks for neuroscientists whereas BindsNet cannot leverage GPU processing on spiking neurons in its current version. In any case, the execution time can be reduced by optimising the trade-off between the presentation time and the accuracy.

6   Conclusions

The growing interest in SNNs has led to the emergence of a wide variety of SNN development libraries aimed at both data scientists and neuroscientists. In this work, seven of the main SNN development libraries have been analysed, in which their flexibility, their variety of spiking neurons and the different learning methods they possess are analysed. In addition, an experimental study is presented in which the ability to adapt a traditional neural network to its spiking version is tested. After analysing the results obtained, the conclusions of the study indicate that the more mature and flexible libraries, such as Brian2 and Nest, require more computation time due to the fact that they are strongly simulation-oriented. In contrast, there are libraries such as ANNarchy, Norse or BindsNet that present a good balance between computation time and quality, but they need more development to be competitive and reach a more general public. In any case, the results extracted are competitive in terms of performance while the energy consumption is significantly reduced due to the characteristics of SNNs. Therefore, developers and practitioners are encouraged to continue the development of such tools and networks in order to provide more sophisticated algorithms with better performance and energy consumption.

Acknowledgments. This work has been supported by the Regional Government of Andalusia, under the program "Personal Investigador Doctor", reference DOC_00235.

References

1. Bekolay, T., et al.: Nengo: a Python tool for building large-scale functional brain models. Front. Neuroinform. **7**, 48 (2014)
2. Bellec, G., Salaj, D., Subramoney, A., Legenstein, R., Maass, W.: Long short-term memory and learning-to-learn in networks of spiking neurons. In: Advances in Neural Information Processing Systems, pp. 787–797 (2018)
3. Bengio, Y., Lee, D.H., Bornschein, J., Mesnard, T., Lin, Z.: Towards biologically plausible deep learning. arXiv preprint arXiv:1502.04156 (2015)
4. Bienenstock, E.L., Cooper, L.N., Munro, P.W.: Theory for the development of neuron selectivity: orientation specificity and binocular interaction in visual cortex. J. Neurosci. **2**(1), 32–48 (1982)
5. Blouw, P., Eliasmith, C.: Event-driven signal processing with neuromorphic computing systems. In: IEEE International Conference on Acoustics, Speech and Signal Processing, pp. 8534–8538 (2020)
6. Bohte, S.M., Kok, J.N., La Poutré, H.: Error-backpropagation in temporally encoded networks of spiking neurons. Neurocomputing **48**(1–4), 17–37 (2002)
7. Diehl, P.U., Neil, D., Binas, J., Cook, M., Liu, S.C., Pfeiffer, M.: Fast-classifying, high-accuracy spiking deep networks through weight and threshold balancing. In: 2015 International Joint Conference on Neural Networks (IJCNN), pp. 1–8 (2015)
8. Eliasmith, C., Anderson, C.H.: Neural Engineering: Computation, Representation, and Dynamics in Neurobiological Systems. MIT Press, Cambridge (2003)
9. Feldman, D.E.: The spike-timing dependence of plasticity. Neuron **75**(4), 556–571 (2012)
10. Gerstner, W., Kistler, W.M.: Spiking Neuron Models: Single Neurons, Populations, Plasticity. Cambridge University Press, Cambridge (2002)
11. Gewaltig, M.O., Diesmann, M.: Nest (neural simulation tool). Scholarpedia **2**(4), 1430 (2007)
12. Hazan, H., et al.: Bindsnet: a machine learning-oriented spiking neural networks library in python. Front. Neuroinform. **12**, 89 (2018)
13. Hebb, D.O.: The Organization of Behavior: A Neuropsycholocigal Theory. Wiley, Hoboken (1949)
14. Hodgkin, A.L., Huxley, A.F.: A quantitative description of membrane current and its application to conduction and excitation in nerve. J. Physiol. **117**(4), 500–544 (1952)
15. Hunsberger, E., Eliasmith, C.: Training spiking deep networks for neuromorphic hardware. arXiv preprint arXiv:1611.05141 (2016)
16. Izhikevich, E.M.: Dynamical Systems in Neuroscience. MIT Press, Cambridge (2007)
17. Krizhevsky, A., Sutskever, I., Hinton, G.E.: ImageNet classification with deep convolutional neural networks. In: Advances in Neural Information Processing Systems 25, pp. 1097–1105 (2012)
18. Liu, C., et al.: Memory-efficient deep learning on a SpiNNaker 2 prototype. Front. Neurosci. **12**, 840 (2018)
19. Maass, W.: Networks of spiking neurons: the third generation of neural network models. Neural Netw. **10**(9), 1659–1671 (1997)

20. Nayak, P., Zhang, D., Chai, S.: Bit efficient quantization for deep neural networks. arXiv preprint arXiv:1910.04877 (2019)
21. Oja, E.: Simplified neuron model as a principal component analyzer. J. Math. Biol. **15**(3), 267–273 (1982). https://doi.org/10.1007/BF00275687
22. Painkras, E., et al.: SpiNNaker: a 1-W 18-core system-on-chip for massively-parallel neural network simulation. IEEE J. Solid-State Circuits **48**(8), 1943–1953 (2013)
23. Pehle, C., Pedersen, J.E.: Norse - a deep learning library for spiking neural networks, January 2021, documentation: https://norse.ai/docs/
24. Pérez-Carrasco, J.A., et al.: Mapping from frame-driven to frame-free event-driven vision systems by low-rate rate coding and coincidence processing-application to feedforward convnets. IEEE Trans. Pattern Anal. Mach. Intell. **35**(11), 2706–2719 (2013)
25. Petro, B., Kasabov, N., Kiss, R.M.: Selection and optimization of temporal spike encoding methods for spiking neural networks. IEEE Trans. Neural Netw. Learn. Syst. **31**(2), 358–370 (2020)
26. Rasmussen, D.: NengoDL: combining deep learning and neuromorphic modelling methods. Neuroinformatics **17**(4), 611–628 (2019). https://doi.org/10.1007/s12021-019-09424-z
27. Rueckauer, B., Lungu, I.A., Hu, Y., Pfeiffer, M., Liu, S.C.: Conversion of continuous-valued deep networks to efficient event-driven networks for image classification. Front. Neurosci. **11**(DEC), 682 (2017)
28. Rumelhart, D.E., Hinton, G.E., Williams, R.J.: Learning representations by back-propagating errors. Nature **323**(6088), 533–536 (1986)
29. Schwartz, R., Dodge, J., Smith, N.A., Etzioni, O.: Green AI. Commun. ACM **63**(12), 54–63 (2020)
30. Silver, D., et al.: Mastering the game of go without human knowledge. Nature **550**(7676), 354–359 (2017)
31. Stimberg, M., Brette, R., Goodman, D.F.: Brian 2, an intuitive and efficient neural simulator. Elife **8**, e47314 (2019)
32. Sze, V., Chen, Y.H., Yang, T.J., Emer, J.S.: Efficient processing of deep neural networks: a tutorial and survey. Proc. IEEE **105**(12), 2295–2329 (2017)
33. Vitay, J., Dinkelbach, H.Ü., Hamker, F.H.: ANNarchy: a code generation approach to neural simulations on parallel hardware. Front. Neuroinform. **9**, 19 (2015)
34. Wang, X., Lin, X., Dang, X.: Supervised learning in spiking neural networks: a review of algorithms and evaluations. Neural Netw. **125**, 258–280 (2020)
35. Zenke, F., Ganguli, S.: SuperSpike: supervised learning in multilayer spiking neural networks. Neural Comput. **30**(6), 1514–1541 (2018)
36. Zhang, Y., Suda, N., Lai, L., Chandra, V.: Hello edge: keyword spotting on micro-controllers. ArXiv preprint ArXiv:1711.07128 (2018)

Edge AI for Covid-19 Detection Using Coughing

J. A. Rincon[✉], V. Julian, and C. Carrascosa

VRAIN, Valencian Research Institute for Artificial Intelligence,
Universitat Politècnica de València, Valencia, Spain
{jrincon,vinglada,carrasco}@dsic.upv.es

Abstract. The emergence of the COVID-19 virus has placed the planet before one of the worst pandemics in 100 years. Early detection of the virus and vaccination have become the main weapons in the fight against the virus. In terms of detection, numerous alternatives have been proposed over the last one and a half years, including the use of artificial intelligence techniques. In this paper we propose the use of such techniques for virus detection using cough. The development of a low-cost device that incorporates the classification model has been proposed, facilitating its use anywhere without the need for connectivity.

Keywords: EDGE AI · COVID-19 · DeepLearning

1 Introduction

The new coronavirus-2019 (COVID-19) is an acute respiratory disease, which is responsible for more than 2.53 million deaths and has infected more than 114 million people, as of 1 March 2021[1]. It has dramatically affected business, economic and social dynamics around the world. All governments have imposed flight restrictions, social distancing and an emphasis on personal hygiene. However, COVID-19 continues to spread at a very rapid rate. COVID-19 presents a series of symptoms associated with viral infections, such as fever, cough, shortness of breath and pneumonia. As a respiratory disease, coronavirus includes acute respiratory distress syndrome (ARDS) or complete respiratory failure. This failure makes it imperative that those people affected by the virus require the support of mechanical ventilation and an intensive care unit (ICU). People with a compromised immune system, such as elderly people or people with chronic diseases, accelerate the occurrence of vital organ failure such as heart and kidney failure, in particular of the kidneys or septic shock [1].

The diagnosis of COVID-19 is performed by a reverse transcription-polymer chain reaction (RT-PCR) test, which is time-consuming. At the same time, it is, in some cases, inaccessible in some regions, either due to lack of medical

---

[1] https://www.who.int/docs/default-source/coronaviruse/situation-reports/20210309_weekly_epi_update_30.pdf.

© Springer Nature Switzerland AG 2021
H. Sanjurjo González et al. (Eds.): HAIS 2021, LNAI 12886, pp. 576–587, 2021.
https://doi.org/10.1007/978-3-030-86271-8_48

supplies, lack of qualified professionals to perform the test or lack of adequate health facilities.

Recently, Artificial Intelligence (AI) has been widely deployed in the health-care sector, allowing the use of different AI techniques to perform diagnostics with high accuracy rates. The use of deep learning as a diagnostic tool is one of the most promising applications. Many of these applications are directly related to image analysis for breast cancer diagnosis [2], retina health diagnosis [3] or the analysis of lung sounds and cough sounds for identification of pneumonia disease [4].

The use of these techniques has accelerated the early detection of different diseases, which could help save a large number of lives. Studies support that the use of breathing, speech and coughing can be used as inputs for machine learning models. These can be used to diagnose different diseases, including COVID-19 [5–7]. These signals could identify whether a person is infected by the virus or not, all using coughing sounds as information. The use of coughing sounds helps to make an early detection of this disease, which is performed by the doctor at the time of the first interaction with the patient. The physician can obtain these sounds using special devices for listening to the sound of the lungs such as stethoscopes or digital stethoscopes, which are able to store the sounds obtained. However, in order to interpret these sounds, it is necessary that the doctor would be able to distinguish a positive case from a negative case of COVID-19. This can be done quickly and inexpensively using appropriate devices and techniques.

Recent advances in embedded systems have led to a reduction in size, increased computing power, integration of WiFi, low energy Bluetooth (BLE) and the ability to integrate MEMS [8] (microelectromechanical systems) micro-phones. These new embedded devices can host small classification models inside them. This new technology is known as Edge-AI [9] or Artificial Intelligence at the edge. This technology allows these devices to perform classification on the fly, without having to send the acquired data to remote computer centres for further analysis. This new way of performing on-device classification decreases latency when sending the data and receiving the response. At the same time, it is very useful in places where there is no good 4G or 5G mobile phone coverage.

This new technology, which makes it possible to host a classification model within devices that are in direct contact with the user, has substantially increased the use of IoMT [10] (Internet of Medical Things) applications and especially in COVID-19 related applications. Raham et al. [11] have developed an affective computing framework that leverages the IoMT deployed in the user's home. Each of the nodes at the edge uses state-of-the-art GPUs to run deep learning applications at the edge, collecting user information and generating alerts for a multitude of COVID-19 related symptoms. Sufian et al. [12] present a study in which they explore the potentials and challenges of Deep Transfer Learning, Edge Computing, as well as related aspects to mitigate the COVID-19 pandemic. At the same time the authors propose a conceptual model of the scope and future challenges of working on critical sites and real data.

As the current pandemic is very new, there are a limited number of peer-reviewed studies and experimental results. Therefore, this paper proposes an early detection system for COVID-19 using cough. Concretely, we consider some previous studies, trying to make some contribution in the area of Edge-AI. The proposed system is able to discriminate COVID-19 positive cases from negative cases or environmental noise. Moreover, it includes an AI algorithm based on deep learning to help the health staff to make the diagnosis. Since the proposed system does not require connection to a WiFi network, it could be of particular interest for using in rural areas or in developing countries.

2   System Proposed

This section describes the proposed architecture for COVID-19 classification using coughing, on an Edge-AI device.

2.1   Hardware Description

The proposed hardware to perform COVID-19 classification using coughing consists of a microcontroller unit (MCU). The ESP32[2] is a feature-rich MCU with Wi-Fi connectivity and Bluetooth connectivity for a wide range of IoT [13] and IoMT [14] applications. The developed device integrates six MEMS microphones[3], which capture the cough signal generated by the user. This signal is captured by the ESP32 using the I2S protocol[4]. I2S, also known as Inter-IC Sound, Integrated Interchip Sound, or IIS, is an electrical serial bus standard used to interconnect digital audio circuits. The sound is captured at a frequency 22050 Hz, 16000 samples of the cough signal are captured. Once the patient's cough is captured, the system scrolls through the buffer where it is stored. The width of the window used by the system is 128, which corresponds to the number of inputs of the trained deep learning model (Fig. 1).

The Fig. 2 shows the designed device, showing the seven MEMS microphones, six of which are spaced at 45 degrees to each other and a seventh microphone in the centre. These microphones are covered by a protective cap to facilitate disinfection of the device after use. Early in the pandemic, one of the most common symptoms of patients diagnosed with COVID-19 was fever. For this reason, the system incorporates an MLX90614 infrared temperature sensor. This sensor measures two temperatures, one is the ambient temperature and the second is the temperature of the object, which in this case is the temperature of the user to be tested.

The Fig. 2 shows how the system presents the results to the user who is using the designed device. To do this, the system presents the results on a screen indicating the patient's temperature, the ambient temperature and finally the diagnosis.

---

[2] https://www.espressif.com/en/products/socs/esp32.

[3] https://www.cuidevices.com/catalog/audio/microphones.

[4] https://prodigytechno.com/i2s-protocol/.

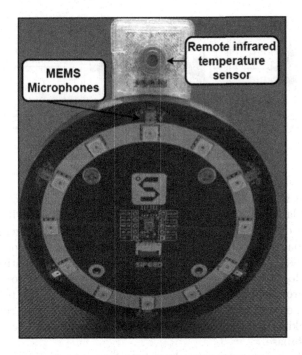

Fig. 1. Device designed for the detection of COVID-19 using cough.

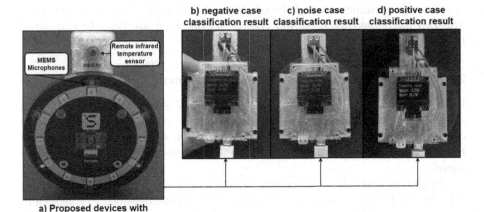

Fig. 2. Device developed for the detection of COVID-19.

2.2 Software Description

To perform the COVID-19 patient classification using cough as an input source, the Virufy COVID-19 Open Cough Dataset [15] was used. The clinical data obtained by the creators of the database are very accurate, as the data collection

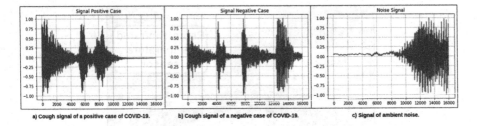

a) Cough signal of a positive case of COVID-19.   b) Cough signal of a negative case of COVID-19.   c) Signal of ambient noise.

Fig. 3. Cough signal in time domain.

was performed in a hospital under the supervision of physicians. The data is processed and labelled with COVID-19 status (supported by PCR tests), at the same time the database provides patient demographic data (age, sex, medical history). The database contains 121 segmented cough samples from 16 patients.

This database provides two classes with audio files for the positive COVID-19 cases and one for the negative cases. However, it was necessary to create a third class with environmental noise, so that the system would be able to discriminate noise from the environment in which the device was used. This third class was acquired by taking into account the sampling frequency parameters of the audio files in the database.

Several experiments were carried out in order to obtain a good classification rate. The signals were approached from the time and frequency domain. All this in order to determine which of the two allowed to obtain the best classification. The two methodologies used and the results obtained are described in detail below.

2.3   Cough Signal Analysis in the Time Domain

In this subsection we have analyse the results obtained by analysing the cough signals in the time domain. The Fig. 3 shows three signals extracted from the database in the time domain, Fig. 3a shows a cough signal from a positive case by COVID-19, Fig. 3b shows a negative case and finally Fig. 3c shows a noise signal.

To classify COVID-19 according to cough sounds, a deep sequential neural network [16] was created, the structure of the model obtained can be seen in Fig. 8. The network has 128 inputs, 16 neurons in the middle layer and three outputs, the hyperparameters of the presented model are described below: *epochs: 100, batch size: 8, optimizers model: adam, loss method: categorical_crossentropy* and *validation split: 0.2.*

For the training of the network, the audio signals were segmented and the first 16000 data were used. This number of data was used, because after this value the signal did not provide any important information for the network. When we had the new signal with a length of 16000, it was segmented into 125 signals with a length of 128 data, thus increasing the number of samples per signal, to perform the training the network (Fig. 4).

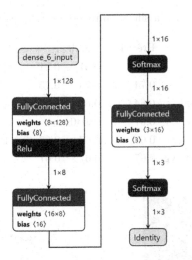

Fig. 4. Deep sequential network model for COVID-19 detection using coughing.

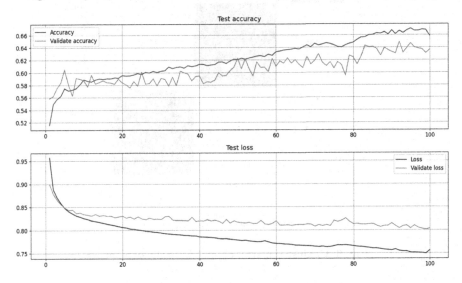

Fig. 5. Loss of accuracy obtained time domain.

The Fig. 5 shows the losses and accuracy of the training and validation processes. It can be seen that the accuracy is around 67%, which is an acceptable percentage, considering the simplicity of the network and the few audio samples.

Moreover, Table 1 shows the scores obtained in the validation phase, we can highlight that the model obtained has an acceptable accuracy percentage with an average accuracy of 67%.

Table 1. Classification report for time domain.

	Precision	Recall	F1-score	Support
Covid	0.62	0.61	0.61	7713
Noise	0.74	0.75	0.75	7681
Not-Covid	0.64	0.65	0.65	7646
Accuracy	0.67	23040		
Macro avg	0.67	0.67	0.67	23040
Weighted avg	0.67	0.67	0.67	23040

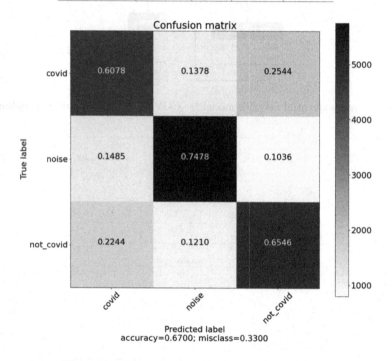

Fig. 6. Confusion matrix obtained in time domain.

Finally, Fig. 6 shows the confusion matrix calculated in the validation phase, it can be seen that the system classifies adequately. However, the system is unable to recognise 33% of the signals used.

2.4   Frequency Domain Cough Signal Analysis

In this subsection we have analyzed the results obtained by analysing the cough signals in the frequency domain.

The Fig. 7 shows three cough signal spectra, the Fig. 7a shows the spectrum obtained from a positive case, the Fig. 7b shows the spectrum of a negative case and finally the Fig. 7c shows a spectrum of ambient noise.

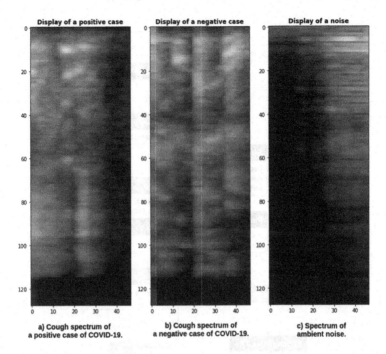

a) Cough spectrum of
a positive case of COVID-19.

b) Cough spectrum of
a negative case of COVID-19.

c) Spectrum of
ambient noise.

Fig. 7. Cough signal in frequency domain.

To classify COVID-19 from the frequency domain, a deep sequential neural network was created [16], the structure of the obtained model can be seen in Fig. 8. The network has 128 inputs, 400 neurons in the middle layer and three outputs, the hyperparameters to optimise the model are described below: *epocs: 300, batch size: 8, optimization model: rmsprop, loss method: mse, validation split: 0.2* and a *learning rate: 0.0001.*

For the training of the network, the same technique used for the analysis over time was used. The signal was segmented and the first 16000 data were used. This number of data was used, because after this value the signal did not provide any important information for the network. Once segmented, the new signal was divided into 125 signals with a length of 128 data, thus increasing the number of samples per signal, to perform the training of the network.

The Fig. 9 shows the losses and accuracy of the training and validation processes. It can be seen that the accuracy is around 77%, which is an acceptable percentage, considering the simplicity of the network and the few audio samples.

The Table 2 shows the scores obtained in the validation phase, we can highlight that the model obtained has an acceptable accuracy percentage with an average accuracy of 77%.

The Fig. 10 shows the confusion matrix calculated in the validation phase, it can be seen that the system classifies adequately. However, the system is unable to recognise 22% of the signals used.

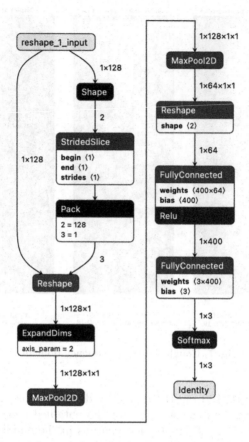

Fig. 8. Deep sequential network model for COVID-19 detection using coughing.

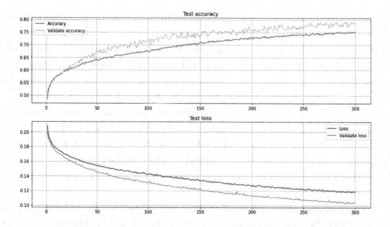

Fig. 9. Loss of accuracy obtained frequency domain.

Table 2. Classification report for frequency domain.

	Precision	Recall	F1-score	Support
Covid	0.70	0.89	0.78	958
Noise	0.92	0.80	0.86	577
Not-Covid	0.79	0.56	0.65	578
Accuracy			0.77	2113
Macro avg	0.81	0.75	0.77	2113
Weighted avg	0.79	0.77	0.77	2113

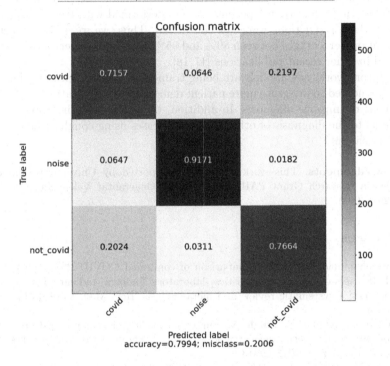

Fig. 10. Confusion matrix obtained in frequency domain.

3   Conclusions and Future Work

In this paper we have presented a COVID-19 classification system using cough. The system was designed taking into account new developments and techniques related to edge artificial intelligence or Edge-AI. The system was designed keeping in mind its simplicity in construction and programming as well as its low cost. As it is completely self-contained and does not require connectivity, its main use is proposed in remote locations to help to obtain a first fast diagnosis of COVID-19. To carry out this diagnosis our system incorporates an array of microphones, which are controlled by a microcontroller ESP-32. This microcontroller uses the signals captured by the microphones to discriminate the positive cases from the negative ones and from the environmental noise. This detection is performed by embedding a deep sequential neural network or SNN model inside the ESP-32. The classification model achieved by this network obtained an average accuracy of 67%. This classification rate is acceptable, considering the simplicity of the neural network and the device in which the model was embedded.

Since the main objective of this work was to embed the model inside an ESP-32 microcontroller, it was necessary to implement some restrictions to perform the training, such as the internal architecture of the network or the number of layers. These restrictions and the limited data made the percentage scores acceptable. However, other experiments were performed with the same data set in which these restrictions were not considered. Thus, we were able to achieve classification percentages between 89% and 95%, being values very close to those achieved in other recent publications [17,18].

As future work, new models are being trained with other datasets that have recently emerged to integrate more patient demographic information, such as age or gender, to improve diagnosis. In addition, we will extend the application of the model to the diagnosis of other types of diseases using cough or lung sounds from the patient.

Acknowledgements. This work was partly supported by Universitat Politecnica de Valencia Research Grant PAID-10-19 and by Generalitat Valenciana (PROME-TEO/2018/002).

References

1. Pormohammad, A., et al.: Comparison of confirmed COVID-19 with SARS and MERS cases-clinical characteristics, laboratory findings, radiographic signs and outcomes: a systematic review and meta-analysis. Rev. Med. Virol. **30**(4), e2112 (2020)
2. Chugh, G., Kumar, S., Singh, N.: Survey on machine learning and deep learning applications in breast cancer diagnosis. Cogn. Comput. 1–20 (2021). https://doi.org/10.1007/s12559-020-09813-6
3. Riaz, H., Park, J., Kim, P.H., Kim, J.: Retinal healthcare diagnosis approaches with deep learning techniques. J. Med. Imaging Health Inform. **11**(3), 846–855 (2021)
4. Muhammad, Y., Alshehri, M.D., Alenazy, W.M., Vinh Hoang, T., Alturki, R.: Identification of pneumonia disease applying an intelligent computational framework based on deep learning and machine learning techniques. Mob. Inf. Syst. **2021** (2021)
5. Deshpande, G., Schuller, B.: An overview on audio, signal, speech, & language processing for COVID-19. arXiv preprint arXiv:2005.08579 (2020)
6. Belkacem, A.N., Ouhbi, S., Lakas, A., Benkhelifa, E., Chen, C.: End-to-end AI-based point-of-care diagnosis system for classifying respiratory illnesses and early detection of COVID-19. arXiv e-prints, arXiv-2006 (2020)
7. Schuller, B.W., Schuller, D.M., Qian, K., Liu, J., Zheng, H., Li, X.: COVID-19 and computer audition: an overview on what speech & sound analysis could contribute in the SARS-CoV-2 corona crisis. arXiv preprint arXiv:2003.11117 (2020)

8. Zawawi, S.A., Hamzah, A.A., Majlis, B.Y., Mohd-Yasin, F.: A review of mems capacitive microphones. Micromachines **11**(5), 484 (2020)
9. Li, W., Liewig, M.: A survey of AI accelerators for edge environment. In: Rocha, Á., Adeli, H., Reis, L.P., Costanzo, S., Orovic, I., Moreira, F. (eds.) WorldCIST 2020. AISC, vol. 1160, pp. 35–44. Springer, Cham (2020). https://doi.org/10.1007/978-3-030-45691-7_4
10. Sun, L., Jiang, X., Ren, H., Guo, Y.: Edge-cloud computing and artificial intelligence in internet of medical things: architecture, technology and application. IEEE Access **8**, 101079–101092 (2020)
11. Rahman, M.A., Hossain, M.S.: An internet of medical things-enabled edge computing framework for tackling COVID-19. IEEE Internet Things J. (2021)
12. Sufian, A., Ghosh, A., Sadiq, A.S., Smarandache, F.: A survey on deep transfer learning to edge computing for mitigating the COVID-19 pandemic. J. Syst. Arch. **108**, 101830 (2020)
13. Ahsan, M., Based, M., Haider, J., Rodrigues, E.M.G., et al.: Smart monitoring and controlling of appliances using LoRa based IoT system. Designs **5**(1), 17 (2021)
14. Rodriguez, F., et al.: IoMT: Rinku's clinical kit applied to collect information related to COVID-19 through medical sensors. IEEE Latin Am. Trans. **19**(6), 1002–1009 (2021)
15. Fakhry, A., Jiang, X., Xiao, J., Chaudhari, G., Han, A., Khanzada, A.: Virufy: a multi-branch deep learning network for automated detection of COVID-19. arXiv preprint arXiv:2103.01806 (2021)
16. Denoyer, L., Gallinari, P.: Deep sequential neural network. arXiv preprint arXiv:1410.0510 (2014)
17. Feng, K., He, F., Steinmann, J., Demirkiran, I.: Deep-learning based approach to identify Covid-19. In: SoutheastCon 2021, pp. 1–4. IEEE (2021)
18. Kumar, L.K., Alphonse, P.J.A.: Automatic diagnosis of COVID-19 disease using deep convolutional neural network with multi-feature channel from respiratory sound data: cough, voice, and breath. Alex. Eng. J. (2021)

Unsupervised Learning Approach for pH Anomaly Detection in Wastewater Treatment Plants

Diogo Gigante[1]([⊠]) [iD], Pedro Oliveira[1] [iD], Bruno Fernandes[1] [iD],
Frederico Lopes[2], and Paulo Novais[1] [iD]

[1] ALGORITMI Centre, University of Minho, Braga, Portugal
{pedro.jose.oliveira,bruno.fernandes}@algoritmi.uminho.pt,
pjon@di.uminho.pt
[2] Águas do Norte, Guimarães, Portugal
frederico.lopes@adp.pt

Abstract. Sustainability has been a concern for society over the past few decades, preserving natural resources being one of the main themes. Among the various natural resources, water was one of them. The treatment of residual waters for future reuse and release to the environment is a fundamental task performed by Wastewater Treatment Plants (WWTP). Hence, to guarantee the quality of the treated effluent in a WWTP, continuous control and monitoring of abnormal events in the substances present in this water resource are necessary. One of the most critical substances is the pH that represents the measurement of the hydrogen ion activity. Therefore, this work presents an approach with a conception, tune and evaluation of several candidate models, based on two Machine Learning algorithms, namely Isolation Forests (iF) and One-Class Support Vector Machines (OCSVM), to detect anomalies in the pH on the effluent of a multi-municipal WWTP. The OCSVM-based model presents better performance than iF-based with an approximate 0.884 of Area Under The Curve - Receiver Operating Characteristics (AUC-ROC).

Keywords: Anomaly detection · Isolation Forest · One-Class Support Vector Machine · pH · Wastewater Treatment Plants

1 Introduction

Sustainability is a very striking area of the last decades, emerged in 1987 as a political concept and went back to a concern with the natural resources and the well-being of future generations [1]. With the world's population of 7.8 billion by 2020 and a projection of 9.8 billion by 2050, the preservation of natural resources becomes an exhaustive and complex task [2].

One of the essential natural resources is water, where the Wastewater Treatment Plants (WWTPs) play a crucial role in removing all microbes and contaminants present in sewers allowing the residual waters to be reused for various

© Springer Nature Switzerland AG 2021
H. Sanjurjo González et al. (Eds.): HAIS 2021, LNAI 12886, pp. 588–599, 2021.
https://doi.org/10.1007/978-3-030-86271-8_49

purposes such as urban, industrial and agricultural [3]. To guarantee the water quality, there are several sensors throughout the multiple phases of treatment in the WWTPs. However, in this facilities, sometimes the existing systems can't efficiently detect anomalies throughout the process.

Among the several substances present in WWTPs, pH is a fundamental substance representing the measurement of hydrogen ion activity. It ranges from 1 to 14, with 7 considered neutral, lower values considered acid and higher values considered alkaline [4]. This substance can be dangerous to the environment if not correctly treated. For example, if the pH value is low, this can indicate the presence of heavy metals, which leads to the environment's ability to foster life being reduced as human, animal and plant health become threatened [5]. So, the control of pH value on the effluent of the WWTP's is a crucial task since after this last phase, the treated residual waters are dumped into the environment. Hence, this work aims to conceive, tune and evaluate Machine Learning (ML) models, namely, Isolation Forests (iF) and One-Class Support Vector Machines (OCSVM), to detect pH anomalies in the treated effluent of the WWTP.

The remainder of this paper is structured as follows: the next section summarises state of the art, focusing on anomaly detection in WWTP. The third section presents the materials and methods, focusing on the collected dataset, the data exploration and pre-processing of the data, a brief review of the algorithms used and evaluation metrics to evaluate the conceived models. The fourth section demonstrates the performed experiences, and the fifth section discusses the obtained results from these experiences. The last section describes the significant conclusions taken from this work and present future work.

2 State of the Art

Several studies have been carried out on the topic of anomaly detection in WWTPs [6,7]. Dairi et al. [6] proposes a data-driven anomaly detection approach in the affluent conditions of a WWTP using deep learning methods and clustering algorithms. The authors present four approaches to the problem, namely, Deep Boltzmann Machine (DBM), Recurrent Temporal Restricted Boltzmann Machines (RBM), Recurrent Neural Networks (RNN) coupled with RBM and standalone Clustering techniques. For the first three types of model, various classifiers were tested to find the best possible solution to solve the problem. The dataset contains seven years (September 1, 2010, to September 1, 2017) with historical data from a WWTP in Thuwal, Saudi Arabia, with 21 features, in which the pH is included. This dataset has more than 150 abnormal events identified, with 9 of them are pH over the limit anomalies. To evaluate and compare the models, various metrics were used, such as True Positive Rate (TPR), False Positive Rate (FPR) and AUC (Area Under The Curve). In the end, the approach involving RNN-RBM produced the best efficiency and accuracy when compared with the other candidate models.

Another study by Muharemi et al. [8] describes several approaches to identify changes or anomalies occurring on water quality data. The authors deal with a dataset that contains 11 features among 122334 records from a German waters company. Even though it is not a uni-variate problem, the pH is present among other substances. Moreover, the data used in the training phase is highly imbalanced, containing less than 2% true events. In this type of problem, maximizing the accuracy is meaningless if we assume that rare class examples are much more important to be identified. In the data preparation phase, missing timesteps were not identified, and the missing values found were adequately treated, filling them out. Hence, the authors use a feature selection approach to measure the importance of each feature. Concluded the data preparation phase, the authors identify anomalies in the dataset with different models conceived, namely, Logistic Regression, Linear Discriminant Analysis, Support Vector Machine (SVM), Artificial Neural Network (ANN), RNN and Long Short-Term Memory (LSTM). For evaluation purposes, the F-score metric was used to compare the conceived models with a k-fold cross-validation technique. In the end, the authors concluded that all algorithms are vulnerable to imbalanced data, although SVM, ANN and logistic regression tend to be less vulnerable with F-score values of 0.36, 0.32 and 0.44, respectively.

Harrou et al. [7] carried out a case study in a WWTP located in Golden City of Colorado in the United States of America to present an innovative and efficient solution for the detection of failures using unsupervised deep learning models. The Golden WWTP aims to produce an effluent suitable for irrigation and, for that, needs to comply with specific standards regulated by local agencies. The data extracted from this WWTP correspond to one month (April 10 from 2010 to May 10, 2010), which results in 4464 observations with 28 variables. During this period, a fault is known to have affected pH and salinity. The authors selected seven variables with the help of expert recommendations from the extracted records, including pH. To achieve the objectives, a hybrid DBN-OCSVM approach was carried out. This approach combines Deep Belief Networks (DBN) and OCSVM, where flawless data is used to build the DBN model and, then, the output features of that model are used by OCSVM to monitor for anomalous events, which allowed the identification of the anomaly five days before the alert triggered for the operators. The authors concluded that the existence of a system with the capacity and performance similar to the presented approach, the operators would identify the anomaly, besides being possible to avoid the degradation caused as well as the two months of repairs that were necessary to normalize the situation.

Hereupon, in the revised articles, only one treated the missing values, used cross-validation in the training phase and performed feature selection. Moreover, none of the revised papers applied hyperparameter tuning to the candidate models.

3   Materials and Methods

The materials and methods used in this paper are described on the subsequent lines. The evaluation metrics are detailed as well as the two chosen ML models.

3.1 Data Collection

The dataset used in this study was provided by a portuguese water company. The data collection started on January 1, 2016, and was maintained until May 28, 2020. The observations represent a real scenario from a multi-municipal WWTP.

3.2 Data Exploration

The collected dataset consists of 1539 timesteps with different time intervals, and each observation consists of 2 features. The *date* feature in each record includes the date and time of the gathering. On the other hand, the *pH* feature consists of the value of pH. Table 1 describes the features of the collected dataset.

Table 1. Features of the collected dataset.

#	Features	Description	Unit
1	*date*	Timestamp	date & time
2	*pH*	Value of pH	Sorensen Scale

After analyzing the data, it is clear that there were missing timesteps that need to be treated. In total, there were 71 missing timesteps, with some of which corresponded to a period of two weeks, between May 4, 2016, to May 20, 2016, and a period of near two months with unrealistic values (e.g. values outside of Sorensen scale), between December 15, 2017, to February 2, 2018. Beyond missing timesteps, there were also missing values regarding *pH* value, with a total of 16 observations. Regarding the *ph* feature, a descriptive statistics is represented in the Table 2. It is possible to verify that the mean *pH* value is 7.047 and the standard deviation of 1.378. Relatively to the skewness, a value of -3.960 represents an asymmetric distribution and a negative inclination in the data distribution. In terms of kurtosis, the value of 14.838 indicates that the data follows a leptokurtic distribution.

Table 2. Descriptive statistics for *ph*

Num. of items	Mean	Median	Std. Deviation	Skewness	Kurtosis
1521	7.047	7.315	1.378	-3.960	14.838

To understand the periodicity of the *pH* in the effluent over the years, the average *pH* per quarter was analyzed, illustrate in Fig. 1. It is possible to verify that the lowest peaks of *pH* are found in the first quarter and, on the other hand, the highest peaks are mostly present in the third quarter. It is also visible that the year with the lowest values is 2018. Also, the first quarter has normal average values in the first two years but, in 2018, the value fell, returning to the usual values in the following years. In addition, the second quarter had a low average value in 2016, and however, in the next 4 years, the average values were around 7.

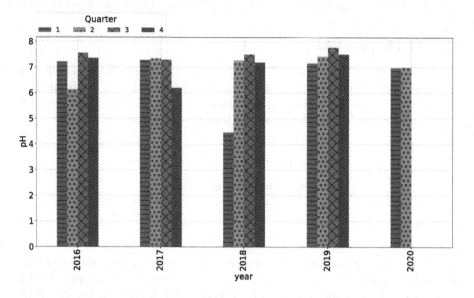

Fig. 1. *pH* variation per quarter and year.

3.3 Data Preparation

Regarding the data preparation, the first step consisted in applying feature engineering to create more features. From the *date* feature, we create three new features, namely, *day of month, month* and *year*. Moreover, with this new features, other five were created, namely, *day of the year, day of the week, season, quarter* and *semester*. Furthermore, taking into account the Kolmogorov-Smirnov test with a $p < 0.05$, it was verified that all the features in the dataset follow an non-Gaussian distribution.

Since the dataset, on average, only exists one record per day, we grouped the records by day, month and year. Therefore, the missing timesteps between the first and last records *date* were inserted. After inserting the missing days, the next step was treating the missing values. As previously mentioned, there were unrealistic values that we replaced with a null value. Hence, the total of missing values increased to 158. When the missing values amounted to four days or more, we removed the corresponding records. The remaining missing values were replaced with the mean of the previous seven days.

Concluded the treatment of missing timesteps and values, we verified the correlation with the target feature, and all features ended up not being used because the non-parametric Spearman's rank correlation coefficient with *pH* obtained a significantly low value. Hereupon, all the features were removed, ending up with a *pH* uni-variate dataset.

Finally, taking into account the use of an unsupervised approach, the data relating to the training phase did not have any type of labeling. However, to be able to evaluate the conceived models comparatively, data labelling is required. Therefore, we used a subset of data for testing where the observations were labelled as abnormal or normal, considering the decreed by the Ministry of the Environment of Portugal, where the pH value (effluent) must be between 6.0 and 9.0 [9]. Nevertheless, a value of pH in the effluent outside of this range, even not decreed, can also be an anomaly, such as, a sudden rise or decrease.

3.4 Evaluation Metrics

Since two ML models were compared, we needed an evaluation metric to obtain the best performance for each one. In this case, the two chosen evaluation metrics were the AUC-ROC (Receiver Operating Characteristics) and the F-score. The first metric is used in the evaluation of binary classification and anomaly detection problems which plots the TPR, also known as Recall (R) or Sensitivity, against FPR where the TPR is on the y-axis and the FPR is on the x-axis.

An excellent model has an AUC near to 1, which means it has a good measure of separability. On the other side, a poor model has AUC near 0, which means it has the worst separability measure [10].

The second metric, F-score, is the weighted average of Precision (P) and R. Hence, this score takes both false positives and false negatives into account. This metric is helpful when the problem contains uneven class distribution [11]. Hereupon, F-score consists in:

$$\text{F-score} = (1 + \beta^2) \times \frac{R \times P}{\beta^2 \times R + P} \tag{1}$$

Where the coefficient β corresponds to the balance between R and P with higher values favouring R.

In this case, F-score is used with $\beta = 1$, also called as F1-score. On the other hand, R equals the TP to divide by the TP plus FN. Lastly, P equals TP to divide by TP plus FP. An F1-score is considered perfect when its value is close to 1, which means that the presence of false positives and false negatives is very low. On the other side, the model is a total failure if the value is near 0.

3.5 Isolation Forest

As it is known, the most frequent techniques used to detect anomalies are based on the creation of a pattern of what is normal data and, consequently, identifying the points that don't match that profile.

However, the iF work on a different methodology consisting on the principle of the Decision Tree algorithm. The process of identifying anomalies is based in the isolation of the outliers from the data [12].

Hence, the architecture of this algorithm is illustrated in Fig. 2 and consists in dividing the data recursively until all instances are isolated. This random partitioning produces shorter paths for anomalies since the smallest number of anomaly occurrences results in a smaller number of partitions. Hereupon, when a forest of iTrees produces shorter path lengths for some specific points, then they are more likely to be abnormal events (anomalies) [13].

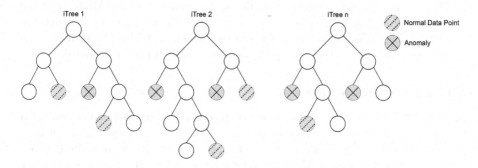

Fig. 2. iF architecture.

Therefore, iF algorithm provides some benefits, such as faster anomaly detection and less need for memory when compared to other anomaly detection algorithms. On the other hand, iF also has some disadvantages, such as, not considering relative importance of different variables, instead, it puts all the variables on equal footing. This drawback most probably degrades the effectiveness of the model [13].

3.6 One-Class Support Vector Machine

Based on the original SVM, the One-Class variation apply two ideas to ensure a good generalization, the maximization of the margin and the mapping of the data to a higher dimensional feature space induced by a kernel function.

Contrary to the iF algorithm, the OCSVM process of identifying anomalies is based on estimating a boundary region that comprises most of the training samples. If a new test sample falls within this boundary, it is classified as a normal record. Otherwise, it is recognized as an anomaly [14].

Regarding the OCSVM architecture, it is represented in Fig. 3 and consists in constructing a hyperplane that has the largest distance to the nearest training-data point of any class in a high or infinite dimensional space which can be used for detecting classification and anomaly detection. Using kernel tricks, this models guarantees to find a hyperplane that provides a good data separation [15].

Therefore, this algorithm has some advantages, such as, working well when there is a clear margin of separation between classes and is memory efficient. Otherwise, it also has some disadvantages, namely, it is not suitable for large datasets and has difficulties when the data has more noise [16].

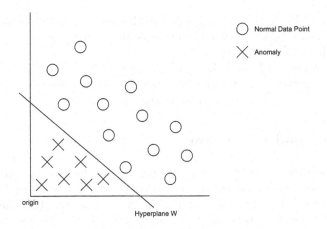

Fig. 3. OCSVM architecture.

4 Experiments

To achieve our goal, several iF and OCSVM candidate models were conceived. Several hyperparameters were tested in all of these models to find the best possible combination for each model.

The search for the best combination of hyperparameters is listed in Table 3 with the values considered.

Table 3. iF vs OCSVM hyperparameters searching space.

iF		
Parameter	**Values**	**Rationale**
Num. Estimators	[50, 100, 150, 200]	Number of base estimators
Max Samples	[100, 150, 200, 250, 300]	Maximum of samples
Contamination	[0.2, 0.25]	Contamination of the dataset
Bootstrap	[True, False]	Replacement of sampled data on fit
OCSVM		
Parameter	**Values**	**Rationale**
Kernel	[linear, poly, rbf]	Kernel type used in the algorithm
Nu	[0.2, 0.25, 0.3, 0.35]	Bound of fractions
Gamma	[auto, 0.001, 0.005, 0.01, 0.02]	Kernel coefficient

To perform this search, a GridSearchCV approach was used, which consists of a dictionary mapping estimator parameters to sequences of allowed values and, for each combination, k dataset splits are tested. On each data split, the

AUC-ROC and F1-score were calculated in the test set, and the average of the k splits for each estimator is used to evaluate the performance of the candidate models, with k value being equal to 5.

Python, version 3.7, was the programming language used for data exploration, data preparation, model conception and evaluation. *Pandas, NumPy, scikit-learn, seaborn* and *matplotlib* were the used libraries.

5   Results and Discussion

After the realization of multiple experiments, the results achieved are presented in Table 4, where the three best iF and OCSVM models are present.

Regarding the best iF-based model, the AUC-ROC was 0.878 and an F1-score of 0.854. On the other hand, the best OCSVM-based model obtained an AUC-ROC of 0.884 and an F1-score of 0.862. Therefore, both metrics over 0.8 is a very positive indicator.

Table 4. iF and OCSVM top-3 models.

	iF						
# Num. Estimators	Max Samples	Contamination	Bootstrap	AUC-ROC	F1-score	Time(s)
25 100	100	0.2	False	**0.878**	**0.854**	**0.343**
32 150	200	0.2	False	0.877	0.853	0.575
9 50	250	0.2	True	0.876	0.852	0.541
	OCSVM						
# Kernel	Nu	Gamma		AUC-ROC	F1-score	Time(s)
20 rbf	0.2	0.001		**0.884**	**0.862**	**0.017**
56 rbf	0.2	0.02		0.883	0.859	0.022
44 rbf	0.2	0.01		0.882	0.859	**0.017**

Overall, the OCSVMs showed better performance results. Moreover, the top-3 OCSVM models' hyperparameters are relatively homogeneous, with only the gamma value changing between the best candidate models. On the other side, iFs hyperparameters differ between the best candidate models, with the number of estimators and maximum samples being always different among them.

Beyond best performance, OCSVMs have a maximum of 0.022 s on the training phase, while iFs took at least 0.343 s to perform the same process.

Furthermore, when analyzing all the results, it was noted that in the iF-based models the performance decreases with higher contamination values. On the other side, the OCSVM-based models a lower performance is correlated with the poly and linear kernels. Concerning the other hyperparameters of the two methods there is no consistent pattern that can correlate with the used evaluation metrics values.

Hence, the best overall model is an OCSVM-based model that uses a rbf kernel, with a *nu* of 0.2 and a *gamma* of 0.02. Figure 4 illustrates the anomalies detected by the best OCSVM candidate model (in circles). For instance, it is also visible that all the values under 6 (minimum regulated value) were identified correctly as an anomaly. Although some records were identified incorrectly as an anomaly (FP). In this scenario, the major problem is determining an abnormal value as normal (FN), which does not happen in this prediction.

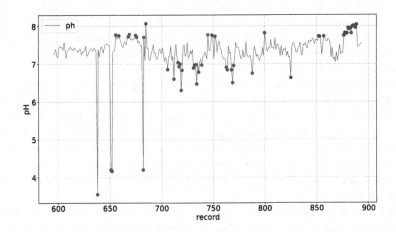

Fig. 4. Detected anomalies by the best OCSVM-based candidate model.

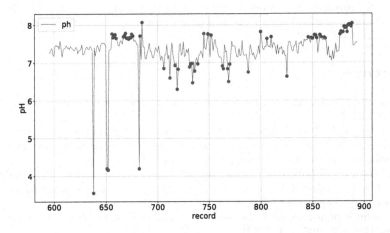

Fig. 5. Detected anomalies by the best iF-based candidate model.

On the other hand, on Fig. 5 are represented the anomalies identified by the best iF-based model. As in the best OCSVM-based model, all the values under 6 were identified as anomaly with the difference being the identification of more FP (normal events identified as anomaly).

These results prove that pH anomaly detection is reliable using uni-variate historical data of a WWTP effluent.

6   Conclusions

The treated effluent values represent how successful the prior treatments were, being this phase, the final goal in the WWTP process. Hence, detecting anomalies in this phase is an important task that can alert the operators about eventual abnormal events in the process. To accomplish this process, ML is used in this study to conceive models, namely, iFs and OCSVMs, to detect pH anomalies in the treated residual waters before they are released.

Multiple experiments were carried out to discover the better hyperparameter combination for each model to find the best candidate model. In the end, the best model was an OCSVM-based model that achieved an AUC-ROC of 0.884 and an F1-score of 0.862. Beyond the better performance, the OCSVM-based models took less time than the iF-based models in the training phase. Furthermore, with the present study it can be seen that it is possible to detect anomalies even inside the decreed limit range that represent a sudden rise or decrease of the pH effluent value with promising results.

With this in mind, the future work aims to conceive models with new algorithms, such as Long Short-Term Memory (LSTM) Autoencoders, and deploying the best candidate model. In addition, the investigation of anomaly detection in other substances of the WWTP effluent, such as Nitrogen (N), Biological Oxygen Demand (BOD), Chemical Oxygen Demand (COD), is planned.

Acknowledgments. This work is financed by National Funds through the Portuguese funding agency, FCT - Fundação para a Ciência e a Tecnologia within project DSAIPA/AI/0099/2019.

References

1. Kuhlman, T., Farrington, J.: What is sustainability? Sustainability **2**(11), 3436–3448 (2010)
2. van Vuuren, D.P., et al.: Pathways to achieve a set of ambitious global sustainability objectives by 2050: explorations using the IMAGE integrated assessment model. Technol. Forecast. Soc. Change **98**, 303–323 (2015)
3. Aghalari, Z., Dahms, H.-U., Sillanpää, M., Sosa-Hernandez, J.E., Parra-Saldívar, R.: Effectiveness of wastewater treatment systems in removing microbial agents: a systematic review. Glob. Health **16**(1), 1–11 (2020)
4. Harashit Kumar Mandal: Influence of wastewater pH on turbidity. Int. J. Environ. Res. Dev **4**(2), 105–114 (2014)
5. Masindi, V., Muedi, K.L.: Environmental contamination by heavy metals. Heavy Metals **10**, 115–132 (2018)
6. Dairi, A., Cheng, T., Harrou, F., Sun, Y., Leiknes, T.O.: Deep learning approach for sustainable WWTP operation: a case study on data-driven influent conditions monitoring. Sustain. Cities Soc. **50**, 101670 (2019)
7. Harrou, F., Dairi, A., Sun, Y., Senouci, M.: Statistical monitoring of a wastewater treatment plant: a case study. J. Environ. Manag. **223**, 807–814 (2018)
8. Muharemi, F., Logofătu, D., Leon, F.: Machine learning approaches for anomaly detection of water quality on a real-world data set. J. Inf. Telecommun. **3**(3), 294–307 (2019)

9. Ministério do Ambiente: Decreto-lei (no. 236/98). https://data.dre.pt/web/guest/pesquisa/-/search/430457/details/maximized. Accessed 12 Apr 2021
10. Narkhede, S.: Understanding AUC-ROC curve. In: Towards Data Science (2018)
11. Derczynski, L.: Complementarity, F-score, and NLP evaluation. In: Proceedings of the Tenth International Conference on Language Resources and Evaluation (LREC 2016), pp. 261–266 (2016)
12. Xu, D., Wang, Y., Meng, Y., Zhang, Z.: An improved data anomaly detection method based on isolation forest. In: 2017 10th International Symposium on Computational Intelligence and Design (ISCID), vol. 2, pp. 287–291. IEEE (2017)
13. Liu, F.T., Ting, K.M., Zhou, Z.-H.: Isolation forest. In: 2008 Eighth IEEE International Conference on Data Mining, pp. 413–422. IEEE (2008)
14. Shin, H.J., Eom, D.-H., Kim, S.-S.: One-class support vector machines-an application in machine fault detection and classification. Comput. Ind. Eng. **48**(2), 395–408 (2005)
15. Zhang, M., Xu, B., Gong, J.: An anomaly detection model based on one-class SVM to detect network intrusions. In: 2015 11th International Conference on Mobile Ad-Hoc and Sensor Networks (MSN), pp. 102–107. IEEE (2015)
16. Senf, A., Chen, X., Zhang, A.: Comparison of one-class SVM and two-class SVM for fold recognition. In: King, I., Wang, J., Chan, L.-W., Wang, D.L. (eds.) ICONIP 2006. LNCS, vol. 4233, pp. 140–149. Springer, Heidelberg (2006). https://doi.org/10.1007/11893257_16

A Hybrid Post Hoc Interpretability Approach for Deep Neural Networks

Flávio Arthur Oliveira Santos[1,3]($\boxtimes$) ⓘ, Cleber Zanchettin[1] ⓘ,
José Vitor Santos Silva[2] ⓘ, Leonardo Nogueira Matos[2] ⓘ, and Paulo Novais[3] ⓘ

[1] Universidade Federal de Pernambuco, Recife, Brazil
{faos,cz}@cin.ufpe.br
[2] Universidade Federal de Sergipe, São Cristóvão, Brazil
{jose.silva,leonardo}@dcomp.ufs.br
[3] University of Minho, Braga, Portugal
pjon@di.uminho.pt, flavio.santos@algoritmi.uminho.pt

Abstract. Every day researchers publish works with state-of-the-art results using deep learning models, however as these models become common even in production, ensuring fairness is a main concern of the deep learning models. One way to analyze the model fairness is based on the model interpretability, obtaining the essential features to the model decision. There are many interpretability methods to produce the deep learning model interpretation, such as Saliency, GradCam, Integrated Gradients, Layer-wise relevance propagation, and others. Although those methods make the feature importance map, different methods have different interpretations, and their evaluation relies on qualitative analysis. In this work, we propose the Iterative post hoc attribution approach, which consists of seeing the interpretability problem as an optimization view guided by two objective definitions of what our solution considers important. We solve the optimization problem with a hybrid approach considering the optimization algorithm and the deep neural network model. The obtained results show that our approach can select the features essential to the model prediction more accurately than the traditional interpretability methods.

Keywords: Deep learning · Optimization · Interpretability · Fairness

1 Introduction

Deep learning [6] models shows state-of-the-art results in a wide range of domains, for example natural language processing [1], computer vision [3], and recommendation systems [23]. Its success is not limited to research projects. Nowadays, deep learning models have become part of people's daily lives through the countless products they use on the web and that are present on their smartphones[1]. Despite this success, these models are hard to be adopted in some

---

[1] https://www.technologyreview.com/2015/02/09/169415/deep-learning-squeezed-onto-a-phone/.

© Springer Nature Switzerland AG 2021
H. Sanjurjo González et al. (Eds.): HAIS 2021, LNAI 12886, pp. 600–610, 2021.
https://doi.org/10.1007/978-3-030-86271-8_50

specific domains, such as healthcare, law systems, and trading systems. This specifics system needs the interpretation of the model decision. The user needs at least an indication of why the model has made the prediction. The difficulty for the deep learning models to be adopted in these domains is that they are black-box models. But why are they black boxes? It's because they generally use high dimensional data, have a large number of processing layers (deeper), and uses a non-linear function. Hence, it becomes challenging to interpret the reason why they returned a certain prediction.

As the black box characteristic is a problem in deep learning models, more research is needed. In the past years, methods were proposed to open the black box and return the interpretability of the model decisions giving us an indication of the feature-degree of importance to the model decision. Some of the interpretability methods proposed are Deconvolution [24], Saliency [17,18], Guided Backpropagation [19], GradCam [15], Integrated Gradients [22], Deep Lift [16], Layer-wise relevance propagation (LRP) [12], Semantic Input Sampling for Explanation (SISE) [14], and Adaptive Semantic Input Sampling for Efficient Explanation of Convolutional Neural Networks (AdaSISE) [21]. Given a trained neural network f and a input vector x, those methods produce a interpretability map (or attribution maps) map whose dimension is equal to x dimension and the map_i value in i position means how much important is the feature x_i to $f(x)$ prediction. Due to the subjectivity of the term *importance*, the evaluation of interpretability methods is mostly qualitative. Sometimes it is hard to evaluate the interpretation because different interpretability methods produce different interpretations from the same model and input. Thus, due to its subjectivity, in this work, we propose a direct hybrid approach combining optimization methods with the deep neural network to select the features which are responsible for producing the model prediction.

The first definition says that if a region r is relevant to the model prediction, the model prediction will decrease if we erase its information. Next, the second definition shows how we can compare the importance of two distinct regions.

- **Definition 1** Given a model $f : R^n \rightarrow \{0,1\}$. A region r is important to f prediction only if $f(x) > f(degrade(x,r))$.
- **Definition 2** A region r_i is more important than r_j to f prediction only if $f(degrade(x,r_i)) < f(degrade(x,r_j))$.

As we propose to combine optimization methods with the deep neural network model, we describe the deep learning interpretability problem as an optimization problem, as follow:

$$\arg\max_{mask} Importance(mask) = f(mask \odot x + (1 - mask) \odot C) \qquad (1)$$

To solve the *argmin* problem in the Eq. 6 we can employ optimization algorithms such as Hill Climbing [4], Ant Colony [2], Genetic Algorithms [5], Particle Swarm [8], and others.

This work is structured in the following way: Sect. 2 present the related work most important to our method. Next, we explain our proposed approach in detail. In Sect. 3, we show the experiments performed in this work and discuss the results obtained. Finally, in Sect. 4, we present the conclusion.

2   Related Works

One of the first works in interpretability is the approach proposed in [20]. It obtains the model interpretability through the computation of Shapley Values. The work as follow First, it selects a permutation from the features in a random way and, gradually, adds each feature in a reference point. Next, it computes the model output before and after the feature is added, then computes the difference between those two outputs to obtain the feature importance. We repeated this process N times, and the final feature importance is the average of the N execution.

As we can see, the method presented before is timing-consuming because it needs to be executed during N times and should use different permutations. In the last years, some works have been proposed in this branch of study; for example, Deconvolution [24] proposed a modified gradient to obtain the model interpretations. It can be seen as a convolutional neural network [11] but in the reverse order. Thus, instead of mapping the input in the output, it maps the output in the input (pixel-space). Saliency (or Vanilla Gradient) [17,18], presented in Eq. 2, performs an execution close to the Deconvolution, but instead, if set the negative gradients to 0, it changes to 0 when the respective neuron output value in the forward step is negative; that is, it follows the derivative of the ReLu [13] function. The method Guided Backpropagation [19] combines both methods, Deconvolution and Vanilla Gradient, and it is set to zero when both the gradients or the respective output neuron value are negatives.

$$Saliency(x, f, c) = \frac{\partial f_c}{\partial x} \qquad (2)$$

Gradient-Weighted Class Activation Mapping (Grad-Cam) [15] uses the gradient information to produce the attribution maps. Its maps highlight the most important parts of the input vector. The most important difference between Grad-Cam and the methods presented so far is because Grad-Cam can point up to the most discriminative parts of the predicted category. In contrast, the others methods point to specifics details of the input vector. To produce the attribution maps, Grad-Cam (Eqs. 3 and 4) computes the gradient of the model output c with relation to the feature map A_k ($\frac{\partial y^c}{\partial A^k}$), compute the weight ($\alpha_k$) of each feature map A_k and calculate the combination of the respective gradients using all α_k as weights. Although the Grad-Cam can find discriminative regions in the input vector, it fails to see important details. Thus, the authors propose to use Grad-Cam jointly with Guided Backpropagation, which is named Guided Grad-Cam.

$$\alpha_k^c = \frac{1}{Z} \sum_i \sum_j \frac{\partial f_c}{\partial A_{ij}^k} \qquad (3)$$

$$L^c_{Grad-CAM} = ReLU(\sum_k \alpha^c_k A^k) \tag{4}$$

Different from the approaches presented so far, [22] proposed an axiomatic method for interpretability called Integrated Gradients (IG). They offered two axioms that all interpretability methods must satisfy (according to the IG authors). The axioms are Sensitivity and Implementation invariance. The sensitivity axiom says that if two images differ only in a single pixel and produce different predictions, the interpretability method must attribute non-zero importance to this pixel. The implementation invariance shows that the interpretability method must be robust to the implementation. In other words, different implementations of the same method must return equal interpretations. Given an input vector x and a baseline vector x, IG obtain the interpretability accumulating the gradients of every point in the straight line from x to x, as presented in the Eq. 5.

$$IntegratedGrads_i ::= (x_i - x'_i) \times \int_{\alpha=0}^{1} \frac{\partial F(x' + \alpha \times (x - x'))}{\partial x_i} d\alpha \tag{5}$$

3   Iterative Post Hoc Attribution

In this work, we proposed the method Iterative Post Hoc Attribution (IPHA). IPHA consists of viewing deep learning interpretability as an optimization problem and using optimization algorithms to find a solution. The Algorithm 1 and Fig. 1 presents the IPHA approach. It receives the deep learning model f and an input vector x and returns two versions of the vector x, one with the important features ($x\_important$) and the other with only the non-important features of x ($x\_non\_important$). Next, we present a more formal definition of our optimization view.

Formally, given a trained deep neural network f and an input image x, we would like to know the pixels from x that most contribute to the $f(x)$ output according to definitions 1 and 2. Thus, we can model this question as an optimization problem present in the Eq. 6.

$$\underset{mask}{\arg\max}\, Importance(mask) = f(mask \odot x + (1 - mask) \odot C) \tag{6}$$

Since we want to select the important pixels in x, the mask parameter in the Eq. 6 is a vector composed of 0 and 1 with the exact dimension of x, that is $mask \in R^n$. The vector C is a constant vector responsible for filling the information in x removed by mask. C is a constant because we do not want to insert any new pattern in x, so it needs to be unbiased. If we search in the space $\{0,1\}^1$ to select the mask which maximizes the $Importance(mask)$ function, we should evaluate 2^n masks. This has a high computational cost for the analysis and can be impractical. However, there is a class of mathematical optimization techniques to mitigate this computational cost, named local search. The local search methods are iterative algorithms that begin with an arbitrary solution to the problem and make small changes to find better solutions.

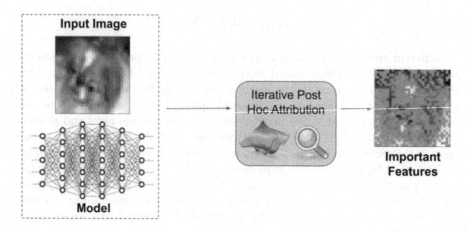

Fig. 1. Overview of the iterative post hoc attribution method.

Algorithm 1: Iterative post hoc attribution.

Input: Given a trained deep learning model f, input vector x, constant vector
        C, and an optimizer algorithm
$P_y \leftarrow f(x)$;
$eval \leftarrow lambda\, x, c, mask : f(mask \odot x + (1 - mask) \odot C)$;
$important\_features \leftarrow optimizer(f, x, c, eval)$;
$x\_important \leftarrow important\_features \odot x + non\_important\_features \odot C$;
$x\_non\_important \leftarrow non\_important\_features \odot x + important\_features \odot C$;
return $x\_important, x\_non\_important$

Using local search methods, we can mitigate the 2^n (n is the number of features) cost and find an approximate solution to our problem. We employ the local search method Hill-Climbing [4] jointly with the neural network f to obtain preliminary results. The Algorithm 2 presents our hill-climbing implementation.

As we can see from the Algorithm 2, the *get_neighbors* function is a significant part of this method because it generates new solutions from the actual best solution. Since our solution is a mask composed of 0 and 1, to create a new neighbor, we randomly select a position in the mask and change to 0 a grid of dimension 2×2 around it. Another important part of our solution is the *eval* function. We have defined it as the *Importance* function present in the Eq. 6.

Algorithm 2: Hill Climbing.

Input: $f, x, C, num\_iterations, num\_neighbors$

$best\_mask \leftarrow random(x.shape, 0, 1)$;

for $i \leftarrow 0; i < num\_iterations$ **do**

    $neighbors \leftarrow get\_neighbors(best\_mask, num\_neighbors)$;

    $next\_eval \leftarrow -INF$;

    $next\_node \leftarrow NULL$;

    for $mask \in neighbors$ **do**

        if $next\_eval < eval(mask)$ **then**

            $next\_eval \leftarrow eval(mask)$;

            $next\_node \leftarrow mask$;

    if $eval(best\_mask) < next\_eval$ **then**

        $best\_mask \leftarrow next\_node$;

return $best\_mask$

---

4 Experiments and Results

To evaluate our proposed method, we have compared the mask obtained from the Hill-Climbing with a wide range of interpretability methods available in the literature. The chosen methods are Saliency, Guided Backprop, GradCam, Guided GradCam, and Integrated Gradients. Since the mask represents the location of the important features, we compute the model outputs in two scenarios: (i) when we use only the pixels of the mask and (ii) when we do not use the features in the mask (i.e., (1 - mask)). In the first scenario, since it obtains the most important features, the model output must be high (or close to when using the full features), while the model output must be low in the second scenario. To perform the experiments, we have trained a ResNet [7] model with the well-known CIFAR-10 [10] dataset. Next, we will explain the dataset and model architecture used.

CIFAR10 [10] is an image classification dataset that is composed of 60000 train and 10000 test images. It has ten categories and is a balanced dataset. That is, each class has 6000 and 1000 images for training and testing, respectively. It is essential to highlight that all images are in RGB format and has three channels with 32×32 dimension. Figure 2 presents a sample of some images present in the CIFAR10 dataset.

Residual neural network (ResNet) [7] is a class of artificial neural networks that employ a residual learning framework to train deeper neural networks. ResNet employs skip-connections to mitigate the vanishing gradient problem and continue learning, achieving better accuracy in image classification tasks. In this work, we used a ResNet-18 architecture and trained the model with the CIFAR-10 dataset during 50 epochs. We have used the Adam [9] optimizer with a learning rate of 0.001

airplane
automobile
bird
cat
deer
dog
frog
horse
ship
truck

Fig. 2. Samples of the CIFAR-10 dataset. Source: [10]

In the Eq. 6 we have presented our optimization problem. The C vector is a constant that must be a neutral value. To be fair, in the experiments, we have tried different types of the constant vector. In the Fig. 3 we present all types of constant used: the noise value is black (0), Gaussian, Normalization mean, and white (1). The rows represent the types of noise, and the columns represent the mask with the most important pattern according to each interpretability method. It is important to highlight that to obtain the most important features from the Saliency, GradCam, Guided Backpropagation, Guided GradCam, and Integrated Gradients, we have computed the interpretability and selected the top-k features, where k is 512 because we have 1024 features in total.

To evaluate our method, we have defined the Feature Impact index (FII) present in Eq. 7. The goal of FII is to compute the absolute distance between the model output with the complete information $f(x)$ and the model output using only the selected features $f(x_{selected})$. The selected feature may be the (i) non-important features or the (ii) important features. Thus, when we use the non-important as the selected feature, we expect that the FII returns a low value. However, when we use the important features, the FII must be higher.

$$FeatureImpactIndex(f, x, x_{selected}) = abs(f(x) - f(x_{selected})) \qquad (7)$$

4.1   Results and Discussion

The Fig. 4 present the results of the analysis from the mask with the less important pixels. As the mask indicates the less important pixels, we expect that the model considering only those pixels is low, so the difference between the original model output and the output based on the non-important features will be higher. Since we believe only the non-important pixels, we need to replace the important with some neutral value. Thus we use different constant values and produce the results. Each y-axis in the graph means the impact of the model when used the respective constant type. From the results, we can see that the Hill Climbing (HC) approach has found the less important features and achieved almost 95% when we consider the 0 value as constant. Besides, all results obtained from the HC method are higher than 80%. Although the others interpretability method are established in the literature, they did not achieve any results above 80%.

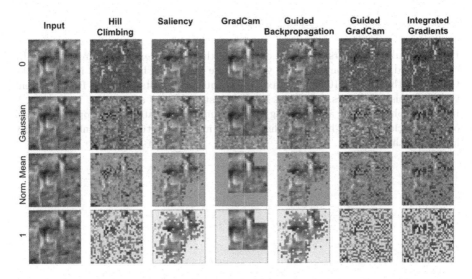

Fig. 3. Samples of noisy types used in the experiments.

From the Fig. 5 we can see the results from the analysis of when we are using the mask pointing to the important features. Since the mask point to the important features, we expect that the model output is high. So, the difference between the original model output and the output based on the important features be lower. As in the scenario before, we also have used different constant values in the evaluation process. The results show that, in our evaluation scenario, the GradCam is the best method to indicate the important feature. Their average difference is lower than 30% in every constant value. When we consider only 0 as the constant value, the Hill-Climbing (HC) approach could find the non-important features and produce results close to the GradCam. However, in others scenarios, their outputs were at least 20% higher than the GradCam.

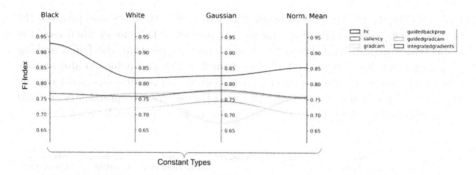

Fig. 4. Results of the first scenario, evaluating the *non-important* features selected by each method. The FII index in the y-axis is the average of between all images in the test set of the CIFAR-10. Each y-axis represents the results obtained from different types of constant values. Examples of the constant values are present in the Fig. 3. In this graph, since we are selecting the non-important features, we hope the model output be close to 0. Thus, as higher the FII index as better is the interpretability method in this scenario.

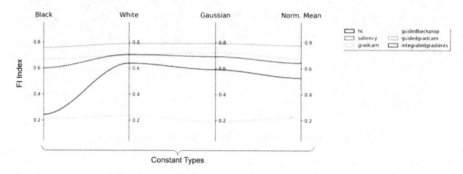

Fig. 5. Results of the second scenario, evaluating the *important* features selected by each method. As this scenario refers to the most important features, it is expected that the model output with the selected features is close to the model output with the full features. Thus, as lower the FII index as better is the interpretability method in this scenario.

The results obtained from the GradCam method in the second scenario are intriguing because it remained consistent in all the different types of constant. We believe that this result is because when we select the lowest top-512 features using the GradCam interpretation, the model selects squared regions, thus erasing a meaning pattern. On the other side, the other method selects specifics pixels.

5   Conclusion and Future Works

In this work, we proposed an optimization view of the deep learning interpretability problem. We have used a hybrid approach combining optimization algorithms with the deep learning model to solve this optimization problem. Our findings show that we can select the less relevant features to the model prediction more accurately than the interpretability methods used in the experiments, such as Saliency, GradCam, Guided Backpropagation, Guided GradCam, and Integrated Gradients. However, the GradCam presented better results in finding the relevant features, but our approach achieved results close to the GradCam in this second scenario.

The preliminary results are exciting, and the overall optimization view has several future works to be done. For example, we can work in the neighbor function of the optimizer HC to generate a more structural importance mask, analyzing the impact of the constant value, adding constraints in the solution such as fixing the number of features that the mask should have. In addition, we can use more sophisticated optimization algorithms (i.e., Genetic Algorithms, Ant Colony, Particle Swam) to solve the optimization problem.

Acknowledgments. This work has been supported by FCT – Fundaçño para a Ciência e Tecnologia within the R&D Units Project Scope: UIDB/00319/2020. The authors also thanks CAPES and CNPq.

References

1. Devlin, J., Chang, M.W., Lee, K., Toutanova, K.: BERT: pre-training of deep bidirectional transformers for language understanding. arXiv preprint arXiv:1810.04805 (2018)
2. Dorigo, M., Birattari, M., Stutzle, T.: Ant colony optimization. IEEE Comput. Intell. Mag. **1**(4), 28–39 (2006)
3. Dosovitskiy, A., et al.: An image is worth 16x16 words: transformers for image recognition at scale. arXiv preprint arXiv:2010.11929 (2020)
4. Gent, I.P., Walsh, T.: Towards an understanding of hill-climbing procedures for sat. In: AAAI, vol. 93, pp. 28–33 (1993)
5. Goldberg, D.E.: Genetic algorithms in search. Optim. Mach. Learn. (1989)
6. Goodfellow, I., Bengio, Y., Courville, A., Bengio, Y.: Deep Learning, vol. 1. MIT Press, Cambridge (2016)
7. He, K., Zhang, X., Ren, S., Sun, J.: Deep residual learning for image recognition. In: Proceedings of the IEEE Conference on Computer Vision and Pattern Recognition, pp. 770–778 (2016)
8. Kennedy, J., Eberhart, R.: Particle swarm optimization. In: Proceedings of ICNN 1995-International Conference on Neural Networks, vol. 4, pp. 1942–1948. IEEE (1995)
9. Kingma, D.P., Ba, J.: Adam: a method for stochastic optimization. arXiv preprint arXiv:1412.6980 (2014)
10. Krizhevsky, A., Hinton, G., et al.: Learning multiple layers of features from tiny images (2009)
11. LeCun, Y., Bengio, Y., et al.: Convolutional networks for images, speech, and time series. Handb. Brain Theory Neural Netw. **3361**(10), 1995 (1995)
12. Montavon, G., Binder, A., Lapuschkin, S., Samek, W., Müller, K.-R.: Layer-wise relevance propagation: an overview. In: Samek, W., Montavon, G., Vedaldi, A., Hansen, L.K., Müller, K.-R. (eds.) Explainable AI: Interpreting, Explaining and Visualizing Deep Learning. LNCS (LNAI), vol. 11700, pp. 193–209. Springer, Cham (2019). https://doi.org/10.1007/978-3-030-28954-6_10

13. Nair, V., Hinton, G.E.: Rectified linear units improve restricted Boltzmann machines. In: Proceedings of the 27th International Conference on Machine Learning (ICML-10), pp. 807–814 (2010)

14. Sattarzadeh, S., et al.: Explaining convolutional neural networks through attribution-based input sampling and block-wise feature aggregation. ArXiv abs/2010.00672 (2020)

15. Selvaraju, R.R., Cogswell, M., Das, A., Vedantam, R., Parikh, D., Batra, D.: Grad-CAM: visual explanations from deep networks via gradient-based localization. In: Proceedings of the IEEE International Conference on Computer Vision, pp. 618–626 (2017)

16. Shrikumar, A., Greenside, P., Kundaje, A.: Learning important features through propagating activation differences. In: Precup, D., Teh, Y.W. (eds.) Proceedings of the 34th International Conference on Machine Learning, ICML 2017, Sydney, NSW, Australia, 6–11 August 2017. Proceedings of Machine Learning Research, vol. 70, pp. 3145–3153. PMLR (2017). http://proceedings.mlr.press/v70/shrikumar17a.html

17. Simon, M., Rodner, E., Denzler, J.: Part detector discovery in deep convolutional neural networks. In: Cremers, D., Reid, I., Saito, H., Yang, M.-H. (eds.) ACCV 2014. LNCS, vol. 9004, pp. 162–177. Springer, Cham (2015). https://doi.org/10.1007/978-3-319-16808-1_12

18. Simonyan, K., Vedaldi, A., Zisserman, A.: Deep inside convolutional networks: visualising image classification models and saliency maps. In: Bengio, Y., LeCun, Y. (eds.) 2nd International Conference on Learning Representations, ICLR 2014, Banff, AB, Canada, 14–16 April 2014. Workshop Track Proceedings (2014). http://arxiv.org/abs/1312.6034

19. Springenberg, J.T., Dosovitskiy, A., Brox, T., Riedmiller, M.A.: Striving for simplicity: the all convolutional net. In: Bengio, Y., LeCun, Y. (eds.) 3rd International Conference on Learning Representations, ICLR 2015, San Diego, CA, USA, 7–9 May 2015. Workshop Track Proceedings (2015). http://arxiv.org/abs/1412.6806

20. Strumbelj, E., Kononenko, I.: An efficient explanation of individual classifications using game theory. J. Mach. Learn. Res. **11**, 1–18 (2010)

21. Sudhakar, M., Sattarzadeh, S., Plataniotis, K.N., Jang, J., Jeong, Y., Kim, H.: Ada-SISE: adaptive semantic input sampling for efficient explanation of convolutional neural networks. In: IEEE International Conference on Acoustics, Speech and Signal Processing (2021)

22. Sundararajan, M., Taly, A., Yan, Q.: Axiomatic attribution for deep networks. In: Precup, D., Teh, Y.W. (eds.) Proceedings of the 34th International Conference on Machine Learning, ICML 2017, Sydney, NSW, Australia, 6–11 August 2017. Proceedings of Machine Learning Research, vol. 70, pp. 3319–3328. PMLR (2017). http://proceedings.mlr.press/v70/sundararajan17a.html

23. Wu, L., Li, S., Hsieh, C.J., Sharpnack, J.: SSE-PT: sequential recommendation via personalized transformer. In: Fourteenth ACM Conference on Recommender Systems, RecSys 2020, p. 328–337. Association for Computing Machinery, New York (2020). https://doi.org/10.1145/3383313.3412258

24. Zeiler, M.D., Fergus, R.: Visualizing and understanding convolutional networks. In: Fleet, D., Pajdla, T., Schiele, B., Tuytelaars, T. (eds.) ECCV 2014. LNCS, vol. 8689, pp. 818–833. Springer, Cham (2014). https://doi.org/10.1007/978-3-319-10590-1_53

Deep Learning Applications on Cybersecurity

Carlos Lago, Rafael Romón, Iker Pastor López[✉], Borja Sanz Urquijo,
Alberto Tellaeche, and Pablo García Bringas

University of Deusto, Avenida de las Universidades 24, 48007 Bilbao, Spain
`iker.pastor@deusto.es`

Abstract. Security has always been one of the biggest challenges faced by computer systems, recent developments in the field of Machine Learning are affecting almost all aspects of computer science and Cybersecurity is no different. In this paper, we have focused on studying the possible application of deep learning techniques to three different problems faced by Cybersecurity: SPAM filtering, malware detection and adult content detection in order to showcase the benefits of applying them. We have tested a wide variety of techniques, we have applied LSTMs for spam filtering, then, we have used DNNs for malware detection and finally, CNNs in combination with Transfer Learning for adult content detection, as well as applying image augmentation techniques to improve our dataset. We have managed to reach an AUC over 0.9 on all cases, demonstrating that it is possible to build cost-effective solutions with excellent performance without complex architectures.

Keywords: Deep learning · Cybersecurity · Transfer learning · Image classification · NLP

1 Introduction

Security has always been one of the biggest challenges faced by computer systems, however in the present era, the world is at the beginning of a fourth industrial revolution based on IoT, this revolution has brought new challenges specially in the field of Cybersecurity. In a threat environment where security products need to be constantly refined or updated to identify the recent exploitation, the challenge is to find a solution that provides a future-proof defense [2]. Recent developments in the field of AI and Machine Learning offer us a new suite of tools to deal with the increasing demand for cybersecurity.

With this research we are aiming to showcase the benefits of applying Deep Learning to Cybersecurity, for this an observation of different projects will be carried out, in order to provide a good variety of examples we have chosen three wildly different use cases: Spam Filtering using LSTMs, Malware Detection with DNNs and Adult Content Filtering using CNNs and Transfer Learning, in the next few paragraphs we will be giving some context for each one of them.

© Springer Nature Switzerland AG 2021
H. Sanjurjo González et al. (Eds.): HAIS 2021, LNAI 12886, pp. 611–621, 2021.
https://doi.org/10.1007/978-3-030-86271-8_51

Digital spam has always been one of the main problems due to its broad scope, with the intention of deceiving or influencing users. Moreover, due to emerging technologies, there is a continuous change which leads to new ways of exploiting it [3].

The detection of malware is essential to improve security, new tools are needed to cope with new malware that is becoming practically impossible to detect, due to existing malware creation tool-kits like Zeus or SpyEye [22], this study performs its detection using deep neural networks.

Content filtering is traditionally defined as using a computer program to scan and block access to web pages o email deemed as suspicious, this type of filtering is often carried out by firewalls or antivirus. For this article, we are going to focus on adult content filtering, mainly because it has been one of the most problematic types of filtering [16], some reasons to filter adult content are:

- Allowing pornographic content in the workplace can generate a hostile work-ing environment and even put the company at risk of receiving a sexual harassment complaint.
- Allowing pornographic content on a platform like a social network can have legal repercussions, for instance in Spain going against article 186 of the penal code, which protects minors from exposure to this type of content [13].

The driving force behind this research is that we believe not enough people are taking advantage of this new set of techniques provided by Machine Learning, we hope that after showing how easy it is to implement a simple yet powerful solution more people will be encouraged to begin using them.

2   State of the Art

2.1   Malware

Previous studies show the amazing range of approaches that can be taken with Machine Learning for malware detection.

Convolutional transformation network for malware classification [19], the author proposes a new way of analyzing malware, converting binary files into images that are then analyzed by a convolutional network. The following image Fig. 1 shows how the files are displayed using different transformation techniques:

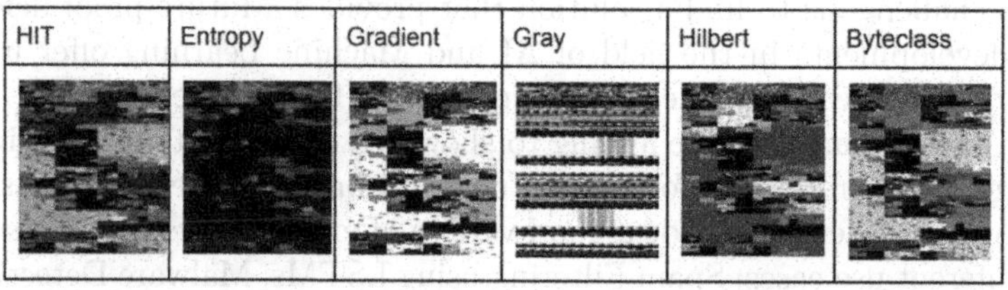

Fig. 1. Visualization by means of different transformations [19]

The best results are obtained through HIT, which extends the entropy color scheme in the previous section to capture more semantic information about PE files [18]. HIT can capture obfuscation information and represent semantic information in file headers as imported functions or libraries, achieving an accuracy of 99% in the validation, these are the average views that are obtained for both benign and malware files Fig. 2.

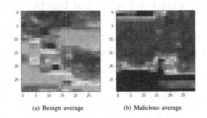

(a) Benign average   (b) Malicious average

Fig. 2. Average file visualization [19]

Ivan et al. [4], performed a comparison of several Machine Learning classifications techniques for malware detection, k-Nearest Neighbor, Naïve Bayes, Support Vector Machine (SVM), J48 decision tree, and Multilayer Perceptron (MLP) neural network. The best results were reached by using Decision Trees, reaching a recall of 95.9%, a precision of 97.3% and an accuracy of 96.8%.

2.2   SPAM

In [6] a comparative study for the detection of SPAM in SMS is carried out, the different techniques compared are:

- **Decision Tree:** this supervised learning algorithm is the preferred method for classification tasks.
- **Logistic Regression:** this technique is considered the best method for classifications with a binary result.
- **Random Forest:** this technique is used to generate decision trees.
- **AdaBoost:** or Adaptive Boosting is a learning algorithm used to increase the performance of a classifier by merging weaker classifiers into a stronger one.
- **Artificial Neural Network:** ANNs are nonlinear statistical data modelling techniques defined over complex relationships between inputs and outputs.
- **Convolutional Neural Network:** this particular type of ANN network uses perceptrons for supervised learning.

The results of this study show that CNNs have the highest performance out of all the techniques presented, this demonstrates that even though CNNs are traditionally associated with image classification this technique is also suited for other kinds of data and it has shown great improvements over more traditional classifiers.

Another example of a previous study is [1], this research is very interesting, having a dataset with tweets from certain microblogs related to the world of finance, it is shown that there is a great relationship in the tweets between NASDAQ stocks and OTCMKTS, which should not really be represented, since OTCMKTS are stocks with very low market capitalization. So spammers were trying to show that they were safer when put together with NASDAQ.

Sang et al. [10] propose a spam detection method using feature selection and parameters optimization, using Random Forest, it is necessary to optimize the number of variables in the random subset at each node (mtry) and the number of trees in the forest (ntree), for example, in the case of ntree values form 0 to 500 were tested.

2.3   Adult Content Filtering

An example of a possible application of Machine Learning for adult content filtering is presented on [21], in this paper J. Wehrmann et al. propose a method of adult content detection using deep neural networks called ACORDE, this method uses CNNs to extract features an LSTMs to classify the end result without needing to adjust or retrains the CNNs, the results of this paper are astounding, even ACORDES weakest version, ACORDE-GN, offers much better performance than it's competitors cutting by half it's false positives and reducing false negatives to a third, it's most powerful version sports an accuracy rate of 94.5% further cementing it as the best available technique of its time.

2.4   Neural Network Architectures

Convolutional Neural Networks. Convolutional neural networks (CNNs) have become the dominant Machine Learning approach for visual object recognition [9], they perform convolution instead of matrix multiplication (as with fully-connected neural networks) [15]. In this way, the number of weights is decreased, thereby reducing the complexity of the network. Furthermore, the images, as raw inputs, can be directly imported to the network, thus avoiding the feature extraction procedure in the standard learning algorithms. It should be noted that CNNs are the first truly successful Deep Learning architecture due to the successful training of the hierarchical layers [11].

In our case we have used them for the detection of adult content, as our filter is an image classifier, so its perfect for the use of CNNs.

LSTMS. Recurrent networks can in principle use their feedback connections to store representations of recent input events in form of activations ("short-term memory", as opposed to "long-term memory" embodied by slowly changing weights). This is potentially significant for many applications, including speech processing, non-Markovian control, and music composition [7].

LSTMs are designed to avoid the long-term dependency problem. They can easily remember information for long periods of time [5] and have a chain like structure, but the repeating module has a different structure. Instead of having a single neural network layer, there are four Fig. 3.

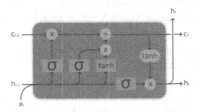

Fig. 3. LSTM architecture

We have used them for the detection of SPAM because of their ability to remember information, which is really convenient for finding semantic relatedness and sentiment analysis [20].

Deep Neural Networks. A deep neural network (DNN) is a conventional multi-layer perceptron with many hidden layers, optionally initialized using the DBN pre-training algorithm [17]. They can represent functions with higher complexity if the numbers of layers and units in a single layer are increased. Given enough labeled training datasets and suitable models, Deep Learning approaches can help humans establish mapping functions for operation convenience [11]. Thanks to the amount of available samples and the accuracies that can be reached thanks to deep neural networks we have used them for the malware detection case.

3   Dataset Preparation

3.1   Data Gathering

Malware. For the case of malware detection a dataset[1] containing 200,000+ executables has been used, the files are divided into malware and secure. Concretely, the 50% of the instances fall into the Malware class, both in train (80%) as in test (20%) sets This separation is done by a scan of each file by VirusTotal. In this tool, when a file is uploaded, it is scanned by different antivirus, in the case of the dataset, the files in which more than 10 detections occur are marked as malware and the others as safe.

The files are coded to have a heterogeneous array of features, the structure would be as follows:

---

[1] https://drive.google.com/file/d/1HIJShr0GvQCUp_0R_kQe_WLG5PippurN/view.

1. Has configuration: a 1 if the file has Load Configuration.
2. Has debug: to 1 if the file has a debug section, which contains debugging information generated by the compiler.
3. Has exceptions: to 1 if the file uses exceptions.
4. Has exports: to 1 if the file has any exported symbols.
5. Has imports: a 1 if the file imports symbols.
6. Has nx: to 1 if you are using the NX bit, which is used in CPUs to segregate memory areas and mark certain parts as non-executable.
7. Has relocations: to 1 if the file has relocation entries, which specify how the section data is to be changed.
8. Has resources: a 1 if the file uses any resources.
9. Has a rich header: at 1 if there is a rich header, it is a PE section that serves as a fingerprint of the Windows executable compilation environment.
10. Has a signature: a 1 if the file is digitally signed.
11. Has tls: a 1 if the file uses TLS.
12. 64 elements: the following elements represent the first bytes of the entry point function, standardized.

SPAM. For this research we decided to treat SMS SPAM separated from Email SPAM, the main reason for this being that unlike emails, the length of text messages is short, that is, less statistically-differentiable information, due to which the availability of number of features required to detect spam SMS are less. Text messages are highly influenced by the presence of informal languages like regional words, idioms, phrases and abbreviations due to which email spam filtering methods fail in the case of SMS [6], both datasets were acquired from kaggle:

– In the case of email SPAM, the dataset[2] contains over 2500 emails, of which 20% are labelled as SPAM and 80% are labelled as HAM, both in train as in test sets.
– For sms SPAM we have used a dataset[3] with more than 5000 sms, already classified where 4457 images are SPAM and the rest we use for validation. In both sets, the distribution of the labels are 13% for SPAM and 87% for HAM.

Adult Content. We decided to build a filter oriented to social networks or instant messaging, for this use case we will define nude photography or other "compromising" media as adult content, for example the infamous "dickpicks" will be labeled as "NSFW", while a normal selfie will be classified as "SFW".

This specific definition meant we had trouble finding an adequate dataset to meet our needs, so in order to generate a dataset with photos that would normally appear on your timeline we turned to Reddit's API, using this API we generated a 2000 photo dataset, in which 50% fall into "NSFW" class, both

---

[2] https://www.kaggle.com/ozlerhakan/spam-or-not-spam-dataset.
[3] https://www.kaggle.com/uciml/sms-spam-collection-dataset.

in train as in test sets. "NFSW" photos were downloaded from /r/GoneWild and /r/GayBrosGoneWild while "SFW" photos were downloaded from /r/Selfies /r/GayBrosGoneMild.

The main reason why we chose two subreddits oriented to the homosexual community is that because of Reddit's upvote system most photographs on neutral subreddits are from women, this also means our dataset is lacking in racial diversity but for the purposes of this research we do not need one.

3.2 Data Preprocessing

SPAM Detection. For spam detection the text messages are first tokenized, transforming the text to sequences. This sequences are then converted to word embedding with word2vec, generating a vector of similarities for each word of the vocabulary, which makes it easier for the network to learn semantic meaning with LSTMs.

Adult Content Filtering. As our data-set is relatively small in order to avoid overfitting we decided to apply Image Augmentation, this strategy allows us to create new images by performing rotation or reflection of the original image, zooming in and out, shifting, applying distortion, changing the color palette, etc. [12].

Although collecting more data will always be better that applying this strategy, data augmentation has been shown to produce promising ways to increase the accuracy of classification tasks [14].

4 Experimentation and Results

The training process and evaluation of the networks has been developed with Tensorflow and Keras, evaluating the accuracy, AUC and F1 score as shown in Table 3.

First, for the SPAM case with SMS, we compared two approaches, first using a deep neural network and then LSTM, which performed better and required less epochs to train (see Table 1), from 100 needed for the first case to 15. The same architecture was then used for the training process of the email case. Training metrics for both networks are shown in Table 1. When comparing our results to the ones presented in [6] both of our models fall short of even the traditional algorithms. Nonetheless, we consider them a success as they showcase that it is possible to produce effective models using deep learning techniques without major modifications.

For adult content, we first tested basic CNN architectures, at first, the network tended to overfit before applying image augmentation techniques, afterwards the results were still poor as shown in Table 2 so we studied the implementation of Transfer Learning with deep convolutional neural networks, from Densenet201 [8], with pre-trained weights from imagenet. Other architectures such as VGG, ResNet were also tested but performed worse.

Table 1. SPAM metrics

Case	Accuracy	AUC	F1
SMS SPAM (DNN)	90.1%	0.92	0.91
SMS SPAM (LSTM)	98%	0.99	0.98

Table 2. Adult content metrics

Case	Accuracy	AUC	F1
Adult content	75.1%	0.66	0.78
Adult content (DenseNet201)	88.51%	0.95	0.88

In the malware case a basic DNNs structure scored amazing results, thanks to the amount of data available and its quality, the confusion matrix obtained on the test set can be seen in Fig. 4.

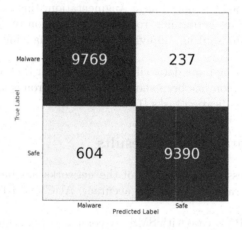

Fig. 4. Malware confusion matrix

4.1 Demo

Once your model is trained it is extremely easy to implement it in your software applications, to illustrate this we have fully integrated all of our filters into a web app powered by Streamlit, Fig. 5. The project is available on github[4].

---

[4] https://github.com/rafaelromon/SecurityML.

Table 3. Network metrics

Case	Accuracy	AUC	F1	Specificity	Sensitivity
SMS SPAM	98%	0.99	0.98	0.90	0.73
Email SPAM	98.17%	0.99	0.93	1.0	0.98
Malware	95.79%	0.98	0.96	0.99	0.73
Adult content	88.51%	0.95	0.88	0.87	0.93

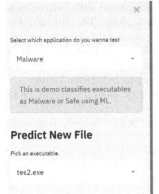

Fig. 5. Demo screenshot

5   Conclusion

In summary, by studying Deep Learning applications to different Cybersecurity problems like content filtering or malware detection we have demonstrated that it is possible to build cost-effective solutions with excellent performance.

A wide range of promising methods were presented, all showing promising results, especially of note is the use of Transfer Learning for image classification, as it has allowed us to achieve really high performance on our tests while also reducing the need for a big dataset. It is also important to mention that one the biggest weaknesses of our research is the quality of some of our datasets, for example our email dataset is quite outdated and our adult content dataset lacks proper multiracial representation, should further research be carried better alternatives must be used. Lastly as our research centered around different problems, we have yet to properly integrate one solution on a real world application, future research should focus on one specific applications or try to apply these Deep Learning techniques alongside traditional solutions in order to provide more robust solution.

References

1. Cresci, S., Lillo, F., Regoli, D., Tardelli, S., Tesconi, M.: Cashtag piggybacking: uncovering spam and bot activity in stock microblogs on Twitter (2018)

2. Ervural, B.C., Ervural, B.: Overview of cyber security in the industry 4.0 era. In: Industry 4.0: Managing The Digital Transformation. SSAM, pp. 267–284. Springer, Cham (2018). https://doi.org/10.1007/978-3-319-57870-5_16
3. Ferrara, E.: The history of digital spam. Commun. ACM **62**(8), 82–91 (2019). https://doi.org/10.1145/3299768
4. Firdausi, I., Erwin, A., Nugroho, A.S., et al.: Analysis of machine learning techniques used in behavior-based malware detection. In: 2010 Second International Conference on Advances in Computing, Control, and Telecommunication Technologies, pp. 201–203. IEEE (2010)
5. Graves, A., Schmidhuber, J.: Framewise phoneme classification with bidirectional LSTM and other neural network architectures. Neural Netw. **18**(5–6), 602–610 (2005)
6. Gupta, M., Bakliwal, A., Agarwal, S., Mehndiratta, P.: A comparative study of spam SMS detection using machine learning classifiers. In: 2018 Eleventh International Conference on Contemporary Computing (IC3), pp. 1–7. IEEE (2018)
7. Hochreiter, S., Schmidhuber, J.: Long short-term memory. Neural Comput. **9**(8), 1735–1780 (1997)
8. Huang, G., Liu, Z., van der Maaten, L., Weinberger, K.Q.: Densely connected convolutional networks (2016)
9. Huang, G., Liu, Z., Van Der Maaten, L., Weinberger, K.Q.: Densely connected convolutional networks. In: Proceedings of the IEEE Conference on Computer Vision and Pattern Recognition, pp. 4700–4708 (2017)
10. Lee, S.M., Kim, D.S., Kim, J.H., Park, J.S.: Spam detection using feature selection and parameters optimization. In: 2010 International Conference on Complex, Intelligent and Software Intensive Systems, pp. 883–888. IEEE (2010)
11. Liu, W., Wang, Z., Liu, X., Zeng, N., Liu, Y., Alsaadi, F.E.: A survey of deep neural network architectures and their applications. Neurocomputing **234**, 11–26 (2017). https://doi.org/10.1016/j.neucom.2016.12.038
12. Mikołajczyk, A., Grochowski, M.: Data augmentation for improving deep learning in image classification problem. In: 2018 International Interdisciplinary PhD Workshop (IIPhDW), pp. 117–122. IEEE (2018)
13. Nogales, I.O., de Vicente Remesal, J., Castañón, J.M.P.: Código penal y legislación complementaria. Editorial Reus (2018)
14. Perez, L., Wang, J.: The effectiveness of data augmentation in image classification using deep learning. arXiv preprint arXiv:1712.04621 (2017)
15. Ronao, C.A., Cho, S.-B.: Deep convolutional neural networks for human activity recognition with smartphone sensors. In: Arik, S., Huang, T., Lai, W.K., Liu, Q. (eds.) ICONIP 2015. LNCS, vol. 9492, pp. 46–53. Springer, Cham (2015). https://doi.org/10.1007/978-3-319-26561-2_6
16. Rowley, H.A., Jing, Y., Baluja, S.: Large scale image-based adult-content filtering (2006)
17. Seide, F., Li, G., Chen, X., Yu, D.: Feature engineering in context-dependent deep neural networks for conversational speech transcription. In: 2011 IEEE Workshop on Automatic Speech Recognition & Understanding (2011). https://doi.org/10.1109/asru.2011.6163899
18. Van De Sande, K., Gevers, T., Snoek, C.: Evaluating color descriptors for object and scene recognition. IEEE Trans. Pattern Anal. Mach. Intell. **32**(9), 1582–1596 (2009)
19. Vu, D.L., Nguyen, T.K., Nguyen, T.V., Nguyen, T.N., Massacci, F., Phung, P.H.: A convolutional transformation network for malware classification (2019)

20. Wang, Y., Huang, M., Zhu, X., Zhao, L.: Attention-based LSTM for aspect-level sentiment classification. In: Proceedings of the 2016 Conference on Empirical Methods in Natural Language Processing, pp. 606–615 (2016)
21. Wehrmann, J., Simões, G.S., Barros, R.C., Cavalcante, V.F.: Adult content detection in videos with convolutional and recurrent neural networks. Neurocomputing **272**, 432–438 (2018)
22. Ye, Y., Li, T., Adjeroh, D., Iyengar, S.S.: A survey on malware detection using data mining techniques. ACM Comput. Surv. (CSUR) **50**(3), 1–40 (2017)

Optimization Problem Applications

A Parallel Optimization Solver
for the Multi-period WDND Problem

Carlos Bermudez[1] (ID), Hugo Alfonso[1] (ID), Gabriela Minetti[1(✉)] (ID),
and Carolina Salto[1,2] (ID)

[1] Facultad de Ingeniería, Universidad Nacional de La Pampa,
General Pico, Argentina
{bermudezc,alfonsoh,minettig,saltoc}@ing.unlpam.edu.ar
[2] CONICET, Buenos Aires, Argentina

Abstract. A secure water distribution system is an essential element
for any city in the world. The importance and huge capital cost of the
system lead to their design optimization. However, the water distribution network design optimization is a multimodal and NP-hard problem,
as a consequence, we propose an intelligent optimization solver based on
a Parallel Hybrid Simulated Annealing (PHSA) to solve it. The parallelism is applied at the algorithmic level, following a cooperative model.
The Markov Chain Length (MCL) is an important Simulated Annealing
control parameter, which represents the number of moves to reach the
equilibrium state at each temperature value. Our main objective is to
analyze the PHSA behavior by considering static and dynamic methods
to compute the MCL. The obtained results by PHSA enhance the found
ones by the published algorithms. This improvement becomes more interesting when PHSA solves a real-world case. Furthermore, the parallel
HSA exhibits efficient scalability to solve the WDND problem.

Keywords: Water distribution network design · Optimization ·
Parallelism · Simulated annealing · Markov chain length

1   Introduction

Public water services provide more than 90% of the water supply in the world
today, and therefore a safe drinking water distribution system is a critical component for any city. This kind of system considers components, such as pipes of
various diameters, valves, service connections, and distribution reservoirs. Furthermore, when diameters are used as the decision variables, the constraints are
implicit functions of the decision variables and require solving conservation of
mass and energy to determine the network's pressure heads. In addition, the
feasible region is nonconvex, and the objective function is multimodal [1]. Water
distribution network design (WDND) belongs to a group of inherently intractable
problems referred to as NP-hard [2].

Hence cost reduction and efficiency have been important reasons for managers to progressively migrate from exact (deterministic) optimization algorithms to the development of approximate (stochastic) optimization solvers. In

© Springer Nature Switzerland AG 2021
H. Sanjurjo González et al. (Eds.): HAIS 2021, LNAI 12886, pp. 625–636, 2021.
https://doi.org/10.1007/978-3-030-86271-8_52

this sense, some research works [3,4] applied solvers based on Simulated Annealing (SA) to optimize water distribution network designs, which were defined as single-period, single-objective, and gravity-fed design optimization problem. Other works expressed it as a multi-objective optimization problem and applied a multi-objective evolutionary algorithm [5] and recently, a multiobjective simulated annealing was proposed [6]. A Genetic Algorithm [7] considered the velocity constraint on the water flowing through the pipes. An Iterative Local Search (ILS) [8] considered that every demand node has 24 h water demand pattern and included a new constraint related to the limit of the maximal water velocity through the pipes. This new WDND formulation was also solved by our hybrid SA solver, named HSA [9], which incorporates a local search procedure to improve the network layout. Moreover, this HSA was enhanced by using three different strategies to configure a control parameter value, the Markov Chain Length (MCL). This HSA was used to solve a new real-world instance of the multi-period WDND optimization problem [10].

The efficiency of the sequential algorithms is highly reduced by solving real-world instances of WDND, which involve very large distribution networks. In this sense, parallel evolutionary algorithms [1,11] were proposed to solve the WDND problem. However, in this field, the parallel SA power is hardly put into practice, then we propose to parallelize the HSA solver at algorithmic level [12]. This model is based on cooperative self-contained HSA solvers, becoming a problem-independent parallelization. This idea is also supported by the fact that the literature presents several solvers based on parallel SA, which improve the performance of their respective sequential versions to solve NP-hard problems in diverse areas. For instance, to solve the vehicle routing problem [13,14], scheduling problems [15,16], and many others [17–19].

The main contribution of our research is to develop an intelligent solver based on a parallel HSA algorithm, named PHSA, which intends to support the decision-making during the design, plan, and management of complex water systems. The focus is to increase the efficiency of this solver when large real-world cases are addressed. Hence, we formulate three research questions (*RQs*): *1)* if the parallelization of HSA increases its efficiency when we deal with real cases, where the WDND optimization includes time-varying demand patterns (extension known as multi-period setting) and the maximum water velocity constraint; *2)* what parametric configurations should be used in the proposed parallel algorithm to enhance the solution quality, and *3)* if our PHSA design is efficient to solve the WDND problem. To answer the *RQs*, we conduct experiments by applying PHSA with different configurations on publicly available benchmarks [20] and real-world instances [10]. Furthermore, we analyze and compare these results considering the published ones in the literature.

The remainder of this article is structured as follows. Section 2 introduces the problem definition. Section 3 explains our algorithmic proposal. Section 4 describes the experimental design and the methodology used. Then, Sect. 5 presents the result analysis of the variants and the comparison with the literature approaches. Finally, Sect. 6 summarizes our conclusions and sketches out our future work.

2   Multi-period Water Distribution Network Design

The mathematical formulation of the WDND is often treated as the least-cost optimization problem. The decision variables are the diameters for each pipe in the network. The problem can be characterized as simple-objective, multi-period, and gravity-fed. Two restrictions are considered: the limit of water speed in each pipe and the demand pattern that varies in time. The network can be modeled by a connected graph, which is described by a set of nodes $N = \{n_1, n_2, ...\}$, a set of pipes $P = \{p_1, p_2, ...\}$, a set of loops $L = \{l_1, l_2, ...\}$, and a set of commercially available pipe types $T = \{t_1, t_2, ...\}$. The objective of the WDND problem is to minimize the Total Investment Cost (TIC) in a water distribution network design. The TIC value is obtained by the formula shown in Eq. 1.

$$\min TIC = \sum_{p \in P} \sum_{t \in T} L_p IC_t x_{p,t} \qquad (1)$$

where IC_t is the cost of a pipe p of type t, L_p is the length of the pipe, and $x_{p,t}$ is the binary decision variable that determines whether the pipe p is of type t or not. The objective function is constrained by: physical laws of mass and energy conservation, minimum pressure demand in the nodes, and the maximum speed in the pipes, for each time $\tau \in T$.

3   HSA Solver for the Multi-period WDND Problem

Simulated Annealing (SA) [21] is based on the principles of statistical thermodynamics, which models the physical process of heating material. SA evolves by a sequence of changes between states generated by transition probabilities, which are calculated involving the current temperature. Therefore, SA can be modeled mathematically by Markov chains.

In this work, we propose an intelligent parallel solver to optimize the Multi-Period WDND problem. This solver is based on the hybrid SA (HSA) introduced in [10]. We begin with the HSA explanation, and then we continue by detailing how the parallel HSA (PHSA) works to solve the problem at hand.

3.1   HSA Solver to Tackle the WDND Problem

As the Algorithm 1 shows, HSA begins with the initialization of the temperature. After that, HSA generates and evaluates a feasible initial solution S_0 applying both HighCost and LowCost mechanisms proposed in [8]. Once the initialization process ends, an iterative process starts. The first step in the iteration involves a hybridization to intensify the exploration into the current search space. In this way, a feasible solution S_1 is obtained by applying the MP-GRASP local

---

Algorithm 1. HSA to solve the WDND Optimization Problem.

1: $k = 0$;
2: initTemp(T)
3: initialize(S_0);
4: TIC_0 = evaluate(S_0);
5: **repeat**
6:   **repeat**
7:     $k = k + 1$;
8:     S_1 = MP-GRASP_LS(S_0);
9:     TIC_1=evaluate(S_1);
10:     **if** $(TIC_1 < TIC_0)$ **or** $(exp^{((TIC_1-TIC_0)/T)} > random(0,1))$ **then**
11:       $S_0 = S_1$; $TIC_0 = TIC_1$
12:     **end if**
13:     S_2 = perturbation_operator(S_0);
14:     TIC_2 = evaluate(S_2);
15:     **if** $(TIC_2 < TIC_0)$ **or** $(exp^{((TIC_2-TIC_0)/T)} > random(0,1))$ **then**
16:       $S_0 = S_2$; $TIC_0 = TIC_2$
17:     **end if**
18:   **until** (k mod MCL == 0)
19:   update(T);
20: **until** *stop criterion is met*
21: **return** S_0;

---

search [8] to S_0, and then a greedy selection mechanism is performed. Therefore, S_0 can be replaced by S_1 if it is better than S_0. In the next step, a perturbation operator is used to obtain a feasible neighbor S_2 from S_0, to explore other areas of the search space. This perturbation randomly changes some pipe diameters. If S_2 is worse than S_0, S_2 can be accepted under the Boltzmann probability. In this way, the search space exploration is strengthened when the temperature (T) is high. In contrast, at low temperatures the algorithm only exploits a promising region of the solution space, intensifying the search. To update T, a random cooling schedule presented in [22] is used. Finally, SA ends the search when the total evaluation number or the temperature equilibrium ($T = 0$) is achieved.

The MCL is the number of required transitions k (moves) to reach the equilibrium state at each T value. This number can be set statically or adaptively. The static MCL (MCLs) assumes that each T value is held constant for a sufficient and fixed number of iterations, defined before the search starts. In this work, each T value is held constant for $k = 30$ iterations, a widely used number in the scientific community. For the adaptive cases, the MCL depends on the characteristics of the search. For instance, Cardoso et al. [23] consider that the equilibrium state is not necessarily attained at each temperature. Here, the cooling schedule is applied as soon as an improved candidate (neighbor) solution is generated. In this way, the computational effort can be drastically reduced without compromising the solution quality. This approach is named MCLa1. Another adaptive strategy is proposed by Ali et al. [24], identified as MCLa2, which uses both the worst and the best solutions found in the Markov chain (inner loop) to compute the next MCL. MCLa2 increases the number of function evaluations at a given temperature if the difference between the worst and the best solutions increases. But if an improved solution is found, the MCL remains unchanged.

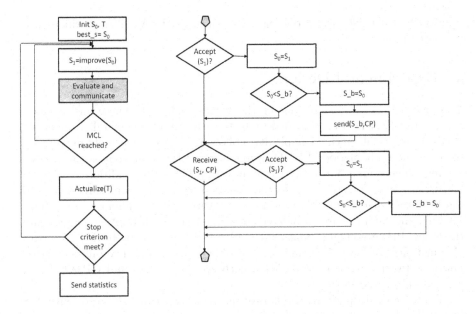

Fig. 1. Scheme of the intelligent PHSA solver.

3.2 Parallel HSA Solver Description

In this work, we use the Algorithmic-Level Parallel Model [12] to parallelize HSA. In this model, many self-contained metaheuristic algorithms are launched in parallel. They may or may not cooperate to solve the target optimization problem. When the cooperation is carried out, the self-contained metaheuristics exchange information to improve the quality solutions and enhance the efficiency. Furthermore, this kind of model offers a problem-independent parallelization.

Hence, the PHSA design follows a cooperative search strategy on a star topology. A central process (CP) launches n HSA solvers in parallel. Each HSA solver is responsible for processing the HSA search and the CP gathers the solutions from the solvers to determine the global best solution. The communication among the HSA solvers and the CP follows an asynchronous behavior. Each HSA solver generates its initial solution S_0 and is allowed to run an independent HSA (see Fig. 1). The HSA solvers exchange information with the CP related to the search to compute better and more robust solutions. This information exchange is decided according to an intelligent adaptive criterion. The decision is related to find an improvement of the best found local solution S_b. In this way, an HSA solver sends a solution to the CP when a better one arises (function send(S_b,CP)). Once the central process retrieves a solution, a comparison with the best global one is carried out. If the incoming solution is best, the CP actualizes their best global one and sends it to the remaining HSA solvers. As soon as the solver receives a new solution from the CP (function receive(S_1,CP)) uses the Bolztman acceptance criteria (function accept(S_1)) to compare the incoming solution S_1 with the current one S_0. After that, the HSA solver continues with its search process.

4   Experimental Design

This new PHSA is used to answer the formulated RQs. These answers need the empirical verification provided by testing this intelligent parallel solver in a WDND test set of varying complexity. Following the experience obtained in [25] to solve the multi-period WDND problem, the PHSA solver uses the random cooling scheme and 100 as seed temperature. The stop condition is to reach 1,500,000 evaluations of the objective function to make a fair comparison with the literature algorithms. Furthermore, the three MCL configurations (MCLs, MCLa1, and MCLa2) are considered (described in Sect. 3.1). As a consequence, three new PHSA configurations arise, which are evaluated by testing 25 Hydro-Gen instances of WDND optimization problem grouped by five different distribution networks, named as HG-MP-i with $i \in \{1; 5\}$ [20], and GP-Z2-2020, a real case presented in [10].

The parallelism performance is assessed regarding three scenarios with 4, 8, and 16 HSA solvers. At least 30 independent executions for each instance and PHSA's configuration and parallel scenarios are necessary due to the stochastic nature of the HSA solver (30 runs × 26 instances × 3 MCL × 3 parallel scenarios). This big experimentation allows us to gather meaningful experimental data and apply statistical confidence metrics to validate our results and conclusions. Before performing the statistical tests, we first check whether the data follow a normal distribution by applying the Shapiro-Wilks test. Where the data are distributed normally, we later apply an ANOVA test. Otherwise, we use the Kruskal–Wallis (KW) test. This statistical study allows us to assess whether or not there are meaningful differences between the compared algorithms with $\alpha = 0.05$. To determine these algorithm pairwise differences is by carrying out a post hoc test, as is the case of the Wilcoxon test if the KW test is used.

5   Result Analysis

The solution quality and efficiency of the proposed parallel HSA are compared against algorithms from the literature, HSA [10] and ILS [8], to show the parallel HSA robustness as an alternative search method. The three MCL configurations of PHSA were run with two different stopping criteria: one is based on a maximum number of evaluations (predefined effort), while the other consists in running the PHSA until a given solution is found or the maximum number of evaluations is reached (predefined solution quality). First, we want to compare the PHSA against HSA and ILS to show the numerical advantages of the proposed parallel algorithm. For a fair comparison, all algorithms used the predefined effort as a cut point. Second, we analyze the parallel performance. We focus in the number of optimum values the PHSA configurations found. Then, we measure the speedup between the serial time (PHSAs using one processor)

Table 1. Minimum of the best TIC values found by each PHSA configuration and the literature algorithms.

	Algorithm	Instances					
		HG-MP-1	HG-MP-2	HG-MP-3	HG-MP-4	HG-MP-5	GP-Z2-2020
	4 HSA solvers	**298000**	245330	310296	589769	**631000**	354374
MCLs	8 HSA solvers	**298000**	245330	310493	596666	**631000**	383753
	16 HSA solvers	**298000**	245391	310322	708778	**631000**	408567
	4 HSA solvers	**298000**	245330	310366	592247	**631000**	340908
PHSA MCLa1	8 HSA solvers	**298000**	245330	310696	595765	**631000**	339004
	16 HSA solvers	**298000**	245330	310704	599148	**631000**	333677
	4 HSA solvers	**298000**	245330	**310111**	596364	**631000**	344364
MCLa2	8 HSA solvers	**298000**	245330	**310111**	595765	**631000**	339271
	16 HSA solvers	**298000**	245330	310696	**588402**	**631000**	**331194**
	HSA	**298000**	245330	310493	590837	**631000**	347596
	ILS	**298000**	**245000**	318000	598000	**631000**	355756

Table 2. Average of the best TIC values found by each PHSA configuration and literature algorithms.

	Algorithm	Instances					
		HG-MP-1	HG-MP-2	HG-MP-3	HG-MP-4	HG-MP-5	GP-Z2-2020
	4 HSA solvers	347776	362058	472524	698116	803742	437638
MCLs	8 HSA solvers	347454	348244	459688	712085	829679	480967
	16 HSA solvers	347262	346742	458758	962414	989804	481771
	4 HSA solvers	347427	346349	471418	691691	802691	377876
PHSA MCLa1	8 HSA solvers	347609	346059	457795	690149	793532	367516
	16 HSA solvers	347689	346742	453274	**688307**	781479	**363614**
	4 HSA solvers	**346784**	349290	470262	691672	799173	400737
MCLa2	8 HSA solvers	347399	338394	457713	691334	789259	372389
	16 HSA solvers	348221	331975	455158	690351	**778605**	371509
	HSA	377467	**299690**	**442055**	736997	830539	401570
	ILS	380000	318000	445000	750000	854000	429948

against the parallel time (PHSAs using 4, 8, and 16 processors) because they run the same underlying algorithm. We assume an orthodox way to report to the research community in parallelism [26]. Thus, general numerical comparisons are always interesting to compare algorithms, while measuring speedup needs a careful definition of the one processor case.

5.1 Results with Predefined Effort

Let us now proceed with the presentation of the results for the six WDND instance groups. Tables 1 and 2 show the minimum and average TIC values, respectively, for the PHSA considering the three MCL strategies and the number of HSA solvers. Moreover, the last two rows present the TIC values corresponding to the HSA and ILS to make a literature comparison about solution qualities. For HG-MP-1 and HG-MP-5, all algorithms obtain the same minimum TIC value. For the rest of the instances, PHSA configurations achieve the minimum TIC value except for HG-MP-2, pointing to the superiority of our devised PHSA approach. Regarding the average TIC values, we can see that the PHSA configurations present lower average values than the HSA and ILS for four of six groups of instances (HG-MP-1, HG-MP-4, HG-MP-5, and GP-Z2-2020). Con-

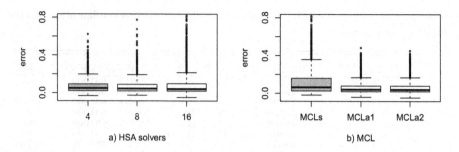

Fig. 2. Dispersion of TIC error values found by each PHSA and MCL strategies for WDND instances.

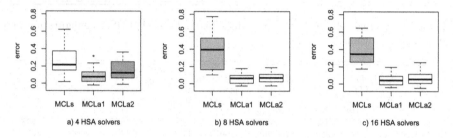

Fig. 3. Dispersion of TIC error values found by each PHSA for WDND instances.

sequently, we prove that PHSA is more efficient than HSA and ILS to solve real-world networks, answering *RQ 1*.

Now, we proceed with the error value, which is the relative distance to the best-known TIC value and the best TIC value of each PHSA configuration. Figures 2 a) and b) show the distribution of the PHSA errors grouped by the number of HSA solvers and MCL strategies, respectively. PHSA with 8 and 16 HSA solvers present lower median errors than the PHSA with 4 HSA solvers. Although the PHSA error dispersion is quite similar between the parallel scenarios, the PHSA with 8 and 16 HSA solvers have similar statistical behavior, but different from the PSHA with 4 HSA solvers. Focusing on the influence of MCL strategies on the error values of the different PHSAs, the adaptive strategies allow the PHSA to find lower error values, and the statistical tests indicate that the no significant differences are present between the PHSAs.

The following analysis is devoted to studying each PHSA configuration with more detail. Figure 3 shows the error dispersion for each MCL strategy and the number of HSA solvers. When 8 and 16 HSA solver are considered, the adaptive MCL strategies present lower error values, but MCLa1 shows less dispersion. Furthermore, the statistical test reveals that the adaptive MCL strategies behave similarly (both boxplots in white). It is important to remark that there are significant differences between the configurations when the PHSA has 4 HSA solvers. In all PHSAs, the static MCL has the worst results. Thus, the PHSA solution quality is enhanced by using adaptive MCL strategies, responding satisfactorily the *RQ 2*.

Therefore, we prove that the proposed PHSA solver tackles the WDND optimization problem finding better solution quality than the algorithms proposed in the literature. As a consequence, the cooperative model presented to parallelize HSA results an excellent option to solve this kind of problem, answering the *RQs 1* and *2*.

5.2   Results with Predefined Quality of Solutions

The following analysis focuses on the study of the PHSA algorithms from a different point of view. Until now, a predefined effort is given to each algorithm to determine the best one. But from now on, a different scenario is faced in which the computational effort of the algorithms is measured to locate a preset solution. Therefore, the same target solution quality is used as the stop criterion in all the algorithms for any given instance. The target TIC value depends on the WDND instance, and the median TIC value obtained in [10] is selected for each one as the stopping value for the PHSA algorithms, to offer a nonbiased scenario produced by an ad-hoc value.

The first study focus on the hit rate of the PHSA algorithms. This measure is the relation between the number of execution that reached the target TIC value and the total number of performed tests. We analyze the results when the algorithms run over n CPUs, and thus we can focus on the effort and hit rate (i.e., numerical effort), which is presented in Fig. 4.a). For all the solved group WDND instances, adaptive MCL strategies obtain a higher hit rate than the static MCL strategies. It can be explained by the enhanced search model that the adaptive MCL strategies present by making a deep search. If we analyze the hit rate when the PHSA run with 16 HSA solvers (Fig. 4.a)), we observe that this PHSA outperforms the remaining parallel algorithms. The PHSA performance is improved as the number of HSA solvers is increased, thus becoming a scalable parallel algorithm and justifying the contribution of this work.

Now, the speedup values of the analyzed PHSA algorithms are discussed. The same PHSA algorithms, executed in both sequential and parallel ways (one versus n processors), are compared following an orthodox speedup definition [26]. Results in Fig. 4.b) show that the speedup is relatively high. In the cases where PHSA uses adaptive MCL strategies, the speedup values are close to linear, showing the efficiency of PHSA to solve the WDND problem and answering affirmatively *RQ 3*.

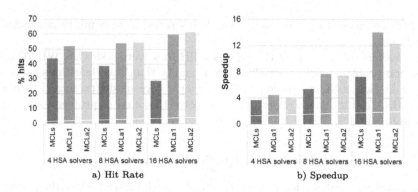

Fig. 4. Hit rate and speedup for each PHSA when solving WDND instances.

6   Conclusions

In this work, we have presented a Parallel Hybrid Simulated Annealing (PHSA) to optimize the design of water distribution networks. The formulation problem includes multi-period settings and the maximum water velocity constraint, which are realistic restrictions. The main feature of our PHSA is to use static and two adaptive MCL strategies. The basic underlying search model consists of a set of HSA solvers connected to a central process by a star topology. The availability of different HSA solvers by cooperating in the process search allows obtaining a powerful parallel algorithm that leads to more accurate solutions. To show the accuracy and efficiency of our PHSA, we compared it with two state-of-the-art algorithms (HSA and ILS).

The whole experimentation was directed at finding the answers to our research questions. The idea behind *RQ 1* was to study the solution quality. After that, we observed that our PHSA solver found minimum TIC values and outperformed the algorithms proposed in the literature, even more in real-world WDND instances. As a consequence, a model based on cooperative self-contained HSA solvers at the algorithmic level resulted in an excellent option to solve this problem. The *RQ 2* suggested an analysis of the effect of several parametric configurations on the solution quality. In this sense, different statistical tests proved that adaptive MCL strategies (MCLa1 and MCLa2) allowed significantly enhancing the solution quality. Finally, the *RQ 3* led us to assess the PHSA efficiency, considering hit rates and speedup measures. Hence, we confirmed that PHSA held the scalability property because its performance was improved by increasing the number of HSA solvers and the speedup values were close to linear.

As future work, we are interested in better analyzing the complex behavior of the PHSA with the adaptive MCL strategies. Furthermore, we plan to use big-data frameworks to implement our PHSA solver, which will allow us to deal with high-dimensional WDND problems.

Acknowledgments. The authors acknowledge the support of Universidad Nacional de La Pampa (Project FI-CD-107/20) and the Incentive Program from MINCyT. The last author is also funded by CONICET.

References

1. Eusuff, M., Lansey, K.: Optimization of water distribution network design using the shuffled frog leaping algorithm. J. Water Resour. Plan. Manag. - ASCE **129**(3), 210–225 (2003)
2. Yates, D., Templeman, A., Boffey, T.: The computational complexity of the problem of determining least capital cost designs for water supply networks. Eng. Optim. **7**(2), 143–155 (1984)
3. Loganathan, G., Greene, J., Ahn, T.: Design heuristic for globally minimum cost water-distribution systems. J. Water Resour. Plan. Manag. **121**(2), 182–192 (1995)
4. Cunha, M., Sousa, J.: Hydraulic infrastructures design using simulated annealing. J. Infrastruct. Syst. **7**(1), 32–39 (2001)
5. Farmani, R., Walters, G.A., Savic, D.A.: Trade-off between total cost and reliability for anytown water distribution network. J. Water Resour. Plan. Manag. **131**(3), 161–171 (2005)
6. Cunha, M., Marques, J.: A new multiobjective simulated annealing algorithm– MOSA-GR: application to the optimal design of water distribution networks. Water Resour. Res. **56**(3), e2019WR025852 (2020). https://agupubs.onlinelibrary.wiley. com/doi/abs/10.1029/2019WR025852
7. Gupta, I., Gupta, A., Khanna, P.: Genetic algorithm for optimization of water distribution systems. Environ. Model. Softw. **14**(5), 437–446 (1999)
8. De Corte, A., Sörensen, K.: An iterated local search algorithm for water distribution network design optimization. Network **67**(3), 187–198 (2016)
9. Bermudez, C., Salto, C., Minetti, G.: Solving the multi-period water distribution network design problem with a hybrid simulated anealling. In: Pesado, P., Aciti, C. (eds.) CACIC 2018. CCIS, vol. 995, pp. 3–16. Springer, Cham (2019). https:// doi.org/10.1007/978-3-030-20787-8_1
10. Bermudez, C., Alfonso, H., Minetti, G., Salto, C.: Hybrid simulated annealing to optimize the water distribution network design: a real case. In: Pesado, P., Eterovic, J. (eds.) CACIC 2020. CCIS, vol. 1409, pp. 19–34. Springer, Cham (2021). https:// doi.org/10.1007/978-3-030-75836-3_2
11. Cruz-Chávez, M.A., Avila-Melgar, E.Y., Serna Barquera, S.A., Juárez-Pérez, F.: General methodology for converting a sequential evolutionary algorithm into parallel algorithm with MPI to water design networks. In: 2010 IEEE Electronics, Robotics and Automotive Mechanics Conference, pp. 149–154 (2010)
12. Talbi, E.-G.: Metaheuristics: From Design to Implementation. Wiley, Hoboken (2009)
13. Czech, Z.J., Mikanik, W., Skinderowicz, R.: Implementing a parallel simulated annealing algorithm. In: Wyrzykowski, R., Dongarra, J., Karczewski, K., Wasniewski, J. (eds.) PPAM 2009. LNCS, vol. 6067, pp. 146–155. Springer, Heidelberg (2010). https://doi.org/10.1007/978-3-642-14390-8_16
14. Mu, D., Wang, C., Zhao, F., Sutherland, J.: Solving vehicle routing problem with simultaneous pickup and delivery using parallel simulated annealing algorithm. Int. J. Shipping Transport Logist. **8**, 81–106 (2016)
15. Cicirello, V.A.: Variable annealing length and parallelism in simulated annealing. CoRR, vol. abs/1709.02877 (2017). http://arxiv.org/abs/1709.02877

16. Gerdovci, P., Boman, S.: Re-scheduling the railway traffic using parallel simulated annealing and tabu search: a comparative study. Master's thesis (2015)
17. Lee, S., Kim, S.B.: Parallel simulated annealing with a greedy algorithm for Bayesian network structure learning. IEEE Trans. Knowl. Data Eng. **32**(6), 1157–1166 (2020)
18. Ding, X., Wu, Y., Wang, Y., Vilseck, J.Z., Brooks, C.L.: Accelerated CDOCKER with GPUs, parallel simulated annealing, and fast Fourier transforms. J. Chem. Theory Comput. **16**(6), 3910–3919 (2020). pMID: 32374996. https://doi.org/10.1021/acs.jctc.0c00145
19. Lincolao-Venegas, I., Rojas-Mora, J.: A centralized solution to the student-school assignment problem in segregated environments via a cuda parallelized simulated annealing algorithm. In: 2020 39th International Conference of the Chilean Computer Science Society (SCCC), pp. 1–8 (2020)
20. De Corte, A., Sörensen, K.: Hydrogen. http://antor.uantwerpen.be/hydrogen. Accessed 27 June 2018
21. Kirkpatrick, S., Gelatt, C.D., Jr., Vecchi, M.P.: Optimization by simulated annealing. Science **220**, 671–680 (1983)
22. Bermudez, C., Minetti, G., Salto, C.: SA to optimize the multi-period water distribution network design. In: XXIX Congreso Argentino de Ciencias de la Computación (CACIC 2018), pp. 12–21 (2018)
23. Cardoso, M., Salcedo, R., de Azevedo, S.: Nonequilibrium simulated annealing: a faster approach to combinatorial minimization. Ind. Eng. Chem. Res. **33**, 1908–1918 (1994)
24. Ali, M., Törn, A., Viitanen, S.: A direct search variant of the simulated annealing algorithm for optimization involving continuous variables. Comput. Oper. Res. **29**(1), 87–102 (2002)
25. Bermudez, C., Salto, C., Minetti, G.: Designing a multi-period water distribution network with a hybrid simulated annealing. In: XLVIII JAIIO: XX Simposio Argentino de Inteligencia Artificial (ASAI 2019), pp. 39–52 (2019)
26. Alba, E.: Parallel evolutionary algorithms can achieve super-linear performance. Inf. Process. Lett. **82**(1), 7–13 (2002)

Waste Collection Vehicle Routing Problem Using Filling Predictions

Haizea Rodriguez, Jon Díaz, Jenny Fajardo-Calderín$^{(\boxtimes)}$, Asier Perallos, and Enrique Onieva

DeustoTech, Faculty of Engineering, University of Deusto,
Av. Universidades, 24, 48007 Bilbao, Spain
{haizearodriguez,jondi}@opendeusto.es,
{fajardo.jenny,perallos,enrique.onieva}@deusto.es

Abstract. With the increase in the global population and rising demand for food and other materials, there has been a rise in the amount of waste being generated daily by each house and locality. Most of the waste that is generated is thrown into the garbage containers, from where it is collected by the area municipalities. Improper management of waste in cities results in a huge loss to any smart city. However, with the importance that this has for the health of the citizens and the environment, there are few tools available to manage the planning of waste collection routes efficiently. In this paper, we present a solution to the planning of waste collection routes considering the existence of monitoring capabilities of the containers. The routing of the trucks for each of the days will be formulated as a Capacitated Vehicle Routing Problem. This research focuses on the minimization of the routing times, as well as avoiding the overflow probability. The paper presents the results of a case study with real data on three municipalities in the Basque Country, Spain.

Keywords: Capacitated vehicle routing problem · Garbage collection · Smart city · Internet of Things

1 Introduction

Technology has disrupted many aspects of our world, creating unlimited paths to solve many complex things in our daily life. Sometimes even taking the next step, not confronting the problem, but rather preventing it from happening, the very basis upon which the smart cities are being built [4]. These cities face a wide variety of challenges, one of them being the accumulation of waste and overfilling of bins in their urban areas. Constant residue collecting routes is the most widespread method of approaching this waste management issue, and it has not evolved in the last century [15].

Residual waste generation is something that technology will not mend in the short term, but it can help to handle it better. One of the approaches to mitigate this problem lies in the proper plan of routes for garbage collection, facing the

© Springer Nature Switzerland AG 2021
H. Sanjurjo González et al. (Eds.): HAIS 2021, LNAI 12886, pp. 637–648, 2021.
https://doi.org/10.1007/978-3-030-86271-8_53

problem as a variant of the well-known Vehicle Routing Problem (VRP). Usually, the problem is represented employing a graph containing a set of locations that have to be visited only once, additionally, restrictions on the capacity of the trucks or specific time windows are considered in the literature [8].

The most classical VRP is the Capacitated VRP (CVRP). In a typical CVRP, a vehicle is allowed to visit and serve each customer on a set of routes exactly once. The vehicle starts and ends its visit at the central depot such that the total travel cost (distance or time) is minimized and the vehicle total capacity is not exceeded [1].

A smart city is an urban area that uses different types of electronic methods and sensors to collect data. Insights gained from that data are used to manage assets, resources and services efficiently; in return, that data is used to improve the operations across the city. The smart city concept integrates information and communication technology (ICT), and various physical devices connected to the IoT (Internet of things) network to optimize the efficiency of city operations and services and connect to citizens. Smart waste management in a smart city is one service that can be provided [3,5].

Smart waste management is mainly compatible with the concept of smart cities. Having waste containers with the capability of measuring their current fill will allow the estimation of future states of filling, which will allow effective planning of resources for their collection. The efficient management of waste has a significant impact on the quality of life of citizens.

This work focuses on the planning of waste collection routes considering the existence of monitoring capabilities of the containers, as well as estimation of the filling of them from one day to the next one. Given so, each container may share its actual filling rate, as well as an estimation of the increase for the next day. In this sense, here we propose an optimization problem considering the collection of garbage from different containers in a scenario in which not all of them have to be collected on the same day.

To achieve it, we propose a formulation of an optimization problem for making the decision of which containers will be collected in different days of a time horizon, and, given that, the routing of the trucks for each of the days will be formulated as a CVRP. The formulation has into account both the minimization of the routing times, as well as avoiding the overflow probability.

This paper is structured as follows. Section 2 presents the related literature review. Section 3 states the formulation of the problem to be solved by optimization methods. Section 4 introduces the optimization method proposed. After that, Sect. 5 presents experimental results obtained. Finally, Sect. 6 states conclusions and future works.

2 Literature Review

Waste management has a great impact on the environment, managing it efficiently and intelligently has a direct impact on the health of society. The waste collection system named Decision Support System, is enhanced with IoT services that enable dynamic scheduling and routing in smart cities. Different works model waste management as a service of an IoT infrastructure.

The waste collection problem has received some attention in recent years. Gutierrez et al. [10] have proposed a waste collection solution based on providing intelligence to trashcans, by using an IoT prototype embedded with sensors, which can read, collect, and transmit trash volume data over the Internet. The system is simulated in a realistic scenario in the city of Copenhagen and using freely available geolocation data of the municipality-owned trashcans as Open Data.

Buhrkal et al. [6] present the study the Waste Collection Vehicle Routing Problem with Time Window, and proposed an adaptive large neighbourhood search algorithm for solving the problem. Huang et al. [11] present the Waste collection problem as a set covering and vehicle routing problem (VRP) involving inter-arrival time constraints, bi-level optimization formula to model the split delivery VRP with several trips to decide the shortest path. Developed an ACO algorithm for the solution obtained.

Reddy et al. [3] proposed work includes a smart dustbin that separates dry and wet waste so that the waste recycling can be done more efficiently. To find the efficient way to find garbage collecting routes, we use shortest path algorithms like Dijkstra's Algorithm.

The idea proposed in [7] considered two garbage bins for waste segregation, and sensors are attached to bins for garbage data collection to avoid overfilling. Overfilling of the bins is prevented using sensors, but no mechanism for waste collection is proposed. Then, Shreyash et al. [12] provides a comparative analysis on different algorithms used for Garbage collection Systems such as Genetic Algorithm Ant Algorithm, Integrated Nearest Neighbor Algorithm and Genetic Algorithm.

Wongsinlatam et al. [16] have proposed the solution to find the minimal total cost for finding the shortest route of waste collection for garbage truck capacity by using metaheuristic techniques. Moreover, Stanković et al. [14] proposed a solution for the collection of waste in a municipality, and in the solution of the model they use the distance-constrained capacitated vehicle routing problem. To solve the problem they used four metaheuristic algorithms, such as: Genetic algorithm, Simulated annealing, Particle swarm optimization, and Ant colony optimization. They develop a case study on the collection and transport of municipal waste in an urban environment in the city of Niš in Serbia.

Wu et al. [17] developed a model for the collection and transportation of urban waste in China, the solution is specific because it is based on the design of algorithms to solve a priority vehicle routing problem. The problem is based on the possibility that waste collection services are immediate for high priority waste containers, for example those containing hospital or medical waste, considering that harmful waste must be collected immediately. They consider the deployment of sensors in the waste containers to make dynamic routes instead of fixed routes. In the solution they used a hybrid search algorithm, which builds the initial solution with the Particle Swarm Optimization, and then a local search is performed on the initial solution, using a Simulated Annealing.

Nurprihatin et al. [13] have proposed a waste collection model using the vehicle routes problem, and add characteristics and constraint, such as: time windows, multiple trips, fractional delivery and heterogeneous fleet. To solve the problem, they use the Nearest Neighbor algorithm, and use data from Cakung District, located in Indonesia, as a case study. Also, Tariq et al. [2] proposed a model IoT-based garbage collection solutions to solid waste management. The proposed system is capable of the effective collection of waste, detection of fire in waste material and forecasting of the future waste generation. The IoT based device performs the controlling and monitoring of the electric bins. Based on the previously collected data from the selected bins, the predictive analytic algorithm namely decision tree and neural network are applied to predict the future waste collection.

After reviewing the literature of various authors, we can see the importance of the Waste Collection Problem for the development of smart cities. We found an increasing need to develop systems that take into account multiple factors, such as capacity of trucks and bins, bin litter level prediction and generates smart routes for the garbage trucks; to reduce the consumption of resources and improve the quality of life of the citizens.

3   Formulation of the Problem

This work faces the problem of the planning of waste collection routes in a dynamic time horizon as a combinatorial optimization problem, where the day in the plan in which a particular container is collected is optimized for solving, and, after that, the particular capacitated VRP for each one of the days is applied. The problem assumes containers having a filling status, as well as a daily increment of it, so the algorithm must ensure that none of the containers overflows. The following subsections explain details of the problem.

3.1   The Concept of the System

To find a better way of handling waste management, the proposed system consists of two steps. The first one will generate a plan for a given amount of days, while the second step will be in charge of optimizing said plan.

A plan is defined as an allocation of the day in which a particular container will be collected, within a given time horizon for generating the plan. Selected containers for collection on a particular day of the plan will be visited by a truck of the fleet, considering each particular day as an instance of a CVRP.

It must be noted that, given this definition, a particular plan may not be suitable, because the amount of garbage to be collected exceeds the capacity of the trucks. On the other hand, given the increasing filling of the containers, a plan can derive to a situation in which one (or more) containers overflow. Both situations are considered through the evaluation function, formally presented in Sect. 3.2, by penalizing those scenarios.

Initially, a feasible plan is created considering the capacity of the trucks, as well as the quantity of garbage to be collected by a greedy procedure. In subsequent iterations of the optimization methods, such an initial plan is slightly modified by a neighbour operator, creating different alternative solutions. The best option among all the possible alternatives is selected for replacing the original one until no more movements can be done, making it so a Hill Climbing Optimization algorithm [9].

Within each day of the plan, an optimization method is in charge of assigning the trucks to locations to visit, as well as establishing the order of the visits, to cover the minimum distance possible. For that procedure, the ORTools[1] framework is used.

The final solution consists of all the information needed to effectively clean all residual waste located in the selected town, which is expressed by a list of routes, a list of arrival times and the load of the truck at each node.

Given this, the plan is done based upon the conditions that every route conforming to a solution must comply with, which are as follows:

1. Both, the starting and ending points are the same for all the vehicles.
2. The total capacity of the trucks may not be exceeded at any point.
3. Each container can only be visited once across the whole plan. After the visit, the filling of the container becomes 0.
4. As every day passed, the load of all the containers is increased by a defined amount.

Finally, in the effort of facilitating the understanding of the outcomes, the routes of each day will be presented in a Folium[2]-generated map, along with a report, addressing specific information, such as the time of the routes and the amount of weight they will manoeuvre.

3.2 Mathematical Model

To proceed with the formulation of the problem, the following nomenclature will be used. A total fleet of N trucks is considered, as well as K containers for collection and D, which will be used to denote the number of days to plan the garbage collection. Each of the trucks has a capacity of N_i $(i = 1 \ldots N)$, while each of the containers has an initial filling of F_j, a daily increment in the filling of Δ_j and a maximum capacity of M_j $(j = 1 \ldots N)$.

Each day of the plan, the non-collected containers increase their filling according to Eq. 1. The overflow of a container is considered when $F_j > M_j$ is fulfilled for any $j \in \{1 \ldots N\}$.

$$F_j = F_{(j-1)} + \Delta_j, \forall (j = 1 \ldots N) \tag{1}$$

Given so, a *plan* to be optimized will be defined and presented as $plan_i = j$, where $i \in \{1 \ldots K\}$ and $j \in \{1 \ldots D\}$. Therefore, the method will optimize the routes among all the containers that have been selected for collection each day.

---

[1] https://developers.google.com/optimization.
[2] https://python-visualization.github.io/folium/.

For the distribution of collection days to each of the K containers ($plan$), the evaluation (or cost function) defined is presented in Algorithm 1. The rationale of the method is as follows for each day in the planning horizon (D):

1. Initially, the cost is initialized to 10.000, which, in this case is used as a very high starting value denoting a extremely bad solution (equivalent to ∞), for each one of the days on the horizon (line 1).
2. If any of the container's overflow (line 5) or the VRP optimizer is not able to find a solution due to capacity constraints (line 12), the cost of this day is calculated as a very high penalty number plus the quantity of garbage to be collected in the day. This means that, for any two plans that make containers overflow, the one with a minor quantity to collect is less penalized. After that, the loop is stopped, and the following days remain with a high cost (the value of 10.000 is used in this case).
3. In the case of the VRP optimizer being able to find a solution for collecting the containers assigned in the day (line 10), the cost for the day is -10.000 plus the distance travelled by the solution.

As a result of the aforementioned rationale, the algorithm will prioritize short routes that do not make containers overflow during the time horizon. Additionally, if overflow occurs, the optimization function will evaluate better solutions with later overflow.

4   Proposal

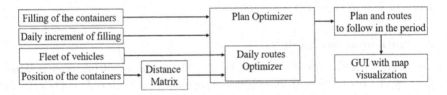

Fig. 1. Overall schema of the solution

To provide an end to end solution, the schema presented in Fig. 1 is followed, where input data and several components interact among themselves:

- The position of the containers is extracted from the Open Data portal of the county of Biscay[3] (Spain).
- Both, the filling and daily increment of the containers are simulated in this study. In Sect. 5, the concrete values used in the experimentation will be specified.
- Distance matrix among the containers is calculated by using Open Source Routing Machine[4].

---

[3] https://opendata.euskadi.eus/catalogo/-/contenedores/.
[4] http://project-osrm.org.

Algorithm 1: Calculation of the cost of a solution (for minimization).

1   $cost_d = +10.000 \; \forall d \in \{1 \ldots D\}$;
2   **for** $d \in \{1 \ldots D\}$ **do**
3      $collect = \{i \mid plan_i == D_c\}$;
4      $noncollect = \{i \mid plan_i! = D_c\}$;
5   **if** $any(F_{collect} > M_{collect})$ **then**
6          $cost_d = +10.000 + \sum(F_{collect})$;
7          $break$;
8   **else**
9          $solution = VRP\_optimizer(collect, F_{collect})$;
10   **if** $solution\ found$ **then**
11             $cost_d = -10.000 + solution\_distance$;
12   **else**
13             $cost_d = +10.000 + \sum(F_{collect})$;
14             $break$;
15   **end**
16          $F_{non\_collect} = F_{non\_collect} + \Delta_{non\_collect}$;
17          $F_{collect} = 0$
18   **end**
19 **end**
20 $return(\sum cost)$

- The plan optimizer is in charge of assigning the collection of different containers to one of the days on the horizon.
- For this, it uses the daily routes optimizer component, which is in charge of searching for the optimal solution to the derived CVRP, given a set of containers and a fleet of vehicles. In this proposal, ORTools[5] is used.
- Finally, the system returns a set of routes for each of the days on the horizon, which are presented to the user in a GUI, presenting visualization of the routes in a map powered by Folium[6].

In this work, a Hill Climber optimization algorithm is used within the Plan Optimizer module, whose working operation procedure is presented in Algorithm 2. The algorithm starts building a plan for the collection of the containers by assigning the collection of each container to a random day on the horizon and evaluating it according to the procedure presented in Algorithm 1. After the initial plan is generated, the algorithm, in an iterative way, creates N_s alternative plans derived from variations of the initial plan, and the cost of each one is subsequently calculated and stored. Once all the alternatives have been evaluated, the plan with the lowest cost replaces the initial one, but only if an improvement is made.

The *modification* operator (line 6 in Algorithm 2) randomly increases or decreases by 1 the assigned day of collection of a randomly selected container,

---

[5] https://developers.google.com/optimization.
[6] https://python-visualization.github.io/folium/.

Algorithm 2: Hill Climber optimization procedure.

---

1 $plan = randomassignation(K, D)$;
2 $cost = cost(plan)$;
3 **for** $i \in \{1 \dots iterations\}$ **do**
4      $bestCost = \infty$
5   **for** $j \in \{1 \dots N_s\}$ **do**
6          $newPlan = modification(plan)$;
7          $newCost = cost(newPlan)$;
8   **if** $newCost < bestCost$ **then**
9              $bestCost = newCost$;
10             $besPlan = newPlan$;
11   **end**
12   **end**
13   **if** $bestCost < cost$ **then**
14         $cost = bestCost$;
15         $plan = bestPlan$;
16   **end**
17 **end**
18 $return(plan)$;

---

always respecting the day of the collection cannot be lower than one or higher than D.

5   Experimentation

5.1   Use Cases

The system includes data from all the municipalities in the Basque Country, Spain, collected from the Open Data portal. For experimentation purposes, a selection of three locations has been done. The selected municipalities ensure a diverse distribution of container, as well as varying size for the problem. The three selected municipalities are:

1. Abadiño: a small municipality with a total of $K = 69$ containers.
2. Durango: a medium-size municipality with a total of $K = 149$ containers.
3. Leioa: a big municipality with a total of $K = 192$ containers.

When generating a solution for any of the locations, parameters such as the number of trucks (N) and the number of days to plan (D) are slightly modified, always taking into consideration that the amount of containers in one location is directly related to the number of trucks needed to collect them. This relationship is of prime importance when defining the values of the parameters because if the value is low, it can lead to failed plans due to the overload of the containers.

The capacity of the containers and trucks are held constant through the entire experimentation. The amount of load a container is capable of bearing is set to $M_j = 400$ kg while trucks can hold up to $N_i = 2000$ kg in each route.

Each container has an initial load and a daily increase. The initial load of the containers (F_j) range from 40 and 200 kg (half of the container capacity), while the daily increase (Δ_j) ranges between 40 and 120 kg. Both of these values have been kept constant across all the experimentation process. Since there are no sensors from which this information can be retrieved, these numbers have been simulated.

5.2   Results and Analysis

Table 1 explores different situations of the chosen municipalities, with slight modifications of the parameters D and N. In all configurations, the number of iterations is kept at a constant of 100. Finally, in each execution, the number of alternative plans generated to explore alternative solutions in the Algorithm 2 is established to $N_s = K/3$. Columns in the table are:

- Load: the sum of the load from all the collected containers across the plan, in tons.
- Time: the total amount of time the trucks have been travelling from one container to another, in hours.
- Distance: the total amount of km the trucks have travelled.
- Overflow: whether the plan has been completed without any bin overflowing or not.

One of the glaring observations we can make is that no 5-day configuration has been able to be resolved without the overflow of one of the containers, even when that same configuration has worked for a lower number of days. This phenomenon occurs because containers can only be assigned one day for collection, which makes the bins collected on the first day susceptible to overfill in later days, meaning that the proposed algorithm is not suitable for long-term or periodic planning operations. The adaptation of the coding of the solution to face this issue will be considered in future work.

We can also see that adding more days to the planning may lead to a reduction in the number of trucks (N) needed to collect the garbage. Taking a closer look at the municipality of Leioa (case #13), with the planning of 3 days, at least 12 trucks are needed to find a solution, but if the planning is meant for 4 days, 10 trucks are enough (case #15). Cases #1 and #3 also reflect this fact, which occurs because the containers can be distributed across more days, leading to a lower number of bins to be collected per day and consequently being able to find a solution with a lower number of trucks.

With regards to the total load collected, in those cases the plan does not end in overflow, we can see minor differences within the same municipality, but higher values in those plans with more days. This is caused to the increment in the load of the containers, which makes the more days a container is not collected, the higher amount of waste to collect.

An example of a possible solution for the configuration of the case #10 can be found in Fig. 2.

Table 1. Experimental results obtained.

#	Location	D	N	Load (tons)	Time (h)	Distance (km)	Overflow
1	Abadiño	3	7	16.95	14.05	118.25	
2	Abadiño.	3	10	17.88	15.4	118.25	
3	Abadiño	4	4	17.42	14.18	119.13	
4	Abadiño	4	7	18.73	13.82	80.74	
5	Abadiño	4	10	20.6	16.55	117.18	
6	Abadiño	5	4	9.38	9.73	71.85	x
7	Abadiño	5	7	9.97	10.15	68.51	x
8	Abadiño	5	10	4.52	10.82	67.36	x
9	Durango	3	10	32.92	22.73	88.85	
10	Durango	4	10	37.62	24.31	93.69	
11	Durango	5	10	29.81	20.02	76.94	x
12	Durango	5	15	27.97	17.67	72.10	x
13	Leioa	3	12	42.83	30.33	93.01	
14	Leioa	3	14	43.8	30.57	140.39	
15	Leioa	4	10	47.37	31.13	167.72	
16	Leioa	5	10	38.32	24.55	95.11	x
17	Leioa	5	20	43.48	27.18	123.17	x

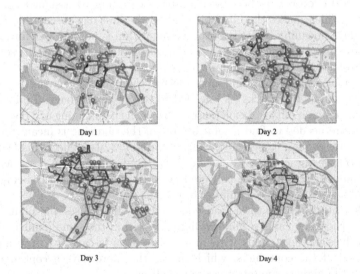

Day 1   Day 2

Day 3   Day 4

Fig. 2. Result of the optimization for the municipality of Durango

6   Conclusions and Future Works

This work has presented a system for the planning of waste collection routes for a fleet of trucks for a specified time horizon, in days. The system is built upon the assumption the containers can be monitored, and an estimation of the increase of the filling each day can be provided. Given that, the system results suitable for Smart Cities scenarios.

For the optimization of the planning, it is proposed a formulation of an optimization problem, where the day in which a particular container is collected is encoded. After that, for each particular day, the specific sequence of containers followed by each one of the trucks in the fleet is subsequently optimized using a CVRP optimizer. The fitness function used in the optimization engine considers the minimization of the distance travelled by the trucks, as well as the avoidance of the overflow of the containers within the time horizon.

Experimental results over three municipalities of the Basque Country, in Spain, show the suitability of the approach for the successful planning of the fleet.

Future works come through the inclusion of the capability of optimizing the collection of the containers in more than a day, to make it suitable for long-term planning of the operations. On the other hand, the improvement of the implemented search heuristic, by using more advanced optimization meta-heuristics are expected to be used in the next stages of the developments. In this regard, the implementation of meta-heuristics able to make use of multiple combination of modification operators, as well as their application and comparative analysis within more use cases will be considered.

Acknowledgements. This work has been supported by The LOGISTAR project, which has received funding from the European Union's Horizon 2020 research and innovation programme under Grant Agreement No. 769142. Also, from the Spanish Government under the project IoTrain (RTI2018-095499-B-C33).

References

1. Adewumi, A.O., Adeleke, O.J.: A survey of recent advances in vehicle routing problems. Int. J. Syst. Assur. Eng. Manage. **9**(1), 155–172 (2016). https://doi.org/10.1007/s13198-016-0493-4
2. Ali, T., Irfan, M., Alwadie, A.S., Glowacz, A.: IoT-based smart waste bin monitoring and municipal solid waste management system for smart cities. Arab. J. Sci. Eng. **45**(12), 10185–10198 (2020). https://doi.org/10.1007/s13369-020-04637-w
3. Amarnadha Reddy, A., Gangadhar, B., Muthukumar, B., Albert Mayan, J.: Advanced garbage collection in smart cities using IoT. In: IOP Conference Series: Materials Science and Engineering, vol. 590. Institute of Physics Publishing, October 2019. https://doi.org/10.1088/1757-899X/590/1/012020
4. Batty, M.: Smart cities, big data (2012)
5. Bhandari, R., Nidhi, S., Swapnil, R., Dhruvi, D., Harsh, K.: Survey on IOT based smart city bin. Int. J. Comput. Appl. **177**(11), 29–33 (2019). https://doi.org/10.5120/ijca2019919528

6. Buhrkal, K., Larsen, A., Ropke, S.: The waste collection vehicle routing problem with time windows in a city logistics context. Proc. - Soc. Behav. Sci. **39**, 241–254 (2012). https://doi.org/10.1016/j.sbspro.2012.03.105

7. Chowdhury, P., Sen, R., Ray, D., Roy, P., Sarkar, S.: Garbage monitoring and disposal system for smart city using Iot. In: Proceedings of the 2nd International Conference on Green Computing and Internet of Things, ICGCIoT 2018, pp. 455–460. Institute of Electrical and Electronics Engineers Inc., August 2018. https://doi.org/10.1109/ICGCIoT.2018.8753060

8. Defalque, C.M., da Silva, A.F., Marins, F.A.S.: Goal programming model applied to waste paper logistics processes. Appl. Math. Model. **98**, 185–206 (2021)

9. Derigs, U., Kaiser, R.: Applying the attribute based hill climber heuristic to the vehicle routing problem. Eur. J. Oper. Res. **177**(2), 719–732 (2007)

10. Gutierrez, J.M., Jensen, M., Henius, M., Riaz, T.: Smart waste collection system based on location intelligence. Proc. Comput. Sci. **61**, 120–127 (2015). https://doi.org/10.1016/j.procs.2015.09.170

11. Huang, S.H., Lin, P.C.: Vehicle routing-scheduling for municipal waste collection system under the "keep trash off the ground" policy. Omega (U.K.) **55**, 24–37 (2015). https://doi.org/10.1016/j.omega.2015.02.004

12. Khandelwal, S., Yadav, R., Singh, S.: IoT based Smart Garbage Management-Optimal Route Search. Int. Res. J. Eng. Technol. **6**, 44–50 (2019)

13. Nurprihatin, F., Lestari, A.: Waste collection vehicle routing problem model with multiple trips, time windows, split delivery, heterogeneous fleet and intermediate facility. Eng. J. **24**(5), 55–64 (2020). https://doi.org/10.4186/ej.2020.24.5.55

14. Stanković, A., Marković, D., Petrović, G., Čojbašić, Ž.: Metaheuristics for the waste collection vehicle routing problem in urban areas. Facta Univ. Ser.: Work. Living Environ. Protect. 001–016 (2020). https://doi.org/10.22190/FUWLEP2001001S

15. Statheropoulos, M., Agapiou, A., Pallis, G.: A study of volatile organic compounds evolved in urban waste disposal bins. Atmos. Environ. **39**(26), 4639–4645 (2005)

16. Wongsinlatam, W., Thanasate-Angkool, A.: Finding the shortest route of waste collection by metaheuristic algorithms using SCILAB. In: Proceedings - 2nd International Conference on Mathematics and Computers in Science and Engineering, MACISE 2020, pp. 20–23. Institute of Electrical and Electronics Engineers Inc., January 2020. https://doi.org/10.1109/MACISE49704.2020.00011

17. Wu, H., Tao, F., Yang, B.: Optimization of vehicle routing for waste collection and transportation. Int. J. Environ. Res. Public Health **17**(14), 4963 (2020). https://doi.org/10.3390/IJERPH17144963

Optimization of Warehouse Layout for the Minimization of Operation Times

Jon Díaz, Haizea Rodriguez, Jenny Fajardo-Calderín[✉], Ignacio Angulo, and Enrique Onieva

DeustoTech, Faculty of Engineering, University of Deusto,
Av. Universidades, 24, 48007 Bilbao, Spain
{jondi,haizearodriguez}@opendeusto.es,
{fajardo.jenny,ignacio.angulo,enrique.onieva}@deusto.es

Abstract. Warehousing management is essential for many companies involved in the supply chain. The optimal arrangement and operation of warehouses play important role in companies, allowing them to maintain and increase their competitiveness. One of the main goals of warehousing is the reduction of costs and improvement of efficiency.

Determination of the ideal warehouse layout is a special optimization case. In the article, a procedure is applied to the optimization of the allocation of bins in an automated warehouse with special characteristics. The initial layout of the warehouse, as well as the automated platforms constraints the search, as well as define the time needed to move goods inside the warehouse. The definition of time needed to move goods, with the analysis of historical data, allows the definition of a mathematical model of the operation of the warehouse. Using that model, an optimization procedure based on the well-known hill-climbing algorithm is defined. Experimental results show increments in the efficiency of the warehousing operations.

Keywords: Warehouse layout design · Optimization of warehouse · Decision support models · Logistics · Optimization

1 Introduction

One of the principal goals of global trends of the market is to put a great effort into trying to distribute goods more rapidly to the customers and consequently reduce the cost of item storage. An initial idea of directly connecting the supplier and the customer has been contemplated, but it is still not a reality[3]. Companies are still in need of a warehouse where goods can be held and organized before distribution in a clever way, and intelligent warehouses are the key.

Storehouses have evolved over the years. At first, all the manhandling of the products was done manually, but in the effort of reducing time and cost, technology has been used to automate the main processes. The machinery installed in the warehouses has proved useful by improving the movement's cost and

© Springer Nature Switzerland AG 2021
H. Sanjurjo González et al. (Eds.): HAIS 2021, LNAI 12886, pp. 649–658, 2021.
https://doi.org/10.1007/978-3-030-86271-8_54

efficiency. Nonetheless, there are some obstacles that even advanced machines cannot solve. One of the main obstacles found at warehouses, while trying to optimize the services provided, is the proper arrangement of goods [1].

Warehouses use to have a static layout that is not supposed to change, but the way products are arranged can be adapted to the overall needs. The main goal is the reduction of time, and placing the most requested items closer to the extraction points can help achieve it. These products are constantly being moved around, thus, by reducing the effort to access them, the overall cost is then reduced. The effect of this is similarly reflected in both, manually managed warehouses (and) automated ones.

In order to find the good's best arrangement, the use of optimization techniques is crucial. The solution provided by the algorithms can accomplish the goal of reducing the cost and improving the overall performance, but it cannot assure a perfect solution is found. Because of this, several algorithms are tested with the intention of finding the arrangement that returns the best warehouse layout.

This article presents a use case of optimization methods applied to the arrangement of the goods within an automated warehouse. The objective of the optimization procedure is to minimize the time the automated platforms inside the warehouse spend moving goods. To achieve so, historical data from a real warehouse is used and processed to build a matrix denoting the flow within each pair of possible locations of goods and then, using the shape of the environment, the optimization module is in charge of re-arranging the goods to minimize the time spent to realize all the movements registered. The module used a model of the warehouse to model the time needed to move a good between two locations of the warehouse. A Hill-Climbing algorithm is implemented and 3 different neighbourhood operators are compared in terms of execution time and performance obtained. This paper is structured as follows. Section 2 presents the related literature review. Section 3 describes the warehouse design and restrictions applicable in the proposed scenario. Section 4 states the formulation of the problem to be solved by the optimization. After that, Sect. 5 presents experimental results obtained. Finally, Sect. 6 states conclusions and future works.

2   Literature Review

The selection of the ideal warehouse layout is a special optimization process, not a typical mathematical optimization. Since the number of layout alternatives is huge or infinite, therefore, the formation and the evaluation of all possible alternatives are impossible. This is one of the reasons that heuristic method and continuous iteration have to be used during the optimization process.

It is a complex problem to design a warehouse. It includes a large number of interrelated decisions involving the functional description of a warehouse, technical specifications, selection of technical equipment and their layout, warehouses processes, warehouses organizations and others. Several authors present general models of warehouse design and planning.

Jaimes et al. [7] presented the evaluation of applying models to improve efficiency in management of warehouses used in shipyards, focused on pick up, packing, and shipping activities. Besides proposing the best physical layout for the storage of goods, the model seeks to minimize three types of costs: costs related to the initial investment (construction and maintenance), shortage costs, and costs associated with storage policies.

Muharni et al. [10] presented an optimization model to obtain a better disposition of a warehouse containing raw materials. For the solution to the design of the arrangement of the elements, the particle swarm algorithm was used and the objective was to minimize the cost of handling materials. Moreover, Ballesteros et al. [4] propose an optimization model to obtain an adequate arrangement of different food products in a warehouse, over different periods of time. They develop a solution that identifies operational areas and the amount of required spaces in which the products must be located, seeking to reduce the costs of maintaining and handling the products. The CPLEX software is used to solve the problem of dynamic allocation of products.

Arif et al. [2] proposed the Genetic Algorithm application in optimizing the arrangement of storage of manufactured goods in warehouses. The goal is to find solutions to minimize the amount of coverage area used by pallets and maximize the number of boxes stored on pallets. The use of Genetic Algorithms in this research aims to find the best fitness value in the allocation of goods in the finished warehouse so that the arrangement and allocation of goods are not done carelessly and can reduce the free space in the warehouse.

Derhami et al. [5] presented an optimal design of the distribution of a warehouse, taking into account: the number of aisles and cross aisles, the depth of the streets and the types of cross aisles. They develop a simulation-based optimization algorithm to find optimal arrangements regarding: material handling cost and space utilization. They use a case study in the beverage industry and show that the resulting layout can save up to 10% of a warehouse's operating costs. Also, Irman et al. [6] developed an integer linear programming model to obtain an optimal layout design that minimizes total travel costs, and they also used the LINGO 17 software.

Sudiarta et al. [11] In this work an analysis of the methods used for the planning of the warehouse distribution is made, they study the following methods: Dedicated Storage, Class-Based Storage, Shared Storage, Random Storage and Fishbone Layout. These methods are used to decrease the handling distance of the materials, so that the distribution reaches an optimal solution.

Kovács et al. [8] presented the characteristics and the detailed procedure of the special optimization process of the warehouse layout design are described. They describe the most common objective functions that can be taken into account to model the problem. In addition, they explain the most important constraints and limitations that the optimization problem has.

Saderova [9] presented warehouse system design methodology that was designed applying the logistics principle-systematic (system) approach. The starting point for designing a warehouse system represents of the process of design logistics systems.

Within literature presented, different specific use cases and problem modeling can be found, with contributions ranging from developments of novel and better performing algorithms to their application to real complex domains. This works is focuses on the application of optimization procedures to a real case, with specific modelling, requirements and constraints, which makes it interesting for the community. The present article proposes an use case of well-known optimization method to specific warehouse layout, which opens the door to future developments in the field.

3   Characteristics of the Warehouse Layout Design

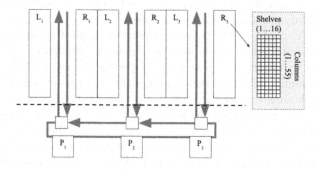

Fig. 1. Layout of the warehouse, composed by three aisles and three packaging posts.

This work is done using data from a real automated warehouse. An schema of the layout of the warehouse can be seen in Fig. 1, which is divided in two sections, the corridor area where the stock is stored ($R_{1,2,3}$ and $L_{1,2,3}$) and three packaging posts ($P_{1,2,3}$), where the shipment is stored in boxes according to the received shipment order. The three corridors are divided into sides, denoted by R or L to denote right of left handed respectively, and each side is divided into 16 shelves and 55 columns.

This makes a total number of 5280 possible locations for the products (3 aisles · 2 sides · 16 shelves · 55 columns). Each location will be identified by variable $R_{\{a,s,c\}}$, $\forall a = 1,2,3$, $\forall s = 1\ldots16$, $\forall c = 1\ldots5$ (similarly for $L_{\{a,s,c\}}$ for left handed sides). In addition, packaging locations will assumed to have shelf and column equal to zero.

A single unidirectional conveyor belt connects both sections of the warehouse, which continuously circles from the corridors to the working posts. An automatic platform operates in each of the aisles, taking the bins from their respective shelves to the belt and vice versa. The platform can only take a single bin at a time, and the belt will completely stop its movement until the bin has either been collected or placed on top of the belt itself.

The platform is also able of interchanging the position of two bins of its corridor. When the bins are on the same side, swapping is almost immediate, but, if that is not the case, then the bin needs to be placed on the belt so that it can be properly placed once it comes back. Therefore, swapping bins of different sides is extremely expensive.

Each of the aisles is equipped with a robotic platform able to pick a bin from a particular location and to move to another in the same aisle, or to put it in the conveyor belt, in order to move it to another one or to one of the three packaging posts. The optimization will consider the minimization of the total time needed for operations of movements between locations inside the warehouse, so in order to calculate this time, following assumptions are considered:

- Each of the robotic platforms moves both, horizontally (between columns) and vertically (between shelves) simultaneously, at speeds S_h and S_v respectively.
- Distances between consecutive columns or shelves are denoted as d_h and d_v, respectively.
- In order to move a bin between two sides of the same aisle (R_i to L_i or vice versa), the platform will first move the bin from the origin to the corresponding packaging point (P_i), and then to the destination.
- To move a bin between different aisles, the platform will move it to the corresponding packaging point, then, the conveyor belt moves it to the destination packaging point and then to the destination. Distance between packaging points is denoted as d_p, while the outer trail of the belt is a total of d_r. The conveyor belt moves bins at speed S_b.

4   Mathematical Model

Given characteristics exposed in previous section, the time needed of moving a bin from a origin ($O_{a1,s1,c1}$) to a destination ($D_{a2,s2,c2}$) is calculated as presented in Algorithm 1.

In this case, it has been considered $S_h = 2.5$, $S_v = 1.5$ and $S_b = 2.5$ for speeds (in m/s), as well as $d_h = 2.5$, $d_v = 1$, $d_p = 2$, $d_r = 10$, in meters, for distances in the layout. The algorithm takes into account four different cases of movements, which are visually explained in Fig. 2:

1. Origin and destination are in the same aisle and side: the robotic platform directly moves the bin. Case (a) in Fig. 2.
2. If both locations are in opposite sides of the same aisle, the bin must be moved to the corresponding packaging point, and then to the destination. This is represented in Case (b) in Fig. 2.
3. If origin and destination are in different aisles, the time will depend if the movement needs to make use of the outer ring of the conveyor belt or not. The first case is represented in case (d) in Fig. 2, while the second in case (c).

The main goal is to accelerate the movement of the bins across the warehouse by decreasing the time spent retrieving and moving them from their respective

---

Algorithm 1: Calculation of the time for moving a bin among two points.

1 **if** $a1 == a2$ **then**
2 | **if** $O == D$ **then**
3 | | time $= max(\frac{|c1-c2|d_h}{S_h}, \frac{|s1-s2|d_v}{S_v})$
4 | **else**
5 | | time $= max(\frac{(56-c1)d_h}{S_h}, \frac{(17-s1)d_v}{S_v}) + max(\frac{(56-c2)d_h}{S_h}, \frac{(17-s2)d_v}{S_v})$
6 **else**
7 | time $= max(\frac{(56-c1)d_h}{S_h}, \frac{(17-s1)d_v}{S_v}) + max(\frac{(56-c2)d_h}{S_h}, \frac{(17-s2)d_v}{S_v})$;
8 | **if** $a1 > a2$ **then**
9 | | time $= time + \frac{(a1-a2)d_r}{S_b}$
10 | **if** $a1 < a2$ **then**
11 | | time $= time + \frac{(a1-1)d_p+(3-a2)d_p+d_r}{S_b}$

---

shelves to the belt. For doing so, optimization methods will be used to minimize the time spent by robotic platforms moving bins by reallocating them along the warehouse.

In order to formulate the optimization algorithm, the fitness function to minimize is presented in Eq. 1, where both M_d and M_f are 5280×5280 matrices denoting the distance and number of movements realized (flow) between 2 locations respectively.

$$F = \sum(M_d \times M_f) \tag{1}$$

M_d is calculated by using the procedure presented in Algorithm 1 over each pair of locations, while, for obtaining the values of M_f, real data coming from a company is used. Data covers the entire historical information of movements in the warehouse from 2019-09-26 to 2020-09-15, having a total of 373933 entries. In order to fill M_f, the data is processed, and the number of movements between each pair of locations is calculated.

4.1   Optimization Procedure

For the minimization of the Eq. 1, a permutation based coding is used. A candidate solution represents a feasible arrangement of the bins, which can be represented as $X = (x_1, x_2, ...x_{5280})$, where x_i represents the product stored in the i-th location.

Initially, $x_i = i$, denoting the initial solution being the original arrangement of the warehouse. With this, we ensure the final solution will be as similar as the original arrangement as possible, avoiding the cost of arranging all the elements within the warehouse from scratch. Interchanging operators will allow to generate neighbour solutions to explore the solutions space. Any change in the arrangement of the positions will derive in a reordering of both the rows and columns in M_f, so result of Eq. 1 will be recalculated.

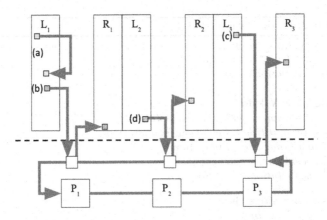

Fig. 2. Visual representation of the possible movements. With origin and destination in: same aisle and side (a), same aisle but different side (b), different aisle without using the external ring (c) or using it (d).

The optimization procedure can be seen in Algorithm 2, a version of the well known Hill-Climber algorithm, where the number of neighbour solutions to explore in each iteration is limited, due the complexity of the solution space. The selection of the Hill-Climber schema is justified by the need of the stakeholder of the solution for using an algorithm able to modify as less as possible the initial arrangement of the warehouse, while guaranteeing as optimal as possible results.

After the initial plan is generated and evaluated using the procedure presented in Algorithm 1, the method, in an iterative way, creates N_s alternative plans derived from variations of the initial solution, and the time of each one is subsequently calculated and stored. Once all the alternatives have been evaluated, the solution with the lowest cost replaces the initial one, but only if an improvement is made.

In this work, three alternatives for the *getNeighbour* operator (line 6 in Algorithm 2) are implemented, which are following:

1. *Swap*: This operator takes randomly two elements of the solution and interchanges them.
2. *Insertion*: The operator selects one of the elements of the solution and a random location in the vector in order to insert it in.
3. *Adjacent*: The operator randomly selects one of the elements in the warehouse and interchanges it with one of the colliding ones (upper or lower shelve or column in the left or right).

Algorithm 2: Optimization procedure.
___

1   $solution = \{1, 2, 3, \ldots 5280\}$;
2   $time = timeCalculation(solution)$;
3   **for** $i \in \{1 \ldots iterations\}$ **do**
4      $bestTime = \infty$
5   **for** $j \in \{1 \ldots N_s\}$ **do**
6         $newSolution = getNeighbour(solution)$;
7         $newTime = timeCalculation(newSolution)$;
8   **if** $newTime < bestTime$ **then**
9            $bestTime = newTime$;
10           $besSolution = newSolution$;
11   **if** $bestTime < time$ **then**
12         $time = bestTime$;
13         $solution = bestSolution$;
14 $return(solution)$;
___

5 Experimentation and Results

This section presents the results of the carried out experiments for comparison of the performance of the algorithm using the three proposed neighbour operators, as well as for different sizes of the explored neighbour of a solution (N_s). During the experimentation, the number of iterations was kept constant at $iterations = 10^5$. Results are presented in Table 1, where the different executions ran are compared in terms of execution time and reduction value of the fitness function with respect to the initial (original) arrangement of the bins.

Table 1. Experimental results obtained. Result is calculated as the percent time reduction over the initial arrangement.

getNeighbour()	N_s	Execution time	Result
Swap	100	00:27:50	1.49%
Swap	500	01:24:13	4.15%
Swap	1000	03:16:24	7.69%
Insertion	100	05:27:21	10.19%
Insertion	500	26:24:06	64.98%
Insertion	1000	–	–
Adjacent	100	00:29:59	22.46%
Adjacent	500	01:40:56	54.39%
Adjacent	1000	03:10:39	68.90%

The *Insertion* algorithm was discarded as it took more than 5 h to perform the 10.000 iterations with the lowest N_s. Since in subsequent executions the neighborhood size was greatly increased, it was expected for the second algorithm to take much longer, which was proven to be true with the neighborhood size of 500. This happened due to the actual implementation of the operator did not allow the use of fast calculation of the fitness function, which was achieved by subtracting and adding values to the previous fitness accordingly to movements performed.

As far as the rest of the algorithms are concerned, their execution times are very close to each other. However, the gap among them increases as the neighborhood becomes bigger.

Based on the result of the experimentation procedure, it can be concluded that the first algorithm does not achieve considerable improvements with the enlargement of the neighborhood. The third algorithm performs better, and it reaches a good optimization values when compared to the other methods while maintaining reasonable amount of time spent.

6   Conclusions and Future Works

This work has presented the modelization of an optimization problem from the historical data of warehousing operations in a real company. The characteristics of the automated platforms operating the warehouse, as well as their movement speeds are used to build a mathematical model of the operation of the warehouse. In addition, historical data regarding past movements of bind within the area are used to model the matrix containing the flow or number of movements between locations in the warehouse.

An optimization procedure is defined to properly arrange the goods in the warehouse for improved efficiency and reduction of costs. The proposal is based on the well-known hill-climbing algorithm, and it is used for the arrangement of products within the warehouse. Three different neighbouring operators are used within the procedure and the performance is measured in terms of both computational time and reduction of operational time.

Future works will be focused on the real-time management of the warehouse, including the optimization procedure within the operation of the warehouse to guarantee the optimal allocation of goods as they arrive for the first time. The exploration of additional operators for modification of the current arrangement, as well as the implementation of more sophisticated optimization algorithms, will be considered in future developments. With this regard, evolutive meta-heuristics able to manage with different recombination or modification methods will be considered as the next step to improve actual results.

Acknowledgement. This work has been supported by The LOGISTAR project, which has received funding from the European Union's Horizon 2020 research and innovation programme under Grant Agreement No. 769142. Also, from the Spanish Government under the project IoTrain (RTI2018-095499-B-C33).

References

1. Arif, F.N., Irawan, B., Setianingsih, C.: Optimization of storage allocation for final goods warehouse using genetic algorithm. In: 2020 2nd International Conference on Electrical, Control and Instrumentation Engineering (ICECIE), pp. 1–6, November 2020. https://doi.org/10.1109/ICECIE50279.2020.9309669
2. Arif, F.N., Irawan, B., Setianingsih, C.: Optimization of storage allocation for final goods warehouse using genetic algorithm. In: 2020 2nd International Conference on Electrical, Control and Instrumentation Engineering, Proceedings, ICECIE 2020. Institute of Electrical and Electronics Engineers Inc., November 2020. https://doi.org/10.1109/ICECIE50279.2020.9309669
3. Baker, P., Canessa, M.: Warehouse design: a structured approach. Eur. J. Oper. Res. **193**(2), 425–436 (2009). https://doi.org/10.1016/j.ejor.2007.11.045
4. Ballesteros-Riveros, F.A., et al.: Storage allocation optimization model in a Colombian company. DYNA **86**(209), 255–260 (2019). https://doi.org/10.15446/DYNA.V86N209.77527
5. Derhami, S., Smith, J.S., Gue, K.R.: A simulation-based optimization approach to design optimal layouts for block stacking warehouses. Int. J. Prod. Econ. **223**, 107525 (2020). https://doi.org/10.1016/J.IJPE.2019.107525
6. Irman, A., Muharni, Y., Yusuf, A.: Design of warehouse model with dedicated policy to minimize total travel costs: a case study in a construction workshop. In: IOP Conference Series: Materials Science and Engineering, vol. 909, no. 1, p. 012088, December 2020. https://doi.org/10.1088/1757-899X/909/1/012088
7. Jaimes, W.A., Otero Pineda, M.A., Quiñones, T.A.R., Tejeda López, L.: Optimization of a warehouse layout used for storage of materials used in ship construction and repair. Ship Sci. Technol. **5**(10) (2012)
8. Kovács, G.: Special optimization process for warehouse layout design. In: Jármai, K., Voith, K. (eds.) VAE 2020. LNME, pp. 194–205. Springer, Singapore (2021). https://doi.org/10.1007/978-981-15-9529-5_17
9. Saderova, J., Rosova, A., Sofranko, M., Kacmary, P.: Example of warehouse system design based on the principle of logistics. Sustainability **13**(8), 4492 (2021). https://doi.org/10.3390/su13084492
10. Septiani, W., Divia, G.A., Adisuwiryo, S.: Warehouse layout designing of cable manufacturing company using dedicated storage and simulation promodel. In: IOP Conference Series: Materials Science and Engineering, vol. 847, no. 1, p. 012054, April 2020. https://doi.org/10.1088/1757-899X/847/1/012054
11. Sudiarta, N., Gozali, L., Marie, I.A., Sukania, I.W.: Comparison study about warehouse layout from some paper case studies. In: IOP Conference Series: Materials Science and Engineering, vol. 852, no. 1, p. 012112, July 2020. https://doi.org/10.1088/1757-899X/852/1/012112

Simple Meta-optimization of the Feature MFCC for Public Emotional Datasets Classification

Enrique de la Cal[1]([✉]), Alberto Gallucci[1], Jose Ramón Villar[1], Kaori Yoshida[2], and Mario Koeppen[2]

[1] Computer Science Department, University of Oviedo, Oviedo, Spain
{delacal,villarjose}@uniovi.es
[2] Departament of Human Intelligence Systems, Kyushu Institute of Technology, Fukuoka, Japan
kaori@brain.kyutech.ac.jp, mkoeppen@ieee.org

Abstract. A Speech Emotion Recognition (SER) system can be defined as a collection of methodologies that process and classify speech signals to detect emotions embedded in them [2]. Among the most critical issues to consider in an SER system are: i) definition of the kind of emotions to classify, ii) look for suitable datasets, iii) selection of the proper input features and iv) optimisation of the convenient features. This work will consider four of the well-known dataset in the literature: EmoDB, TESS, SAVEE and RAVDSS. Thus, this study focuses on designing a low-power SER algorithm based on combining one prosodic feature with six spectral features to capture the rhythm and frequency. The proposal compares eleven low-power Classical classification Machine Learning techniques (CML), where the main novelty is optimising the two main parameters of the MFCC spectral feature through the meta-heuristic technique SA: the n_mfcc and the hop_length.

The resulting algorithm could be deployed on low-cost embedded systems with limited computational power like a smart speaker. In addition, the proposed SER algorithm will be validated for four well-known SER datasets. The obtained models for the eleven CML techniques with the optimised MFCC features outperforms clearly (more than a 10%) the baseline models obtained with the not-optimised MFCC for the studied datasets.

Keywords: Speech Emotion Recognition · SER · Segmental features · MFCC · Emotional speech datasets · SER classification algorithms · Optimization

1 Introduction and Motivation

A Speech Emotion Recognition (SER) system can be defined as a collection of methodologies that process and classify speech signals to detect emotions embedded in them [2].

© Springer Nature Switzerland AG 2021
H. Sanjurjo González et al. (Eds.): HAIS 2021, LNAI 12886, pp. 659–670, 2021.
https://doi.org/10.1007/978-3-030-86271-8_55

Some of the current most brilliant applications of SER are [8]: i) robotics: to design intelligent collaborative or service robots which can interact with humans, ii) marketing: to create specialised adverts, based on the emotional state of the potential customer, iii) in education: used for improving learning processes, knowledge transfer, and perception methodologies, iv)entertainment industries: to propose the most appropriate entertainment for the target audience and v) health, to gather real-time emotional information from the patients to make decisions to improve their lives.

Among the most critical issues to consider in an SER system are: i) Definition of the kind of emotions to classify, ii) Look for the suitable datasets, iii) Selection of the suitable input features, and iv) Definition of the strategy of the proposal.

1.1 Affective State Taxonomy

According to the latter classification, the affective states can be clustered in four main concepts [8]:

- "emotion" is a response of the organism to a particular stimulus (person, situation or event). Usually, it is an intense, short-duration experience, and the person is typically well aware of it;
- "affect" is a result of the effect caused by emotion and includes their dynamic interaction;
- "feeling" is always experienced with a particular object of which the person is aware; its duration depends on the length of time that the representation of the object remains active in the person's mind;
- "mood" tends to be subtler, longer-lasting, less intensive, and more in the background, but it can affect a person's affective state in a positive or negative direction.

Usually, the applications of SER systems are focused on "emotions" recognition since his concept provides instant and a piece of valuable information for the decision module. Besides, it is easier to obtain affective state "emotion" from the voice speech than the remaining affective states, so most SER datasets available in the literature label the data with "emotion" classes. Hence, current work will be focused on **emotions** identification.

1.2 Election of the Emotion Speech Dataset

Since the classification process is dependent on labelled data, databases are an integral component of SER. Furthermore, the success of the recognition process is influenced by the size and quality of the data. Data that is incomplete, low-quality, or faulty may result in incorrect predictions; thus, data should be carefully designed and collected [2]. So, there are three types of Speech Emotion Datasets (SED) for Speech Emotion Recognition: a) Simulated SEDs, b) Elicited (Induced) SEDs, and c) Natural SEDs. Thus, SED Utterances in simulated SEDs are played and recorded in soundproof studios by professional or semi-professional actors; elicited SEDs are produced in a simulated emotional environment that can induce various emotions. In this case, emotions are not entirely evoked, similar to actual emotions; Natural language datasets are compiled chiefly from talk shows, call centre interviews, radio conversations, and related media. This kind of data is more challenging to get because collecting and distributing private data entails ethical and legal challenges.

1.3   Emotional Speech Features

Speech is a variable-length signal that carries both information and emotion, so global or local features can be extracted depending on the required goal. Global features represent the gross statistics such as mean, minimum and maximum values, and standard deviation. Local features represent the temporal dynamics, where the purpose is to approximate a stationary state. These stationary states are essential because emotional features are not uniformly distributed over all positions of the speech signal [24]. For example, emotions such as anger are predominant at the beginning of utterances, whereas surprise is overwhelmingly conveyed at the end of it. Hence, to capture the temporal information from the speech, local features are used. These local and global features of SER systems can be categorised mainly under four groups [2]: prosodic, spectral, voice quality, and features based on Teager energy operator. Nevertheless, classifiers can be improved by incorporating additional features from other modalities, such as visual or linguistic depending on the application and availability. Commonly, prosodic and spectral are the typical features used in SER, although in practice are combined to obtain better performance [2]. Prosodic features are those that can be perceived by humans, such as intonation and rhythm [28], while spectral features capture the vocal tract shape of the person [17]. Characteristics of the vocal tract are well represented in the frequency domain, so usually, spectral features are obtained by transforming the time domain signal into the frequency domain signal using the Fourier transform. One of the more useful spectral features used in SER is MFCC (Mel Frequency Cepstral Coefficient).

1.4   The Goal

Thus, this study focuses on designing a low-power SER algorithm based on combining one prosodic feature with six spectral features to capture the rhythm and frequency. The proposal compares eleven low-power Classical classification Machine Learning techniques (CML), where the main novelty is the optimisation of the two main parameters of the MFCC [1, 7, 26] spectral feature through the meta-heuristic technique SA: the n_mfcc and the hop_length.

The resulting algorithm should be deployed on low-cost embedded systems with limited computational power like smart speakers or smartwatches. Finally, the proposed SER algorithm will be validated with four of the most well-known SEDs.

The structure of the paper is as follows. The following section deals with the description of the proposal, together with the combination and description of the input features. Section 3 will include the details of the selected SED, the experimental setup and discussion of the results. Finally, conclusions and future work are depicted.

2   The Proposal

Typically, the stages that comprise an SER system are: Pre-processing, Feature computation, Feature selection and Classification [2, 23]. However, it is not common in this field to optimise each feature individually after or before the feature selection stage. So this paper proposes including a Feature Optimisation stage (in Green) where the key features will be optimised (see Fig. 1).

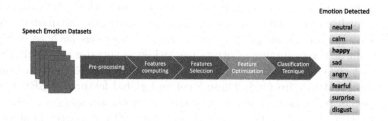

Fig. 1. Overall steps of the proposed SER algorithm

There are two essential issues concerning this new stage: including this step after or before calculating the features and which features are suitable to be optimised. Thus, after the feature selection, the optimisation stage is included to not compute the optimisation on non-significative features. Instead, the features that will be optimised will be the ones with more critical parameters.

Algorithm 1 details the stages defined in Fig. 1 for one generic speech dataset D composed of NF records:

1. L1-7: The sampling rate of all the dataset D files has been set to 16kHz, and the stereo channels have been unified in mono. As the range of values of all the sound wave files is typical, it was not updated.
2. L9-14: As this is a preliminary study of an optimal SER algorithm, segmental transformations have been considered, avoiding spectrographic ones. Thus, two groups of prosodic and spectral features have been selected to capture the rhythm and frequency:
   a) Prosodic features [27]
      - Root-mean-square (RMS): value for each frame, either from the audio samples or from a spectrogram S. It represents the energy of the signal and can be calculated by taking the root average of the square of the amplitude.
   b) Spectral features.

- Mel-Frequency Cepstral Coefficients (MFCCs): are a compact representation of the spectrum(When a waveform is represented by a summation of a possibly infinite number of sinusoids) of an audio signal.
- Chroma_stft: Compute a chromagram from a waveform or power spectrogram.
- Spectral_centroid (spec_cent): Compute the spectral centroid.
- Spectral_bandwidth (spec_bw): Compute p'th-order spectral bandwidth.
- Spectral_rolloff (rolloff): Compute roll-off frequency.
- Zero_crossing_rate (zcr): It calculates how many times a signal is crossing its zero axes. Due to a change in peoples' biological and psychological behaviour with a change in their emotion, it also changed how many times a signal crossed its zero axis.

The seven features are calculated for each Window of WS secs.

3. L16-18: After computing the features, a PCA analysis is carried out discarding the features above the 90 of representativity.
4. L20-28: Each problem and feature must be analysed carefully to decide which features are suitable for optimisation. In the case of our problem, the MFCC feature will be the one.
5. L30-34: Once the features have been optimised, the best model can be obtained through a cross-validation process to be deployed in the proper device.
6. L33-34: The Model obtained in the previous step has to be adapted to the embedded device to be deployed in a real context.

2.1 MFCC Optimization

As one of the most common features and powerful [3,21], in the absence of noise, in the SER field is the MFCC feature, it has been selected to be optimised. The steps to calculate MFCCs are [26]:

1. Frame the signal into short frames.
2. Take the Fourier transform of each frame.
3. Map the powers of the spectrum obtained above onto the mel scale, using overlapping triangular windows or alternatively, cosine overlapping windows.
4. Take the logs of the powers at each of the mel-frequencies.
5. Take the discrete cosine transform of the list of mel log powers as if it were a signal.
6. The MFCCs are the amplitudes of the resulting spectrum.

Hence, we have two relevant issues when calculating MFCC: the number of coefficients that determines the precision of the representation of the original signal (n_mfcc) and the overlapping size (hop_length) between frames(see Fig. 2). According to [16,18], the frame size has been set to 2048 tics as standard, and the n_mfcc and hop_length have taken values in the ranges 20–128 and 4–2048,

Algorithm 1. SER_SYSTEM(D: Dataset, NF: Number of Files in D, WS: Window Size, WO: Window Overlapping, S: Significance value for PCA, Features: Features, FP: Features Parameters, NE: Number of Emotions, NP: Number of Participants, SR: SamplingRate)

```
1:  Step1: Preprocessing
2:  W ← []
3:  for f in 1:NF do
4:      D[f]← SetStandardSamplingRate(D[f], SR)
5:      D[f]← ConvertToMono(D[f])
6:      W← W + WindowFraming(D[f], WS, WO)
7:  end for
8:
9:  Step2: Feature calculation
10: for w in 1:Size(W) do
11:     for ft in 1:Size(Features) do
12:         FEATURES[w, ft] ← ComputeFeature(W, ft, FP[ft])
13:     end for
14: end for
15:
16: Step3: Feature selection
17: PCAContributions← PCA(D, FEATURES, S, FP)
18: FEATURES'[] ← DiscardFeatures(FEATURES, PCAContributions)
19:
20: Step4: Feature optimization
21: for ft in 1:Size(FEATURES') do
22:     if ft should be OPTIMIZED then
23:         OPTIMISED_FP ← Meta-Heuristic(FEATURES')
24:         for w in 1:Size(W) do
25:             OPTIMISED_FEATURES[w, ft] ← ComputeFeature(W, ft, OPTI-
                MISED_FP)
26:         end for
27:     end if
28: end for
29:
30: Step5: Train Clasification Model
31: MODEL ← CrossValidationRun(OPTIMISED_FEATURES, NF)
32:
33: Step6: Test Clasification Model
34: OUTPUT_TEST ← ModelDeployment(MODEL, OPTIMISED_FEATURES, NF)
```

respectively, since each complete sound wave will be 3 secs (16Khz). Between the meta-heuristics available in the literature, it has been selected one of the simplest and effective, Simulated Annealing (SA), [25]. The parameters used have been: i) Stopping criteria: FunctionTolerance under 1.0e-4 and individuals with two variables: number of parameters of MFCC and hop_length.

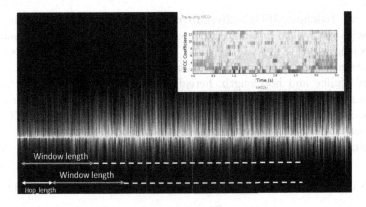

Fig. 2. MFCC coefficients example and Window framing (hop_length)

3   Numerical Results

3.1   Materials and Methods

In order to validate the proposal presented in this work three public Simulated SEDs, and one Elicited SED have been selected: EmoDB [5], TESS [22], SAVEE [11–13] and RAVDESS [19] (see Table 1).

Table 1. Summary of the Speech Emotion Dataset details

Dataset	Participants	Language	Type	Emotions	#Utterances
EmoDB	5 actresses and 5 actors (mixed gender)	German	Simulated	anger, disgust, fear, happiness,sadness, surprise and neutral (7)	535
TESS	2 non professional actresses (female gender)	English	Simulated	anger, disgust, fear, happiness, pleasant surprise, sadness, and neutral (7)	2800
SAVEE	4 actors (male gender)	English	Simulated	anger, disgust, fear, happiness, sadness, surprise and neutral (8)	480
RAVDESS	12 actresses and 12 actors (mixed gender)	English	Elicited	neutral, calm, happy, sad, angry, fearful, surprise, and disgust (8)	1440

Eleven CML techniques has been selected in order to compare the effect of the MFCC optimization using the SA, hereafter MFCCSA: BernoulliNB (Ber) [20], DecisionTree (DT) [14], RandomForestClassifier (RF) [4], ExtraTrees (XT) [10], KNeighbors (KN), RidgeClassifierCV (RC), SVC [6], AdaBoost (AB) [29], GradientBoosting (GB) [14], MLP (Multi Layer Perceptron) [15] and XGB [9].

666   E. de la Cal et al.

And two baseline MFCCs alternatives have been considered: i) the MFCC feature with the typical parameters by default and no optimisation, n_mfcc = 20 and hop_length = 512 [16,18], hereafter MFCCDef, and ii) the MFCC feature optimisation using the exhaustive enumeration method combining all potential values of n_mfcc and hop_length, hereafter MFCCExh. Considering the CML techniques and the three MFCC setups, two studies have been carried out: i) comparing the performance between the selected CML using MFCCSA and MFCCDef, and ii) a time computing cost comparison between the run of the CML using the MFCCSA and MFCCExh.

3.2 Comparison Between CML with MFCCSA and MFCCDef

Table 2 shows the accuracy for each model after optimising the MFCC parameters with SA 500 epochs (MFCCSA) for the four studied datasets, including their combination (All). The baseline results have been calculated for the models using MFCC with the typical parameters by default (MFCCDef). It can be stated that all the models run with the optimised MFCC (*-SA) outperform the corresponding ones with MFCCDef (*-Def). ExtraTrees (XT) outperforms clearly the remaining models for all the datasets. Moreover, the dataset TESS is the one with the best results since it comprises just two participants.

Table 2. Accuracy (the greater, the better) for eleven CML methods carried out with MFCC-SA (*-SA) and MFCC-Def (*-Def) on datasets RAVNESS, SAVEE, TESS, EMO-DB and All.

	Ber	DT	**XT**	RF	MLP	KM	RC	SVC	AB	XGB	GB	Mean
All-Def	35,6	62,5	81,4	79,1	37,6	43,6	54,7	23,8	28,8	80,3	71,1	54,4
All-SA	49,3	66,4	84,5	82,5	62,2	48,8	67,4	24,2	35,2	84,1	77,4	62,0
RAV-Def	24,0	44,2	69,5	65,8	29,3	29,0	46,3	23,2	34,4	66,3	56,3	44,4
RAV-SA	38,1	49,7	75,3	71,7	44,3	30,5	60,2	23,3	36,8	73,6	65,1	51,7
SAV-Def	31,5	46,0	57,1	60,2	39,2	36,7	59,0	25,4	32,3	60,0	57,7	45,9
SAV-SA	46,5	58,3	70,0	69,4	58,3	37,9	70,8	25,4	45,4	70,4	70,8	56,7
TESS-Def	60,7	87,7	98,8	98,1	90,4	58,9	92,0	24,8	40,7	98,2	97,7	77,1
TESS-SA	88,7	92,1	100,0	99,9	99,0	69,3	99,7	26,4	67,6	99,6	99,5	**85,6**
EMO-Def	44,1	53,9	71,8	69,0	49,55	43,0	66,9	29,2	8,0	70,5	68,4	54,5
EMO-SA	49,6	57,6	77,4	73,7	63,8	44,9	72,4	29,2	33,1	74,6	72,4	58,9
MFCCDef	39,2	58,9	75,7	74,4	49,1	42,2	63,8	25,3	32,9	75,1	70,3	
MFCCSA	54,4	64,8	**81,4**	79,4	65,5	46,3	74,1	25,7	43,6	80,4	77,0	

Figure 3 shows the evolution of the accuracy for the metaheuristic SA carrying the ExtraTree CML technique out. For the sake of space, just the curve corresponding to the ExtraTrees technique has been included. As a result, it can be observed that the accuracy of all the datasets has risen approximately the same.

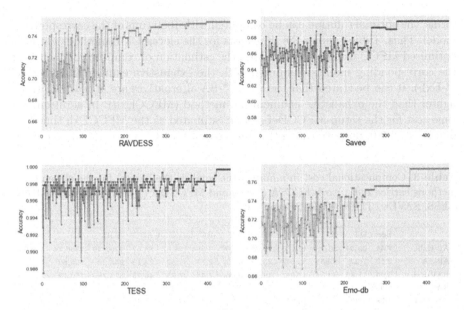

Fig. 3. Evolution of the accuracy for the metaheristic SA for the model ExtraTrees (RAVDESS, Savee, TESS and EmoDB train-datasets)

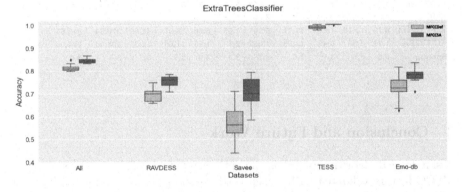

Fig. 4. Comparison of 10 k-fold boxplot with base and optimized parameters

Figure 4 includes the boxplot for the baseline models obtained with the parameters by default for MFCC (MFCCDef) compared with the models trained with optimised MFCC parameters (MFCCSA). Thus, it can be stated that the optimised models outperform all the baseline models for all the datasets.

3.3 Computational Cost Study

The potential value-range of the parameters optimised with the meta-heuristic SA are [20–128] for n_mfcc (108 different values) and $[2^2-2^10]$ for hop_length (10 different values), so the whole possible combinations are: $108*10 = 1080$. In these terms, the optimal solution can be computed with 1080 runs of the corresponding model. Thus, Table 3 shows the time costs for the eleven CML methods with the optimised MFCC using SA (*-SA) and the estimation of computational cost for the corresponding models computed with the exhaustive enumeration method (*-Exh). It can be stated that the whole *-SA alternatives are at least 3,3 times lighter than the exhaustive enumeration method (MFCCExh). In addition, the time cost for the setup MFCCDef can be estimated as the MFCCExh time cost divided by 1080, outperforming the time cost of MFCCSA but not the accuracy.

Table 3. Computational cost in minutes (the smaller, the better) for the eleven CML methods carried out with MFCCSA (*-SA) and MFCCDef (*-Exh) on datasets RAV-NESS, SAVEE, TESS, EMO-DB and All

	Ber	DT	XT	RF	MLP	KM	RC	SVC	AB	XGB	GB	Mean
All-Exh	7,41	25,54	33,38	117,51	90,65	14,09	9,58	124,67	88,57	1202,39	2366,76	370,96
All-SA	2,13	7,22	8,85	32,67	25,12	4,43	2,73	37,52	24,13	355,20	666,60	106,05
RAV-Exh	6,90	11,82	23,54	47,91	48,63	9,06	7,48	27,63	40,73	374,78	750,12	122,60
RAV-SA	1,95	3,35	6,28	12,97	12,98	2,72	2,13	8,42	10,80	106,93	212,65	34,65
SAV-Exh	4,78	5,32	10,12	16,05	22,93	4,82	4,77	5,05	14,35	38,63	132,62	23,58
SAV-SA	1,27	1,42	2,60	4,43	6,50	1,53	1,32	1,58	3,82	10,10	35,65	6,38
TES-Exh	6,99	12,86	21,49	52,08	75,60	8,74	7,85	32,49	49,94	196,06	725,08	108,11
TES-SA	1,87	3,55	6,47	15,05	20,18	2,77	2,28	9,15	13,88	58,82	220,18	32,20
EMO-Exh	4,86	5,12	10,75	15,02	12,53	5,07	4,33	5,03	14,85	39,83	124,69	22,01
EMO-SA	1,35	1,37	2,90	4,05	3,30	1,48	1,20	1,58	4,10	10,63	35,23	6,11
$\overline{MFCCExh}$	6,19	12,13	19,86	49,71	50,07	8,35	6,80	38,97	41,69	370,34	819,85	
$\overline{MFCCSA}$	1,71	3,38	5,42	13,83	13,62	2,59	1,93	11,65	11,35	108,34	234,06	
Speedup	3,6	3,6	3,7	3,6	3,7	3,2	3,5	3,3	3,7	3,4	3,5	

4   Conclusion and Future Work

A simple SER algorithm based on the optimising of two of the parameters of the MFCC feature validated with eleven CML methods as well as four well-known SER datasets. The best model obtained, ExtraTress, enhances all the baseline models accuracy and especially the dataset SAVEE with an improvement of 10%. Attending that one dataset corresponds to german speakers and the three to English native speakers. The results are invariant to obtaining good results for all the models and dataset event the fusion of datasets (All). As this is preliminary work, we think that the optimisation of the remaining feature would improve the results. Moreover, the comparison of more meta-heuristics methods is needed and a more powerful hybrid technique like a Bag of Models.

Moreover, we think that the kind of records of the dataset are gathered, simulated, elicited or natural, and should be analysed to study how this information affects the transfer learning process. More datasets corresponding to other languages must be included in the study.

Acknowledgement. This research has been funded partially by the Spanish Ministry of Economy, Industry and Competitiveness (MINECO) under grant TIN2017-84804-R/PID2020-112726RB-I00.

References

1. Ahsan, M., Kumari, M.: Physical features based speech emotion recognition using predictive classification. Int. J. Comput. Sci. Inf. Technol. **8**(2), 63–74 (2016). https://doi.org/10.5121/ijcsit.2016.8205
2. Akçay, M.B., Oğuz, K.: Speech emotion recognition: emotional models, databases, features, preprocessing methods, supporting modalities, and classifiers. Speech Commun. **116**(October 2019), 56–76 (2020). https://doi.org/10.1016/j.specom.2019.12.001
3. Anagnostopoulos, C.N., Iliou, T., Giannoukos, I.: Features and classifiers for emotion recognition from speech: a survey from 2000 to 2011. Artif. Intell. Rev. **43**(2), 155–177 (2012). https://doi.org/10.1007/s10462-012-9368-5
4. Breiman, L.: Random forests. Mach. Learn. **45**(1), 5–32 (2001). https://doi.org/10.1023/A:1010933404324
5. Burkhardt, F., Paeschke, A., Rolfes, M., Sendlmeier, W., Weiss, B.: A database of German emotional speech. In: 9th European Conference on Speech Communication and Technology, pp. 1517–1520 (2005)
6. Chang, C.C., Lin, C.J.: LIBSVM: a library for support vector machines. ACM Trans. Intell. Syst. Technol. **2**(3) (2011). https://doi.org/10.1145/1961189.1961199
7. Chatterjee, S., Koniaris, C., Kleijn, W.B.: Auditory model based optimization of MFCCs improves automatic speech recognition performance. In: Proceedings of the Annual Conference of the International Speech Communication Association, INTERSPEECH (January), pp. 2987–2990 (2009)
8. Dzedzickis, A., Kaklauskas, A., Bucinskas, V.: Human emotion recognition: review of sensors and methods. Sensors (Switzerland) **20**(3) (2020). https://doi.org/10.3390/s20030592
9. Friedman, J.H.: Greedy function approximation: a gradient boosting machine. Ann. Stat. **29**(5), 1189–1232 (2001). https://doi.org/10.1214/aos/1013203451
10. Geurts, P., Ernst, D., Wehenkel, L.: Extremely randomized trees. Mach. Learn. **63**(1), 3–42 (2006). https://doi.org/10.1007/s10994-006-6226-1
11. Haq, S., Jackson, P.J.B.: Speaker-dependent audio-visual emotion recognition. In: Proceedings of the International Conference on Auditory-Visual Speech Processing (AVSP 2008), Norwich, UK (2009)
12. Haq, S., Jackson, P.J.B.: Machine Audition: Principles, Algorithms and Systems. chap. Multimodal, pp. 398–423. IGI Global, Hershey (2010)
13. Haq, S., Jackson, P., Edge, J.: Audio-visual feature selection and reduction for emotion classification. Expert Syst. Appl. **39**, 7420–7431 (2008)
14. Hastie, T., Tibshirani, R., Friedman, J.: Springer Series in Statistics The Elements of Statistical Learning Data Mining, Inference, and Prediction. Technical report
15. Kingma, D.P., Ba, J.L.: Adam: a method for stochastic optimization. In: 3rd International Conference on Learning Representations, ICLR 2015 - Conference Track Proceedings. International Conference on Learning Representations, ICLR (2015)
16. Klapuri, A., Davy, M.: Signal Processing Methods for Music Transcription. Springer, Heidelberg (2007)
17. Koolagudi, S.G., Rao, K.S.: Emotion recognition from speech: a review. Int. J. Speech Technol. **15**(2), 99–117 (2012). https://doi.org/10.1007/s10772-011-9125-1

18. Librosa.org: MFCC implementation (2021). https://librosa.org/doc/main/modules/librosa/feature/spectral.html#mfcc

19. Livingstone, S.R., Russo, F.A.: The Ryerson audio-visual database of emotional speech and song (RAVDESS): a dynamic, multimodal set of facial and vocal expressions in north American English. PLoS ONE **13**(5), e0196391 (2018). https://doi.org/10.1371/journal.pone.0196391

20. Manning, C.D., Raghavan, P., Schuetze, H.: The Bernoulli model. In: Introduction to Information Retrieval, pp. 234–265 (2009)

21. Pandey, S.K., Shekhawat, H.S., Prasanna, S.R.: Deep learning techniques for speech emotion recognition: a review. In: 2019 29th International Conference Radioelektronika, RADIOELEKTRONIKA 2019 - Microwave and Radio Electronics Week, MAREW 2019 (2019). https://doi.org/10.1109/RADIOELEK.2019.8733432

22. Pichora-Fuller, M.K., Dupuis, K.: Toronto emotional speech set (TESS) (2020). https://doi.org/10.5683/SP2/E8H2MF

23. Rahi, P.K.: Speech emotion recognition systems: review. Int. J. Res. Appl. Sci. Eng. Technol. **8**(1), 45–50 (2020). https://doi.org/10.22214/ijraset.2020.1007

24. Rao, K.S., Koolagudi, S.G., Vempada, R.R.: Emotion recognition from speech using global and local prosodic features. Int. J. Speech Technol. **16**(2), 143–160 (2013). https://doi.org/10.1007/s10772-012-9172-2

25. Rutenbar, R.A.: Simulated annealing algorithms: an overview. IEEE Circuits Dev. Mag. **5**(1), 19–26 (1989). https://doi.org/10.1109/101.17235

26. Sahidullah, M., Saha, G.: Design, analysis and experimental evaluation of block based transformation in MFCC computation for speaker recognition. Speech Commun. **54**(4), 543–565 (2012). https://doi.org/10.1016/j.specom.2011.11.004

27. Väyrynen, E.: Emotion recognition from speech using prosodic features. Ph.D. thesis (2014)

28. Zeng, Z., Pantic, M., Roisman, G.I., Huang, T.S.: A survey of affect recognition methods: audio, visual, and spontaneous expressions. IEEE Trans. Pattern Anal. Mach. Intell. **31**(1), 39–58 (2009). https://doi.org/10.1109/TPAMI.2008.52

29. Zhu, J., Zou, H., Rosset, S., Hastie, T.: Multi-class AdaBoost*. Technical report (2009)

Correction to: Hybrid Model to Calculate the State of Charge of a Battery

María Teresa García Ordás, David Yeregui Marcos del Blanco, José Aveleira-Mata, Francisco Zayas-Gato, Esteban Jove, José-Luis Casteleiro-Roca, Héctor Quintián, José Luis Calvo-Rolle, and Héctor Alaiz-Moretón

Correction to:
Chapter "Hybrid Model to Calculate the State of Charge of a Battery" in: H. Sanjurjo González et al. (Eds.): *Hybrid Artificial Intelligent Systems*, LNAI 12886, https://doi.org/10.1007/978-3-030-86271-8_32

In an older version of this paper, the "s" was missing from the first name of Francisco Zayas-Gato. This has been corrected.

The updated version of this chapter can be found at
https://doi.org/10.1007/978-3-030-86271-8_32

© Springer Nature Switzerland AG 2021
H. Sanjurjo González et al. (Eds.): HAIS 2021, LNAI 12886, p. C1, 2021.
https://doi.org/10.1007/978-3-030-86271-8_56

Author Index

Printed in the United States
by Baker & Taylor Publisher Services

Printed in the United States
by Baker & Taylor Publisher Services